Springer Series in Computational Physics

Springer Series in Computational Physics

A Computational Method in Plasma Physics
F. Bauer, O. Betancourt, P. Garabedian

Implementation of Finite Element Methods for Navier-Stokes Equations
T. Thomasset

Finite-Difference Techniques for Vectorized Fluid Dynamics Calculations
Edited by D. Book

Unsteady Viscous Flows. D. P. Telionis

Computational Methods for Fluid Flow. R. Peyret, T. D. Taylor

Computational Methods in Bifurcation Theory and Dissipative Structures
M. Kubicek, M. Marek

Optimal Shape Design for Elliptic Systems. O. Pironneau

The Method of Differential Approximation. Yu. I. Shokin

Computational Galerkin Methods. C. A. J. Fletcher

Numerical Methods for Nonlinear Variational Problems
R. Glowinski

Numerical Methods in Fluid Dynamics. Second Edition
M. Holt

Computer Studies of Phase Transitions and Critical Phenomena
O. G. Mouritsen

Finite Element Methods in Linear Ideal Magnetohydrodynamics
R. Gruber, J. Rappaz

Numerical Simulation of Plasmas. Y. N. Dnestrovskii, D. P. Kostomarov

Computational Methods for Kinetic Models of Magnetically Confined Plasmas
J. Killeen, G. D. Kerbel, M. C. McCoy, A. A. Mirin

Spectral Methods in Fluid Dynamics
C. Canuto, M. Y. Hussaini, A. Quarteroni, T. A. Zang

Computational Techniques for Fluid Dynamics 1
Fundamental and General Techniques. C. A. J. Fletcher

Computational Techniques for Fluid Dynamics 2
Specific Techniques for Different Flow Categories. C. A. J. Fletcher

Methods for the Localization of Singularities in Numerical Solutions of Gas Dynamics Problems. E. V. Vorozhtsov and N. N. Yanenko

Classical Orthogonal Polynomials of a Discrete Variable
A. F. Nikiforov, S. K. Suslov, and V. B. Uvarov

Flux Coordinates and Magnetic Field Structure: A Guide to a Fundamental Tool of Plasma Theory
W. D. D'haeseleer, W. N. G. Hitchon, J. D. Callen, J. L. Shohet

Monte Carlo Methods in Boundary Value Problems. K. K. Sabelfeld

Optimization of Weighted Monte Carlo Methods. G. A. Mikhailov

Claudio Canuto M. Yousuff Hussaini
Alfio Quarteroni Thomas A. Zang

Spectral Methods in Fluid Dynamics

With 88 Illustrations

Springer-Verlag
Berlin Heidelberg New York London
Paris Tokyo Hong Kong Barcelona

Authors

Professor Claudio Canuto
Dipartimento di Matematica
Politecnico di Torino
Corso Duca degli Abruzzi, 24
10129 Torino,
Italy

Professor Alfio Quarteroni
Dipartimento di Matematica
Politecnico di Milano
Via Bonardi 9
20133 Milano
Italy

Dr. M. Yousuff Hussaini
ICASE. Mail Stop 132C
NASA Langley Research Center
Hampton, VA 23681-0001

Dr. Thomas A. Zang, Jr.
Mail Stop 159
Computational Sciences Branch
NASA Langley Research Center
Hampton, VA 23681-0001

Editors

Library of Congress Cataloging-in-Publication Data
Spectral methods in fluid dynamics.
(Springer series in computational physics)
Bibliography: p.
Includes index.
1. Differential equations, Partial—Numerical solutions. 2. Numerical analysis
3. Fluid dynamics. I. Canuto, C. II. Title. III. Series.
QA377.S676 1987 515.3′53 87-12855

Typeset by Asco Trade Typesetting Ltd., Hong Kong.

9 8 7 6 5 4 (Corrected Third Printing)

ISBN-13:978-3-540-52205-8 e-ISBN-13:978-3-642-84108-8
DOI: 10.1007/978-3-642-84108-8

Preface

This is a book about spectral methods for partial differential equations: when to use them, how to implement them, and what can be learned from their rigorous theory. The computational side of spectral methods has evolved vigorously since the early 1970s, especially in computationally intensive applications in fluid dynamics. Some of the more spectacular applications are discussed here, first in general terms as examples of the power of these methods and later in great detail after the specifics of the methods have been covered. This book pays special attention to those algorithmic details which are essential to successful implementation of spectral methods. The focus is on algorithms for fluid dynamical problems in transition, turbulence, and aerodynamics. This book does not address specific applications in meteorology, partly because of the lack of experience of the authors in this field and partly because of the coverage provided by Haltiner and Williams (1980).

The success of spectral methods in practical computations has led to an increasing interest in their theoretical aspects, especially since the mid-1970s. Although the theory does not yet cover the complete spectrum of applications, the analytical techniques which have been developed in recent years have facilitated the examination of an increasing number of problems of practical interest. In this book we present a unified theory of the mathematical analysis of spectral methods and apply it to many of the algorithms in current use. The attention of the reader will be focused mainly on those aspects of the techniques which are typical of the analysis of spectral methods but which do not reduce to the straightforward application of well-known results from the analysis of finite difference methods. Some significant proofs will be given as examples of how the analysis is carried out.

No numerical method is suitable for all problems. This book aims to guide the reader towards those applications for which spectral methods are preferable and not just workable. It is addressed both to computational fluid dynamicists who wish to use spectral methods and to numerical analysts who are interested in their rigorous analysis. It is directed towards research workers in these fields and presumes an elementary knowledge of linear algebra, differential equations, numerical analysis, and fluid dynamics. A complete appreciation of the mathematical analysis does require the reader to be familiar with Hilbert space theory, the modern formulation of partial differential equations, and distribution theory. Nevertheless, the book is self-contained in the sense that these subjects are reviewed in an appendix.

During the ICASE Workshop on Spectral Methods in August 1982, it became apparent to the authors that the time had come for a thorough book on spectral methods which would bring some coherence to the rapidly emerging developments on both algorithms and theory. The interaction between engineers and theorists was beginning to bear fruit in the form that theoretical developments were influencing the course of algorithms and that the real problems of interest to engineers were being analyzed rigorously.

Our objective is to reach both audiences. The algorithm descriptions are given in a mathematically unambiguous form. The theoretical discussions are presented in such a way that the mathematically oriented, but nonspecialist, reader can follow them. Details which are only technical are relegated to the original references.

The outline of the book is as follows. Chapter 1 consists of a colloquial introduction to spectral methods and a motivation for their use in fluid dynamical problems. A classical presentation of the material used for the approximation of a function and its derivatives by orthogonal polynomials is given in Chapter 2. This is followed in Chapter 3 by an illustration on a simple nonlinear evolution equation of how to construct various spectral discretizations. This chapter also deals with certain algorithmic details, such as the proper implementation of boundary conditions and the efficient evaluation of convolution sums. Chapter 4 consists of a discussion of the time discretizations which are appropriately coupled with spectral spatial discretizations. A description of direct and iterative methods for solving implicit spectral equations is provided in Chapter 5. Then in Chapter 6 we turn to the application of spectral methods to some relatively simple one- and two-dimensional problems in incompressible flow. The focal point of the applications part is Chapter 7. Here we describe in detail the algorithms for unsteady incompressible problems in transition and turbulence. This is the one field in which spectral methods are unchallenged. Chapter 8 discusses the more recent (and sometimes controversial) use of spectral methods for compressible flow. Chapter 9 consists of a review of those results from approximation theory which are pertinent to the theoretical analysis of spectral methods. Chapter 10 is the other focal point of the book. In it, the fundamental stability and convergence theory for spectral approximations to linear partial differential equations is presented. Chapters 11 and 12 deal with the specific applications of the general theory to steady and unsteady problems, respectively. Chapter 11 places particular emphasis on the steady Navier-Stokes equations, whereas Chapter 12 stresses hyperbolic problems. Chapter 13 is devoted to the recent developments, both algorithmical and theoretical, in applying spectral methods to more general geometries.

For organizational purposes and to accommodate those readers interested in only the algorithms or the analysis alone, we have split the detailed description of these two subjects into separate parts of the book—Chapters 4 to 8 for the algorithms and Chapters 9 to 12 for the analysis. Nevertheless, and despite the particular research interests of the authors, virtually every chapter in the book is the product of a vigorous interaction.

The authors are grateful to Professors Holt and Beiglböck who encouraged us throughout this project. The writing of this book has benefited enormously from the scientific atmosphere and the active support furnished by the Institute for Computer Applications in Science and Engineering, NASA Langley Research Center, Hampton, Virginia and the Istituto di Analisi Numerica del Consiglio Nazionale delle Ricerche, Pavia, Italy. The first and third authors are particularly grateful to Professors Magenes and Brezzi for their encouragement and sage advice on this project. The authors have profited from discussions and technical advice from virtually all the active researchers on spectral methods. Special thanks are due to those who advised us on particular sections of the book or supplied some of the figures: Jill Dahlburg, Daniele Funaro, John Kim, David Kopriva, Michele Macaraeg, Yvon Maday, Ralph Metcalfe, Tony Patera, Robert Rogallo, Phillippe Spalart, and Craig Streett.

We wish to take this opportunity to initiate the canonization procedure for Barbara Kraft. She mastered the authors' hieroglyphics, accurately interpreted their convoluted editing instructions, suffered their innumerable changes of mind, met their unreasonable deadlines, and still produced nearly perfect drafts. It was truly a miraculous performance. Special thanks are also due to Barbara Stewart who prepared the final manuscript.

Hampton, Virginia
Pavia, Italy
September, 1986

C. Canuto
M. Y. Hussaini
A. Quarteroni
T. A. Zang

Preface to Second Printing

In this, the second printing of the book, we have taken the opportunity to correct a number of typographical errors and to furnish a supplementary bibliography which contains new work which has appeared since early 1987 as well as the completion of many citations that were still in press at the time of the first printing. Moreover, we have added a new section to Chapter 13 which summarizes some of the recent key developments in the rapidly evolving subject of spectral domain decomposition methods.

Although a number of significant developments have also occurred for conventional spectral methods, we have made no attempt to incorporate them in the text. The interested reader, however, can find many of them by a careful perusal of the supplementary bibliography. More substantial changes to the text will have to await a possible future edition.

Preface to Third Printing

The third printing of the book gave us the opportunity to correct a few typographical errors and to complete those citations in the supplementary bibliography that were still in press at the time of the second printing. No further changes have been carried out in the text although there have been a number of significant developments in spectral methods. A possible future edition will incorporate substantial changes in the text.

Contents

Authors' Affiliations

Claudio Canuto Dipartimento di Matematica, Politecnico di Torino, Corso Duca degli Abruzzi, 24, 10129 Torino, Italy

M. Yousuff Hussaini Institute for Computer Applications in Science and Engineering, Mail Stop 132C, NASA Langley Research Center, Hampton, VA 23681-0001, U.S.A.

Alfio Quarteroni Dipartimento di Matematica, Politecnico di Milano, Via Bonardi 9, 20133 Milano, Italy

Thomas A. Zang Computational Sciences Branch, Mail Stop 159, NASA Langley Research Center, Hampton, VA 23681-0001, U.S.A.

CHAPTER 1

Introduction

1.1. Historical Background

Spectral methods may be viewed as an extreme development of the class of discretization schemes for differential equations known generically as the method of weighted residuals (MWR) (Finlayson and Scriven (1966)). The key elements of the MWR are the trial functions (also called the expansion or approximating functions) and the test functions (also known as weight functions). The trial functions are used as the basis functions for a truncated series expansion of the solution. The test functions are used to ensure that the differential equation is satisfied as closely as possible by the truncated series expansion. This is achieved by minimizing the residual, i.e., the error in the differential equation produced by using the truncated expansion instead of the exact solution, with respect to a suitable norm. An equivalent requirement is that the residual satisfy a suitable orthogonality condition with respect to each of the test functions.

The choice of trial functions is one of the features which distinguish spectral methods from finite-element and finite-difference methods. The trial functions for spectral methods are infinitely differentiable global functions. (Typically they are tensor products of the eigenfunctions of singular Sturm–Liouville problems.) In the case of finite-element methods, the domain is divided into small elements, and a trial function is specified in each element. The trial functions are thus local in character, and well suited for handling complex geometries. The finite-difference trial functions are likewise local.

The choice of test functions distinguishes between the three most commonly used spectral schemes, namely, the Galerkin, collocation, and tau versions. In the Galerkin approach, the test functions are the same as the trial functions. They are, therefore, infinitely smooth functions which individually satisfy the boundary conditions. The differential equation is enforced by requiring that the integral of the residual times each test function be zero. In the collocation approach the test functions are translated Dirac delta functions centered at special, so-called collocation points. This approach requires the differential equation to be satisfied exactly at the collocation points. Spectral tau methods are similar to Galerkin methods in the way that the differential equation is enforced. However, none of the test functions need satisfy the boundary conditions. Hence, a supplementary set of equations is used to apply the boundary conditions.

The collocation approach is the simplest of the MWR, and appears to have been first used by Slater (1934) and by Kantorovic (1934) in specific applications. Frazer, Jones and Skan (1937) developed it as a general method for solving ordinary differential equations. They used a variety of trial functions and an arbitrary distribution of collocation points. The work of Lanczos (1938) established for the first time that a proper choice of trial functions and distribution of collocation points is crucial to the accuracy of the solution. Perhaps he should be credited with laying down the foundation of the orthogonal collocation method. This method was revived by Clenshaw (1957), Clenshaw and Norton (1963) and Wright (1964). These studies involved application of Chebyshev polynomial expansions to initial value problems. Villadsen and Stewart (1967) developed this method for boundary value problems.

The earliest applications of the spectral collocation method to partial differential equations were made for spatially periodic problems by Kreiss and Oliger (1972) (who called it the Fourier method) and Orszag (1972) (who termed it pseudospectral). This approach is especially attractive because of the ease with which it can be applied to variable-coefficient and even non-linear problems. The essential details will be furnished below.

The Galerkin approach is perhaps the most esthetically pleasing of the methods of weighted residuals since the trial functions and the test functions are the same and the physical problem can be discretized in terms of a variational principle. Finite-element methods customarily use this approach. Moreover, the first serious application of spectral methods to PDE's—that of Silberman (1954) for meteorological modeling—was a Galerkin method. However, spectral Galerkin methods only became practical for high resolution calculations of such non-linear problems after Orszag (1969, 1970) and Eliasen, Machenhauer and Rasmussen (1970) developed transform methods for evaluating convolution sums arising from quadratic non-linearities. (Non-linear terms also increase the cost of finite-element methods, but not nearly as much as they do for spectral Galerkin methods). For problems containing more complicated non-linear terms, high resolution spectral Galerkin methods remain impractical.

The tau approach is a modification of the Galerkin method that is applicable to problems with non-periodic boundary conditions. It may be viewed as a special case of the so-called Petrov–Galerkin method. Lanczos (1938) developed the spectral tau method and, although it too is difficult to apply to non-linear problems, it has proven quite useful for constant-coefficient problems or subproblems, e.g., for semi-implicit time-stepping algorithms.

The first unifying mathematical assessment of the theory of spectral methods was contained in the monograph by Gottlieb and Orszag (1977). Since then, the theory has been extended to cover a large variety of problems, such as variable-coefficient and non-linear equations. A sound approximation theory for the polynomial families used in spectral methods has been developed. Stability and convergence analyses for spectral methods have been based on several kinds of approaches. The interpretation of spectral methods as MWR methods (or, in mathematical terms, as variational methods) has proven very successful in the theoretical investigation. As a matter of fact, it has opened the road to the use of techniques of functional analysis to handle complex problems and to obtain the sharpest results.

In the middle of the last decade Gottlieb and Orszag (1977) summarized the state of the art in the theory and application of spectral methods. Developments of the following five years were reviewed in the symposium proceedings edited by Voigt, Gottlieb and Hussaini (1984). These references emphasized applications in fluid dynamics. Applications in meteorology have been covered in the review by Jarraud and Baede (1985) and in Haltiner and Williams (1980). Numerous other introductory or review articles have appeared recently, e.g., Mercier (1981), Hussaini, Salas and Zang (1985), Zang and Hussaini (1985c), Deville (1984), Gottlieb (1985) and Hussaini and Zang (1987).

1.2. Some Examples of Spectral Methods

Spectral methods are distinguished not only by the type of the method (Galerkin, collocation, or tau), but also by the particular choice of the trial functions. The most frequently used trial functions are trigonometric polynomials, Chebyshev polynomials, and Legendre polynomials. In this section we shall illustrate the basic principles of each method and the basic properties of each set of polynomials by examining in detail one particular spectral method on each of the fundamental types of equations. Each of these examples will be reconsidered in Chap. 10 from a rigorous theoretical basis.

1.2.1. A Fourier Galerkin Method for the Wave Equation

Many evolution equations can be written as

$$\frac{\partial u}{\partial t} = M(u), \tag{1.2.1}$$

where $u(\mathbf{x}, t)$ is the solution and $M(u)$ is an operator which contains all the spatial derivatives of u. Equation (1.2.1) must be coupled with an initial condition $u(\mathbf{x}, 0)$ and suitable boundary conditions.

For simplicity suppose that there is only one spatial dimension, that the spatial domain is $(0, 2\pi)$, and that the boundary conditions are periodic. Most often the MWR is used only for the spatial discretization. The approximate solution is represented as

$$u^N(x, t) = \sum_{k=-N/2}^{N/2} a_k(t)\phi_k(x). \tag{1.2.2}$$

The ϕ_k are the trial functions, whereas the a_k are the expansion coefficients. In general, u^N will not satisfy (1.2.1), i.e., the residual

$$\frac{\partial u^N}{\partial t} - M(u^N)$$

will not vanish everywhere. The MWR approximation results by demanding that

$$\int_0^{2\pi} \left[\frac{\partial u^N}{\partial t} - M(u^N)\right]\psi_k(x)\,dx = 0 \tag{1.2.3}$$

for $k = -N/2, \ldots, N/2$, where the test functions ψ_k determine the weights of the residual.

The most straightforward spectral method for this problem is based on trigonometric polynomials:

$$\phi_k(x) = e^{ikx} \tag{1.2.4}$$

$$\psi_k(x) = \frac{1}{2\pi} e^{-ikx}. \tag{1.2.5}$$

Note that the trial functions and the test functions are essentially the same and that they satisfy the orthonormality condition

$$\int_0^{2\pi} \phi_k(x)\psi_l(x)\,dx = \delta_{kl}. \tag{1.2.6}$$

If this were merely an approximation problem, then (1.2.2) would be the truncated Fourier series of the known function $u(x, t)$ with

$$a_k(t) = \int_0^{2\pi} u(x, t)\psi_k(x)\,dx \tag{1.2.7}$$

being simply the familiar Fourier coefficients. For the partial differential equation (PDE), however, $u(x, t)$ is not known; the approximation (1.2.2) is determined by (1.2.3).

For the linear hyperbolic problem,

$$M(u) = \frac{\partial u}{\partial x}, \tag{1.2.8}$$

i.e.,

$$\frac{\partial u}{\partial t} - \frac{\partial u}{\partial x} = 0, \tag{1.2.9}$$

this condition becomes

$$\frac{1}{2\pi}\int_0^{2\pi}\left[\left(\frac{\partial}{\partial t}-\frac{\partial}{\partial x}\right)\sum_{l=-N/2}^{N/2} a_l(t)e^{ilx}\right]e^{-ikx}\,dx = 0.$$

The next two steps are the analytical (spatial) differentiation of the trial functions,

$$\frac{1}{2\pi}\int_0^{2\pi}\left[\sum_{l=-N/2}^{N/2}\left(\frac{da_l}{dt} - ila_l\right)e^{ilx}\right]e^{-ikx}\,dx = 0,$$

and the analytical integration of this expression, which produces the dynamical equations

$$\frac{da_k}{dt} - ika_k = 0 \qquad k = -N/2, \ldots, N/2. \tag{1.2.10}$$

The initial conditions for this system of ordinary differential equations (ODEs) are the coefficients for the expansion of the initial condition. For this Galerkin approximation

$$a_k(0) = \int_0^{2\pi} u(x,0)\psi_k(x)\,dx. \tag{1.2.11}$$

For more complicated problems the analytical component of spectral Galerkin methods is carried as far as these last two equations. In general, numerical quadratures are performed both to obtain the initial values of the expansion coefficients and to advance the dynamical equations in time.

We shall use the initial condition

$$u(x,0) = \sin(\pi\cos x) \tag{1.2.12}$$

to illustrate the accuracy of the Fourier Galerkin method for (1.2.9). The exact solution

$$u(x,t) = \sin[\pi\cos(x+t)] \tag{1.2.13}$$

has the Fourier expansion

$$u(x,t) = \sum_{k=-\infty}^{\infty} a_k(t)e^{ikx}, \tag{1.2.14}$$

where the Fourier coefficients

$$a_k(t) = \sin\left(\frac{k\pi}{2}\right)J_k(\pi)e^{ikt} \tag{1.2.15}$$

and $J_k(t)$ is the Bessel function of order k. The asymptotic properties of the Bessel functions imply that

$$k^p a_k(t) \to 0 \qquad \text{as } k \to \infty \tag{1.2.16}$$

for all positive integers p. As a result, the truncated Fourier series

$$u^N(x,t) = \sum_{k=-N/2}^{N/2} a_k(t)e^{ikx} \tag{1.2.17}$$

converges faster than any finite power of $1/N$. This property is often referred to as exponential convergence.

An illustration of the superior accuracy available from the spectral method for this problem is provided in Table 1.1 and Fig. 1.1. Shown in the table are the maximum errors after one period at $t = 2\pi$ for the spectral Galerkin method as well as for second-order and fourth-order finite-difference methods. The time-discretization was the classical fourth-order Runge–Kutta method and the exact initial Fourier coefficients were used for the spectral

Table 1.1. Maximum error for a one-dimensional periodic problem

N	Fourier Galerkin	Second-order finite-difference	Fourth-order finite-difference
8	9.87 (−2)	1.11 (0)	9.62 (−1)
16	2.55 (−4)	6.13 (−1)	2.36 (−1)
32	1.05 (−11)	1.99 (−1)	2.67 (−2)
64	6.22 (−13)	5.42 (−2)	1.85 (−3)
128		1.37 (−2)	1.18 (−4)

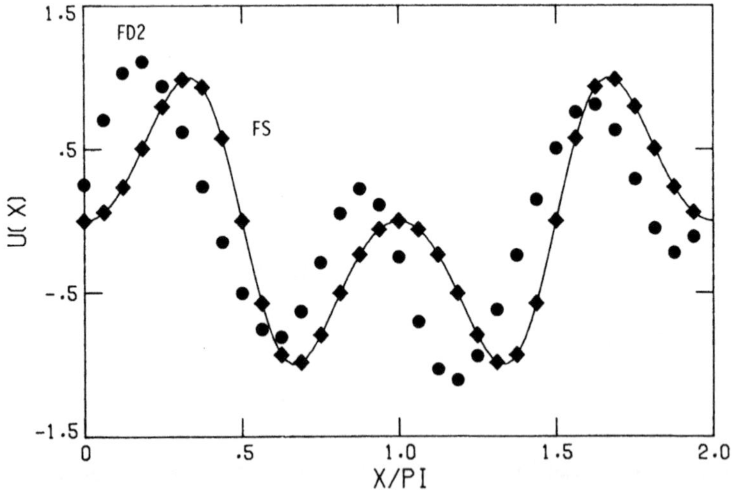

Figure 1.1. Second-order finite-difference (FD2) and Fourier spectral (FS) solutions to a periodic wave equation. The solid line is the exact solution after one full period.

method. In all cases the time-step was chosen so small that the temporal discretization error was negligible. The figure compares the second-order finite-difference and spectral Galerkin solutions for $N = 32$ with the exact answer. Note that the major error in the finite-difference solution is one of *phase* rather than *amplitude*. In many problems the very low phase errors of spectral methods is a significant advantage.

Because the solution is infinitely smooth, the convergence of the spectral method on this problem is more rapid than any finite power of $1/N$. (The error for the $N = 64$ spectral result is so small that it is swamped by the round-off error of these single-precision CDC Cyber 175 calculations.) In most practical applications the benefit of the spectral method is not the extraordinary accuracy available for large N but rather the small size of N necessary for a moderately accurate solution.

1.2.2. A Chebyshev Collocation Method for the Heat Equation

Fourier series, despite their simplicity and familiarity, are not always a good choice for the trial functions. In fact, for reasons that will be explored in the next chapter, Fourier series are only advisable for problems with periodic boundary conditions. A more versatile set of trial functions is composed of the Chebyshev polynomials. These are defined on $[-1, 1]$ by

$$T_k(x) = \cos(k \cos^{-1} x) \tag{1.2.18}$$

for $k = 0, 1, \ldots$.

Let us focus on the linear heat equation

$$M(u) = \frac{\partial^2 u}{\partial x^2}, \tag{1.2.19}$$

i.e.,

$$\frac{\partial u}{\partial t} - \frac{\partial^2 u}{\partial x^2} = 0 \tag{1.2.20}$$

on $(-1, 1)$ with homogeneous Dirichlet boundary conditions

$$\begin{aligned} u(1, t) &= 0 \\ u(-1, t) &= 0. \end{aligned} \tag{1.2.21}$$

Choosing the trial functions

$$\phi_k(x) = T_k(x) \qquad k = 0, 1, \ldots, N, \tag{1.2.22}$$

the approximate solution has the representation

$$u^N(x, t) = \sum_{k=0}^{N} a_k(t) \phi_k(x). \tag{1.2.23}$$

In the collocation version of the MWR the test functions are the shifted Dirac delta-functions

$$\psi_j(x) = \delta(x - x_j) \qquad j = 1, \ldots, N-1, \tag{1.2.24}$$

where the x_j are distinct collocation points in $(-1, 1)$. The standard MWR condition

$$\int_{-1}^{1} \left[\frac{\partial u^N}{\partial t} - M(u^N) \right] \psi_j(x)\, dx = 0 \qquad j = 1, \ldots, N-1 \tag{1.2.25}$$

reduces to the requirement that (1.2.20) be satisfied exactly by (1.2.23) at each of the x_j:

$$\left. \frac{\partial u^N}{\partial t} - M(u^N) \right|_{x=x_j} = 0 \qquad j = 1, \ldots, N-1. \tag{1.2.26}$$

The boundary conditions

$$\begin{aligned} u^N(1, t) &= 0 \\ u^N(-1, t) &= 0 \end{aligned} \tag{1.2.27}$$

and the initial condition

$$u^N(x_k, 0) = u(x_k, 0) \qquad k = 0, \ldots, N \tag{1.2.28}$$

accompany (1.2.26).

A particularly convenient choice for the collocation points x_j is

$$x_j = \cos\frac{\pi j}{N}. \tag{1.2.29}$$

Not only does this choice produce highly accurate approximations, but it also is economical. Note that

$$\phi_k(x_j) = \cos\frac{\pi j k}{N}. \tag{1.2.30}$$

This enables the Fast Fourier Transform to be employed in the evaluation of $M(u_N)|_{x=x_j}$.

For the particular initial condition

$$u(x, 0) = \sin \pi x, \tag{1.2.31}$$

the exact solution is

$$u(x, t) = e^{-\pi^2 t} \sin \pi x. \tag{1.2.32}$$

It has the infinite Chebyshev expansion

$$u(x, t) = \sum_{n=0}^{\infty} b_n(t) T_n(x), \tag{1.2.33}$$

where

$$b_n(t) = \frac{2}{c_n} \sin\left(\frac{n\pi}{2}\right) J_n(\pi) e^{-\pi^2 t}, \tag{1.2.34}$$

with

$$c_n = \begin{cases} 2 & n = 0 \\ 1 & n \geqslant 1 \end{cases}. \tag{1.2.35}$$

The truncated series converges at an exponential rate. A well-designed collocation method will do the same. (Since the finite series (1.2.23) is not simply the truncation of the infinite series (1.2.33) at order N, the expansion coefficients $a_n(t)$ and $b_k(t)$ are not identical.)

Unlike a Galerkin method which is implemented in terms of the expansion coefficients $b_k(t)$, a collocation method uses the expansion coefficients in an intermediate step, namely, in the analytic differentiation (with respect to x) of (1.2.23). The details of this step, which will be derived in Sec. 2.4, follow.

Let $u_j(t)$ denote $u^N(x_j, t)$. The expansion coefficients are given by

$$a_k(t) = \frac{2}{N\bar{c}_k} \sum_{j=0}^{N} \bar{c}_j^{-1} u_j \cos\frac{\pi jk}{N} \qquad k = 0, 1, \ldots, N, \tag{1.2.36}$$

where

$$\bar{c}_j = \begin{cases} 2 & j = 0 \text{ or } N \\ 1 & 1 \leqslant j \leqslant N-1 \end{cases}. \tag{1.2.37}$$

The analytic derivative of (1.2.23) is

$$\frac{\partial^2 u^N}{\partial x^2} = \sum_{k=0}^{N} a_k^{(2)}(t) T_k(x), \tag{1.2.38}$$

where

$$\begin{aligned} a_{N+1}^{(1)}(t) &= 0 \\ a_N^{(1)}(t) &= 0 \\ \bar{c}_k a_k^{(1)}(t) &= a_{k+2}^{(1)}(t) + 2(k+1)a_{k+1} \qquad k = N-1, N-2, \ldots, 0, \end{aligned} \tag{1.2.39}$$

and

$$\begin{aligned} a_{N+1}^{(2)}(t) &= 0 \\ a_N^{(2)}(t) &= 0 \\ \bar{c}_k a_k^{(2)}(t) &= a_{k+2}^{(2)}(t) + 2(k+1)a_{k+1}^{(1)}(t) \qquad k = N-1, N-2, \ldots, 0. \end{aligned} \tag{1.2.40}$$

Finally, (1.2.26) becomes

$$\left.\frac{\partial u^N}{\partial t}\right|_{x=x_j} = \sum_{k=0}^{N} a_k^{(2)}(t) \cos\frac{\pi jk}{N}. \tag{1.2.41}$$

Results pertaining to $t = 1$ for a Chebyshev collocation method and a second-order finite-difference method are given in Table 1.2.

Table 1.2. Maximum error for the one-dimensional heat equation

N	Chebyshev collocation	Second-order finite-difference
8	4.58 (−4)	6.44 (−1)
10	8.25 (−6)	3.59 (−1)
12	1.01 (−7)	2.50 (−1)
14	1.10 (−9)	1.74 (−1)
16	2.09 (−11)	1.35 (−1)

1.2.3. A Legendre Tau Method for the Poisson Equation

Spectral methods are also applicable to non-evolution equations. The Poisson equation is perhaps the simplest example. Here the general equation is

$$M(u) = f, \tag{1.2.42}$$

along with the boundary conditions

$$\mathbf{B}(u) = 0. \tag{1.2.43}$$

For the Poisson equation on $(-1, 1) \times (-1, 1)$ with homogeneous Dirichlet boundary conditions these are

$$M(u) = \frac{\partial^2 u}{\partial x^2} + \frac{\partial^2 u}{\partial y^2} \tag{1.2.44}$$

$$B_1(u) = u(x, -1) \tag{1.2.45a}$$

$$B_2(u) = u(x, +1) \tag{1.2.45b}$$

$$B_3(u) = u(-1, y) \tag{1.2.45c}$$

$$B_4(u) = u(+1, y). \tag{1.2.45d}$$

Both Legendre and Chebyshev polynomials are suitable trial functions. A two-dimensional Legendre expansion is produced by the tensor product choice

$$\phi_{kl}(x, y) = L_k(x)L_l(y) \qquad k, l = 0, 1, \ldots, N, \tag{1.2.46}$$

where L_k is the Legendre polynomial of degree k. The expansion is

$$u^N(x, y) = \sum_{k=0}^{N} \sum_{l=0}^{N} a_{kl} L_k(x) L_l(y). \tag{1.2.47}$$

The properties of Legendre polynomials are well known; a detailed discussion of them is furnished in Sec. 2.3.

Note that the trial functions do not satisfy the boundary conditions individually. (In most Galerkin methods the trial functions do satisfy the boundary conditions.) Thus, it is necessary to have weighted residual conditions for both the PDE and the boundary conditions. Two sets of test functions are used. For the PDE they are

$$\psi_{kl}(x, y) = Q_k(x)Q_l(y) \qquad k = 0, 1, \ldots, N-2, \tag{1.2.48}$$

where

$$Q_k(x) = \frac{2k+1}{2} L_k(x); \tag{1.2.49}$$

for the boundary conditions they are

$$\chi_k^i(x) = Q_k(x) \qquad \begin{array}{l} i = 1, 2 \\ k = 0, 1, \ldots, N \end{array} \tag{1.2.50a}$$

$$\chi_l^i(y) = Q_l(y) \qquad \begin{array}{l} i = 3, 4 \\ l = 0, 1, \ldots, N. \end{array} \tag{1.2.50b}$$

The MWR conditions are

$$\int_{-1}^{1} dy \int_{-1}^{1} M(u^N)\psi_{kl}(x, y)\,dx = 0 \qquad k, l = 0, 1, \ldots, N-2, \tag{1.2.51}$$

and

$$\int_{-1}^{1} B_i(u^N)\chi_k^i(x)\,dx = 0 \qquad \begin{array}{l} i = 1, 2 \\ k = 0, 1, \ldots, N \end{array} \tag{1.2.52a}$$

$$\int_{-1}^{1} B_i(u^N)\chi_l^i(y)\,dy = 0 \qquad \begin{array}{l} i = 3, 4 \\ l = 0, 1, \ldots, N. \end{array} \tag{1.2.52b}$$

Four of the conditions in (1.2.52) are linearly dependent upon the others; in effect the boundary conditions at each of the four corner points have been applied twice. For the Poisson equation the above integrals may be performed analytically. The result is

$$a_{kl}^{(2,0)} + a_{kl}^{(0,2)} = f_{kl} \qquad k, l = 0, 1, \ldots, N-2 \tag{1.2.53}$$

$$\left.\begin{array}{l} \sum_{k=0}^{N} a_{kl} = 0 \\ \sum_{k=0}^{N} (-1)^k a_{kl} = 0 \end{array}\right. \qquad l = 0, 1, \ldots, N \tag{1.2.54a}$$

$$\left.\begin{array}{l} \sum_{l=0}^{N} a_{kl} = 0 \\ \sum_{l=0}^{N} (-1)^l a_{kl} = 0 \end{array}\right. \qquad k = 0, 1, \ldots, N \tag{1.2.54b}$$

Table 1.3. Maximum error for the two-dimensional Poisson equation

N	Legendre tau	Second-order finite-difference
8	1.55 (−3)	5.30 (−2)
10	3.40 (−5)	3.36 (−2)
12	6.05 (−7)	2.32 (−2)
14	6.98 (−9)	1.70 (−2)
16	6.37 (−11)	1.30 (−2)

where

$$f_{kl} = \int_{-1}^{1} dy \int_{-1}^{1} f(x,y)\psi_{kl}(x,y)\,dx \tag{1.2.55}$$

$$a_{kl}^{(2,0)} = (k+\tfrac{1}{2}) \sum_{\substack{p=k+2 \\ p+k \text{ even}}}^{N} [p(p+1) - k(k+1)]a_{pl} \tag{1.2.56a}$$

$$a_{kl}^{(0,2)} = (l+\tfrac{1}{2}) \sum_{\substack{q=l+2 \\ q+l \text{ even}}}^{N} [q(q+1) - l(l+1)]a_{kq}. \tag{1.2.56b}$$

These last two expressions represent the expansions of $\partial^2 u^N/\partial x^2$ and $\partial^2 u^N/\partial y^2$, respectively, in terms of the trial functions.

The Legendre tau approximation to the Poisson equation consists of (1.2.53) and (1.2.54). A scheme for the solution of these equations is provided in Sec. 5.1.

The specific example that will be used to illustrate the accuracy of this method is

$$f(x,y) = -2\pi^2 \sin \pi x \sin \pi y, \tag{1.2.57}$$

which corresponds to the analytic solution

$$u(x,y) = \sin \pi x \sin \pi y. \tag{1.2.58}$$

The results are given in Table 1.3 along with results for a second-order finite-difference scheme.

1.2.4. Basic Aspects of Galerkin, Tau and Collocation Methods

The Galerkin, tau and collocation methods are more general than suggested by any of the above examples. In broad terms, Galerkin and tau methods are implemented in terms of the expansion coefficients, whereas collocation methods are implemented in terms of the physical space values of the unknown function. The first example illustrated only one of the key aspects of Galerkin methods—the test functions are the same as the trial functions. The other important aspect is that the trial functions must individually satisfy the boundary conditions. In the case of periodic boundary conditions the trigonometric polynomials automatically satisfy this condition. Otherwise, simple linear combinations of the orthogonal polynomials will usually suffice. For example, a Chebyshev Galerkin approximation to the third example would use the trial functions

$$\phi_k(x) = \begin{cases} T_k(x) - T_0(x) & k \text{ even} \\ T_k(x) - T_1(x) & k \text{ odd} \end{cases}.$$

On the other hand, for the tau method the trial functions do not individually satisfy the boundary conditions. Thus, some equations are needed to ensure that the global expansion satisfies the boundary conditions. Some of the highest order MWR equations are then dropped in favor of these boundary condition equations.

The collocation method uses the values of the function at certain physical points as the fundamental representation; the expansion functions are employed solely for evaluating derivatives. The collocation points for both the differential equations and the boundary conditions are usually the same as the physical grid points. The most effective choice for the grid points are those that correspond to quadrature formulas of maximum precision.

1.3. The Equations of Fluid Dynamics

The purpose of this book is to describe spectral methods, not only for simple model problems such as those described in the proceding section, but also for large scale applications in fluid dynamics. In this section we summarize the basic equations of fluid dynamics (in Eulerian form). Those interested in detailed derivations and physical interpretations of these equations should consult standard references such as Batchelor (1967), Liepmann and Roshko (1957), Rosenhead (1963) and Schlichting (1979). The mathematical theory of these equations has been summarized by Ladyzhenskaya (1969) and Temam (1977).

1.3.1. Compressible Navier-Stokes

A complete description of a fluid is available if one knows, as a function of space and time, the velocity $\mathbf{u} = (u, v, w)$, any two thermodynamic variables, and an equation-of-state. The thermodynamic variables of most interest in fluid dynamics are the density ρ, temperature T, pressure p, specific internal energy e, specific enthalpy $h = e + p/\rho$ and specific entropy s. The latter is defined by

$$ds = \frac{1}{T}\left[de + pd\left(\frac{1}{\rho}\right)\right].$$

Two important quantities are the specific heat at constant pressure

$$C_p = \left.\frac{\partial h}{\partial T}\right|_p$$

and the specific heat at constant volume (or density)

$$C_v = \left.\frac{\partial e}{\partial T}\right|_\rho.$$

Their difference gives the gas constant

$$\mathcal{R} = C_p - C_v,$$

and their ratio the adiabatic index

$$\gamma = C_p/C_u.$$

We shall assume here that we are dealing with a perfect gas for which the specific heats are constant (independent of T) and for which the equation-of-state is

$$p = \rho\mathcal{R}T. \tag{1.3.1}$$

The sound speed c is given by

$$c^2 = \left.\frac{\partial p}{\partial \rho}\right|_s$$

and for an adiabatic gas this is

$$c^2 = \frac{\gamma p}{\rho}. \tag{1.3.2}$$

The Navier–Stokes equations are a differential form of the conservation laws which govern fluid motion. The equation of mass conservation, known as the continuity equation, is

$$\frac{\partial \rho}{\partial t} + \nabla \cdot (\rho \mathbf{u}) = 0. \tag{1.3.3}$$

The momentum equation is

$$\frac{\partial}{\partial t}(\rho \mathbf{u}) + \nabla \cdot (\rho \mathbf{u}\mathbf{u}) = -\nabla p + \nabla \cdot \underline{\tau} \tag{1.3.4}$$

where the stress tensor

$$\underline{\tau} = \mu(\nabla \mathbf{u} + (\nabla \mathbf{u})^T) - \tfrac{2}{3}\mu(\nabla \cdot \mathbf{u})\underline{\mathbf{I}}, \tag{1.3.5}$$

where $\underline{\mathbf{I}}$ is the unit tensor and μ is the fluid viscosity. The equation of energy conservation takes many forms. Let k be the thermal conductivity and define the dissipation function Φ by

$$\Phi = \tau_{ij}\frac{\partial u_i}{\partial x_j} \tag{1.3.6}$$

with the summation convention employed for repeated indices. These alternative forms of the energy equation are

$$\rho C_v\left(\frac{\partial T}{\partial t} + \mathbf{u}\cdot\nabla T\right) + p\nabla\cdot\mathbf{u} = \nabla\cdot(k\nabla T) + \Phi \tag{1.3.7}$$

$$\rho\left(\frac{\partial e}{\partial t} + \mathbf{u}\cdot\nabla e\right) + p\nabla\cdot\mathbf{u} = \nabla\cdot(k\nabla T) + \Phi \tag{1.3.8}$$

$$\frac{\partial p}{\partial t} + \mathbf{u}\cdot\nabla p + \gamma p\nabla\cdot\mathbf{u} = (\gamma - 1)[\nabla\cdot(k\nabla T) + \Phi] \tag{1.3.9}$$

$$\rho T\left(\frac{\partial s}{\partial t} + \mathbf{u}\cdot\nabla s\right) = \nabla\cdot(k\nabla T) + \Phi \tag{1.3.10}$$

$$\rho\left(\frac{\partial h}{\partial t} + \mathbf{u}\cdot\nabla h\right) + \gamma p\nabla\cdot\mathbf{u} = \gamma\nabla\cdot(k\nabla T) + \gamma\Phi. \tag{1.3.11}$$

These equations have been written in dimensional form. They comprise an incompletely parabolic system (Belov and Yanenko (1971)). The absence of a diffusion term in the continuity equation prevents them from being fully parabolic.

On solid, stationary walls the boundary conditions are no-slip for velocity

$$\mathbf{u} = 0$$

and either constant temperature

$$T = T_{\text{wall}}$$

or adiabatic conditions

$$\frac{\partial T}{\partial n} = 0,$$

where $\hat{\mathbf{n}}$ is the unit vector normal to the wall.

1.3.2. Compressible Euler

If the problem is inviscid, i.e., $\mu = 0$, then the Navier–Stokes equations reduce to the Euler equations

$$\frac{\partial \rho}{\partial t} + \nabla \cdot (\rho \mathbf{u}) = 0 \tag{1.3.12}$$

$$\frac{\partial}{\partial t}(\rho \mathbf{u}) + \nabla \cdot (\rho \mathbf{u}\mathbf{u}) + \nabla p = 0 \tag{1.3.13}$$

$$\frac{\partial p}{\partial t} + \mathbf{u} \cdot \nabla p + \gamma p \nabla \cdot \mathbf{u} = 0. \tag{1.3.14}$$

Appropriate boundary conditions are discussed in Sec. 8.2.

This system is hyperbolic. It admits weak or discontinuous solutions (Courant and Friedrichs (1948)). The boundary conditions at solid walls are no-flux,

$$\mathbf{u} \cdot \hat{\mathbf{n}} = 0.$$

The conditions at inflow or outflow boundaries are dependent upon the characteristic directions there. The flow is called subsonic if $|\mathbf{u}| < c$ and supersonic if $|\mathbf{u}| > c$. The Mach number M is given by

$$M = |\mathbf{u}|/c.$$

In order for the flow to change from a supersonic state to a subsonic one, it must (except in rare circumstances) undergo a shock wave. Here the flow variables are discontinuous and the differential equations themselves do not apply, although the more basic integral conservation laws still hold.

Even in problems which are not strictly inviscid, the viscous effects are generally confined to thin layers adjacent to the boundary. In these circumstances the Euler equations are useful for describing the gross features of the flow.

1.3.3. Compressible Potential

In many practical applications, the flow may be assumed to be irrotational, even in the presence of weak shocks. In this case, the velocity is derivable from a potential, i.e.,

$$\mathbf{u} = \nabla \phi, \tag{1.3.15}$$

and (1.3.12) becomes

$$\frac{\partial \rho}{\partial t} + \nabla \cdot (\rho \nabla \phi) = 0, \tag{1.3.16}$$

where ρ is related to $\nabla \phi$ by the isentropic relation

$$\frac{\rho_0}{\rho} = \left[1 + \frac{\gamma - 1}{2} \frac{|\nabla \phi|^2}{c^2}\right]^{1/(\gamma-1)}, \tag{1.3.17}$$

where ρ_0 is the stagnation density. The momentum equation (1.3.13) is automatically satisfied.

1.3.4. Incompressible Flow

There are essentially four different formulations of the incompressible Navier–Stokes equations—primitive variable (velocity and pressure), streamfunction-vorticity, streamfunction and velocity-vorticity formulations. The primitive-variable formulation can be found in any text on fluid dynamics (e.g., Batchelor (1967)). It has been the formulation most extensively employed in three-dimensional spectral calculations.

The incompressible Navier–Stokes equations on a domain Ω are usually written as

$$\frac{\partial \mathbf{u}}{\partial t} + \mathbf{u}\cdot\nabla\mathbf{u} = -\nabla p + \nu\Delta\mathbf{u} \tag{1.3.18}$$

$$\nabla\cdot\mathbf{u} = 0, \tag{1.3.19}$$

where $\mathbf{u}$ is the velocity vector, p the pressure, and $\nu = \mu/\rho$ the kinematic viscosity, ρ is constant (and taken to be unity), and Δ is the Laplacian, sometimes denoted by ∇^2. Equation (1.3.18) is the momentum equation and (1.3.19) is the continuity constraint.

The pressure in the incompressible Navier–Stokes equations is not a thermodynamic variable satisfying an equation-of-state, but it is an implicit dynamic variable which adjusts itself instantaneously in a time-dependent flow to satisfy the incompressibility or divergence-free condition. From the mathematical point of view, it may be considered as a Lagrange multiplier that ensures the kinematical constraint of incompressibility (i.e., solenoidity of the velocity field). Notice that no initial or boundary conditions are required for the pressure. Various results on the existence and uniqueness of solutions to the Navier–Stokes equations are furnished in the treatises on the mathematical analysis of these equations (Lions (1969), Ladyzhenskaya (1969), Temam (1977)).

One of the most interesting characteristics of a flow is its vorticity. This is denoted by ω and is given by

$$\boldsymbol{\omega} = \nabla \times \mathbf{u}. \tag{1.3.20}$$

It represents (half) the local rotation rate of the fluid. A dynamical equation for the vorticity is derived by taking the curl of (1.3.18)

$$\frac{\partial \boldsymbol{\omega}}{\partial t} + \mathbf{u}\cdot\nabla\boldsymbol{\omega} = \boldsymbol{\omega}\cdot\nabla\mathbf{u} + \nu\Delta\boldsymbol{\omega}. \tag{1.3.21}$$

This is an advection-diffusion equation with the addition of the term $\boldsymbol{\omega}\cdot\nabla\mathbf{u}$, which represents the effects of vortex stretching. This term is identically zero for two-dimensional flows. It is responsible for many of the interesting aspects of three-dimensional flows.

The vorticity can be combined with the streamfunction ψ to yield a concise description of two-dimensional flows. The streamfunction is related to the velocity by

$$u = \frac{\partial \psi}{\partial y}$$
$$v = -\frac{\partial \psi}{\partial x} \tag{1.3.22}$$

and to the vorticity by

$$\Delta \psi = -\omega, \tag{1.3.23}$$

where $\mathbf{u} = (u, v, 0) = \nabla \times (\psi \hat{\mathbf{k}})$ and $\boldsymbol{\omega} = (0, 0, \omega)$. Equation (1.3.18) reduces to

$$\frac{\partial \omega}{\partial t} + \frac{\partial \psi}{\partial y}\frac{\partial \omega}{\partial x} - \frac{\partial \psi}{\partial x}\frac{\partial \omega}{\partial y} = \nu \Delta \omega. \tag{1.3.24}$$

The flow is parallel to curves of constant ψ—the streamlines. The boundary conditions that accompany (1.3.22)—(1.3.24) are

$$\psi = 0$$
$$\frac{\partial \psi}{\partial n} = 0. \tag{1.3.25}$$

Equations (1.3.22)—(1.3.25) provide a complete description of a two-dimensional incompressible flow. Note that the pressure is not needed. The subtlety of the streamfunction-vorticity formulation is that there are no physical boundary conditions for the vorticity.

The elimination of the vorticity leads to the pure streamfunction formulation

$$\frac{\partial}{\partial t}(\Delta \psi) + \frac{\partial \psi}{\partial y}\frac{\partial}{\partial x}(\Delta \psi) - \frac{\partial \psi}{\partial x}\frac{\partial}{\partial y}(\Delta \psi) = \nu \Delta^2 \psi. \tag{1.3.26}$$

The extension of these approaches to three-dimensional flows requires the introduction of a second streamfunction. The appropriate equations are given by Murdock (1986).

1.3.5. Boundary Layer

Prandtl (1904) introduced the concept that the effects of viscosity are often confined to a thin layer—the boundary layer—adjacent to solid walls. One of the simplest boundary layers is the one sketched in Fig. 1.2 for two-dimensional flow over a flat plate. The boundary-layer thickness δ is taken to

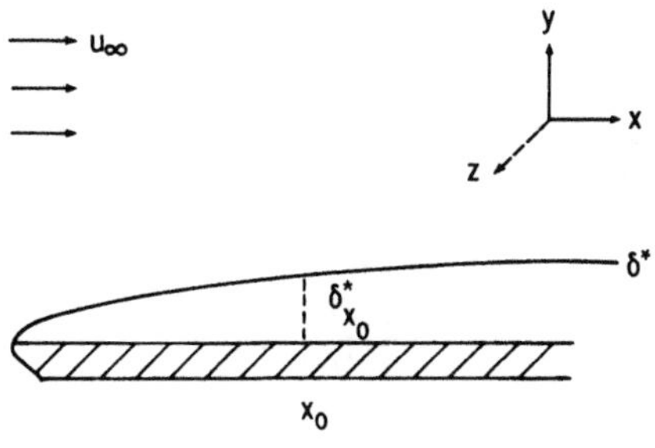

Figure 1.2. Flow over a flat plate. The x, y, and z directions are called the streamwise, normal, and spanwise directions, respectively.

be the height y at which the streamwise velocity u reaches 99% of the freestream velocity u_∞. The displacement thickness δ^* measures the deflection of the incident streamlines in the direction normal to the wall. It is typically three times the boundary-layer thickness δ. Within the boundary layer u/u_∞ is order 1 and v/u_∞ is of order δ, where derivatives with respect to y are of order x_0/δ larger than those with respect to x. Moreover, the pressure variation between $y = 0$ and $y = \delta$ is very small, and p is assumed to be independent of y (and known as a function of x). The lowest order terms from (1.3.18) and (1.3.19) are

$$u\frac{\partial u}{\partial x} + v\frac{\partial u}{\partial y} = -\frac{\partial p}{\partial x} + \nu\frac{\partial^2 u}{\partial y^2} \tag{1.3.27}$$

$$\frac{\partial u}{\partial x} + \frac{\partial v}{\partial y} = 0. \tag{1.3.28}$$

The boundary conditions are

$$\begin{aligned} u = v &= 0 \qquad \text{at } y = 0 \\ u &= u_\infty \qquad \text{at } y = \infty, \end{aligned} \tag{1.3.29}$$

plus appropriate inflow conditions at some $x = x_0$ and the prescribed pressure gradient.

1.4. Spectral Accuracy for a Two-Dimensional Fluid Calculation

A problem of long-standing interest is incompressible flow in a straight channel, commonly referred to as plane Poiseuille flow (Fig. 1.3). Let the upper and lower boundaries in the direction normal to the walls be given by $y = +1$ and $y = -1$, respectively, and let the centerline velocity be unity, i.e., scale distances by h and velocities by the mean flow u_0 at $y = 0$.

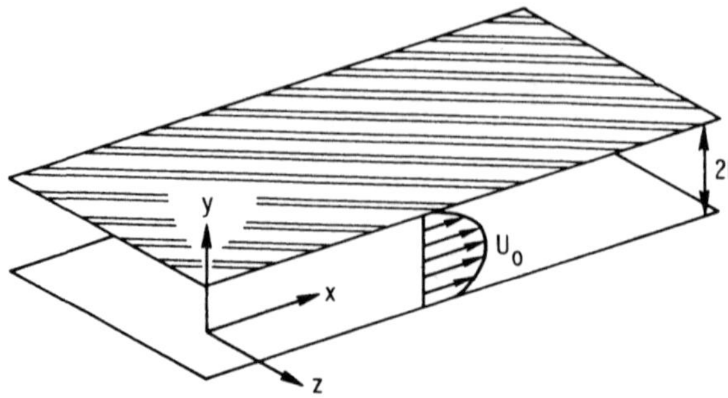

Figure 1.3. Plane channel flow. The x, y, and z directions are called the streamwise, normal, and spanwise directions, respectively.

In two dimensions (1.3.18)—(1.3.19) reduce to

$$\frac{\partial u}{\partial t} + u\frac{\partial u}{\partial x} + v\frac{\partial u}{\partial y} = -\frac{\partial p}{\partial x} + \nu\left(\frac{\partial^2 u}{\partial x^2} + \frac{\partial^2 u}{\partial y^2}\right) \tag{1.4.1a}$$

$$\frac{\partial v}{\partial t} + u\frac{\partial v}{\partial x} + v\frac{\partial v}{\partial y} = -\frac{\partial p}{\partial y} + \nu\left(\frac{\partial^2 v}{\partial x^2} + \frac{\partial^2 v}{\partial y^2}\right) \tag{1.4.1b}$$

$$\frac{\partial u}{\partial x} + \frac{\partial v}{\partial y} = 0. \tag{1.4.1c}$$

The boundary conditions at these impermeable and no-slip walls are

$$\begin{aligned} u(x, +1, t) &= v(x, +1, t) = 0 \\ u(x, -1, t) &= v(x, -1, t) = 0. \end{aligned} \tag{1.4.2}$$

There is a class of problems in which the relevant boundary conditions in the x (streamwise) direction are periodic. The selection of the streamwise period and the initial conditions for velocity and pressure complete the specification of a particular problem.

An equilibrium (time-independent) solution is

$$\begin{aligned} u(x, y, t) &= 1 - y^2 \\ v(x, y, t) &= 0 \\ p(x, y, t) &= -2\nu x. \end{aligned} \tag{1.4.3}$$

The evolution of small perturbations from this state has been studied extensively by analytical means. It has been customary to focus on a particular wavenumber α in the streamwise direction and to look for solutions of the form

$$u(x, y, t) = (1 - y^2) + \varepsilon\,\mathrm{Re}\{\phi(y)e^{i\alpha x - i\omega t}\}. \tag{1.4.4}$$

The function $\phi(y)$ and the temporal frequency ω come from solutions to the Orr–Sommerfeld eigenvalue problem (Sec. 6.4). One would expect the numeri-

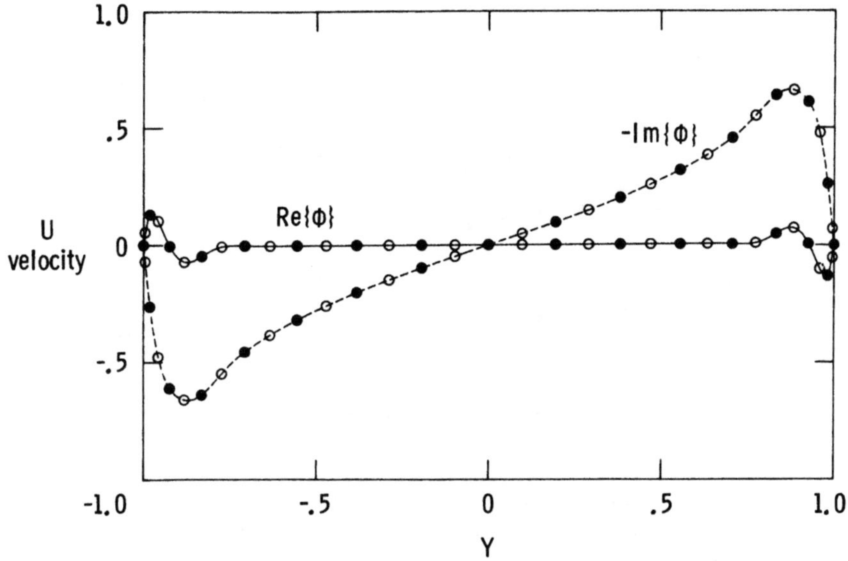

Figure 1.4. Streamwise velocity eigenfunction of an unstable mode in plane channel flow. The circles denote the Chebyshev collocation points for $N_y = 32$. The filled circles are the points for $N_y = 16$.

cal solutions to (1.4.1), subject to initial conditions taken from (1.4.4) at $t = 0$ for small amplitude ε, to be approximated closely by (1.4.4).

This is the sort of problem that will be used as a more elaborate illustration of the power of spectral methods. The particular problem chosen for study has $\nu = (7500)^{-1}$ and $\alpha = 1$; the eigenfrequency of the only growing mode is $\omega = \omega_r + i\omega_i$, where $\omega_r = .24989154$ and $\omega_i = .00223497$. The eigenfunction ϕ for this mode is shown in Fig. 1.4. The amplitude parameter $\varepsilon = 0.0001$. The natural choice of the streamwise periodicity length is $L_x = 2\pi/\alpha = 2\pi$. Three discretizations in x were used: (1) Fourier collocation (FS), (2) second-order finite-difference (FD2) and (3) fourth-order finite-difference (FD4). In y the options were: (1) Chebyshev collocation (CB) and (2) second-order finite-difference (FD2). All grids were uniform in x and used the Chebyshev collocation distribution in y. The time-stepping was a second-order semi-implicit scheme with the time-step chosen so small that the spatial errors predominated. All runs were made from $t = 0$ to $t = 4\pi/\omega_r$. This is twice the length of time T_0 $(=2\pi/\omega_r)$ that it takes for the perturbation wave to propagate through the streamwise computational domain.

A useful integral property of the solution is the perturbation kinetic energy

$$E = \frac{1}{2}\int_{-1}^{1} dy \int_{0}^{L_x} [(u - (1 - y^2))^2 + v^2]\,dx. \tag{1.4.5}$$

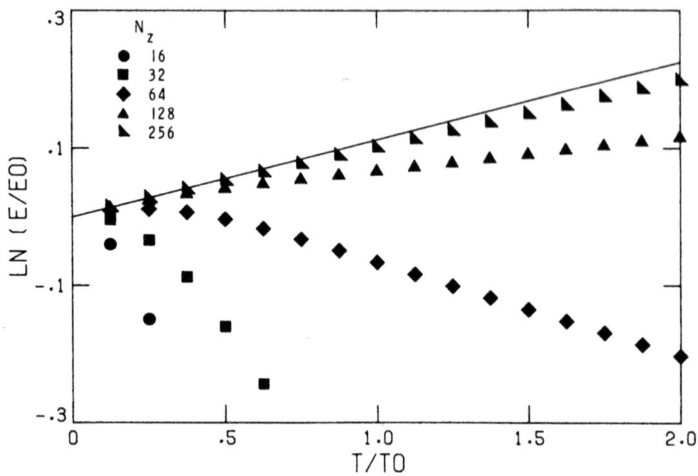

Figure 1.5. Computed perturbation energy for a Fourier spectral method in x and a second-order finite-difference method in y. Results are shown for various grids in y and for a four point grid in x. The solid line is the correct result.

For sufficiently low amplitude waves this ought to exhibit the exponential behavior

$$E(t) = E_0 e^{2\omega_i t}, \tag{1.4.6}$$

where E_0 is the perturbation energy at $t = 0$.

The performance of various discretizations of this problem is illustrated in Figs. 1.5 to 1.8. These diagrams display the time dependence of the logarithm of the computed perturbation energy divided by the initial perturbation energy. In each figure the discretization in one of the coordinate directions is fixed (and extremely accurate), and data points are provided for several grids in the other direction. The linear theory result is shown for comparison as a straight line.

Figures 1.5 and 1.6 contrast the finite-difference and Chebyshev discretizations in the normal direction. Figure 1.4 indicates that a thirty two point grid resolves the key features of the eigenfunction, whereas a sixteen point grid has inadequate resolution for the real part near $y = \pm 1$. As indicated in Fig. 1.6, the computed fully-spectral results for this unresolved case are exceedingly unreliable, but thirty two Chebyshev polynomials provide a solution which is correct to within graphical accuracy. Even 256 finite-difference points are still a long way from achieving this accuracy level. Table 1.4 provides a numerical comparison of these methods. It lists the error in the logarithm of the energy ratio at $t = 2T_0$, i.e., after two wave periods. The sixty four mode Chebyshev result is so accurate that it is contaminated by nonlinear effects and time-

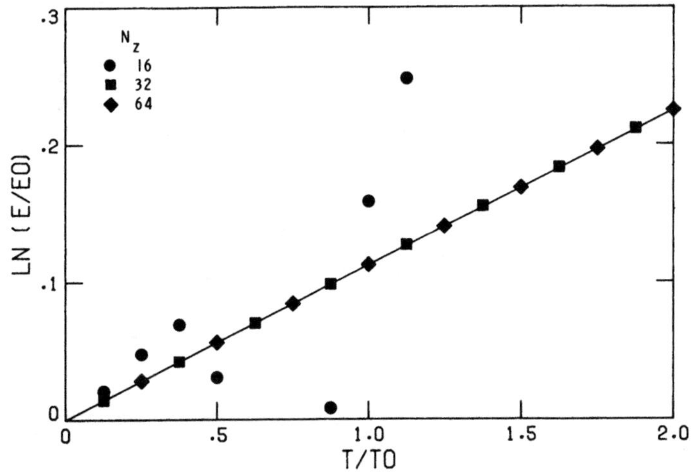

Figure 1.6. Computed perturbation energy for a Fourier spectral method in x and a Chebyshev spectral method in y. Results are shown for various grids in y and for a four point grid in x. The solid line is the correct result.

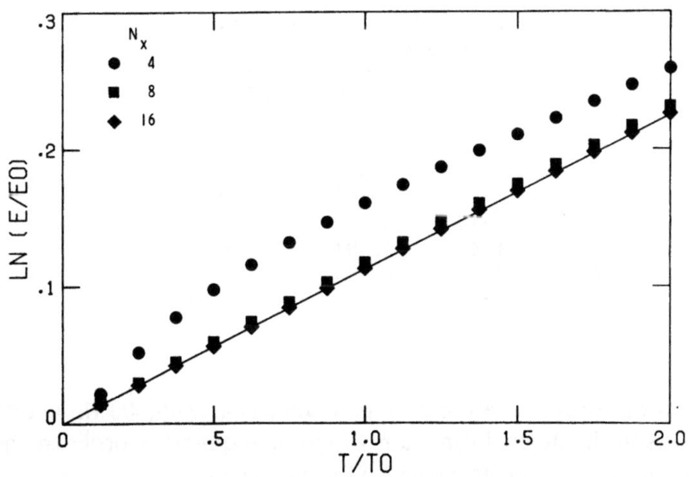

Figure 1.7. Computed perturbation energy for a fourth-order finite-difference method in x and a Chebyshev spectral method in y. Results are shown for three grids in x and for a thirty two point grid in y. The solid line is the correct result.

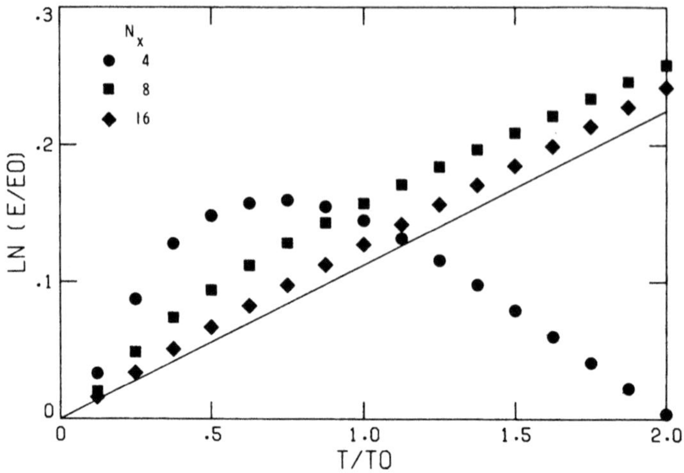

Figure 1.8. Computed perturbation energy for a second-order finite-difference method in x and a Chebyshev spectral method in y. Results are shown for three grids in x and for a thirty two point grid in y. The solid line is the correct result.

Table 1.4. Accuracy of vertical discretization for $N_x = 4$

N_y	FS-FD2 error	FS-CB error
16	−.79322	−1.11895
32	−.98480	.00087
64	−.43564	.00009
128	−.12983	
256	−.03384	

discretization errors. The table suggests that nearly 10,000 finite-difference grid points in the normal direction would be required in order to equal the accuracy of thirty two Chebyshev collocation points.

A similar comparison of the streamwise discretization is provided in Figs. 1.6, 1.7 and 1.8 along with Table 1.5. Since the perturbation is a trigonometric function, four Fourier collocation points in the streamwise direction suffice for essentially perfect resolution. The source of error in the last column of Table 1.5 is the normal discretization (and to a lesser extent, of course, non-linear effects and time-discretization errors). Note that even the fourth-order streamwise discretization is much more demanding than the spectral version.

Table 1.5. Accuracy of horizontal discretization for $N_y = 32$

N_x	FD2-CB error	FD4-CB error	FS-CB error
4	−.24835	.04334	.00087
8	.04246	.00804	.00087
16	.02122	.00135	.00087

1.5. Three-Dimensional Applications in Fluids

The fluid flows illustrated in Figs. 1.2 and 1.3 are characteristic of the simplest class of flows, termed laminar flow, in which the motion is quite regular and predictable, even though possibly unsteady. Many flows fall into the far more complex category of turbulent flow which is described by Hinze (1959) as:

> Turbulent fluid motion is an irregular condition of flow in which the various quantities show a random variation with time and space coordinates, so that statistically distinct average values can be discerned.

Figure 1.9, taken from a calculation of three-dimensional isotropic turbulence by Erlebacher, Hussaini, Speziale and Zang (1987), is typical of the results produced by numerical simulation of turbulence. The resolution requirements for the simulation of wall-bounded turbulent flows are more

Figure 1.9. Typical cross section of a vorticity component in isotropic turbulent flow.

stringent that the requirements for simulating the more tractable problem of homogeneous turbulence (a flow which can be studied with periodic boundary conditions).

Turbulent flows contain a wide range of length scales, bounded above by the geometric dimension of the flow field and bounded below by the dissipative action of the molecular viscosity (see, for instance, Tennekes and Lumley (1972, Chap. 3)). The ratio of the macroscopic (largest) length scale L to the microscopic (smallest) length l (usually known as Kolmogorov scale) is

$$\frac{L}{l} = R^{3/4},$$

where

$$R = \frac{uL}{\nu} \qquad u = (\tfrac{1}{3}\overline{u'^2})^{1/2},$$

u' in the fluctuating velocity, and bar denotes time averaging. To resolve these scales N mesh points would be needed in each direction, where

$$N = c_1 \frac{L}{l}.$$

Thus, for the simulation of homogeneous turbulence if a spectral method could be used, it is sufficient to take $c_1 = 1$; for a fourth-order scheme c_1 could be between 4 to 7 (Kreiss and Oliger (1979), see also Figs. 1.5—1.8 of Sec. 1.4). The ratio of the time scales of the macroscopic and microscopic motions is

$$T/t = \sqrt{R}.$$

Consequently, the number of time-steps required to describe the flow during the characteristic period of the physically significant events is proportional to $\sqrt{R}$. Now, the number of operations required to update the solution per time step is

$$c_2 N^3 \log N + c_3 N^3,$$

where for the fourth-order method $c_2 = 15$, $c_3 = 90$, and for the spectral method, $c_2 = 75$, $c_3 = 100$. Thus, for homogeneous turbulence simulation, the storage requirement is roughly proportional to

$$4c_1^3 R^{9/4},$$

and the total number of operations is proportional to

$$c_1^3 R^{11/4}[c_2 \log_2(c_1 R^{3/4}) + c_3].$$

Assuming a performance of 100 MFLOPS, typical of Cray 1X or CDC Cyber-205 with careful programming, the computer time required for one realization of homogeneous turbulence by a spectral method is three minutes for R equal to 100, and nine hours for R equal to 400. Spectral methods have been singularly successful for this problem since the corresponding requirements for a fourth-order finite-difference method are one or two orders of magnitude larger. Besides its obvious storage advantage, spectral methods for this class of problems typically have a substantial speed advantage as well. Although a finite-difference method is faster than a spectral method which uses the same number of grid points, this advantage is overridden by the necessity to use many more grid points in a finite-difference method to produce a comparably accurate solution. Moreover, Fourier functions arise naturally in the theoretical analysis of homogeneous turbulence, and they are the natural choice of trial functions for spectral methods. Thus, the spectral methods, apart from their computational efficiency, have the added advantage of readily permitting one to monitor and diagnose non-linear interactions which contribute to resonance effects, energy transfer, dissipation and other dynamic features. Furthermore, if there are any symmetries underlying a problem, and symmetry-breaking phenomena are precluded, spectral methods permit unique exploitation of these symmetries.

The focus of theoretical interest in homogeneous turbulence is the experimental result on the existence of a range of scales of motion, called the inertial range, which are not directly affected by the energy maintenance and dissipation mechanisms (Mestayer et al. (1976)) and possess an energy spectrum exhibiting a scaling behavior (Grant, Stewart, and Moillet (1962)):

$$E(k, t) = k^{-m}$$

where k is the magnitude of the wavenumber vector and m is close to 5/3. The spectrum with $m = 5/3$ is the famous Kolmogorov spectrum. The huge Reynolds numbers required to produce an extended inertial range are experimentally accessible only in geophysical flows such as planetary boundary layers and tidal channels.

The pioneering work on homogeneous isotropic turbulence by Orszag and Patterson (1972b) has, as algorithm refinements and increased computational power developed, been followed by fruitful investigations of intermittency effects (Siggia and Patterson (1978), of constant strain effects (Rogallo (1977, 1981)), magnetic field effects (Schumann (1976)), Pouquet and Patterson (1978)), compression (Wu, Ferziger and Chapman (1985)) and passive scalar transport (Kerr (1985)). The early work was on 32^3 grids for (Taylor microscale) Reynolds numbers of 20–40. Now, calculations are practical on 128^3 grids for Reynolds numbers as large as 80.

Perhaps the most spectacular application to date has been the first numerically computed three-dimensional inertial range (Brachet et al. (1983)). The Reynolds number was 3000 and, of course, crude by experimental standards. This calculation of the Taylor–Green vortex was feasible only because the symmetries of the problem were fully exploitable with the spectral method. Among the salient results of this study is the physical insight gained into the behavior of turbulence at high Reynolds number, including the formation of an inertial range and the geometry of the regions of high vorticity. Various features characteristic of fully developed turbulence which are observed in this study include the key result that the energy spectrum and the dissipation spectrum in this range are proportional to k^{-n} and $k^{-1+\mu}$, respectively, where n lies in the range 1.6 to 2.2, and μ in the range 0.3 to 0.7.

The study of Orszag and Pao (1974) is the first instance of the direct simulation of turbulent free shear layers by spectral methods. More recently, Riley and Metcalfe (1980) and Metcalfe et al. (1987) have made significant progress in this area. They use a spectral collocation method with the Fourier series representation in the longitudinal and lateral directions, and sine and cosine series representation in the transverse direction. Figure 1.10 compares a typical experimental flow field with a typical simulation of the mixing layer. The experimental flow visualization due to Bernal (1981) suggests the existence of mushroom shaped features in the flow. The numerical simulation, which includes simple chemical reactions, exhibits the same type of structure in the concentration of one of the species.

The applications cited above were all for problems with no physical boundaries. Spectral algorithms for problems with solid boundaries are more subtle, largely because a pure Fourier method is no longer appropriate. It was not until the late 1970s that reliable Fourier–Chebyshev algorithms were applied to the simplest wall-bounded flows. The principal advantage of such spectral methods over finite-difference methods is their minimal phase errors (Sec. 1.2.1). This is especially important in numerical simulations of instability and transition to turbulence, because such simulations must follow the evolution and non-linear interaction of waves through several characteristic periods. Since phase errors are cumulative, a method which admits phase errors of even a few percent per period is unacceptable.

Orszag and Kells (1980), Kleiser and Schumann (1984), Orszag and Patera (1983), Herbért (1983a) and Metcalfe et al. (1987) have used spectral methods to isolate and study in detail a prototype of three-dimensional instability leading to transition. Similar studies for simple boundary layers have been carried out by Zang and Hussaini (1985b) and by Spalart and Yang (1987). The development of strong detached shear layers, i.e., regions which are at some remove from a wall and which have large values of $\partial u/\partial y$, appears to be a crucial aspect of the transition phenomenon. Figure 1.11, taken from the work of Zang and Hussaini (1985b), presents a comparison of the most intense cross-section of a shear layer computed by spectral methods with one observed experimentally by Kovasznay, Komoda and Vasudeva (1962). The two results are in close correspondence, even on the quantitative level. (The relative displacement in x is due to choice of origin.)

Herbert (1983b, 1984) has unravelled various subharmonic mechanisms of instability in wall-bounded shear flows with the use of spectral methods. Marcus (1984a, 1984b), and Marcus and Tuckermann (1987a, 1987b) have

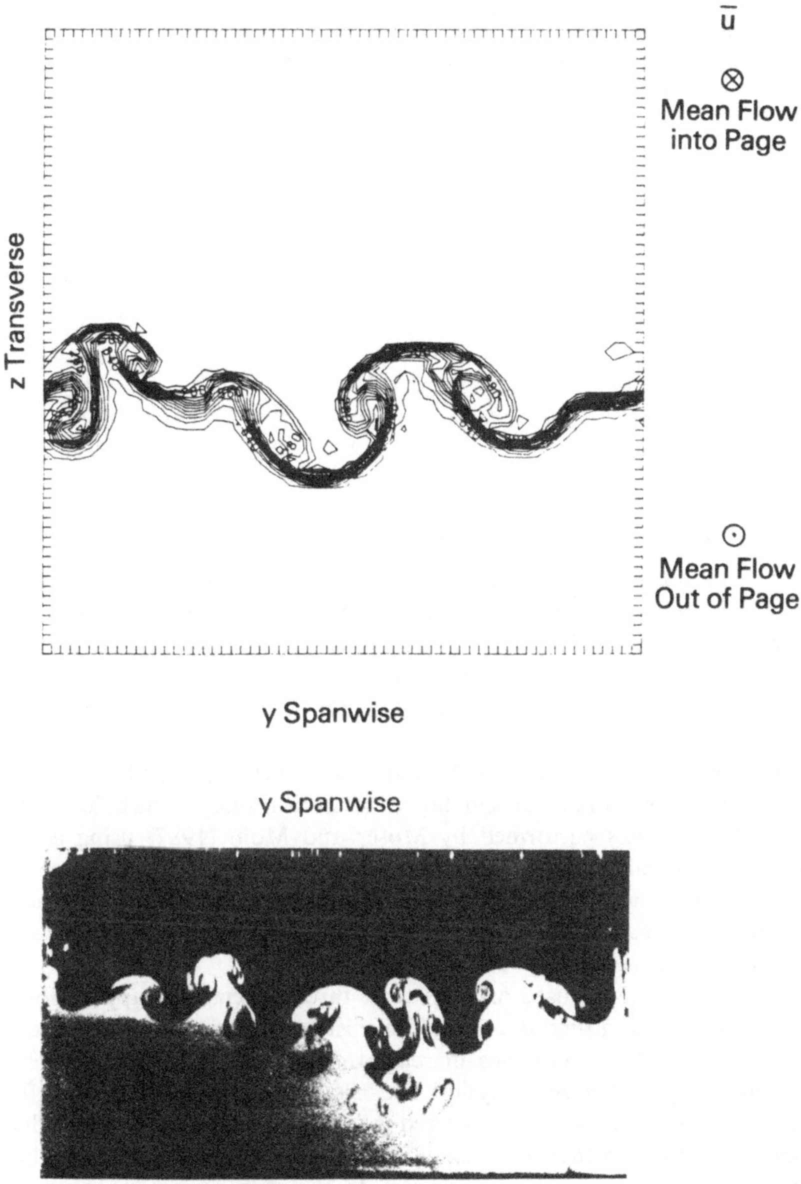

Figure 1.10. Comparison of experimental (bottom) and computed (top) vortex structures in a plane mixing layer. (Courtesy of R. Metcalfe and L. Bernal)

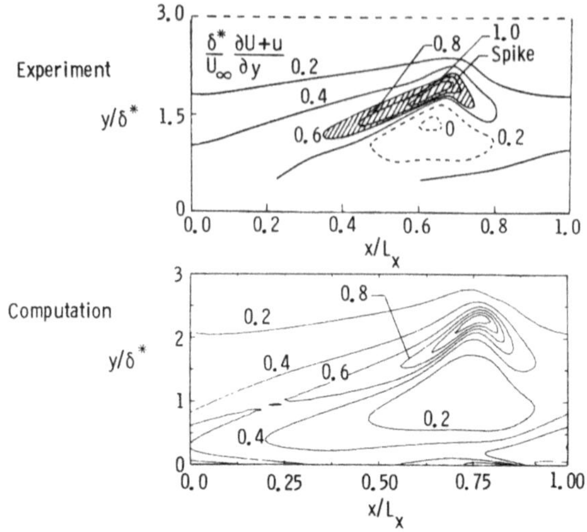

Figure 1.11. Comparison of experimental (top) and computed (bottom) detached shear layer in boundary layer transition.

used spectral methods to study the transition process of the Taylor–Couette flow in the cylindrical and spherical geometries respectively. A large amount of research exists on the Rayleigh–Benard convection problem based on the application of spectral methods. Some of the notable studies are those of Siggia and Zippelius (1981), McLaughlin and Orszag (1982) and Curry et al. (1983).

A few brief results on turbulent channel flows were reported by Orszag and Patera (1983). A more thorough, higher resolution study of turbulent curved channel flows was performed by Moser and Moin (1987) using a novel weighted residual technique. Kim (1985) has used the data from this simulation to produce a numerical flow field visualization of the vortical structure near the wall. Spalart and Leonard (1985) have performed a systematic study of turbulence in a simple boundary layer.

This section has focused on large scale three-dimensional applications of spectral methods in fluid dynamics. This list is by no means exhaustive and certainly neglects applications in related disciplines such as meteorology, oceanography and plasma physics. The calculations, dating from the 1970s, were so impressive that they triggered many other applications and induced some of the more recent sophisticated numerical analyses of spectral methods. The numerical analysis has progressed to the point where rigorous error estimates are available for some Navier–Stokes algorithms. These results will be reviewed in Sec. 11.3.

Algorithms for the Navier–Stokes equations will be described in detail in Chap. 7. Other applications will be reviewed in Chaps. 6 and 8. The essential elements of the numerical analysis are provided in Chaps. 9–12.

CHAPTER 2

Spectral Approximation

The expansion of a function u in terms of an infinite sequence of orthogonal functions $\{\phi_k\}$, $u = \sum_{k=-\infty}^{\infty} \hat{u}_k \phi_k$, underlies many numerical methods of approximation. The accuracy of the approximations and the efficiency of their implementation influence decisively the domain of applicability of these methods in scientific computations.

The most familiar approximation results are those for periodic functions expanded in Fourier series. The k-th coefficient of the expansion decays faster than any inverse power of k when the function is infinitely smooth and all its derivatives are periodic as well. In practice this decay is not exhibited until there are enough coefficients to represent all the essential structures of the function. The subsequent rapid decay of the coefficients implies that the Fourier series truncated after just a few more terms represents an exceedingly good approximation of the function. This characteristic is usually referred to as "spectral accuracy" of the Fourier method.

The property of spectral accuracy is also attainable for smooth but non-periodic functions provided that the expansion functions are chosen properly. It is not necessarily true that the coefficients of the expansion of a smooth function in terms of any orthogonal smooth basis decay faster than algebraically—usually spectral accuracy is attained only when the function exhibits very special boundary behavior. However, the eigenfunctions of a singular Sturm–Liouville operator allow spectral accuracy in the expansion of any smooth function. No a priori restriction on the boundary behavior is required. Moreover, since the eigenfunctions of the most common singular Sturm–Liouville problems are polynomials, such systems are a natural extension of the Fourier system for the approximation of non-periodic functions.

The expansion in terms of an orthogonal system introduces a linear transformation between u and the sequence of its expansion coefficients $\{\hat{u}_k\}$. This is usually called the *finite transform* of u between physical space and transform space. If the system is complete in a suitable Hilbert space, this transform can be inverted. Hence, functions can be described both through their values in physical space and through their coefficients in transform space.

The expansion coefficients depend on all the values of u in physical space; hence, they can rarely be computed exactly. A finite number of approximate expansion coefficients can be easily computed using the values of u at a finite number of selected points, usually the nodes of high precision quadrature formulas. This procedure defines a *discrete transform* between the set of values of u at the quadrature points and the set of approximate, or *discrete* coefficients. With a proper choice of the quadrature formulas, the finite series defined by the discrete transform is actually the interpolant of u at the quadrature nodes. If the properties of accuracy (in particular the spectral accuracy) are retained in replacing the finite transform with the discrete transform, then the interpolant series can be used instead of the truncated series in approximating functions.

For some of the most common orthogonal systems (Fourier and Chebyshev polynomials) the discrete transform can be computed in a "fast" way, i.e., with an operation count $(5/2)\,N\log_2 N$, where N is the number of polynomials, rather than with the $2N^2$ operations required by a matrix-vector multiplication. Fast discrete transforms for other orthogonal systems have been suggested (Orszag (1986)), but their utility in practical computations is, at present, unproven.

In this chapter we shall describe in detail those orthogonal systems which guarantee spectral accuracy. Some of their approximation properties will be surveyed, and practical indications on how to use the approximating functions will be given. A rigorous description of the approximation properties is postponed to Chap. 9. The description in this chapter is confined to one-dimensional approximation. Multidimensional approximations on a Cartesian domain are constructed by the familiar tensor product approach. The formulas given in the present chapter extend in an obvious way to multidimensional problems. Some specific formulas are given in Sec. 9.7.

The technical definitions of the integrals, Hilbert spaces, and norms used on the analysis of spectral methods are provided in Appendix A. They are referenced within the text by the label of that section in the appendix in which they are discussed.

2.1. The Fourier System

2.1.1. The Continuous Fourier Expansion

The set of functions

$$\phi_k(x) = e^{ikx} \tag{2.1.1}$$

is an orthogonal system over the interval $(0, 2\pi)$:

$$\int_0^{2\pi} \phi_k(x)\,\overline{\phi_l(x)}\,dx = 2\pi\delta_{kl} = \begin{cases} 0 & \text{if } k \neq l \\ 2\pi & \text{if } k = l \end{cases}. \tag{2.1.2}$$

For a complex-valued function u defined on $(0, 2\pi)$, we introduce the *Fourier coefficients* of u:

$$\hat{u}_k = \frac{1}{2\pi}\int_0^{2\pi} u(x)e^{-ikx}\,dx \qquad k = 0, \pm 1, \pm 2, \ldots. \tag{2.1.3}$$

The integrals in (2.1.3) exist if u is Riemann integrable (see (A.8), i.e., Section A.8 of Appendix A), which is ensured, for instance, if u is bounded and piecewise continuous in $(0, 2\pi)$. More generally, the Fourier coefficients are defined for any function which is integrable in the sense of Lebesgue (see (A.9)).

The relation (2.1.3) associates with u a sequence of complex numbers called the *Fourier transform* of u. It is possible as well to introduce a *Fourier cosine transform* and a *Fourier sine transform* of u, respectively, through the formulas

$$a_k = \frac{1}{2\pi}\int_0^{2\pi} u(x)\cos kx\,dx \qquad k = 0, \pm 1, \pm 2, \ldots \tag{2.1.4}$$

and

$$b_k = \frac{1}{2\pi}\int_0^{2\pi} u(x)\sin kx\,dx \qquad k = 0, \pm 1, \pm 2, \ldots. \tag{2.1.5}$$

The three Fourier transforms of u are related by the formula

$\hat{u}_k = a_{|k|} - ib_{|k|}$ for $k = 0, \pm 1, \pm 2, \ldots$. Moreover, if u is a real valued function, a_k and b_k are real numbers and $\hat{u}_{-k} = \overline{\hat{u}_k}$.

The Fourier series of the function u is defined as

$$Su = \sum_{k=-\infty}^{\infty} \hat{u}_k \phi_k. \tag{2.1.6}$$

It represents the formal expansion of u in terms of the Fourier orthogonal system. In order to make this expansion rigorous, one has to cope with three problems:

(i) when and in what sense is the series convergent;
(ii) what is the relation between the series and the function u;
(iii) how rapidly does the series converge.

The basic issue is how u is approximated by the sequence of trigonometric polynomials

$$P_N u(x) = \sum_{k=-N/2}^{N/2-1} \hat{u}_k e^{ikx} \tag{2.1.7}$$

as N tends to ∞. Theoretical discussions of truncated (or finite) Fourier series are customarily given for

$$P_N u(x) = \sum_{k=-N}^{N} \hat{u}_k e^{ikx} \tag{2.1.8}$$

rather than for (2.1.7). We have chosen to use the (mathematically unconventional) form (2.1.7) because it corresponds directly to the way spectral methods are actually programmed. In most cases, the most important characterization of the approximation is the number of degrees of freedom. Equation (2.1.7) corresponds to N degrees of freedom and is preferred by us for this reason. We shall refer to $P_N u$ as the N-th order truncated Fourier series of u.

Points (i), (ii) and (iii) have been subjected to a thorough mathematical investigation. We review here only those basic results relevant to the application of spectral methods to partial differential equations.

We recall the following results about the convergence of the Fourier series. Hereafter, a function u defined in $(0, 2\pi)$ will be called periodic if $u(0^+)$ and $u(2\pi^-)$ exist and are equal.

(a) If u is continuous, periodic, and of bounded variation on $[0, 2\pi]$ (see (A.8)), then Su is uniformly convergent to u, i.e.,

$$\max_{x \in [0, 2\pi]} |u(x) - P_N u(x)| \to 0, \qquad \text{as } N \to \infty.$$

(b) If u is of bounded variation on $[0, 2\pi]$, then $P_N u(x)$ converges pointwise to $(u(x^+) + u(x^-))/2$ for any $x \in [0, 2\pi]$ (here $u(0^-) = u(2\pi^-)$).

(c) If u is continuous and periodic, its Fourier series does not necessarily converge at every point $x \in [0, 2\pi]$.

A full characterization of the functions for which the Fourier series is everywhere pointwise convergent is not known. However, a full characterization is available within the framework of Lebesgue integration for convergence in the mean. The series Su is said to be convergent in the mean (or L^2-convergent) to u if

$$\int_0^{2\pi} |u(x) - P_N u(x)|^2 \, dx \to 0, \qquad \text{as } N \to \infty. \tag{2.1.9}$$

Clearly, the convergence in the mean can be defined for square integrable functions. Integrability can be intended in the Riemann sense, but the most general results require that the integral in (2.1.9) be defined according to Lebesgue. Henceforth, we assume that $u \in L^2(0, 2\pi)$, where $L^2(0, 2\pi)$ is the space of (classes) of the Lebesgue—measurable functions $u\colon (0, 2\pi) \to \mathbb{C}$ such that $|u|^2$ is Lebesgue—integrable over $(0, 2\pi)$ (see (A.9)). $L^2(0, 2\pi)$ is a complex Hilbert space (see (A.1)) with inner product

$$(u, v) = \int_0^{2\pi} u(x)\overline{v(x)} \, dx \tag{2.1.10}$$

and norm

$$\|u\| = \left(\int_0^{2\pi} |u(x)|^2 \, dx \right)^{1/2}. \tag{2.1.11}$$

Let S_N be the space of the trigonometric polynomials of degree $N/2$, defined as

$$S_N = \text{span}\{e^{ikx} | -N/2 \leqslant k \leqslant N/2 - 1\}. \tag{2.1.12}$$

Then by the orthogonality relation (2.1.2) one has

$$(P_N u, v) = (u, v) \qquad \text{for all } v \in S_N. \tag{2.1.13}$$

This shows that $P_N u$ is the orthogonal projection of u upon the space of the trigonometric polynomials of degree $N/2$. Equivalently, $P_N u$ is the closest element to u in S_N with respect to the norm (2.1.11).

Functions in $L^2(0, 2\pi)$ can be characterized in terms of their Fourier coefficients, according to the Riesz theorem, in the following sense. If $u \in L^2(0, 2\pi)$, then its Fourier series converges to u in the sense of (2.1.9), and

$$\|u\|^2 = 2\pi \sum_{k=-\infty}^{\infty} |\hat{u}_k|^2 \qquad \text{(Parseval identity).} \tag{2.1.14}$$

(In particular, the numerical series on the right-hand side is convergent.) Conversely, for any complex sequence $\{c_k\}$, $k = 0, \pm 1, \ldots$ such that $\sum_{k=-\infty}^{\infty} |c_k|^2 < \infty$, there exists a unique function $u \in L^2(0, 2\pi)$ such that its Fourier coefficients are precisely the c_k's for any k. Thus, for any function $u \in L^2(0, 2\pi)$ we can write

$$u = \sum_{k=-\infty}^{\infty} \hat{u}_k \phi_k, \tag{2.1.15}$$

where the equality has to be intended between two functions in $L^2(0, 2\pi)$. The Riesz theorem states that the finite Fourier transform is an isomorphism between $L^2(0, 2\pi)$ and the space l^2 of complex sequences $\{c_k\}$, $k = 0, \pm 1, \pm 2, \ldots$, such that $\sum_{k=-\infty}^{\infty} |c_k|^2 < \infty$.

The L^2-convergence does not imply the pointwise convergence of $P_N u$ to u at all points of $[0, 2\pi]$. However, a nontrivial result by Carleson (1966) asserts that $P_N u(x)$ converges to $u(x)$ as $N \to \infty$ for any x outside a set of zero measure in $[0, 2\pi]$.

We deal now with the problem of the speed of convergence of the Fourier series. Hereafter, we set

$$\sum_{|k| \gtrsim N/2} = \sum_{\substack{k < -N/2 \\ k \geqslant N/2}}$$

First of all, note that by the Parseval identity one has

$$\|u - P_N u\| = \left(2\pi \sum_{|k| \gtrsim N/2} |\hat{u}_k|^2 \right)^{1/2} \tag{2.1.16}$$

On the other hand, if u is sufficiently smooth,

$$\max_{0 \leqslant x \leqslant 2\pi} |u(x) - P_N u(x)| \leqslant \sum_{|k| \gtrsim N/2} |\hat{u}_k|. \tag{2.1.17}$$

This shows that the size of the error created by replacing u with its N-th order truncated Fourier series depends upon how fast the Fourier coefficients of u decay to zero. This in turn depends on the regularity of u in the domain

$(0, 2\pi)$ and on the periodicity properties of u. Indeed, if u is continuously differentiable in $[0, 2\pi]$, then for $k \neq 0$

$$\begin{aligned} 2\pi \hat{u}_k &= \int_0^{2\pi} u(x) e^{-ikx}\, dx \\ &= \frac{-1}{ik}(u(2\pi^-) - u(0^+)) + \frac{1}{ik}\int_0^{2\pi} u'(x) e^{-ikx}\, dx. \end{aligned} \tag{2.1.18}$$

Hence,

$$\hat{u}_k = O(k^{-1}). \tag{2.1.19}$$

If now u' is itself continuously differentiable in $[0, 2\pi]$, the last integral in (2.1.18) is 2π times the k-th Fourier coefficient of u'; hence, it decays like k^{-1}. It follows that $\hat{u}_k = O(k^{-2})$ if and only if $u(2\pi^-) = u(0^+)$. Iterating this argument, one proves that *if u is m-times continuously differentiable in* $[0, 2\pi]$ $(m \geqslant 1)$ *and if $u^{(j)}$ is periodic for all $j \leqslant m - 2$, then*

$$\hat{u}_k = O(k^{-m}) \qquad k = \pm 1, \pm 2, \ldots . \tag{2.1.20}$$

(The symbol $u^{(j)}$ denotes the jth derivative of u.) The same result holds if u is $(m-1)$-times differentiable almost everywhere in $(0, 2\pi)$, with $(m-1)$th derivative of bounded variation in $[0, 2\pi]$, and $u^{(j)}$ is periodic for all $j \leqslant m - 2$. In this case the integral on the right-hand side of (2.1.18) has to be replaced by the Riemann-Stieltjes integral $\int_0^{2\pi} e^{-ikx}\, du(x)$ (see (A.8)).

As a corollary of (2.1.20), we conclude that *the k-th Fourier coefficient of a function which is infinitely differentiable and periodic with all its derivatives on* $[0, 2\pi]$ *decays faster than any negative power of k.*

EXAMPLES.
(1) The function

$$u(x) = \begin{cases} 1 & \dfrac{\pi}{2} < x \leqslant \dfrac{3\pi}{2} \\ 0 & 0 < x \leqslant \dfrac{\pi}{2}, \dfrac{3\pi}{2} < x \leqslant 2\pi \end{cases}$$

is of bounded variation in $[0, 2\pi]$. Its Fourier coefficients are:

$$\hat{u}_k = \begin{cases} \pi & \text{if } k = 0 \\ 0 & \text{if } k \neq 0, \text{ even} \\ \dfrac{(-1)^{(k-1)/2}}{k} & \text{if } k \neq 0, \text{ odd.} \end{cases}$$

The truncated Fourier series for this function are illustrated in Fig. 2.1 (a). The pointwise convergence is linear and the series is not uniformly convergent. A more detailed discussion of the convergence is given in Sec. 2.1.4.

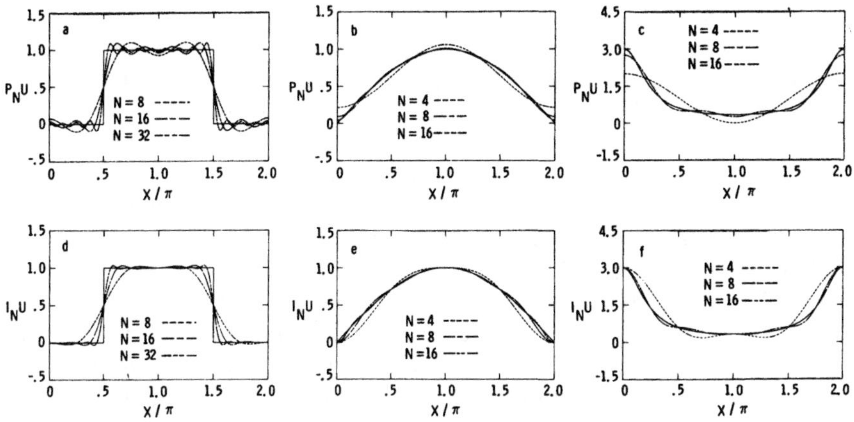

Figure 2.1. Trigonometric approximations to the square wave ((a) and (d)), to $u(x) = \sin(x/2)$ ((b) and (e)) and to $u(x) = 3/(5 - 4\cos x)$ ((c) and (f)). Parts (a), (b), and (c) display truncated Fourier series. Parts (d), (e), and (f) display Fourier interpolating polynomials. The exact function is denoted by the solid curve.

(2) The function $u(x) = \sin(x/2)$ is infinitely differentiable in $[0, 2\pi]$, but $u'(0^+) \neq u'(2\pi^-)$. Its Fourier coefficients are

$$\hat{u}_k = \frac{2}{\pi} \frac{1}{1 - 4k^2}.$$

The truncated series for this function are shown in Fig. 2.1(b). The convergence is quadratic except at the endpoints. Here it is linear and monotonic, which is an obvious consequence of the coefficients decaying quadratically with the same sign.

(3) The function

$$u(x) = \frac{3}{5 - 4\cos x}$$

is infinitely differentiable and periodic with all its derivatives in $[0, 2\pi]$. Its Fourier coefficients are

$$\hat{u}_k = 2^{-|k|} \qquad k = 0, \pm 1, \ldots.$$

Note that u actually is real analytic on the real-axis. This results in the exponential decay of its Fourier coefficients. The resulting rapid convergence is evident in Fig. 2.1(c). Note that the truncated series for $N = 16$ is virtually indistinguishable from the function itself.

We should stress that the asymptotic rate of decay of the Fourier coefficients does not convey the whole story of the error made in a given approximation. If a series has a finite rate of decay, $\hat{u}_k = O(k^{-m})$, then this decay is observed only for $k >$ some k_0. Should the series be truncated below k_0, then the approximation will be quite bad indeed. Even for an infinitely differentiable function there is some minimum acceptable k_0, and truncations below this level yield thoroughly unacceptable approximations.

Estimates (2.1.16) and (2.1.17) show that the error between u and its N-th order truncated Fourier series decays faster than algebraically in $1/N$, when u is infinitely smooth and periodic with all its derivatives. As noted above, this property is commonly termed *spectral accuracy*. It is also known as *exponential convergence* and *infinite-order accuracy*. However, in the analysis of spectral methods for PDEs, one is often interested in estimating global errors like (2.1.16) or (2.1.17) for those functions u having finite regularity. In such cases, using (2.1.20) in (2.1.16) or (2.1.17) will result in a non-optimal rate of convergence of $P_N u$ to u. A different approach is then required, and it will be the subject of Sec. 9.1.2.

2.1.2. The Discrete Fourier Expansion

In many practical applications, numerical methods based upon Fourier series cannot be implemented in precisely the way suggested by the standard treatment of Fourier series that was reviewed in the previous subsection. Some of the difficulties are: the Fourier coefficients of an arbitrary function are not known in closed form and must therefore be approximated in some way, there needs to be an efficient way to recover in physical space the information that is calculated in transform space; and all but the simplest nonlinearities lead to extreme complications. The key to overcoming these difficulties is the use of the discrete Fourier transform and the related discrete Fourier series.

For any integer $N > 0$, consider the set of points

$$x_j = \frac{2\pi j}{N} \qquad j = 0, \ldots, N-1, \tag{2.1.21}$$

referred to as nodes or grid points or knots. The *discrete Fourier coefficients* of a complex-valued function u in $[0, 2\pi]$ with respect to these points are

$$\tilde{u}_k = \frac{1}{N} \sum_{j=0}^{N-1} u(x_j)\, e^{-ikx_j} \qquad -N/2 \leqslant k \leqslant N/2 - 1. \tag{2.1.22}$$

Due to the orthogonality relation

$$\frac{1}{N} \sum_{j=0}^{N-1} e^{ipx_j} = \begin{cases} 1 & \text{if } p = Nm,\ m = 0, \pm 1, \pm 2, \ldots \\ 0 & \text{otherwise,} \end{cases} \tag{2.1.23}$$

we have the inversion formula

$$u(x_j) = \sum_{k=-N/2}^{N/2-1} \tilde{u}_k e^{ikx_j} \qquad j = 0, \ldots, N-1. \tag{2.1.24}$$

Consequently, the polynomial

$$I_N u(x) = \sum_{k=-N/2}^{N/2-1} \tilde{u}_k e^{ikx} \tag{2.1.25}$$

is the $N/2$-degree trigonometric interpolant of u at the nodes (2.1.21), i.e., $I_N u(x_j) = u(x_j)$, $j = 0, \ldots, N-1$. This polynomial is also known as the discrete Fourier series of u. Three examples of such series are provided in Fig. 2.1(d), (e), (f).

The $\tilde{u}_k$'s depend only on the N values of u at the nodes (2.1.21). The discrete Fourier transform (DFT) is the mapping between the N complex numbers $u(x_j)$, $j = 0, \ldots, N-1$ and the N complex numbers $\tilde{u}_k$, $k = -N/2, \ldots, N/2-1$. The two conventional forms for the DFT are given in (2.1.22) and (2.1.24), with the latter sometimes referred to as the inverse DFT. They show that the discrete Fourier transform is an orthogonal transformation in $\mathbb{C}^N$. From a computational point of view, it can be accomplished by the Fast Fourier Transform algorithm (Cooley and Tukey (1965)).

The simplest Fast Fourier Transform (FFT) requires N to be a power of 2. If the data are fully complex it requires $5N \log_2 N - 6N$ real operations, where addition and multiplication are counted as separate operations. In most applications, u is real and $\tilde{u}_{-k} = \bar{\tilde{u}}_k$. In this case the operation count is halved. Fast Fourier Transforms which allow factors of 2, 3, 4, 5, and 6 are widely available (Temperton (1983a)) and offer a 10–20% reduction in the operation count over the basic power of 2 FFT. For simplicity, we shall use just $5N \log_2 N$ as the operation count for a complex FFT. A more complete discussion of FFT's (including several FORTRAN programs) is contained in Appendix B.

Note that the continuous Fourier coefficients of the interpolant are precisely the values computed via the discrete Fourier transform (2.1.22). On the other hand, $\tilde{u}_k$ can be regarded as an approximation to $\hat{u}_k$ using the trapezoidal rule to evaluate the integral in (2.1.3). For infinitely differentiable, periodic functions the trapezoidal rule is the quadrature formula of Lagrange type with maximum precision.

The interpolation operator I_N can be regarded as an orthogonal projection upon the space S_N of the trigonometric polynomials of degree $N/2$, with respect to the discrete approximation of the inner product (2.1.10). Actually, the bilinear form

$$(u, v)_N = \frac{2\pi}{N} \sum_{j=0}^{N-1} u(x_j)\overline{v(x_j)} \tag{2.1.26}$$

coincides with the inner product (2.1.10) if u and v are polynomials of degree $N/2$, due to (2.1.23):

$$(u, v)_N = (u, v) \qquad \text{for all } u, v \in S_N. \tag{2.1.27}$$

As a consequence, (2.1.26) is an inner product on S_N. The interpolant $I_N u$ of a continuous function u satisfies trivially the identity

$$(I_N u, v)_N = (u, v)_N \qquad \text{for all } v \in S_N. \tag{2.1.28}$$

The discrete Fourier coefficients can be expressed also in terms of the exact Fourier coefficients of u. If Su converges to u at every node (2.1.21), then by (2.1.22) one gets

$$\tilde{u}_k = \hat{u}_k + \sum_{\substack{m=-\infty \\ m \neq 0}}^{+\infty} \hat{u}_{k+Nm} \qquad k = -N/2, \ldots, N/2 - 1. \tag{2.1.29}$$

Formula (2.1.29) shows that the k-th mode of the trigonometric interpolant of u depends not only on the k-th mode of u, but also on all the modes of u which "alias" the k-th one on the discrete grid. The $(k + Nm)^{\text{th}}$ frequency aliases the kth frequency on the grid; they are indistinguishable at the nodes since $\phi_{k+Nm}(x_j) = \phi_k(x_j)$. The phenomenon is illustrated in Fig. 2.2. Shown there are three sine waves with frequencies $k = 6$, -2, and -10. Superimposed upon each wave are the eight grid point values of the function. In each case these grid point values coincide with the $k = -2$ wave.

An equivalent formulation of (2.1.29) is

$$I_N u = P_N u + R_N u, \tag{2.1.30}$$

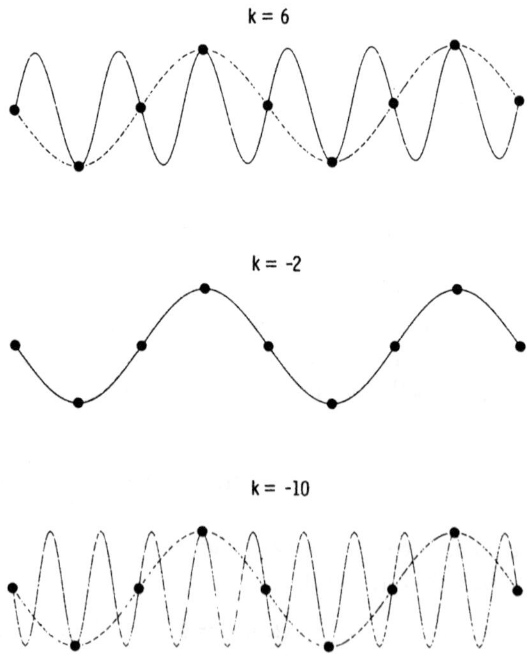

Figure 2.2. Three sine waves which have the same $k = -2$ interpretation on an eight point grid. The nodal values are denoted by the filled circles. The actual sine waves are denoted by the solid curves. Both the $k = 6$ and the $k = -10$ waves are misinterpreted as a $k = -2$ wave (dashed curves) on the coarse grid.

with

$$R_N u = \sum_{k=-N/2}^{N/2-1} \left(\sum_{\substack{m=-\infty \\ m \neq 0}}^{\infty} \hat{u}_{k+Nm} \right) \phi_k. \tag{2.1.31}$$

The error $R_N u$ between the interpolation polynomial and the truncated Fourier series is called "aliasing error." It is orthogonal to the truncation error $u - P_N u$, so that

$$\|u - I_N u\|^2 = \|u - P_N u\|^2 + \|R_N u\|^2. \tag{2.1.32}$$

Hence, the error due to the interpolation is actually always larger than the error due to the truncation of the Fourier series.

In the past years, there has been an abundant literature on the role of aliasing errors in spectral methods. The debate has concerned the influence of these errors on both the stability and the accuracy of the methods. Clever methods have been proposed to remove or control the aliasing effects on spectral calculations (Secs. 3.2, 7.2). At the present time, it has been proven that the influence of aliasing on the accuracy of spectral methods is asymptotically of the same order as the truncation error (Kreiss and Oliger (1979)). Indeed, error estimates (9.1.10) and (9.1.18) show that the truncation and interpolation errors decay at the same rate. This implies similar behavior of the approximation errors for a Galerkin and a collocation scheme. The influence of aliasing on the stability and accuracy of actual spectral solutions of PDEs will be discussed in Secs. 7.2.4 and 7.3.6. Rigorous analyses of aliasing errors in steady Navier–Stokes algorithms are given in Sec. 11.3.

The sequence of interpolating polynomials exhibits convergence properties similar to those of the sequence of truncated Fourier series; furthermore, the continuous and the discrete Fourier coefficients share the same asymptotic behavior. More precisely, when $N \to \infty$, we have:

(a) if u is continuous, periodic and of bounded variation on $[0, 2\pi]$, $I_N u$ converges to u uniformly on $[0, 2\pi]$;
(b) if u is of bounded variation on $[0, 2\pi]$, $I_N u$ is uniformly bounded on $[0, 2\pi]$ and converges pointwise to u at every continuity point for u;
(c) if u is Riemann integrable, $I_N u$ converges to u in the mean.

Concerning the discrete Fourier coefficients, we have:

(d) for any integer $k \neq 0$, and any positive N such that $N/2 > |k|$, let $\tilde{u}_k = \tilde{u}_k^{(N)}$ be the k-th Fourier coefficient of $I_N u$. If u is infinitely smooth and periodic with all its derivatives, formula (2.1.29) shows that $|\tilde{u}_k^{(N)}|$ decays faster than algebraically in k^{-1}, uniformly in N. More generally, if u satisfies the hypotheses for which (2.1.20) holds, the same asymptotic behavior holds for $\tilde{u}_k^{(N)}$, uniformly in N.

2.1.3. Differentiation

The manner in which differentiation is accomplished in a spectral method depends upon whether one is working with a representation of the function in transform space or in physical space. Differentiation in transform space consists of simply multiplying each Fourier coefficient by the imaginary unit times the corresponding wavenumber. If $Su = \sum_{k=-\infty}^{\infty} \hat{u}_k \phi_k$ is the Fourier series of a function u, then

$$Su' = \sum_{k=-\infty}^{\infty} ik\, \hat{u}_k \phi_k \tag{2.1.33}$$

is the Fourier series of the derivative of u. Consequently,

$$(P_N u)' = P_N u', \tag{2.1.34}$$

i.e., truncation and differentiation commute. The series (2.1.33) is L^2-convergent provided that the derivative of u (in the sense of distributions) is a function in $L^2(0, 2\pi)$.

Differentiation in physical space is based upon the values of the function u at the Fourier nodes (2.1.21). These are used in the evaluation of the discrete Fourier coefficients of u according to (2.1.22), these coefficients are multiplied by ik, and the resulting Fourier coefficients are then transformed back to physical space according to (2.1.24). The values $(\mathscr{D}_N u)_l$ of the approximate derivative at the grid points are thus given by

$$(\mathscr{D}_N u)_l = \sum_{k=-N/2}^{N/2-1} a_k e^{2ikl\pi/N} \qquad (l = 0, 1, \ldots, N-1), \tag{2.1.35}$$

where

$$a_k = \frac{ik}{N} \sum_{j=0}^{N-1} u_j e^{-2ikj\pi/N}. \tag{2.1.36}$$

This procedure amounts to computing the grid values of the derivative of the discrete Fourier series of u, i.e.,

$$\mathscr{D}_N u = (I_N u)', \tag{2.1.37}$$

where $I_N u$ is defined in (2.1.25). Since, in general,

$$\mathscr{D}_N u \neq P_N u',$$

the function $\mathscr{D}_N u$ is called the *Fourier collocation derivative* of u, to distinguish it from the true spectral derivative of u, which we refer to as the *Fourier Galerkin derivative*.

Interpolation and differentiation do not commute, i.e.,

$$(I_N u)' \neq I_N(u'), \tag{2.1.38}$$

unless $u \in S_N$. However, we shall prove in Sec. 9.1.3 that the error

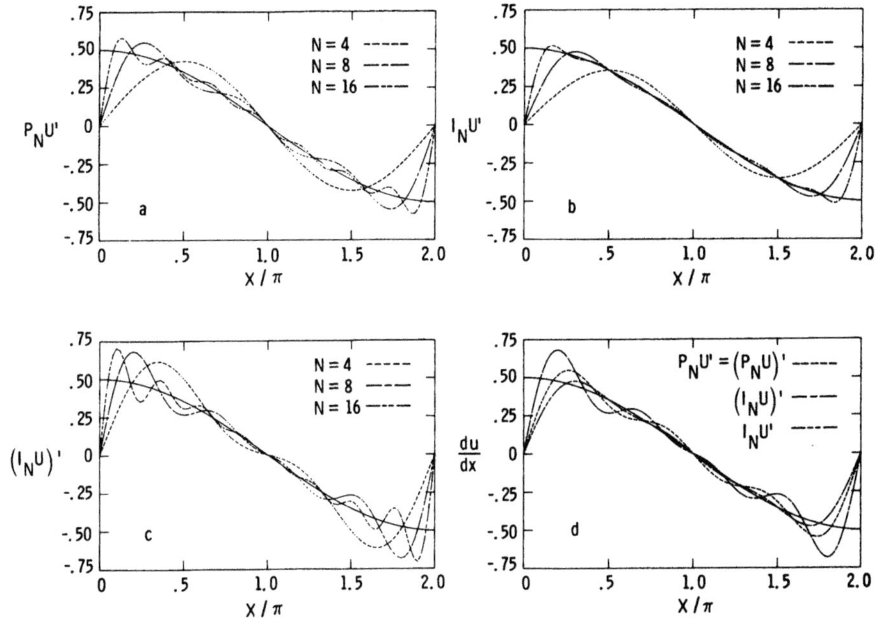

Figure 2.3. Several versions of Fourier differentiation for $u(x) = \sin(x/2)$. The exact result is indicated by the solid curves and the approximate results for $N = 4, 8$ and 16 are indicated by the dashed curves. (a) $P_N u'$ and $(P_N u)'$; (b) $I_N u'$; (c) $(I_N u)'$. Part (d) shows all versions for $N = 8$.

$$(I_N u)' - I_N(u')$$

is of the same order as the truncation error for the derivative

$$u' - P_N u'.$$

It follows that collocation differentiation is spectrally accurate.

These various Fourier differentiation procedures are illustrated in Fig. 2.3 for the function $u(x) = \sin(x/2)$. Part (a) shows both $P_N u'$ and $(P_N u)'$, which are identical. Part (b) displays $I_N u'$ and part (c) shows $(I_N u)'$. The function u' has a discontinuity of the same character as the square wave. The characteristic oscillations arising from a discontinuity, known as the Gibbs phenomenon, will be discussed at length in Sec. 2.1.4. The difference between $(I_N u')$ and $(I_N u)'$ is apparent in parts (b) and (c). Although the truncation errors of both have the same asymptotic behavior, in this example at least, the constant is much larger for $(I_N u)'$.

If $u \in S_N$ then $\mathscr{D}_N u = u'$. Thus, due to (2.1.27), $\mathscr{D}_N$ is a skew-symmetric operator on S_N:

$$(\mathscr{D}_N u, v)_N = -(u, \mathscr{D}_N v)_N \qquad \text{for all } u, v \in S_N. \tag{2.1.39}$$

From a computational point of view, the Fourier collocation derivative can be evaluated according to (2.1.36) and (2.1.35). These require N multiplications and two discrete Fourier transforms. The total operation count is $(5\log_2 N - 5)N$ real multiplications or additions, provided that the discrete Fourier transforms are accomplished by an FFT which takes advantage of the reality of u.

Fourier collocation differentiation can be represented by a matrix. Equations (2.1.35) and (2.1.36) can be combined to yield

$$(\mathscr{D}_N u)_l = \sum_{j=0}^{N-1} (D_N)_{lj} u_j, \tag{2.1.40}$$

where

$$(D_N)_{lj} = \frac{1}{N} \sum_{k=-N/2}^{N/2-1} ik e^{2ik(l-j)\pi/N}.$$

This sum may be evaluated in closed form:

$$(D_N)_{lj} = \begin{cases} \frac{1}{2}(-1)^{l+j} \cot\left[\frac{(l-j)\pi}{N}\right], & l \neq j \\ 0, & l = j. \end{cases} \tag{2.1.41}$$

The skew-symmetry of this real matrix is evident. Its eigenvalues are ik, $k = -N/2 + 1, \ldots, N/2 - 1$. The eigenvalue 0 has double multiplicity. It corresponds to the eigenfunctions 1 and $\cos Nx/2$. Note that central difference operators for the first derivative also have a double zero eigenvalue. Peyret (1986) has given an explicit expression for the second derivative matrix.

If a Fourier collocation method is based on an odd number of points rather than an even number, then the derivative matrix has a zero eigenvalue of single multiplicity. This alternative version of the Fourier method uses the collocation points

$$x_j = \frac{2j}{N+1}\pi \qquad j = 0, \ldots, N \tag{2.1.42}$$

and keeps both the $\cos Nx/2$ and $\sin Nx/2$ terms in the discrete real Fourier series. The derivatives of these terms are both nonzero at the collocation points. Most applications use FFTs where N is a multiple of 2. For this reason we have chosen to present Fourier methods for an even number of collocation points.

There is a way to retain the information in the $\cos Nx/2$ mode for a diffusion operator of the form

$$\frac{d}{dx}\left(a(x)\frac{du}{dx}\right).$$

Table 2.1. Time (in msec) for collocation derivatives

	Cyber 855			Cyber 205		
N	Matrix multiply	Fourier transform	Chebyshev transform	Matrix multiply	Fourier transform	Chebyshev transform
8	0.11	0.12	0.18	.0010	.0010	.0018
16	0.38	0.29	0.39	.0032	.0028	.0044
32	1.54	0.65	0.91	.0119	.0058	.0091
64	5.74	1.56	1.92	.0456	.0151	.0215
128	24.02	3.21	4.20	.1782	.0317	.0442
256	93.49	7.87	9.40	.7061	.0775	.1015

The trick is to evaluate du/dx not at $x_j = 2\pi j/N$ but at $x_{j+1/2} = 2\pi(j + \frac{1}{2})/N$, to form the product $a(x_{j+1/2})du/dx|_{j+1/2}$, and to evaluate the final result at x_j. This approach was suggested by Brandt, Fulton and Taylor (1985). They note that it can be implemented by standard FFTs and that it does lead to more accurate solutions.

In principle, it is possible to accomplish Fourier collocation differentiation by simply performing the matrix multiplication implied by (2.1.40) rather than resorting to Fourier transforms. This requires $2N^2$ operations. This operation count is lower than the operation count for transforms for $N \leqslant 8$. In practice, the exact crossover point will depend on the details of implementation on a given computer. This is especially true for vector computers.

Table 2.1 lists some timings for both methods obtained on a scalar machine (CDC Cyber 855) and a vector machine (CDC Cyber 205 with two pipes). All routines were coded in FORTRAN and the FFTs did not take advantage of the extra 10–15% efficiency available from using radix 4 rather than just radix 2 (see Appendix B). The times cited in the table for the Cyber 205 are the time per derivative for codes which compute a large number of derivatives in pipeline fashion. The transform method is at least twice as fast for $N \geqslant 32$. For $N \geqslant 256$ it is at least an order of magnitude faster. The transform method has the additional advantage of lesser contamination by round-off error.

2.1.4. The Gibbs Phenomenon

The Gibbs phenomenon describes the characteristic oscillatory behavior of the truncated Fourier series or the discrete Fourier series of a function of bounded variation in the neighborhood of a point of discontinuity. Figures 2.1(a), (b) and (c) furnish an interesting contrast. Each truncated Fourier series exhibits some oscillations about the exact function. However, the oscillations for the square wave example have some distinguishing features. The maximum amplitude of the oscillation nearest the discontinuity (the overshoot) tends to a finite limit and the location of the overshoot tends toward the point of discontinuity as the number of retained frequencies is increased. The truncated series for the other two examples are uniformly convergent over $[0, 2\pi]$. They do not exhibit a finite limiting overshoot.

The behavior represented in Fig. 2.1(a) can be easily explained in terms of the singular integral representation of a truncated Fourier series. We assume here that the truncation is symmetric with respect to N, i.e., we set

$$P_N u = \sum_{|k| \leq N/2} \hat{u}_k e^{ikx}. \tag{2.1.43}$$

By (2.1.3) we have

$$\begin{aligned} P_N u(x) &= \sum_{|k| \leq N/2} \frac{1}{2\pi} \int_0^{2\pi} u(y) e^{-iky}\, dy\, e^{ikx} \\ &= \frac{1}{2\pi} \int_0^{2\pi} \left[\sum_{|k| \leq N/2} e^{ik(x-y)} \right] u(y)\, dy. \end{aligned}$$

The integral representation of $P_N u$ is therefore

$$P_N u(x) = \frac{1}{2\pi} \int_0^{2\pi} D_N(x-y) u(y)\, dy, \tag{2.1.44}$$

where $D_N(\xi)$ is the Dirichlet kernel (where in keeping with our notational convention for this part of the book we use D_N for what is classically denoted by $D_{N/2}$)

$$\begin{aligned} D_N(\xi) &= 1 + 2 \sum_{k=1}^{N/2} \cos k\xi \\ &= \begin{cases} \dfrac{\sin((N+1)\xi/2)}{\sin(\xi/2)} & \xi \neq 2j\pi \\ N+1 & \xi = 2j\pi. \end{cases} \qquad j \in \mathbb{Z} \end{aligned} \tag{2.1.45}$$

It is illustrated in Fig. 2.4, where it is shown, for esthetic reasons, on the interval $[-\pi, \pi]$. The Dirichlet kernel can be considered as the orthogonal projection of the delta function upon the space of trigonometric polynomials of degree $N/2$, in the L^2-inner product. D_N is an even function which changes sign at the points $\xi_j = 2j\pi/(N+1)$ and which satisfies

$$\frac{1}{2\pi} \int_0^{2\pi} D_N(\xi)\, d\xi = 1, \tag{2.1.46}$$

as can be seen by making $u = 1$ in (2.1.44). Moreover, as $N \to \infty$, D_N tends to zero uniformly on every closed interval excluding the singular points $\xi = 2j\pi, j \in \mathbb{Z}$. This means that for all $\delta > 0$ and all $\varepsilon > 0$ there exists an integer $N(\delta, \varepsilon) > 0$ such that

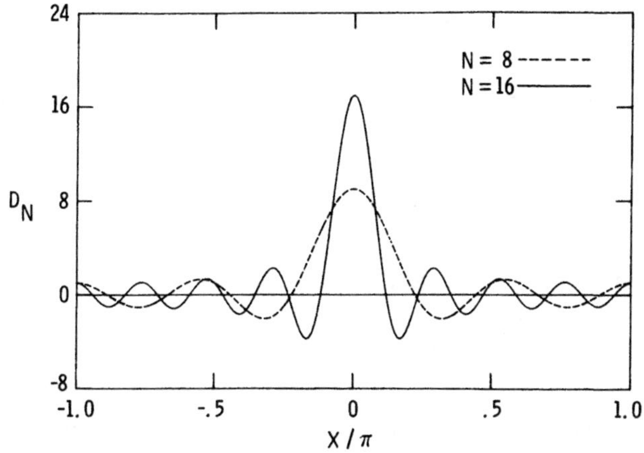

Figure 2.4. The Dirichlet kernel for $N = 8$ and $N = 16$.

$$|D_N(\xi)| < \varepsilon \quad \text{if } N > N(\varepsilon, \delta) \text{ and } \delta \leqslant \xi \leqslant 2\pi - \delta. \tag{2.1.47}$$

We return now to the square wave represented in Fig. 2.1(a). For simplicity we shift the origin to the point of discontinuity, i.e., we consider the periodic function

$$\phi(x) = \begin{cases} 1 & 0 \leqslant x < \pi \\ 0 & \pi \leqslant x < 2\pi. \end{cases} \tag{2.1.48}$$

Its truncated Fourier series is

$$\begin{aligned} P_N\phi(x) &= \frac{1}{2\pi} \int_{x-\pi}^{x} D_N(y)\,dy \\ &= \frac{1}{2\pi} \left[\int_0^x D_N(y)\,dy + \int_{-\pi}^{0} D_N(y)\,dy + \int_{x-\pi}^{-\pi} D_N(y)\,dy \right]. \end{aligned} \tag{2.1.49}$$

As soon as x is not close to π, the last integral on the right-hand side is arbitrarily small, provided N is large enough, by (2.1.47). The middle integral equals π by (2.1.46); hence,

$$P_N\phi(x) \simeq \frac{1}{2} + \frac{1}{2\pi} \int_0^x D_N(y)\,dy \qquad \text{as } N \to \infty. \tag{2.1.50}$$

This formula explains the Gibbs phenomenon for the square wave. If $x > 0$ is far enough from 0, then $1/2\pi \int_0^x D_N(y)\,dy \simeq 1/2\pi \int_0^\pi D_N(y)\,dy = 1/2$ by (2.1.46) and (2.1.47); hence $P_N\phi(x)$ is close to 1. But the function $x \to 1/2\pi \int_0^x D_N(y)\,dy$ has alternating maxima and minima at the points where D_N vanishes, $\xi_j = 2j\pi/(N+1)$; this accounts for its oscillatory behavior. The absolute maximum occurs at $\xi_1 = 2\pi/(N+1)$, where for large enough N

$$\frac{1}{2\pi}\int_0^{2\pi/(N+1)} D_N(y)\,dy \simeq \frac{1}{\pi}\int_0^{\pi}\frac{\sin t}{t} = 0.58949\ldots. \tag{2.1.51}$$

Thus, the sequence $(P_N\phi)(2\pi/(N+1))$ tends to $1.08949\ldots > 1 = \phi(0^+)$, as $N \to \infty$. Equivalently

$$\limsup_{\substack{N\to\infty\\ x\to 0^+}} (P_N\phi)(x) > \phi(0^+). \tag{2.1.52}$$

Similarly, for x negative one has

$$\liminf_{\substack{N\to\infty\\ x\to 0^-}} (P_N\phi)(x) < \phi(0^-).$$

This is a mathematical characterization of the Gibbs phenomenon.

If now $u = u(x)$ is any function of bounded variation in $[0, 2\pi]$ which has at $x = x_0$ an isolated jump discontinuity, we can write

$$u(x) = \tilde{u}(x) + j(u; x_0)\phi(x - x_0),$$

where $j(u; x_0) = u(x_0^+) - u(x_0^-)$ is the jump of u at x_0. The function $\tilde{u}(x) = u(x) - j(u; x_0)\phi(x - x_0)$ has at most a removable singularity at $x = x_0$. Hence, its Fourier series converges uniformly in a neighborhood of x_0. Thus, by (2.1.50)

$$\begin{aligned} P_N u(x) &\simeq \frac{1}{2}[u(x_0^+) + u(x_0^-)] \\ &\quad + \frac{1}{2\pi}[u(x_0^+) - u(x_0^-)]\int_0^{x-x_0} D_N(y)\,dy \quad \text{as } N \to \infty. \end{aligned} \tag{2.1.53}$$

This shows that the sequence $\{P_N u\}$ undergoes a Gibbs phenomenon at $x = x_0$ with the same structure as the Gibbs phenomenon for the square wave (2.1.48).

From a mathematical point of view it is worthwhile to observe that truncation does not preserve the boundedness of the total variation of a function. This means that even if the total variation of u is finite, the total variation of $P_N u$ is not bounded independently of N. For the square wave (2.1.48), formula (2.1.50) shows that the total variation $V_N(\phi; a)$ of $P_N\phi$ in the neighborhood $[-a, a]$ of the origin is approximately

$$V_N(\phi; a) \simeq \frac{1}{\pi}\int_0^a |D_N(y)|\,dy.$$

Since $D_N(y) = \sin(\frac{1}{2}(N+1)y)/y$ for y close to 0, and $\int_0^{+\infty} |\sin t/t|\,dt = \infty$, $V_N(\phi; a)$ diverges as $N \to \infty$.

The Gibbs phenomenon influences the behavior of the truncated Fourier series not only in the neighborhood of the point of singularity, but over the entire interval $[0, 2\pi]$. The convergence rate of the truncated series is linear in N^{-1} at a given non-singular point. This asymptotic behavior is evident in Fig. 2.1(a) for the square wave. The point $x_0 = \pi/2$ is the farthest from all the singularity points. There one has

$$P_N\phi\left(\frac{\pi}{2}\right) = \frac{1}{2\pi}\int_{-\pi/2}^{\pi/2} D_N(y)\,dy,$$

or

$$1 - P_N\phi\left(\frac{\pi}{2}\right) = \frac{1}{\pi}\int_{\pi/2}^{\pi} D_N(y)\,dy.$$

A primitive of the Dirichlet kernel is $(\int D_N)(x) = x + 2\sum_{k=1}^{N/2} (\sin kx)/k$, whence

$$1 - P_N\phi\left(\frac{\pi}{2}\right) = \frac{2}{\pi}\sum_{p \geqslant N/4} \frac{(-1)^p}{2p+1} \simeq \frac{2}{N} \qquad \text{as } N \to \infty.$$

The Gibbs phenomenon also occurs for the sequence $\{I_N u\}$ of the trigonometric interpolating polynomials of u. If the points

$$x_j = \frac{2j\pi}{N+1}, \qquad j = 0, \ldots, N,$$

already introduced in Sec. 2.1.3, are used in the interpolation process, then the interpolating polynomial has the following discrete integral representation:

$$I_N u(x) = \frac{1}{N+1}\sum_{j=0}^{N} D_N(x - x_j)u(x_j). \tag{2.1.54}$$

Note that $D_N(x - x_j)/(N + 1)$ is the Lagrange polynomial of degree $N/2$ at the nodes (2.1.42), i.e., a trigonometric polynomial of degree $N/2$ such that

$$\frac{1}{N+1}D_N(x_l - x_j) = \delta_{lj}, \qquad 0 \leqslant l, j \leqslant N.$$

The representation (2.1.54) for the discrete Fourier series can be related to the representation (2.1.44) for the truncated Fourier series via the use of the trapezoidal quadrature rule for evaluating the singular integral. This accounts, at least heuristically, for the similarity of the Gibbs phenomenon arising in the truncation and interpolation processes. Figure 2.1(d) shows the Gibbs phenomenon for the sequence of the discrete Fourier series of the square wave. The qualitative behavior is the same as for the truncated series, although quantitatively the oscillations appear here less pronounced. (Compare also Fig. 2.3(b), 2.3(d).)

We have seen so far how the Gibbs phenomenon occurs in the two most common trigonometric approximations of a discontinuous function: truncation and interpolation. The capability of constructing alternative trigonometric approximations which avoid or at least reduce the Gibbs phenomenon near the discontinuity points while producing a faithful representation of the function elsewhere in physical space is desirable both theoretically and practically. To be of any practical use this "smoothing" process ought to employ only that information which is available from a finite approximation to the function, namely a finite number of its Fourier coefficients or else its values at the gridpoints.

Since the Gibbs phenomenon is related to the slow decay of the Fourier coefficients of a discontinuous function (as seen in Sec. 2.1.1), it is natural to use smoothing procedures which attenuate the higher order coefficients. Thus, the oscillations associated with the higher modes in the trigonometric approximant are damped. On the other hand, the intrinsic structure of the coefficients carries information about the discontinuities, and this information should not be wasted. Too strong a smoothing procedure may result in excessively smeared approximations, which are again unfaithful representations of the true function. Therefore, the smoothing method has to be suitably tuned.

Let us now focus on smoothing for truncated Fourier series. A straightforward way to attenuate the higher order Fourier coefficients is to multiply each Fourier coefficient $\hat{u}_k$ by a factor σ_k. Thus, the truncated Fourier series $P_N u$ is replaced by the smoothed series

$$\mathscr{S}_N u = \sum_{k=-N/2}^{N/2} \sigma_k \hat{u}_k e^{ikx}. \tag{2.1.55}$$

Typically, the σ_k are required to be real non-negative numbers such that $\sigma_0 = 1$, $\sigma_k = \sigma_{-k}$ and $\sigma_{|k|}$ is a decreasing function of $|k|$.

The Cesáro sums are a classical way of smoothing the truncated Fourier series. They consist of taking the arithmetic means of the truncated series, i.e.,

$$\mathscr{S}_N u = \frac{1}{N/2+1} \sum_{k=0}^{N/2} P_k u = \sum_{k=-N/2}^{N/2} \left(1 - \frac{|k|}{N/2+1}\right) \hat{u}_k e^{ikx}. \tag{2.1.56}$$

In this case the smoothing factors are $\sigma_k = 1 - |k|/(N/2+1)$; they decay linearly in $|k|$.

Other simple smoothing methods are the Lanczos smoothing and the raised cosine smoothing. The factors which define the Lanczos smoothing are

$$\sigma_k = \frac{\sin 2k\pi/N}{2k\pi/N} \qquad k = -N/2, \ldots, N/2. \tag{2.1.57}$$

These are flat near $k = 0$ and approach 0 linearly as $k \to N/2$. The factors for the raised cosine smoothing are

$$\sigma_k = \frac{1 + \cos \dfrac{2k\pi}{N}}{2} \qquad k = -N/2, \ldots, N/2. \tag{2.1.58}$$

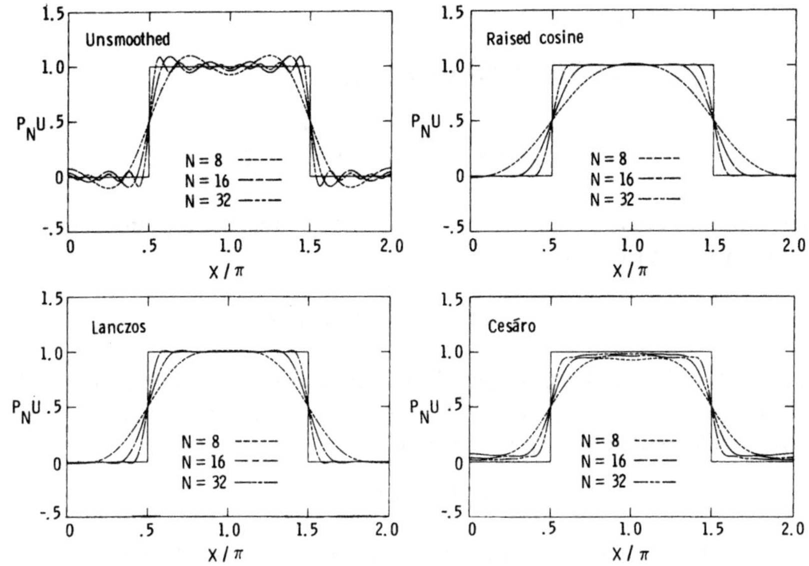

Figure 2.5. Several smoothings for the square wave.

These are flat at $k = N/2$ as well as $k = 0$. The effect of each of these three smoothings upon the square wave is represented in Fig. 2.5.

The smoothed series (2.1.55) can be represented in terms of a singular integral, as

$$\mathcal{S}_N u(x) = \frac{1}{2\pi} \int_0^{2\pi} K_N(x - y) u(y)\, dy, \tag{2.1.59}$$

where the kernel $K_N(\xi)$ is given by

$$K_N(\xi) = 1 + 2 \sum_{k=1}^{N/2} \sigma_k \cos k\xi. \tag{2.1.60}$$

The representation (2.1.59) allows one to describe more general forms of smoothing than (2.1.55). The kernel $K_N(\xi)$ need not have the particular form (2.1.60). The only requirement is that K_N be an approximate polynomial delta function, i.e., a trigonometric polynomial of degree $N/2$ such that

$$\frac{1}{2\pi} \int_0^{2\pi} K_N(\xi)\, d\xi = 1, \tag{2.1.61}$$

and such that for $\delta > 0$ and all $\varepsilon > 0$ there exists an integer $N(\delta, \varepsilon) > 0$ for which

$$|K_N(\xi)| < \varepsilon \quad \text{if } N > N(\delta, \varepsilon) \text{ and } \delta \leqslant \xi \leqslant 2\pi - \delta. \tag{2.1.62}$$

Under these assumptions, one can repeat the arguments used in deriving

(2.1.53) and obtain the asymptotic formula

$$\begin{aligned} \mathscr{S}_N u(x) &\simeq \frac{1}{2}[u(x_0^+) + u(x_0^-)] \\ &\quad + \frac{1}{2\pi}[u(x_0^+) - u(x_0^-)] \int_0^{x-x_0} K_N(y)\, dy, \end{aligned} \tag{2.1.63}$$

near a point of discontinuity for u. Thus, the behavior of $\mathscr{S}_N u$ depends on the behavior of the function

$$\psi_N(z) = \frac{1}{2\pi} \int_0^z K_N(y)\, dy \tag{2.1.64}$$

in a neighborhood of the origin. There will be a Gibbs phenomenon if there exists a point $z_N > 0$, with $z_N \to 0$ as $N \to \infty$, at which $\psi_N(z_N) \geqslant \alpha > \frac{1}{2}$ (for some α independent of N); in this case

$$\lim_{N\to\infty} \mathscr{S}_N u(z_N) > u(x_0^+).$$

The kernel K_N^F generated by the Cesáro sums is known as the Fejer kernel. Its analytic expression is

$$\begin{aligned} K_N^F(\xi) &= 1 + 2 \sum_{k=1}^{N/2} \left(1 - \frac{k}{N/2+1}\right) \cos k\xi \\ &= \begin{cases} \dfrac{1}{N/2+1}\left[\dfrac{\sin((N/2+1)\xi/2)}{\sin \xi/2}\right] & \xi \neq 2j\pi \\ N/2+1 & \xi = 2j\pi. \end{cases} \quad j \in \mathbb{Z} \end{aligned} \tag{2.1.65}$$

This kernel is plotted in Fig. 2.6(a).

Since K_N^F is non-negative and $1/2\pi \int_{-\pi}^{\pi} K_N^F(y)\, dy = 1$, the corresponding function $\psi_N(z)$ is monotonically increasing and satisfies $0 < \psi_N(z) < \frac{1}{2}$ in the interval $(0, \pi)$. It follows that the Cesáro sums do not exhibit the Gibbs phenomenon near a discontinuity point (see Fig. 2.5). The Cesáro sums have several useful theoretical properties of approximation: if u is a continuous function in $[0, 2\pi]$, then the sequence $\mathscr{S}_N u$ converges to u uniformly in the interval as $N \to \infty$. Moreover, the Cesáro sums are bounded variations preserving,

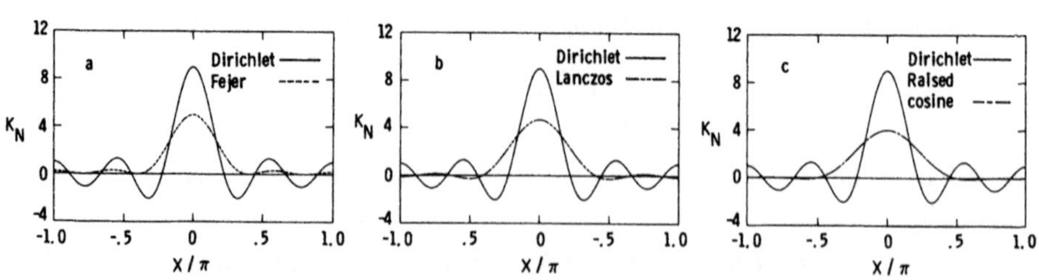

Figure 2.6. Comparison of the Dirichlet kernel with the smoothed kernels for $N = 8$. The Dirichlet kernel is denoted by the solid curves and the smoothed kernels by the dashed curves.

in the sense that if u is of bounded variation in $[0, 2\pi]$, then the total variation of $\mathscr{S}_N u$ can be bounded independently of N. However, as Fig. 2.5 shows, the Cesáro sums produce a heavy smearing of the function near a singularity point. In most applications it is desirable to have a sharper representation of the function, at the expense of retaining some oscillations. For this reason, other forms of smoothing, such as Lanczos' or the raised cosine, are preferred.

The kernel K_N^L corresponding to Lanczos' smoothing is given by

$$K_N^L(\xi) = 1 + \sum_{k=1}^{N/2} \frac{\sin\left(k\left(x + \frac{2\pi}{N}\right)\right) + \sin\left(k\left(x - \frac{2\pi}{N}\right)\right)}{\frac{2k\pi}{N}} \tag{2.1.66}$$

(see Fig. 2.6(b)), while the kernel $K_N^R(\xi)$ associated with the raised cosine smoothing is

$$K_N^R(\xi) = \frac{1}{4}\left[D_N\left(\xi - \frac{2\pi}{N}\right) + 2D_N(\xi) + D_N\left(\xi + \frac{2\pi}{N}\right)\right] \tag{2.1.67}$$

(see Fig. 2.6(c)). Thus, the raised cosine kernel can be considered as a smoothing of the Dirichlet kernel by local averages. Both K_N^L and K_N^R change sign away from the origin. Thus, the associated functions ψ_N defined in (2.1.64) exhibit an oscillatory behavior there. Since the first maximum value attained is larger than 1/2, both the Lanczos' and the raised cosine smoothing produce the Gibbs phenomenon near a discontinuity point. However, Figs. 2.6(b) and (c) show that the oscillations of K_N^L and K_N^R away from the origin are considerably less pronounced than the oscillations of the Dirichlet kernel; hence the overshooting is dramatically reduced. Moreover, K_N^R is better behaved then K_N^L from the point of view of the oscillations. Consequently, the raised cosine smoothing is the most effective among those considered in this discussion.

Similar smoothing procedures can be implemented for discrete Fourier series by applying the smoothing factors to the discrete coefficients. An alternative procedure is to represent the function as the sum of discontinuities of a simple type, such as a step-function, and a smoother component. This has been proposed by Abarbanel, Gottlieb and Tadmor (1986). Examples and further discussion are provided in Chaps. 8 and 12 in the context of Fourier collocation solutions to partial differential equations.

2.2. Orthogonal Polynomials in $(-1, 1)$

2.2.1. Sturm-Liouville Problems

The importance of Sturm-Liouville problems for spectral methods lies in the fact that the spectral approximation of the solution of a differential problem is usually regarded as a finite expansion of eigenfunctions of a suitable Sturm-Liouville problem. We recall that a Sturm-Liouville problem is an eigenvalue problem of the form

$$\begin{cases} -(pu')' + qu = \lambda wu & \text{in the interval } (-1,1) \\ + \text{ suitable boundary conditions for } u. \end{cases} \tag{2.2.1}$$

The coefficients p, q and w are three given, real-valued functions such that: p is continuously differentiable, strictly positive in $(-1,1)$ and continuous at $x = \pm 1$; q is continuous, non-negative and bounded in $(-1,1)$; the weight function w is continuous, non-negative and integrable over $(-1,1)$.

The Sturm-Liouville problems of interest in spectral methods are such that the expansion of an infinitely smooth function in terms of their eigenfunctions guarantees spectral accuracy. This means that the "Fourier" coefficients according to this basis decay faster than algebraically in the inverse of the wavenumber. As pointed out in Gottlieb and Orszag (1977, Sec. 3) not all the Sturm-Liouville problems ensure this property. For instance, the Strum-Liouville problem

$$u'' + \lambda u = 0$$

$$u'(-1) = u'(1) = 0$$

has eigenvalues $\lambda_k = (\pi k)^2/2$ and corresponding eigenfunctions $\phi_k(x) = \cos(\pi/2)k(x+1)$. A smooth function can be approximated by the cosine series on $(-1,1)$ with spectral accuracy if and only if all its odd derivatives vanish at the boundary. This is due to the fact that the coefficient $p(x)$ in the operator does not vanish at the boundary in this case, i.e., the Sturm-Liouville problem is *regular*. Conversely, spectral accuracy is ensured if the problem is *singular*, i.e., p vanishes at the boundary. A mathematical proof of these facts is given in Sec. 9.2.

Among the singular Sturm-Liouville problems, particular importance rests with those problems whose eigenfunctions are algebraic polynomials because of the efficiency with which they can be evaluated and differentiated numerically. It is also proven in Sec. 9.2 that the Jacobi polynomials are precisely the only polynomials arising as eigenfunctions of a singular Sturm-Liouville problem.

2.2.2. Orthogonal Systems of Polynomials

We shall consider here from a general point of view the problem of the expansion of a function in terms of a system of orthogonal polynomials. We denote by $\mathbb{P}_N$ the space of all polynomials of degree $\leqslant N$. Assume that $\{p_k\}_{k=0,1\ldots}$ is a system of algebraic polynomials (with degree of $p_k = k$) which are mutually orthogonal over the interval $(-1,1)$ with respect to a weight function w:

$$\int_{-1}^{1} p_k(x)p_m(x)w(x)\,dx = 0 \qquad \text{whenever } m \neq k. \tag{2.2.2}$$

The classical Weierstrass theorem implies that such a system is complete in the space $L^2_w(-1, 1)$. This is the space of functions v such that the norm

$$\|v\|_w = \left(\int_{-1}^{1} |v(x)|^2 w(x)\,dx\right)^{1/2} \tag{2.2.3}$$

is finite. The associated inner product is

$$(u, v)_w = \int_{-1}^{1} u(x)v(x)w(x)\,dx. \tag{2.2.4}$$

The formal series of a function $u \in L^2_w(-1, 1)$ in terms of the system $\{p_k\}$ is

$$Su = \sum_{k=0}^{\infty} \hat{u}_k p_k,$$

where the expansion coefficients $\hat{u}_k$ are defined as

$$\hat{u}_k = \frac{1}{\|p_k\|_w^2} \int_{-1}^{1} u(x)p_k(x)w(x)\,dx. \tag{2.2.5}$$

For an integer $N > 0$, the truncated series of u of order N is the polynomial

$$P_N u = \sum_{k=0}^{N} \hat{u}_k p_k. \tag{2.2.6}$$

Due to (2.2.2), $P_N u$ is the orthogonal projection of u upon $\mathbb{P}_N$ in the inner product (2.2.4), i.e.,

$$(P_N u, v)_w = (u, v)_w \qquad \text{for all } v \in \mathbb{P}_N. \tag{2.2.7}$$

The completeness of the system $\{p_k\}$ is equivalent to the property that for all $u \in L^2_w(-1, 1)$

$$\|u - P_N u\|_w \to 0 \qquad \text{as } N \to \infty. \tag{2.2.8}$$

2.2.3. Gauss-Type Quadratures and Discrete Polynomial Transforms

We discuss here the close relation between orthogonal polynomials and Gauss-type integration formulas on the interval $[-1, 1]$. The material of this subsection includes the interpolation formulas and discrete transforms pertinent to finite polynomial expansions.

First, we review Gaussian integration formulas, including those with preassigned abscissas. The first result can be found in any textbook on numerical analysis. For completeness we report the proofs concerning Gauss–Radau and Gauss–Lobatto formulas (see also Mercier (1981)).

Gauss integration. *Let $x_0, \ldots, x_N$ be the roots of the $(N+1)$-th orthogonal polynomial p_{N+1}, and let $w_0, \ldots, w_N$ be the solution of the linear system*

$$\sum_{j=0}^{N} (x_j)^k w_j = \int_{-1}^{1} x^k w(x)\,dx \qquad 0 \leqslant k \leqslant N. \tag{2.2.9}$$

Then

(i) *$w_j > 0$ for $j = 0, \ldots, N$ and*

$$\sum_{j=0}^{N} p(x_j) w_j = \int_{-1}^{1} p(x) w(x)\,dx \qquad \text{for all } p \in \mathbb{P}_{2N+1}. \tag{2.2.10}$$

The positive numbers w_j are called "weights".

(ii) *It is not possible to find x_j, w_j, $j = 0, \ldots, N$ such that* (2.2.10) *holds for all polynomials $p \in \mathbb{P}_{2N+2}$.*

This version of Gauss integration is quite well known. However, the roots, which correspond to the collocation points, are all in the interior of $(-1, 1)$. The requirement of imposing boundary conditions at one or both end points creates the need for the generalized Gauss integration formulas which include these points.

To obtain the Gauss–Radau formula let us consider the polynomial

$$q(x) = p_{N+1}(x) + a p_N(x), \tag{2.2.11}$$

where a is chosen to produce $q(-1) = 0$ (hence $a = -p_{N+1}(-1)/p_N(-1)$).

Gauss–Radau integration. *Let $-1 = x_0, x_1, \ldots, x_N$ be the $N+1$ roots of the polynomial* (2.2.11), *and let $w_0, \ldots, w_N$ be the solution of the linear system*

$$\sum_{j=0}^{N} (x_j)^k w_j = \int_{-1}^{1} x^k w(x)\,dx \qquad 0 \leqslant k \leqslant N. \tag{2.2.12}$$

Then

$$\sum_{j=0}^{N} p(x_j) w_j = \int_{-1}^{1} p(x) w(x)\,dx \qquad \text{for all } p \in \mathbb{P}_{2N}. \tag{2.2.13}$$

Proof: From the definition of q and the orthogonality of the polynomials, it follows that

$$(q, \phi)_w = 0 \qquad \text{for all } \phi \in \mathbb{P}_{N-1}. \tag{2.2.14}$$

For any $p \in \mathbb{P}_{2N}$ there exist $r \in \mathbb{P}_{N-1}$ and $s \in \mathbb{P}_N$ such that

$$p(x) = q(x) r(x) + s(x).$$

Since $q(x_j) = 0$, $0 \leqslant j \leqslant N$, we have $p(x_j) = s(x_j)$, $0 \leqslant j \leqslant N$. It follows that

$$\sum_{j=0}^{N} p(x_j)w_j = \sum_{j=0}^{N} s(x_j)w_j = \int_{-1}^{1} s(x)w(x)\,dx$$
$$= \int_{-1}^{1} p(x)w(x)\,dx - \int_{-1}^{1} q(x)r(x)w(x)\,dx.$$

Now (2.2.13) is a consequence of (2.2.14).

To obtain the Gauss–Radau formula including the right-hand point $x = +1$, one has to take a in (2.2.11) in such a way that $q(1) = 0$. If $x_0, x_1, \ldots, x_N$ $(=1)$ are the roots of $q(x)$, and $w_0, \ldots, w_N$ is the solution of the system (2.2.12) relative to these new points x_j, then (2.2.13) holds.

The Gauss–Lobatto formula is obtained in a similar way. We consider now

$$q(x) = p_{N+1}(x) + ap_N(x) + bp_{N-1}(x), \tag{2.2.15}$$

where a and b are chosen so that $q(-1) = q(1) = 0$. Then we have

Gauss–Lobatto integration. *Let $-1 = x_0, x_1, \ldots, x_N = 1$ be the $N + 1$ roots of the polynomial* (2.2.15), *and let $w_0, \ldots, w_N$ be the solution of the linear system*

$$\sum_{j=0}^{N} (x_j)^k w_j = \int_{-1}^{1} x^k w(x)\,dx \qquad 0 \leqslant k \leqslant N. \tag{2.2.16}$$

Then

$$\sum_{j=0}^{N} p(x_j)w_j = \int_{-1}^{1} p(x)w(x)\,dx \qquad \textit{for all } p \in \mathbb{P}_{2N-1}. \tag{2.2.17}$$

The proof of this result is similar to the previous one: here the decomposition $p = qr + s$ holds with $r \in \mathbb{P}_{N-2}$ and $s \in \mathbb{P}_N$.

In the important special case of a Jacobi weight (see Sec. 2.5.1) there is an alternative characterization of the Gauss–Lobatto points, namely they are the points -1, $+1$ and the roots of the polynomial

$$q(x) = p_N'(x). \tag{2.2.18}$$

In fact, each $p \in \mathbb{P}_{2N-1}$ can be represented in the form

$$p(x) = (1 - x^2)p_N'(x)r(x) + s(x)$$

with $r \in \mathbb{P}_{N-2}$ and $s \in \mathbb{P}_N$. By partial integration we have

$$\int_{-1}^{1} p_N'(x)(1 - x^2)r(x)w(x)\,dx$$
$$= -\int_{-1}^{1} p_N(x)[(1 - x^2)r(x)]'w(x)\,dx$$
$$- \int_{-1}^{1} p_N(x)r(x)(1 - x^2)\frac{w'(x)}{w(x)}w(x)\,dx.$$

Figure 2.7. The Gauss–Lobatto points for $N = 8$ for the Jacobi polynomials with the weight function $w(x) = (1-x)^\alpha(1+x)^\alpha$.

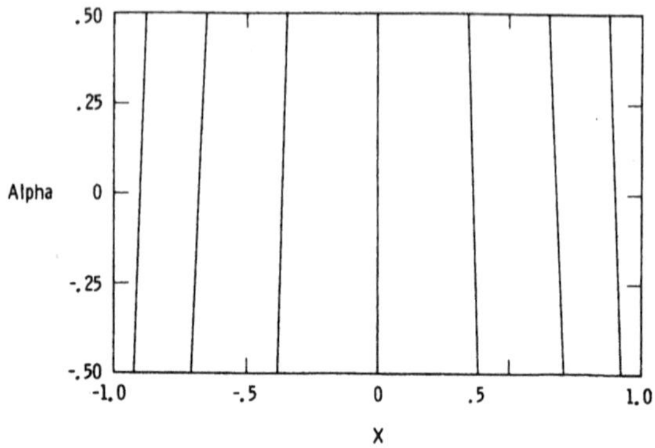

If $w(x) = (1-x)^\alpha(1+x)^\beta\,(-\frac{1}{2} \leqslant \alpha, \beta \leqslant \frac{1}{2})$ is a Jacobi weight, then $(1-x^2)(w'(x)/w(x))$ is a polynomial of degree 1. It follows that p_N' is orthogonal to $(1-x^2)r(x)$; hence (2.2.17) holds when the interior quadrature knots are the zeroes of p_N' and the weights are defined by (2.2.16).

The Gauss–Lobatto points for the particular Jacobi polynomials corresponding to the weights $w(x) = (1-x)^\alpha(1+x)^\alpha$ are illustrated in Fig. 2.7 for $N = 8$.

As observed at the beginning of this section, the nodes of the Gauss-type formulas play an important role in collocation approximations—they are precisely the collocation points at which the differential equations are enforced (see Sec. 9.4.3). We assume here that a weight function w is given, together with the corresponding sequence of orthogonal polynomials p_k, $k = 0, 1, 2, \ldots$. For a given $N \geqslant 0$, we denote by $x_0, x_1, \ldots, x_N$ the nodes of the $N+1$-point integration formula of Gauss, Gauss–Radau or Gauss–Lobatto type, and by $w_0, w_1, \ldots, w_N$ the corresponding weights.

In a collocation method the fundamental representation of a smooth function u on $(-1, 1)$ is in terms of its values at the discrete Gauss-type points. Derivatives of the function are approximated by analytic derivatives of the interpolating polynomial. The interpolating polynomial is denoted by $I_N u$. It is an element of $\mathbb{P}_N$ and satisfies

$$I_N u(x_j) = u(x_j) \qquad 0 \leqslant j \leqslant N. \tag{2.2.19}$$

$I_N u$ is uniquely defined since the x_j's are distinct. Since it is a polynomial of degree N, it admits an expression of the form

$$I_N u = \sum_{k=0}^{N} \tilde{u}_k p_k. \tag{2.2.20}$$

Obviously,

$$u(x_j) = \sum_{k=0}^{N} \tilde{u}_k p_k(x_j). \tag{2.2.21}$$

The $\tilde{u}_k$ are called the *discrete polynomial coefficients* of u. They are sometimes referred to as discrete expansion coefficients. The inverse relationship is

$$\tilde{u}_k = \frac{1}{\gamma_k} \sum_{j=0}^{N} u(x_j) p_k(x_j) w_j, \tag{2.2.22}$$

where

$$\gamma_k = \sum_{j=0}^{N} p_k^2(x_j) w_j. \tag{2.2.23}$$

Equation (2.2.22) will be derived below. Explicit formulas for γ_k for the more common orthogonal polynomials are supplied in Secs. 2.3 and 2.4.

Equations (2.2.21) and (2.2.22) enable one to transform freely between physical space $\{u(x_j)\}$ and transform space $\{\tilde{u}_k\}$. Such a transformation for orthogonal polynomials is the analogue of the transformation (2.1.24) and (2.1.22) for trigonometric polynomials. We shall call it the *discrete polynomial transform* associated with the weight w and the nodes $x_0, \ldots, x_N$.

For any u, v continuous on $[-1, 1]$, we set

$$(u, v)_N = \sum_{j=0}^{N} u(x_j) v(x_j) w_j. \tag{2.2.24}$$

The Gauss integration formulas imply that

$$(u, v)_N = (u, v)_w \qquad \text{if } uv \in \mathbb{P}_{2N+\delta}, \tag{2.2.25}$$

where $\delta = 1, 0, -1$ for Gauss, Gauss–Radau or Gauss–Lobatto integration, respectively. In particular, $(u, v)_N$ is an inner product on $\mathbb{P}_N$. For any continuous v, (2.2.19) gives

$$(I_N u, v)_N = (u, v)_N. \tag{2.2.26}$$

This shows that, as for the trigonometric systems, the interpolant $I_N u$ is the projection of u upon $\mathbb{P}_N$ with respect to the discrete inner product (2.2.24).

The orthogonality of the p_m's, together with (2.2.25) give

$$(p_m, p_k)_N = \gamma_k \delta_{km} \qquad \text{if } 0 \leqslant k \leqslant N, \tag{2.2.27}$$

where γ_k is defined in (2.2.23).

From (2.2.26) and (2.2.27) we obtain

$$(u, p_k)_N = (I_N u, p_k)_N = \sum_{m=0}^{N} \tilde{u}_m (p_m, p_k)_N = \gamma_k \tilde{u}_k \qquad 0 \leqslant k \leqslant N,$$

and (2.2.22) follows directly. In terms of the discrete inner product this is just

$$\tilde{u}_k = \frac{1}{\gamma_k} (u, p_k)_N \qquad 0 \leqslant k \leqslant N. \tag{2.2.28}$$

The discrete polynomial coefficients $\tilde{u}_k$ can be expressed in terms of the continuous coefficients $\hat{u}_k$ as follows:

$$\tilde{u}_k = \hat{u}_k + \frac{1}{\gamma_k} \sum_{l>N} (p_l, p_k)_N \hat{u}_l. \tag{2.2.29}$$

This formula is an easy consequence of (2.2.28) and (2.2.27). Equivalently, one can write

$$I_N u = P_N u + R_N u, \tag{2.2.30}$$

where

$$R_N u = \sum_{k=0}^{N} \left(\frac{1}{\gamma_k} \sum_{l>N} (p_l, p_k)_N \hat{u}_l \right) p_k \tag{2.2.31}$$

can be considered the *aliasing error* due to interpolation (compare with (2.1.31)). The aliasing error is orthogonal to the truncation error, $u - P_N u$, so that

$$\|u - I_N u\|_w^2 = \|u - P_N u\|_w^2 + \|R_N u\|_w^2. \tag{2.2.32}$$

In general, $(p_l, p_k)_N \neq 0$ for all $l > N$. Thus the k-th mode of the algebraic interpolant of u depends on the k-th mode of u and *all* the modes whose wavenumber is larger than N. The aliasing error has a simpler expression for the Chebyshev interpolation points (see (2.4.20)).

2.3. Legendre Polynomials

2.3.1. Basic Formulas

We present here a collection of the essential formulas for Legendre polynomials. For proofs, the reader may refer to Szegö (1939). The Legendre polynomials $\{L_k(x), k = 0, 1, \ldots,\}$ are the eigenfunctions of the singular Sturm–Liouville problem

$$((1 - x^2)L_k'(x))' + k(k+1)L_k(x) = 0, \tag{2.3.1}$$

which is (2.2.1) with $p(x) = 1 - x^2$, $q(x) = 0$ and $w(x) = 1$. $L_k(x)$ is even if k is even, and odd if k is odd. If $L_k(x)$ is normalized so that $L_k(1) = 1$, then for any k:

$$L_k(x) = \frac{1}{2^k} \sum_{l=0}^{[k/2]} (-1)^l \binom{k}{l} \binom{2k-2l}{k} x^{k-2l}, \tag{2.3.2}$$

where $[k/2]$ denotes the integral part of $k/2$. The Legendre polynomials satisfy the recurrence relation

$$L_{k+1}(x) = \frac{2k+1}{k+1} x L_k(x) - \frac{k}{k+1} L_{k-1}(x), \tag{2.3.3}$$

where $L_0(x) = 1$ and $L_1(x) = x$. Relevant properties are

$$|L_k(x)| \leqslant 1, \qquad -1 \leqslant x \leqslant 1, \tag{2.3.4}$$

$$L_k(\pm 1) = (\pm 1)^k \tag{2.3.5}$$

$$|L_k'(x)| \leqslant \tfrac{1}{2}k(k+1) \qquad -1 \leqslant x \leqslant 1 \tag{2.3.6}$$

$$L_k'(\pm 1) = (\pm 1)^k \tfrac{1}{2}k(k+1) \tag{2.3.7}$$

$$\int_{-1}^{1} L_k^2(x)\, dx = (k + \tfrac{1}{2})^{-1}. \tag{2.3.8}$$

The expansion of any $u \in L^2(-1, 1)$ in terms of the L_k's is

$$u(x) = \sum_{k=0}^{\infty} \hat{u}_k L_k(x), \quad \hat{u}_k = (k + \tfrac{1}{2}) \int_{-1}^{1} u(x) L_k(x)\, dx. \tag{2.3.9}$$

We consider now discrete Legendre series. Since explicit formulas for the quadrature nodes are not known, such points have to be computed numerically as zeroes of appropriate polynomials. In Appendix C a routine to compute the Legendre–Gauss–Lobatto points is supplied. The quadrature weights can be expressed in closed form in terms of the nodes, as indicated in the following formulas (see, e.g., Davis and Rabinowitz (1984)):

Legendre–Gauss

$$\begin{aligned} &x_j (j = 0, \ldots, N) \text{ zeroes of } L_{N+1}; \\ &w_j = \frac{2}{(1 - x_j^2)[L_{N+1}'(x_j)]^2} \qquad j = 0, \ldots, N. \end{aligned} \tag{2.3.10}$$

Legendre–Gauss–Radau

$$\begin{aligned} &x_j (j = 0, \ldots, N) \text{ zeroes of } L_N + L_{N+1}; \\ &w_0 = \frac{2}{(N+1)^2} \\ &w_j = \frac{1}{(N+1)^2} \frac{1 - x_j}{[L_N(x_j)]^2} \qquad j = 1, \ldots, N. \end{aligned} \tag{2.3.11}$$

Legendre–Gauss–Lobatto

$$\begin{aligned} &x_0 = -1,\ x_N = 1,\ x_j (j = 1, \ldots, N-1) \text{ zeroes of } L_N'; \\ &w_j = \frac{2}{N(N+1)} \frac{1}{[L_N(x_j)]^2} \qquad j = 0, \ldots, N. \end{aligned} \tag{2.3.12}$$

The normalization factors γ_k introduced in (2.2.23) are given by

$$\gamma_k = (k + \tfrac{1}{2})^{-1} \qquad \text{for } k < N$$
$$\gamma_N = \begin{cases} (N + \tfrac{1}{2})^{-1} & \text{for Gauss and Gauss–Radau formulas} \\ 2/N & \text{for the Gauss–Lobatto formula.} \end{cases} \tag{2.3.13}$$

2.3.2. Differentiation

As for the Fourier expansion, differentiation can be accomplished in transform space or in physical space, according to the representation of the function.

Differentiation in transform space consists of computing the Legendre expansion of the derivative of a function in terms of the Legendre expansion of the function itself. If $u = \sum_{k=0}^{\infty} \hat{u}_k L_k$, u' can be (formally) represented as

$$u' = \sum_{k=0}^{\infty} \hat{u}_m^{(1)} L_m, \tag{2.3.14}$$

where

$$\hat{u}_m^{(1)} = (2m + 1) \sum_{\substack{p=m+1 \\ p+m \text{ odd}}}^{\infty} \hat{u}_p. \tag{2.3.15}$$

The key to proving this formula is the relation

$$(2k + 1)L_k(x) = L'_{k+1}(x) - L'_{k-1}(x) \qquad k \geqslant 0. \tag{2.3.16}$$

This, in turn, is an easy consequence of the identity (see, e.g., Abramowitz and Stegun (1972, Chap. 22))

$$(1 - x^2)L'_k(x) = kL_{k-1}(x) - kxL_k(x) \tag{2.3.17}$$

and the recurrence relation (2.3.3). By (2.3.16),

$$\begin{aligned} u'(x) &= \sum_{k=0}^{\infty} \frac{\hat{u}_k^{(1)}}{2k + 1} L'_{k+1}(x) - \sum_{k=0}^{\infty} \frac{\hat{u}_k^{(1)}}{2k + 1} L'_{k-1}(x) \\ &= \sum_{k=1}^{\infty} \frac{\hat{u}_{k-1}^{(1)}}{2k - 1} L'_k(x) - \sum_{k=-1}^{\infty} \frac{\hat{u}_{k+1}^{(1)}}{2k + 3} L'_k(x) \\ &= \sum_{k=1}^{\infty} \left[\frac{\hat{u}_{k-1}^{(1)}}{2k - 1} - \frac{\hat{u}_{k+1}^{(1)}}{2k + 3} \right] L'_k(x). \end{aligned}$$

On the other hand,

$$u'(x) = \sum_{k=0}^{\infty} \hat{u}_k L'_k(x)$$

and since the L'_k are linearly independent

$$\hat{u}_k = \frac{\hat{u}_{k-1}^{(1)}}{2k-1} - \frac{\hat{u}_{k+1}^{(1)}}{2k+3} \qquad k \geqslant 1, \tag{2.3.18}$$

which imply (2.3.15). The previous identity generalizes, with obvious notation, to

$$\hat{u}_k^{(q-1)} = \frac{\hat{u}_{k-1}^{(q)}}{2k-1} - \frac{\hat{u}_{k+1}^{(q)}}{2k+3} \qquad k \geqslant 1, \tag{2.3.19}$$

from which it is possible to get explicit expressions for the Legendre coefficients of higher derivatives. For the second derivative we have

$$\hat{u}_m^{(2)} = (m + \tfrac{1}{2}) \sum_{\substack{p=m+2 \\ p+m \text{ even}}} [p(p+1) - m(m+1)]\hat{u}_p. \tag{2.3.20}$$

The previous expansions are not merely formal provided u is smooth enough. For instance, the series (2.3.14) is convergent in the mean if the derivative of u (in the sense of distributions) is a function in $L^2(-1,1)$.

Unlike for the Fourier system, differentiation and Legendre truncation do not commute, i.e., in general

$$(P_N u)' \neq P_{N-1} u'. \tag{2.3.21}$$

This is an immediate consequence of (2.3.15). It is the quantity on the left which is referred to as the *Legendre Galerkin derivative*. The error $(P_N u)' - P_{N-1} u'$ decays spectrally for infinitely smooth solutions. However, if u has finite regularity then this difference decays at a slower rate than the truncation error for the derivative $u' - P_{N-1} u'$. This means that $(P_N u)'$ is asymptotically a worse approximation to u' than $P_{N-1} u'$. This topic is discussed in Sec. 9.4.2.

The function $u(x) = |x|^{3/2}$ will serve as an illustration of the results produced by Legendre differentiation procedures. It has the Legendre coefficients

$$\hat{u}_k = \begin{cases} 0 & k \text{ odd} \\ 1/(a+1) & k = 0 \\ \dfrac{(2k+1)a(a-2)\ldots(a-k+2)}{(a+1)(a+3)\ldots(a+k+1)} & \text{otherwise,} \end{cases}$$

where $a = 3/2$. A comparison between $P_{N-1}u'$ and $(P_N u)'$ is furnished in Figs. 2.8(a) and (b). (Only the right half of the approximation interval $[-1, 1]$ is displayed.) Both approximations yield the expected slow convergence near the singularity at $x = 0$. The global nature of the approximation leads to additional problems caused by the singularity which are most apparent at $x = \pm 1$. Further discussion of this behavior will be given in Sec. 9.4.2 after we have presented the general results on the error between u' and $(P_N u)'$ in terms of N and the regularity of u.

Let us consider now differentiation in physical space. If the function u is

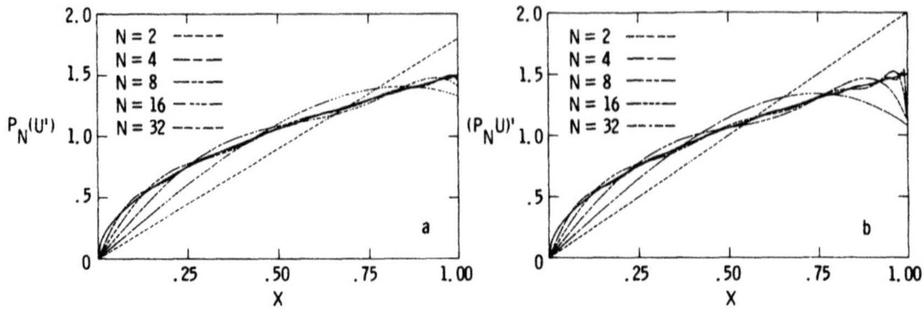

Figure 2.8. Several versions of Legendre differentiation for $u(x) = |x|^{3/2}$ on $[-1, 1]$. The exact result is indicated by the solid curves and the approximate results for $N = 2, 4, 8, 16$, and 32 are indicated by the dashed curves. Only the right half of the interval is shown. (a) $P_N u'$; (b) $(P_N u)'$.

known at one set of quadrature points (2.3.10), (2.3.11) or (2.3.12), one can compute an approximate derivative of u by differentiating the interpolant $I_N u$ (as defined in (2.2.20)) and evaluating it at the same nodes. The polynomial of degree $N - 1$,

$$\mathscr{D}_N u = (I_N u)', \tag{2.3.22}$$

is called the *Legendre collocation derivative of u* relative to the chosen set of quadrature nodes, since in general, it is different from the Galerkin derivative $(P_N u)'$.

The error between u' and the Legendre collocation derivative of u can be estimated in terms of N and the regularity of u. This is done in Sec. 9.4.3 (see (9.4.26)).

In order to compute the values $(\mathscr{D}_N u)(x_l)$, $(l = 0, \ldots, N)$ from the values $u(x_j)$, $(j = 0, \ldots, N)$, one could use formula (2.2.22) to get the discrete Legendre coefficients of u, then use (2.3.15) to differentiate in transform space and finally compute $(\mathscr{D}_N u)_l$ through (2.2.21). However, this procedure is not efficient for N of practical interest in the absence of a fast transform method for the Legendre expansion. Therefore, it is preferable to obtain the collocation derivative at the nodes through matrix multiplication, namely

$$(\mathscr{D}_N u)(x_l) = \sum_{j=0}^{N} (D_N)_{lj} u(x_j) \qquad (l = 0, \ldots, N). \tag{2.3.23}$$

The entries $(D_N)_{lj}$ can be computed by differentiating the Lagrange polynomials ψ_j which are 1 at x_j and 0 at all the other collocation points. For the commonly used Gauss–Lobatto points (2.3.12) one has

$$\psi_j(x) = \frac{1}{N(N+1)L_N(x_j)} \frac{(1 - x^2)L_N'(x)}{x - x_j}, \tag{2.3.24}$$

so that

$$
(D_N)_{lj} = \begin{cases} \dfrac{L_N(x_l)}{L_N(x_j)} \dfrac{1}{x_l - x_j} & l \neq j \\[2ex] \dfrac{(N+1)N}{4} & l = j = 0 \\[2ex] -\dfrac{(N+1)N}{4} & l = j = N \\[2ex] 0, & \text{otherwise.} \end{cases} \tag{2.3.25}
$$

Solomonoff and Turkel (1986) have furnished a detailed derivation of (2.3.24).

The matrix of the collocation derivative can be obtained by a similarity transformation from the matrix of the spectral derivative, which is associated with the linear transformation (2.3.15) with the summation truncated to $p \leqslant N$. Thus they both have 0 as generalized eigenvalue of order $N + 1$; the only eigenvector is $L_0(x)$, while each $L_k(x)$, $1 \leqslant k \leqslant N$, is a generalized eigenvector, i.e., a function f for which $f^{(k)}$ is zero.

In spectral methods of Legendre type, differentiation is usually associated with suitable boundary conditions. In this case, the spectra of the related operators may exhibit different behavior. This topic is discussed in Secs. 4.2 and 11.4.

2.4. Chebyshev Polynomials

2.4.1. Basic Formulas

Classical references on the Chebyshev polynomials are Fox and Parker (1968) and Rivlin (1974). The Chebyshev polynomials of first kind $\{T_k(x), k = 0, 1, \dots\}$ are the eigenfunctions of the singular Sturm–Liouville problem

$$
(\sqrt{1 - x^2}\, T_k'(x))' + \frac{k^2}{\sqrt{1 - x^2}} T_k(x) = 0, \tag{2.4.1}
$$

which is (2.2.1) with $p(x) = (1 - x^2)^{1/2}$, $q(x) = 0$ and $w(x) = (1 - x^2)^{-1/2}$. For any k, $T_k(x)$ is even if k is even, and odd if k is odd. If T_k is such that $T_k(1) = 1$, then

$$
T_k(x) = \cos k\theta \qquad \theta = \arccos x. \tag{2.4.2}
$$

Thus, the Chebyshev polynomials are nothing but cosine functions after a change of independent variable. This property is the origin of their widespread popularity in the numerical approximation of non-periodic boundary value problems. The transformation $x = \cos\theta$ enables many mathematical relations as well as theoretical results concerning the Fourier system to be adapted readily to the Chebyshev system.

The Chebyshev polynomials can be expanded in power series as

$$T_k(x) = \frac{k}{2} \sum_{l=0}^{[k/2]} (-1)^k \frac{(k-l-1)!}{l!(k-2l)!} (2x)^{k-2l}, \tag{2.4.3}$$

where $[k/2]$ denotes the integral part of $k/2$. Moreover, the trigonometric relation $\cos(k+1)\theta + \cos(k-1)\theta = 2\cos\theta\cos k\theta$ gives the recurrence relation

$$T_{k+1}(x) = 2xT_k(x) - T_{k-1}(x) \tag{2.4.4}$$

with $T_0(x) \equiv 1$ and $T_1(x) \equiv x$.

Some properties of the Chebyshev polynomials are:

$$|T_k(x)| \leqslant 1 \qquad -1 \leqslant x \leqslant 1 \tag{2.4.5}$$

$$T_k(\pm 1) = (\pm 1)^k \tag{2.4.6}$$

$$|T_k'(x)| \leqslant k^2 \qquad -1 \leqslant x \leqslant 1 \tag{2.4.7}$$

$$T_k'(\pm 1) = (\pm 1)^k k^2 \tag{2.4.8}$$

$$\int_{-1}^{1} T_k^2(x) \frac{dx}{\sqrt{1-x^2}} = c_k \frac{\pi}{2}, \tag{2.4.9}$$

where

$$c_k = \begin{cases} 2 & \text{if } k = 0 \\ 1 & \text{if } k \geqslant 1. \end{cases} \tag{2.4.10}$$

The Chebyshev expansion of a function $u \in L_w^2(-1, 1)$ is

$$u(x) = \sum_{k=0}^{\infty} \hat{u}_k T_k(x) \qquad \hat{u}_k = \frac{2}{\pi c_k} \int_{-1}^{1} u(x) T_k(x) w(x)\, dx. \tag{2.4.11}$$

If we define the even periodic function $\tilde{u}$ by $\tilde{u}(\theta) = u(\cos\theta)$, then

$$\tilde{u}(\theta) = \sum_{k=0}^{\infty} \hat{u}_k \cos k\theta;$$

hence, the Chebyshev series for u corresponds to a cosine series for $\tilde{u}$. It is easy to verify that if $u(x)$ is infinitely differentiable on $[-1, 1]$, then $\tilde{u}(\theta)$ is infinitely differentiable and periodic with all its derivatives on $[0, 2\pi]$. According to the integration-by-parts argument for Fourier series developed in Sec. 2.2.1, the Chebyshev coefficients are guaranteed to decay faster than algebraically.

Turning now to relations of interest for discrete Chebyshev series, explicit formulas for the quadrature points and weights are:

Chebyshev–Gauss

$$x_j = \cos\frac{(2j+1)\pi}{2N+2} \qquad w_j = \frac{\pi}{N+1} \qquad j = 0, \ldots, N. \tag{2.4.12}$$

Chebyshev–Gauss–Radau

$$x_j = \cos\frac{2\pi j}{2N+1} \qquad w_j = \begin{cases} \dfrac{\pi}{2N+1} & j = 0 \\[2ex] \dfrac{2\pi}{2N+2} & 1 \leqslant j \leqslant N \end{cases} \tag{2.4.13}$$

Chebyshev–Gauss–Lobatto

$$x_j = \cos\frac{\pi j}{N} \qquad w_j = \begin{cases} \dfrac{\pi}{2N} & j = 0, N \\[2ex] \dfrac{\pi}{N} & 1 \leqslant j \leqslant N-1 \end{cases} \tag{2.4.14}$$

The most commonly used points are those for the Gauss–Lobatto case which we consider in detail hereafter. The matrix representing the transformation from physical space to Chebyshev transform space (see (2.2.22)) is available in the simple form

$$C_{kj} = \frac{2}{N\bar{c}_j\bar{c}_k}\cos\frac{\pi jk}{N}, \tag{2.4.15}$$

where

$$\bar{c}_j = \begin{cases} 2 & j = 0, N \\ 1 & 1 \leqslant j \leqslant N-1 \end{cases}. \tag{2.4.16}$$

Likewise, the inverse transformation (see (2.2.21)) is represented by

$$(C^{-1})_{jk} = \cos\frac{\pi jk}{N}. \tag{2.4.17}$$

Both transforms may be evaluated by the Fast Fourier Transform (Appendix B).

The normalization factors γ_k introduced in (2.2.23) are here given by:

$$\gamma_k = \frac{\pi}{2}c_k \qquad \text{for } k < N$$

$$\gamma_N = \begin{cases} \dfrac{\pi}{2} & \text{for Gauss and Gauss–Radau formulas} \\[2ex] \pi & \text{for the Gauss–Lobatto formula} \end{cases} \tag{2.4.18}$$

The structure of the aliasing error (2.2.31) due to interpolation takes a very simple form for the Chebyshev–Gauss–Lobatto points. Recalling (2.4.2) and using the identity (2.1.23) with N replaced by $2N$, one gets for $k = 0, \ldots, N$

$$(T_k, T_l)_N = \begin{cases} (T_k, T_k)_N & \text{if } l = 2mN \pm k,\ m \geqslant 0 \\ 0 & \text{otherwise;} \end{cases} \tag{2.4.19}$$

hence, (2.2.29) becomes

$$\tilde{u}_k = \hat{u}_k + \sum_{\substack{j=2mN\pm k \\ j>N}} \hat{u}_j. \tag{2.4.20}$$

As for the Fourier points, the k-th Chebyshev mode of the interpolant polynomial depends upon all the Chebyshev modes which alias $T_k(x)$ on the grid.

2.4.2. Differentiation

The derivative of a function u expanded in Chebyshev polynomials according to (2.4.11) can be represented formally as

$$u' = \sum_{m=0}^{\infty} \hat{u}_m^{(1)} T_m, \tag{2.4.21}$$

where

$$\hat{u}_m^{(1)} = \frac{2}{c_m} \sum_{\substack{p=m+1 \\ p+m \text{ odd}}}^{\infty} p\hat{u}_p. \tag{2.4.22}$$

This expression is a consequence of the relation

$$2T_k(x) = \frac{1}{k+1} T'_{k+1}(x) - \frac{1}{k-1} T'_{k-1}(x) \qquad k \geqslant 1, \tag{2.4.23}$$

which, due to (2.4.2), is a different form of the trigonometric identity

$$2\sin\theta\cos k\theta = \sin(k+1)\theta - \sin(k-1)\theta.$$

From (2.4.23) one has

$$2k\hat{u}_k = c_{k-1}\hat{u}_{k-1}^{(1)} - \hat{u}_{k+1}^{(1)} \qquad k \geqslant 1; \tag{2.4.24}$$

whence (2.4.22) follows. Note that the last relation suggests an efficient way of differentiating a polynomial of degree N in Chebyshev space. Since $\hat{u}_k^{(1)} = 0$ for $k \geqslant N$, the non-zero coefficients are computed in decreasing order by the recurrence relation

$$c_k\hat{u}_k^{(1)} = \hat{u}_{k+2}^{(1)} + 2(k+1)\hat{u}_{k+1} \qquad 0 \leqslant k \leqslant N-1, \tag{2.4.25}$$

in $2N$ multiplications or additions. The generalization of this relation is

$$c_k\hat{u}_k^{(q)} = \hat{u}_{k+2}^{(q)} + 2(k+1)\hat{u}_{k+1}^{(q-1)}. \tag{2.4.26}$$

The coefficients of the second derivative are

$$\hat{u}_k^{(2)} = \frac{1}{c_k} \sum_{\substack{p=k+2 \\ p+k \text{ even}}}^{\infty} p(p^2 - k^2)\hat{u}_p. \tag{2.4.27}$$

The *Chebyshev Galerkin derivative* is just $(P_N u)'$. The *Chebyshev collocation derivative* of a function u known at one set of quadrature nodes—(2.4.12), (2.4.13) or (2.4.14)—is defined as the derivative of the discrete Chebyshev series of u at the same nodes,

$$\mathscr{D}_N u = (I_N u)'. \tag{2.4.28}$$

As for Legendre polynomials, Chebyshev truncation and interpolation do not commute with differentiation. $(P_N u)'$ or $(I_N u)'$ are asymptotically worse approximations of u' than $P_{N-1} u'$ and $I_{N-1} u'$ respectively, for functions with finite regularity. These results are made more precise in Sec. 9.5.2.

Chebyshev collocation differentiation can be accomplished efficiently by means of a transform method. The discrete Chebyshev coefficients of u are computed according to (2.2.22), then (2.4.25) is used to differentiate in transform space, and finally the values of $\mathscr{D}_N u$ at the grid points are obtained by transforming back to physical space. If the discrete Chebyshev transforms are computed by an FFT algorithm which takes advantage of the reality and the parity of the function $\tilde{u}(\theta) = u(\cos\theta)$, the total number of operations required to differentiate in physical space is $(5\log_2 N + 8 + 2q)N$, where q is the order of the derivative. A FORTRAN program is furnished in Appendix B.

The Chebyshev collocation derivative can also be represented in matrix form as

$$(\mathscr{D}_N u)(x_l) = \sum_{j=0}^{N} (D_N)_{lj} u(x_j) \qquad (l = 0, \ldots, N). \tag{2.4.29}$$

The entries $(D_N)_{lj}$ can be computed by differentiating the Lagrange polynomials ψ_j which are 1 at x_j and 0 at all the other collocation points. For the popular Gauss–Lobatto points (2.4.14) one has

$$\psi_j(x) = \frac{(-1)^{j+1}(1 - x^2)T_N'(x)}{\bar{c}_j N^2 (x - x_j)} \tag{2.4.30}$$

$$(D_N)_{lj} = \begin{cases} \dfrac{\bar{c}_l}{\bar{c}_j} \dfrac{(-1)^{l+j}}{x_l - x_j} & l \neq j \\[2ex] \dfrac{-x_j}{2(1 - x_j^2)} & 1 \leqslant l = j \leqslant N - 1 \\[2ex] \dfrac{2N^2 + 1}{6} & l = j = 0 \\[2ex] -\dfrac{2N^2 + 1}{6} & l = j = N. \end{cases} \tag{2.4.31}$$

(See Gottlieb, Hussaini and Orszag (1984) and Solomonoff and Turkel (1986).) Peyret (1986) has given an explicit expression for the second derivative matrix.

The matrix (2.4.31) is not skew-symmetric, as opposed to the matrix (2.1.41) of the Fourier differentiation. Since it is obtained by a similarity transformation from the matrix of differentiation in transform space (see (2.4.22)), it is immediate that the only eigenvalue is 0 with algebraic multiplicity $N + 1$. Clearly, introducing boundary conditions results in a different structure of the spectrum, as discussed in Secs. 4.2 and 11.4.

If the collocation derivative is computed by matrix multiplication, the total number of operations is $2N^2$. Table 2.1 (p. 45) also provides a timing comparison of matrix-multiply and transform-based Chebyshev derivatives. The extra overhead in the cosine transform (reflected in the $(8 + 2q)N$ term in the Chebyshev derivative operation count) compared with the Fourier transform leads to a 20–40% greater cost for a Chebyshev derivative compared with a Fourier derivative.

2.5. Generalizations

2.5.1. Jacobi Polynomials

As noted in Sec. 2.2.1, the class of Jacobi polynomials comprises all the polynomial solutions to singular Sturm–Liouville problems on $(-1, 1)$. They are solutions to (2.2.1) with $p(x) = (1 - x)^{1+\alpha}(1 + x)^{1+\beta}$ for α, $\beta > -1$, $q(x) = 0$, and $w(x) = (1 - x)^\alpha(1 + x)^\beta$. They are denoted by $P_k^{(\alpha,\beta)}(x)$. Under the normalization $P_k^{(\alpha,\beta)}(1) = \binom{k+\alpha}{k}$,

$$P_k^{(\alpha,\beta)}(x) = \frac{1}{2^k} \sum_{l=0}^{k} \binom{k+\alpha}{l} \binom{k+\beta}{k-l} (x-1)^l (x+1)^{k-l}. \qquad (2.5.1)$$

They are related to the Legendre polynomials by

$$L_k(x) = P_k^{(0,0)}(x) \qquad (2.5.2)$$

and to the Chebyshev polynomials by

$$T_k(x) = \frac{2^{2k}(k!)^2}{(2k)!} P_k^{(-1/2,-1/2)}(x). \qquad (2.5.3)$$

They satisfy the two recursion relations

$$\begin{aligned}
P_0^{(\alpha,\beta)}(x) &= 1 \\
P_1^{(\alpha,\beta)}(x) &= (1 + \alpha)x \\
a_{1,k} P_{k+1}^{(\alpha,\beta)}(x) &= a_{2,k}(x) P_k^{(\alpha,\beta)}(x) - a_{3,k} P_{k-1}^{(\alpha,\beta)}(x) \\
a_{1,k} &= 2(k+1)(k+\alpha+\beta+1)(2k+\alpha+\beta) \qquad (2.5.4) \\
a_{2,k}(x) &= (2k+\alpha+\beta+1)(\alpha^2-\beta^2) + x\Gamma(2k+\alpha+\beta+3)/\Gamma(2k+\alpha+\beta) \\
a_{3,k} &= 2(k+\alpha)(k+\beta)(2k+\alpha+\beta+2)
\end{aligned}$$

and

$$b_{1,k}(x)\frac{d}{dx}P_k^{(\alpha,\beta)}(x) = b_{2,k}(x)P_k^{(\alpha,\beta)}(x) + b_{3,k}P_{k-1}^{(\alpha,\beta)}(x)$$

$$\begin{aligned} b_{1,k}(x) &= (2k+\alpha+\beta)(1-x^2) \\ b_{2,k}(x) &= k[\alpha-\beta-(2k+\alpha+\beta)x] \\ b_{3,k} &= 2(k+\alpha)(k+\beta) \end{aligned} \tag{2.5.5}$$

(see Abramowitz and Stegun (1972, Chap. 22)).

Jacobi series are given by

$$\begin{aligned} u(x) &= \sum_{k=0}^{\infty} \hat{u}_k P_k^{(\alpha,\beta)}(x) \\ \hat{u}_k &= \frac{2k+\alpha+\beta+1}{2^{\alpha+\beta+1}} \frac{k!\Gamma(k+\alpha+\beta+1)}{\Gamma(k+\alpha+1)\Gamma(k+\beta+1)} \\ &\quad \times \int_{-1}^{1} u(x)P_k^{(\alpha,\beta)})x)(1-x)^\alpha(1+x)^\beta\,dx. \end{aligned} \tag{2.5.6}$$

The discussion in Sec. 2.2.3 contains the general formulas for the Jacobi nodes and discrete quadrature weights. Appendix C contains a routine for computing the Gauss–Lobatto nodes.

Although spectral accuracy can be achieved for expansion in Jacobi polynomials (see Sec. 9.6.1), they have seen comparatively little use, aside, of course, from the special cases of Chebyshev ($\alpha = \beta = -1/2$) and Legendre ($\alpha = \beta = 0$) polynomials. The lack of a fast transform is a major handicap in large problems. Their principal use to date has been in some special Galerkin methods for wall-bounded incompressible flows (see Sec. 7.3.3).

2.5.2. Mapping

We focus here on expansions in Chebyshev polynomials on $[-1, 1]$. Problems on other finite intervals may trivially be mapped onto the standard interval. Mappings of $[-1, 1]$ onto itself can often be useful in improving the accuracy of a Chebyshev expansion. A comparison of the location of the Gauss points with the function itself gives a good indication of the accuracy of the series representation: a rule of thumb is that there needs to be at least three nodes for every significant feature (or "wavelength") in the function (see Gottlieb and Orszag (1977, Sec. 3)). For functions whose structure is localized, mapping can lead to much more efficient approximations.

Let $x = \phi(\xi)$ denote a mapping of the computational coordinate ξ into the natural coordinate x. The relation

$$dx = \phi'(\xi)\,d\xi \tag{2.5.7}$$

indicates that the nodes (in x) are compressed together when $\phi'(\xi)$ is small and dilated when $\phi'(\xi)$ is large. This serves as a rough guide in the choice of a mapping which will cluster points as desired. The expansion is thus

$$u(x) = \sum_{k=0}^{N} \hat{u}_k T_n(\xi) \tag{2.5.8}$$

and derivatives are evaluated via

$$\frac{du}{dx} = \frac{d\psi}{dx} \sum_{k=0}^{N} \hat{u}_k^{(1)} T_n(\xi) \tag{2.5.9}$$

where $\psi = \phi^{-1}$.

The function which is actually being expanded is $u(\phi(\xi))$. In order for spectral accuracy to be retained, we need $\phi(\xi)$ and all of its derivatives to be sufficiently well-behaved that the standard integration-by-parts argument goes through. This fails if, say, $\phi(\xi)$ is singular at the endpoints. The mapping $x = \phi(\xi) = -2/\pi \cos^{-1}\xi + 1$ leads to a uniform distribution of the nodes in x. (One might be tempted to use this mapping in order to alleviate the severe restriction on the time-step in evolution problems which results from the standard Chebyshev distribution (see Sec. 4.2).) Unfortunately, $\phi'(\pm 1)$ is singular. We expect the resulting approximation to be quite poorly behaved. In fact, it exhibits the well-known Runge phenomenon for polynomial interpolation on a uniformly spaced grid.

2.5.3. Semi-Infinite Intervals

There are three basic ways to construct global approximations to functions defined on the semi-infinite interval $[0, \infty)$: (1) expand in Laguerre functions, (2) map the semi-infinite interval into a finite one and then expand in Chebyshev polynomials, and (3) truncate the domain to $[0, x_{\max}]$ and use a Chebyshev expansion.

The two most common accuracy criteria by which the approximations are judged are weighted mean errors and maximum errors. Some approximation results for Laguerre expansions are cited in Sec. 9.6.2. In the present subsection we shall present some guidelines for selecting global approximations which yield faster than algebraic decay of the maximum error.

The Laguerre functions $\phi_n(x)$ are defined by $\phi_n(x) = e^{-x/2}\ell_n(x)$, where $\ell_n(x)$ is the classical Laguerre polynomial of degree n. Maday, Pernaud–Thomas and Vandeven (1985) summarize their recursion formulas and other useful properties. Boyd (1987b) has noticed that (uniform) spectral accuracy of Laguerre approximations requires at least exponential decay of the function as $x \to \infty$. No fast transform is available for Laguerre functions.

The combination of the mapping $x = \phi(\xi)$, for $\xi \in [-1, 1]$, with a Chebyshev polynomial expansion in ξ is appealing because it allows the FFT to be employed for many of the requisite series manipulations. The convergence properties of the approximation to $u(x)$ can be determined from the behavior of the function $v(\xi) = u(\phi(\xi))$. Exponential convergence is expected when $v(\xi)$ is infinitely differentiable on $[-1, 1]$. Assuming that $u(x)$ itself is infinitely differentiable on $[0, \infty)$, the critical issue is the behavior of the derivatives of $v(\xi)$ at $\xi = \pm 1$. Loosely put, uniform spectral accuracy can be achieved provided the derivatives of $u(x)$ decay fast enough and oscillate slowly enough as $x \to \infty$.

The most frequently used mappings are algebraic, e.g.,

$$x = L\frac{1+\xi}{1-\xi} \qquad \xi = \frac{x-L}{x+L} \tag{2.5.10}$$

and exponential, e.g.,

$$x = -L\ln\left(\frac{1-\xi}{2}\right) \qquad \xi = 1 - 2e^{-x/L}, \tag{2.5.11}$$

where the constant L sets the length scale of the mapping. Neither mapping induces any singularities at $x = 0$, and they both guarantee spectral accuracy for infinitely differentiable functions which decay exponentially as $x \to \infty$.

A mapping which is useful when combined with domain truncation is the logarithmic map

$$\begin{aligned} x &= x_{\max}\frac{e^{a(1-\xi)} - e^{2a}}{1 - e^{2a}} \\ \xi &= 1 - \frac{1}{a}\ln[(1 - e^{2a})(x/x_{\max}) + e^{2a}]. \end{aligned} \tag{2.5.12}$$

The stretching of the grid (near $x = 0$) is strongest for the exponential map and weakest for the logarithmic one.

Boyd (1987b) has discussed the algebraic map in detail. Uniform spectral accuracy can be achieved with this mapping for algebraically decaying functions which are analytic at infinity, i.e., which admit convergent power series in $1/x$. Numerous authors (Grosch and Orszag (1977), Boyd (1982), Herbert (1984)) have found that, in practice, algebraic mappings are more accurate and more robust (less sensitive to the scale factor L) than exponential ones. For functions which decay only algebraically, the logarithmic map is the most robust.

Boyd (1987b) has termed the basis functions (in x) which result from a Chebyshev expansion in terms of the mapped variable ξ given by (2.5.10) the "rational Chebyshev functions on the semi-infinite interval." He has catalogued their basic properties and provides numerous examples of their usefulness for approximations to ordinary differential equations.

Spalart (1984) observed that the use of the exponential mapping (2.5.11) for a function which decays faster than exponentially (as a Gaussian, for example) results in an inefficient distribution of grid points. Because of the clustering of nodes at $\xi = -1$ and $\xi = 1$, there will be more nodes for large x than are required to resolve the function. Spalart proposed replacing (2.5.11) with

$$x = -L \ln \xi \qquad \xi = e^{-x/L} \tag{2.5.13}$$

for $\xi \in [0, 1]$. The function $v(\xi)$ and all of its derivatives are zero at $\xi = 0$. Hence, $v(\xi)$ may be extended smoothly to a function on $[-1, 1]$. (In some cases, just the odd or just the even Chebyshev polynomials become appropriate expansion functions.) The grid points are clustered near $\xi = 1$ $(x = 0)$ and are coarsely distributed near $\xi = 0$ $(x = \infty)$. Likewise, for an exponentially decaying function, Chebyshev expansions may be combined with the map

$$x = L\frac{\xi}{1-\xi} \qquad \xi = \frac{x}{x+L}. \tag{2.5.14}$$

When the infinite interval is handled by truncating the domain to $[0, x_{max}]$, exponential convergence can only be achieved by increasing x_{max} as the number of terms in the series is increased. Boyd (1982) provides some guidance on how x_{max} should increase with N.

2.5.4. Infinite Intervals

Similar considerations apply to expansions on infinite intervals as on semi-infinite ones. The classical preference is for expansions in Hermite functions. However, there is no fast transform for them and exponential convergence requires that the function decay at least exponentially fast as $|x| \to \infty$ (Boyd (1984)).

Cain, Ferziger, and Reynolds (1984) suggested the use of the mapping

$$x = -L \cot(\xi/2) \qquad \xi \in [0, 2\pi] \tag{2.5.15}$$

in conjunction with Fourier series. Exponential convergence is only achieved if the function $u(x)$ and all of its derivatives exist and match at $x = -\infty$ and $x = +\infty$. The reason is that the function $v(\xi) = u(\phi(\xi))$ is implicitly extended periodically by the use of Fourier series, and continuity of $v(\xi)$ and all its derivatives is required for spectral accuracy.

A simple function such as $u(x) = \tanh x$ does not meet those conditions, since $u(-\infty) \neq u(+\infty)$. For functions such as this, which approach different limits (but exponentially fast) at $x = \pm\infty$, Cain et al. proposed the mapping

$$x = -L \cot(\xi) \qquad \xi \in [0, \pi] \tag{2.5.16}$$

with $v(\xi)$ extended to $\xi \in [\pi, 2\pi]$ by reflection. When coupled with Fourier series, this yields exponential convergence.

Boyd (1987a) has discussed the use of the mapping (2.5.16) on just $[0, \pi]$ in conjunction with a sine and cosine expansion (as opposed to the complex

Fourier series on $[0, 2\pi]$). He noted that if just the cosine expansion is used, then $u(x)$ must at least have exponential decay (or special symmetries). If the decay is only algebraic and no special symmetries are present, then only algebraic convergence is possible with the cosine expansion.

An alternative approximation couples either the algebraic map

$$x = L\frac{\xi}{\sqrt{1-\xi^2}} \qquad \xi \in [-1, 1] \tag{2.5.17}$$

or else the exponential map

$$x = L \tanh^{-1}\xi \qquad \xi \in [-1, 1] \tag{2.5.18}$$

with an expansion in Chebyshev polynomials. One expects exponential convergence, even if $u(-\infty) \neq u(+\infty)$, provided that the derivatives of u decay sufficiently fast, i.e., algebraic decay with (2.5.17) and exponential decay with (2.5.18), and of, course, provided that $u(x)$ is analytic at $x = \pm\infty$.

CHAPTER 3

Fundamentals of Spectral Methods for PDEs

For the remainder of this book we shall be concerned with the use of spectral methods to approximate solutions to partial differential equations (PDEs). Our concern in this chapter is to illustrate how spectral methods are actually implemented for PDEs. We start by deriving the semi-discrete (continuous in time) ODE equations which are satisfied by various spectral approximations to Burgers equation. This will involve a discussion of non-linear terms, boundary conditions, projection operators, and different spectral discretizations. The second section provides a detailed discussion of transform methods for evaluating convolution sums. Next, we discuss Neumann, Robin and radiation boundary conditions. Finally, we remark on the treatment of coordinate singularities and the use of mapping techniques in two-dimensional problems.

3.1. Spectral Projection of the Burgers Equation

The discretization by spectral methods of the non-linear Burgers equation

$$\frac{\partial u}{\partial t} + u\frac{\partial u}{\partial x} - \nu\frac{\partial^2 u}{\partial x^2} = 0, \qquad (3.1.1)$$

where ν is a positive constant, raises many issues that occur for much more complicated problems. Of course, (3.1.1) must be supplemented with an initial condition

$$u(x,0) = u_0(x) \qquad (3.1.2)$$

and appropriate boundary conditions.

Equation (3.1.1) can be written as

$$\frac{\partial u}{\partial t} + G(u) + Lu = 0 \qquad (3.1.3)$$

where the non-linear operator G is defined by $G(u) = u(\partial u/\partial x)$ and the linear operator L is just $-\nu(\partial^2/\partial x^2)$. The discretization process consists of defining a space X_N of trial functions, a space Y_N of test functions, discrete approximations G_N and L_N of the operators G and L, respectively, and an orthogonal projection operator Q_N from a suitable Hilbert space which contains X_N onto the space Y_N. The weighted residual minimization statement is equivalent to the application of the orthogonal projection operator. It may be expressed as

$$u^N \in X_N$$
$$Q_N\left(\frac{\partial u^N}{\partial t} + G_N(u^N) + L_N u^N\right) = 0 \tag{3.1.4}$$

or, in variational form as

$$u^N \in X_N$$
$$\left(\frac{\partial u^N}{\partial t} + G_N(u^N) + L_N u^N, v\right) = 0 \quad \text{for all } v \in Y_N. \tag{3.1.5}$$

The rigorous discussion of this discretization process is given in Chap. 10. This section will illustrate it for several spectral approximations to Burgers equation. We consider here different treatments of the non-linear and linear terms as well as different treatments of the boundary conditions.

3.1.1. Fourier Galerkin

We look for a solution which is periodic in space on the interval $(0, 2\pi)$. The trial space X_N is S_N, the set of all trigonometric polynomials of degree $\leqslant N/2$. The approximate function u^N is represented as the truncated Fourier series

$$u^N(x,t) = \sum_{k=-N/2}^{N/2-1} \hat{u}_k(t)e^{ikx}. \tag{3.1.6}$$

In this method the fundamental unknowns are the coefficients $\hat{u}_k(t)$, $k = -N/2, \ldots, N/2-1$. A set of ODEs for the $\hat{u}_k$ are obtained by requiring that the residual of (3.1.1) be orthogonal to all the test functions in $Y_N = S_N$:

$$\int_0^{2\pi}\left(\frac{\partial u^N}{\partial t} + u^N\frac{\partial u^N}{\partial x} - \nu\frac{\partial^2 u^N}{\partial x^2}\right)e^{-ikx}\,dx = 0$$
$$k = -\frac{N}{2}, \ldots, \frac{N}{2} - 1. \tag{3.1.7}$$

Due to the orthogonality property of the test and trial functions,

$$\frac{\partial \hat{u}_k}{\partial t} + \widehat{\left(u^N\frac{\partial u^N}{\partial x}\right)}_k + k^2\nu\hat{u}_k = 0, \quad k = -\frac{N}{2}, \ldots, \frac{N}{2} - 1, \tag{3.1.8}$$

where

$$\widehat{\left(u^N\frac{\partial u^N}{\partial x}\right)}_k = \frac{1}{2\pi}\int_0^{2\pi} u^N\frac{\partial u^N}{\partial x}e^{-ikx}\,dx. \tag{3.1.9}$$

The initial conditions are clearly

$$\hat{u}_k(0) = \frac{1}{2\pi}\int_0^{2\pi} u(x,0)e^{-ikx}\,dx. \tag{3.1.10}$$

This ODE system is typically discretized in time by a method which is explicit for the non-linear term and implicit for the linear one.

The operator L_N is defined by $L_N u^N = -\nu(\partial^2 u^N/\partial x^2)$, whereas the discrete non-linear operator G_N is defined by $G_N(u^N) = u^N(\partial u^N/\partial x)$. The operator Q_N is the orthogonal projection operator, with respect to the continuous inner product (2.1.10), from $L^2(0,2\pi)$ to S_N. Equation (3.1.7) is the variational expression of this projection.

The wavenumber $k = -N/2$ appears unsymmetrically in this approximation. If $\hat{u}_{-N/2}$ has a non-zero imaginary part, then $u^N(t)$ is not a real-valued function. This can lead to a number of difficulties and it is advisable in practice simply to enforce the condition that $\hat{u}_{-N/2}$ is zero. This nuisance would, of course, be avoided if the approximation contained an odd rather than an even number of modes. However, the most widely used FFTs require an even number of modes. Our objective is to describe spectral methods in a way that corresponds directly to the way they are implemented. This problem of the $k = -N/2$ mode arises, in practice, for all Galerkin methods in which the FFT is employed. We assume, throughout this and the next five chapters, that these modes are always set to zero.

Equation (3.1.9) is a particular case of the general quadratic non-linear term

$$\widehat{(uv)}_k = \frac{1}{2\pi}\int_0^{2\pi} uv e^{-ikx}\,dx, \tag{3.1.11}$$

where u and v denote generic trigonometric polynomials of degree $N/2$, i.e., elements of S_N (see (2.1.12)). They have expansions similar to (3.1.6). When these are inserted into (3.1.11) and the orthogonality property (2.1.2) is invoked, the expression

$$\widehat{(uv)}_k = \sum_{p+q=k} \hat{u}_p \hat{v}_q \tag{3.1.12}$$

results. This is a convolution sum. The straightforward evaluation of (3.1.12) requires $O(N^2)$ operations. Fortunately, transform methods allow this term to be evaluated in only $O(N\log_2 N)$ operations (see Sec. 3.2). Hence, a single time-step for this Fourier Galerkin method takes only $O(N\log_2 N)$ operations.

3.1.2. Fourier Collocation

We again presume periodicity on $(0, 2\pi)$ and take $X_N = S_N$, but now think of the approximate solution u^N as represented by its values at the grid points $x_j = 2\pi j/N, j = 0, \ldots, N-1$. Recall that the grid values of u^N are related to its discrete Fourier coefficients by (2.1.22) and (2.1.24). For the collocation method we require that (3.1.1) be satisfied at these points, i.e.,

$$\left.\frac{\partial u^N}{\partial t} + u^N \frac{\partial u^N}{\partial x} - \nu \frac{\partial^2 u^N}{\partial x^2}\right|_{x=x_j} = 0 \qquad j = 0, 1, \ldots, N-1. \quad (3.1.13)$$

Initial conditions here are obviously

$$u^N(x_j, 0) = u_0(x_j). \quad (3.1.14)$$

The linear operator $L_N u^N$ is $-\nu \mathscr{D}_N^2 u^N$, where $\mathscr{D}_N$ is the Fourier collocation differentiation operator (see Sec. 2.1.3). Note that $\mathscr{D}_N^2 u^N$ is just $\partial^2 u^N/\partial x^2$. The non-linear term is discretized as $G_N(u^N) = u^N \mathscr{D}_N u^N$. The orthogonal projection is already expressed by (3.1.13). It is taken from $C^0([0, 2\pi])$ into S_N with respect to the discrete inner product (2.1.26).

In vector form, with $U = (u^N(x_0, t), u^N(x_1, t), \ldots, u^N(x_{N-1}, t))$, (3.1.13) is

$$\frac{\partial U}{\partial t} + U \boxtimes D_N U - \nu D_N^2 U = 0, \quad (3.1.15)$$

where D_N is the matrix, given by (2.1.41), which represents Fourier collocation differentiation, and $U \boxtimes V$ is the pointwise product of two vectors U and V.

Assume again that an explicit/implicit time-discretization is employed. The derivative $\partial u^N/\partial x$ is most efficiently evaluated by the transform differentiation procedure described in Sec. 2.1.3. An efficient solution procedure for the implicit term is discussed in Sec. 5.1.1. It, too, resorts to transform methods. A single time-step, then, can be performed in $O(N \log_2 N)$ operations.

The PDE itself can be written in the equivalent form

$$\frac{\partial u}{\partial t} + \frac{1}{2}\frac{\partial}{\partial x}(u^2) - \nu \frac{\partial^2 u}{\partial x^2} = 0. \quad (3.1.16)$$

The non-linear operator is approximated as $G_N(u^N) = (1/2)\mathscr{D}_N[(u^N)^2]$. The collocation discretization of (3.1.16) is

$$\frac{\partial U}{\partial t} + \frac{1}{2} D_N(U \boxtimes U) - \nu D_N^2 U = 0. \quad (3.1.17)$$

Note that the non-linear term is evaluated by first taking the pointwise square of U and then differentiating. The set of equations (3.1.17) is *not* equivalent to (3.1.15). In contrast, the Galerkin method produces the same discrete equations regardless of the precise form used for the PDE.

3.1.3. Chebyshev Tau

We now seek a solution to (3.1.1) on $(-1, 1)$ which satisfies the Dirichlet boundary conditions

$$u(-1,t) = u(1,t) = 0. \tag{3.1.18}$$

The trial space X_N consists of all the members of $\mathbb{P}_N$ (the set of polynomials of degree $\leqslant N$) which vanish for $x = \pm 1$. The discrete solution is expressed as the Chebyshev series

$$u^N(x,t) = \sum_{k=0}^{N} \hat{u}_k(t) T_k(x), \tag{3.1.19}$$

with the Chebyshev coefficients comprising the fundamental representation of the approximation. The equation (3.1.1) is enforced by insisting that the residual be orthogonal to the test functions in $Y_N = \mathbb{P}_{N-2}$:

$$\int_{-1}^{1} \left(\frac{\partial u^N}{\partial t} + u^N \frac{\partial u^N}{\partial x} - \nu \frac{\partial^2 u^N}{\partial x^2} \right) T_k(x)(1-x^2)^{-1/2}\, dx = 0 \qquad k = 0, \ldots, N-2. \tag{3.1.20}$$

Note that the weight function $w(x) = (1-x^2)^{-1/2}$ appropriate to the Chebyshev polynomials is used in the orthogonality condition. The boundary conditions (3.1.18) impose the additional constraints

$$u^N(-1,t) = u^N(1,t) = 0. \tag{3.1.21}$$

Equation (3.1.20) reduces to

$$\frac{\partial \hat{u}_k}{\partial t} + \widehat{\left(u^N \frac{\partial u^N}{\partial x} \right)}_k - \nu \hat{u}_k^{(2)} = 0, \quad k = 0, 1, \ldots, N-2, \tag{3.1.22}$$

where $\hat{u}_k^{(2)}$ is given by (2.4.27) and

$$\widehat{\left(u^N \frac{\partial u^N}{\partial x} \right)}_k = \frac{2}{\pi c_k} \int_{-1}^{1} \left(u^N \frac{\partial u^N}{\partial x} \right) T_k(x)(1-x^2)^{-1/2}\, dx, \tag{3.1.23}$$

where the c_k are given by (2.4.10). In terms of the Chebyshev coefficients, the boundary conditions (3.1.21) become, through the use of (2.4.6),

$$\sum_{k=0}^{N} \hat{u}_k = 0 \tag{3.1.24}$$

$$\sum_{k=0}^{N} (-1)^k \hat{u}_k = 0. \tag{3.1.25}$$

The initial conditions are

$$\hat{u}_k(0) = \frac{2}{\pi c_k} \int_{-1}^{1} u_0(x) T_k(x)(1-x^2)^{-1/2}\, dx \quad k = 0, 1, \ldots, N. \tag{3.1.26}$$

Equations (3.1.22), (3.1.24), (3.1.25) and (3.1.26) form a complete set of ODEs for this approximation.

The operator L_N is defined by $L_N u^N = -\nu(\partial^2 u^N / \partial x^2)$ and G_N is defined by

$G_N(u^N) = u^N(\partial u^N/\partial x)$. The operator Q_N is the orthogonal projection operator, with respect to the continuous inner product (2.2.4) for the weight $w(x) = (1 - x^2)^{-1/2}$, from $L^2_w(-1, 1)$ into $\mathbb{P}_{N-2}$. Equation (3.1.20) is the variational expression of this projection.

The expression in (3.1.23) is a special case of

$$\widehat{(uv)}_k = \frac{2}{\pi c_k} \int_{-1}^{1} uvT_k(x)(1 - x^2)^{-1/2}\, dx, \tag{3.1.27}$$

which is equal to

$$\widehat{(uv)}_k = \frac{1}{2} \sum_{p+q=k} \hat{u}_p \hat{v}_q + \sum_{|p-q|=k} \hat{u}_p \hat{v}_q. \tag{3.1.28}$$

A typical time-discretization is explicit for the non-linear term and implicit for the linear one. Transform methods (see Sec. 3.2) are an efficient means of evaluating the non-linear term. The implicit terms (including the boundary conditions) can be solved in $O(N)$ operations by the method described in Sec. 5.1.2.

If the boundary conditions are Neumann, then (3.1.24) and (3.1.25) are replaced by

$$\begin{aligned} \sum_{k=0}^{N} k^2 \hat{u}_k &= 0 \\ \sum_{k=0}^{N} (-1)^k k^2 \hat{u}_k &= 0. \end{aligned} \tag{3.1.29}$$

3.1.4. Chebyshev Collocation

For a collocation approximation to the Dirichlet problem the trial space X_N is the same as for the previous example and the solution u^N is represented by its values at the grid points $x_j = \cos \pi j/N$, $j = 0, 1, \ldots, N$. The grid values of u^N are related to the discrete Chebyshev coefficients by (2.4.15) and (2.4.17). The discretization of the PDE is

$$\left. \frac{\partial u^N}{\partial t} + u^N \frac{\partial u^N}{\partial x} - \nu \frac{\partial^2 u^N}{\partial x^2} \right|_{x=x_j} = 0 \qquad j = 1, \ldots, N-1, \tag{3.1.30}$$

with

$$u^N(-1, t) = u^N(1, t) = 0 \tag{3.1.31}$$

$$u^N(x_j, 0) = u_0(x_j) \qquad j = 0, \ldots, N. \tag{3.1.32}$$

The linear operator $L_N u^N$ is $-\nu \mathscr{D}^2 u^N$, where $\mathscr{D}_N$ is the Chebyshev collocation differentiation operator and $G_N(u^N) = u^N \mathscr{D}_N u^N$. The orthogonal projection is expressed by (3.1.30)—(3.1.31). It maps $C^0([-1, 1])$ into Y_N (which is the same as X_N) with respect to the discrete inner product (2.2.24). Let

$U = (u^N(x_0, t), \ldots, u^N(x_N, t))$. Then, (3.1.30) can be written

$$Z_N\left(\frac{\partial U}{\partial t} + U \boxtimes D_N U - \nu D_N^2 U\right) = 0, \tag{3.1.33}$$

where D_N is the Chebyshev collocation differentiation matrix given by (2.4.31) and Z_N is the matrix which represents setting the first and last points of a vector to zero. The boundary conditions (3.1.31) are enforced by directly setting the first and last entries of U to 0. The numerical analysis of this Chebyshev collocation method is reviewed in Sec. 12.3.

The non-linear term can be evaluated efficiently by transform methods. The best direct solution method for the implicit term (see Sec. 5.1.3), however, takes $O(N^2)$ operations. Iterative solution methods (see Chap. 5) are an alternative.

If a Chebyshev pseudospectral transform method (see Sec. 3.2) is used for the non-linear term and a Chebyshev tau method for the implicit terms, then a single time-step takes only $O(N \log_2 N)$ operations. Such a mixed discretization scheme is, in fact, typical of most large scale algorithms.

3.2. Convolution Sums

A principal algorithmic component of efficient Galerkin methods for non-linear or variable-coefficient problems is the evaluation of convolution sums. Consider, however, the Fourier Galerkin treatment of the quadratic term

$$w(x) = u(x)v(x). \tag{3.2.1}$$

In the case of an infinite series expansion, we have the familiar convolution sum

$$\hat{w}_k = \sum_{m+n=k} \hat{u}_m \hat{v}_n, \tag{3.2.2}$$

where

$$u(x) = \sum_{m=-\infty}^{\infty} \hat{u}_m e^{imx} \tag{3.2.3}$$

$$v(x) = \sum_{n=-\infty}^{\infty} \hat{v}_n e^{inx}$$

$$\hat{w}_k = \frac{1}{2\pi} \int_0^{2\pi} w(x) e^{-ikx}\, dx. \tag{3.2.4}$$

When u, v, and w are approximated by their respective truncated Fourier series of degree $N/2$, (3.2.2) becomes

$$\hat{w}_k = \sum_{\substack{m+n=k \\ |m|,|n| \leqslant N/2}} \hat{u}_m \hat{v}_n, \tag{3.2.5}$$

where $|k| \leqslant N/2$. The direct summation implied by (3.2.5) takes $O(N^2)$ operations. (In three dimensions, the cost is $O(N^4)$, provided, as discussed in Orszag (1980), that one utilizes the tensor product nature of multidimensional spectral approximations.) This is prohibitively expensive, especially when one considers that for a non-linear term a finite-difference algorithm takes $O(N)$ operations in one dimension (and $O(N^3)$ in three). However, the use of transform methods enables (3.2.5) to be evaluated in $O(N \log_2 N)$ operations (and the three-dimensional generalization in $O(N^3 \log_2 N)$ operations). This technique was developed independently by Orszag (1969, 1970) and Eliasen, Machenhauer and Rasmussen (1970). It was the single most important development which made spectral Galerkin methods practical for large scale computations.

3.2.1. Pseudospectral Transform Methods

The approach taken in the transform method for evaluating (3.2.5) is to use the inverse discrete Fourier transform (DFT) to transform $\hat{u}_m$ and $\hat{v}_n$ to physical space, to perform there a multiplication similar to (3.2.1), and then to use the DFT to determine $\hat{w}_k$. This must be done carefully, however. To illustrate the subtle point involved, we introduce the discrete transforms (Sec. 2.1.2)

$$\begin{aligned} U_j &= \sum_{k=-N/2}^{N/2-1} \hat{u}_k e^{ikx_j} \\ V_j &= \sum_{k=-N/2}^{N/2-1} \hat{v}_k e^{ikx_j} \end{aligned} \qquad j = 0, 1, \ldots, N-1 \tag{3.2.6}$$

and define

$$W_j = U_j V_j \qquad j = 0, 1, \ldots, N-1 \tag{3.2.7}$$

and

$$\hat{W}_k = \frac{1}{N} \sum_{j=0}^{N-1} W_j e^{-ikx_j} \qquad k = -\frac{N}{2}, \ldots, \frac{N}{2} - 1, \tag{3.2.8}$$

where

$$x_j = 2\pi j/N.$$

Use of the discrete transform orthogonality relation (2.1.23) leads to

$$\begin{aligned} \hat{W}_k &= \sum_{m+n=k} \hat{u}_m \hat{v}_n + \sum_{m+n=k\pm N} \hat{u}_m \hat{v}_n \\ &= \hat{w}_k + \sum_{m+n=k\pm N} \hat{u}_m \hat{v}_n. \end{aligned} \tag{3.2.9}$$

The second term on the right-hand side is the aliasing error. If the convolution sums are evaluated as described above, then the method is not a true spectral Galerkin method. Orszag (1971d) termed it a pseudospectral method. The convolution sum (3.2.5) in the pseudospectral method is evaluated at the cost of 3 FFTs and N multiplications. The total operation count is $(15/2)N\log_2 N$ multiplications. The generalization of the pseudospectral evaluation of convolution sums to more than one dimension is straightforward.

There are two basic techniques for removing the aliasing error from (3.2.9). They are discussed in the following two subsections.

3.2.2. Aliasing Removal by Padding or Truncation

The key to this "de-aliasing" technique is the use of a discrete transform with M rather than N points, where $M \geqslant 3N/2$. Let

$$\begin{aligned} y_j &= 2\pi j/M \\ U_j &= \sum_{k=-M/2}^{M/2-1} \tilde{u}_k e^{iky_j} \qquad j = 0, 1, \ldots, M-1 \\ V_j &= \sum_{k=-M/2}^{M/2-1} \tilde{v}_k e^{iky_j} \end{aligned} \tag{3.2.10}$$

$$W_j = U_j V_j, \tag{3.2.11}$$

where

$$\tilde{u}_k = \begin{cases} \hat{u}_k & |k| \lesssim N/2 \\ 0 & \text{otherwise} \end{cases}. \tag{3.2.12}$$

Thus, the $\tilde{u}_k$ coefficients are the $\hat{u}_k$ coefficients padded with zeros for the additional wavenumbers. Likewise, let

$$\tilde{W}_k = \frac{1}{M} \sum_{j=0}^{M-1} W_j e^{-iky_j} \qquad k = -\frac{M}{2}, \ldots, \frac{M}{2} - 1. \tag{3.2.13}$$

Then

$$\tilde{W}_k = \sum_{m+n=k} \tilde{u}_m \tilde{v}_n + \sum_{m+n=k\pm M} \tilde{u}_m \tilde{v}_n. \tag{3.2.14}$$

We are only interested in $\tilde{W}_k$ for $|k| \leqslant N/2$, and choose M such that the second term on the right-hand side vanishes for these k. Since $\tilde{u}_m$ and $\tilde{v}_m$ are zero for $|m| > N/2$, the worst case condition is

$$-\frac{N}{2} - \frac{N}{2} \leqslant \frac{N}{2} - 1 - M$$

or

$$M \geqslant \frac{3N}{2} - 1. \tag{3.2.15}$$

The operation count for this transform method is $(45/4)N\log_2(\frac{3}{2}N)$, which is roughly 50% larger than the simpler, but aliased method discussed earlier. For obvious reasons this technique is sometimes referred to as the 3/2-rule. As described here it requires an FFT which can handle prime factors of 3. If only a prime factor 2 FFT is available, then this de-aliasing technique can be implemented by choosing $(3/2)N$ to be a power of 2. This de-aliasing technique is also termed truncation and is sometimes referred to as the 2/3-rule.

3.2.3. Aliasing Removal by Phase Shifts

A second method to remove the aliasing terms employs phase shifts. In this case, (3.2.6) is replaced with

$$\begin{aligned} \tilde{U}_j' &= \sum_{k=-N/2}^{N/2-1} \hat{u}_k e^{ik(x_j+\Delta)} \\ \tilde{V}_j' &= \sum_{k=-N/2}^{N/2-1} \hat{v}_k e^{ik(x_j+\Delta)}, \end{aligned} \tag{3.2.16}$$

which are just the transforms on a grid shifted by the factor Δ in physical space. One then computes

$$W_j' = U_j' V_j' \tag{3.2.17}$$

$$\hat{W}_k' = \frac{1}{N} \sum_{j=0}^{N-1} W_k' e^{-ik(x_j+\Delta)}. \tag{3.2.18}$$

This last quantity is just

$$\hat{W}_k' = \sum_{m+n=k} \hat{u}_m \hat{v}_n + e^{\pm iN\Delta} \sum_{m+n=k\pm N} \hat{u}_m \hat{v}_n. \tag{3.2.19}$$

If one chooses $\Delta = \pi/N$, i.e., one shifts by half a grid cell, then

$$\hat{w}_k = \tfrac{1}{2}[\hat{W}_k + \hat{W}_k']. \tag{3.2.20}$$

Thus, the aliasing contributions to the non-linear term can be eliminated completely at the cost of two evaluations of the convolution sum. The cost here is $15N\log_2 N$. This is greater than the cost of the padding technique. However, if only a power of 2 FFT is available, then the padding technique requires the use of $M = 2N$ points rather than $(3/2)N$. Its cost then increases to $15N\log_2 N$.

The phase shift technique and the padding method can both be extended to two and three dimensions. This discussion is postponed until Sec. 7.2, where it is given in the context of applications to simulations of incompressible, homogeneous turbulence.

Rogallo (1977) observed how the phase shifting strategy can be incorporated at no extra cost into an otherwise pseudospectral algorithm to produce a method which has greatly reduced aliasing errors. Suppose that the time-differencing scheme is second-order Runge–Kutta (see Sec. 4.3.2). At the first stage, the convolution sum is evaluated by the pseudospectral transform method described in Sec. 3.2.1 except that U_j and V_j are computed not by (3.2.6) but rather by (3.2.16), where Δ is a random number in $(0, 2\pi/N)$. In the second stage, (3.2.16) is again used for U_j and V_j, but now with Δ replaced by $\Delta + \pi/N$. As a result the aliasing errors at the end of the full Runge–Kutta step are reduced to $O(\Delta t^2)$ times the pure pseudospectral aliasing errors, where Δt is the size of the time-step. The use of a random shift Δ ensures that the remaining aliasing errors are uncorrelated from step to step.

3.2.4. Convolution Sums in Chebyshev Methods

Quadratic non-linearities also produce convolution-type sums in Chebyshev Galerkin and tau methods. A typical sum is given in (3.1.28). The 3/2-rule, but for Chebyshev rather than Fourier transforms, provides a straightforward method of evaluating this sum without aliasing.

It can also be evaluated pseudospectrally by transforming $\hat{u}_k$ and $\hat{v}_k$ to physical space with an N-mode Chebyshev transform, forming the product $U_j V_j$ there and then transforming back. This, of course, introduces aliasing errors. This modification to the algorithm discussed in Sec. 3.1.3 produces a pseudospectral Chebyshev tau method.

3.2.5. Relation Between Collocation and Pseudospectral Methods

In most cases Fourier pseudospectral methods are algebraically equivalent to collocation methods. Consider again the simple Burgers equation (3.1.1) periodic on $(0, 2\pi)$. The Galerkin approximation is

$$\frac{d\hat{u}_k}{dt} + \sum_{m+n=k} \hat{u}_m \hat{v}_n + \nu k^2 \hat{u}_k = 0 \qquad k = -\frac{N}{2}, \ldots, \frac{N}{2} - 1, \quad (3.2.21)$$

where $\hat{v}_k = ik\hat{u}_k$.

The pseudospectral approximation uses a fully-aliased transform method to evaluate the convolution sum. Equation (3.2.21) is, in effect, replaced by

$$\frac{d\hat{u}_k}{dt} + \sum_{m+n=k} \hat{u}_m \hat{v}_n + \sum_{m+n=k\pm N} \hat{u}_m \hat{v}_n + \nu k^2 \hat{u}_k = 0 \qquad (3.2.22)$$

(see (3.2.9)).

The collocation approximation may be written

$$\left.\frac{\partial u^N}{\partial t} + u^N(x) v^N(x) - \nu \frac{\partial^2 u^N}{\partial x^2}\right|_{x=x_j} = 0 \qquad j = 0, \ldots, N-1, \quad (3.2.23)$$

where $v^N(x) = du^N/dx$. Resorting to the discrete Fourier series representation of u and v at the grid points we have that (3.2.23) is

$$\sum_{l=-N/2}^{N/2-1} \frac{d\tilde{u}_l}{dt} e^{ilx_j} + \left(\sum_{m=-N/2}^{N/2-1} \tilde{u}_m e^{imx_j}\right)\left(\sum_{n=-N/2}^{N/2-1} \tilde{v}_n e^{inx_j}\right) + \nu \sum_{l=-N/2}^{N/2-1} l^2 \tilde{u}_l e^{ilx_j} = 0 \qquad j = 0, 1, \ldots, N-1. \tag{3.2.24}$$

Applying the inverse DFT to (3.2.24) and using the orthogonality relation (2.1.23), we find

$$\frac{d\tilde{u}_k}{dt} + \sum_{m+n=k} \tilde{u}_m \tilde{v}_n + \sum_{m+n=k\pm N} \tilde{u}_m \tilde{v}_n + \nu k^2 \tilde{u}_k = 0 \qquad k = -\frac{N}{2}, \ldots, \frac{N}{2} - 1. \tag{3.2.25}$$

This is identical to (3.2.22). Thus, except for round-off error and the precise choice of initial condition, the pseudospectral and collocation discretizations of (3.1.1) are identical. So are the pseudospectral and collocation discretizations of the Burgers equation in the form (3.1.16). The same equivalence occurs for more complicated systems of equations such as incompressible Navier–Stokes (see Sec. 7.2.4).

It should be obvious that a scheme for the Burgers equation implemented as a standard collocation method can be de-aliased, if desired, by a truncation method. If at every time-step one sets to zero the discrete Fourier coefficients for which $|k| \geqslant (1/3)N$, the aliasing term in (3.2.22) vanishes. The collocation scheme then becomes algebraically equivalent to a Galerkin method. In this context the truncation method is known as the 2/3-rule. For the Burgers equation, this truncation can be accomplished as part of the solution of the implicit part of the equation. This is solved in transform space (see Sec. 5.1.1) and the unwanted Fourier coefficients are easily discarded.

The algebraic equivalence between pseudospectral and collocation methods does not extend to Chebyshev tau discretizations. Here one must be careful to distinguish between the two approaches.

We might add that in some quarters the term pseudospectral method is used to refer to what we call in this book a collocation method. We use the adjective pseudospectral solely in terms of otherwise Galerkin or tau methods in which the non-linear terms are subjected to a pseudospectral evaluation.

3.3. Boundary Conditions

We have seen that there are several ways of imposing numerically the given mathematical boundary conditions. Galerkin and tau methods have uniquely defined numerical boundary conditions. Collocation methods, on the other hand, allow a great deal of freedom. For instance, Neumann and Robin boundary conditions may be imposed directly or they may be incorporated indirectly into the collocation operator.

Consider, for simplicity, the one-dimensional problem with Robin boundary conditions

$$\begin{aligned} -\frac{d^2u}{dx^2}+u&=f \qquad -1<x<1 \\ B_+u \equiv \beta\frac{du}{dx}+\alpha u &= 0, \qquad \text{at } x=1 \\ B_-u \equiv \delta\frac{du}{dx}+\gamma u &= 0, \qquad \text{at } x=-1, \end{aligned} \tag{3.3.1}$$

where α, β, γ, and δ are real constants such that

$$\alpha^2+\beta^2 \neq 0 \qquad \alpha\beta \geqslant 0$$
$$\gamma^2+\delta^2 \neq 0 \qquad \gamma\delta \leqslant 0.$$

This problem has a unique solution. Assume that we are given $N+1$ collocation points $x_j, j=0,\ldots,N$ where $x_0=+1$ and $x_N=-1$.

One formulation of this problem is

$$\begin{aligned} -\frac{d^2u^N}{dx^2}+u^N-f\bigg|_{x=x_j} &= 0 \qquad j=1,\ldots,N-1 \\ (B_+u^N)(x_0) &= 0 \\ (B_-u^N)(x_N) &= 0, \end{aligned} \tag{3.3.2}$$

where u^N is a polynomial of degree N. This is clearly the appropriate formulation for Dirichlet boundary conditions ($\beta=\delta=0$), and if used with Neumann or Robin boundary conditions constitutes a direct imposition of these boundary conditions.

An alternative formulation is

$$-\frac{d}{dx}\left[I_N\left(B\frac{du^N}{dx}\right)\right]+u^N-f\bigg|_{x=x_j}=0 \qquad j=0,\ldots,N. \tag{3.3.3}$$

The polynomial $I_N(B(dv/dx))$ is defined for any $v\in\mathbb{P}_N$ by

$$I_N\left(B\frac{dv}{dx}\right)\in\mathbb{P}_N$$

$$\begin{aligned} I_N\left(B\frac{dv}{dx}\right)(x_0) &= -\gamma v(x_0) \\ I_N\left(B\frac{dv}{dx}\right)(x_j) &= \frac{dv}{dx}(x_j), \qquad j=1,\ldots,N-1 \\ I_N\left(B\frac{dv}{dx}\right)(x_N) &= -\alpha v(x_N). \end{aligned} \tag{3.3.4}$$

Thus, $I_N(B(dv/dx))$ is the interpolant of dv/dx at the interior nodes. However, at the boundary points x_0 and x_N, it modifies the values of dv/dx in accordance with (3.3.1). In this case, the boundary conditions are incorporated into the differential operator and thus are enforced indirectly.

Clearly, the solution to (3.3.2) satisfies the boundary conditions exactly, whereas the solution to (3.3.3) does not. In the latter case it has been shown that the error in the boundary conditions decays spectrally. The stability and convergence of this approximation can be deduced from the general theory presented in Chap. 10. See, for instance, Example 4 of Sec. 10.5.1. The details are available in Canuto (1986).

An alternative scheme for imposing the boundary conditions indirectly is given by (Streett (1983))

$$\begin{aligned} -\frac{d^2u^N}{dx^2} + u^N - f\bigg|_{x=x_j} &= 0 \qquad j = 1, \ldots, N-1 \\ -\frac{d}{dx}\left[I_N\left(B\frac{du^N}{dx}\right)\right] + u^N - f\bigg|_{x=x_j} &= 0 \qquad j = 0, N. \end{aligned} \tag{3.3.5}$$

The equation is enforced as usual in the interior and the boundary conditions are included in the differential operator only at the boundary. This formulation is advantageous in the context of iterative solution procedures for the implicit equations (see Sec. 5.2.3).

We also note that the technique of incorporating Robin boundary conditions indirectly into the operator provides a convenient means of handling the boundary conditions of fourth-order problems which have both Dirichlet and Neumann data at the boundaries.

Problems in an infinite domain can be reduced to problems in a bounded domain by introducing an artificial boundary at a finite distance. A boundary condition (a so-called "radiation" condition) has to be enforced on the artificial boundary. It should simulate the behavior of the solution at infinity. When a spectral method is used on the finite domain, the radiation condition has to be of the highest accuracy, in order to reduce the effects of the truncation of the domain. For the Poisson equation, Canuto, Hariharan and Lustman (1985) have proposed a non-local radiation condition for which the formal truncation error is $O(r_\infty^{-N})$, where r_∞ is the distance of the artificial boundary Γ_∞ and N is the degree of the polynomials used. Assuming Γ_∞ to be a circle, the radiation condition has the form

$$\frac{\partial u}{\partial n} + Ku = g, \tag{3.3.6}$$

where Ku is an integral operator on Γ_∞ which can be efficiently computed by the FFT. Canuto et al. recommend the implicit implementation of this boundary condition, and indicate a way of preconditioning the corresponding algebraic system (see Sec. 5.2.3). The use of a similar radiation condition for more general Helmholtz problems has been discussed by Hariharan (1986).

3.4. Coordinate Singularities

3.4.1. Polar Coordinates

Poisson's equation in a disk,

$$\begin{aligned} -\Delta u &= f \qquad 0 < r < 1,\ 0 \leqslant \theta < 2\pi \\ u &= 0 \qquad r = 1 \end{aligned} \tag{3.4.1}$$

is an obvious example of a problem with a coordinate singularity. A standard Fourier expansion in θ, either Galerkin or collocation, is clearly in order. The numerical solution may be written

$$u(r, \theta) = \sum_{m=-M/2}^{M/2-1} \tilde{u}_m(r) e^{im\theta}. \tag{3.4.2}$$

There have been several proposals for Chebyshev expansions in radius. One of these is

$$\tilde{u}_m(r) = \sum_{\substack{n=0 \\ n+m \text{ even}}}^{N} a_{mn} T_n(r). \tag{3.4.3}$$

Thus, the numerical solution to (3.4.1) will have the same parity, $\tilde{u}_m(-r) = (-1)^m \tilde{u}_m(r)$, as the analytic one. A further refinement (Orszag and Patera (1983)) is to incorporate the decay of $u(r, \theta)$ near the origin by using

$$\tilde{u}_m(r) = r^m \sum_{\substack{n=0 \\ n+m \text{ even}}}^{N} a_{mn} T_n(r). \tag{3.4.4}$$

Both of these expansions have better resolution near the outer edge than near the origin, as is evident from the concentration of the zeroes of $T_n(r)$ near the edge. Improved center resolution can be achieved by expanding in

$$x = 2r - 1, \tag{3.4.5}$$

and using all of the Chebyshev polynomials.

These expansions must satisfy the condition

$$\frac{\partial u}{\partial \theta} = 0 \tag{3.4.6}$$

at the origin. This expresses the requirement that the solution be single-valued. Obviously, this requires

$$\tilde{u}_m(r) = 0 \qquad \text{at } r = 0 \text{ for } m \neq 0. \tag{3.4.7}$$

The appropriate condition on the remaining component is

$$\frac{d\tilde{u}_0}{dr} = 0 \qquad \text{at } r = 0. \tag{3.4.8}$$

These latter two conditions are readily applied in a tau approach. Note that the expansion (3.4.4) automatically satisfies (3.4.7) and (3.4.8).

When $\mathbf{u}$ is a vector quantity, such as velocity, the necessary condition at the origin is

$$\frac{\partial \mathbf{u}}{\partial \theta} = 0. \tag{3.4.9}$$

In polar coordinates, $\mathbf{u} = u_r \hat{\mathbf{r}} + u_\theta \hat{\boldsymbol{\theta}}$, where $\hat{\mathbf{r}}$ and $\hat{\boldsymbol{\theta}}$ are the unit vectors in the radial and azimuthal directions, and u_r and u_θ are the respective velocity components. These unit vectors depend upon θ, and this dependence must be included in applying (3.4.9). The result is

$$\begin{aligned} u_{r,m} = u_{\theta,m} &= 0 \qquad \text{for } |m| \neq 1 \\ u_{r,m} + imu_{\theta,m} &= 0 \qquad \text{for } |m| = 1. \end{aligned} \tag{3.4.10}$$

These types of boundary conditions at the origin have been used (for mixed spectral/finite-difference calculations) by Schnack and Killeen (1980) and by Aydemir and Barnes (1985). They have been justified theoretically (for mixed spectral/finite-element calculations) by Mercier and Raugel (1982).

The expansions (3.4.3) and (3.4.4) are not well suited to pure collocation methods because there would need to be different collocation points in r for the even m and odd m components. One needs a Fourier Galerkin–Chebyshev collocation method.

Suppose now that a standard Chebyshev expansion is combined with the mapping (3.4.5). If the Gauss–Lobatto points are used, then conditions such as (3.4.6) and (3.4.9) need to be imposed at $r = 0$ (or $x = -1$). Alternatively, one can use the Gauss–Radau points which include the point $r = 1$ (or $x = 1$) but exclude the origin. There is then no need to impose a boundary condition at $r = 0$.

3.4.2. Spherical Polar Coordinates

In addition to a singularity at the origin, spherical coordinates also exhibit singularities at the poles. The natural expansion functions are spherical harmonics. These automatically account for the proper behavior at the poles and are heavily used in meteorological computations (Haltiner and Williams (1980), Jarraud and Baede (1985)). Non-linear terms can be calculated by transform methods (Orszag (1970), Eliasen, Machenhauer and Rasmussen (1970)). However, since spherical harmonics use associated Legendre functions, the FFT cannot be employed for the transform in latitude.

Orszag (1974) suggested the use of special Fourier series in latitude (coupled with regular Fourier series in longitude). Boyd (1978a, 1978b) has also considered these expansions. They allow the use of the FFT for all the series manipulations. However, great care is needed at the poles and no rigorous approximation theory results are available for these expansions. Furthermore, severe time-step limitations can arise in time-dependent problems.

3.5. Two-Dimensional Mapping

If one wishes to solve a two-dimensional problem by spectral methods and the geometry is not directly conducive to the use of a tensor product expansion, then one might be able to map the domain of interest onto a more standard computational domain such as a square or a circle. (This might not always be possible or even desirable. One must then resort to the domain decomposition methods discussed in Chap. 13.)

One of the standard mapping techniques is based on conformal transformations. These are discussed in most elementary texts on complex variables, e.g., in Carrier, Krook and Pearson (1966) or Ahlfors (1966). Among their advantages are the preservation of orthogonality and of simple operators such as the divergence and the gradient. Conformal mappings are widely used in two-dimensional fluid dynamical problems. The book by Milne-Thomson (1966) contains an extensive discussion. More recently, several numerical methods have been devised for generating conformal mappings; see, for example, Meiron, Orszag and Israeli (1981) and Trefethen (1980).

A fairly simple procedure exists for mapping a square $\hat{\Omega}$ into a quadrilateral Ω with curved boundaries. The basic geometry is illustrated in Fig. 3.1. Let the four sides of the quadrilateral be denoted by Γ_i, for $i = 1, 2, 3, 4$ and those of the square by $\hat{\Gamma}_i$. One uses mappings π_i from $\hat{\Gamma}_i$ to Γ_i to construct the mapping Ψ from $\hat{\Omega}$ to Ω. Following Gordon and Hall (1973a, 1973b), the

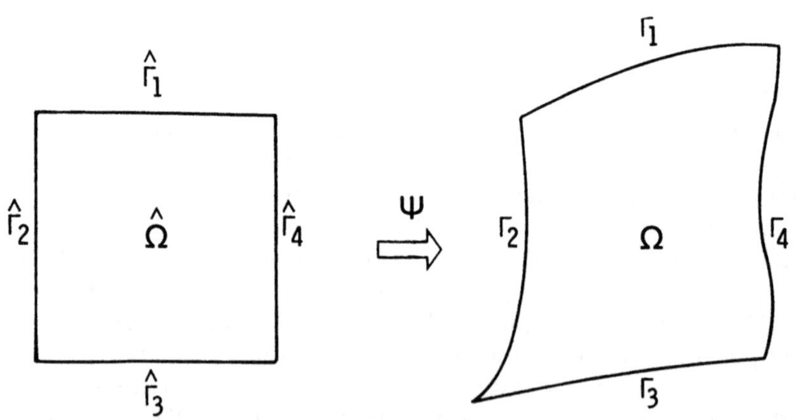

Figure 3.1. Mapping of the unit rectangle $\hat{\Omega} = [-1, 1]^2$ into a quadrilateral Ω with curved boundaries.

mapping Ψ can be expressed in terms of the π_i as

$$\begin{aligned}\Psi(\xi,\eta) &= \frac{1-\eta}{2}\pi_1(\xi) + \frac{1+\eta}{2}\pi_3(\xi)\\ &\quad + \frac{1-\xi}{2}\left[\pi_2(\eta) - \frac{1+\eta}{2}\pi_2(1) - \frac{1-\eta}{2}\pi_2(-1)\right] \qquad (3.5.1)\\ &\quad + \frac{1+\xi}{2}\left[\pi_4(\eta) - \frac{1+\eta}{2}\pi_4(1) - \frac{1-\eta}{2}\pi_4(-1)\right].\end{aligned}$$

(We assume that the arcs Γ_1 and Γ_3 are oriented from left to right and the arcs Γ_2 and Γ_4 from bottom to top.)

In the event that the domain Ω is actually a subdomain in a domain decomposition method (see Chap. 13), the use of an isoparametric description of the curves Γ_i may be desirable. Here one chooses the curves Γ_i so that they are exactly parametrizable by polynomials of the same order as the discretization within Ω. This approach is common in finite-element methods (see, e.g., Ciarlet (1978)) and has been used in spectral-element methods (see Sec. 13.3) by Korczak and Patera (1986).

CHAPTER 4

Temporal Discretization

4.1. Introduction

In most applications of spectral methods to partial differential equations the spatial discretization is spectral but the temporal discretization uses conventional finite-differences. The typical evolution equation can be written

$$\frac{\partial u}{\partial t} = f(u, t) \qquad t > 0$$
$$u(0) = 0, \tag{4.1.1}$$

where the (generally) non-linear operator f contains the spatial part of the PDE. (The dependence of f upon t will not be indicated explicitly hereafter except when it is essential.) Following the general formulation of Chap. 3 the semi-discrete version is

$$Q_N \frac{du^N}{dt} = Q_N f_N(u^N),$$

where u^N is the spectral approximation to u, f_N denotes the spectral approximation to the operator f, and Q_N is the projection operator which characterizes the scheme. Let us set $U(t) = Q_N u^N(t)$. For example, in a collocation method for a Dirichlet boundary value problem, $U(t)$ represents the set of the interior grid values of $u^N(t)$. Then, the previous discrete problem can be written in the form

$$\frac{dU}{dt} = F(U). \tag{4.1.2}$$

This is also called a method-of-lines approach or a continuous-in-time discretization. We shall often confine our discussion of time-discretizations to the linearized version of (4.1.2), i.e., we consider

$$\frac{dU}{dt} = LU, \tag{4.1.3}$$

where L is the matrix resulting from the linearization of (4.1.2). For simplicity, we assume that L is a diagonalizable matrix. We can now draw upon the extensive literature on numerical methods for ODEs to examine alternative time discretizations of the full PDE. Some standard references are the books by Gear (1971), Lambert (1973), and Shampine and Gordon (1975), and review articles by Shampine, Watts, and Davenport (1976) and Gear (1981). The most crucial property of L is its spectrum, for this will determine the stability of the time-discretization.

If the spatial discretization is presumed fixed, then we use the term stability in its ODE context. Let U^n denote the computed solution at the time $t_n = n\Delta t$. The time-discretization is said to be stable if there exist constants δ, C, and K, independent of Δt, such that, for all $T > 0$,

$$\| U^n \| \leqslant Ce^{KT} \| U^0 \| \tag{4.1.4}$$

for $0 \leqslant t_n \leqslant T$ and for all $0 \leqslant \Delta t < \delta$, where $\|\cdot\|$ is some spatial norm of U^n.

In many problems the solution is bounded in some norm for all $t > 0$. In these cases a method which produces the exponential growth allowed by the estimate (4.1.4) is not practical for long-time integrations. For such problems the notion of asymptotic (or absolute) stability is useful. The region of absolute stability of a numerical method is defined for the scalar model problem

$$\frac{dU}{dt} = \lambda U \tag{4.1.5}$$

to be the set of all $\lambda \Delta t$ such that $\| U^n \|$ is bounded as $t \to \infty$. Furthermore, a method is called A-stable if the region of absolute stability includes the region $\mathrm{Re}\{\lambda \Delta t\} \leqslant 0$. Finally, we say that a numerical method is asymptotically stable for a particular problem if, for sufficiently small $\Delta t > 0$, the product of Δt times every eigenvalue of L lies within the region of absolute stability.

On the other hand, we may be interested in the behavior of the computed solution as both the spatial and temporal discretizations are refined. We now define stability by an estimate of the form (4.1.4) where C and K are independent of both Δt and the spatial discretization parameter N, the norm is independent of N, but δ will in general be a function of N. The functional dependence of δ upon N which is necessary to obtain an estimate of the form (4.1.4) is termed the stability limit of the numerical method. If δ is in fact independent of N, then the method is called unconditionally stable. Clearly, a necessary condition for the fully discrete problem to be stable is that the semi-discrete problem be stable in the sense to be discussed in Sec. 10.5.

In this chapter we will summarize what is known about the spectra of the basic spatial operators. We will then discuss several standard time-discretizations and their stability limits on various problems. Next, we shall mention some recent work on obtaining spectral accuracy for the time-discretization as well. Finally, we will address the significance of conservation laws and aliasing in evolution problems.

4.2. The Eigenvalues of Basic Spectral Operators

The spatial eigenfunctions of Fourier approximations to constant-coefficient problems are just e^{ikx} for $-N/2 \leqslant k < N/2$. The eigenvalues of such problems are apparent. We shall discuss here the qualitative behavior of the eigenvalues of Chebyshev and Legendre spectral approximations of the first-order hyperbolic operator $Lu = du/dx$ and the second-order diffusion operator $Lu = d^2u/dx^2$. The theoretical discussion of the spectra of these and slightly more general operators is postponed until Sec. 11.4.

4.2.1. The First-Derivative Operator

We consider here the advection operator

$$Lu = \frac{du}{dx} \qquad \text{on } (-1, 1) \tag{4.2.1}$$

subject to the boundary condition

$$u(1) = 0. \tag{4.2.2}$$

The eigenvalues of tau approximations are complex numbers. Their real parts are all strictly negative. One has the estimate

$$|\lambda| \leqslant O(N^2) \qquad \text{as } N \to \infty. \tag{4.2.3}$$

As shown by Trefethen and Trummer (1987), round-off errors have a significant effect upon numerical computations of first-derivative eigenvalues. They explain that the source of the problem is the exponentially decaying character of the eigenfunctions: these behave roughly as $e^{x\,\mathrm{Re}\{\lambda\}}$. Once $e^{+2\,\mathrm{Re}\{\lambda\}}$ falls below the machine precision, the eigenfunctions cannot be approximated in any meaningful sense. Since the real part of λ becomes increasingly negative as N increases, there will be a value of N beyond which the eigenvalues can no longer be computed reliably (with fixed precision arithmetic).

The eigenvalues of the Chebyshev tau method computed with fourteen significant digits are plotted in Fig. 4.1. The results for $N = 64$ are seriously contaminated by round-off errors. Consequently, the entire computed eigenvalue pattern is markedly different from the true spectrum. As noted above, the real parts (of the reliably calculated eigenvalues) are all negative. Moreover, there is numerical evidence that both the real and imaginary parts of the largest eigenvalue (in modulus) grow like $O(N^2)$ as $N \to \infty$. However, the smallest 7/8 of the eigenvalues grow only linearly with N.

The eigenvalues of the Legendre tau method differ qualitatively from those of the Chebyshev tau method, in that their largest modulus satisfies an

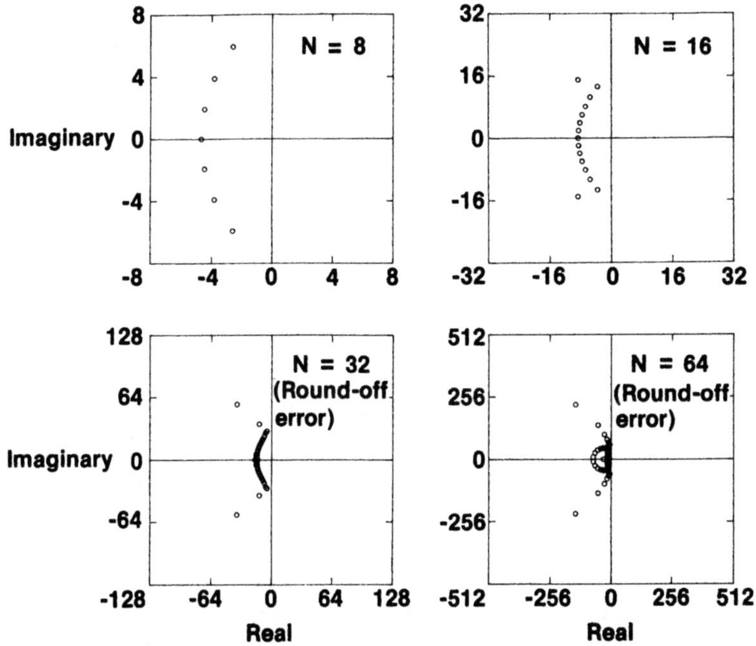

Figure 4.1. Chebyshev tau first-derivative eigenvalues computed with fourteen digit precision. Results contaminated by round-off error are indicated.

estimate of the form

$$|\lambda| = O(N) \qquad \text{as } N \to \infty, \tag{4.2.4}$$

instead of (4.2.3). On the other hand, when the Legendre tau method is applied to a system of hyperbolic equations, the corresponding eigenvalues grow again at the rate of $O(N^2)$. In Fig. 4.2 we report the Legendre tau eigenvalues for the first-derivative operator. Indeed, the largest eigenvalue is only $O(N)$, but round-off error is again a concern.

We consider now collocation methods using the Gauss–Lobatto points. The eigenvalues of the collocation operator are defined by the set of equations

$$\begin{aligned} &\frac{dU}{dx}(x_j) = \lambda U(x_j), \qquad j = 1, \ldots, N, \\ &U(x_0) = 0, \end{aligned} \tag{4.2.5}$$

provided U is a non-trivial polynomial of degree N.

For the Chebyshev points, the real parts of λ are strictly negative, while the modulus satisfies a bound of the form

$$|\lambda| = O(N^2). \tag{4.2.6}$$

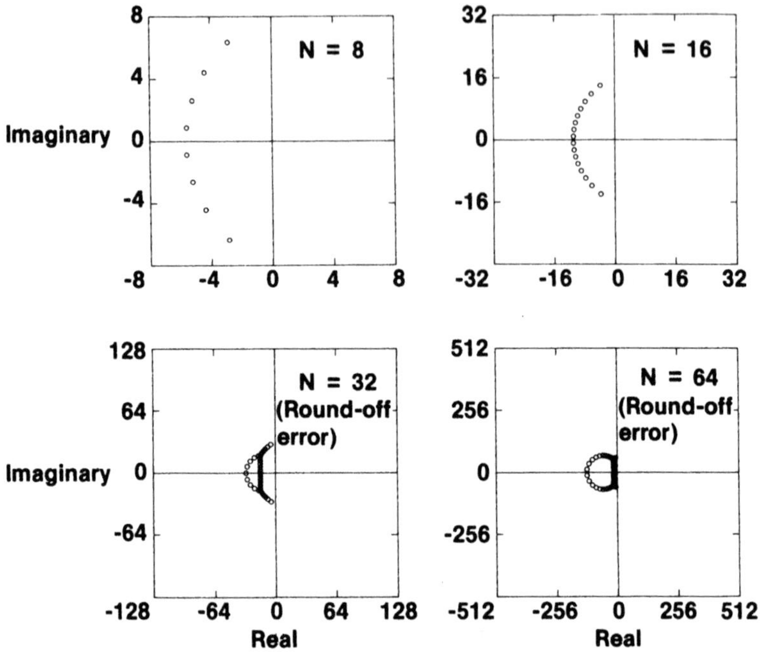

Figure 4.2. Legendre tau first-derivative eigenvalues computed with fourteen digit precision. Results contaminated by round-off error are indicated.

The estimate (4.2.6) is sharp, as confirmed by numerical experiments (see Fig. 4.3). As for the tau method, it is seen that the largest real part of λ grows like $O(N^2)$. The eigenvalues of the Legendre collocation operator are similar (see Fig. 4.4). As with the tau methods, one must worry about round-off error.

The recent results of Trefethen and Trummer on the effects of round-off error on the computed eigenvalues are troubling and their full significance has not yet been assessed. They present a numerical example which suggests that in actual computations the stability limit depends upon the numerical eigenvalues appropriate to the precision of the computations. No results are yet available on how this round-off error affects the accuracy of the computation.

4.2.2. The Second-Derivative Operator

The operator of interest here is the diffusion operator

$$Lu = \frac{d^2u}{dx^2} \qquad \text{on } (-1, 1). \tag{4.2.7}$$

The boundary conditions we shall impose are of Dirichlet type

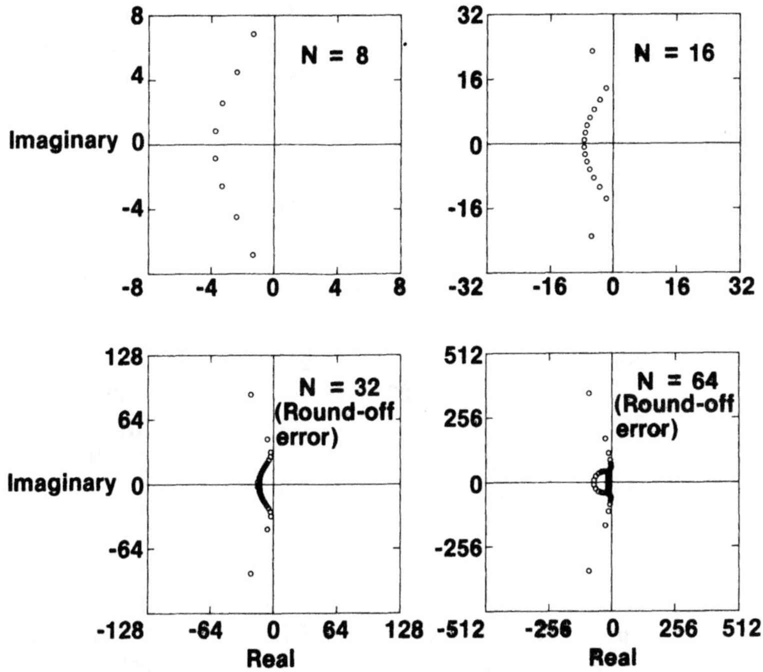

Figure 4.3. Chebyshev collocation first-derivative eigenvalues computed with fourteen digit precision. Results contaminated by round-off error are indicated.

$$u(+1) = u(-1) = 0 \tag{4.2.8a}$$

or of Neumann type

$$\frac{du}{dx}(-1) = \frac{du}{dx}(1) = 0. \tag{4.2.8b}$$

We shall focus on the collocation method for these boundary value problems which uses the Gauss–Lobatto points. We show in Sec. 11.4 that, except for the zero eigenvalue of the Neumann problem, there exist two positive constants c_1, c_2 independent of N such that

$$0 < c_1 \leqslant -\lambda \leqslant c_2 N^4. \tag{4.2.9}$$

Numerical evidence shows that the largest several discrete eigenvalues with Dirichlet or Neumann boundary conditions grow like $O(N^4)$ as $N \to \infty$. (No insidious round-off effects occur for the eigenvalues of the second-derivative operator.) Of course, the smaller discrete eigenvalues are good approximations to the eigenvalues of the corresponding analytic problem. It is only

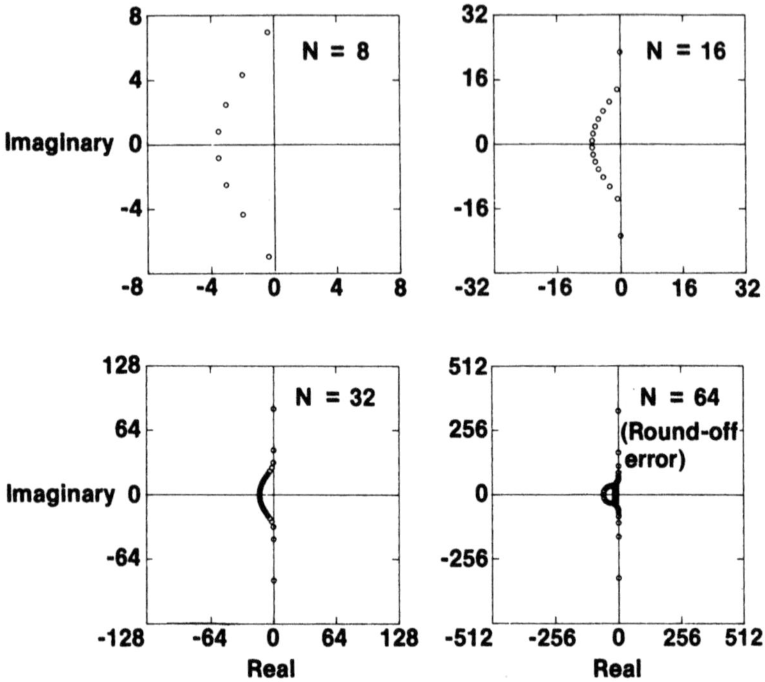

Figure 4.4. Legendre collocation first-derivative eigenvalues computed with fourteen digit precision. Results contaminated by round-off error are indicated.

the upper-third of the discrete eigenvalue spectrum which differs from the analytic eigenvalues by more than 10%. Tau approximations have eigenvalues with similar properties. The asymptotic constants of the largest eigenvalue are given in Table 4.1. (The ratio λ_{max}/N^4 is a decreasing function of N. The constants given in the table are good to two digits for $N \geqslant 64$.)

The Neumann boundary conditions (4.2.8b) can also be enforced indirectly in a collocation approximation, as described in Sec. 3.3. These eigenvalues, too, are negative, distinct, and bounded by CN^4 as $N \to \infty$.

Table 4.1. Asymptotic growth of second-derivative eigenvalues

Approximation	$-\lambda_{max}$(Dirichlet)$/N^4$	$-\lambda_{max}$(Neumann)$/N^4$
Chebyshev tau	.30	.047
Chebyshev collocation	.047	.014
Legendre tau	.11	.026
Legendre collocation	.026	.0064

4.3. Some Standard Schemes

Among the factors which influence the choice of a time-discretization are the accuracy, stability, storage requirements, and work demands of the methods. The storage and work requirements of a method can be deduced in a straightforward manner from the definition of the method and the nature of the PDE. The accuracy of a method follows from a truncation error analysis and the stability for a given problem is intimately connected with the spectrum of the spatial discretization. In this section we will describe some of the standard methods for ODEs and relate their stability regions to the spectra of the advection and diffusion operators. Bear in mind that in many problems different time-discretizations are used for different spatial terms in the equation.

4.3.1. Multistep Schemes

Leap Frog

This is a second-order, two-step scheme given by

$$U^{n+1} = U^{n-1} + 2\Delta t F^n, \tag{4.3.1}$$

where the superscripts denote the time-level at which the term is evaluated. The stability condition for the linear model problem is that $\lambda\Delta t$ be on the imaginary axis and that $|\lambda\Delta t| \leqslant 1$. Thus, leap frog is a suitable explicit scheme for problems with purely imaginary eigenvalues. It also is a time-reversible or non-dissipative method. However, since it is only stable on a line in the complex $\lambda\Delta t$-plane for the model problem, extra care is needed in practical situations.

The most obvious application is to periodic advection problems, for the eigenvalues of the Fourier approximation to $\partial/\partial x$ are imaginary. The difficulty with the leap frog method is that the solution is subject to a temporal oscillation with period $2\Delta t$. This arises from the extraneous solution to the temporal difference equations. The oscillations can be controlled by every so often averaging the solution at two consecutive time levels.

Leap frog is quite inappropriate for problems whose spatial eigenvalues have non-zero real parts. This certainly includes diffusion operators. Leap frog is also not viable for advection operators with non-periodic boundary conditions. Figures 4.1–4.4 indicate clearly that the discrete spectra of Chebyshev and Legendre approximations to the standard advection operator have appreciable real parts.

Adams–Bashforth Methods

This is a class of explicit methods which includes the simple forward Euler (FE) method

Figure 4.5. Absolute stability regions of Adams–Bashforth methods. The stability boundaries along the imaginary and the negative real axis are marked.

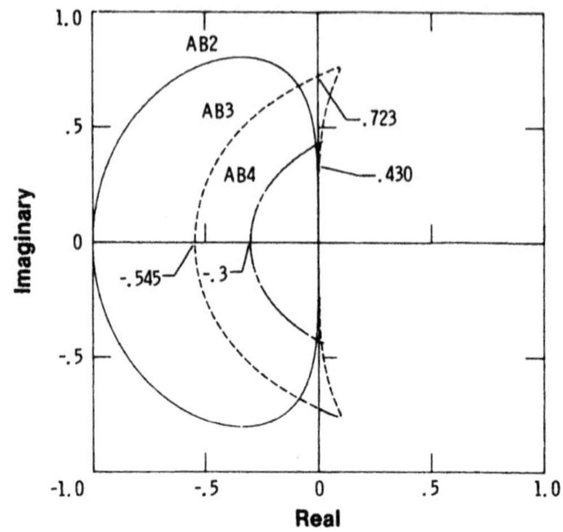

$$U^{n+1} = U^n + \Delta t F^n, \tag{4.3.2}$$

the popular second-order Adams–Bashforth (AB2) method

$$U^{n+1} = U^n + \Delta t[\tfrac{3}{2}F^n - \tfrac{1}{2}F^{n-1}], \tag{4.3.3}$$

the still more accurate third-order (AB3) method

$$U^{n+1} = U^n + \Delta t[\tfrac{23}{12}F^n - \tfrac{16}{12}F^{n-1} + \tfrac{5}{12}F^{n-2}], \tag{4.3.4}$$

and even higher-order versions. The absolute stability regions of these methods are shown in Fig. 4.5. Note that the size of the stability region decreases as the order of the method increases. Note also that except for the origin, no portion of the imaginary axis is included in the asymptotic stability region of the first- and second-order methods, whereas the third- and fourth-order versions are stable along some portion of the imaginary-axis.

Let us consider how the AB2 method performs for Fourier approximations to the periodic advection problem

$$\frac{\partial u}{\partial t} = \frac{\partial u}{\partial x},$$

on $(0, 2\pi)$. The fundamental solution to the temporal difference equation for the Fourier component with wavenumber k is

$$r_k = 1 + ik\Delta t - \tfrac{1}{2}(k\Delta t)^2 + \tfrac{1}{4}i(k\Delta t)^3 - \tfrac{1}{8}(k\Delta t)^4. \tag{4.3.5}$$

Thus,

$$|r_k| = 1 + \tfrac{1}{4}(k\Delta t)^4 + O(k^5\Delta t^5).$$

For fixed and sufficiently small Δt, the growth in time of the k-th component of the numerical solution can be bounded by

$$\begin{aligned} |r_k|^{t/\Delta t} &\leqslant [1 + k^4 \Delta t^4]^{t/\Delta t} \\ &\leqslant e^{(k^4 \Delta t^3)t}. \end{aligned} \tag{4.3.6}$$

Thus, the full solution satisfies

$$\| U_N(t) \| \leqslant C e^{(N^4 \Delta t^3/16)t}. \tag{4.3.7}$$

One can show that this method is stable, but that it is not asymptotically stable. Of course, the true solution exhibits no growth at all. Ultimately, the growth of the numerical solution will render it meaningless. Nevertheless, the AB2 method is quite useful for integration over a fixed time interval $[0, T]$. Since the growth rate of the numerical solution is proportional to Δt^3, one need merely choose Δt sufficiently small so that the spurious, numerical growth of the solution is insignificant. The slow growth of the AB2 method for advection problems is often referred to as weak instability.

Equation (4.3.7) can also be used to derive the relation between Δt and N that leads to stability in the full PDE sense. It is a relationship of the form $\Delta t = O(1/N^{4/3})$ as the spatial grid is refined. Note that this stability limit is more strict than the usual criterion, as occurs for example with leap frog, that $\Delta t = O(1/N)$.

As is evident from Fig. 4.5, higher order AB methods are asymptotically stable for Fourier approximations to periodic advection problems. The stability bound along the imaginary axis has been marked in the figure. If this bound is denoted by c, then the stability limit is

$$\frac{N}{2} \Delta t \leqslant c$$

or

$$\Delta t \leqslant \left(\frac{1}{\pi} c \right) \Delta x. \tag{4.3.8}$$

The limit on Δt is smaller by a factor of π than the corresponding limit for a second-order finite-difference approximation in space. The Fourier spectral approximation is more accurate in space because it represents the high frequency components much more accurately than the finite-difference method. The artificial damping of the high frequency components which is produced by finite-difference methods enables the stability restriction on the time-step to be relaxed.

Chebyshev approximations to advection problems appear to be asymptotically stable under all AB methods for sufficiently small Δt. As discussed in Sec. 4.2.1, the spatial eigenvalues all have negative real parts. Thus, the failure of the AB2 method to include the imaginary axis in its stability region does not preclude asymptotic stability. The Chebyshev limit for large N ($N \geqslant 32$) is essentially (Zakaria (1985))

$$\Delta t \lesssim 9/N^2, \tag{4.3.9}$$

which is equivalent to

$$\Delta t \leqslant \tfrac{9}{4}(\Delta x_{\text{avg}})^2$$

and (4.3.10)

$$\Delta t \leqslant 2(\Delta x_{\text{min}}).$$

The stability limits for AB methods for both Fourier and Chebyshev approximations to diffusion equations are easy to deduce since their spatial eigenvalues are real and negative (limited as indicated in Table 4.1) and the stability bounds along the negative real-axis are given in Fig. 4.5.

Adams–Moulton Methods

A related set of implicit multi-step methods are the Adams–Moulton methods. They include backward Euler (BE)

$$U^{n+1} = U^n + \Delta t F^{n+1}, \tag{4.3.11}$$

the Crank–Nicolson (CN) method

$$U^{n+1} = U^n + \tfrac{1}{2}\Delta t[F^{n+1} + F^n], \tag{4.3.12}$$

the third-order Adams–Moulton (AM3) method

$$U^{n+1} = U^n + \tfrac{1}{12}\Delta t[5F^{n+1} + 8F^n - F^{n-1}], \tag{4.3.13}$$

and higher order versions. The absolute stability regions of these methods are displayed in Fig. 4.6. Both BE and CN are A-stable, i.e., they are absolutely stable in the entire left-half plane.

In comparison with the explicit Adams–Bashforth method of the same order, an Adams–Moulton method has a smaller truncation error (by factors of five and nine for second and third-order versions), a larger stability region, and requires one less level of storage. However, it does require the solution of an implicit set of equations.

The CN method is commonly used for diffusion problems. In Navier–Stokes calculations, it is frequently applied to the viscous and pressure gradient components. Although CN is absolutely stable for such terms, it has the disadvantage that it damps high frequency components very weakly, whereas these components in reality decay very rapidly. Deville, Kleiser and Montigny–Rannou (1984) have noted that this is undesirable in Navier–Stokes applications for which the solution itself decays rapidly. One remedy is to resort to BE—it damps the high frequency components rapidly. An alter-

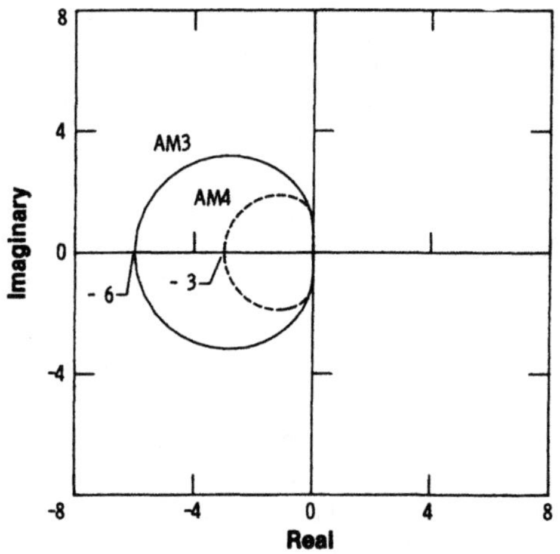

Figure 4.6. Absolute stability regions of Adams–Moulton methods. The stability boundaries along the negative real axis are marked.

native approach is to use the θ-method

$$U^{n+1} = U^n + \Delta t[\theta F^{n+1} + (1 - \theta)F^n] \tag{4.3.14}$$

for $\theta = 1/2 + \alpha\Delta t$, where α is a small positive constant. This method damps all components of the solution and is formally second-order accurate in time.

The Adams–Moulton methods of third and higher order are only conditionally stable for advection and diffusion problems. The stability limits marked on Figs. 4.5 and 4.6 indicate that the stability limit of a high order Adams–Moulton method is roughly ten times as large for a diffusion problem as the stability limit of the corresponding Adams–Bashforth method. In addition, both Adams–Moulton methods are weakly unstable for Fourier advection problems, since the origin is the only part of the imaginary axis which is included in their stability regions.

For an advection-diffusion equation, or indeed for the full Navier–Stokes system, a very popular time-discretization scheme uses Adams–Bashforth for advection and Crank–Nicolson for diffusion. Ouazzani, Peyret and Zakaria (1985) have made a detailed experimental study of the stability limits for Chebyshev approximations to such problems. More generally, one can couple an explicit method for advection with an implicit method for diffusion. An experimental analysis of such a strategy can be found in Basdevant, et al. (1986). With this kind of approach, a stability estimate like (4.1.4) can be obtained for a δ independent of N. (This is proved in Sec. 12.3 for a model problem.) Moreover, absolute stability can be obtained by such methods with a stability limit depending on the convective term only.

Predictor-Corrector Methods

A wide variety of predictor-corrector schemes are in use for ODEs, most commonly combining an Adams–Bashforth method for the predictor and an

Adams–Moulton method of the same or one higher order for the corrector. Such methods have been used on PDEs, but usually with the corrector employed explicitly, i.e., with the predicted value of the solution used to approximate the F^{n+1} term. We focus here on fully explicit predictor-corrector schemes in which the Adams–Moulton formula for the corrector stage is of order s and the Adams–Bashforth formula for the predictor stage has order $s-1$. Such a scheme, which we denote by APCs, has order s. The stability regions of these methods are displayed in Fig. 4.7. Only the APC3 method is stable along the imaginary-axis.

Compared with an ABs scheme, the APCs scheme has a lower truncation error, greater stability, the same storage requirement, but involves two evaluations of F per (full) step rather than just one. In many problems, including incompressible Navier–Stokes algorithms in which the viscous and pressure gradient terms are treated implicitly at each stage, the CPU time for a full step is essentially proportional to the number of stages. When operated at the Fourier advective stability limit, AB3 is more efficient than APC3.

An alternative class of predictor-corrector methods has been devised by Zlatev, Berkowicz and Prahm (1984) for use in Fourier collocation approximations to multidimensional problems. Their methods have the form

$$U^* = \alpha^* U^n + (1-\alpha^*)U^{n-1} + \Delta t \sum_{k=1}^{s-1} \beta^*_{sk} F^{n+1-k} \tag{4.3.15}$$

$$U^{n+1} = \alpha U^n + (1-\alpha)U^{n-1} + \Delta t \beta_{s0} F^* + \Delta t \sum_{k=1}^{s-1} \beta_{sk} F^{n+1-k}, \tag{4.3.16}$$

where $\alpha^* \in \mathbb{R}$, $\alpha \in [0, 2)$, the β^*_{sk} are chosen so that (4.3.15) is an explicit method of order $s-1$, and the β_{sk} so that (4.3.16) is an implicit method of order s. These schemes are denoted by (α^*, α)PCs. They have order s and the stability

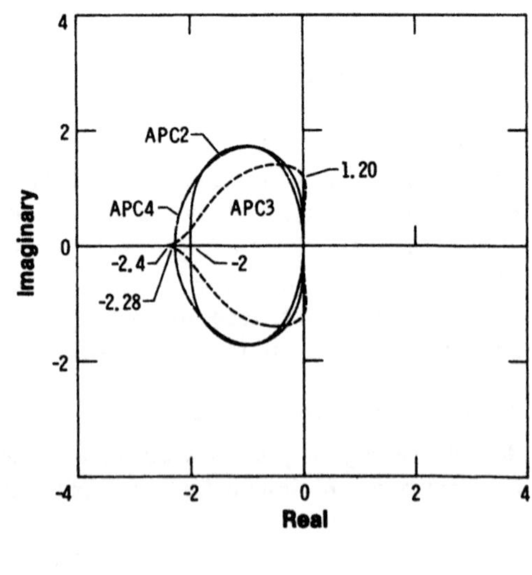

Figure 4.7. Absolute stability regions of Adams predictor-corrector methods. The stability boundaries along the imaginary and the negative real axes are marked.

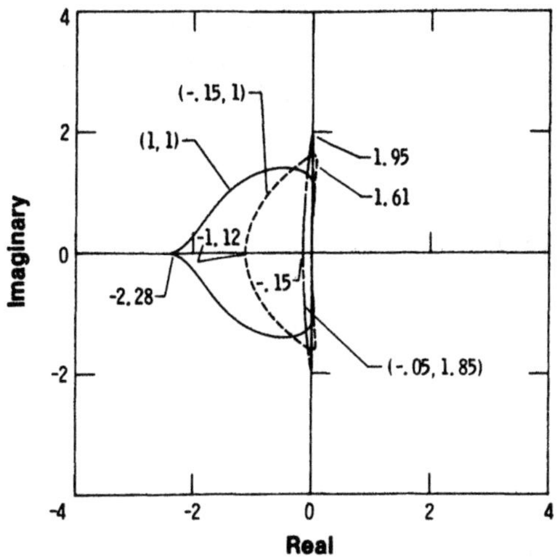

Figure 4.8. Absolute stability regions of a class of Adams predictor-corrector methods proposed by Zlatev, Berkowicz, and Prahm. The stability boundaries along the imaginary and the negative real axes are marked.

regions of several of these schemes are sketched in Fig. 4.8. These methods require one more level of storage than either the ABs or APCs schemes. They do, however, have large stability regions near the imaginary-axis and may be more efficient in terms of CPU time. The review by Zlatev et al. contains several comparisons of these methods. However, the enhanced stability along the imaginary axis is achieved at the cost of reduced stability elsewhere. Note how narrow the stability region for the $(-.15, 1.85)$PC3 method is. This version is clearly unsuitable for non-periodic advection problems and for diffusion problems.

4.3.2. Runge–Kutta Methods

Numerous Runge–Kutta time discretizations have been applied in combination with spectral methods for PDE's. The general second-order Runge–Kutta (RK2) method can be written

$$\begin{aligned} U &= U^n \\ G &= F(U, t_n) \\ U &= U + \alpha \Delta t G \\ G &= aG + F(U, t_n + \alpha \Delta t) \\ U^{n+1} &= U + \Delta t \frac{1}{2\alpha} G, \end{aligned} \tag{4.3.17}$$

where

$$a = -1 + 2\alpha - 2\alpha^2.$$

The choice $\alpha = \frac{1}{2}$ produces the modified Euler method and $\alpha = 1$ corresponds to the Heun method. Note that only two levels of storage are required for (4.3.17). The classical fourth-order Runge–Kutta (RK4) method is

$$\begin{aligned}
K_1 &= F(U^n, t_n) \\
K_2 &= F(U^n + \tfrac{1}{2}\Delta t K_1, t_n + \tfrac{1}{2}\Delta t) \\
K_3 &= F(U^n + \tfrac{1}{2}\Delta t K_2, t_n + \tfrac{1}{2}\Delta t) \\
K_4 &= F(U^n + \Delta t K_3, t_n + \Delta t) \\
U^{n+1} &= U^n + \tfrac{1}{6}\Delta t[K_1 + 2K_2 + 2K_3 + K_4].
\end{aligned} \tag{4.3.18}$$

An equivalent formulation is

$$\begin{aligned}
U &= U^n \\
G &= U \\
P &= F(U, t_n) \\[1ex]
U &= U + \tfrac{1}{2}\Delta t P \\
G &= P \\
P &= F(U, t_n + \tfrac{1}{2}\Delta t) \\[1ex]
U &= U + \tfrac{1}{2}\Delta t(P - G) \\
G &= \tfrac{1}{6}G \\
P &= F(U, t_n + \tfrac{1}{2}\Delta t) - \tfrac{1}{2}P \\[1ex]
U &= U + \Delta t P \\
G &= G - P \\
P &= F(U, t_n + \Delta t) + 2P \\
U^{n+1} &= U + \Delta t(G + \tfrac{1}{6}P).
\end{aligned} \tag{4.3.19}$$

This version, which requires only three levels of storage, was proposed by Blum (1962).

Williamson (1980) has listed numerous low-storage versions of Runge–Kutta methods. An example of an RK3 method which requires only two levels of storage is

$$\begin{aligned}
U &= U^n \\
G &= F(U, t_n) \\[1ex]
U &= U + \tfrac{1}{3}\Delta t G \\
G &= -\tfrac{5}{9}G + F(U, t_n + \tfrac{1}{3}\Delta t)
\end{aligned}$$

$$\begin{aligned} U &= U + \tfrac{15}{16}\Delta t G \\ G &= -\tfrac{153}{128}G + F(U, t_n + \tfrac{3}{4}\Delta t) \qquad (4.3.20) \\ U^{n+1} &= U + \tfrac{8}{15}G. \end{aligned}$$

In the event that F contains no explicit dependence upon t, the following formulation, due to Jameson, Schmidt and Turkel (1981) applies:

Set

$$U = U^n$$

For $k = s, 1, -1$

$$U = U^n + \frac{1}{k}\Delta t F(U) \qquad (4.3.21)$$

End for

$$U^{n+1} = U.$$

It yields a Runge–Kutta method of order s (for linear problems) and requires at most three levels of storage.

All Runge–Kutta methods of a given order have the same stability properties. The regions of absolute stability are given in Fig. 4.9. Note that the stability regions expand as the order increases. Note also that RK2 methods

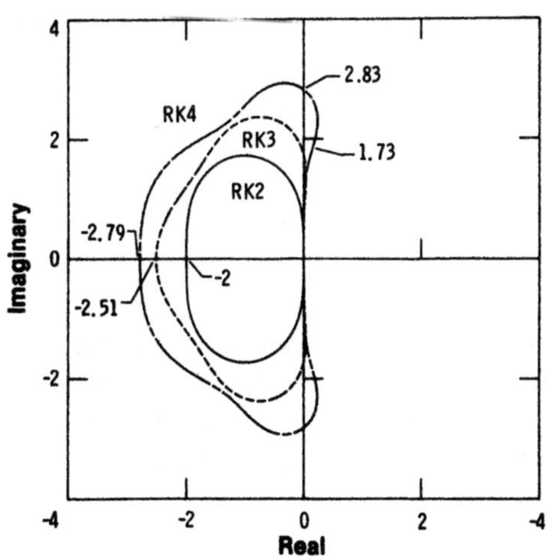

Figure 4.9. Absolute stability regions of Runge–Kutta methods. The stability boundaries along the imaginary and negative real axes are marked.

are afflicted with the same weak instability as the AB2 scheme. Funaro (1985) finds experimentally that the stability limit of an RK2 method for a Chebyshev collocation scheme for an advection equation is (for large N)

$$\Delta t \leqslant 16/N^2. \tag{4.3.22}$$

For scalar equations the recently developed rational Runge–Kutta methods (Wambecq (1978), Dekker and Verwer (1984)) are attractive because they can yield unconditional stability with an explicit method. Zakaria (1985) has investigated these for problems in which the spatial discretization is spectral. The utility of these methods for systems is not clear since the stability analysis cannot be carried out by the standard diagonalization technique.

4.4. Special Purpose Schemes

The temporal integration schemes which were discussed in the preceding section are classical methods for ODEs. In this section we describe some integration techniques, most of which are at least semi-implicit, that are designed especially for PDEs.

4.4.1. High Resolution Temporal Schemes

The methods described in the preceding section are infinite-order accurate in space, but only of finite-order in time. Especially for second-order time discretizations, one expects that the error of the fully discretized method will be dominated by the temporal errors. The following example, due to Tal–Ezer (1986a), is instructive. Consider the scalar, hyperbolic problem for $x \in (0, 2\pi)$ and $t \in [0, T]$

$$\begin{aligned} \frac{\partial u}{\partial t} &= a\frac{\partial u}{\partial x} \\ u(x,0) &= u_0(x) \\ u(x+2\pi, t) &= u(x,t), \end{aligned} \tag{4.4.1}$$

where a is a constant. Use Fourier collocation in space and leapfrog in time. The errors of the leapfrog method reside entirely in the phases of the individual Fourier components. The amount of phase error increases with the Fourier wavenumber of the component and with the length of the time interval. In this example, it equals

$$\Delta\phi_{\max} = \frac{1}{6}(\Delta t)^2\left(\frac{a\pi}{\Delta x}\right)^3.$$

If one desires the phase error in all components to be less than ε, then one needs

$$\Delta t < \sqrt{\frac{3\varepsilon}{4\pi^3 T}} \left(\frac{\Delta x}{a}\right)^{3/2} \tag{4.4.2}$$

The time-step needn't be quite this small for an accuracy ε in the solution itself, since the high-wavenumber components, which have the greatest phase errors, have much lower amplitudes than the lower-wavenumber components. Nevertheless, the true accuracy limit on Δt should differ from the result (4.4.2) by only a constant factor. In general, for a time-discretization of order s, an accuracy condition of the form

$$\Delta t < C \left(\frac{\varepsilon}{T}\right)^{1/s} \left(\frac{\Delta x}{a}\right)^{(s+1)/s} \tag{4.4.3}$$

is plausible.

The accuracy condition for the leapfrog method is *more* demanding than the stability condition

$$\Delta t < \frac{1}{\pi} \left(\frac{\Delta x}{a}\right). \tag{4.4.4}$$

Thus, if extremely accurate solutions to an evolution equation are desired, either very small time-steps or else high-order time-differencing methods are required.

Morchoisne (1979, 1981a) and Tal–Ezer (1986a, 1987b) have developed time-differencing methods which have infinite-order accuracy. The integration from $t = 0$ to $t = T$ is accomplished by successive integration over subintervals of length Δt. On each of these subintervals the time dependence of the solution is expressed as a series of Chebyshev polynomials of order M. According to the usual philosophy of spectral methods, increased resolution is achieved not by decreasing Δt, but rather by increasing M.

Morchoisne's method is a (necessarily implicit) collocation scheme which resorts to the iterative techniques described in Chap. 5 to obtain the solution of the implicit system. An important consideration is that storage for $M + 1$ time-levels is required. In practice, especially for three-dimensional problems, the size of M is constrained by the storage demands. A comparison of this method with more conventional time-differencing methods is given by Deville, Montigny–Rannou and Kleiser (1984). Some examples of two-dimensional and three-dimensional Navier–Stokes calculations are furnished by Morchoisne (1981a).

Tal–Ezer's method is more restricted in applicability, but when it does apply it is substantially more efficient than the alternative method, both in storage and computation. The problems to which it applies are linear ones with known bounds in the complex plane on the spectrum of the spatial operator. In these cases the desired Chebyshev polynomial in the time variable can be generated by a three-term recursion formula. Tal–Ezer has given a detailed discussion of constant-coefficient hyperbolic and parabolic problems with periodic boundary conditions.

4.4.2. Special Integration Techniques

In some problems with periodic boundary conditions the preferred technique for handling constant-coefficient linear terms is exact integration. The Burgers equation (3.1.1) will serve as an illustration. The semi-discrete Fourier Galerkin formulation of this is given by (3.1.8), which we write here as

$$\frac{d\hat{U}_k}{dt} = -\nu k^2 \hat{U}_k + \hat{G}_k(\hat{U}) \qquad k = -\frac{N}{2}, \ldots, \frac{N}{2} - 1, \tag{4.4.5}$$

where $\hat{G}_k(\hat{U})$ is given by (3.1.9). Equation (4.4.5) can be written

$$\frac{d}{dt}[e^{\nu k^2 t}\hat{U}_k] = e^{\nu k^2 t}\hat{G}_k(\hat{U}).$$

The forward Euler approximation reduces to

$$\hat{U}_k^{n+1} = e^{-\nu k^2 \Delta t}[\hat{U}_k^n + \Delta t \hat{G}_k(\hat{U}^n)]. \tag{4.4.6}$$

The treatment of the linear term is both unconditionally stable and exact. The accuracy and stability restrictions of the method arise solely from the non-linear term.

The Fourier collocation method can be handled in a similar, but not equivalent manner:

$$U^{n+1} = C^{-1}\Lambda C U^n + \Delta t G(U^n), \tag{4.4.7}$$

where C represents the discrete Fourier transform matrix (see (2.1.22) and (5.1.9)) and

$$\Lambda = \operatorname{diag}\{e^{-\nu k^2 \Delta t}\}. \tag{4.4.8}$$

This integrating-factor technique has found extensive use in Fourier Galerkin simulations of homogeneous turbulence (Rogallo (1977)) and has also been used for the horizontal diffusion terms in calculations of parallel boundary layers (Spalart (1986)). The integrating factors are especially useful in these Navier–Stokes applications because they do not suffer from the weak or non-existent damping of the high-frequency components that arise in backward Euler or Crank–Nicolson discretizations of the viscous terms.

Fornberg and Whitham (1978) employed exact integration of the linear term of the Korteweg-de Vries equation

$$\frac{\partial u}{\partial t} = -\frac{\partial^3 u}{\partial x^3} - u\frac{\partial u}{\partial x} \tag{4.4.9}$$

in their Fourier collocation-leap frog calculations. In this application, exact integration enables the stability limit to be increased from $\Delta t < (1/\pi^3)\Delta x^3$ to $\Delta t < (3/2\pi^2)\Delta x^3$. This is a fivefold increase. Note, however, that the $O(\Delta x^3)$ limit does not disappear entirely in favor of an $O(\Delta x)$ limit, as it would for a Fourier Galerkin method applied in conjunction with exact integration. Chan and Kerkhoven (1985) discuss alternative time-discretization of the Korteweg-de Vries equations. They show that, with the leap frog method for the advection term and the Crank–Nicolson method for the linear term, the stability limit is independent of Δx for any finite time interval.

4.4.3. Lerat Schemes

We close this discussion by mentioning a recently developed class of implicit methods which may prove useful if asymptotic stability rather than temporal accuracy is the primary concern. This is the case, for example, in methods which achieve a steady-state solution by a time-marching approach. Lerat (1979, 1983) developed a methodology which has proven useful in finite-difference calculations.

Consider the linear problem (4.1.3) and write

$$U^{n+1} = U^n + W^n. \tag{4.4.10}$$

The first step is to adopt the Lax–Wendroff approach of expanding about $t = t_n$ to second-order in Δt. Next, substitute $U^n + \beta W^n$ for U^n in the second-order term. The result is

$$U^{n+1} + \tfrac{1}{2}\beta\Delta t^2 L^2 U^{n+1} = U^n + \Delta t L U^n + \tfrac{1}{2}(1 + \beta)\Delta t^2 L^2 U^n. \tag{4.4.11}$$

For the model problem (4.1.5) we have

$$U^{n+1} = \frac{1 - \lambda\Delta t + \frac{1}{2}(1 + \beta)(\lambda\Delta t)^2}{1 + \frac{1}{2}\beta(\lambda\Delta t)^2} U^n. \tag{4.4.12}$$

For $\beta = -\frac{1}{2}$, A-stability is apparent. Furthermore, if λ is imaginary, then asymptotic stability is achieved provided $\beta \leqslant -\frac{1}{2}$. This is the case for a periodic, constant-coefficient advection problem. Canuto and Quarteroni (1987) have discussed this method for a non-periodic advection problem. They note that the operator L^2 is *not* the standard diffusion operator, but rather the square of the first-derivative operator which incorporates the boundary condition. Unconditional stability is only achieved if the latter selection is made for L^2. This makes the inversion of the left-hand side of (4.4.11) non-trivial. The extension to non-linear problems is given in the above references.

Lerat (1986, private communication) reports that when this implicit time-discretization scheme is employed in finite-difference calculations in aerodynamics, it has proven sufficient to use the standard diffusion operator (with its customary boundary conditions) in place of L^2 on the left-hand side of (4.4.11). This is but one of many instances in which spectral methods are far more sensitive to boundary conditions than finite-difference or finite-element methods.

4.5. Conservation Forms

Often, the solution to the PDE satisfies one or more conservation properties. The independent variables themselves are conserved (except for boundary effects) if the spatial operator is in divergence form, i.e.,

$$\mathbf{f}(\mathbf{u}) = \nabla \cdot f, \tag{4.5.1}$$

where the tensor f is called the flux function. Gauss' Theorem implies that the solution to the evolution equation (4.1.1) satisfies

$$\frac{d}{dt}\int_\Omega \mathbf{u} = \int_{\partial\Omega} f \cdot \hat{\mathbf{n}}. \tag{4.5.2}$$

Hence, the only integral changes in $\mathbf{u}$ are those due to fluxes through the boundaries. If the spatial operator is orthogonal to the solution, i.e.,

$$(\mathbf{u}, \mathbf{f}(\mathbf{u})) = 0, \tag{4.5.3}$$

then the quadratic conservation law

$$\frac{d}{dt}(\mathbf{u}, \mathbf{u}) = \frac{d}{dt}\|\mathbf{u}\|^2 = 0 \tag{4.5.4}$$

holds.

Naturally, it is desirable for the discrete solution to satisfy analogous conservation laws. The conservation laws which apply are influenced by both the spatial and temporal discretizations. We focus first on the spatial discretization and consider the semi-discrete evolution equation. We assume here that both the solution and its approximation are real-valued functions.

For a Galerkin spectral method the quadratic conservation law (4.5.4) follows directly from (4.5.3), provided that these inner products coincide with the one used in defining the Galerkin method. Since the Galerkin projection is self-adjoint, we have that

$$\begin{aligned}\frac{1}{2}\frac{d}{dt}(\mathbf{u}^N, \mathbf{u}^N) &= (\mathbf{u}^N, Q_N\mathbf{f}(\mathbf{u}^N)) \\ &= (Q_N\mathbf{u}^N, \mathbf{f}(\mathbf{u}^N)) \\ &= (\mathbf{u}^N, \mathbf{f}(\mathbf{u}^N)) \\ &= 0.\end{aligned} \tag{4.5.5}$$

(In this section we use $\mathbf{u}^N$ rather than $\mathbf{U}$ and denote the projection operator explicitly.)

An important special case arises when the operator f is linear and skew-symmetric, i.e.,

$$\mathbf{f}(\mathbf{u}) = L\mathbf{u} \tag{4.5.6}$$

with

$$L^* = -L. \tag{4.5.7}$$

In this case

$$\begin{aligned}(\mathbf{u}, \mathbf{f}(\mathbf{u})) &= (\mathbf{u}, L\mathbf{u}) \\ &= \tfrac{1}{2}(\mathbf{u}, L\mathbf{u}) + \tfrac{1}{2}(\mathbf{u}, L\mathbf{u}) \\ &= \tfrac{1}{2}(\mathbf{u}, L\mathbf{u}) - \tfrac{1}{2}(L\mathbf{u}, \mathbf{u}) \\ &= 0.\end{aligned}$$

Note that, for periodic boundary conditions, $f(u) = \partial u/\partial x$ satisfies these conditions, but that $f(u) = a(x)(\partial u/\partial x)$ does not when $da/dx \not\equiv 0$. In more than one space dimension, the corresponding conditions are $f(u) = \mathbf{a}\cdot\nabla u$ with $\nabla\cdot\mathbf{a} \equiv 0$ and $\nabla\cdot\mathbf{a} \not\equiv 0$, respectively. In the former case, we have $f(u) = \frac{1}{2}\mathbf{a}\cdot\nabla u + \frac{1}{2}\nabla\cdot(u\mathbf{a})$ and the skew-symmetry is apparent.

An illustrative non-linear example is

$$f(u) = -u\frac{\partial u}{\partial x}, \tag{4.5.8}$$

corresponding to the inviscid Burgers equation. Condition (4.5.1) is satisfied since $u(\partial u/\partial x) = \frac{1}{2}(\partial(u^2)/\partial x)$, and (4.5.2) holds for both the Fourier and Legendre inner products.

In the periodic case, if one looks for real-valued solutions, the integrals $\int u^k$ for all integers k are conserved. Fourier Galerkin approximations also conserve $\int u^N$ and $\int (u^N)^2$, but they do not conserve $\int (u^N)^k$ for $k \geqslant 3$. In the nonperiodic case, the integrals $\int u^k$ are conserved up to a boundary term. In particular, for Legendre Galerkin approximations this property holds for $\int u^N$ and $\int (u^N)^2$. Thus, not all the invariants of the fully continuous problem are conserved by the spatial discretization. This behavior is typical of conservation properties of systems.

The incompressible Navier–Stokes equations are an application of considerable importance. In the absence of viscosity (see (7.3.1))

$$\mathbf{f}(\mathbf{u}) = \mathbf{u} \times \boldsymbol{\omega} - \nabla P, \tag{4.5.9}$$

with $\nabla \cdot \mathbf{u} = 0$, where $\boldsymbol{\omega} = \nabla \times \mathbf{u}$ and $P = p + \frac{1}{2}|\mathbf{u}|^2$.

Since $\mathbf{u} \cdot (\mathbf{u} \times \boldsymbol{\omega}) \equiv 0$, we have, for either periodic or no-slip boundary conditions

$$\begin{aligned}(\mathbf{u}, \mathbf{f}(\mathbf{u})) &= -(\mathbf{u}, \nabla P)\\ &= (\nabla \cdot \mathbf{u}, P) = 0.\end{aligned}$$

Thus, Galerkin approximations to these equations in the Fourier and Legendre norms are quadratically conservative. Note that any passive scalar that is associated with the inviscid Navier–Stokes equations is also quadratically conserved, i.e., if

$$f(a) = -\mathbf{u} \cdot \nabla a \tag{4.5.10}$$

with

$$\nabla \cdot \mathbf{u} = 0,$$

then (a, a) is conserved. The same property holds for the Galerkin approximation a^N.

The conservation properties of Fourier collocation approximations to periodic problems can be readily analyzed. On the space S_N the bilinear form $(u, v)_N$, defined by (2.1.26), is an inner product. Moreover, the differentiation operator D_N is skew-symmetric. For multidimensional problems, the discrete divergence and gradient operators are obvious. Denote these by $\mathbb{D}_N$ and $\mathbb{G}_N$, respectively and note that $\mathbb{G}_N^* = -\mathbb{D}_N$, where the $*$ denotes the adjoint operator.

One can easily show that Fourier collocation approximations to evolution equations which satisfy (4.5.1) satisfy the discrete conservation law

$$\frac{d}{dt} \sum_{j=0}^{N-1} \mathbf{u}_j^N = 0. \tag{4.5.11}$$

However, unlike Galerkin methods, the condition (4.5.3) does not guarantee the discrete quadratic conservation law

$$\frac{d}{dt}(\mathbf{u}^N, \mathbf{u}^N)_N = 0. \tag{4.5.12}$$

Consider the case for which

$$f(u) = -\mathbf{a} \cdot \nabla u \quad \text{with} \quad \nabla \cdot \mathbf{a} = 0. \tag{4.5.13}$$

The semi-discrete method

$$\frac{\partial u^N}{\partial t} = \mathbf{a}^N \cdot \mathbb{G}_N u^N \tag{4.5.14}$$

with $\mathbb{D}_N \cdot \mathbf{a}^N = 0$ is not conservative, since

$$(u^N, \mathbf{a}^N \cdot \mathbb{G}_N u^N)_N \equiv 0 \tag{4.5.15}$$

does not, in general, hold. However, the method

$$\frac{\partial u^N}{\partial t} = -\frac{1}{2}\mathbf{a}^N \cdot \mathbb{G}_N u^N - \frac{1}{2}\mathbb{D}_N \cdot (u^N \mathbf{a}^N) \tag{4.5.16}$$

with

$$\mathbb{D}_N \cdot \mathbf{a}^N = 0$$

does imply

$$\left(u^N, \frac{1}{2}\mathbf{a}^N \cdot \mathbb{G}_N u^N + \frac{1}{2}\mathbb{D}_N \cdot (u^N \mathbf{a}^N)\right)_N = 0 \tag{4.5.17}$$

since

$$\begin{aligned}(u^N, \mathbb{D}_N \cdot (u^N \mathbf{a}^N))_N &= -\,(\mathbb{G}_N u^N, u^N \mathbf{a}^N)_N \\ &= -\,(u^N, \mathbf{a}^N \cdot \mathbb{G}_N u^N).\end{aligned}$$

Likewise, for the non-linear problem (4.5.8), the quadratically conservative collocation version is

$$\frac{\partial u^N}{\partial t} = -\frac{1}{3}D_N((u^N)^2) - \frac{1}{3}u^N D_N u^N. \tag{4.5.18}$$

This form, however, does not conserve $\int u^N$. The Galerkin method, on the other hand, conserves both $\int u^N$ and $\int (u^N)^2$. This result, too, is typical: collocation methods may not satisfy as many conservation properties as Galerkin ones.

Conservation laws for Legendre and Chebyshev collocation methods are more subtle than those for Fourier methods. In these cases the first-derivative operator is not skew-symmetric. Moreover, for Chebyshev approximations, the inner product of the approximation does not correspond to the physical inner product in which the conservation property holds.

For problems of the form (4.5.1) the discrete Legendre counterpart of (4.5.2) holds, by exactness of the quadrature formula. For Chebyshev approximations, however, one must resort to the Clenshaw–Curtis rule. This has degree of precision N. (See Davis and Rabinowitz (1984) or (13.2.18).)

The semi-discrete conservation laws are not satisfied by the fully discrete solution unless the time-discretization is symmetric. The leap frog and Crank–Nicolson methods are symmetric. However, the departure from conservation is small for unsymmetric time-discretization schemes, such as Adams–Bashforth and Runge–Kutta, and the departure decreases as the time-step is reduced.

4.6. Aliasing

In Sec. 2.1.2 we noted that the discrete Fourier coefficients of a function are not identical to the continuous ones (see (2.1.29)). The difference is attributable to the aliasing phenomenon (see Fig. 2.2). Hence, the principal difference between Galerkin and collocation (or pseudospectral) methods is the presence of truncation error alone in the former versus the presence of both truncation and aliasing errors in the latter. The question of whether the additional aliasing errors in the collocation methods are indeed serious has been highly controversial. The two most pertinent issues are the effects of aliasing upon the accuracy and, in evolution problems, the temporal numerical stability of the calculation.

Here, we both review the available theoretical results on this subject and also summarize the vast numerical experience that has been accumulated in the last two decades. Many approximation theory results are presented in Chap. 9, for Fourier, Legendre and Chebyshev series. Compare, for example, the Fourier Galerkin (truncation) and collocation (interpolation) approximation error bounds given in the L^2 norm by (9.1.9) and (9.1.15), respectively. These imply that, although for fixed N the collocation error will be larger than the Galerkin error, both errors exhibit the same asymptotic decay rate for large N. The Legendre Galerkin and collocation estimates are furnished in (9.4.6) and (9.4.24); the Chebyshev ones are (9.5.6) and (9.5.19). In the case of the Legendre polynomial approximation, the collocation approximation has an asymptotic error decay rate which is slower, by a factor of $\sqrt{N}$, than the rate of the Gakerkin approximation. If the function has m derivatives, then the Galerkin error decays as N^{-m}, whereas the collocation error decays as $N^{1/2-m}$. For smooth functions, this should be a very minor difference, once there are enough polynomials to resolve the essential structure. Nevertheless for marginally resolved cases we do anticipate more difficulty with aliasing in spectral approximations to non-periodic problems than for periodic ones.

In the late 1970s, some results appeared on aliasing effects in Fourier approximations to simple partial differential equations with periodic boundary conditions. Kreiss and Oliger (1979) proved that the aliasing error decays at the same rate as the truncation error in Fourier approximations for the one-dimensional, linear wave equation. The theory of spectral approximations to the steady three-dimensional, Navier–Stokes equations has matured to the point that Galerkin (de-aliased) and collocation (aliased) approximations have been proven to have the same asymptotic error decay rate. This holds for both Fourier and Chebyshev approximations. The details of this analysis are supplied in Sec. 11.3.

Thus, there is reasonable support for the claim that, for any given problem, an aliased calculation will yield just as acceptable an answer as a de-aliased one, once sufficient resolution has been achieved.

For evolution problems, however, one must still address the issue of the temporal numerical stability of the calculation. Collocation approximations must be formulated with more care than Galerkin approximations. The reason is that for evolution problems with quadratic conservation properties, the Galerkin formulation will automatically yield semi-discrete quadratic conservation laws. This ensures that, at least for symmetric time-discretizations, the numerical solution will be bounded as $t \to \infty$. However, as discussed in Sec. 4.5, collocation approximations do not necessarily yield semi-discrete quadratic conservation properties.

A dramatic example of the importance of the proper collocation formulation is furnished by the ideal two-dimensional MHD problem (Dahlburg (1985))

$$\frac{\partial \omega}{\partial t} = \nabla \cdot (\mathbf{b}j - \mathbf{u}\omega) \tag{4.6.1}$$

$$\frac{\partial a}{\partial t} = -\mathbf{u} \cdot \nabla a \tag{4.6.2}$$

$$\omega = -\Delta\psi \tag{4.6.3}$$

$$j = -\Delta a \tag{4.6.4}$$

$$\nabla\psi = (-v, u) \tag{4.6.5}$$

$$\nabla a = (-b_y, b_x), \tag{4.6.6}$$

where $\mathbf{b}$, a, and j are the magnetic field, the magnetic potential and the current, respectively. The boundary conditions are taken to be periodic. The quantities

$$\mathscr{E} = \tfrac{1}{2}(\nabla\psi, \nabla\psi) + \tfrac{1}{2}(\nabla a, \nabla a) \tag{4.6.7}$$

$$\mathscr{H} = (\omega, a) \tag{4.6.8}$$

$$\mathscr{A} = (a, a) \tag{4.6.9}$$

are conserved. Semi-discrete Galerkin approximations also conserve each of these quantities.

Equation (4.6.2) may also be written

$$\frac{\partial a}{\partial t} = -\nabla \cdot (\mathbf{u}a). \tag{4.6.2b}$$

Using the principles discussed in Sec. 4.5, one can show that the collocation approximation (returning in this section to the convention of denoting the discrete variables by upper case symbols)

$$\begin{aligned} \frac{\partial \Omega}{\partial t} &= \mathbb{D}_N \cdot (\mathbf{B}J - \mathbf{U}\Omega) \\ \frac{\partial A}{\partial t} &= -\mathbf{U} \cdot \mathbb{G}_N A \end{aligned} \tag{4.6.10}$$

conserves

$$\mathscr{E}_N = \tfrac{1}{2}(\mathbb{G}_N\Psi, \mathbb{G}_N\Psi)_N + \tfrac{1}{2}(\mathbb{G}_N A, \mathbb{G}_N A)_N \tag{4.6.11}$$

and

$$\mathscr{H}_N = (\Omega, A)_N, \tag{4.6.12}$$

but not

$$\mathscr{A}_N = (A, A)_N, \tag{4.6.13}$$

whereas the collocation approximation

$$\begin{aligned} \frac{\partial \Omega}{\partial t} &= \mathbb{D}_N \cdot (\mathbf{B}J - \mathbf{U}\Omega) \\ \frac{\partial A}{\partial t} &= -\mathbb{D}_N \cdot (\mathbf{U}A) \end{aligned} \tag{4.6.14}$$

conserves none of these quantities. Because formulation (4.6.10) conserves $\mathscr{E}_N$, we are assured that aliasing instabilities are not present. However, formulation (4.6.14) may succumb to numerical instability.

Figure 4.10 summarizes the results of some calculations performed by Dahlburg for these MHD equations. A sine series of degree $N = 16$ was used to approximate ω and a and random initial conditions were employed. The solid line in Fig. 4.10(a) represents $\mathscr{E}_N$ for the Galerkin approximation of one such calculation. The total energy is conserved to a high degree of accuracy. There is a slight increase in $\mathscr{E}_N$ in the collocation formulation (4.6.10). This arises from the time-differencing errors. (An RK2 method was employed. This is not a time symmetric scheme.) On the other hand, $\mathscr{E}_N$ grows catastrophically for the collocation formulation (4.6.14). Figures 4.10(b) and (c) illustrate the separate contributions of the kinetic $-\tfrac{1}{2}(\mathbb{G}_N\Psi, \mathbb{G}_N\Psi)_N$— and magnetic $-\tfrac{1}{2}(\mathbb{G}_N A, \mathbb{G}_N A)_N$—energies. Note that the collocation and Galerkin results do not agree. The resolution for this calculation is inadequate. Since there is no dissipation in (4.6.1) and (4.6.2), the high frequency components saturate rapidly. Figure 4.10(d) gives the time history of $\mathscr{E}_N$ for a similar calculation which includes a dissipative term of sufficient size that the numerical resolution is adequate. There is very little difference between the aliased and de-aliased results.

Additional collocation formulations of (4.6.1) and (4.6.2) are possible. The version

$$\begin{aligned} \frac{\partial \Omega}{\partial t} &= \mathbb{D}_N \cdot (\mathbf{B}J - \mathbf{U}\Omega) \\ \frac{\partial A}{\partial t} &= -\frac{1}{2}\mathbf{U} \cdot \mathbb{G}_N A - \frac{1}{2}\mathbb{D}_N \cdot (\mathbf{U}A) \end{aligned} \tag{4.6.15}$$

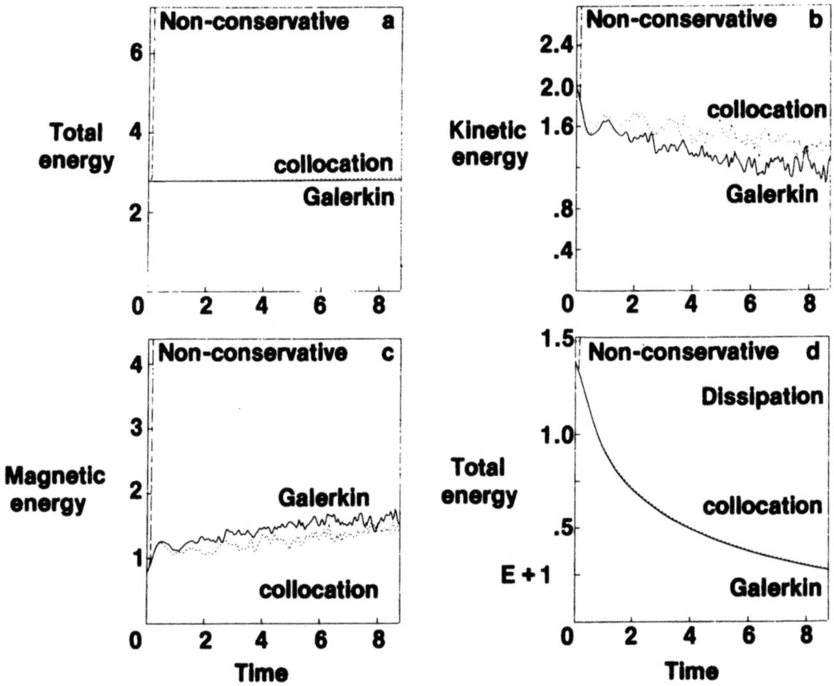

Figure 4.10. Galerkin and collocation solutions to a two-dimensional MHD problem. Results from a nonconservative as well as a correct (conservative) collocation formulation are given. (Courtesy of J. Dahlburg.)

conserves only $\mathscr{A}_N$. The version

$$\begin{aligned}\frac{\partial \Omega}{\partial t} &= -\frac{1}{2}\mathbf{U}\cdot\mathbb{G}_N\Omega - \frac{1}{2}\mathbb{D}_N\cdot(\mathbf{U}\Omega) + \mathbb{D}_N\cdot(\mathbf{B}J)\\ \frac{\partial A}{\partial t} &= -\frac{1}{2}\mathbf{U}\cdot\mathbb{G}_N A - \frac{1}{2}\mathbb{D}_N\cdot(\mathbf{U}A)\end{aligned} \tag{4.6.16}$$

conserves $\mathscr{A}_N$ and $\mathscr{H}_N$, but not $\mathscr{E}_N$. From the point of view of numerical stability, conservation of $\mathscr{E}_N$ is the most important law, so that (4.6.10) is the preferred collocation formulation.

Numerous comparisons have been performed for aliased and de-aliased calculations of the periodic, multidimensional Navier–Stokes equations. Useful discussions may be found in Orszag (1972), Fox and Orszag (1973), Montigny–Rannou (1982) and Kerr (1985). All of these authors conclude that with sufficient resolution, aliased calculations are quite acceptable.

An early report which was highly critical of aliased spectral calculations was published by Schamel and Elsässer (1976). Among their two examples of supposedly unacceptable results was a Korteweg-de Vries equation. Fornberg and Whitham (1978) performed extensive Fourier collocation computations for a related Korteweg-de Vries equation and experienced no difficulty.

We ourselves have performed Fourier collocation calculations for precisely the example cited by Schamel and Elsässer. The problem we solved was

$$\frac{\partial u}{\partial t} = -2\pi \frac{\partial}{\partial x}\left(u + \frac{1}{2}u^2\right) - \frac{1}{2}\lambda_D^2 (2\pi)^3 \frac{\partial^3 u}{\partial x^3} \tag{4.6.17}$$

with the exact solution

$$\begin{aligned} u(x,t) &= u_0 + \Delta u \operatorname{sech}^2\left[\frac{1}{2\pi\lambda_D}\sqrt{\frac{\Delta u}{6}}(x - ct - \pi)\right] \\ c &= 2\pi\left(1 + u_0 + \frac{1}{3}\Delta u\right) \\ u_0 &= -2\lambda_D\sqrt{6\Delta u}\tanh\left[\sqrt{\frac{\Delta u}{24}}\Big/\lambda_D\right] \end{aligned} \tag{4.6.18}$$

on $(-\infty, \infty)$. Schamel and Elsässer chose $\lambda_D = 0.01$, $\Delta u = 0.2$, and imposed periodicity on $(0, 2\pi)$. (The true solution to the problem with initial condition

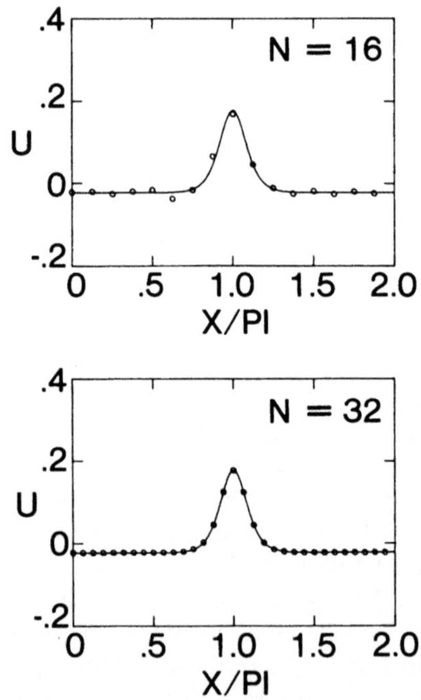

Figure 4.11. Collocation solutions to the Korteweg-de Vries equation at $t = 1$.

given by (4.6.18) with $t = 0$ and imposed periodicity on $(0, 2\pi)$ differs from the expression (4.6.18) for $t > 0$ by an exponentially small amount.) Our solutions for fully aliased calculations are shown in Fig. 4.11. The higher resolution results are very good approximations and display nothing at all like the errors reported by Schamel and Elsässer. In view of our own results, and especially those of Fornberg and Whitham, the conclusions of Schamel and Elsässer must be viewed with suspicion.

Recently, Moser, Moin and Leonard (1983) have also cautioned against aliased calculations. Their results are for axisymmetric Taylor–Couette flow between rotating cylinders of infinite length. This problem has one periodic and one non-periodic direction. They present a single, poorly resolved, aliased calculation and compare it with three de-aliased calculations, one poorly resolved, one moderately resolved and one well resolved. Their single aliased result is certainly much worse than their well resolved, de-aliased case, but their poorly resolved, de-aliased case is no better than the aliased one. Hence, their conclusion is not supported by their evidence.

Additional comparisons of aliased versus de-aliased calculations for the three-dimensional Navier–Stokes equations are given in Secs. 7.2.4 and 7.3.5. The conclusions are the same as those reached here.

This discussion has presumed that the solution to the problem is well behaved. Some investigators, e.g., Sulem, Sulem and Frisch (1983), have used spectral methods on problems for which the solution develops a singularity in a finite time. They have used the evolution of the spectra as a diagnostic for the onset of the singularity. If a problem such as this is simulated with a fixed resolution, then a de-aliasing procedure is advisible. Moreover, if a fully turbulent flow is simulated with marginal resolution, then de-aliasing may also be useful (Spalart (1986)).

CHAPTER 5

Solution Techniques for Implicit Spectral Equations

The solution of implicit equations is an important component of many spectral algorithms. For steady problems this task is unavoidable, while spectral algorithms for many unsteady problems are only feasible if they incorporate implicit (or semi-implicit) time discretizations (see Sec. 3.1 and Chap. 7). We concentrate on linear systems, assuming that non-linear ones are attacked by standard linearization techniques.

We focus (though not exclusively) on the elliptic equation

$$\Delta u - \lambda u = f, \tag{5.1}$$

where f is a function of $\mathbf{x}$ and $\lambda > 0$ is a constant. The simplest generalization is to the self-adjoint form

$$\nabla \cdot (a \nabla u) - \lambda u = f, \tag{5.2}$$

where $a > 0$ is a function of $\mathbf{x}$. Another generalization which arises when mappings are employed is

$$\sum_{i=1}^{d} b_i \frac{\partial}{\partial x_i} \left[a b_i \frac{\partial}{\partial x_i} \right] - \lambda u = f, \tag{5.3}$$

where b_i is a function of x_i and is the inverse of the Jacobian of the mapping in the coordinate x_i, $i = 1, \ldots, d$. All equations are, of course, subject to appropriate boundary conditions. Equation (5.1) contains as special cases the steady incompressible potential equation ((1.3.16) with constant ρ) and implicit temporal discretizations of the heat equation.

Spectral collocation approximations lead to a linear system

$$LU = F, \tag{5.4}$$

where U and F are vectors consisting of the grid point values of u, f and any boundary data, and L is a matrix (constructed as a tensor product matrix in two or more spatial dimensions). A similar linear system is obtained for Galerkin and tau approximations, but now U and F are vectors consisting of the expansion coefficients of u, f and the boundary data, and L is the appropriate matrix in transform space.

The linear systems arising from (5.2) or (5.3) are usually full. Gaussian elimination may, in principle, be applied to (5.4), but it requires $\frac{2}{3}N^{3d}$

operations (where addition, subtraction, multiplication and division are counted as separate and equal operations) and $O(N^{2d})$ storage, where d is the dimension of the problem. (We assume, for simplicity, that the number of degrees of freedom in *each* spatial dimension is N.) In the first section of this chapter we discuss some direct solution techniques which can yield the solution to (5.4) in $O(N^d)$, $O(N^d \log_2 N)$ or at worst $O(N^{d+1})$ operations with at most $O(N^d)$ additional storage. The remainder of the chapter is devoted to a discussion of iterative techniques. These require $O(N^d \log_2 N)$ operations per iteration and $O(N^d)$ additional storage.

5.1. Direct Methods

Our objectives in this section are to explain the principles underlying the basic direct solution techniques, to illustrate these on some specific problems that arise in practice, and to summarize the literature on more specialized applications.

We shall call a solution "efficient" if it enables the solution to (5.4) to be obtained in at most $O(N^d \log_2 N)$ operations. This makes the cost of solving (5.4) comparable, even for N large, to the cost of typical explicit spectral operations such as differentiation and the evaluation of convolution sums. In many cases, a solution cost of $O(N^{d+1})$ is still acceptable in the sense that it only overwhelms the cost of other spectral operations for values of N of 128 or so.

An important consideration is whether only a few or else a large number of solutions to (5.4) with different data F are sought. The latter case is typical of semi-implicit methods for unsteady problems: hundreds, or even thousands of solutions to a single linear system are required. In such situations, it is reasonable to invest a substantial amount of calculations on a pre-processing stage that greatly reduces the subsequent cost of solving (5.4). The matrix-diagonalization techniques presented in Sec. 5.1.3 belong to this category. The discussion of Fourier and Chebyshev methods in Secs. 5.1.1 and 5.1.2 is concerned with techniques for furnishing a solution to a single implicit equation. Naturally, they may also be employed in unsteady algorithms as well.

5.1.1. Fourier Approximations

The discussion will open with the simplest case—a one-dimensional, constant-coefficient, periodic problem:

$$\frac{d^2u}{dx^2} - \lambda u = f. \tag{5.1.1}$$

A Galerkin approximation takes the form

$$-k^2\hat{u}_k - \lambda\hat{u}_k = \hat{f}_k \qquad k = -\frac{N}{2}, \ldots, \frac{N}{2} - 1, \tag{5.1.2}$$

where the Fourier coefficients $\hat{u}_k$ are defined by (2.1.3) and the corresponding truncated Fourier series by (2.1.7). The solution to (5.1.2) is trivially

$$\begin{matrix} \hat{u}_k = -\hat{f}_k/(k^2 + \lambda) \\ (\hat{u}_0 \text{ arbitrary for } \lambda = 0) \end{matrix} \qquad k = -\frac{N}{2}, \ldots, \frac{N}{2} - 1, \tag{5.1.3}$$

with an operation count of $3N$, presuming u is real, so that $\hat{u}_{-k} = \bar{\hat{u}}_k$.

A collocation approximation is (with x_j given by (2.1.21))

$$\left.\frac{d^2u}{dx^2} - \lambda u - f\right|_{x=x_j} = 0 \qquad j = 0, \ldots, N-1. \tag{5.1.4}$$

This may be solved by using the discrete Fourier transform (DFT) to diagonalize (5.1.4):

$$-k^2\tilde{u}_k - \lambda\tilde{u}_k = \tilde{f}_k \qquad k = -\frac{N}{2}, \ldots, \frac{N}{2} - 1, \tag{5.1.5}$$

where the discrete Fourier coefficients $\tilde{u}_k$ are defined by (2.1.22), then solving for $\tilde{u}_k$ as in (5.1.3), and finally reversing the discrete Fourier Transform to recover u_j for $j = 0, 1, \ldots, N-1$. The operation count for the direct solution of (5.1.4) is $5N\log_2 N$ real operations, with the FFT used to accomplish the discrete Fourier transform. (Lower order terms in the operation counts, such as those linear in N in this case, are ignored unless they have especially large coefficients.)

The problem

$$\frac{d}{dx}\left[a(x)\frac{du}{dx}\right] - \lambda u = f \tag{5.1.6}$$

represents the next level of complexity. The collocation approximation to (5.1.6) may be written in the form (5.4) with

$$L = DAD - \lambda I, \tag{5.1.7}$$

where D is given explicitly by (2.1.41), A is the diagonal matrix representing multiplication by $a(x)$ in physical space, and I is the identity matrix. An alternative expression to (2.1.41) for D is

$$D = C^{-1}KC, \tag{5.1.8}$$

where

$$C_{kj} = \frac{1}{N}e^{-ikx_j} \qquad \begin{cases} k = -\frac{N}{2}, \ldots, \frac{N}{2} - 1 \\ j = 0, \ldots, N-1 \end{cases} \tag{5.1.9}$$

represents the DFT and

$$K = \operatorname{diag}\{ik'\} \qquad k = -\frac{N}{2}, \ldots, \frac{N}{2} - 1$$

$$k' = \begin{cases} k & k = -\frac{N}{2} + 1, \ldots, \frac{N}{2} - 1 \\ 0 & k = -\frac{N}{2} \end{cases} \tag{5.1.10}$$

represents differentiation in transform space. Equation (5.1.6) admits an efficient direct solution only if $\lambda = 0$ or if the Fourier series of $a(x)$ contains just a few low order terms.

In the former case we have for (5.4)

$$C^{-1}KCAC^{-1}KCU = F, \tag{5.1.11}$$

which is equivalent to

$$U = C^{-1}K^{-1}CA^{-1}C^{-1}K^{-1}CF. \tag{5.1.12}$$

Although K is technically singular—because of the $k = 0$ and $k = -(N/2)$ components—this merely reflects the non-uniqueness of the problem. The offending Fourier components may be assigned arbitrary values. The solution procedure described by (5.1.12) involves four FFTs and three multiplications, for a total cost of $10N\log_2 N$.

The latter condition on $a(x)$ is trivially satisfied by $a(x) = 1$. A more interesting example is

$$a(x) = \sin^2(x/2) = \tfrac{1}{2} - \tfrac{1}{4}(e^{ix} + e^{-ix}). \tag{5.1.13}$$

The collocation approximation to (5.1.6) can be expressed as

$$-\frac{1}{2}\sum_{k=-(N/2)+1}^{(N/2)-1}\left[k^2\tilde{u}_k - \frac{k(k-1)}{2}\alpha_{k-1}\tilde{u}_{k-1} - \frac{k(k+1)}{2}\alpha_{k+1}\tilde{u}_{k+1} + 2\lambda\tilde{u}_k\right]e^{ikx_j} = f_j \quad j = 0, \ldots, N-1, \tag{5.1.14}$$

where we have ignored the contributions of the $\tilde{u}_{-(N/2)}$ term, and where

$$\alpha_k = \begin{cases} 1 & |k| \leqslant \frac{N}{2} - 1 \\ 0 & |k| > \frac{N}{2} - 1 \end{cases}. \tag{5.1.15}$$

The solution procedure clearly requires two FFTs and one tridiagonal solution. Since the cost of the latter is minor, the entire solution requires $5N\log_2 N$ operations.

A closely related problem arises for the mapping (2.5.15) introduced by Cain, Ferziger, and Reynolds (1984) for problems on $(-\infty, \infty)$ with solutions which tend to the same constant at $\pm\infty$. In this case, the Poisson problem of interest is really not (5.1.6) but rather a one-dimensional version of (5.3):

$$b(x)\frac{d}{dx}\left[b(x)\frac{du}{dx}\right] - \lambda u = f, \tag{5.1.16}$$

where $b(x) = 1/h'(x)$ (and is given by (5.1.13)) is essentially the inverse of the Jacobian of the mapping $z = h(x) = -\cot(x/2)$ which was discussed in Sec. 2.5.4. The relevant collocation approximation is now expressible as

$$\begin{aligned} -\sum_{k=-(N/2)+1}^{(N/2)-1} \Bigg[&\frac{1}{4}(k-1)(k-2)\alpha_{k-2}\tilde{u}_{k-2} - \frac{1}{2}(k-1)(2k-1)\alpha_{k-1}\tilde{u}_{k-1} \\ &+\left(\frac{3}{2}k^2 + \lambda\right)\tilde{u}_k - \frac{1}{2}(k+1)(2k+1)\alpha_{k+1}\tilde{u}_{k+1} \\ &+\frac{1}{4}(k+1)(k+2)\alpha_{k+2}\tilde{u}_{k+2}\Bigg] e^{ikx_j} = f_j \qquad j = 0, 1, \ldots N-1, \end{aligned} \tag{5.1.17}$$

assuming $\tilde{u}_k = 0$ for $|k| = (N/2) - 1,\ N/2$. The solution to (5.1.17) requires two FFTs and a pentadiagonal solution. The linear system needs $19N$ operations, whereas the two FFTs together require $5N\log_2 N$ operations.

The mapping (2.5.16) leads to the following approximation to (5.1.16):

$$\begin{aligned} -\frac{1}{4}\sum_{k=-(N/2)+1}^{(N/2)-1} \Bigg[&\frac{1}{4}(k-2)(k-4)\alpha_{k-4}\tilde{u}_{k-4} - (k-1)(k-2)\alpha_{k-2}\tilde{u}_{k-2} \\ &+\left(\frac{3}{2}k^2 + 4\lambda\right)\tilde{u}_k - (k+1)(k+2)\alpha_{k+2}\tilde{u}_{k+2} \\ &+\frac{1}{4}(k+2)(k+4)\alpha_{k+4}\tilde{u}_{k+4}\Bigg] e^{ikx_j} = f_j \qquad j = 0, 1, \ldots, N-1. \end{aligned} \tag{5.1.18}$$

This requires the same number of operations to solve as (5.1.17); since the odd modes decouple from the even ones, two pentadiagonal solutions of length $N/2$ suffice for the linear equations.

As a rule, the generality of efficient direct methods decreases as the dimensionality of the problem increases. Clearly, the generalization of techniques for (5.1.1) are straightforward. The operation count of a Galerkin solution to (5.1) is $(2d + 1)N^d$ and that of a collocation approximation is $5dN^d\log_2 N$.

Two-dimensional versions of (5.2) and (5.3) are equally straightforward if, for (5.2) the coefficient a depends only on x, and for (5.3) b_1 depends only on x and a and b_2 are constant. In this case, a Fourier transform in y produces uncoupled sets of equations in x, which are of the form (5.1.6) and (5.1.16) with λ replaced with $\lambda + k_y^2$. If, however, $a(\mathbf{x})$ in (5.1.6) contains a general dependence on x and y, then even for $\lambda = 0$, no efficient direct solution is available. The prospects for problems of the type (5.1.16) arising from the use of trigonometric mappings in two directions are almost as bad. In this case the matrix L is banded, with half-bandwidth $2N$. Banded Gaussian elimination methods require $4N^4$ operations, which is quite expensive for a two-dimensional problem. Similar considerations apply to a third dimension.

5.1.2. Chebyshev Tau Approximations

Efficient solution processes are available for a limited class of Chebyshev and Legendre tau approximations to one-dimensional problems. An example of considerable importance is (5.1.1) with homogeneous Dirichlet boundary conditions. We write the Chebyshev tau approximation as

$$\hat{u}_n^{(2)} - \lambda \hat{u}_n = \hat{f}_n \qquad n = 0, 1, \ldots, N-2, \tag{5.1.19}$$

$$\sum_{n=0}^{N} \hat{u}_n = 0$$
$$\sum_{n=0}^{N} (-1)^n \hat{u}_n = 0. \tag{5.1.20a}$$

The boundary conditions may also be written as

$$\sum_{\substack{n=0\\ n \text{ even}}}^{N} \hat{u}_n = 0$$
$$\sum_{\substack{n=1\\ n \text{ odd}}}^{N} \hat{u}_n = 0. \tag{5.1.20b}$$

Equation (5.1.19) may be expressed as (see (2.4.27))

$$\frac{1}{c_n} \sum_{\substack{p=n+2\\ p+n \text{ even}}}^{N} p(p^2 - n^2)\hat{u}_p - \lambda \hat{u}_n = \hat{f}_n \qquad n = 0, 1, \ldots, N-2. \tag{5.1.21}$$

Using (5.1.20b) and (5.1.21), we arrive at a linear system of the form (5.4) in which L is upper-triangular. The solution process requires N^2 operations. A far more efficient solution procedure is obtained by rearranging the equations. We invoke the recursion relation (2.4.26) with $q = 2$:

$$2n\hat{u}_n^{(1)} = c_{n-1}\hat{u}_{n-1}^{(2)} - \hat{u}_{n+1}^{(2)},$$

and use (5.1.19) to obtain

$$2n\hat{u}_n^{(1)} = c_{n-1}(\hat{f}_{n-1} + \lambda \hat{u}_{n-1}) - (\hat{f}_{n+1} + \lambda \hat{u}_{n+1}) \qquad n = 1, \ldots, N-3. \tag{5.1.22}$$

Next use (2.4.26) with $q = 1$ in combination with (5.1.22):

$$2n\hat{u}_n = \frac{c_{n-1}}{2(n-1)}[c_{n-2}(\hat{f}_{n-2} + \lambda \hat{u}_{n-2}) - (\hat{f}_n + \lambda \hat{u}_n)]$$
$$- \frac{1}{2(n+1)}[c_n(\hat{f}_n + \lambda \hat{u}_n) - (\hat{f}_{n+2} + \lambda \hat{u}_{n+2})] \qquad n = 2, \ldots, N-4.$$

This simplifies to

$$\frac{c_{n-2}}{4n(n-1)}\lambda\hat{u}_{n-2} + \left(1 - \frac{\lambda}{2(n^2-1)}\right)\hat{u}_n + \frac{\lambda}{4n(n+1)}\hat{u}_{n+2}$$
$$= \frac{c_{n-2}}{4n(n-1)}\hat{f}_{n-2} - \frac{1}{2(n^2-1)}\hat{f}_n + \frac{1}{4n(n+1)}\hat{f}_{n+2}, \quad n = 2, \ldots, N-4. \tag{5.1.23}$$

By accounting carefully for the four equations which were dropped in going from (5.1.19) to (5.1.23), we can write (5.1.19) as

$$\frac{c_{n-2}\lambda}{4n(n-1)}\hat{u}_{n-2} + \left[1 - \frac{\lambda\beta_n}{2(n^2-1)}\right]\hat{u}_n + \frac{\lambda\beta_{n+2}}{4n(n+1)}\hat{u}_{n+2}$$
$$= \frac{c_{n-2}}{4n(n-1)}\hat{f}_{n-2} - \frac{\beta_n}{2(n^2-1)}\hat{f}_n + \frac{\beta_{n+2}}{4n(n+1)}\hat{f}_{n+2}, \quad n = 2, \ldots, N, \tag{5.1.24}$$

where

$$\beta_n = \begin{cases} 1 & 0 \leqslant n \leqslant N-2 \\ 0 & n > N-2 \end{cases}. \tag{5.1.25}$$

Note that the even and odd coefficients are uncoupled in (5.1.24) and (5.1.20b). The structure of the linear system for the even coefficients is quasi-tridiagonal, namely

$$\begin{pmatrix} 1 & 1 & 1 & \cdots & & & & 1 \\ x & x & x & & & & & \\ & x & x & x & & & & \\ & & x & x & x & & & \\ & & & & \vdots & & & \\ & & & & x & x & x & \\ & & & & & x & x & \\ & & & & & & x & x \end{pmatrix} \begin{pmatrix} \hat{u}_0 \\ \hat{u}_2 \\ \hat{u}_4 \\ \\ \vdots \\ \hat{u}_{N-4} \\ \hat{u}_{N-2} \\ \hat{u}_N \end{pmatrix} = \begin{pmatrix} 0 \\ \hat{g}_0 \\ \hat{g}_2 \\ \\ \vdots \\ \hat{g}_{N-6} \\ \hat{g}_{N-4} \\ \hat{g}_{N-2} \end{pmatrix}, \tag{5.1.26}$$

where x's denote the non-zero coefficients from (5.1.24) and $\hat{g}_n$ is the right-hand side of (5.1.24). This ordering has been chosen to minimize the round-off errors arising from a specially tailored Gauss elimination procedure for (5.1.26) which performs no pivoting (and works from the "bottom up" rather than the more customary "top down"). Assuming that the coefficients in (5.1.24) have already been calculated, the cost of solving for both the even and odd coefficients is $16N$.

The coefficient of $\hat{u}_n$ in (5.1.24) is the largest coefficient and it is desirable for it to be on the main diagonal. The system (5.1.26) is not diagonally dominant and, in practice, round-off errors are a mild problem: typically four digits are lost for $N = 128$. The accuracy may be increased through iterative improvement (see Golub and Van Loan (1983), Chap. 4) or double-precision.

Dennis and Quartapelle (1985) have shown that if a constant-coefficient first-derivative term is included in (5.1.19), then a similar set of equations can be derived. They are, however, quasi-pentadiagonal, since the even and odd modes do not decouple.

The solution process for a mixed collocation/tau approximation to (5.1.1) is: (1) perform a discrete Chebyshev transform on the grid point values f_j; (2) solve the quasi-tridiagonal system (5.1.24), (5.1.20b); (3) perform an inverse Chebyshev transform on $\tilde{u}_n$ to produce the u_j. Step 1 prevents this from being a pure tau method, since the Chebyshev coefficients are computed by quadrature rather than exact integration. This solution requires $5N\log_2 N + 24N$ operations, where we include the latter term because of its large coefficient.

The Neumann problem may be solved just as efficiently. In this case, (5.1.20) is replaced by

$$\sum_{n=1}^{N} n^2\hat{u}_n = 0 \qquad \sum_{n=1}^{N} (-1)^n n^2\hat{u}_n = 0 \tag{5.1.27a}$$

or equivalently

$$\sum_{\substack{n=2 \\ n \text{ even}}}^{N} n^2\hat{u}_n = 0 \qquad \sum_{\substack{n=1 \\ n \text{ odd}}}^{N} n^2\hat{u}_n = 0. \tag{5.1.27b}$$

The even and odd coefficients decouple, so that the cost is the same as that of the Dirichlet problem. If $\lambda = 0$, then the compatibility condition

$$\sum_{\substack{n=0 \\ n \text{ even}}}^{N-2} \frac{-2}{n^2-1}\hat{f}_n = 0$$

is required by the algebraic problem. This is the discrete analog of the compatibility condition

$$\int_{-1}^{1} f(x)\,dx = 0$$

for the continuous problem.

This procedure can be generalized to problems with inhomogeneous boundary conditions of Robin type. Haldenwang et al. (1984) have catalogued the relevant formulas. The solution process, however, is more costly for the mixed boundary conditions since the even and odd modes do not decouple.

The procedure used in the derivation of (5.1.24) may be used for some other simple differential operators. Consider the equation

$$\frac{1}{r}\frac{d}{dr}\left(r\frac{du}{dr}\right) - \frac{\lambda}{r^2}u = f, \tag{5.1.28}$$

which has obvious applications to the Poisson equation on the unit circle. We rewrite (5.1.28) as

$$r\frac{d}{dr}\left(r\frac{du}{dr}\right) - \lambda u = r^2 f = g, \tag{5.1.29}$$

and invoke (Gottlieb and Orszag (1977, Appendix)) the property that if $v = r(du/dr)$, then

$$c_n \hat{v}_n = n\hat{u}_n + 2 \sum_{\substack{p=n+2 \\ p+n \text{ even}}}^{N} p\hat{u}_p; \tag{5.1.30}$$

hence, we have the recursion relation

$$c_{n-1}\hat{v}_{n-1} - \hat{v}_{n+1} = (n-1)\hat{u}_{n-1} + (n+1)\hat{u}_{n+1}. \tag{5.1.31}$$

After considerable algebra we arrive at

$$-\lambda\hat{u}_0 + \frac{8+4\lambda}{6}\hat{u}_2 + \frac{16-\lambda}{6}\hat{u}_4 = \hat{g}_0 - \frac{2}{3}\hat{g}_2 + \frac{1}{6}\hat{g}_4$$

$$\frac{(n-2)^2-\lambda}{2(n-1)}\hat{u}_{n-2} + \frac{n(n^2-2+\lambda)}{n^2-1}\hat{u}_n + \frac{(n+2)^2-\lambda}{2(n+1)}\hat{u}_{n+2}$$

$$= \frac{1}{2(n-1)}\hat{g}_{n-2} - \frac{n}{n^2-1}\hat{g}_n + \frac{1}{2(n+1)}\hat{g}_{n+2}, \quad n = 3, \ldots, N-2,$$

$$\frac{(N-3)^2+\lambda}{2(N-4)}\hat{u}_{N-3} + \frac{3N^2-10N+7-\lambda}{2(N-4)}\hat{u}_{N-1} \tag{5.1.32}$$

$$= -\frac{1}{2(N-4)}\hat{g}_{N-3} + \frac{1}{2(N-4)}\hat{g}_{N-1},$$

$$\frac{(N-2)^2+\lambda}{2(N-3)}\hat{u}_{N-2} + \frac{3N^2-4N-\lambda}{2(N-3)}\hat{u}_N = -\frac{1}{2(N-3)}\hat{g}_{N-2} + \frac{1}{2(N-3)}\hat{g}_N$$

$$\frac{(N-1)^2+\lambda}{2(N-2)}\hat{u}_{N-1} = \frac{1}{2(N-2)}\hat{g}_{N-1},$$

plus the boundary condition

$$\sum_{n=0}^{N} \hat{u}_n = 0. \tag{5.1.33}$$

In the specific application to the Poisson equation in a disk the appropriate expansion is

$$u(r,\theta) = \sum_{n=0}^{N} \sum_{m=-N/2}^{N/2-1} \hat{u}_{nm} T_n(r) e^{im\theta}, \tag{5.1.34}$$

where only the even Chebyshev polynomials are retained for m even and only the odd polynomials are used for m odd. Note that the procedures described here permit solutions to two-dimensional Chebyshev–Fourier approximations to the Poisson equation in Cartesian and polar coordinates to be obtained in $O(N^2 \log_2 N)$ operations.

Let us now consider a first-order problem

$$\frac{du}{dx} + \lambda u = f \tag{5.1.35}$$

with

$$u(-1) = 0.$$

(An implicit time-discretization of $\partial u/\partial t + c(\partial u/\partial x) = 0$, with $c > 0$, leads to such an equation.) The tau approximation is

$$\begin{aligned} &\hat{u}_n^{(1)} + \lambda \hat{u}_n = \hat{f}_n \qquad n = 0, 1, \ldots, N-1 \\ &\sum_{n=0}^{N} (-1)^n \hat{u}_n = 0. \end{aligned} \tag{5.1.36}$$

When combined with the recursion relation (2.4.25) this yields

$$\begin{aligned} &\hat{u}_{n-1} + \frac{2}{\lambda} n\hat{u}_n - \beta_{n+1}\hat{u}_{n+1} = \hat{f}_{n-1} - \beta_{n+1}\hat{f}_{n+1}, \quad n = 1, \ldots, N \\ &\sum_{n=0}^{N} (-1)^n \hat{u}_n = 0. \end{aligned} \tag{5.1.37}$$

Legendre tau methods are quite similar. Here, of course, one uses the recursion relation (2.3.19) in place of (2.4.26).

Unfortunately, tau approximations in two dimensions require considerably more work. In fact, the best known direct methods require $O(N^3)$ operations. These are discussed in the next subsection.

5.1.3. Schur-Decomposition and Matrix-Diagonalization

Let us consider the Poisson equation in a square

$$\Delta u - \lambda u = f \qquad \text{on } (-1, 1)^2, \tag{5.1.38}$$

with homogeneous Dirichlet boundary conditions. The collocation approximation to this can be written

$$A\mathscr{U} + \mathscr{U}B - \lambda\mathscr{U} = \mathscr{F}, \tag{5.1.39}$$

where $\mathscr{U}$ is the $(N-1) \times (M-1)$ matrix (u_{ij}) for $i = 1, \ldots, N-1, j = 1, \ldots, M-1$, $\mathscr{F}$ is defined similarly, A is the second-derivative operator (in x) in which the boundary conditions have been incorporated, and B is the transpose of the second-derivative operator (in y).

Systems of the form (5.1.39) are solvable by Schur-decomposition (Bartels and Stewart (1972)). An orthogonal transformation is used to reduce A to block-lower-triangular form with blocks of size at most two. Similarly, B is reduced to block-upper-triangular form. If P and Q denote the respective orthogonal transformations, then (5.1.39) is equivalent to

$$A'\mathscr{U}' + \mathscr{U}'B' - \lambda\mathscr{U}' = \mathscr{F}', \tag{5.1.40}$$

where

$$\begin{aligned} A' &= P^{\mathrm{T}}AP \\ B' &= Q^{\mathrm{T}}BQ \\ \mathscr{U}' &= P^{\mathrm{T}}\mathscr{U}Q \\ \mathscr{F}' &= P^{\mathrm{T}}\mathscr{F}Q. \end{aligned} \tag{5.1.41}$$

The solution process has four steps: (1) reduction of A and B to real Schur form (and determination of P and Q); (2) construction of $\mathscr{F}'$ via (5.1.41); (3) solution of (5.1.40) for $\mathscr{U}'$; and (4) transformation of $\mathscr{U}'$ to $\mathscr{U}$ via (5.1.41).

The first step can be accomplished via the QR algorithm (Wilkinson (1965)) in $(4 + 8\alpha)(N^3 + M^3)$ operations, where α is the average number of QR steps. Step (3) requires $NM(N + M)$ operations and steps (2) and (4) take $2NM(N + M)$ operations apiece. Assuming $\alpha = 2$, a single solution requires $20(N^3 + M^3) + 5NM(N + M)$ operations. Hence, step (1) is the most time-consuming. When the same problem must be solved repeatedly, then step (1) need only be performed once, in a pre-processing stage. The matrices A', B', P and Q may then be stored and used as needed. In this case a complete solution takes $5NM(N + M)$ operations, or $10N^3$ operations when $M = N$.

Bartels and Stewart supply a complete FORTRAN program for this solution technique. To date, however, this method has seen little use in spectral methods, in part because of the matrix-diagonalization technique described next. It would, however, be the method of choice for solving one equation of the form (5.1.39).

The matrix-diagonalization approach is similar to the Schur-decomposition method. The difference is that the matrices A and B in (5.1.39) are diagonalized rather than merely reduced to block-triangular form. An algebraic problem of the form (5.1.40) is obtained with (5.1.41) replaced by

$$\begin{aligned} A' &= P^{-1}AP = \Lambda_A \\ B' &= Q^{-1}BQ = \Lambda_B \\ \mathscr{U}' &= P^{-1}\mathscr{U}Q \\ \mathscr{F}' &= P^{-1}\mathscr{F}Q, \end{aligned} \tag{5.1.42}$$

where Λ_A is the diagonal matrix with the eigenvalues of A on the diagonal. Thus, we have

$$\Lambda_A\mathscr{U}' + \mathscr{U}'\Lambda_B - \lambda\mathscr{U}' = \mathscr{F}'. \tag{5.1.43}$$

The matrices P and Q are not necessarily orthogonal and their columns consist of the eigenvectors of A and B, respectively.

The matrix-diagonalization scheme for (5.1.39) consists of the same four steps as the Schur-decomposition method except that the first, pre-processing stage also requires that the eigenvectors and the inverse transformations be computed. This takes an additional $4(N^3 + M^3)$ operations (Golub and Van Loan (1983, Algorithm 7.6-3)). Step (3) takes only $3NM$ operations since the system is diagonal, and steps (2) and (4) require $2NM(N + M)$ operations apiece, as before.

For collocation problems requiring multiple solutions, the matrix-diagonalization method has the advantage of taking only 80% of the solution time of the Schur-decomposition method: $8N^3$ operations when $M = N$. Moreover, the entire solution process—steps 2, 3 and 4—is extremely simple and can be optimized readily. In fact, most computer libraries contain assembly-language routines for all the requisite operations. The third stage of the Schur-decomposition method is more complicated.

This solution strategy is an application of the tensor product approach devised by Lynch, Rice and Thomas (1964) for finite-difference approximations to Poisson's equation. For second-order approximations to (5.1.38) on a rectangular grid, the pre-processing stage can be performed analytically.

In the case of tau approximations to (5.1.38), further gains in efficiency are possible. The discrete problem may be written in the form (5.1.39) where $\mathscr{U}$ is the $N - 1$ by $M - 1$ matrix $(\hat{u}_{nm})$ consisting of the Chebyshev coefficients of u (minus those used to enforce the boundary conditions). $\mathscr{F}$ is defined similarly, and A and B are the representations in transform space of the second-derivative operator (with the boundary conditions used to eliminate the two highest-order coefficients in each direction).

In the case of Dirichlet (or Neumann) boundary conditions, the even and odd modes decouple. Thus, A, B, P and Q contain alternating zero and non-zero elements. This property may be exploited to reduce the cost of both the pre-processing step (by a factor of 4) and the matrix multiplies (by a factor of 2). The cost of steps (2) through (4) is, thus, $2NM(N + M)$, or $4N^3$ when $M = N$.

The cost of the solution stages may be halved again by performing the diagonalization in only one direction and resorting to a standard tau solution in the other. Thus, (5.1.39) is reduced to

$$A\mathscr{U}' + \mathscr{U}'\Lambda_B - \lambda\mathscr{U}' = \mathscr{F}', \tag{5.1.44}$$

where

$$\begin{aligned} \mathscr{U}' &= \mathscr{U}Q \\ \mathscr{F}' &= \mathscr{F}Q, \end{aligned} \tag{5.1.45}$$

instead of to (5.1.43). The system (5.1.44) decouples into $M - 1$ systems of the form (5.1.19). Each of these may be reduced to a system like (5.1.24) and solved accordingly in $16N$ operations. The cost of the solution process is essentially halved, to $2NM(4 + M)$ operations, since the number of matrix multiplies is cut in two. Note that if $N \neq M$, then it is preferable to apply diagonalization to B if $M < N$ and to A otherwise.

This particular algorithm has come to be known as the Haidvogel-Zang algorithm after the paper by Haidvogel and Zang (1979) in which the method was explained in detail and compared with finite-difference methods for the Poisson equation. The method had been used earlier by both Murdock (1977) and Haidvogel (1977) in computations of the Navier–Stokes equations with two non-periodic directions.

Murdock applied the method on the domain $(-1, 1) \times (0, \infty)$, using an exponential mapping in y. A tau discretization is still applicable in y, but the diagonalization must be performed in this direction in order to perform a standard tau quasi-tridiagonal solution in the other.

In these algorithms, as indeed with matrix computations in general, the accumulation of round-off error is a concern. Haidvogel and Zang reported the loss of three to four digits (for N between 16 and 64) with the Schur-decomposition method. These were recovered through iterative improvement. Since the computation of eigenvectors can be a sensitive process, double-precision is advisable for the pre-processing stage of the matrix-diagonalization method.

Both methods can be generalized. The use of Neumann or Robin boundary conditions is straightforward. However, with Robin boundary conditions the even and odd modes do not decouple, and hence, some of the economies of the tau method are lost. These methods can be applied to separable equations of the form (5.3). A third, periodic direction is trivial to include in (5.1.38) since, after Fourier transforming in this direction, one simply has an independent set of equations with different λ. The pre-processing is independent of λ and hence of the third, periodic direction. A third, non-periodic direction may be treated by diagonalizing in that direction and then using whichever of the preceding methods is most convenient. Haldenwang et al. (1984) discuss several alternatives. Of course, both algorithms may be applied to separable, variable-coefficient periodic problems.

As it happens, these methods are *more* attractive in three-dimensional problems than in two-dimensional ones. Suppose that the number of degrees of freedom in each direction is N. The pre-processing cost is some large multiple of N^3. In two dimensions, the solution cost is a small multiple of N^3, and typical explicit spectral calculations take $O(N^2 \log_2 N)$ operations. Thus, the pre-processing cost is substantially larger than the cost of a single solution. In three dimensions, the solution cost is a small multiple of N^4 and typical explicit spectral calculations require $O(N^3 \log_2 N)$ operations. Thus, the pre-processing cost may even be smaller than the cost of the solution phase. Similarly, the extra memory required for A', P, and its inverse is proportionally smaller in three dimensions than in two.

The situation for vector and parallel computers is less obvious. Steps (2) through (4) can easily be vectorized. Vectorization of the decomposition phase is more difficult, since the QR algorithm used in the Schur-decomposition and diagonalization procedures is intrinsically scalar. A recent parallelizable algorithm devised by Stewart (1985) for Schur-decomposition may be useful in this regard. Furthermore, the use of double-precision (for the decomposition and eigenvector calculations) can be extremely costly on current vector machines.

5.2. Fundamentals of Iterative Methods

5.2.1. Richardson Iteration

The fundamentals of iterative methods for spectral equations are perhaps easiest to grasp for the simple model problem

$$-\frac{d^2u}{dx^2} = f \tag{5.2.1}$$

on $(0, 2\pi)$ with periodic boundary conditions. The Fourier approximation to the left-hand side of (5.2.1) at the collocation points $x_j = 2\pi j/N$ for $j = 0, \ldots, N-1$ is

$$\sum_{p=-N/2+1}^{N/2-1} p^2 \tilde{u}_p e^{ipx_j}, \tag{5.2.2}$$

where $\tilde{u}_p$ are the discrete Fourier coefficients of u.

This may be represented by the linear system (5.4) with $U = (u_0, u_1, \ldots, u_{N-1})$, $F = (f_0, f_1, \ldots, f_{N-1})$, and $L = -D_N^2 = -C^{-1}K^2C$ where D_N is given by (2.1.41), C by (5.1.9) and K by (5.1.10). The eigenfunctions of this approximation are

$$\xi_j(p) = e^{2\pi ijp/N}, \tag{5.2.3}$$

with the corresponding eigenvalues

$$\lambda(p) = p^2, \tag{5.2.4}$$

where $j = 0, 1, \ldots, N-1$ and $p = -N/2 + 1, \ldots, N/2 - 1$. The index p has a natural interpretation as the frequency of the eigenfunction. The $p = 0$ eigenfunction corresponds to the mean level of the solution. Since it is at one's disposal for this problem, it can essentially be ignored.

A particularly simple iterative scheme is the Richardson (1910) method. Given an initial guess V^0 to U, subsequent approximations are obtained via

$$V^{n+1} = V^n + \omega(F - LV^n), \tag{5.2.5}$$

where ω is a relaxation parameter. The error obeys the relation

$$(V^{n+1} - U) = G(V^n - U), \tag{5.2.6}$$

where the iteration matrix, G, of the Richardson scheme is given by

$$G = I - \omega L. \tag{5.2.7}$$

The iterative scheme is convergent if the spectral radius, ρ, of G is less than 1. In the case of the Richardson scheme this condition is equivalent to

$$|1 - \omega\lambda| < 1, \tag{5.2.8}$$

for all the eigenvalues λ of L. In the present case all the eigenvalues are positive (ignoring the eigenvalue for $p = 0$) and lie in the interval $[\lambda_{\min}, \lambda_{\max}]$, where $\lambda_{\min} = 1$ and $\lambda_{\max} = N^2/4$. Condition (5.2.8) is satisfied for $0 < \omega < \omega_{\max}$, where

$$\omega_{\max} = 2/\lambda_{\max}. \tag{5.2.9}$$

The best choice of ω is that which minimizes ρ. It is obtained from the relation

$$(1 - \omega\lambda_{\max}) = -(1 - \omega\lambda_{\min}), \tag{5.2.10}$$

for then the largest values of $1 - \omega\lambda$ are equal in magnitude and have opposite sign (see Fox and Parker (1968)). The optimal relaxation parameter is thus

$$\omega_{\text{opt}} = \frac{2}{\lambda_{\max} + \lambda_{\min}}. \tag{5.2.11}$$

It produces the spectral radius

$$\rho = \frac{\lambda_{\max} - \lambda_{\min}}{\lambda_{\max} + \lambda_{\min}}. \tag{5.2.12}$$

Note that the dependence upon the extreme eigenvalues enters only in the combination

$$\kappa = \frac{\lambda_{\max}}{\lambda_{\min}}. \tag{5.2.13}$$

We shall call this ratio the spectral condition number to distinguish it from the customary condition number defined by

$$\kappa(L) = \left[\frac{\lambda_{\max}(L^T L)}{\lambda_{\min}(L^T L)}\right]^{1/2}$$

In terms of this ratio,

$$\rho = \frac{\kappa - 1}{\kappa + 1}. \tag{5.2.14}$$

Define the rate of convergence $\mathscr{R}$ to be

$$\mathscr{R} = -\log \rho \tag{5.2.15}$$

and denote its reciprocal by $\mathscr{I}$. The latter quantity measures the number of iterations required to reduce the error by a factor of e. Clearly, the larger the convergence rate that a method has for a problem, the fewer iterations that are required to obtain a solution to a given accuracy.

For the Richardson method described above, the necessary number of iterations increases as

$$\mathscr{I} \cong \tfrac{1}{2}\kappa. \tag{5.2.16}$$

For the problem (5.2.1),

$$\kappa = \tfrac{1}{4}N^2. \tag{5.2.17}$$

The major expense in Richardson iteration is the evaluation of LV^n. This requires $5N\log_2 N$ operations via transform methods. To reduce the error by a single order of magnitude takes $1.4\,N^3\log_2 N$ operations. Fortunately, even the simple Richardson method can be made far more efficient by the technique described in the next section.

5.2.2. Preconditioning

The primary cause of the inefficiency of the straightforward Richardson method is the rapid increase with N of the spectral condition number. This can be alleviated by "preconditioning" the problem, in effect solving

$$H^{-1}LU = H^{-1}F$$

rather than (5.4). A preconditioned version of (5.2.5) is

$$V^{n+1} = V^n + \omega H^{-1}(F - LV^n). \tag{5.2.18}$$

In practice the inverse of the preconditioning matrix H is never explicitly required; instead one solves

$$H(V^{n+1} - V^n) = \omega(F - LV^n). \tag{5.2.19}$$

One obvious requirement for H is that this equation can be solved inexpensively, i.e., in fewer operations than are required to evaluate LU^n. The effective iteration matrix is now

$$G = I - \omega H^{-1}L. \tag{5.2.20}$$

The second requirement on the preconditioning matrix is that H^{-1} be a good approximation to L^{-1}, i.e., that the spectral condition number of $H^{-1}L$ be small.

Preconditioning techniques have been investigated extensively for finite-difference and finite-element methods (see Evans (1983)). Orszag (1980) proposed a preconditioning for spectral methods which amounts to using a low-order finite-difference approximation as H. Let $H^{(2)}$, $H^{(4)}$ and L denote second-order, fourth-order and spectral discretizations of the operator $-d^2/dx^2$. For example, the second-order approximation to (5.2.1) is given by

$$-\frac{u_{j+1} - 2u_j + u_{j-1}}{(\Delta x)^2} = f_j \qquad j = 0, 1, \ldots, N-1, \tag{5.2.21}$$

where $\Delta x = 2\pi/N$. The inversion of (5.2.21) requires the solution of a cyclic tridiagonal system. The fourth-order approximation is equally straightforward and requires the solution of a cyclic pentadiagonal system. Both types of systems can be inverted far more quickly than the computation of LV^n. The eigenfunctions of these discretizations are all given by (5.2.3) and the eigenvalues are

$$\lambda_p^{(2)} = 4\frac{\sin^2\left(\dfrac{p\Delta x}{2}\right)}{(\Delta x)^2}, \tag{5.2.22}$$

$$\lambda_p^{(4)} = \frac{\cos(2p\Delta x) - 16\cos(p\Delta x) + 15}{6(\Delta x)^2}, \tag{5.2.23}$$

$$\lambda_p^{(\infty)} = p^2. \tag{5.2.24}$$

Since the spectral operator and the finite-difference operator have the same eigenfunctions, it is clear that the effective eigenvalues of the preconditioned iterations based on $(H^{(2)})^{-1}L$ and $(H^{(4)})^{-1}L$ are then given by

$$\begin{aligned}\Lambda_p^{(2)} &= (p^2)(\lambda_p^{(2)})^{-1}\\ &= \frac{(p\Delta x/2)^2}{\sin^2(p\Delta x/2)}\end{aligned} \tag{5.2.25}$$

$$\begin{aligned}\Lambda_p^{(4)} &= (p^2)(\lambda_p^{(4)})^{-1}\\ &= \frac{6(p\Delta x)^2}{\cos(2p\Delta x) - 16\cos(p\Delta x) + 15}.\end{aligned} \tag{5.2.26}$$

The argument $p\Delta x$ lies in $[-\pi, \pi]$ and, in fact, only $[0, \pi]$ need be considered due to symmetry.

Similar results for even higher order finite-difference preconditionings are straightforward but tedious to obtain. The key properties of this class of preconditioning are given in Table 5.1. Unlike the original system, which has a spectral condition number scaling as N^2, the preconditioned system has a spectral condition number which is independent of N. The fourth-order finite-difference operator offers around a 40% improvement in convergence rate over the second-order operator. This is partially offset by the additional cost of inverting the finite-difference operator. The sixth-order finite-difference operator requires still more work for even less of a relative reduction in the eigenvalue range. The higher order preconditionings are thus of doubtful utility.

Some additional issues are raised by the problem

Table 5.1. Properties of finite difference preconditioning for the model problem (5.2.1)

Finite difference order	$\Lambda_{\min}$	$\Lambda_{\max}$	ρ
2	1.00	2.47	0.424
4	1.00	1.85	0.298
6	1.00	1.63	0.240

$$\frac{du}{dx} = f. \tag{5.2.27}$$

We continue to presume periodic boundary conditions. Using the second-order, central-difference approximation

$$\frac{u_{j+1} - u_{j-1}}{2\Delta x} = f_j \qquad j = 0, 1, \ldots, N-1, \tag{5.2.28}$$

we arrive at

$$\Lambda_p^{(2)} = \frac{p\Delta x}{\sin(p\Delta x)}, \tag{5.2.29}$$

for $|p\Delta x| \in [0, \pi]$. The obvious difficulty is that $\Lambda_{\max}^{(2)}$ is unbounded. No iterative scheme can overcome this property.

Orszag suggested one way around this difficulty: in the Fourier collocation evaluation of du/dx, simply set the upper third or so of the frequency spectrum to zero. The prescription for this is first to compute

$$\tilde{u}_k = \frac{1}{N} \sum_{j=0}^{N-1} u_j e^{-ikx_j} \qquad k = -\frac{N}{2}, -\frac{N}{2} + 1, \ldots, \frac{N}{2} - 1 \tag{5.2.30}$$

as usual; then to set, for example,

$$\tilde{u}_k^{(1)} = \begin{cases} ik\,\tilde{u}_k & |k| \leqslant N/3 \\ 0 & \dfrac{N}{3} < |k| \leqslant \dfrac{N}{2}; \end{cases} \tag{5.2.31}$$

and finally to use

$$\left.\frac{du}{dx}\right|_j = \sum_{k=-N/2+1}^{N/2-1} \tilde{u}_k^{(1)} e^{ikx_j} \qquad j = 0, 1, \ldots, N-1. \tag{5.2.32}$$

The relevant range of $|p\Delta x|$ is $[0, 2\pi/3]$. The upper bound on $\Lambda_p^{(2)}$ is 2.42; the lower bound is still 1. In addition to the loss of accuracy which this method produces, there is also the need to remove the upper third of the spectrum of f so that the residual may be used to monitor the convergence of the scheme.

Another approach is to use a first-order, one-sided finite-difference ap-

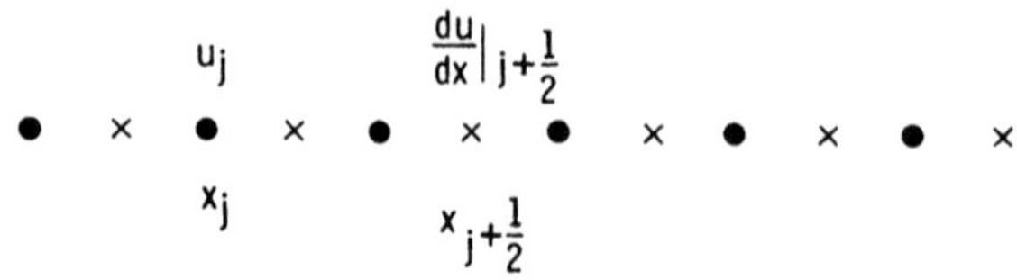

Figure 5.1. The staggered Fourier grid. The standard collocation points are denoted by the circles and the shifted points by the x's.

proximation such as

$$\frac{u_{j+1} - u_j}{\Delta x} = f_j \qquad j = 0, 1, \ldots, N - 1. \tag{5.2.33}$$

The eigenvalues resulting from this preconditioning are

$$\Lambda_p^{(1)} = \frac{\dfrac{p\Delta x}{2}}{\sin\left(\dfrac{p\Delta x}{2}\right)} e^{-i(p\Delta x/2)}. \tag{5.2.34}$$

These eigenvalues are bounded in absolute value, but are complex. Since the entire frequency spectrum has been retained, there is no loss of accuracy. However, the iterative scheme must be able to handle complex eigenvalues.

Yet another alternative is to shift, or stagger, the grid on which the derivative is evaluated with respect to the grid on which the function itself is defined. This is illustrated in Fig. 5.1. Fourier derivative evaluations are performed by computing $\tilde{u}_k$ as usual and then using

$$\left.\frac{du}{dx}\right|_{j+1/2} = \sum_{k=-N/2}^{N/2-1} ik\tilde{u}_k e^{ik(x_j + (\pi/N))}. \tag{5.2.35}$$

The finite-difference eigenvalues on this staggered grid are

$$\lambda_p^{(s)} = ik \frac{\sin\left(\dfrac{p\Delta x}{2}\right)}{\dfrac{p\Delta x}{2}} e^{i(p\Delta x/2)}, \tag{5.2.36}$$

but the spectral eigenvalues have a similar complex phase shift. Thus, the preconditioned eigenvalues are

$$\Lambda_p^{(s)} = \frac{\dfrac{p\Delta x}{2}}{\sin\left(\dfrac{p\Delta x}{2}\right)}. \tag{5.2.37}$$

These are real and confined to the narrow interval $[1, \pi/2]$. Even the simple

Table 5.2. Preconditioned eigenvalues for a one-dimensional first-derivative model problem

Preconditioning	Eigenvalues
Central differences	$\dfrac{p\Delta x}{\sin(p\Delta x)}$
One-sided differences	$e^{-i(p\Delta x/2)}\dfrac{p\Delta x/2}{\sin(p\Delta x/2)}$
High mode cut-off	$\begin{cases} \dfrac{p\Delta x}{\sin(p\Delta x)} & 0 \leqslant \lvert p\Delta x\rvert \leqslant (2\pi/3) \\ 0 & (2\pi/3) < \lvert p\Delta x\rvert \leqslant \pi \end{cases}$
Staggered grid	$\dfrac{p\Delta x/2}{\sin(p\Delta x/2)}$

Richardson method will perform well with the staggered grid preconditioning. These alternative first-order preconditionings are summarized in Table 5.2.

Another difficulty is posed by the traditional Helmholtz equation

$$-\frac{d^2u}{dx^2} - \lambda u = f, \tag{5.2.38}$$

where $\lambda > 0$. This problem is indefinite with

$$\lambda_p^{(\infty)} = p^2 - \lambda, \tag{5.2.39}$$

but has a well-defined solution so long as $\lambda \neq p^2$ for some integer p. Second-order finite-difference preconditioning leads to

$$\lambda_p^{(2)} = \frac{4\sin^2\left(\dfrac{p\Delta x}{2}\right)}{\Delta x^2} - \lambda \tag{5.2.40}$$

and

$$\Lambda_p^{(2)} = \frac{p^2 - \lambda}{\dfrac{p^2 \sin^2\left(\dfrac{p\Delta x}{2}\right)}{\left(\dfrac{p\Delta x}{2}\right)^2} - \lambda}. \tag{5.2.41}$$

There is likely to be a range of p for which $\Lambda_p^{(2)} < 0$. Thus, a preconditioned version of this Helmholtz problem will have both positive and negative eigenvalues.

As a final example of the complications which can arise in practice, let us take a simple model of a high Reynolds number flow:

$$-\varepsilon\frac{d^2u}{dx^2}+\frac{du}{dx}=f. \tag{5.2.42}$$

Second-order finite-difference preconditioning leads to

$$\begin{aligned}\Lambda_p^{(2)} &= \frac{\varepsilon p^2+ip}{\varepsilon p^2\dfrac{\sin^2(p\Delta x/2)}{(p\Delta x/2)^2}+ip\dfrac{\sin(p\Delta x)}{(p\Delta x)}}\\ &= \frac{\varepsilon^4p^4\dfrac{\sin^2(p\Delta x/2)}{(p\Delta x/2)^2}+p^2\sin(p\Delta x)/(p\Delta x)}{\left(\varepsilon^2p^4\dfrac{\sin^4(p\Delta x/2)}{(p\Delta x/2)^4}+p^2\dfrac{\sin^2(p\Delta x)}{(p\Delta x)^2}\right)}\\ &\quad+i\frac{\varepsilon p^3\sin^2(p\Delta x/2)/(p\Delta x/2)^2-\varepsilon p^3\sin(p\Delta x)/(p\Delta x)}{\left(\varepsilon^2p^4\dfrac{\sin^4(p\Delta x/2)}{(p\Delta x/2)^4}+p^2\dfrac{\sin^2(p\Delta x)}{(p\Delta x)^2}\right)}.\end{aligned} \tag{5.2.43}$$

The eigenvalues are complex and although the real parts are positive, there are some real parts which are close to zero for small ε. The staggered grid preconditioning produces complex eigenvalues as well, but their real parts are safely bounded greater than zero.

5.2.3. Non-Periodic Problems

These simple estimates of the eigenvalue range of the preconditioned Fourier operator are a good guide to the range of the preconditioned Chebyshev one as well. Chebyshev polynomials, of course, would be employed in place of trigonometric functions for problems with Dirichlet or Neumann boundary conditions. The appropriate preconditioning is a second-order finite-difference approximation on the non-uniform, Chebyshev grid. For (5.2.1) on $(-1, 1)$ with homogeneous Dirichlet boundary conditions, this preconditioning is

$$\begin{gathered}\frac{-2}{h_{j-1}(h_j+h_{j-1})}u_{j-1}+\frac{2}{h_jh_{j-1}}u_j+\frac{-2}{h_j(h_j+h_{j-1})}u_{j+1}=f_j\\ j=1,\ldots,N-1\\ u_0=0\\ u_N=0,\end{gathered} \tag{5.2.44}$$

where $h_j = x_j - x_{j+1}$ with $x_j = \cos \pi j/N$. Haldenwang et al. (1984) have shown analytically that the eigenvalues of the preconditioned matrix $H^{-1}L$ are given by

$$\Lambda_p^{(2)} = \frac{p(p-1)\sin^2\dfrac{\pi}{2N}\cos\dfrac{\pi}{2N}}{\sin\dfrac{(p-1)\pi}{2N}\sin\dfrac{\pi}{2N}} \qquad p = 2, 3, \ldots, N. \tag{5.2.45}$$

Hence,

$$\Lambda_{\min}^{(2)} = 1 \tag{5.2.46}$$

and

$$\Lambda_{\max}^{(2)} = N(N-1)\sin^2\frac{\pi}{2N}. \tag{5.2.47}$$

Note that $\Lambda_{\max}^{(2)} \leqslant \pi^2/4$, which is the same upper bound that applies to the second-order preconditioned Fourier operator.

Numerical eigenvalue calculations by Phillips, Zang and Hussaini (1986) indicate that the largest eigenvalue for the fourth-order finite-difference preconditioning of the Chebyshev operator is bounded by 1.85. Once again, the estimate from the preconditioned Fourier operator is reliable for the more complicated Chebyshev case. Even for the periodic problem, fourth-order preconditioning seemed not worthwhile. The case is even more compelling for non-periodic problems since (1) special difference formulas are needed at points adjacent to a boundary and (2) fourth-order finite-difference approximations on a non-uniform grid to variable-coefficient problems are extremely tedious to obtain.

Preconditionings for problems with Neumann or Robin boundary conditions are less clear-cut. Since the Dirichlet problem has already been discussed, we concentrate here on a Robin boundary condition at, say $x = +1$, of the form

$$\frac{du}{dx} + \alpha u = g. \tag{5.2.48}$$

As discussed in Sec. 3.3, there are several ways to enforce this condition. The most straightforward way is to collocate (5.2.48) directly at $x = +1$ (see (3.3.2)). The obvious preconditioning involves using a first-order one-sided derivative at $x = +1$:

$$-\frac{u_1 - u_0}{h_0} + \alpha u_0 = 0. \tag{5.2.49}$$

Table 5.3 presents the eigenvalue range for this preconditioning on the second-

Table 5.3. Extreme eigenvalues for preconditioned Dirichlet/Neumann problem

	$(H^{-1}L)_{\text{direct}}$		$(H^{-1}L)_{\text{indirect},1}$		$(H^{-1}L)_{\text{indirect},2}$	
N	$\lambda_{\min}$	$\lambda_{\max}$	$\lambda_{\min}$	$\lambda_{\max}$	$\lambda_{\min}$	$\lambda_{\max}$
4	1.000	1.757	0.329	1.609	0.805	1.757
8	1.000	2.131	0.180	2.000	0.818	2.131
16	1.000	2.306	0.093	2.219	0.821	2.306
32	1.000	2.388	0.047	2.340	0.822	2.388
64	1.000	2.428	0.024	2.403	0.822	2.428

derivative operator with Dirichlet conditions at $x = -1$ and Neumann ones at $x = +1$. The preconditioned problem has all of its eigenvalues in the interval $[1, \pi^2/4]$, just as the Dirichlet problem does.

There are two ways to impose the Robin boundary condition indirectly. The preconditioning at the boundary $x_0 = 1$ for both indirect approaches can be derived by the use of a "ghost point" $x_{-1} = 1 + \cos \pi/N$. Its value is determined by a central-difference approximation to (5.2.48) at x_0 and this ghost point is then eliminated from the second-order, finite-difference approximation to (5.2.1) at x_0. This yields

$$\left[\frac{2(1 + \alpha h_0)}{h_0^2}\right] u_0 - \frac{2}{h_0^2} u_1 = f_0. \tag{5.2.50}$$

The preconditioning in the interior is clearly given by (5.2.44). The middle two columns of Table 5.3 indicate the preconditioned eigenvalues resulting from this method. The smallest eigenvalue (and only it) decreases as N^{-1}. The remaining eigenvalues are confined to $[1, \pi^2/4]$. The second way to impose the boundary conditions implicitly alters the spectral operator only at the boundary. It is described by (3.3.5). The resulting eigenvalues of $H^{-1}L$ for large N are confined to $[0.82, \pi^2/4]$, as indicated in the last two columns of Table 3.2.

Preconditionings for the Helmholtz problem with Robin boundary conditions at both endpoints described in Sec. 3.3 present an additional difficulty that has been addressed by Canuto (1986). If the preconditioning (5.2.50) is modified to include the additional term in the ODE and is applied at both boundaries along with the appropriate modification of (5.2.44) in the interior, then although the largest eigenvalues of $H^{-1}L$ remain bounded, the smallest eigenvalue tends to zero as N^{-3}. This occurs because 1 is an eigenvalue of double multiplicity for L (the eigenfunctions are 1 and $T_N(x)$) but only single multiplicity for H. Canuto presented an alternative preconditioning for (3.3.1) with $\beta = \delta = 1$ that ameliorates this problem. It is given by

$$H = -\begin{pmatrix} \frac{1}{h_0} & -\frac{1}{h_0} & & & & \\ \frac{2}{h_0+h_1} & 0 & -\frac{2}{h_0+h_1} & & & \\ & \frac{2}{h_1+h_2} & 0 & -\frac{2}{h_1+h_2} & & \\ & & & & & \\ & & & \frac{2}{h_{N-2}+h_{N-1}} & 0 & -\frac{2}{h_{N-2}+h_{N-1}} \\ & & & & \frac{1}{h_{N-1}} & -\frac{1}{h_{N-1}} \end{pmatrix}$$

$$\times \begin{pmatrix} -\alpha & 0 & & & & \\ \frac{2}{h_0+h_1} & 0 & -\frac{2}{h_0+h_1} & & & \\ & \frac{2}{h_1+h_2} & 0 & -\frac{2}{h_1+h_2} & & \\ & & & & & \\ & & & \frac{2}{h_{N-2}+h_{N-1}} & 0 & -\frac{2}{h_{N-2}+h_{N-1}} \\ & & & & 0 & \gamma \end{pmatrix}$$

$$+I. \tag{5.2.51}$$

In this case, $\lambda_{\min}(H^{-1}L)$ is 1 and $|\lambda_{\max}(H^{-1}L)|$ grows as $N^{3/2}$, and the two largest eigenvalues are complex conjugates. In addition, H is now pentadiagonal rather than tridiagonal.

For the Richardson method, and its generalization via Chebyshev acceleration (discussed below in Sec. 5.3.3), the eigenvalues of $H^{-1}L$ are what matter for the iterative methods. For the descent (or variational) methods discussed in Secs. 5.3.1 and 5.3.2, it is the eigenvalues of $(LH^{-1}) + (LH^{-1})^T$ which matter. None of the finite-difference preconditioned Chebyshev methods for Robin boundary conditions are well-behaved in this sense, as will be discussed in Sec. 5.3.2.

Funaro (1987b) has analyzed the staggered grid preconditioning for the non-periodic first-order problem (5.2.27) with Dirichlet boundary conditions at $x = +1$. He has shown that the preconditioned eigenvalues are

$$\Lambda_p^{(s)} = \frac{p \sin \dfrac{\pi}{2N}}{\sin \dfrac{p\pi}{N}} \qquad p = 1, \ldots, N. \tag{5.2.52}$$

These are confined to the interval $[1, \pi/2]$, just as they are for the periodic problem. Funaro also presents some theoretical and numerical results for preconditioned, one-dimensional first-order systems.

5.2.4. Finite-Element Preconditioning

An alternative type of preconditioning is based on finite elements rather than finite differences. This was originally proposed by Canuto and Quarteroni (1985) and by Deville and Mund (1985). A clear discussion of its merits has recently been provided by Canuto and Pietra (1990). Deville and co-workers have applied it extensively in Navier–Stokes calculations (in a yet to be published work). Some of its advantages are: (1) the preconditioned operator has a smaller spread of eigenvalues and hence a reduced condition number; (2) improved treatment of Neumann and Robin boundary conditions; and (3) simpler implementation of higher order preconditioning.

In practical terms, the principal difference between finite-difference and finite-element preconditioning is that the latter includes a weighting of the residuals. In effect, (5.2.19) is replaced by

$$H(V^{n+1} - V^n) = \omega M(F - LV^n). \tag{5.2.53}$$

In finite-element terminology the matrix M is called the mass matrix and H the stiffness matrix. For the Fourier model problem with linear finite-element preconditioning, (5.2.21) is replaced by

$$-\frac{u_{j+1} - 2u_j + u_{j-1}}{\Delta x} = \frac{1}{6}\Delta x[f_{j+1} + 4f_j + f_{j-1}]. \tag{5.2.54}$$

The matrix which determines the effective eigenvalues of the preconditioned operator is $H^{-1}ML$ rather than just $H^{-1}L$. For the model problem

$$\Lambda_p^{(FE)} = \frac{(p\Delta x/2)^2}{\sin^2(p\Delta x/2)} \frac{2 + \cos(p\Delta x)}{3}. \tag{5.2.55}$$

These eigenvalues are confined to the interval [.693,1], giving a spectral condition number of 1.44 rather than the 2.47 which finite-difference preconditioning produces. This yields a spectral radius of 0.18. With finite-difference preconditioning, the spectral radius is 0.42; hence, a single iteration with the finite-element preconditioning produces the same reduction in the error that follows from two iterations with finite-difference preconditioning.

The proper formulation of the finite-element, preconditioning equations can be found in standard texts on finite-element methods, such as those by

Strang and Fix (1973), Glowinski (1984), and Fletcher (1984). The use of finite-element preconditioning is a recent development. Although the discussion of preconditioning in the remainder of the book is limited to finite-difference methods, the reader should keep in mind the potential advantages of finite-element preconditioning.

5.3. Conventional Iterative Methods

The past three decades have witnessed extensive research into iterative schemes for linear equations. Some standard references are the books by Varga (1962), Young (1971), and Hageman and Young (1981). The most thorough analyses are available for symmetric, positive-definite systems. The descent methods, discussed in the first subsection, are simple, robust and efficient schemes for such systems. Unfortunately, they are not strictly applicable to spectral equations, with the exception of Fourier approximations to self-adjoint problems. Of course, a non-symmetric system of the form (5.4) can always be transformed into a positive-definite system given by the normal equation

$$L^T L U = L^T F. \tag{5.3.1}$$

But the normal equation generally has a condition number that is the square of that for the original system, and the operator L must be applied twice. In most cases, effective alternatives to the normal equation approach are extremely useful. The second subsection is devoted to descent methods for non-symmetric problems. The simple Richardson method can be improved via Chebyshev acceleration. Furthermore, this method can be applied to some problems with complex eigenvalues. This technique is treated in the third subsection.

5.3.1. Descent Methods for Symmetric, Positive-Definite Systems

The method of steepest descent (SD) is a classical scheme for solving (5.4). (The name arises from the geometrical interpretation of the scheme for solving the minimization problem equivalent to (5.4).) The principle is to adjust the current guess V^n via

$$V^{n+1} = V^n + \alpha_n R^n, \tag{5.3.2}$$

where R^n is the residual

$$R^n = F - LV^n \tag{5.3.3}$$

and the scalar α_n is chosen to minimize the inner product

$$(E^{n+1}, LE^{n+1}) = \|E^{n+1}\|_L^2, \tag{5.3.4}$$

where E^n denotes the error

$$E^n = V^n - U. \tag{5.3.5}$$

It is a simple matter to show that

$$\alpha_n = \frac{(R^n, R^n)}{(R^n, LR^n)}. \tag{5.3.6}$$

Furthermore,

$$\|E^{n+1}\|_L^2 = \|E^n\|_L^2 - \frac{(R^n, R^n)^2}{(R^n, LR^n)}. \tag{5.3.7}$$

Thus, the energy error is strictly reduced at each step. The following bound holds:

$$\|E^n\|_L \leqslant \left(\frac{\kappa - 1}{\kappa + 1}\right)^n \|E^0\|_L. \tag{5.3.8}$$

(Since we have assumed L to be symmetric, κ is the true condition number of L.)

Like the steepest descent method, the minimum residual (MR) method adjusts the approximation in the direction of the current residual, as given by (5.3.3). However, the criterion for choosing the scalar α_n is that the Euclidean norm of the new residual should be minimized. This leads to the choice

$$\alpha_n = \frac{(R^n, LR^n)}{(LR^n, LR^n)}. \tag{5.3.9}$$

It follows that

$$\|R^{n+1}\|^2 = \|R^n\|^2 - \frac{(R^n, LR^n)^2}{(LR^n, LR^n)}. \tag{5.3.10}$$

Since L is symmetric, positive-definite, the residual is strictly reduced at every step. The bound for the iterates is

$$\|R^n\| \leqslant \left(\frac{\kappa - 1}{\kappa + 1}\right)^n \|R^0\|. \tag{5.3.11}$$

This is the same sort of bound as (5.3.8) for steepest descent, except that here the bound is on the residual norm rather than the energy norm of the error. Note that the number of iterations required for convergence (in their respective norms) of both descent methods is proportional to

$$\mathscr{I} \cong \tfrac{1}{2}\kappa. \tag{5.3.12}$$

The methods described in this section have the advantage of not requiring the specification of any parameter such as the parameter ω used in the Richardson iteration. In fact, the parameter α_n in the SD and MR methods is analogous to the ω in the Richardson method, but it is calculated dynamically.

A substantial improvement in convergence rate can be achieved by using conjugate direction methods in place of SD or MR. The two most common conjugate direction methods are known as the conjugate gradient (CG) method and the conjugate residual (CR) method. These methods were proposed by Hestenes and Stiefel (1952) as a direct method for solving symmetric, positive-definite linear systems. For such problems the conjugate direction methods produce the exact answer (in the absence of round-off errors) in a finite number of steps. In the late 1960s and early 1970s these methods began to be considered seriously as iterative rather than direct solution schemes which can produce a very accurate result in a small number of iterations. The papers by Reid (1971) and by Concus, Golub and O'Leary (1976) were particularly influential.

In a conjugate direction method the update of the iterate is generalized from (5.3.2) to

$$V^{n+1} = V^n + \alpha_n P^n. \tag{5.3.13}$$

In the conjugate gradient version, the directions satisfy the orthogonality property

$$(P^{n+1}, LP^n) = 0. \tag{5.3.14}$$

The scheme is initialized with an initial guess V^0. The initial direction vector is chosen to be

$$P^0 = R^0, \tag{5.3.15}$$

where R^0 is the initial residual. Subsequent iterations are made according to

$$\alpha_n = \frac{(R^n, R^n)}{(P^n, LP^n)} \tag{5.3.16a}$$

$$V^{n+1} = V^n + \alpha_n P^n \tag{5.3.16b}$$

$$R^{n+1} = R^n - \alpha_n LP^n \tag{5.3.16c}$$

$$\beta_n = \frac{(R^{n+1}, R^{n+1})}{(R^n, R^n)} \tag{5.3.16d}$$

$$P^{n+1} = R^{n+1} + \beta_n P^n. \tag{5.3.16e}$$

The formula (5.3.16a) for the familiar scalar α_n results from the requirement that V^{n+1} minimize the energy norm of the error, and the formula (5.3.16d) for the additional scalar β_n follows from the requirement (5.3.14).

The following orthogonality properties hold:

$$(R^k, R^l) = 0 \qquad \text{for } k \neq l \tag{5.3.17}$$

and

$$(P^k, LP^l) = 0 \qquad \text{for } k \neq l. \tag{5.3.18}$$

The first of these implies that $R^m = 0$ for some $m \leqslant M$, where M is the order of the matrix L. This explains the claim that the exact solution is obtained in a finite number of iterations. However, the presence of rounding errors leads to some contamination of the residual and direction vectors. The second orthogonality relation shows that the CG method does far more than the original requirement (5.3.14).

The favorable convergence properties of this method are reflected by the estimate for the energy error

$$\|E^n\|_L \leqslant 2\left(\frac{\sqrt{\kappa}-1}{\sqrt{\kappa}+1}\right)^n \|E^0\|_L. \tag{5.3.19}$$

The number of iterations required for convergence is therefore proportional to

$$\mathscr{I} = \tfrac{1}{2}\sqrt{\kappa}. \tag{5.3.20}$$

This is a decided improvement over the SD result (5.3.12). Of course, the CG method is more costly per iteration, both in CPU time and storage.

The orthogonality property for the conjugate residual version is

$$(LP^{n+1}, LP^n) = 0, \tag{5.3.21}$$

rather than (5.3.14). The requirement on V^{n+1} is now that it minimize the Euclidean norm of the residual. The algorithm is the same as (5.3.15)–(5.3.16) except that (5.3.16a) and (5.3.16d) are replaced by

$$\alpha_n = \frac{(R^n, LP^n)}{(LP^n, LP^n)} \tag{5.3.22a}$$

$$\beta_n = -\frac{(LR^{n+1}, LP^n)}{(LP^n, LP^n)} \tag{5.3.22d}$$

and the recursion relation

$$LP^{n+1} = LR^{n+1} + \beta_n LP^n \tag{5.3.22f}$$

is added.

The relevant error estimate is

$$\|R^n\| \leqslant 2\left(\frac{\sqrt{\kappa}-1}{\sqrt{\kappa}+1}\right)^n \|R^0\| \tag{5.3.23}$$

and (5.3.20) applies here as well.

Let us now include a symmetric preconditioning, denoted as usual by H, in these descent methods. It is tempting to write (5.4) as

$$\tilde{L}\tilde{U} = F, \tag{5.3.24a}$$

where

$$\tilde{L} = LH^{-1} \tag{5.3.24b}$$

and

$$\tilde{U} = HU, \tag{5.3.24c}$$

and to apply the preceding formulas to (5.3.24a). However, LH^{-1} is not necessarily symmetric, positive-definite (unless L and H^{-1} commute). We can, however, choose Q such that

$$H = QQ^T \tag{5.3.25}$$

and use

$$\tilde{L}\tilde{U} = \tilde{F} \tag{5.3.26a}$$

with

$$\tilde{L} = Q^{-1}LQ^{-T} \tag{5.3.26b}$$

$$\tilde{F} = Q^{-1}F \tag{5.3.26c}$$

$$\tilde{U} = Q^T U. \tag{5.3.26d}$$

Also use

$$\tilde{V} = Q^T V \tag{5.3.27a}$$

$$\tilde{P} = Q^T P \tag{5.3.27b}$$

$$\tilde{R} = Q^{-1}R. \tag{5.3.27c}$$

The matrix $\tilde{L}$ is symmetric, positive-definite. After inserting (5.3.26b) into the preceding schemes and then manipulating the expressions into computationally convenient forms we arrive at the following:

Preconditioned Steepest Descent (PSD)

Initialize

$$\begin{aligned} &V^0 \\ R^0 &= F - LV^0 \\ HP^0 &= R^0. \end{aligned}$$

Iterate (5.3.28)

$$\begin{aligned} \alpha_n &= \frac{(R^n, P^n)}{(P^n, LP^n)} \\ V^{n+1} &= V^n + \alpha_n P^n \\ R^{n+1} &= R^n - \alpha_n LP^n \\ HP^{n+1} &= R^{n+1}. \end{aligned}$$

Preconditioned Minimum Residual (PMR)

Initialize

$$
\begin{aligned}
&V^0 \\
R^0 &= F - LV^0 \\
HP^0 &= R^0.
\end{aligned}
$$

Iterate (5.3.29)

$$
\begin{aligned}
\alpha_n &= \frac{(R^n, LP^n)}{(LP^n, LP^n)} \\
V^{n+1} &= V^n + \alpha_n P^n \\
R^{n+1} &= R^n - \alpha_n LP^n \\
HP^{n+1} &= R^{n+1}.
\end{aligned}
$$

Preconditioned Conjugate Gradient (PCG)

Initialize

$$
\begin{aligned}
&V^0 \\
R^0 &= F - LV^0 \\
HZ^0 &= R^0 \\
P^0 &= Z^0.
\end{aligned}
$$

Iterate (5.3.30)

$$
\begin{aligned}
\alpha_n &= \frac{(R^n, Z^n)}{(P^n, LP^n)} \\
V^{n+1} &= V^n + \alpha_n P^n \\
R^{n+1} &= R^n - \alpha_n LP^n \\
HZ^{n+1} &= R^{n+1} \\
\beta_n &= \frac{(R^{n+1}, Z^{n+1})}{(R^n, Z^n)} \\
P^{n+1} &= Z^{n+1} + \beta_n P^n.
\end{aligned}
$$

Preconditioned Conjugate Residual (PCR)

Initialize

$$
\begin{aligned}
&V^0 \\
R^0 &= F - LV^0
\end{aligned}
$$

$$HZ^0 = R^0$$
$$P^0 = Z^0.$$

Iterate (5.3.31)

$$\alpha_n = \frac{(R^n, LP^n)}{(LP^n, LP^n)}$$
$$V^{n+1} = V^n + \alpha_n P^n$$
$$R^{n+1} = R^n - \alpha_n LP^n$$
$$HZ^{n+1} = R^{n+1}$$
$$\beta_n = -\frac{(LZ^{n+1}, LP^n)}{(LP^n, LP^n)}$$
$$P^{n+1} = Z^{n+1} + \beta_n P^n$$
$$LP^{n+1} = LZ^{n+1} + \beta_n LP^n.$$

The steepest descent and conjugate gradient methods minimize the L norm of the error, whereas the minimum residual and conjugate residual methods minimize the H^{-1} norm of the residual. The relevant condition number is that of $Q^{-1}LQ^{-T}$ rather than that of L.

The methods described in this section are, with few exceptions, strictly applicable only to Fourier approximations to self-adjoint problems. Standard Chebyshev collocation approximations do not yield symmetric discretizations, even for self-adjoint problems. (However, it is possible to perform collocation on a symmetric, weak formulation of the problem (5.2)—see Spalart (1986) for the weak formulation of the corresponding Fourier/Jacobi Galerkin approximation.) The methods described in the following two subsections are applicable to non-symmetric problems.

5.3.2. Descent Methods for Non-Symmetric Problems

There is not now, and may never be, a totally satisfactory iterative scheme for this class of problems: one that is guaranteed to converge and that requires no more additional storage and work than, say, the PCR method. The subject of iterative schemes for non-symmetric problems is currently receiving much attention. The descent methods that we discuss in this subsection are but a small subset of the schemes that have been proposed. Practitioners of iterative methods for spectral equations will need to keep abreast of the literature.

Since the matrix L is not symmetric, the transformation (5.3.24) rather than (5.3.25)—(5.3.26) is appropriate here. The preconditioned matrix $\tilde{L} = LH^{-1}$ determines the performance of descent methods. In fact, since these descent methods employ inner products, it is really

$$\hat{L} = WLH^{-1}, \tag{5.3.32}$$

which is relevant, where W is the diagonal matrix which represents the weights used in the inner product. These may be unity, the weights (2.4.14) for the standard Chebyshev inner product, or the Clenshaw–Curtis weights (13.2.18) for mimicking the L_2-inner product.

The performance of descent methods depends upon the symmetric part of $\hat{L}$, defined by

$$\hat{L}_S = \tfrac{1}{2}(\hat{L} + \hat{L}^T), \tag{5.3.33}$$

Similarly, the anti-symmetric part is given by

$$\hat{L}_A = \tfrac{1}{2}(\hat{L} - \hat{L}^T). \tag{5.3.34}$$

Eisenstat, Elman, and Schultz (1983), hereafter referred to as EES, considered a class of methods which are generalizations of the MR and CR schemes. They proved that the iterative methods defined by (5.3.29) and (5.3.31) converge provided that $\hat{L}_S$ is positive-definite. Furthermore, they furnished several estimates of the convergence rates, such as

$$\|R^n\| \leqslant \left[1 - \frac{\lambda_{\min}(\hat{L}_S)^2}{\lambda_{\max}(\hat{L}^T\hat{L})}\right]^{n/2} \|R^0\|. \tag{5.3.35}$$

One key property of the CR method which is lost when $\hat{L}$ is not symmetric is that the orthogonality property (5.3.21) does not hold when P^{n+1} is computed via (5.3.16e) and (5.3.22d). EES devised a method that they termed the generalized conjugate residual (GCR) method in which the relevant equations in (5.3.31) are replaced by

$$\beta_j^{(n)} = -\frac{(LZ^{n+1}, LP^n)}{(LP^n, LP^n)}. \tag{5.3.36a}$$

$$P^{n+1} = Z^{n+1} + \sum_{j=0}^{n} \beta_j^{(n)} P^j \tag{5.3.36b}$$

$$LP^{n+1} = LZ^{n+1} + \sum_{j=0}^{n} \beta_j^{(n)} LP^j \tag{5.3.36c}$$

(We have listed the forms appropriate for the preconditioned version.) This restores the orthogonality property of the direction vectors at the price of considerable increase in storage and computation. EES discuss an alternative version of GCR in which the algorithm is restarted every k steps. This restarted GCR method is similar to an earlier algorithm developed by Vinsome (1976), known as Orthomin (k) in which (5.3.36b) is replaced by

$$P^{n+1} = Z^{n+1} + \sum_{j=n-k+1}^{n} \beta_j^{(n)} P^j. \tag{5.3.37}$$

Orthomin (0) is equivalent to MR and Orthomin (1) will also be referred to as the truncated conjugate residual (TCR) method. The estimate (5.3.35) applies to all of these methods as does the requirement that $\hat{L}_S$ be positive-definite.

Some related methods are Orthodir (Young and Jea (1980)), Orthores (Hageman and Young (1981)), and GMRES (Saad and Schultz (1983)). The latter method works even if $\hat{L}_S$ is not positive-definite, but it requires considerable computation and the storage of all the P^j's to achieve this.

Scaling can be crucial for these descent methods. Suppose that the rows of L are scaled by Q_1 and the columns by Q_2, and likewise, for H. Then we are interested in

$$\begin{aligned} L_Q &= Q_1 L Q_2 \\ H_Q &= Q_1 H Q_2. \end{aligned} \tag{5.3.38}$$

We have that

$$L_Q H_Q^{-1} = Q_1 L H^{-1} Q_1^{-1}.$$

Although the spectrum of $L_Q H_Q^{-1}$ corresponds to that of LH^{-1}, the same is not true of their symmetric parts. An example of the crucial role that scaling can play is furnished in Sec. 5.6.

There is one fairly simple problem for which these descent methods have difficulty—(5.2) with Neumann (or Robin) boundary conditions. Canuto (1986) presented numerical computations of the eigenvalues of $\hat{L}_S$ for several ways of imposing and preconditioning the Neumann boundary condition. In Sec. 5.2.3 we saw that the best-behaved eigenvalues of $H^{-1}L$ occurred for the direct imposition of the boundary conditions. For the corresponding $\hat{L}_S$, however, the eigenvalue range grows as N^4. Moreover, $\hat{L}_S$ is not positive-definite. Both methods of enforcing the boundary conditions implicitly also produce an indefinite $\hat{L}_S$.

5.3.3. Chebyshev Acceleration

The basic Richardson method can be improved and extended in several ways. The discussion in Sec. 5.2.1 concerned only the stationary Richardson method. In a non-stationary Richardson method, the parameter ω in (5.2.18) is allowed to depend on n, usually by cycling through a fixed number k of parameters. Using the minimax property of Chebyshev polynomials, one derives the following expression for the optimal parameters (Young (1954))

$$\omega_j = \frac{2/\lambda_{\min}}{(\kappa - 1)\cos\dfrac{(2j-1)\pi}{2k} + (\kappa + 1)} \qquad j = 1, \ldots, k \tag{5.3.39}$$

and the effective spectral radius

$$\rho = \frac{1}{T_k\left(\dfrac{\kappa + 1}{\kappa - 1}\right)^{1/k}}, \tag{5.3.40}$$

where κ is the spectral condition number of $H^{-1}L$.

Richardson's method may be submitted to Chebyshev acceleration (Varga (1962)). For the preconditioned problem this is

Chebyshev Iteration

Initialize

$$V^0$$

$$V^1 = V^0 + \omega H^{-1}(F - LV^1).$$

Iterate

$$\beta_n = \frac{2T_n(1/\sigma)}{\sigma T_{n+1}(1/\sigma)} \tag{5.3.41}$$

$$V^{n+1} = [\beta_n V^n + (1 - \beta_n) V^{n-1}] + \omega\beta_n H^{-1}[F - LV^n],$$

where $\sigma = (\kappa - 1)/(\kappa + 1)$ is the spectral radius of the basic Richardson method.

The first k steps of Chebyshev iteration produce the same results as the first full cycle of non-stationary Richardson with k parameters. Chebyshev iteration, however, has more favorable round-off error properties as well as a smaller spectral radius, given by (5.3.40) after k steps and

$$\rho = \frac{\sqrt{\kappa} - 1}{\sqrt{\kappa} + 1} \tag{5.3.42}$$

in the limit $k \to \infty$. If $\kappa = \pi^2/4$, then $\rho = 0.225$. Hence, it takes 1.9 iterations of stationary Richardson to produce the same reduction in error that one obtains from a single Chebyshev iteration. However, Chebyshev iteration does require an extra level of storage.

The discussion so far has presumed that the eigenvalues of $H^{-1}L$ are confined to the interval $[\lambda_{\min}, \lambda_{\max}]$ on the positive real-axis. Both the Richardson and Chebyshev iteration schemes can work on problems for which the eigenvalues are complex but have positive real parts. In the case of Richardson's method, this involves the use of a complex parameter ω. The iteration can be performed entirely in real arithmetic according to

$$V^{n+1} = V^n + 2\,\mathrm{Re}\,\{\omega\} H^{-1}(F - LV^n) + |\omega|^2 H^{-1}LH^{-1}(F - LV^n). \tag{5.3.43}$$

The parameter ω is obtained by solving a complex minimax problem.

A similar complex minimax problem arises for the more efficient Chebyshev iteration. Manteuffel (1977) developed a method that is based on finding an ellipse which encloses the spectrum of $H^{-1}L$. Let the foci of the ellipse be located at $d + c$ and $d - c$. Then the method is

Complex Chebyshev Iteration

Initialize

$$V^0$$

$$P^0 = \frac{1}{d} H^{-1}(F - LV^0).$$

Iterate

$$\alpha_n = \frac{2}{c} \frac{T_n(d/c)}{T_{n+1}(d/c)}$$

$$\beta_n = \frac{T_{n-1}(d/c)}{T_{n+1}(d/c)} \tag{5.3.44}$$

$$P^n = \alpha_n H^{-1}(F - LV^n) + \beta_n P^{n-1}$$

$$V^{n+1} = V^n + P^n.$$

Manteuffel (1978) has developed an adaptive version of this method which dynamically estimates the crucial parameters d and c. Ashby (1985) has incorporated this algorithm into publicly available software.

5.4. Multidimensional Preconditioning

For one-dimensional problems, finite-difference preconditionings of the spectral operator (see Sec. 5.2.2) are a quite inexpensive part of the iterative scheme. The finite-difference inversion part of the algorithm typically costs $O(N)$ operations compared with the $O(N \log_2 N)$ cost for the application of the spectral operator. In higher dimensions, however, the finite-difference inversion becomes relatively expensive and/or complicated. The best that one can do in terms of a direct solution is for separable problems, for which the cost is $O(N^d(\log_2 N)^{d-1})$. This compares unfavorably to the $O(N^d \log_2 N)$ cost of the spectral operator. For non-separable problems, direct solution of the finite-difference equations is even more expensive. In this section we first review methods for solving the full finite-difference equations and then consider the alternative of using more readily inverted, but less effective preconditioners.

5.4.1. Finite-Difference Solvers

Remarkably efficient direct solvers for separable elliptic equations have been available for over a decade. Dorr (1970) and Swarztrauber (1977) have provided reviews of various direct and iterative methods for such problems. The best direct methods are termed cyclic reduction (CR) and Fourier analysis/cyclic reduction (FACR). The CR method has an operation count of $5N^2 \log_2 N$ in two dimensions and $5N^3(\log_2 N)^2$ in three dimensions. The FACR method has smaller operation counts, but it requires that the coefficients of the separable equation depend on at most one coordinate and that a uniform grid is used in the remaining directions. (Note that a finite-difference approximation to a Chebyshev collocation operator produces what is, in effect, a variable-coefficient problem.) The FACR algorithm is not useful in the present context, for a Fourier transform in the constant-coefficient directions reduces the problem to a set of uncoupled one-dimensional problems and the only remaining finite-difference equations are one-dimensional. Hence, CR is the recommended procedure for inverting second-order finite-difference approximations to separable problems. A software package (Swarztrauber and Sweet (1975)) is readily available. No corresponding efficient software is yet available for higher-order finite-difference or finite-element methods. Software is, however, available for the more costly banded Gauss elimination and sparse matrix algorithms (Dongarra et al. (1978), Duff (1981), Eisenstat et al. (1982)).

Outside of the realm of some special separable equations, iterative methods are the preferred strategy for inverting the finite-difference preconditioner to the spectral operator. The two strongest entries in this field are conjugate direction methods (and their generalizations to non-symmetric problems) and multigrid methods. There is a vast literature on conjugate direction methods for finite-difference discretizations. The classic work was cited in Sec. 5.3.1. In practice, preconditioning of the *finite-difference* approximation is essential. This continues to be an active field of research. The volume edited by Evans (1983) contains numerous articles on the subject. Some of the flavor of multigrid methods is conveyed in Sec. 5.5. The article by Stuben and Trottenberg (1982) is a thorough introduction to the method, as is the book by Hackbusch (1985). At the present time, the general two-dimensional problem may be considered solved, but there are still some three-dimensional problems which remain challenging.

The use of iterative methods for the solution of the preconditioned spectral equations requires an inner iteration (for the finite-difference equations) imbedded within an outer iteration (for the spectral equation itself).

5.4.2. Modified Finite-Difference Preconditioners

The discussion here will focus on methods for the two-dimensional Poisson equation (5.1.38). We begin with a description of several types of preconditioning based on incomplete-LU decomposition (Meijerink and Van der Vorst (1981)) of the matrix H_{FD} which represents the standard five-point

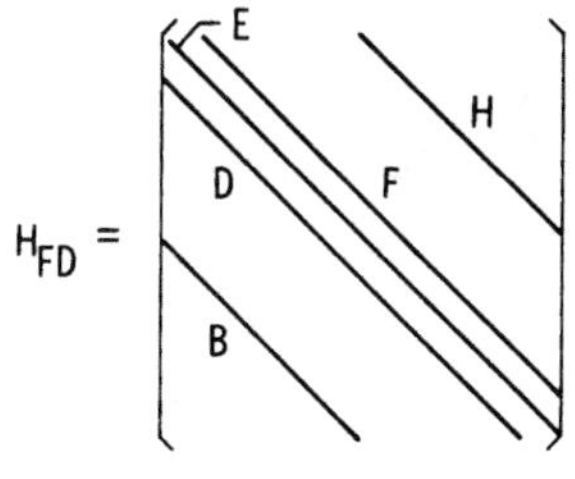

Figure 5.2. Structure of the full finite-difference preconditioning for a two-dimensional problem.

second-order finite-difference approximation to the differential equation (5.1.38). The first preconditioning is given by

$$H_{LU} = \mathscr{L}\mathscr{U} \tag{5.4.1}$$

where $\mathscr{L}$ is identical to the lower-triangular portion of H_{FD}, and $\mathscr{U}$ is chosen so that the two super diagonals of $\mathscr{L}\mathscr{U}$ agree with those of H_{FD}. The second preconditioning, denoted by H_{RS}, is similar but the diagonal elements of $\mathscr{L}$ are altered from those of H_{FD} so as to ensure that the row-sums of H_{RS} and H_{FD} are identical.

A five-point approximation on a Chebyshev grid to (5.1.38) may be written as

$$\begin{aligned}(H_{FD}U)_{i,j} = E_{i,j}U_{i,j} + D_{i,j}U_{i-1,j} + F_{i,j}U_{i+1,j} \\ + H_{i,j}U_{i,j+1} + B_{i,j}U_{i,j-1}.\end{aligned} \tag{5.4.2}$$

Figure 5.2 shows the structure of the matrix H_{FD}. A five-diagonal incomplete-LU factorization is given by (5.4.2) where

$$(\mathscr{L}U)_{i,j} = v_{i,j}U_{i,j} + t_{i,j}U_{i-1,j} + g_{i,j}U_{i,j-1} \tag{5.4.3}$$

and

$$(\mathscr{U}U)_{i,j} = U_{i,j} + e_{i,j}U_{i+1,j} + f_{i,j}U_{i,j+1}. \tag{5.4.4}$$

Figure 5.3 shows the structure of the factors $\mathscr{L}$ and $\mathscr{U}$. The coefficients in (5.4.3) and (5.4.4) are related to those in (5.4.1) by

$$\begin{aligned} t_{i,j} &= D_{i,j} \\ g_{i,j} &= B_{i,j} \\ v_{i,j} &= E_{i,j} - t_{i,j}f_{i,j-1} - g_{i,j}e_{i-1,j} \\ &\quad - \alpha[t_{i,j}e_{i,j-1} + g_{i,j}f_{i-1,j}] \\ e_{i,j} &= F_{i,j}/v_{i,j} \\ f_{i,j} &= H_{i,j}/v_{i,j}. \end{aligned} \tag{5.4.5}$$

The choice of $\alpha = 0$ gives the standard LU (H_{LU5}) result and $\alpha = 1$ gives the

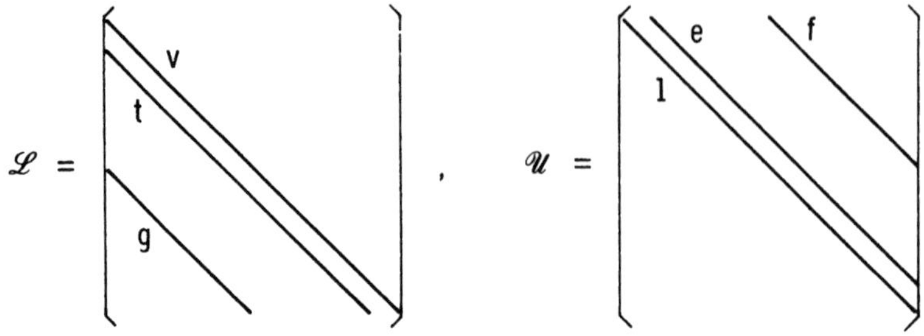

Figure 5.3. Structure of the five-diagonal incomplete-LU preconditioning.

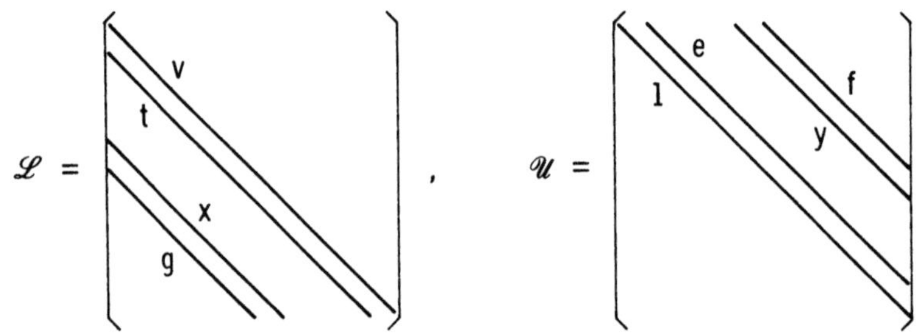

Figure 5.4. Structure of the seven-diagonal incomplete-LU preconditioning.

row-sum-equivalence version (H_{RS5}). Since neither version is an exact factorization of the original finite-difference matrix, some error is inevitable. Roughly speaking, the standard LU case does better on the high frequency components and the row-sum-equivalence alternative is more accurate on the low frequency end.

A more accurate factorization can be achieved by including one extra nonzero diagonal in $\mathscr{L}$ and $\mathscr{U}$ as indicated in Fig. 5.4. The seven-diagonal LU factorization is given by

$$(\mathscr{L}U)_{i,j} = v_{i,j}U_{i,j} + t_{i,j}U_{i-1,j} + g_{i,j}U_{i,j-1} + x_{i,j}U_{i+1,j-1} \tag{5.4.6}$$

$$(\mathscr{U}U)_{i,j} = U_{i,j} + e_{i,j}U_{i+1,j} + f_{i,j}U_{i,j+1} + y_{i,j}U_{i-1,j+1}. \tag{5.4.7}$$

These coefficients can be calculated from

$$g_{i,j} = B_{i,j}$$

$$x_{i,j} = -\, g_{i,j}e_{i,j-1}$$

$$t_{i,j} = D_{i,j} - g_{i,j}y_{i,j-1}$$

$$\begin{aligned} v_{i,j} &= E_{i,j} - g_{i,j}f_{i,j-1} - x_{i,j}y_{i+1,j-1} - t_{i,j}e_{i-1,j} \\ &\quad - \alpha[e_{i+1,j-1}x_{i,j} - y_{i-1,j}t_{i,j}] \\ e_{i,j} &= [F_{i,j} - f_{i+1,j-1}x_{i,j}]/v_{i,j} \\ y_{i,j} &= -t_{i,j}f_{i-1,j}/v_{i,j} \\ f_{i,j} &= H_{i,j}/v_{i,j}. \end{aligned} \tag{5.4.8}$$

Once again, $\alpha = 0$ gives the H_{LU7} version and $\alpha = 1$ the H_{RS7} case.

A good indication of the effectiveness of these preconditionings is provided by their eigenvalue distribution. Let us consider the case of a 16×16 grid. The fully finite-difference preconditioning produces eigenvalues which are purely real and confined to the interval $[1, 2.31]$. All but two of the eigenvalues resulting from the standard LU preconditioning are real; the imaginary parts of the two complex eigenvalues are only of order 10^{-3}. The real parts are in $[.22, 2.4]$. The row-sum-equivalence preconditioned eigenvalues have real parts in $[1, 2.7]$ and the imaginary parts of the only two complex ones are of order 10^{-3} as well.

Table 5.4 summarizes how the extreme eigenvalues depend on the $N \times N$ grid. Note that the two-dimensional $H_{FD}^{-1}L$ eigenvalues obtained here computationally agree exactly with the one-dimensional analytic formula (5.2.45). As N increases beyond 16, a few more complex eigenvalues arise for the factored preconditionings, but their imaginary parts are still very small. Table 5.5 gives the spectral condition numbers for these cases. Since all the iterative methods discussed in Sec. 5.3 perform better for smaller κ, the choice there lies between H_{FD} and H_{RS7}. For the small values of N contained in the tables, H_{RS7} is clearly preferable since, in practice the inversion of H_{RS7} takes only a few percent of the time of the evaluation of LV and its condition number is nearly as good as the much more expensive H_{FD}. Wong, Zang and Hussaini (1986) present several numerical examples of these incomplete-LU preconditionings coupled with the MR and TCR iterative schemes discussed in Sec. 5.3. The balance may change for large N, however, since the condition number of $H_{RS7}^{-1}L$ grows as $\sqrt{N}$ whereas that of $H_{FD}^{-1}L$ remains bounded by 2.47. In this

Table 5.4. Extreme eigenvalues for the preconditioned Chebyshev Laplace operator

	$H_{FD}^{-1}L$		$H_{LU5}^{-1}L$		$H_{LU7}^{-1}L$		$H_{RS5}^{-1}L$		$H_{RS7}^{-1}L$	
N	λ_{min}	λ_{max}	λ_{min}	λ_{max}	λ_{min}	λ_{max}	λ_{min}	λ_{max}	λ_{min}	λ_{max}
4	1.000	1.757	0.929	1.717	0.997	1.761	1.037	1.781	1.001	1.514
8	1.000	2.131	0.582	2.273	0.851	2.160	1.061	2.877	1.039	2.209
16	1.000	2.305	0.224	2.603	0.455	2.376	1.043	4.241	1.036	2.672
24	1.000	2.363	0.111	2.737	0.252	2.467	1.031	5.379	1.028	3.061

Table 5.5. Condition number for the preconditioned Chebyshev Laplace operator

N	$H_{FD}^{-1}L$	$H_{LU5}^{-1}L$	$H_{LU7}^{-1}L$	$H_{RS5}^{-1}L$	$H_{RS7}^{-1}L$
4	1.757	1.848	1.766	1.717	1.512
8	2.131	3.905	2.538	2.712	2.126
16	2.305	11.621	5.222	4.066	2.579
24	2.363	24.658	9.790	5.217	2.978

case the best approach seems to be the multigrid schemes described in Sec. 5.5.

The incomplete-LU preconditionings involve a lot of recursive operations. Hence, they are less attractive on vector and parallel computers. A more vectorizable preconditioning is given by approximate factorization schemes. In light of some of its applications, it is convenient to consider the non-linear version of (5.4), which we write as

$$M(U) = 0, \tag{5.4.9}$$

where

$$M(U) = L(U) - F. \tag{5.4.10}$$

Next, view U not as the solution to (5.4.9), but rather as the steady-state solution to the evolution equation

$$\frac{\partial U}{\partial t} = M(U). \tag{5.4.11}$$

This is surely sensible if the original PDE is elliptic, for then (5.4.11) represents the spatial discretization of a parabolic problem. Semi-implicit time-stepping procedures are desirable for such problems because of the severe explicit time-step limitations. (This is especially acute for collocation discretizations employing Chebyshev series because of the very small spacing between the collocation points near the boundary.) The simplest practical time-discretization of (5.4.11) is

$$\frac{U^{n+1} - U^n}{\Delta t} = M(U^n) + J(U^n)(U^{n+1} - U^n), \tag{5.4.12}$$

where

$$J(U) = \frac{\partial M}{\partial U}(U), \tag{5.4.13}$$

and a superscript refers to a time-level. Let

$$\alpha = \frac{1}{\Delta t} \tag{5.4.14}$$

and

$$\Delta U^n = U^{n+1} - U^n, \tag{5.4.15}$$

and then rewrite (5.4.12) as

$$[\alpha I - J(U^n)]\Delta U^n = M(U^n). \tag{5.4.16}$$

This motivates the relaxation scheme

$$V^{n+1} = V^n + \omega \Delta V^n, \tag{5.4.17}$$

where ΔV^n is the solution to

$$[\alpha_n I - J(V^n)]\Delta V^n = M(V^n). \tag{5.4.18}$$

In many cases the Jacobian $J(V)$ can be split into the sum of two operators, $J_x(V)$ and $J_y(V)$, each involving derivatives in only the one coordinate direction indicated by the subscript. Approximate factorization methods encompass various approximations to the left-hand side of (5.4.18). The most straightforward of these is

$$[\alpha_n I - J_x(V^n)][\alpha_n I - J_y(V^n)]\Delta V^n = \alpha_n M(V^n), \tag{5.4.19}$$

in combination with (5.4.17). This is just the Douglas and Gunn (1964) version of ADI. For second-order spatial discretizations, the term $[\alpha_n I - J_x(V^n)]$ leads to a set of tridiagonal systems, one for each value of y. The second left-hand side factor produces another set of tridiagonal systems. For collocation discretizations, however, these systems are full; hence, (5.4.19) is still relatively expensive to invert. A compromise analogous to the one invoked in the incomplete-*LU* decomposition preconditioning is to replace J_x and J_y with their second-order finite-difference analogs, denoted by H_x and H_y, respectively:

$$[\alpha_n I - H_x(V^n)][\alpha_n I - H_y(V^n)]\Delta V^n = \alpha_n M(V^n). \tag{5.4.20}$$

(The solution algorithm for these equations may include vectorization over the y-direction for the inversion of the matrix in the first brackets and over the x-direction for the other matrix.)

An essential part of this type of preconditioning is the choice of the parameters α_n and ω_n. A brief discussion is provided by Streett, Zang and Hussaini (1985). At present, trial and error is still a major component of the selection process.

5.5. Spectral Multigrid Methods

The iterative methods described above for two-dimensional problems are not entirely satisfactory. Those which employ the incomplete-*LU* factorization, for example, have rather inexpensive iterations but the requisite number of iterations increases with the number of unknowns. Although the methods which employ the full finite-difference preconditioning need a fixed number of iterations (regardless of the number of unknowns), the cost of a single iteration is quite high and it increases with the number of unknowns faster than does the cost of evaluating the spectral residual. A more desirable situation would be to use a cheap preconditioning such as incomplete-*LU* factorization but within an iterative scheme for which the number of iterations is independent of the number of unknowns. Spectral multigrid (SMG) techniques are a promising way to achieve this goal.

5.5.1. Model Problem Discussion

We begin the discussion of spectral multigrid techniques by reverting to the unpreconditioned Richardson scheme with which Sec. 5.2 began. The condition number of this method increases as N^2. The resulting slow convergence was the outcome of balancing the damping of the lowest frequency eigenfunction with that of the highest frequency one in (5.2.10). The multigrid approach takes advantage of the fact that the low frequency modes ($|p| < N/4$) can be represented just as well on coarser grids. It settles for balancing the middle frequency one ($|p| = N/4$) with the highest frequency one ($|p| = N/2$), and hence damps effectively only those modes which cannot be resolved on coarser grids. In (5.2.11) and (5.2.12), $\lambda_{\min}$ is replaced by $\lambda_{\text{mid}} = \lambda(N/4)$. The optimal relaxation parameter in this context is

$$\omega_{MG} = \frac{2}{\lambda_{\max} + \lambda_{\text{mid}}}. \tag{5.5.1}$$

The multigrid smoothing factor

$$\mu_{MG} = \frac{\lambda_{\max} - \lambda_{\text{mid}}}{\lambda_{\max} + \lambda_{\text{mid}}} \tag{5.5.2}$$

measures the damping rate of the high frequency modes. Alternatively, we may write

$$\mu_{MG} = \frac{\kappa_{MG} - 1}{\kappa_{MG} + 1}, \tag{5.5.3}$$

where $\kappa_{MG} = \lambda_{\max}/\lambda_{\text{mid}}$ is known as the multigrid condition number. In this

Table 5.6. Damping factors for $N = 64$

p	Single grid	Multigrid
1	0.9980	0.9984
2	0.9922	0.9938
4	0.9688	0.9750
8	0.8751	0.9000
12	0.7190	0.7750
16	0.5005	0.6000
20	0.2195	0.3750
24	0.1239	0.1000
28	0.5298	0.2250
32	0.9980	0.6000

example $\mu_{MG} = 0.60$, independent of N. The price of this effective damping of the high frequency errors is that the low frequency errors are hardly damped at all. Table 5.6 compares the single grid and multigrid damping factors for $N = 64$. However, on a grid with $N/2$ collocation points, the modes for $|p| \in [N/8, N/4]$ are now the high frequency ones. They get damped on this grid. Still coarser grids can be used until relaxations are so cheap that one can afford to damp all the remaining modes, or even to solve the discrete equations exactly. For the case illustrated in Table 5.6 the high frequency error reduction in the multigrid context is roughly 250 times as fast as the single grid reduction for $N = 64$.

We describe the multigrid process by considering the interplay between two grids. The fine grid problem can be written in the form

$$L^f U^f = F^f. \tag{5.5.4}$$

The decision to switch to the coarse grid is made after the fine grid approximation V^f has been sufficiently smoothed by the relaxation process, i.e., after the high frequency content of the error $V^f - U^f$ has been sufficiently reduced. For the model problem, three relaxations on a grid reduce the error by a factor of $(.60)^3$, which is roughly an order of magnitude. The auxiliary equation on the coarse grid is

$$L^c U^c = F^c, \tag{5.5.5}$$

where

$$F^c = R[F^f - L^f V^f]. \tag{5.5.6}$$

The restriction operator R interpolates a function from the fine grid to the coarse grid. The coarse grid operator and the correction are denoted by L^c and U^c, respectively. After an adequate approximation V^c to the coarse grid

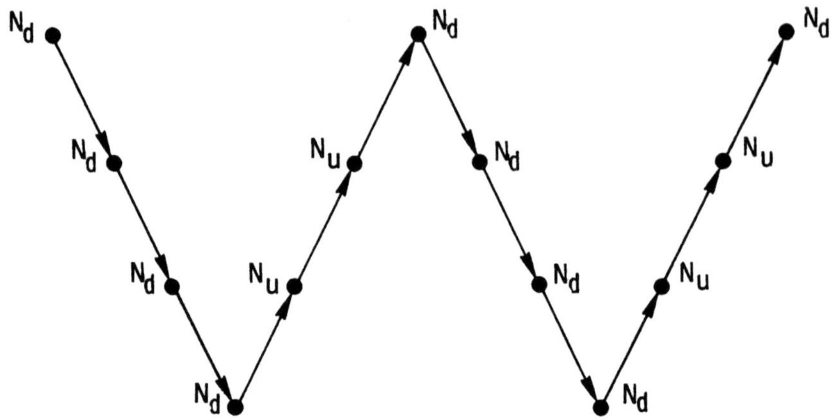

Figure 5.5. The multigrid V-cycle. The number of relaxations after restriction is denoted by N_d and the number of relaxations before prolongation is denoted by N_u.

problem has been obtained, the fine grid approximation is updated using

$$V^f \leftarrow V^f + PV^c. \tag{5.5.7}$$

The prolongation operator P interpolates a function from the coarse grid to the fine grid. Figure 5.5 shows one possible control structure. This fixed algorithm is known as a V-cycle.

For the model problem it is clear that the ideal interpolation operators—both restriction and prolongation—are those which transfer the eigenfunctions intact and without contamination. Trigonometric interpolation accomplishes precisely this and can be implemented efficiently by the FFT.

Finite-difference preconditioning applied to the one-dimensional Fourier model problem yields a single grid condition number which is independent of N. In fact the spectral radius is 0.43, which is lower than the smoothing rate of 0.60 exhibited by the unpreconditioned multigrid scheme. If this same preconditioning is used in a multigrid setting, then the multigrid smoothing rate is reduced to 0.33 since $\lambda_{\text{mid}} = 1.23$ and $\kappa_{MG} = 2$. Even for the one-dimensional Dirichlet model problem multigrid offers only this same modest improvement in convergence rate. Thus, in one dimension spectral multigrid does not seem worth the trouble. In two dimensions, however, it has distinct advantages. These are explored in the next section.

5.5.2. Two-Dimensional Problems

We turn now to an assessment of various finite-difference preconditionings for multigrid solutions to the Dirichlet problem for (5.1) with $\lambda = 0$. Tables 5.4 and 5.5 give the essential features of the spectra of the preconditioned

Table 5.7. Largest eigenvalue for the preconditioned Chebyshev Laplace operator

N	$H_{FD}^{-1}L$	$H_{LU5}^{-1}L$	$H_{LU7}^{-1}L$	$H_{RS5}^{-1}L$	$H_{RS7}^{-1}L$
4	1.757	1.717	1.761	1.781	1.514
8	2.131	2.273	2.160	2.877	2.209
16	2.305	2.603	2.376	4.421	2.672
32	2.388	2.835	2.527	6.416	3.932
64	2.428	3.081	2.695	12.579	7.513
128	2.448	3.115	2.838		

Table 5.8. Multigrid condition numbers for the preconditioned Laplace operator

N	$H_{FD}^{-1}L$	$H_{LU5}^{-1}L$	$H_{LU7}^{-1}L$	$H_{RS5}^{-1}L$	$H_{RS7}^{-1}L$
8	1.733	1.863	1.756	1.984	1.796
16	1.874	2.134	1.932	2.925	2.172
32	1.941	2.324	2.054	4.425	2.934
64	1.974	2.525	2.191	8.675	5.607
128	1.990	2.763	2.307		

Table 5.9. Multigrid smoothing rates for the preconditioned Laplace operator

N	$H_{FD}^{-1}L$	$H_{LU5}^{-1}L$	$H_{LU7}^{-1}L$
8	0.268	0.301	0.274
16	0.304	0.362	0.318
32	0.320	0.398	0.345
64	0.328	0.433	0.373
128	0.331	0.469	0.395

Chebyshev Laplace operator. The lowest high frequency eigenvalue (λ_{mid}) turns out to be insensitive to N, having the value 1.23 from all but the row sum cases, for which it is 1.45 for the five-diagonal version and 1.34 for the seven-diagonal one. Using this information as well as the results of eigenvalue calculations, the results for $N \leqslant 64$ in Tables 5.7, 5.8 and 5.9 are obtained. The $N = 128$ results are based on extrapolations from the empirical formulas (based on the $N \leqslant 64$ results):

$$\begin{aligned} \lambda_{\max} &\sim 1.381\, N^{1/8} \qquad \text{for } H_{LU5}^{-1}L \\ \lambda_{\max} &\sim 1.894\, N^{1/12} \qquad \text{for } H_{LU7}^{-1}L. \end{aligned} \tag{5.5.8}$$

Note that the standard incomplete-*LU* factorization is superior to the row-sum-equivalence alternative in the multigrid context. The former evidently does a better job on the high frequency components of the solution. Although it performs far worse on the low frequency components, this is immaterial for a multigrid scheme.

Another obvious conclusion is that the seven-diagonal standard incomplete-*LU* factorization has a multigrid smoothing rate which is nearly as good as that of the original finite-difference preconditioning. Since the incomplete-*LU* decomposition usually costs far less than the evaluation of the spectral residual, whereas the inversion of the original finite-difference matrix is more expensive than the residual evaluation, the usefulness of the multigrid approach in two dimensions is evident.

5.5.3. Interpolation Operators

We shall focus on the one-dimensional problem

$$\frac{d}{dx}\left\{a(x)\frac{du}{dx}\right\} = f(x) \tag{5.5.9}$$

on either $(0, 2\pi)$, as in the periodic case, or on $(-1, 1)$, as in the Dirichlet case.

Fourier Series

The natural interpolation operators represent trigonometric interpolation. Consider first the prolongation process: given a function on a coarse grid (with N_c points), compute the discrete Fourier coefficients and then use the resulting discrete Fourier series to construct the interpolated function on the fine grid (with N_f points). This may be accomplished by performing two FFTs.

On the coarse grid, the discrete Fourier coefficients of the corrections u_j at the coarse-grid collocation points $x_j, j = 0, 1, \ldots, N_c - 1$, are computed using

$$\tilde{u}_p = \frac{1}{N_c}\sum_{j=0}^{N_c-1} u_j e^{-ipx_j} \qquad p = -N_c/2, \ldots, N_c/2 - 1. \tag{5.5.10}$$

The fine-grid approximation is then updated using

$$u_j \leftarrow u_j + \sum_{p=-N_c/2}^{N_c/2-1} \tilde{u}_p e^{ip\bar{x}_j}, \tag{5.5.11}$$

where $\bar{x}_j, j = 0, 1, \ldots, N_f - 1$, are the fine-grid collocation points.

The restriction operator is constructed in a similar fashion. It turns out that except for a factor of N_f/N_c, P and R are adjoint. An explicit representation of the prolongation operator is

$$P_{jk} = \frac{1}{N_c}\sum_{l=-N_c/2+1}^{N_c/2-1} e^{2\pi i l((j/N_f)-(k/N_c))}, \tag{5.5.12}$$

which sums to yield

$$P_{jk} = \frac{1}{N_c} S\left(\frac{j}{N_f} - \frac{k}{N_c}\right), \tag{5.5.13}$$

where

$$S(r) = \begin{cases} N_c - 1 & r \text{ integer} \\ \sin(\pi r N_c)\cot(\pi r) - \cos(\pi r N_c) & \text{otherwise} \end{cases}. \tag{5.5.14}$$

The corresponding restriction operator is essentially the adjoint of this:

$$R_{jk} = \frac{1}{N_f} S\left(\frac{j}{N_c} - \frac{k}{N_f}\right). \tag{5.5.15}$$

Chebyshev Series

Interpolation for non-periodic coordinates employs Chebyshev series in an analogous fashion. The prolongation operator is

$$P_{jk} = \frac{2}{\bar{c}_k N_c} \sum_{l=0}^{N_c} \bar{c}_l^{-1} \cos\frac{\pi l j}{N_f} \cos\frac{\pi l k}{N_c}, \tag{5.5.16}$$

where $\bar{c}_k$ is defined by (2.4.16) with $N = N_c$. This sums to

$$P_{jk} = \frac{2}{\bar{c}_k N_c}\left[Q\left(\frac{j}{N_f} - \frac{k}{N_c}\right) + Q\left(\frac{j}{N_f} + \frac{k}{N_c}\right)\right], \tag{5.5.17}$$

where

$$Q(r) = \begin{cases} \dfrac{N_c}{2} & r \text{ integer} \\ \dfrac{1}{4} - \dfrac{1}{4}\cos(\pi r N_c) + \dfrac{1}{2}\cos\left(\dfrac{\pi r}{2}(N_c + 1)\right)\sin\left(\dfrac{\pi r N_c}{2}\right)\csc\left(\dfrac{\pi r}{2}\right) & \text{otherwise} \end{cases} \tag{5.5.18}$$

We will have occasion to use two distinct restriction operators. One is sometimes used in forming the coarse-grid operator and is obtained by applying Chebyshev restriction in the obvious fashion. It will be denoted by R and is given by

$$R_{jk} = \frac{2}{\bar{c}_k N_f}\left[\bar{Q}\left(\frac{j}{N_c} - \frac{k}{N_f}\right) + \bar{Q}\left(\frac{j}{N_c} + \frac{k}{N_f}\right)\right]. \tag{5.5.19}$$

where $\bar{c}_k$ is defined by (2.4.16) with $N = N_f$ and

$$\bar{Q}(r) = \begin{cases} \dfrac{1}{4} + \dfrac{N_c}{2} & r \text{ integer} \\ \dfrac{1}{4} + \dfrac{1}{2}\cos\left(\dfrac{\pi r}{2}(N_c + 1)\right)\sin\left(\dfrac{\pi r N_c}{2}\right)\csc\left(\dfrac{\pi r}{2}\right) & \text{otherwise} \end{cases} \tag{5.5.20}$$

The other is used for interpolation, is denoted by $R^{(i)}$, and is defined by the adjoint requirement:

$$R^{(i)}_{jk} = \frac{2}{\bar{c}_k N_c}\left[Q\left(\frac{j}{N_c} - \frac{k}{N_f}\right) + Q\left(\frac{j}{N_c} + \frac{k}{N_f}\right)\right], \tag{5.5.21}$$

where $\bar{c}_k$ is defined by (2.4.16) with $N = N_c$.

5.5.4. Coarse-Grid Operators

A typical term in the class of problems considered here is given by the left-hand side of (5.5.9). The discrete operator which represents its fine-grid collocation approximation is

$$L^f = D_{N_f} A D_{N_f}, \tag{5.5.22}$$

where D_{N_f} is given by (2.1.41) or (2.4.31) and A is the diagonal matrix

$$A_{jk} = a(x_j)\delta_{j,k}. \tag{5.5.23}$$

Many multigrid investigators have advocated choosing the coarse-grid operator so that

$$L^c = RL^f P. \tag{5.5.24}$$

Both the Fourier and the Chebyshev first-derivative operators, defined by (2.1.41) and (2.4.31), satisfy

$$D_{N_c} = RD_{N_f}P. \tag{5.2.25}$$

However, (5.5.24) itself is not satisfied if the coarse grid analog of (5.5.22) is used to define L^c, except in the trivial case for which $a(x)$ is a constant. On the other hand, much of the efficiency of the collocation method is lost if (5.5.24) is used to define the coarse-grid operator. Some compromises were suggested by Zang, Wong, and Hussaini (1984). The most satisfactory one seems to be using (5.5.22) but with the restricted values of $a(x_j)$ in place of the pointwise values. The Chebyshev restrictions should be performed with R.

5.5.5. Relaxation Schemes

Non-stationary Richardson relaxation (see (5.3.39)) using, say, $k = 3$ relaxation parameters is a reasonable scheme for spectral multigrid. Then the smoothing rate of the unpreconditioned one-dimensional Fourier model problem is reduced from 0.60 (for $k = 1$) to 0.42, and the two-dimensional problem is improved to 0.60 from 0.78. For the $N = 32$, two-dimensional, Chebyshev model problem with standard seven-diagonal incomplete-LU factorization as a preconditioner, the smoothing rate improves to 0.22 from 0.37.

Additional improvements to Richardson relaxation for the Poisson problem have been suggested by Brandt, Fulton and Taylor (1985). For variable-coefficient operators such as (5.5.9) with periodic boundary conditions, one can obtain essentially the smoothing rate that is indicated by the model problem analysis by letting the relaxation parameter depend on position according to

$$\omega(x) = \frac{\omega_*}{a(x)}, \tag{5.5.26}$$

where ω_* is the parameter appropriate for the $a(x) \equiv 1$ case. Moreover, by weighting the residuals, one can reduce the smoothing rate from 0.60 to 0.20. Instead of using r_i, one should use $\alpha r_{i-1} + \beta r_i + \alpha r_{i+1}$, where α and β can be chosen to maximize the smoothing.

Both of these devices also work in two dimensions provided that the problem is isotropic, i.e., $a(\mathbf{x})/\Delta x = a(\mathbf{x})/\Delta y$. As shown by Brandt et al., the smoothing rate is reduced from 0.78 to 0.11. Erlebacher, Zang and Hussaini (1987) have examined the periodic, isotropic three-dimensional Poisson problem and demonstrated that residual weighting reduces the smoothing rate for stationary Richardson iteration from 0.85 to 0.20. If the problem is not isotropic, however, these refinements are little help. Moreover, they also pointed out that residual weighting is not very effective for the Helmholtz problem (5.1) with $\lambda \neq 0$. Neither the position-dependent relaxation parameter nor the residual weighting are likely to help much for Chebyshev SMG, even in one dimension. The finite-difference preconditioning, in effect, already takes both of these tricks into account.

For the non-isotropic Fourier SMG problem in two dimensions, Brandt et al. resorted to finite-difference preconditioning. They used alternating line Gauss–Seidel (Zebra) and achieved smoothing rates for the purely Fourier problem which were comparable to the smoothing rates obtained by Zang, Wong and Hussaini (1984) for the purely Chebyshev problem.

With periodic boundary conditions, Zebra preconditioning seems preferable to incomplete-LU, because the enforcement of periodicity is much simpler. Moreover, this relaxation scheme is more economical of storage than incomplete-LU because one needs auxiliary storage for one-dimensional vectors rather than for many two-dimensional vectors.

The approximate factorization scheme described in Sec. 5.4.2 is also suitable for multigrid. Indeed, it has been used for non-linear problems by Streett, Zang and Hussaini (1985). One simply has to choose the parameter α_l to be the lowest high frequency eigenvalue.

Finally, we should mention parameter-free relaxation schemes. For Dirichlet problems the minimum residual method can work as a smoother. One change in the control structure is advisable, however. Instead of using the V-cycle as illustrated in Fig. 5.5, one should use the sawtooth cycle shown in Fig. 5.6. This is the special case of the V-cycle in which no relaxations are

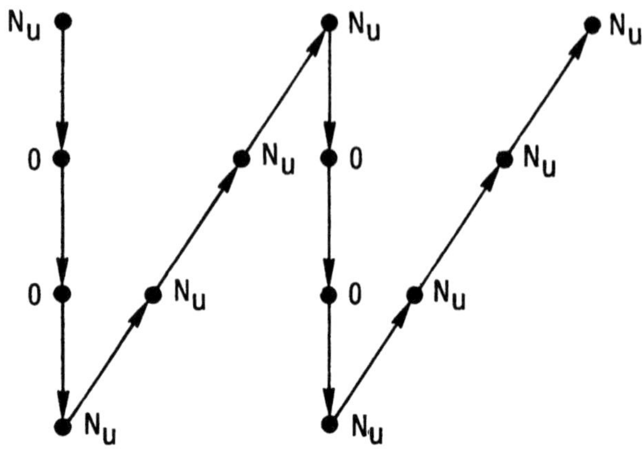

Figure 5.6. The multi grid sawtooth cycle. No relaxations are performed after restriction and N_u relaxations are performed before prolongation.

performed after the restrictions. The minimum residual method chooses the parameter to optimize the damping of those components which are most significant in the current approximation. One would like the low frequency components to be very small. Relaxations on the way down are likely to be inefficient because of the presence of substantial low frequency error components. The relaxations on the way up are much more effective because the low frequency components have already been greatly reduced.

5.6. A Semi-Implicit Method for the Navier–Stokes Equations

We close this chapter with a fairly complex application of iterative methods in a semi-implicit spectral algorithm for the incompressible Navier–Stokes equations. The algorithm was developed by Malik, Zang and Hussaini (1985) and extended to three-dimensional flows as well as analyzed in some detail by Zang and Hussaini (1985a). The details of the discretization are furnished in Sec. 7.3.1. Here, we consider the version for flow in a channel (Fig. 1.3) which is periodic on $(0, 2\pi)$ in x and z, with no-slip walls at $y = \pm 1$. The kinematic viscosity ν is presumed to be a function of $\mathbf{x}$ and t.

The spatial discretization is Fourier collocation in x and z and Chebyshev collocation in y. The temporal discretization is backward Euler for the pressure, Crank–Nicolson for normal diffusion, and Adams–Bashforth or Runge–Kutta for the remaining terms. The continuity equation is enforced as a constraint at the new time level.

The primitive variables have two series representations which will be useful in this discussion. The first is

$$u(x, y, z, t) = \sum_{k_x=-N_x/2}^{N_x/2-1} \sum_{k_z=-N_z/2}^{N_z/2-1} \hat{u}_{k_x,k_z}(y, t) e^{i(k_x x + k_z z)}. \tag{5.6.1}$$

(In this section we depart from our usual custom of denoting continuous coefficients by ^'s and discrete ones by ~'s.) The second involves the additional series

$$\hat{u}_{k_x,k_z}(y, t) = \sum_{m=0}^{N_y} \tilde{u}_{k_x,m,k_z}(t) T_m(y). \tag{5.6.2}$$

Henceforth, the subscripts k_x, m, and k_z will not be written explicitly unless necessary. The collocation points in the periodic directions are

$$\begin{aligned} x_i &= \frac{2\pi i}{N_x} \qquad i = 0, 1, \ldots, N_x - 1 \\ z_k &= \frac{2\pi k}{N_z} \qquad k = 0, 1, \ldots, N_z - 1. \end{aligned} \tag{5.6.3}$$

A staggered grid (see Fig. 7.2) is employed in the normal direction. Velocities are defined at the points

$$y_j = \cos\frac{\pi j}{N_y} \qquad j = 0, 1, \ldots, N_y \tag{5.6.4}$$

and the pressures at

$$y_{j+1/2} = \cos\frac{\pi(j + 1/2)}{N_y} \qquad j = 0, 1, \ldots, N_y - 1. \tag{5.6.5}$$

The continuity equation is enforced at the latter points and the remaining equations at the former ones. The virtues of the staggered grid are discussed in Chaps. 7 and 11.

Chebyshev interpolation is the natural process for transferring variables between the grids of (5.6.4) and (5.6.5). For example, consider the velocity component u. Let u_j, for $j = 0, 1, \ldots, N_y$, denote its values at the points (5.6.4). The Chebyshev coefficients are given by the usual quadrature rule

$$\begin{aligned} \tilde{u}_m &= \frac{2}{N_y \bar{c}_m} \sum_{j=0}^{N_y} \bar{c}_j^{-1} u_j T_m(y_j) \\ &= \frac{2}{N_y \bar{c}_m} \sum_{j=0}^{N_y} \bar{c}_j^{-1} u_j \cos\frac{m\pi j}{N_y} \qquad m = 0, 1, \ldots, N_y. \end{aligned} \tag{5.6.6}$$

The interpolated values of u are

$$\begin{aligned} u_{j+1/2} &= \sum_{m=0}^{N_y-1} \tilde{u}_m T_m(y_{j+1/2}) \\ &= \sum_{m=0}^{N_y-1} \tilde{u}_m \cos\frac{(j + 1/2)\pi m}{N_y} \qquad j = 0, 1, \ldots, N_y - 1. \end{aligned} \tag{5.6.7}$$

(Note that $T_{N_y}(y_{j+1/2}) = 0$ for $j = 0, 1, \ldots, N_y - 1$.) The Fast Fourier Transform may be used to evaluate both sums (5.6.6) and (5.6.7). The less familiar sum in (5.6.7) may be handled by the technique presented in Appendix B.

The temporal discretization of (1.3.18)—(1.3.19) leads, after a Fourier transform in x and z, to an implicit system of the form

$$\left.\begin{aligned} \hat{u}^{n+1} - \partial\left(b\frac{\partial \hat{u}^{n+1}}{\partial y}\right)\Big/\partial y + ik_x\gamma\hat{q}^{n+1} &= \hat{u}_e \\ \hat{v}^{n+1} - \partial\left(b\frac{\partial \hat{v}^{n+1}}{\partial y}\right)\Big/\partial y + \gamma\partial\hat{q}^{n+1}/\partial y &= \hat{v}_e \\ \hat{w}^{n+1} - \partial\left(b\frac{\partial \hat{w}^{n+1}}{\partial y}\right)\Big/\partial y + ik_z\gamma\hat{q}^{n+1} &= \hat{w}_e, \end{aligned}\right\} \quad \text{at } y_j \tag{5.6.8}$$

$$-ik_x\bar{\gamma}\hat{u}^{n+1} - \bar{\gamma}\partial\hat{v}^{n+1}/\partial y - ik_z\bar{\gamma}w^{n+1} = 0, \qquad \text{at } y_{j+1/2} \tag{5.6.9}$$

along with Dirichlet boundary conditions on the velocities. An overbar denotes the complex conjugate. The coefficient $b = \frac{1}{2}\Delta t v^n_{\text{avg}}(y, t)$, where the last term is the average value of $v^n(x, y, z, t)$ at fixed y and t. The pressure has been included in terms of the scaled variable $\hat{q} = (\Delta t/\gamma)\hat{P}$, where γ is a complex constant whose role is explained below.

The right-hand sides of (5.6.8) contain the explicit terms in the temporal discretization. More details are provided in Sec. 7.3.1. Let

$$U = (\hat{u}_0^{n+1}, \hat{u}_1^{n+1}, \ldots, \hat{u}_{N_y}^{n+1})$$
$$Q = (\hat{q}_{1/2}^{n+1}, \hat{q}_{3/2}^{n+1}, \ldots, \hat{q}_{N_y-1/2}^{n+1}),$$

and define V and W similarly to U. Let B be the diagonal matrix with the elements of b on its diagonal. Let the effect of (5.6.6) and (5.6.7) on U be denoted by A_+ and the reverse interpolation procedure (for Q) by A_0. Finally, let D denote the matrix which represents Chebyshev differentiation in the y-direction. Then (5.6.8) and (5.6.9) reduce to the algebraic set

$$\begin{aligned} (I - DBD)U + ik_x\gamma A_0 Q &= U_e \\ (I - DBD)V + \gamma D A_0 Q &= V_e \\ (I - DBD)W + ik_z\gamma A_0 Q &= W_e \end{aligned} \tag{5.6.10}$$

$$-ik_x\bar{\gamma}A_+U - \bar{\gamma}A_+DV - ik_z\bar{\gamma}A_+W = 0, \tag{5.6.11}$$

where the first and last rows of (5.6.10) are replaced by the boundary conditions. Clearly, the equations for each pair (k_x, k_z) are independent. The matrices A_0, A_+, and D are full. Except in special cases the direct solution of these equations is not practical.

The key to this algorithm is the solution of the system (5.6.10)—(5.6.11). A simplified model problem is instructive. Suppose that the boundary conditions in the normal direction are periodic instead of Dirichlet and that the viscosity, i.e., b, is constant. Replace the Chebyshev discretization with a Fourier one, on, say $(0, 2\pi)$. Then the vertical collocation points are

$$y_j = \frac{2\pi j}{N_y} \qquad j = 0, 1, \ldots, N_y - 1 \tag{5.6.12}$$

$$y_{j+1/2} = \frac{2\pi(j+1/2)}{N_y} \qquad j = 0, 1, \ldots, N_y - 1 \tag{5.6.13}$$

(see Fig. 5.1). The fully discrete equations may be cast in a form analogous to (5.6.10)—(5.6.11) where now

$$\begin{aligned} D &= C_0^* K C_0 \\ DBD &= b C_0^* K^2 C_0 \\ A_0 &= C_0^* C_+ \\ A_+ &= C_+^* C_0 \end{aligned} \tag{5.6.14}$$

and

$$(C_0)_{k,j} = \frac{1}{\sqrt{N_y}} e^{-iky_j} \qquad \begin{aligned} &j = 0, 1, \ldots, N_y - 1 \\ &k = -N_y/2, -N_y/2 + 1, \ldots, N_y/2 - 1 \end{aligned} \tag{5.6.15a}$$

$$(C_+)_{k,j} = \frac{1}{\sqrt{N_y}} e^{iky_{j+1/2}} \tag{5.6.15b}$$

$$K_{k,l} = ik\delta_{k,l} \qquad k, l = -N_y/2, -N_y/2 + 1, \ldots, N_y/2 - 1 \tag{5.6.16}$$

and the Hermitian transpose of a matrix is denoted by an asterisk. Thus, we have for the spectral equations

$$\begin{aligned} (I - bC_0^* K^2 C_0)U + ik_x \gamma C_0^* C_+ Q &= U_e \\ (I - bC_0^* K^2 C_0)V + \gamma C_0^* K C_+ Q &= V_e \\ (I - bC_0^* K^2 C_0)W + ik_z \gamma C_0^* C_+ Q &= W_e \end{aligned} \tag{5.6.17}$$

$$-ik_x \bar{\gamma} C_+^* C_0 U - \bar{\gamma} C_+^* K C_0 V - ik_z \bar{\gamma} C_+^* C_0 W = 0. \tag{5.6.18}$$

This can be written as the system

$$LX = F, \tag{5.6.19}$$

where, for instance, $X = (U, V, W, Q)$. Now let $\tilde{U}_k = C_0 U$, $\tilde{Q}_k = C_+ Q$, $\tilde{X} = RX$, and $\tilde{L} = RLR^*$, where

$$R = \begin{pmatrix} C_0 & 0 & 0 & 0 \\ 0 & C_0 & 0 & 0 \\ 0 & 0 & C_0 & 0 \\ 0 & 0 & 0 & C_+ \end{pmatrix}$$

After a permutation of the rows and columns of $\tilde{L}$, we obtain a block-diagonal matrix with blocks

$$\tilde{L}_k \tilde{X}_k = \tilde{B}_k \tag{5.6.20}$$

$$\tilde{X}_k = (\tilde{u}_k, \tilde{v}_k, \tilde{w}_k, \tilde{q}_k)$$

$$\tilde{L}_k = \begin{pmatrix} 1+bk^2 & 0 & 0 & ik_x\gamma \\ 0 & 1+bk^2 & 0 & ik\gamma \\ 0 & 0 & 1+bk^2 & ik_z\gamma \\ -ik_x\bar{\gamma} & -ik\bar{\gamma} & -ik_z\bar{\gamma} & 0 \end{pmatrix}. \tag{5.6.21}$$

Consider now a finite-difference approximation to this model problem. Let E denote the forward shift operator subject to periodic boundary conditions. Then (5.6.8) and (5.6.9) become

$$\begin{aligned} \left[I - \frac{b}{(\Delta y)^2}(E - 2I + E^{-1})\right]U + ik_x\gamma\frac{1}{2}(E+I)Q &= U_e \\ \left[I - \frac{b}{(\Delta y)^2}(E - 2I + E^{-1})\right]V + \frac{\gamma}{\Delta y}(E-I)Q &= V_e \\ \left[I - \frac{b}{(\Delta y)^2}(E - 2I + E^{-1})\right]W + ik_z\gamma\frac{1}{2}(E+I)Q &= W_e \end{aligned} \tag{5.6.22}$$

$$-ik_x\bar{\gamma}\frac{1}{2}(I + E^{-1})U - \frac{\bar{\gamma}}{\Delta y}(I - E^{-1})V - ik_z\bar{\gamma}\frac{1}{2}(I + E^{-1})W = 0. \tag{5.6.23}$$

Denote the matrix which represents the left-hand side by H. This system can be reduced to block-diagonal form by the same transformation that was used for the spectral operator. The result can be written

$$\tilde{H}_k \tilde{X}_k = \tilde{B}_k, \tag{5.6.24}$$

$$\tilde{H}_k = \begin{pmatrix} 1+bk^2s^2 & 0 & 0 & ik_x\gamma a \\ 0 & 1+bk^2s^2 & 0 & ik\gamma s \\ 0 & 0 & 1+bk^2s^2 & ik_z\gamma a \\ -ik_x\bar{\gamma}a & -ik\bar{\gamma}s & -ik_z\bar{\gamma}a & 0 \end{pmatrix}, \tag{5.6.25}$$

where

$$s = \frac{\sin(k\Delta y/2)}{(k\Delta y/2)} \tag{5.6.26}$$

$$a = \cos\left(\frac{k\Delta y}{2}\right). \tag{5.6.27}$$

The relevant range is $|k\Delta y| \leqslant \pi$.

If a and s were identically one, then the preconditioning would be perfect. In any case, the derivative terms cause no serious problem for $(2/\pi) \leqslant s \leqslant 1$.

It is the averaging operator a which is a source of potential difficulty. As $|k\Delta y| \to \pi$, the averaging becomes useless. We anticipate difficulty only in circumstances for which k_x and/or k_z are large relative to the reciprocal of the grid spacing in y.

Consider first Richardson or Chebyshev iteration for (5.6.19). The convergence properties of these schemes depend upon the eigenvalues of $H^{-1}L$. The eigenvalues of the model problem are especially easy to obtain: since H and L were block-diagonalized by the same transformation, we need only compute the eigenvalues of $\tilde{H}_k^{-1}\tilde{L}_k$ for $k = 1, 2, \ldots, N_y/2$. Some results are shown in Fig. 5.7. In these calculations k_z has been set to zero. Similar calculations for both k_x and k_z non-zero lead to qualitatively similar results. Figure 5.7(a) portrays the easiest of these six cases for the iterative scheme. Most of the eigenvalues are near unity, and they are located near the real axis between 1 and $\pi/2$. The eigenvalue spread in part (b) for $k_x = 30$ is much larger. Nevertheless, the real parts of the eigenvalues are safely greater than zero. A comparison of parts (b), (c) and (d) reveals that for fixed k_x the eigenvalue spread is reduced as the resolution in y is increased. The heuristic explanation for this welcome behavior is that as N_y increases, the eigenvalues of the first-derivative operator become more important than those of the averaging operator. In actual numerical simulations the number of points in the x and y directions is likely to be nearly the same. Part (d) corresponds to the worst case that would arise in a 64^3 calculation. Parts (e) and (f) show the eigenvalue spectrum for a situation in which the viscosity is significant. The major difference from the previous cases is the presence of additional eigenvalues along the real-axis as large as $\pi^2/4$. This is characteristic of preconditionings of second-derivative spectral operators.

Clearly, the major shortcoming of this particular preconditioning is its treatment of the averaging operator. If the 2/3-rule is used to de-alias the collocation approximation in the normal direction (see Sec. 3.2.2), then the averaging operator is well-behaved. In this event, $|k\Delta y| \leqslant 2\pi/3$, so that $1/2 \leqslant a \leqslant 1$. For the case shown in Fig. 5.7(c), the largest six eigenvalues disappear. With the use of the 2/3-rule, this case becomes quite manageable.

The actual system that must be solved is given by (5.6.10) and (5.6.11). It clearly can be written in the form of (5.6.19) by an obvious adaptation of the previous notation. Likewise, let H represent the finite-difference counterpart of L on the Chebyshev staggered grid.

The eigenvalues of some channel flow cases are shown in Fig. 5.8. There are no significant differences between these results and those for the model problem. This numerical evidence leaves little doubt that Chebyshev iteration will succeed for this problem. Especially in problems for which ν varies with y and t, an adaptive version is recommended.

A sufficient condition for convergence of descent methods is that the symmetric part of LH^{-1} be positive-definite. The constant γ that appears in (5.6.10) and (5.6.11) is used to ensure that this condition is met.

Let us return momentarily to the model problem. For $\gamma = 1$, the extreme

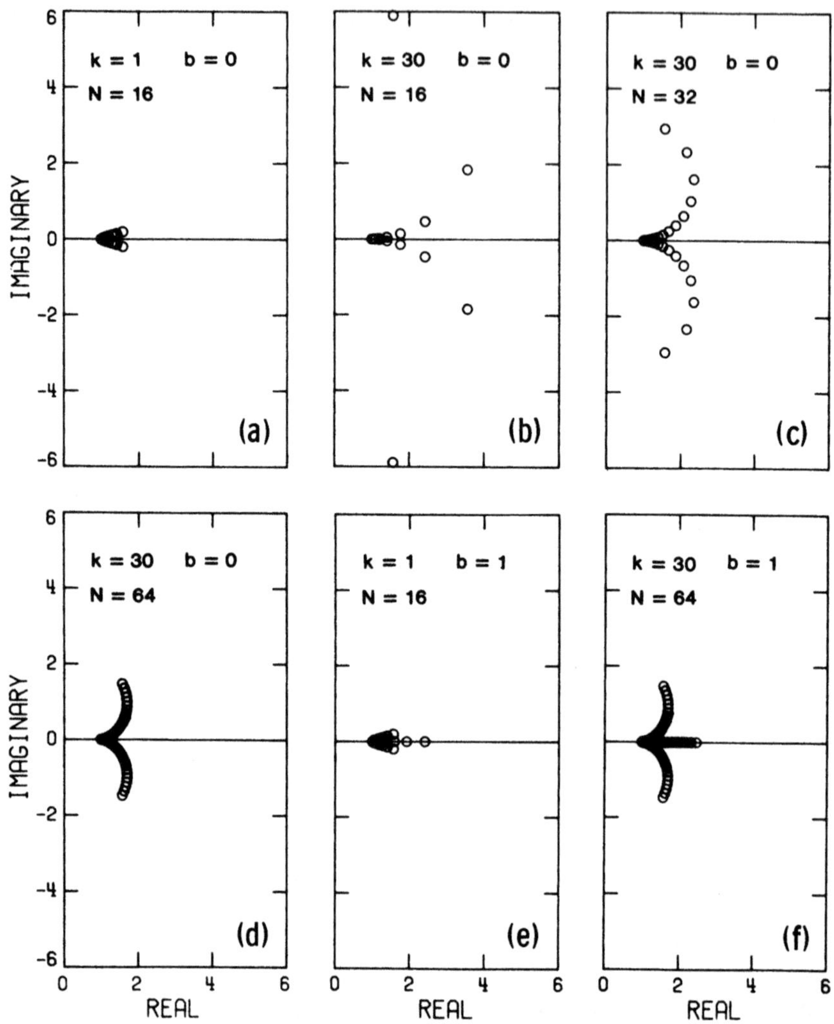

Figure 5.7. Eigenvalues of $H^{-1}L$ for the model problem for incompressible flow. The z-wavenumber k_z is zero and the x-wavenumber k_x is denoted here simply by k. The coefficient b is $\frac{1}{2}\Delta t v_{avg}$, and N denotes N_y, the number of points in the y-direction.

eigenvalues of the symmetric part of LH^{-1} are 1.59 and 0.71 for the case shown in Figure 5.7(a) whereas they are 18.8 and -15.6 for part (b). A descent method will clearly fail for the latter case. However, for $\gamma = 1/\sqrt{k_x^2 + k_z^2}$, these latter eigenvalues improve to 2.55 and 0.43. The importance of γ is evident. It is even more important when there is appreciable diffusion. The

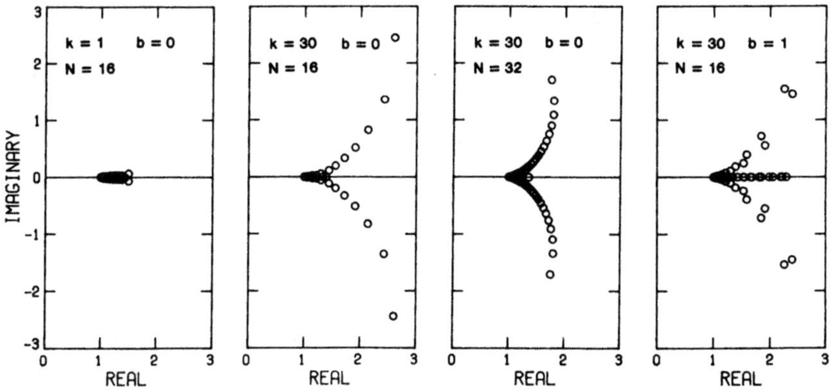

Figure 5.8. Eigenvalues of $H^{-1}L$ for the real channel flow problem. The z-wavenumber k_z is zero and the x-wavenumber k_x is denoted here simply by k. The coefficient b is $\frac{1}{2}\Delta t v_{avg}$, and N denotes N_y, the number of points in the y-direction.

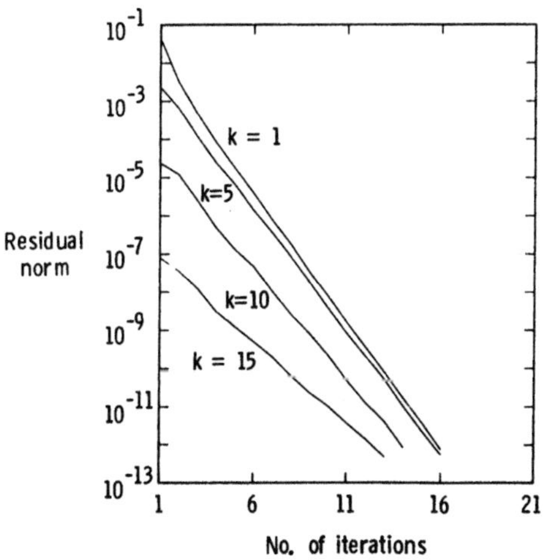

Figure 5.9. Convergence history of the minimum residual method for a two-dimensional channel flow problem. The symbol k refers to the Fourier component in the x direction. The normal direction resolution is $N_y = 32$.

$\gamma = 1$ extreme eigenvalues are 474 and -471 for the case of Fig. 5.7(e). The choice

$$\gamma = [1 + dk_{\max}^2]/\sqrt{k_x^2 + k_z^2 + k_{\max}^2},$$

where $k_{\max} = N_y/2$ leads to 2.98 and 0.98. This scaling is based on balancing the norms of the diffusion and gradient operators. The principle applies to the actual channel problem as well.

There is clearly the potential for improvements in the preconditioning and scaling. One intriguing scheme is suggested by the observation that if a non-staggered grid were used for the pressure, then the preconditioning problems would shift from the averaging operator (which would become the identity) to the first-derivative operator. Perhaps one should employ a scheme which alternates iterations on a staggered grid with ones on a non-staggered grid.

The preconditioning matrix H is block-tridiagonal. Note that (5.6.19) can be separated into two independent real systems. Zang and Hussaini (1985a) reported numerical evidence that this matrix can be inverted without pivoting.

Figure 5.9 contains the residual histories of this iterative method for the channel flow application reported by Malik et al. This is a two-dimensional problem for which $k_z = 0$ and $N_y = 32$. Note that the convergence rates for the higher x harmonics, denoted by k, are slower than those for the lower harmonics. Zang and Hussaini (1985a, 1985b) and Krist and Zang (1987) have reported the results of numerous simulations of channel and boundary layer flows which incorporate this iterative technique. They have used grids as large as 96^3 for channel flow and $48 \times 96 \times 48$ for boundary layers. In these time-dependent calculations typically five to ten iterations suffice for even the highest order modes.

CHAPTER 6

Simple Incompressible Flows

Some simple one- and two-dimensional problems from incompressible fluid dynamics will now be used to illustrate the spectral discretization and solution techniques which have been discussed in the preceding chapters.

6.1. Burgers Equation

The quasi-linear parabolic equation

$$\frac{\partial u}{\partial t} = -u\frac{\partial u}{\partial x} + \nu\frac{\partial^2 u}{\partial x^2}, \tag{6.1.1}$$

known as Burgers equation, has been of considerable physical interest because of its statistical properties and because of its role in the hierarchy of approximations to the Navier–Stokes equations. It first appeared in a paper by Bateman (1915) and is named after Burgers (1948, 1974), who studied it extensively as a mathematical model of turbulence. Burgers equation successfully models certain gas dynamic (Lighthill (1956)), acoustic (Blackstock (1966)) and turbulence phenomena (Burgers (1948)). Solutions to (6.1.1) exhibit a delicate balance between (non-linear) advection and diffusion. Moreover, it is one of the few non-linear PDEs for which exact and complete solutions are known in terms of the initial values (Cole (1951), Hopf (1950)). Thus, the Burgers equation is a convenient and useful test problem for numerical schemes.

Whitham (1974) has provided a solution to (6.1.1) with periodic boundary conditions:

$$u(x,t) = -2\nu\frac{\dfrac{\partial\phi}{\partial x}(x-ct,t+1)}{\phi(x-ct,t+1)} \tag{6.1.2}$$

$$\phi(x,t) = \sum_{n=-\infty}^{\infty} e^{-(x-(2n+1)\pi)^2/4\nu t}.$$

Initial conditions with $c = 4$ for several values of ν are sketched in Fig. 6.1.

Figure 6.1. Exact solutions at $t = 0$ to the periodic Burgers equation problem.

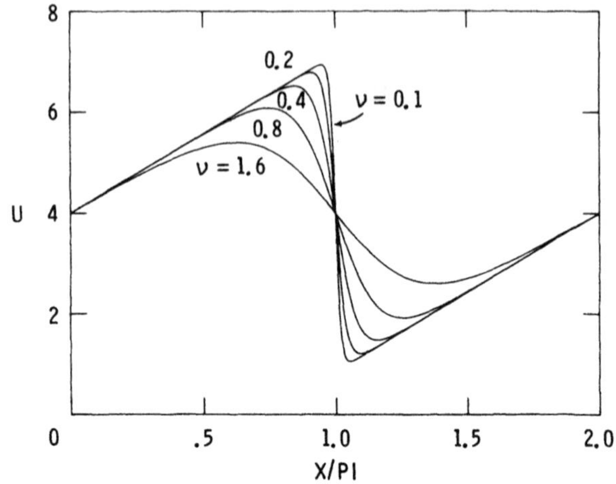

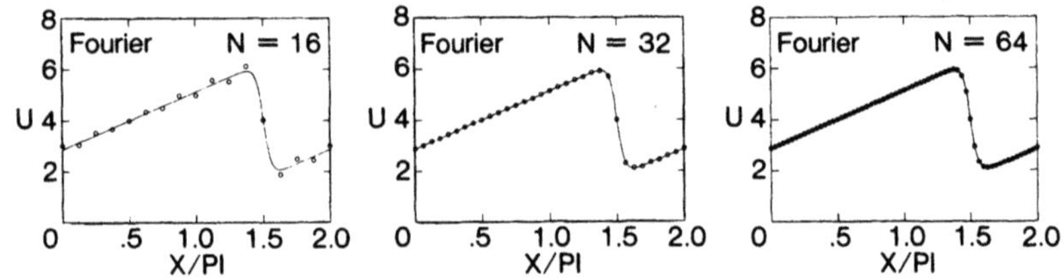

Figure 6.2. Fourier spectral solutions to the periodic Burgers equation problem at $t = \pi/8$.

These are smoothed sawtooth waves—linear except for a transition region of width $\nu(t + 1)$ near $x = \pi + ct$. They have a mean height of c, propagate to the right with speed c, and gradually diffuse to a uniform state. The standard test case uses $c = 4$ and $\nu = 0.20$.

Fourier collocation methods for this problem were discussed in Sec. 3.1.2. We use the spatial discretization given by (3.1.13) coupled with an explicit fourth-order Runge–Kutta method in time. The Runge–Kutta scheme provides the high temporal accuracy needed to demonstrate spectral accuracy in space. Figure 6.2 presents the computed solutions at $t = \pi/8$ for $N = 16$, 32, and 64. The approximation with only sixteen collocation points is unable to resolve the transition zone, and severe oscillations ensue. Once the transition zone has been resolved—with two or more points—these oscillations disappear and the numerical approximation exhibits the expected rapid convergence, as documented by the maximum errors listed in Table 6.1.

In Sec. 2.1.4 we discussed the Gibbs phenomenon, which arises in approximations to functions with discontinuities. The present example illustrates that similar oscillations arise whenever the solution contains gradients which are

Table 6.1. Maximum errors for the periodic Burgers equation

N	Fourier spectral	Second-order finite-difference	Fourth-order finite-difference
16	2.1 (−1)	2.9 (−2)	2.0 (−1)
32	2.5 (−2)	8.2 (−2)	2.5 (−2)
64	3.6 (−4)	2.6 (−2)	5.2 (−4)
128	6.1 (−8)	3.3 (−3)	2.5 (−5)

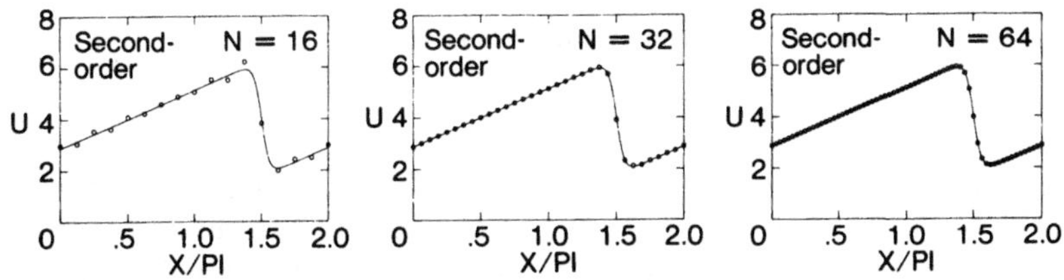

Figure 6.3. Second-order central difference solutions to the periodic Burgers equation problem at $t = \pi/8$.

too steep for the grid to resolve. In principle, oscillations arising from solutions with finite gradients can always be avoided by increasing the spatial resolution. The use of spectral approximations on partial differential equations with genuinely discontinuous solutions is considered in detail in Chap. 8.

A comparison with finite-difference schemes is instructive. Figure 6.3 displays second-order central-difference solutions. The oscillations for the under-resolved, $N = 16$ case are similar to those of the spectral solution. Table 6.1 also lists the maximum errors for these two approximations. The spectral scheme becomes superior to the second-order method for $N = 32$ and to the fourth-order method for $N = 64$. This is a fairly easy problem for a finite-difference method since the solution is essentially linear (and thus represented exactly even by a second-order finite-difference method) over all but the transition region. The real superiority of spectral methods emerges for problems with more structure in the solution—see the examples in Secs. 1.2 and 1.4.

The above examples were geared towards illustrating the spatial accuracy of the method. The time-steps were typically well below the stability limit of the RK4 method. For the $N = 128$ spectral case, $\Delta t = .0008$ was needed in order to push the temporal errors below the spatial ones.

In large scale applications, such as the Navier–Stokes equations, very high accuracy is rarely desired. In such instances, an implicit discretization of the diffusion term is useful.

6.2. Shear Flow Past a Circle

The streamfunction-vorticity formulation of viscous, two-dimensional flow was described in Sec. 1.3.4. Here, we assume that the fluid is inviscid and we consider the steady flow which develops when a circular cylinder is inserted into a uniform shear flow. This problem is illustrated in Fig. 6.4. The unperturbed flow is described by the streamfunction

$$\psi_0 = \tfrac{1}{2}\omega y^2. \tag{6.2.1}$$

Its vorticity is uniform and is given by $-\omega$. In an inviscid, incompressible flow there is no physical mechanism for generating additional vorticity. Therefore, even in the presence of the circle the vorticity will be uniform. Hence, the streamfunction ψ of the perturbed flow satisfies

$$\Delta\psi = \omega \tag{6.2.2}$$

$$\begin{aligned} \psi &= 0 && \text{at } r = 1 \\ \psi &\to \tfrac{1}{2}\omega y^2 && \text{as } r \to \infty, \end{aligned} \tag{6.2.3}$$

(see (1.3.23) and (1.3.25)). We write

$$\psi = \Psi + \tfrac{1}{2}\omega y^2$$

and use (6.2.2) in the form

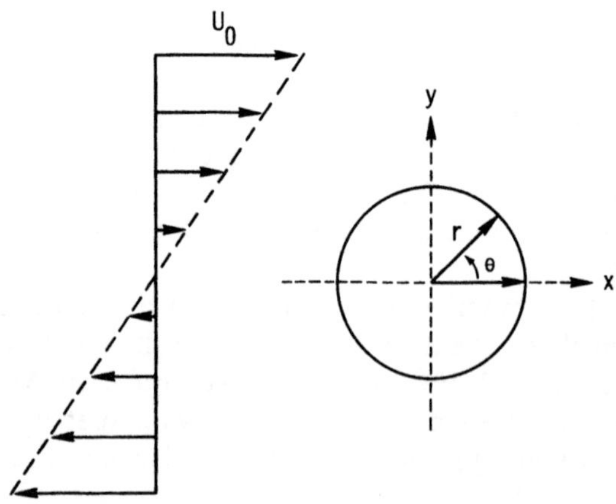

Figure 6.4. Uniform shear flow past a circle.

$$\Delta\Psi = 0 \tag{6.2.4}$$

$$\begin{aligned} \Psi &= -\tfrac{1}{2}\omega y^2 && \text{at } r = 1 \\ \Psi &\to 0 && \text{as } r \to \infty. \end{aligned} \tag{6.2.5}$$

The exact solution to this problem can be derived from the Second Circle Theorem (Milne–Thomson (1966)). It is simply

$$\Psi(x, y) = \frac{1}{4}\omega\frac{(x^2 - y^2)}{(x^2 - y^2)^2 + 4x^2y^2}. \tag{6.2.6}$$

The use of polar coordinates for (6.2.4)—(6.2.5) is natural. The semi-infinite domain in r is handled by the logarithmic mapping

$$r = 1 + (r_{\max} - 1)\left[\frac{e^{a(1-\xi)} - e^{2a}}{1 - e^{2a}}\right], \tag{6.2.7}$$

where $r_{\max}$ is the outer boundary of the computational domain and a is a stretching parameter (see (2.5.12)). Thus, both domain truncation and mapping are employed. The logarithmic map was chosen because of the slow, algebraic decay of the solution. The solution is represented as

$$\Psi(r, \theta) = \sum_{n=0}^{N} \sum_{m=-M/2}^{M/2-1} \Psi_{nm} T_n(\xi) e^{im\theta}. \tag{6.2.8}$$

The solution to (6.2.4)—(6.2.5) has been obtained by the spectral multigrid methods described in Sec. 5.5. The finest grid was 16×32 and the coarser grids were 14×24, 12×16, and 8×12. The multigrid iterations were driven by the approximate factorization preconditioner discussed in Sec. 5.4. Choosing $r_{\max} = 300$ and $a = 7$ suffices for six digit accuracy on the 16×32 grid. (Values of a between 4 and 10 all produce reasonably accurate results.) The velocity on the surface of the circle is illustrated in Fig. 6.5. The solid line is the exact value and the symbols denote the computed solution on the 8×12

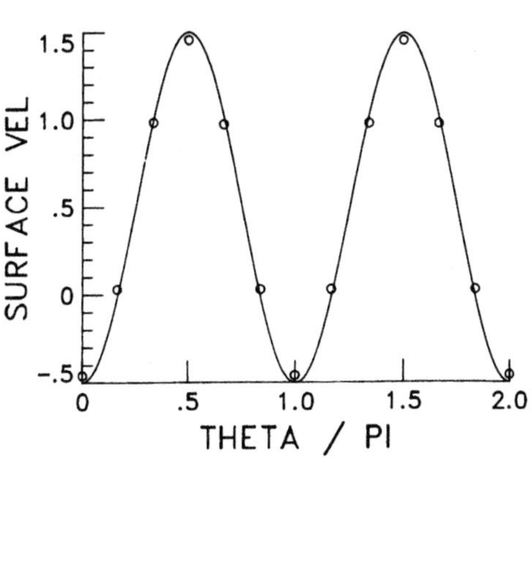

Figure 6.5. Surface velocity for uniform shear flow past a circle. The solid line is the exact solution and the circles represent the solution on a coarse 8×12 grid. (Courtesy of C. Streett.)

Table 6.2. Error in surface velocity for shear flow problem

$N \times M$	Maximum relative error
8×12	9.74 (−2)
12×16	1.17 (−4)
14×24	3.98 (−5)
16×32	1.38 (−6)

grid. The computed solutions on finer grids are indistinguishable from the solid line. Table 6.2 lists the maximum error in surface velocity as a function of the grid size. Note that spectral accuracy is indeed attained.

The spectral method illustrated above for the Poisson equation can also be applied to fluid dynamical problems which are steady, incompressible, and irrotational. These flows are necessarily inviscid. The irrotationality condition implies that a velocity potential ϕ exists:

$$\mathbf{u} = \nabla\phi. \tag{6.2.9}$$

When combined with the incompressibility constraint, this produces the Laplace equation

$$\Delta\phi = 0. \tag{6.2.10}$$

Neumann conditions apply to solid boundaries and for external flows appropriate free stream conditions are enforced at ∞. Spectral methods have actually been applied to the compressible version of this potential flow equation. They are discussed in Sec. 8.5.1.

6.3. Boundary-Layer Flows

The two-dimensional boundary-layer equations were presented in Sec. 1.3.5. For the special case of constant-viscosity flow over a flat plate subjected to a pressure gradient, a similarity solution to (1.3.27)—(1.3.29) exists. The transformation

$$\begin{aligned} \eta &= y\sqrt{\frac{m+1}{2}\frac{u_\infty}{\nu x}} \\ \psi &= \sqrt{\frac{2}{m+1}}\,\nu u_\infty x f(\eta), \end{aligned} \tag{6.3.1}$$

where

$$u_\infty = \alpha x^m$$

$$u = \frac{\partial \psi}{\partial y} = u_\infty f'(\eta) \tag{6.3.2}$$

$$v = -\frac{\partial \psi}{\partial x} = \frac{1}{2}\sqrt{\nu u_\infty / x}[\eta f'(\eta) - f(\eta)]$$

leads to the Falkner–Skan similarity equation (Schlichting (1979))

$$f''' + ff'' + \beta(1 - f'^2) = 0, \tag{6.3.3}$$

where $\beta = 2m/(m + 1)$ is the pressure gradient parameter. The boundary conditions are

$$\begin{aligned} f = f' &= 0 \qquad \text{at } \eta = 0 \\ f' &\to 1 \qquad \text{as } \eta \to \infty. \end{aligned} \tag{6.3.4}$$

Solutions to the Falkner–Skan equation can be written as

$$f(\eta) = \eta + g(\eta), \tag{6.3.5}$$

where $g(\eta)$ decays faster than exponentially as $\eta \to \infty$. As discussed in Sec. 2.5.3, Chebyshev spectral methods can be applied to the function $g(\eta)$ in terms of a computational coordinate ζ. Here we combine mapping with domain truncation, namely, $\zeta \in [-1, 1]$ is mapped into $\eta \in [0, \eta_{\max}]$ with an algebraic map such as

$$\eta = L\frac{1 + \zeta}{b - \zeta}, \tag{6.3.6}$$

where L is the scale length and

$$b = 1 + \frac{2L}{\eta_{\max}}.$$

Thus, we write

$$g(\eta) = \sum_{n=0}^{N} a_n T_n(\zeta) \tag{6.3.7}$$

and use $d/d\eta = (d\zeta/d\eta)(d/d\zeta)$. The algebraic map clusters the collocation points near $\eta = 0$. Exponential and algebraic stretchings are both feasible (since $g(\eta)$ decays faster than exponentially), although algebraic stretchings appear to be more robust. Another, and probably more efficient, approach is to map $\zeta \in [0, 1]$ into $\eta \in [0, \infty)$ and to use (6.3.7). As noted in Sec. 2.5.3, this alternative representation will deliver spectral accuracy because $g(\eta)$ decays faster than exponentially as $\eta \to \infty$. Most of the subsequent discussion, however, presumes that the standard approach has been adopted.

One of the complications of the Falkner–Skan equation is the double boundary condition at $\eta = 0$. This is straightforward to enforce in a tau method—simply apply an equation similar to (5.1.20a) for the Dirichlet condition and an equation similar to (5.1.27a) for the Neumann condition. There are several options for a collocation method. One is to enforce $g(0) = 0$ and $dg/d\eta(0) = 0$ and to drop the differential equation at the collocation point closest to the wall. Another is to build the Neumann boundary condition into the differential operator (as discussed in Sec. 3.3) and to require $g(0) = 0$ at the wall.

Equation (6.3.3) is non-linear and must be solved by iteration. Newton's method can, in principle, be employed. However, both tau and collocation discretizations require considerable algebra for implementation in a Newton's iteration (since the matrices representing differentiation are full), and a single iteration is relatively expensive (for the same reason).

Preconditioned iterative schemes (with a local linearization at each step) of the type discussed in the preceding chapter are certainly easier to implement than a full Newton's method. The subtle part of applying such schemes to a third-order equation is the proper preconditioning of the odd-order derivatives. Following the discussion of Chap. 5, we write

$$G(f) = f''' + ff'' + \beta(1 - f'^2) \tag{6.3.8}$$

and update the current guess f via

$$f \leftarrow f + \omega\Delta f, \tag{6.3.9}$$

where Δf satisfies

$$(\Delta f)''' + f(\Delta f)'' - 2\beta f'(\Delta f)' + f''\Delta f = -G(f) \tag{6.3.10}$$

and ω is a relaxation parameter. The essence of the preconditioning is to replace the spectral operator on the left-hand side of (6.3.10) with a low-order finite-difference operator. One must, however, use a finite-difference grid, with, say, 50% more points than the spectral grid, since the differential operator has odd order (see Sec. 5.2.2). There is no problem in evaluating $G(f)$ on a grid with 3/2 as many points. The delicate part is projecting the solution on the finite-difference grid onto the coarser, spectral grid.

An effective strategy for avoiding these complications was devised by Streett, Zang and Hussaini (1984). The key element is the abandonment of the single equation (6.3.3) description of the boundary-layer problem in favor of a system consisting of a second-order differential equation and an integral equation. This system arises most naturally as a special case of the general boundary-layer equations for a non-similar problem. In this context the Görtler variables (Cebeci and Bradshaw (1977)) are convenient. The independent variables are

$$\begin{aligned} \xi &= \int_0^x u_\infty(x')\,dx' \\ \eta &= u_\infty(x)y/\sqrt{2\xi\nu}, \end{aligned} \tag{6.3.11}$$

and the dependent ones are

$$\bar{u} = u/u_\infty(x)$$
$$\bar{v} = \frac{\sqrt{2\xi/\nu}}{u_\infty(x)} v + \frac{2\xi u}{\sqrt{\nu u_\infty(x)}} \frac{\partial \eta}{\partial x}. \tag{6.3.12}$$

The corresponding boundary-layer equations are

$$\frac{\partial^2 \bar{u}}{\partial \eta^2} - \bar{v}\frac{\partial \bar{u}}{\partial \eta} + \beta(1 - \bar{u}^2) - 2\xi\bar{u}\frac{\partial \bar{u}}{\partial \xi} = 0 \tag{6.3.13}$$

$$\frac{\partial \bar{v}}{\partial \eta} + \bar{u} + 2\xi\frac{\partial \bar{u}}{\partial \xi} = 0, \tag{6.3.14}$$

and the boundary conditions are

$$\begin{aligned} &\bar{u} = \bar{v} = 0 && \text{at } \eta = 0 \\ &\bar{u} \to 1 && \text{as } \eta \to \infty, \end{aligned} \tag{6.3.15}$$

plus suitable inflow conditions at some $\xi = \xi_0$. The pressure gradient parameter $\beta = (2\xi/u_\infty(x))(\partial u_\infty/\partial \xi)$.

These equations reduce to an alternative form of the similar boundary-layer equations when $\partial/\partial\xi = 0$:

$$\frac{\partial^2 \bar{u}}{\partial \eta^2} - \bar{v}\frac{\partial \bar{u}}{\partial \eta} + \beta(1 - \bar{u}^2) = 0 \tag{6.3.16}$$

$$\frac{\partial \bar{v}}{\partial \eta} + \bar{u} = 0, \tag{6.3.17}$$

with boundary conditions again given by (6.3.15). (Note that (6.3.16) and (6.3.17) combine into a single equation for $\bar{v}$ that is similar to (6.3.3).) The robust scheme developed by Streett et al. is as follows. Let the collocation points in η (after a suitable mapping) be denoted by η_i. Then (6.3.16) is collocated in standard form, whereas (6.3.17) is collocated in integral form:

$$\left.\frac{\partial^2 \bar{u}}{\partial \eta^2} - \bar{v}\frac{\partial \bar{u}}{\partial \eta} + \beta(1 - \bar{u}^2)\right|_{\eta=\eta_i} = 0 \qquad i = 1, \ldots, N-1 \tag{6.3.18}$$

$$\bar{v}_i - \bar{v}_{i-1} = -\int_{\eta_{i-1}}^{\eta_i} \bar{u}(\eta')\, d\eta' \qquad i = 1, \ldots, N, \tag{6.3.19}$$

$$\begin{aligned} &\bar{u}_N = \bar{v}_N = 0, \\ &\bar{u}_0 = 1. \end{aligned} \tag{6.3.20}$$

Equation (6.3.18) can be preconditioned effectively by a standard second-order finite-difference scheme and the trapezoidal rule is an obvious approximation to (6.3.19). Simple Fourier analysis, of the type used in Chap. 5, suggests that the preconditioned operator is well-behaved.

Streett et al. obtained efficient solutions of the similar boundary-layer

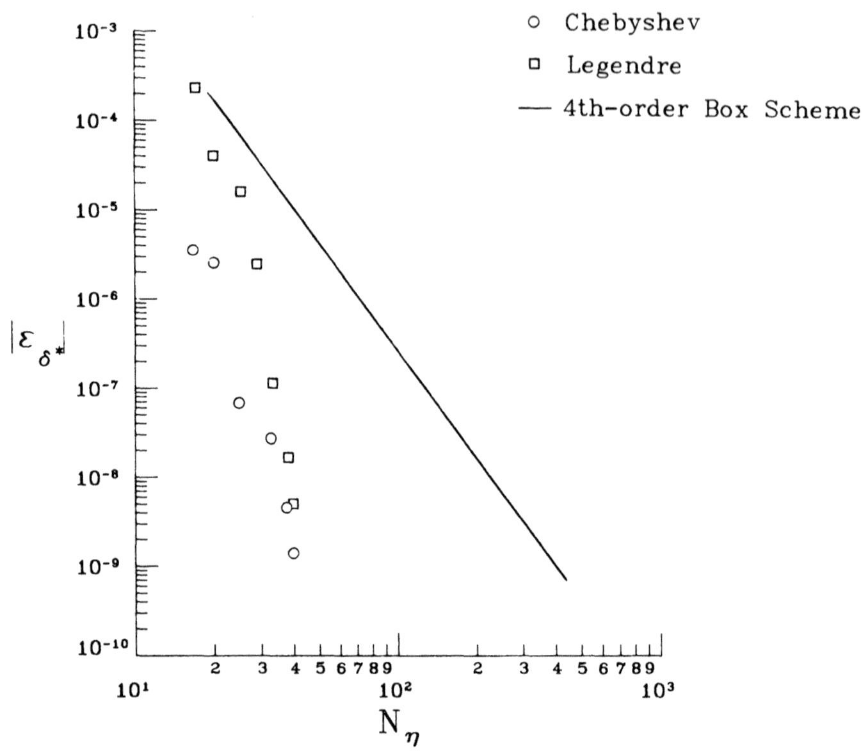

Figure 6.6. Error in displacement thickness as a function of resolution for Chebyshev spectral, Legendre spectral, and fourth-order finite-difference solutions to the similar boundary-layer equations.

equations by this method. Placing the outer boundary at $\eta_{max} = 35$ permitted nine digit accuracy. Both Chebyshev and Legendre solutions were obtained. Figure 6.6 provides a grid refinement study (in terms of error in the displacement thickness) for these spectral solutions as well as for a fourth-order finite-difference code (Wornom (1978)). Notice that 0.1% accuracy is achievable from the spectral methods with a mere twenty collocation points.

Streett et al. solved the non-similar boundary-layer equations (6.3.13)—(6.3.15) in an analogous fashion. The preconditioning chosen for the streamwise direction ξ was a three point (second-order) upwind scheme. (This is the standard discretization in a finite-difference, marching scheme.)

Since the spectral discretization is global in ξ, it is capable of handling separated flow. Figure 6.7 illustrates such a flow calculated by this spectral method. Even for this challenging case, four digit accuracy in skin friction ($\nu(\partial u/\partial y)$) was obtained with a 26×40 grid, as opposed to the 240×200 grid

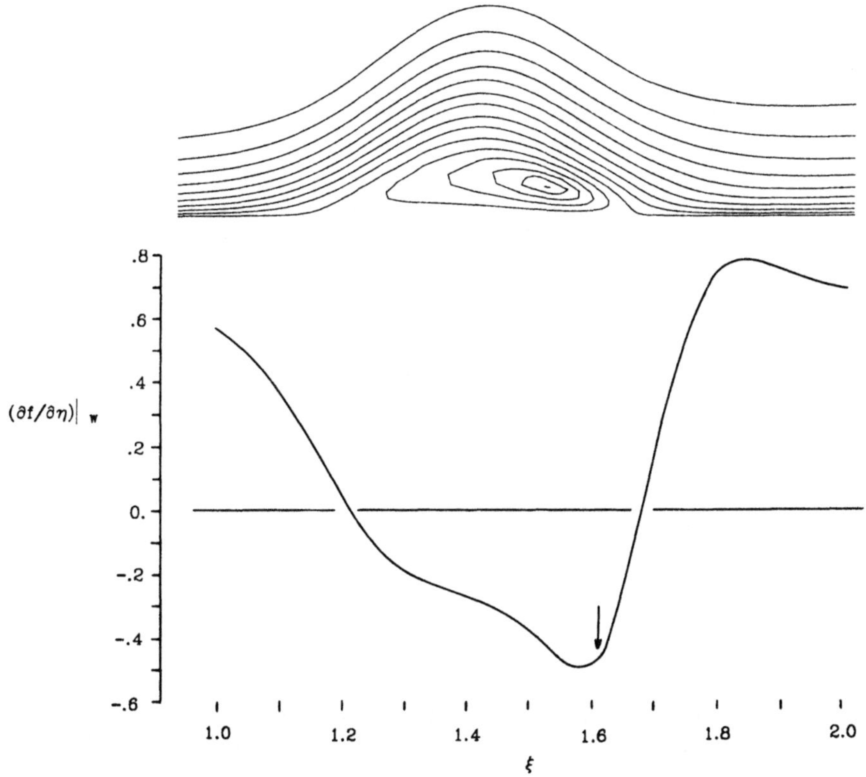

Figure 6.7. Streamlines (top) and skin friction (bottom) from a Chebyshev solution to the non-similar boundary-layer equations. The arrow marks the region of the flow which is most sensitive to the numerical resolution. (Courtesy of C. Streett.)

(and an order of magnitude greater computer time) required by a standard second-order finite-difference scheme.

6.4. Linear Stability

The Navier–Stokes equations admit many equilibrium solutions. However, a given equilibrium solution may not be physically realizable due to instability of the flow to small disturbances. The question of whether a given equilibrium solution is stable or unstable is crucial to many applications. This issue is often resolved within the framework of linear stability theory (Lin (1955),

Drazin and Reid (1981)). Chebyshev spectral methods have proven to be a useful technique for supplying accurate, efficient answers to linear stability problems.

Let $\mathbf{u}_0$ and p_0 denote the velocity and pressure of an equilibrium solution to the incompressible Navier–Stokes equations. Then write $\mathbf{u} = \mathbf{u}_0 + \mathbf{u}'$ and $p = p_0 + p'$, where $\mathbf{u}'$ and p' are perturbations to the mean flow. If $\mathbf{u}'$ and p' are presumed small and quadratic terms are neglected, then (1.3.18)—(1.3.19) become

$$\frac{\partial \mathbf{u}}{\partial t} + \mathbf{u}_0 \cdot \nabla \mathbf{u} + \mathbf{u} \cdot \nabla \mathbf{u}_0 = -\nabla p + \nu \Delta \mathbf{u} \tag{6.4.1}$$

$$\nabla \cdot \mathbf{u} = 0, \tag{6.4.2}$$

where the primes have been dropped.

A simple stability problem of long-standing interest is flow in a channel (Fig. 1.3). The mean flow is

$$\begin{aligned} \mathbf{u}_0 &= (u_0(y), 0, 0) \\ u_0(y) &= 1 - y^2 \\ p_0(x) &= -\frac{1}{R}x^2, \end{aligned} \tag{6.4.3}$$

where distances have been scaled by the half-channel width h, velocities by the centerline velocity $u_c = u_0(0)$, and the Reynolds number R is given by $R = u_c h/\nu$.

The stability of this flow is assessed by studying perturbations of the form

$$\mathbf{u}(\mathbf{x}, t) = \hat{\mathbf{u}}(y) e^{i(\alpha x + \beta z) - i\omega t}, \tag{6.4.4}$$

where α, β, and ω are complex constants. Equations (6.4.1) and (6.4.2) become, in component form,

$$\begin{aligned} \{D^2 - (\alpha^2 + \beta^2) - i\alpha R u_0\}\hat{u} - R(Du_0)\hat{v} - i\alpha R\hat{p} &= -i\omega R\hat{u} \\ \{D^2 - (\alpha^2 + \beta^2) - i\alpha R u_0\}\hat{v} - RD\hat{p} &= -i\omega R\hat{v} \\ \{D^2 - (\alpha^2 + \beta^2) - i\alpha R u_0\}\hat{w} - i\beta R\hat{p} &= -i\omega R\hat{w} \end{aligned} \tag{6.4.5}$$

$$i\alpha\hat{u} + D\hat{v} + i\beta\hat{w} = 0, \tag{6.4.6}$$

where $D = d/dy$. The boundary conditions are

$$\hat{u} = \hat{v} = \hat{w} = 0, \qquad \text{at } y = \pm 1. \tag{6.4.7}$$

The system (6.4.5)—(6.4.7) describes a dispersion relation between α, β and ω with R as a parameter. If four real quantities out of α, β and ω are prescribed, then the dispersion relation constitutes an eigenvalue problem for the remaining two real quantities. If α and β are fixed, real quantities, then ω

is the complex eigenvalue. When approached in this manner, the problem is one of temporal stability. If $\text{Im}\{\omega\} > 0$, then the corresponding mode grows in time and the mean flow will be disrupted. The equilibrium solution, then, is unstable if a growing mode exists for any real α and β. An alternative approach to this problem is one of spatial stability. Here, ω is real and fixed, and two relations are imposed upon α and β to complete the specification of the problem (Nayfeh (1980), Cebeci and Stewartson (1980)). If $\text{Im}\{\alpha\} > 0$ or $\text{Im}\{\beta\} > 0$, then the mode grows in space. If such growing modes exist for any real ω and for any orientations of the waves, then the flow is spatially unstable. Gaster (1962) has given a procedure for relating the results of temporal and spatial stability analyses.

Both the temporal and spatial stability formulations reduce the dispersion relation to a generalized eigenvalue problem. However, in the spatial version, the eigenvalue enters non-linearly and hence, this is a more difficult problem. We consider first numerical approaches to the temporal stability problem.

By manipulating (6.4.5) and (6.4.6) we arrive at

$$\begin{aligned}[D^2 - (\alpha^2 + \beta^2)]^2\hat{v} - i\alpha R u_0[D^2 - (\alpha^2 + \beta^2)]\hat{v} + i\alpha R(D^2 u_0)\hat{v} \\ = -i\omega R[D^2 - (\alpha^2 + \beta^2)]\hat{v}\end{aligned} \tag{6.4.8}$$

and

$$\begin{aligned}[D^2 - (\alpha^2 + \beta^2)](\alpha\hat{w} - \beta\hat{u}) - i\alpha R u_0(\alpha\hat{w} - \beta\hat{u}) \\ = -i\omega R(\alpha\hat{w} - \beta\hat{u}) - \beta R u_0'\hat{v}.\end{aligned} \tag{6.4.9}$$

The first of these is the celebrated Orr–Sommerfeld equation and it is subjected to the boundary conditions

$$\hat{v} = D\hat{v} = 0, \qquad \text{at } y = \pm 1. \tag{6.4.10}$$

(The condition on $D\hat{v}$ follows from (6.4.6) and (6.4.7).) The quantity $\alpha\hat{w} - \beta\hat{u}$ is the normal component of the perturbation vorticity. It satisfies

$$\alpha\hat{w} - \beta\hat{u} = 0, \qquad \text{at } y = \pm 1. \tag{6.4.11}$$

For this reason (6.4.9) is often referred to as the vertical vorticity equation (although Herbert (1983b) has taken to calling it the Squire equation). Hence, there are two distinct classes of solutions to the sixth-order system (6.4.5)—(6.4.7). The first class comprises the eigenmodes of (6.4.8), (6.4.10), with (6.4.9) serving merely to determine the normal vorticity of this mode. The second class has $\hat{v} \equiv 0$ and contains the eigenmodes of (6.4.9), (6.4.11). Squire (1933) showed that all solutions of the second class were damped modes. Thus, until recently attention has been focused on the Orr–Sommerfeld solutions.

Chebyshev tau approximations to the Orr–Sommerfeld equation are obtained in a straightforward manner. The dependent variable $\hat{v}$ is expanded as

$$\hat{v}(y) = \sum_{n=0}^{N} a_n T_n(y), \tag{6.4.12}$$

and the mean flow is written as

$$u_0(y) = \sum_{n=0}^{N} b_n T_n(y), \tag{6.4.13}$$

where, for channel flow,

$$\begin{aligned} b_0 &= \tfrac{1}{2} \\ b_2 &= \tfrac{1}{2} \\ b_n &= 0, \qquad \text{otherwise.} \end{aligned}$$

One enforces (6.4.8) in Chebyshev space for $n = 0, 1, \ldots, N - 4$ and then adds four equations for the boundary conditions

$$\sum_{\substack{n=0 \\ n \text{ even}}}^{N} a_n = \sum_{\substack{n=1 \\ n \text{ odd}}}^{N} a_n = \sum_{\substack{n=2 \\ n \text{ even}}}^{N} n^2 a_n = \sum_{\substack{n=1 \\ n \text{ odd}}}^{N} n^2 a_n = 0. \tag{6.4.14}$$

A generalized eigenvalue problem of the form

$$Aa = \omega Ba \tag{6.4.15}$$

results.

This discretization of the Orr–Sommerfeld equation was proposed by Orszag (1971e). His influential paper contains a detailed description of how to compute the elements of the matrices A and B for general mean flow profiles $u_0(y)$. Herbert (1977a) observed that one must be careful to avoid round-off error in the computation of the matrix elements. This application of the Chebyshev tau method requires an expression for the Chebyshev coefficients of $d^4\hat{v}/dy^4$:

$$\hat{v}_n^{(4)} = \frac{1}{c_n} \sum_{\substack{p=n+4 \\ p+n \text{ even}}}^{N} p[p^2(p^2-4)^2 - 3n^2p^4 + 3n^4p^2 - n^2(n^2-4)^2]a_p. \tag{6.4.16}$$

For n and p large, there can be substantial loss of significance in the bracketed term. Herbert suggested a straightforward re-arrangement of the terms within the bracket using $p = q + n$:

$$[q^4(q^2-8) + nq^3(6q^2-32) + (12n^2q^2 + 8n^3q)(q^2-4) + 16q^2 + 32nq]$$

The round-off errors for this expression are much lower.

In some applications one requires the complete spectrum of (6.4.15). The QR algorithm (Wilkinson (1965)) is the standard technique for this. In other cases, one is interested in only a few, or perhaps just one, eigenvalue of (6.4.15). Often, a good guess for ω is available. This occurs, for example, when computing a neutral curve—the locus (α, R) for which $\mathrm{Im}\{\omega\} = 0$. In these situations an inverse Rayleigh iteration is appropriate. Suppose that ω_0 is an approximate value of ω, and that x^n and y^n are current approximations to the eigenvectors of (6.4.15) and to the adjoint problem, respectively. These approximations are updated via

$$\begin{aligned}(A - \omega_0 B)x^{n+1} &= Bx^n \\ (A - \omega_0 B)^* y^{n+1} &= B^* y^n.\end{aligned} \tag{6.4.17}$$

The eigenvalue ω is then approximated by

$$\omega \cong \frac{(y, Ax)}{(y, Bx)}, \tag{6.4.18}$$

and, of course, x^{n+1} and y^{n+1} are improved approximations to the eigenvectors corresponding to ω.

The Chebyshev tau approach can be applied to many eigenvalue problems. Chaves and Ortiz (1968) applied it to a second-order eigenvalue problem with polynomial coefficients. Ortiz and collaborators have pursued this alternative implementation for many years. See Ortiz and Samara (1983) for some recent work and a detailed list of references.

Collocation approximations to the Orr–Sommerfeld equations are less straightforward. The problem is the double boundary condition at the walls. Suppose that the standard set of collocation points is employed

$$y_j = \cos(\pi j/N) \qquad j = 0, \ldots, N. \tag{6.4.19}$$

If one enforces (6.4.8) for $j = 1, \ldots, N - 1$ and (6.4.10) for $j = 0$ and N, then the problem is overdetermined. One way to remove the overdeterminacy is to enforce (6.4.8) for $j = 2, \ldots, N - 2$ only, dropping the differential equation condition at $j = 1$ and $j = N - 1$, the interior points nearest the wall. The matrices A and B can be constructed readily from the differentiation matrix D_N (see (2.4.31)), its powers, and diagonal matrices representing the constant terms, u_0 and u_0''.

Alternatively, one can use the collocation points

$$y_j = \cos\frac{\pi j}{N-2} \qquad j = 0, \ldots, N - 2 \tag{6.4.20}$$

imposing the differential equation for $j = 1, \ldots, N - 1$ and the four boundary conditions at $j = 0$ and $j = N$. Maday (1986, private communication) noted that the latter choice achieves spectral accuracy with an optimal-order error estimate for solutions with finite regularity. The former choice also produces spectral accuracy but with non-optimal order. Moreover, the order of the former method deteriorates even more as the neglected collocation points are moved further away from the boundary.

Herbert (1977a) has devised another collocation method. He replaces the collocation points (6.4.19) with

$$y_j = \cos\frac{\pi j}{N-4} \qquad j = 0, 1, \ldots, N-4. \tag{6.4.21}$$

The $N + 1$ coefficients a_N in the expansion (6.4.12) are determined by enforcing (6.4.8) at $j = 0, 1, \ldots, N-4$ and adding in the conditions (6.4.14). The result is a linear system similar to (6.4.12) in which the a_n are the unknowns, but for which the rows arising from the differential equations come from a physical space rather than a spectral space condition.

A third approach is to impose the Neumann boundary condition indirectly, as has been discussed in Sec. 3.3. Here, one uses the standard collocation points and computes $D\hat{v}$ as usual, then defines

$$\hat{D}\hat{v} = \begin{cases} D\hat{v} & j = 1, \ldots, N-1 \\ 0 & j = 0, N, \end{cases} \tag{6.4.22}$$

and uses $D^2\hat{v} = D(\hat{D}\hat{v})$ and $D^4\hat{v} = D^3(\hat{D}\hat{v})$ as needed. The differential equation is imposed for $j = 1, \ldots, N-1$ and $\hat{v} = 0$ is required at $j = 0, N$.

Herbert's collocation method is the only one of these that has been used in hydrodynamic stability calculations. He finds that it is comparable in accuracy to the tau method.

Linear stability investigations of channel flow are fairly tractable because the mean flow is strictly parallel, i.e., $\mathbf{u}_0$ is parallel to the wall and is independent of x. Even such a simple flow as that over a flat plate is non-parallel. As we saw in the last section, v_0, as well as u_0, are non-zero (and both depend on x in addition to y). However, $|v_0| \ll u_0$, and the x dependence is much weaker than that on y. A reasonable approximation to boundary layer stability is to make the "parallel flow assumption." Here, one analyzes the stability in the vicinity of some point x_0 by supposing that $u_0(x, y)$ is given by $u_0(x_0, y)$ for all x and that v_0 is negligible. This approximation is illustrated in Fig. 6.8. Within this approximation it is customary to take for $u_0(y)$ the solution to the similar boundary-layer equations. This stability problem is also governed by (6.4.8)—(6.4.11) except that the boundary conditions are imposed at $y = 0$ and $y = \infty$. The same strategies that were discussed in Sec. 6.3 for the Chebyshev spectral solution of the mean flow are also available here—a suitable mapping in y (bearing in mind that the solution to the stability problem decays exponentially) and either the standard Chebyshev expansion over $[-1, 1]$ in the computational coordinate or else an expansion over $[0, 1]$.

The spatial stability problem can also be attacked by Chebyshev methods. Bridges and Morris (1984a, 1984b) have given an extensive discussion of this problem in the context of the similar boundary layer. The discretization methodology for the vertical direction is the same. The principal algorithmic

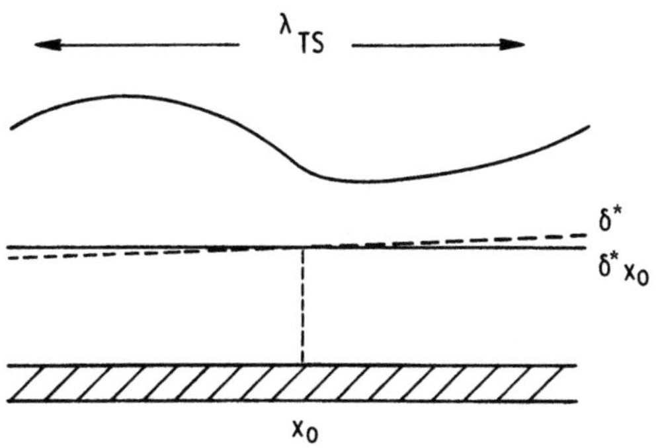

Figure 6.8. The parallel boundary layer: the variation of the displacement thickness δ^* over one wave length λ_{TS} of a perturbation is neglected.

difficulty is the solution of the eigenvalue problem: ω is fixed and α is the eigenvalue (if $\beta = 0$). This enters non-linearly in (6.4.8). They advocate the use of a companion matrix method for determining the eigenvalues.

Several other applications of spectral methods to linear stability problems are worthy of note. Von Kerczek (1982) has applied a Chebyshev tau method to the linear stability problem of oscillatory channel flow. In this flow the pressure gradient is given not by (6.4.3) but rather by

$$p_0(x, t) = -\frac{2x}{R}[1 + \Lambda \cos \Omega t] \tag{6.4.23}$$

and the mean velocity $u_0(y, t)$ is also periodic in time. The appropriate Orr–Sommerfeld equation includes a time dependence. Floquet theory guarantees that the perturbations may be written as

$$\mathbf{u}(\mathbf{x}, t) = \hat{\mathbf{u}}(y, t)e^{i(\alpha x + \beta z) + \lambda t}, \tag{6.4.24}$$

where $\hat{\mathbf{u}}(y, t)$ is periodic in t with period $2\pi/\Omega$. The Floquet exponent λ determines the stability of infinitesimal perturbations (Davis (1976)). Von Kerczek discusses several techniques for computing these Floquet exponents in Chebyshev approximations.

Herbert (1977b, 1983a) and Orszag and Patera (1983) have used Chebyshev methods to determine finite amplitude solutions to channel flow perturbations. Their approach is to write two-dimensional perturbations as

$$u(\mathbf{x}, t) = A \sum_{m=-M}^{M} \hat{u}_m(y)e^{im\alpha(x-ct)}, \tag{6.4.25}$$

where c is a real phase speed and A is an amplitude. Here one makes a two-dimensional approximation—Fourier in x and Chebyshev in y. The result is a neutral stability *surface* defined by a characteristic equation involving R, A, and α. Spectral methods permit this surface, and the corresponding finite-amplitude solutions, to be computed with great precision. An example for channel flow is provided in Fig. 6.9, where E denotes the perturbation kinetic energy divided by the mean kinetic energy.

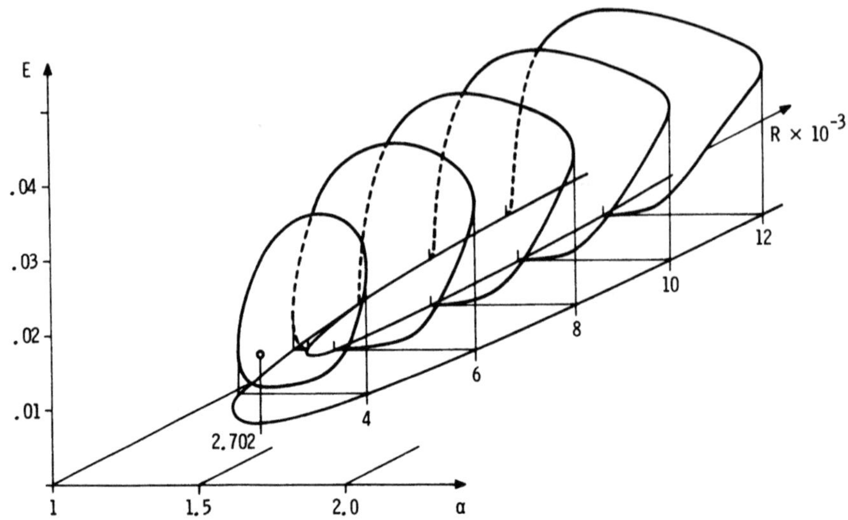

Figure 6.9. Neutral surface for finite amplitude perturbations to channel flow. (Courtesy of T. Herbert.)

It has long been known experimentally that channel flow is strongly unstable to three-dimensional perturbations at Reynolds numbers far below the critical Reynolds number for linear instability. The key phenomenon appears to be the non-linear interaction of a two-dimensional and a three-dimensional wave. The initial stages of this secondary instability may be assessed by performing a linear stability analysis of the mean flow plus a finite amplitude two-dimensional wave such as (6.4.25). If one changes coordinates to $x' = x - ct$, then the mean flow is independent of t. It does, however, depend upon x'. One can look for solutions of the form

$$\mathbf{u}(\mathbf{x}, t) = \sum_{m=-M}^{M} \hat{\mathbf{u}}_m(y) e^{im\alpha x'} e^{i\beta z - i\omega t}, \tag{6.4.26}$$

to the appropriate Orr–Sommerfeld equation. Methods similar to those applied to the purely linear problem may be used here. (Of course, the algebraic problem is more complex because of the coupling in x.) Numerous results have been obtained by Orszag and Patera (1983) and especially by Herbert (1983a, 1983b, 1984, 1985).

CHAPTER 7

Some Algorithms for Unsteady Navier–Stokes Equations

7.1. Introduction

There are essentially four different formulations of the incompressible Navier–Stokes equations—the primitive-variable (velocity and pressure), streamfunction-vorticity, streamfunction and velocity-vorticity formulations. The primitive-variable formulation can be found in any text on fluid dynamics (e.g., Batchelor (1967)), as can the two-dimensional version of both streamfunction formulations. The three-dimensional streamfunction equations are given by Murdock (1986). Both streamfunction formulations avoid the complications of dealing with the pressure, as does the straightforward velocity-vorticity approach. Murdock (1977, 1986) and Vanel, Peyret and Bontoux (1985) have developed spectral algorithms based on the two-dimensional streamfunction-vorticity formulation. Murdock (1986) has extended this to three dimensions. Gottlieb and Orszag (1977) and Maday and Métivet (1986) discuss spectral methods for the pure streamfunction version in two-dimensional flows. The velocity-vorticity formulation (Dennis, Ingham and Cook (1979)) has not yet been employed in spectral calculations. The primitive-variable formulation has been the one most extensively employed in three-dimensional spectral calculations and spectral methods based on it will be the focus of this chapter.

The Navier–Stokes equations on a domain Ω are usually written as (see (1.3.18) and (1.3.19))

$$\frac{\partial \mathbf{u}}{\partial t} + \mathbf{u}\cdot\nabla\mathbf{u} = -\nabla p + \nu\Delta\mathbf{u} \qquad \text{in } \Omega \tag{7.1.1}$$

$$\nabla\cdot\mathbf{u} = 0 \qquad \text{in } \Omega, \tag{7.1.2}$$

where $\mathbf{u}$ is the velocity vector, p the pressure, and ν the kinematic viscosity. Equation (7.1.1) is the momentum equation and (7.1.2) is the continuity equation. The boundary conditions are that

$$\mathbf{u}(\mathbf{x}, t) = 0 \qquad \text{on } \partial\Omega, \tag{7.1.3}$$

and the initial conditions are

$$\mathbf{u}(\mathbf{x}, 0) = \mathbf{u}_0(\mathbf{x}) \qquad \text{in } \Omega. \tag{7.1.4}$$

As a rule, the algebraic structure of the discretized Navier–Stokes equations should mimic as many of the key properties of the continuous system as possible. These include conservation and invariance (symmetry) properties as well as the consistency relation between the divergence, gradient, and Laplacian operators. The inviscid part of the Navier–Stokes equations conserves linear and angular momentum, and kinetic energy. Ideally, one should ensure that the discrete system conserves all these quantities. Often it is not possible to satisfy all of these properties simultaneously. Then, one must assign priority to those properties which are most important to the physics of the problem under consideration.

Analytically, the divergence operator is the adjoint of the (negative) gradient operator, and the Laplacian is the composition of the divergence and the gradient. Both properties are virtually automatic for Fourier methods, as is the latter for Chebyshev methods. The former property is violated by most Chebyshev methods, since the differentiation operator is not skew-symmetric.

The choice of the initial velocity field is not arbitrary. Clearly, it must be divergence-free; otherwise, the continuous problem itself fails to have a classical solution (see Heywood and Rannacher (1986)). Moreover, an initial velocity field, even though divergence-free, may induce transients (due to incompatibility with the momentum equation) which overwhelm delicate time-dependent stability mechanisms (see Deville, Kleiser and Montigny-Rannou (1984)).

The physical boundaries are usually formed by solid bodies with no-slip boundary conditions. However, in some cases the domain Ω is unbounded and it can be mapped onto a bounded computational domain (see Sec. 2.5). In other cases, artificial boundaries are required. These artificial boundaries are often of non-characteristic type. Boundary conditions need to be placed on the primitive variables at these artificial boundaries, but no rigorous theory exists with regard to the implications of these boundary conditions for the existence and uniqueness of the Navier–Stokes solutions.

For any discretization of the incompressible Navier–Stokes equations, the principal conceptual difficulty is the treatment of the pressure. Unlike the velocity, there is no evolution equation for the pressure. Instead, it is determined by the constraint (7.1.2). Some numerical methods, known as penalty methods or artificial-compressibility methods, have circumvented this problem by introducing a time derivative of the pressure into the continuity equation (Temam (1977, Chap. 3)). This approach has produced satisfactory solutions to steady-state problems, but has not yet furnished highly accurate solutions to time-dependent cases. We will not discuss these approaches any further (but see Deville et al. (1984)). The focus in this chapter will be on spectral methods for unsteady problems, as typified by turbulence and transition.

7.2. Homogeneous Flows

Von Neumann (1949) and Emmons (1949) proposed numerical simulation of turbulence as early as the late 40s. Nevertheless, nearly two decades elapsed before this vision became a reality. It materialized at the National Center for Atmospheric Research in Boulder, Colorado in the late 60s, due in large part to the availability of the CDC 6600. Deardorff (1970) combined a finite-difference method with a subgrid-scale model to compute turbulent channel flow on a $24 \times 20 \times 14$ grid. Orszag and Patterson (1972b) performed the first direct simulation of homogeneous, isotropic turbulence on a 32^3 grid using a spectral Galerkin method. This algorithm remains to this day the workhorse for numerical simulations of homogeneous turbulence. Several refinements of their basic algorithm have been developed, notably those by Rogallo (1977, 1981) and Basdevant (1983), and by Brachet et al. (1983) for flows with special symmetries.

For homogeneous flows, periodic boundary conditions in all three directions may be justified on physical grounds. In this case (7.1.1)—(7.1.4) become

$$\frac{\partial \mathbf{u}}{\partial t} + \mathbf{u}\cdot\nabla\mathbf{u} = -\nabla p + \nu\Delta\mathbf{u} \tag{7.2.1}$$

$$\nabla\cdot\mathbf{u} = 0 \tag{7.2.2}$$

$$\mathbf{u}(\mathbf{x} + 2\pi\mathbf{e}_j, t) = \mathbf{u}(\mathbf{x}, t) \qquad j = 1, 2, 3 \tag{7.2.3}$$

$$\mathbf{u}(\mathbf{x}, 0) = \mathbf{u}_0(\mathbf{x}), \tag{7.2.4}$$

where $\mathbf{e}_j$ is a unit vector in the jth direction. We have assumed for simplicity that the periodicity lengths in all three directions are 2π. Rigorous theories for this problem can be found in Temam (1983).

7.2.1. A Spectral Galerkin Solution Technique

The solution to the problem has the Fourier series representation

$$\mathbf{u}(\mathbf{x}, t) = \sum_{\mathbf{k}} \hat{\mathbf{u}}_{\mathbf{k}}(t) e^{i\mathbf{k}\cdot\mathbf{x}} \tag{7.2.5}$$

$$p(\mathbf{x}, t) = \sum_{\mathbf{k}} \hat{p}_{\mathbf{k}}(t) e^{i\mathbf{k}\cdot\mathbf{x}}. \tag{7.2.6}$$

In Fourier space (7.2.1) and (7.2.2) become

$$\left(\frac{d}{dt} + \nu|\mathbf{k}|^2\right)\hat{\mathbf{u}}_{\mathbf{k}} = -i\mathbf{k}\hat{p}_{\mathbf{k}} - \widehat{(\mathbf{u}\cdot\nabla\mathbf{u})}_{\mathbf{k}} \tag{7.2.7}$$

$$i\mathbf{k}\cdot\hat{\mathbf{u}}_{\mathbf{k}} = 0. \tag{7.2.8}$$

The term

$$\hat{\mathbf{f}}_{\mathbf{k}} = -\widehat{(\mathbf{u}\cdot\nabla\mathbf{u})}_{\mathbf{k}} \tag{7.2.9}$$

is non-linear and responsible for most of the algorithmic complexities of the problem. Its numerical treatment is discussed extensively below. The pressure may be eliminated by taking $i\mathbf{k}$ dotted into (7.2.7) and using (7.2.8). Hence,

$$\hat{p}_{\mathbf{k}} = -\frac{1}{|\mathbf{k}|^2} i\mathbf{k}\cdot\hat{\mathbf{f}}_{\mathbf{k}} \tag{7.2.10}$$

$$\left(\frac{d}{dt} + \nu|\mathbf{k}|^2\right)\hat{\mathbf{u}}_k = \hat{\mathbf{f}}_{\mathbf{k}} - \mathbf{k}\frac{(\mathbf{k}\cdot\hat{\mathbf{f}}_{\mathbf{k}})}{|\mathbf{k}|^2}. \tag{7.2.11}$$

The Fourier Galerkin approximation consists of truncating the sums at $|k_1|$, $|k_2|$, $|k_3|$, $< N/2$. (For the reasons given in Sec. 3.1.1, the modes for which any $k_\alpha = -N/2$, $\alpha = 1$, 2, or 3 are not retained.) Note that in (7.2.7) and (7.2.8) the negative gradient operator is the adjoint of the divergence operator and the Laplacian operator is the composition of the divergence and the gradient.

The original Orszag–Patterson algorithm (Orszag (1969), Patterson and Orszag (1971), Orszag (1971d), Orszag and Patterson (1972b)) employed leap frog time-differencing for the non-linear term and Crank–Nicolson for the viscous term. (In most applications, though, an explicit scheme for the viscous term will suffice since the stability restriction arising from the convection terms can be more severe than the viscous stability limit.) Other implementations, for example, that by Rogallo (1977), have used fourth-order Runge–Kutta for the non-linear terms and an integrating factor technique on diffusion.

7.2.2. Treatment of the Nonlinear Terms

Let $\mathbf{u} = (u, v, w) = (u_1, u_2, u_3)$. In component form the non-linear term is

$$\hat{\mathbf{f}}_{\alpha,\mathbf{k}} = -ik_\beta \sum_{\mathbf{m}+\mathbf{n}=\mathbf{k}} \hat{u}_{\beta,\mathbf{m}}\hat{u}_{\alpha,\mathbf{n}} \tag{7.2.12}$$

(where repeated indices—in this case β—imply summation). This convolution sum is the standard Galerkin approximation to the non-linear term $-\mathbf{u}\cdot\nabla\mathbf{u}$. Let us focus on a typical term in the triple convolution sum, namely

$$\hat{w}_{\mathbf{k}} = \sum_{\mathbf{m}+\mathbf{n}=\mathbf{k}} \hat{u}_{\mathbf{m}}\hat{v}_{\mathbf{n}}. \tag{7.2.13}$$

Efficient transform methods for the one-dimensional version of (7.2.13) are discussed in Sec. 3.2. The three-dimensional pseudospectral method is apparent. The basic techniques for de-aliasing are truncation and phase shifts. The 3/2-rule, in which the discrete transforms are evaluated with $3N/2$ points, extends to three dimensions in a straightforward manner. The phase shift technique, however, becomes much more involved. In three dimensions the analog of (3.2.9) is

$$\begin{aligned}\hat{W}_{\mathbf{k}} = &\sum_{\mathbf{m}+\mathbf{n}=\mathbf{k}} \hat{u}_{\mathbf{m}}\hat{v}_{\mathbf{n}} + \sum_{\mathbf{m}+\mathbf{n}=\mathbf{k}\pm N\mathbf{e}_1} \hat{u}_{\mathbf{m}}\hat{v}_{\mathbf{n}} + \sum_{\mathbf{m}+\mathbf{n}=\mathbf{k}\pm N\mathbf{e}_2} \hat{u}_{\mathbf{m}}\hat{v}_{\mathbf{n}} + \sum_{\mathbf{m}+\mathbf{n}=\mathbf{k}\pm N\mathbf{e}_3} \hat{u}_{\mathbf{m}}\hat{v}_{\mathbf{n}} \\ &+ \sum_{\mathbf{m}+\mathbf{n}=\mathbf{k}\pm N\mathbf{e}_1\pm N\mathbf{e}_2} \hat{u}_{\mathbf{m}}\hat{v}_{\mathbf{n}} + \sum_{\mathbf{m}+\mathbf{n}=\mathbf{k}\pm N\mathbf{e}_1\pm N\mathbf{e}_3} \hat{u}_{\mathbf{m}}\hat{v}_{\mathbf{n}} \\ &+ \sum_{\mathbf{m}+\mathbf{n}=\mathbf{k}\pm N\mathbf{e}_2\pm N\mathbf{e}_3} \hat{u}_{\mathbf{m}}\hat{v}_{\mathbf{n}} + \sum_{\mathbf{m}+\mathbf{n}=\mathbf{k}\pm N\mathbf{e}_1\pm N\mathbf{e}_2\pm N\mathbf{e}_3} \hat{u}_{\mathbf{m}}\hat{v}_{\mathbf{n}}.\end{aligned} \tag{7.2.14}$$

The second, third, and fourth terms on the right-hand side are the singly-aliased contributions; the fifth, sixth and seventh are the doubly-aliased ones; and the last term is the triply-aliased contribution. Aliasing removal by phase shifts thus requires eight separate evaluations of the convolution terms. (See Rogallo (1981) for details.)

Another option is combining truncation with just two shifted grids, the usual one with

$$\mathbf{x}_{\mathbf{j}} = \frac{2\pi}{N}(j_1, j_2, j_3) \qquad j_\alpha = -\frac{N}{2}, -\frac{N}{2}+1, \ldots, \frac{N}{2}-1 \tag{7.2.15}$$

and the grid offset from this by a half-cell in each direction

$$\mathbf{x}'_{\mathbf{j}} = \mathbf{x}_{\mathbf{j}} + \left(\frac{\pi}{N}, \frac{\pi}{N}, \frac{\pi}{N}\right). \tag{7.2.16}$$

This eliminates both the singly- and triply-aliased contributions. The doubly- and triply-aliased terms vanish if the spherical truncation (Patterson and Orszag (1971))

$$k_1^2 + k_2^2 + k_3^2 \leqslant \left(\frac{\sqrt{2}}{3}N\right)^2 \tag{7.2.17}$$

or the polyhedral truncation (Orszag (1971d))

$$\begin{aligned} |k_\alpha| &\leqslant N/2 \qquad & \alpha &= 1, 2, 3 \\ |k_\alpha \pm k_\beta| &\leqslant 2N/3 \qquad & \alpha, \beta &= 1, 2, 3, \quad \alpha \neq \beta \end{aligned} \tag{7.2.18}$$

is used. The latter truncation has the advantage of retaining 27% more modes than the former one.

A pseudospectral method employing random shifts at each step was discussed at the end of Sec. 3.2.3. This technique was in fact developed by Rogallo (1977) for the present application of homogeneous turbulence simulations. The spherical truncation (7.2.17) removes the double-aliasing and triple-aliasing errors. At the first stage of a second-order Runge–Kutta method, the convolution sum is evaluated by a pseudospectral transform method on the grid $\mathbf{x}_j$ given by (7.2.15), except that on the right-hand side each j_α is replaced by $j_\alpha + \Delta_\alpha$, where the Δ_α are random numbers in $(0, 2\pi/N)$. At the second stage the grid (7.2.16) is used in the pseudospectral transform method. This reduces the single-aliasing errors to $O(\Delta t^2)$ times their ordinary pseudospectral values. The random shifts at each time-step inhibit these remaining aliasing errors from accumulating over many time-steps. This strategy can be adapted to most time-discretization schemes.

The truncations described by (7.2.17) and (7.2.18) are both isotropic (in addition to eliminating the double- and triple-aliasing errors). Since homogeneous turbulence appears to be isotropic on the smallest scales, one may wish to incorporate the isotropic truncation

$$k_1^2 + k_2^2 + k_3^2 < (N/2)^2 \tag{7.2.19}$$

in those methods which do not employ a truncation for de-aliasing purposes.

Table 7.1 summarizes the various transform methods available for three-dimensional convolution sums. The operation counts are for the evaluation of one triple convolution sum and they presume that a single full three-dimensional FFT on an N^3 grid takes $(15/2) N^3 \log_2 N$ operations and that the three-dimensional FFT used for the 3/2-rule methods takes only $(235/16) N^3 \log_2 N$ operations because of the large number of zeros in the data. (Addition and multiplication count as separate operations and the simple radix 2 transform is assumed; more general and more efficient transforms are discussed in Appendix B.) The active modes are those which survive the truncation. The last column indicates which, if any, aliasing errors are present.

Rogallo (1981) proposed combining the truncation

$$|k_\alpha| \leqslant N/3 \quad \text{for at least one } \alpha = 1, 2, 3$$

Table 7.1. Transform methods for a triple convolution sum

Method	Operations	Active Modes	Operations/ Mode	Aliasing Errors
pseudospectral	$22.5\, N^3 \log_2 N$	N^3	$22.5 \log_2 N$	single, double and triple
$\frac{3}{2}$-rule	$44\, N^3 \log_2(\frac{3}{2}N)$	N^3	$44 \log_2(\frac{3}{2}N)$	none
8 shifted grids	$180\, N^3 \log_2 N$	N^3	$180 \log_2 N$	none
$\frac{3}{2}$-rule + (7.2.19)	$44\, N^3 \log_2(\frac{3}{2}N)$	$\frac{\pi}{6} N^3$	$84 \log_2(\frac{3}{2}N)$	none
random pseudospectral + (7.2.17)	$22.5\, N^3 \log_2 N$	$\frac{8\pi\sqrt{2}}{81} N^3$	$51 \log_2 N$	$O(\Delta t^2)$ single
2 shifted grids + (7.2.17)	$45\, N^3 \log_2 N$	$\frac{8\pi\sqrt{2}}{81} N^3$	$103 \log_2 N$	none
2 shifted grids + (7.2.18)	$45\, N^3 \log_2 N$	$\frac{5}{9} N^3$	$81 \log_2 N$	none

with four shifted grids, but he later abandoned this method since not all of the doubly-aliased terms were in fact eliminated (Rogallo (1986, private communication)).

7.2.3. Refinements

The original Orszag–Patterson algorithm was suitable only for isotropic turbulent flows. Rogallo (1977, 1981) extended it to all homogeneous turbulent flows. These have the form $\mathbf{u} = \mathbf{u}_0 + \mathbf{u}'$ and $p = p_0 + p'$ with the mean flow

$$\mathbf{u}_0 = A\mathbf{x} \tag{7.2.20}$$

$$p_0 = \mathbf{x}^T B\mathbf{x}, \tag{7.2.21}$$

where the tensors A and B depend only on t (and B is symmetric). (Any more general form of the mean flow necessarily leads to inhomogeneous turbulence.) The computations are done in a moving coordinate system $\mathbf{x}'$ described by

$$\mathbf{x}' = C\mathbf{x}, \tag{7.2.22}$$

where C also depends only on t. The tensor A may be written

$$A = S + \Omega, \tag{7.2.23}$$

where S and Ω are the symmetric and antisymmetric parts and represent the strain rate and vorticity of the mean flow. Rogallo (1981) shows that, subject to

$$\sum_{i=1}^{3} S_{ii} = 0, \tag{7.2.24}$$

S is an arbitrary function of time and the other tensors are related by

$$\frac{d\Omega}{dt} + S\Omega + \Omega S = 0 \tag{7.2.25}$$

$$\frac{dS}{dt} + S^2 + \Omega^2 + 2B = 0 \tag{7.2.26}$$

$$\frac{dC}{dt} + CA = 0. \tag{7.2.27}$$

The equation governing the fluctuating part of the flow is, with tensor notation employed,

$$\frac{\partial u_i'}{\partial t} + A_{ij}u_j' + C_{kj}(u_i'u_j')_{,k} = -C_{ji}p'_{,j} + \nu C_{kj}C_{lj}u'_{i,kl} \tag{7.2.28}$$

$$C_{ji}u'_{i,j} = 0, \tag{7.2.29}$$

where the derivatives are taken with respect to $\mathbf{x}'$ and derivatives with respect to x'_k are denoted by $,k$. Since the coefficients of (7.2.28) and (7.2.29) are independent of $\mathbf{x}'$, Fourier approximations are suitable.

There are several refinements which can profitably be employed to reduce the storage requirements, CPU time, or I/O cost for an out-of-core simulation. Rogallo (1981) and Basdevant (1983) provide extended discussions. Basdevant, for example, shows how isotropic simulations may be performed (with the 3/2-rule) at a cost of roughly eight three-dimensional FFT's per step. The generation of divergence-free initial conditions has been discussed by Rogallo (1981) and Schumann (1985). The paper by Brachet et al. (1983) describes how symmetries in special flows such as the Taylor–Green vortex may be exploited to provide a wider dynamic range.

7.2.4. Pseudospectral and Collocation Methods

The pseudospectral version of the Orszag–Patterson algorithm uses (7.2.9) and (7.2.11), but approximates the non-linear terms $\hat{\mathbf{f}}_{\mathbf{k}}$ pseudospectrally, as discussed in Sec. 3.2.1. In this approach the fully-aliased convolution sum (7.2.14) is used in place of the correct sum (7.2.13). The operation count is clearly $(45/2)\,N^3 \log_2 N$. This is one half the cost of a true Galerkin method implemented via the 3/2-rule. If, for physical or esthetical reasons, an isotropic truncation is desired, then the effective cost per mode of a pseudospectral method rises to 60% of the Galerkin cost.

Collocation algorithms for homogeneous flows are based on the Navier–Stokes equations in their physical space, rotation form:

$$\frac{\partial \mathbf{u}}{\partial t} + \boldsymbol{\omega} \times \mathbf{u} = -\nabla P + \nabla\cdot(\nu \nabla \mathbf{u}) \tag{7.2.30}$$

$$\nabla\cdot\mathbf{u} = 0, \tag{7.2.31}$$

where $\boldsymbol{\omega} = \nabla \times \mathbf{u}$ and $P = p + \frac{1}{2}|\mathbf{u}|^2$. As noted in Sec. 4.5, the rotation form (7.2.30) semi-conserves kinetic energy for inviscid flow, but the standard form (7.2.1) does not. In practice, the rotation form in a standard collocation scheme (without randomly shifted grids) must be used to ensure numerical stability.

Consider first the case in which the viscosity ν is a constant. Assume, for simplicity, that backward Euler is applied to the pressure gradient and forward Euler to the non-linear and viscous terms in (7.2.30) and that the incompressibility constraint is enforced at the new time-level. Then the fully discrete approximation to (7.2.30) and (7.2.31) is, without bothering to adopt new symbols for the discrete variables,

$$\mathbf{u}^{n+1} = \mathbf{u}^n - \Delta t(\boldsymbol{\omega}^n \times \mathbf{u}^n) + \Delta t \nu \mathbb{L}_N \mathbf{u}^n - \Delta t \mathbb{G}_N P^{n+1} \tag{7.2.32}$$

$$\mathbb{D}_N \cdot \mathbf{u}^{n+1} = 0, \tag{7.2.33}$$

where $\mathbb{L}_N = \mathbb{D}_N \cdot \mathbb{G}_N$ is the discrete Laplacian. Taking the discrete divergence of (7.2.32) and using (7.2.33), which applies to $\mathbf{u}^n$ as well, plus the property that $\mathbb{D}_N$ and $\mathbb{L}_N$ commute for Fourier methods, we obtain

$$\mathbb{L}_N P^{n+1} = \mathbb{D}_N \cdot (\boldsymbol{\omega}^n \times \mathbf{u}^n). \tag{7.2.34}$$

In Fourier space, (7.2.32)—(7.2.34) become

$$\hat{\mathbf{u}}_{\mathbf{k}}^{n+1} = \hat{\mathbf{u}}_{\mathbf{k}}^n - \Delta t \nu k^2 \hat{\mathbf{u}}_{\mathbf{k}}^n - \Delta t \widehat{(\boldsymbol{\omega}^n \times \mathbf{u}^n)}_{\mathbf{k}} + \Delta t \frac{\mathbf{k}}{|\mathbf{k}|^2} [\mathbf{k} \cdot \widehat{(\boldsymbol{\omega}^n \times \mathbf{u}^n)}_{\mathbf{k}}]. \tag{7.2.35}$$

(Note that the viscous term may be treated implicitly at essentially no extra cost.) The collocation method may be identified with a particular pseudospectral method. Unlike Galerkin methods, pseudospectral methods which write the non-linear term in different (but equivalent) forms for the continuous problem, will not be equivalent. Thus, the algorithm which uses $(\mathbf{u}^n \cdot \mathbb{G}_N \mathbf{u}^n)$ in place of $(\boldsymbol{\omega}^n \times \mathbf{u}^n)$ in (7.2.35) will yield a different result. In fact, since it does not semi-conserve energy, it is susceptible to numerical instabilities.

In the event that the fluid has variable transport properties (as in computations which include a turbulence model), the viscosity will depend on $\mathbf{x}$ and t. The following splitting method is then appropriate:

$$\mathbf{u}^{n+1/2} = \mathbf{u}^n - \Delta t(\boldsymbol{\omega}^n \times \mathbf{u}^n) + \Delta t \mathbb{D}_N \cdot (\nu \mathbb{G}_N \mathbf{u}^n) \tag{7.2.36}$$

$$\mathbf{u}^{n+1} = \mathbf{u}^{n+1/2} - \Delta t \mathbb{G}_N P^{n+1} \tag{7.2.37}$$

$$\mathbb{D}_N \cdot \mathbf{u}^{n+1} = 0. \tag{7.2.38}$$

The first step, given by (7.2.36), is straightforward. If an implicit treatment of the viscous term is required, then the Fourier spectral multigrid methods discussed in Sec. 5.5 are advisable (see Erlebacher, Zang and Hussaini (1987)). For the second step, (7.2.37) and (7.2.38) imply that the pressure satisfies

$$\mathbb{L}_N P^{n+1} = \frac{1}{\Delta t} \mathbb{D}_N \cdot \mathbf{u}^{n+1/2}. \tag{7.2.39}$$

This equation is readily solved in Fourier space.

The principal difference between the collocation and Galerkin methods is the inclusion of aliasing interactions in the former. A crucial issue is whether, in practical calculations, it is necessary to remove the aliasing terms in order to obtain a reliable result. This issue is addressed for smaller scale calculations in Sec. 4.6, and in numerous papers on simulations of homogeneous turbulence, most recently by Kerr (1985). Here we compare some full three-dimensional turbulence simulations performed both with and without aliasing, taken from the report by Erlebacher, Hussaini, Speziale and Zang (1987).

The initial conditions were chosen to match the experimental data of Comte–Bellot and Corrsin (1971) for isotropic turbulence in a wind tunnel.

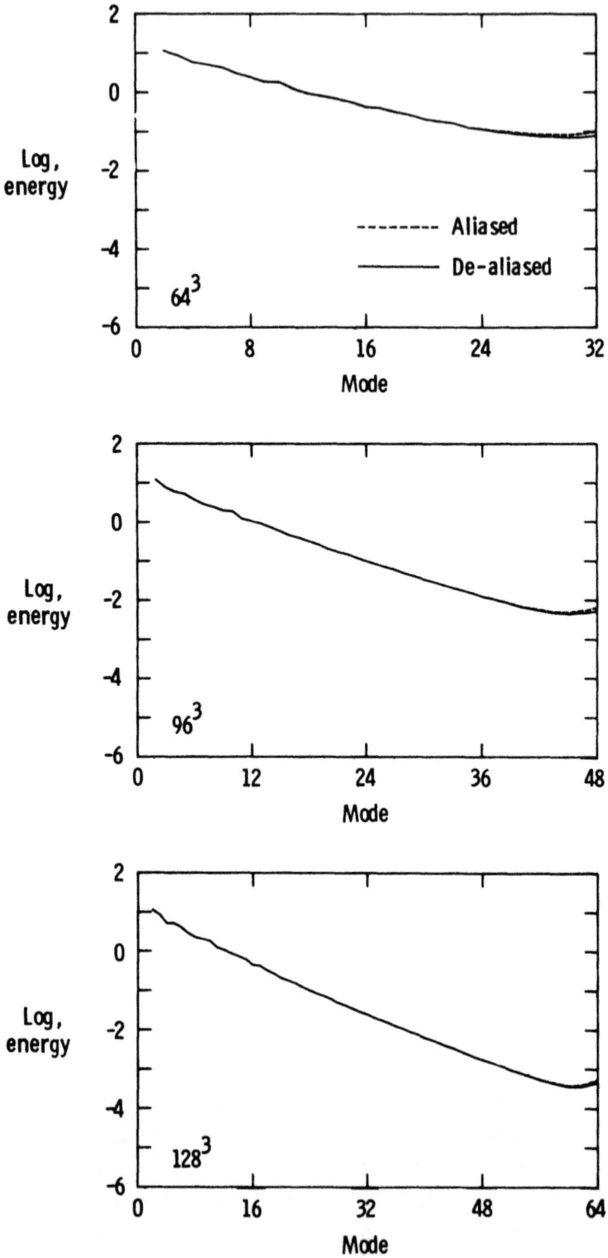

Figure 7.1. Energy spectra for aliased and de-aliased simulations of incompressible, isotropic turbulence on 64^3, 96^3 and 128^3 grids. The energy $\varepsilon(k)$ is plotted as a function of the wavenumber (or mode) k.

Table 7.2. Aliased vs. de-aliased calculations of isotropic turbulence

Quantity	64^3 aliased	64^3 de-aliased	96^3 aliased	96^3 de-aliased	128^3 aliased	128^3 de-aliased
kinetic energy	51.369	51.604	51.144	51.180	51.079	51.083
dissipation rate	238.94	233.74	230.27	229.86	231.20	231.17
Taylor micro-scale Reynolds number	40.657	40.944	37.150	37.160	38.789	38.787
skewness	−.40005	−.42289	−.48952	−.49192	−.52202	−.52220
flatness	3.5305	3.5069	3.7839	3.7760	3.9045	3.9030

(The specific case chosen for these examples was the one used by Clark, Ferziger and Reynolds (1979) in their 64^3, fourth-order finite-difference simulations. The present simulations were run for half of the physical time considered by Clark et al.) Calculations were performed on 64^3, 96^3 and 128^3 grids, with an isotropic truncation applied to modes for which $|\mathbf{k}|^2 \geqslant (N/2)^2$. De-aliasing was implemented via the 3/2-rule. The 64^3 runs suffer from inadequate resolution—the initial state captures only 81% of the dissipation. The 96^3 and 128^3 runs capture 94% and 97% of it, respectively.

Figure 7.1 displays the three-dimensional energy spectra at the end of the simulations and Table 7.2 summarizes the final turbulent states. Since the initial conditions for turbulence simulations are based on random initial fields, subject to the divergence-free condition and a desired energy spectrum, there will be statistical fluctuations (in the 1%–10% range) between turbulence quantities measured from distinct realizations. The realizations for the 64^3, 96^3 and 128^3 cases are different. Thus, one should not expect these three cases to exhibit convergence with grid refinement in the same sense as a deterministic problem does. Of course, on a given grid the same realization was used for the aliased and de-aliased cases.

The differences between the aliased and de-aliased runs diminish rapidly as the resolution increases. For the 128^3 case they occur in the fourth digit and are not perceptible on the plot of the energy spectra. The aliasing errors are more substantial for the 64^3 case, but they are obviously much smaller than the truncation errors. In Sec. 11.3 we prove that aliased and de-aliased algorithms for the corresponding steady problem have the same asymptotic rate of error decay. In our opinion, accuracy considerations do not yield a compelling reason to eliminate the aliasing errors.

The importance of the rotation form in collocation methods has been checked by us. We were able to obtain a stable solution with the standard form (7.2.1) of the momentum equation on a 64^3 grid under the conditions used in the previous example. However, if the Reynolds number were only twice as large, the calculation was unstable. In contrast, the collocation solutions based on the rotation form were stable for arbitrarily large Reynolds numbers. The non-rotation form calculation was only stable at low enough Reynolds numbers for viscous damping to be strong enough to counteract the numerical instability of the advection terms.

Very recent results (Zang 1991) indicate that a skew-symmetric form for the nonlinear terms [see (4.5.13) and (4.5.16) with $\mathbf{a}$ and u replaced by $\mathbf{u}$] is superior to the rotation form; it is more accurate and also semi-conserves kinetic energy.

7.3. Inhomogeneous Flows

To date, the inhomogeneous flows which have been fruitfully investigated by three-dimensional spectral solutions of the incompressible Navier–Stokes equations include channel flow, pipe flow, the parallel boundary layer, Taylor–Couette flow, the Rayleigh–Benard problem, and free shear layers. In each of these applications the boundary conditions are periodic in two directions, and non-periodic in one. In all but the last case, no-slip conditions are required at one or more walls. Problems with two or more non-periodic directions are more complex and will be addressed in Sec. 7.4.

The channel flow problem illustrates the salient features of all the wall-bounded algorithms cited above, even in a two-dimensional setting, and this is used here to describe the different spectral numerical methods. Then, the unique features of spectral algorithms for the other problems are discussed.

We consider plane Poiseuille flow (Fig. 1.3) within the framework of the two-dimensional Navier–Stokes equations

$$\frac{\partial \mathbf{u}}{\partial t} + \boldsymbol{\omega} \times \mathbf{u} = -\nabla P + \nabla \cdot [\nu(\nabla \mathbf{u} + \nabla \mathbf{u}^T)] \qquad \text{in } \Omega \tag{7.3.1}$$

$$\nabla \cdot \mathbf{u} = 0 \qquad \text{in } \Omega \tag{7.3.2}$$

$$\mathbf{u} = 0 \qquad \text{on } y = \pm 1, \tag{7.3.3}$$

where $\mathbf{u}$ is the velocity vector with components u and v in the x and y directions. The domain Ω is $(0, L) \times (-1, 1)$. In the laminar case, the kinematic viscosity ν may depend upon temperature. In some simple models of turbulent flows, ν represents the combination of the kinematic viscosity and an eddy viscosity. Consequently, ν may be a function of both position and time. Periodic boundary conditions in x and no-slip boundary conditions at $y = \pm 1$ are assumed. A Fourier representation in x and a Chebyshev representation in y are customary. Note that the momentum equation (7.3.1) is again used in rotation form.

Three classes of spectral methods will be discussed here in detail: (i) coupled methods in which a single system combining the continuity equation with the implicit contributions of the pressure gradient and viscous terms of the momentum equations is inverted at each time-step; (ii) splitting methods which decouple the viscous and non-linear terms from the pressure gradient and the divergence-free constraint; and (iii) Galerkin methods with velocity trial functions which are divergence-free in addition to complying with the appropriate boundary conditions. All of the algorithms described in the next three subsections have been applied to three-dimensional flows. They use a Fourier representation in z. We are focusing on the two-dimensional version solely to simplify the discussion.

Since the pressure adjusts itself instantaneously to changes in the velocity according to the divergence-free constraint, an implicit treatment of the pressure gradient term is imperative. (In fact, since the pressure acts as an advection term with infinite wave speed, its time-discretization must be A-stable. This limits one to backward Euler, Crank–Nicolson or an A-stable θ-method (see Chap. 4).) An explicit time-discretization of the viscous term is possible, although rarely practical for realistic viscosities and grids. The explicit viscous stability limit scales as $1/N_x^2$ in x and $1/N_y^4$ in y (see Chap. 4), where N_x and N_y are the number of modes in the x and y directions, respectively. The latter restriction is the most severe and virtually mandates an implicit treatment of the normal diffusion term. The principal algorithmic complexity for these problems arises from the need to invert the operators which represent the implicit treatment of the pressure and normal diffusion terms. In general, a direct solution method is impractical in terms of total storage and computational time. Thus, iterative techniques such as those discussed in Chap. 5 must be employed for the solution of the implicit equations. In the description of the three classes of spectral methods which follows we shall discuss both general iterative techniques as well as some efficient direct methods which can be applied in special (and very useful) cases.

7.3.1. Coupled Methods

The most general iterative method yet proposed was described by Malik, Zang and Hussaini (1985) and by Zang and Hussaini (1985a). This section will furnish a detailed description of this method and then will discuss some coupled, direct solution methods which are quite efficient in some important special cases.

The spectral collocation method is used for spatial discretization of (7.3.1) and (7.3.2). The collocation points in the x-direction are

$$x_j = jL/N_x, \qquad j = 0, 1, \ldots, N_x - 1, \tag{7.3.4}$$

where L is the periodicity length in the streamwise direction and N_x is the number of intervals in the x-direction. The dependent variables have discrete Fourier representations of the form

$$\mathbf{u}(x, y, t) = \sum_{k=-N_x/2}^{N_x/2-1} \hat{\mathbf{u}}_k(y, t) e^{2\pi i k x/L}. \tag{7.3.5}$$

Figure 7.2. The staggered grid for the normal (y) direction in channel flow.

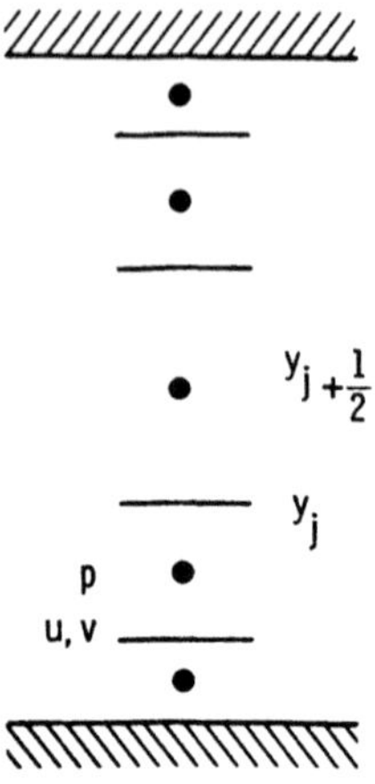

We will use the same symbols to denote the continuous variables **u** and P, and their discrete physical space approximations. The collocation points in the y-direction are

$$y_j = \cos\frac{\pi j}{N_y}, \qquad j = 0, 1, \ldots, N_y \tag{7.3.6}$$

$$y_{j+1/2} = \cos\frac{\pi(j+\frac{1}{2})}{N_y}, \qquad j = 0, 1, \ldots, N_y - 1, \tag{7.3.7}$$

for the velocity and the pressure respectively. They are illustrated in Fig. 7.2. The use of Chebyshev interpolation to transfer variables between these two staggered grids is described in Sec. 5.6.

In the y-direction, the velocity has the discrete Chebyshev polynomial representation

$$\hat{\mathbf{u}}_k(y,t) = \sum_{m=0}^{N_y} \tilde{\mathbf{u}}_{k,m}(t) T_m(y), \tag{7.3.8}$$

and the pressure representation is of the form

$$\hat{P}_k(y,t) = \sum_{m=0}^{N_y-1} \tilde{P}_{k,m}(t) T_m(y). \tag{7.3.9}$$

(This section is an exception to our usual practice of using ^'s for exact coefficients and ~'s for discrete ones.) The continuity equation is enforced at the half-points given by (7.3.7). Note that the maximum order of the trial functions for the pressure is one less than that for the velocity (see Sec. 11.3). If the same collocation points are used for the pressure and the continuity equation as for the velocity and momentum equation, then the linear system for the discrete dependent variables is underdetermined. Hence, some extra condition consistent with the incompressibility constraint must be imposed at the wall.

Moreover, the pressure mode $\tilde{P}_{0,N_y}(t)T_{N_y}(y)$ has no effect upon the velocity. The reason for this is that the pressure affects the velocity only through the momentum equation. At the interior Gauss–Lobatto points (2.4.14) the gradient of this pressure mode vanishes. For this reason it is called a spurious (or parasitic) mode. The mode $\tilde{P}_{0,0}(t)$ also has no effect upon the velocity. Since it represents the mean value of the pressure and has a counterpart in the continuous formulation of the problem, it is not usually called spurious. The issue of spurious pressure modes is addressed again in Sec. 7.4 for problems with more than one non-periodic direction and also in Sec. 11.3 from the theoretical point of view.

Various time-discretizations are possible (see Chap. 4). A simple one employs Crank–Nicolson on the implicit terms and second-order Adams–Bashforth on the remaining terms. In many cases, the x-dependence of the viscosity is much weaker than its y-dependence. Thus, a semi-implicit treatment of the normal diffusion term (in which only the y- and t-dependence of the viscosity is included) still overcomes the severe stability restriction arising from this term. After a discrete Fourier transform in x (with $L = 2\pi$ and the subscript k suppressed), but with y still continuous, (7.3.1) and (7.3.2) become

$$\frac{\partial}{\partial y}\left(b\frac{\partial \hat{\mathbf{u}}^{n+1}}{\partial y}\right) - (1+\hat{k}^2)\hat{\mathbf{u}}^{n+1} - c\hat{\nabla}\hat{p}^{n+1} = -\hat{\mathbf{S}}^n \tag{7.3.10}$$

$$\hat{\nabla}\cdot\hat{\mathbf{u}}^{n+1} = 0, \tag{7.3.11}$$

and the boundary conditions reduce to

$$\hat{\mathbf{u}}^{n+1}(\pm 1) = 0, \tag{7.3.12}$$

where

$$\begin{cases} \hat{\mathbf{u}}^{n+1} = (\hat{u}^{n+1}, \hat{v}^{n+1}) \\ b = \dfrac{\Delta t}{2}\nu_{\text{avg}}(y,t) \\ c = \dfrac{\Delta t}{2} \\ \hat{k}^2 = \dfrac{1}{2}\Delta t k^2 \nu_{\text{avg}}(y,t). \end{cases}$$

The symbol $\nu_{\text{avg}}(y,t)$ denotes the average value of $\nu(x,y,t)$ at fixed y and t, and k is the wavenumber. The superscript $n+1$ represents the time-level, the x-discrete, y-continuous operators are

$$\begin{cases} \hat{\nabla}\hat{p}^{n+1} = \left(ik\hat{p}^{n+1}, \dfrac{\partial \hat{p}^{n+1}}{\partial y}\right) & (7.3.13) \\ \hat{\nabla}\cdot\hat{\mathbf{u}}^{n+1} = ik\hat{u}^{n+1} + \dfrac{\partial \hat{v}^{n+1}}{\partial y}, & (7.3.14) \end{cases}$$

and

$$\hat{\mathbf{S}}^n = \hat{\mathbf{u}}^n + \frac{\Delta t}{2}(3\hat{\mathbf{H}}^n - \hat{\mathbf{H}}^{n-1}) - c\hat{\nabla}\hat{P}^n + \frac{\partial}{\partial y}\left(b\frac{\partial \hat{\mathbf{u}}^n}{\partial y}\right) - \hat{k}^2\hat{\mathbf{u}}^n \tag{7.3.15}$$

$$\mathbf{H} = \mathbf{u} \times \boldsymbol{\omega} + \nabla\cdot[(v - v_{\text{avg}})(\nabla u + \nabla u^T)] + \frac{\partial v_{\text{avg}}}{\partial y}\nabla v - \mathbf{e}_1 P_{x,\text{mean}}. \tag{7.3.16}$$

The last term in (7.3.16) is the mean streamwise pressure gradient term. After the y-derivatives are discretized by spectral collocation (Secs. 2.1.3 and 2.4.2) the system (7.3.10)—(7.3.12) reduces to (for each wave number k)

$$LU = F, \tag{7.3.17}$$

where $U = [\hat{u}^{n+1}, \hat{v}^{n+1}, \hat{P}^{n+1}]$ and F is the known right-hand side. The matrix L is a full $M \times M$ matrix where $M \cong 3N$. A direct solution of (7.3.17) by Gauss elimination methods would require $O(M^2)$ storage and $O(M^3)$ arithmetic operations. An iterative solution, on the other hand, requires only $O(M)$ storage and $O(M \log_2 M)$ operations per iteration. A detailed description of iterative methods for solving (7.3.17) is furnished in Sec. 5.6.

For the constant-viscosity case, (7.3.10) to (7.3.12) reduce to

$$v\hat{\mathbf{u}}'' - \lambda\hat{\mathbf{u}} - \hat{\nabla}\hat{P} = -\hat{\mathbf{R}} \tag{7.3.18}$$

$$\hat{\nabla}\cdot\hat{\mathbf{u}} = 0, \tag{7.3.19}$$

$$\hat{\mathbf{u}}(\pm 1) = 0, \tag{7.3.20}$$

where $\lambda = 2/(\Delta t) + vk^2$, $\hat{\mathbf{R}} = (2/\Delta t)\hat{\mathbf{S}}$, primes denote derivatives with respect to y, and the superscripts denoting time-level are omitted for convenience. Moin and Kim (1980) and Kleiser and Schumann (1980) have devised particular methods for solving these equations. Moin and Kim use a Chebyshev tau discretization for the y-derivative terms in (7.3.18) and (7.3.19). However, they formulate the tau equations for (7.3.18)–(7.3.19) unconventionally—they truncate the equations in Chebyshev space after applying the recursion formula (2.4.26) to obtain a quasi-tridiagonal form. The truncation should be performed first—see Sec. 5.1.2. The resulting system of equations is nearly block-tridiagonal in either case. Moin and Kim used a direct inversion.

Kleiser and Schumann obtain the solution to the system (7.3.18)—(7.3.20) by solving a sequence of one-dimensional scalar Helmholtz equations. Let us first consider the continuous (in y) version of the problem (7.3.18)—(7.3.20). Taking the divergence of (7.3.18) yields the equation for pressure

$$\hat{P}'' - k^2\hat{P} = \hat{\nabla}\cdot\hat{\mathbf{R}}, \tag{7.3.21}$$

and the boundary condition is

$$\hat{\nabla}\cdot\hat{\mathbf{u}}(\pm 1) = 0, \qquad \text{i.e., } \hat{v}'(\pm 1) = 0. \tag{7.3.22}$$

The equation for $\hat{v}$ is

$$\nu\hat{v}'' - \lambda\hat{v} - \hat{P}' = -\hat{R}_y \qquad \hat{v}(\pm 1) = 0, \tag{7.3.23}$$

and the equation for $\hat{u}$ reads

$$\nu\hat{u}'' - \lambda\hat{u} - ik\hat{P} = -\hat{R}_x \qquad \hat{u}(\pm 1) = 0. \tag{7.3.24}$$

Equations (7.3.21)–(7.3.23) form a complete set for $\hat{v}$ and $\hat{P}$. Let us call it the "A-Problem." To solve this, we consider the inhomogeneous "B-problem":

$$\hat{P}'' - k^2\hat{P} = \hat{\nabla}\cdot\hat{\mathbf{R}} \qquad \hat{P}(\pm 1) = \hat{P}_{b\pm} \tag{7.3.25}$$

$$\nu\hat{v}'' - \lambda\hat{v} - \hat{P}' = -\hat{R}_y \qquad \hat{v}(\pm 1) = 0. \tag{7.3.26}$$

The pressure $\hat{P}_{b\pm}$ at the walls is unknown a priori, but it is required to be consistent with the conditions $\hat{v}'(\pm 1) = 0$. Let $(\hat{P}_p, \hat{v}_p)$ be the solution of (7.3.25)—(7.3.26) but with homogeneous Dirichlet boundary conditions on $\hat{P}$. Let $(\hat{P}_+, \hat{v}_+)$ and $(\hat{P}_-, \hat{v}_-)$ be the solutions of the homogeneous B-Problems, i.e., (7.3.25)—(7.3.26) with zero on the right-hand sides of the differential equations, with boundary conditions $\hat{P}_+(-1)=0$, $\hat{P}_+(+1)=1$, and $\hat{P}_-(-1)=1$, $\hat{P}_-(+1) = 0$, respectively. Write the solution of the A-Problem as

$$\begin{pmatrix}\hat{P}\\ \hat{v}\end{pmatrix} = \begin{pmatrix}\hat{P}_p\\ \hat{v}_p\end{pmatrix} + \delta_+\begin{pmatrix}\hat{P}_+\\ \hat{v}_+\end{pmatrix} + \delta_-\begin{pmatrix}\hat{P}_-\\ \hat{v}_-\end{pmatrix}. \tag{7.3.27}$$

The boundary conditions of the A-Problem require

$$\begin{pmatrix}\hat{v}'_+(+1) & \hat{v}'_-(+1)\\ \hat{v}'_+(-1) & \hat{v}'_-(-1)\end{pmatrix}\begin{pmatrix}\delta_+\\ \delta_-\end{pmatrix} = -\begin{pmatrix}\hat{v}'_p(+1)\\ \hat{v}'_p(-1)\end{pmatrix}. \tag{7.3.28}$$

This determines δ_+ and δ_-, and hence the correct pressure boundary condition

$$\hat{P}(\pm 1) = \delta_\pm. \tag{7.3.29}$$

The 2×2 coefficient matrix in (7.3.28) is called the *influence-matrix*, and is calculated once initially for each k and stored. In summary, for each time-step the B-Problem is solved with homogeneous pressure boundary conditions, the correct pressure boundary conditions are derived, and then the B-Problem is solved with the correct boundary condition, thus yielding the solution to the A-Problem. Having now the final $\hat{P}$ and $\hat{v}$, then $\hat{u}$ can be obtained either from (7.3.24) or else from the continuity equation.

The discretization in y may be handled by either tau or collocation methods. Regardless of the choice, one must keep in mind that the derivation of (7.3.25) was based on the properties of the continuous differentiation operators and that discretization effects introduce additional terms into its right-hand side.

The tau approximation to (7.3.18)—(7.3.20) can be written

$$\nu\tilde{u}_m^{(2)} - \lambda\tilde{u}_m - ik\tilde{P}_m = -\tilde{R}_{x,m} - \tilde{\tau}_{x,m} \qquad m = 0, \ldots, N \tag{7.3.30}$$

$$\hat{u}(\pm 1) = 0 \tag{7.3.31}$$

$$\nu\tilde{v}_m^{(2)} - \lambda\tilde{v}_m - \tilde{P}_m^{(1)} = -\tilde{R}_{y,m} - \tilde{\tau}_{y,m} \qquad m = 0, \ldots, N \tag{7.3.32}$$

$$\hat{v}(\pm 1) = 0 \tag{7.3.33}$$

$$\tilde{d}_m \equiv ik\tilde{u}_m + \tilde{v}_m^{(1)} = 0 \qquad m = 0, \ldots, N. \tag{7.3.34}$$

The tau terms $\tilde{\tau}_{x,m}$ and $\tilde{\tau}_{y,m}$ vanish for $0 \leqslant m \leqslant N-2$. The application of the discrete divergence to (7.3.30) and (7.3.32) yields

$$\nu\tilde{d}_m^{(2)} - \lambda\tilde{d}_m - \tilde{P}_m^{(2)} + k^2\tilde{P}_m = -\tilde{r}_m - (ik\tilde{\tau}_{x,m} + \tilde{\tau}_{y,m}^{(1)}) \qquad m = 0, \ldots, N, \tag{7.3.35}$$

where

$$\tilde{r}_m = ik\tilde{R}_{x,m} + \tilde{R}_{y,m}^{(1)} \qquad m = 0, \ldots, N. \tag{7.3.36}$$

But, (7.3.34) is equivalent to

$$\begin{aligned} \tilde{d}_m &= 0 \qquad m = 0, \ldots, N-2 \\ \hat{d}(\pm 1) &= 0. \end{aligned} \tag{7.3.37}$$

Hence, from (7.3.35) the discrete A-problem is

$$\begin{aligned} \tilde{P}_m^{(2)} - k^2\tilde{P}_m &= \tilde{r}_m + \tilde{\sigma}_m^{(1)} \qquad m = 0, \ldots, N-2 \\ \hat{v}'(\pm 1) &= 0 \\ \nu\tilde{v}_m^{(2)} - \lambda\tilde{v}_m - \tilde{P}_m^{(1)} &= -\tilde{R}_{y,m} - \tilde{\sigma}_m \qquad m = 0, \ldots, N \\ \hat{v}(\pm 1) &= 0, \end{aligned} \tag{7.3.38}$$

and the discrete B-problem is

$$\begin{aligned} \tilde{P}_m^{(2)} - k^2\tilde{P}_m &= \tilde{r}_m + \tilde{\sigma}_m^{(1)} \qquad m = 0, \ldots, N-2 \\ \hat{P}(\pm 1) &= \hat{P}_\pm \\ \nu\tilde{v}_m^{(2)} - \lambda\tilde{v}_m - \tilde{P}_m^{(1)} &= -\tilde{R}_{y,m} - \tilde{\sigma}_m \qquad m = 0, \ldots, N \\ \hat{v}(\pm 1) &= 0, \end{aligned} \tag{7.3.39}$$

where we use $\tilde{\sigma}_m = \tilde{\tau}_{y,m}$. If not for the "tau correction" embodied by the $\tilde{\sigma}_n$ and $\tilde{\sigma}_n^{(1)}$ terms, the influence-matrix solution procedure would be a straightforward application of the Helmholtz equation techniques discussed in Sec. 5.1.2. Kleiser and Schumann (1980) describe how to solve the discrete B-problem. Define the $\bar{B}_1$-problem by

$$\begin{aligned} \tilde{P}_m^{(2)} - k^2\tilde{P}_m &= \tilde{r}_m \qquad m = 0, \ldots, N-2 \\ \hat{P}(\pm 1) &= \hat{P}_{b\pm} \\ \nu\tilde{v}_m^{(2)} - \lambda\tilde{v}_m - \tilde{P}_m^{(1)} &= -\tilde{R}_{y,m} \qquad m = 0, \ldots, N-2 \\ \hat{v}(\pm 1) &= 0, \end{aligned} \tag{7.3.40}$$

and the $\bar{B}_0$-problem by

$$\begin{aligned}
\tilde{P}_m^{(2)} - k^2\tilde{P}_m &= \frac{2}{\bar{c}_m} m' \qquad m = 0, \ldots, N-2 \\
\hat{P}(\pm 1) &= 0 \\
\nu\tilde{v}_m^{(2)} - \lambda\tilde{v}_m &= \tilde{P}_m^{(1)} \qquad m = 0, \ldots, N-2 \\
\hat{v}(\pm 1) &= 0,
\end{aligned} \tag{7.3.41}$$

where

$$m' = \begin{cases} N-1 & m \text{ even} \\ N & m \text{ odd} \end{cases} \tag{7.3.42}$$

(assuming N is even). Furthermore, define $\tilde{\sigma}_{1_m}$ and $\tilde{\sigma}_{0_m}$ for $m = N-1, N$ as the tau terms that must be added to the v-momentum equations in (7.3.40) and (7.3.41), respectively, for them to hold for $m = N-1, N$. One can show that

$$\tilde{\sigma}_m = \tilde{\sigma}_{1_m}/(1 - \tilde{\sigma}_{0_m}) \qquad m = N-1, N, \tag{7.3.43}$$

and that

$$\begin{aligned}
\tilde{P}_m &= \tilde{P}_{1_m} + \tilde{\sigma}_{m'}\tilde{P}_{0_m} \\
\tilde{v}_m &= \tilde{v}_{1_m} + \tilde{\sigma}_{(m+1)'}\tilde{v}_{0_m}.
\end{aligned} \qquad m = 0, \ldots, N \tag{7.3.44}$$

The solution to the original B-problem is achieved by:

(1) Solving (7.3.41) for $\tilde{P}_0$, $\tilde{v}_0$ and evaluating $\tilde{\sigma}_{0_m}$ for $m = N-1, N$ from the v-momentum equation of (7.3.41).
(2) Solving (7.3.40) for $\tilde{P}_1$, $\tilde{v}_1$ and evaluating $\tilde{\sigma}_{1_m}$ for $m = N-1, N$ from the v-momentum equation of (7.3.40).
(3) Determining $\tilde{\sigma}_m$ from (7.3.43) and $\tilde{\sigma}_m^{(1)}$ from the standard recurrence relation (2.4.25).
(4) Determining $\tilde{P}$, $\tilde{v}$ from (7.3.44).

Step (1) is redundant for the second B-problem in the influence method calculation. Hence, for each wavenumber k the tau solution to the A-problem can be found at the cost of five complex Helmholtz equation solutions. (This can be reduced to four if one wishes to store $\tilde{P}_0$ and $\tilde{v}_0$.) To this cost must be added the cost of solving for $\tilde{u}$ from either (7.3.30) and (7.3.31) or (7.3.34). The cost is negligible in the latter case.

If the tau correction is simply ignored, then the computed solution will not satisfy all of (7.3.30)—(7.3.34). If (7.3.30)—(7.3.31) is used to determine $\tilde{u}$, then the solution will have a non-zero divergence in the interior. If (7.3.34) is used instead, then the momentum equation will not be satisfied, and the numerical experience is that catastrophic numerical instability occurs (Kleiser (1986, private communication)).

An important question is whether one need bother with the tau correction. Kleiser and Schumann (1980) estimate the magnitude of the error that would be made in the interior divergence as

$$\frac{N}{\nu \Delta t}\tilde{u}_m \qquad m = N-1, N. \tag{7.3.45}$$

Since $\tilde{u}_{N-1}$ and $\tilde{u}_N$ decrease rapidly with N, this error should be small. A rigorous proof of the convergence of this method (for the steady problem) is available and is summarized in Sec. 11.3.3. The theory states that spectral accuracy is achieved for both the corrected and uncorrected versions. However, Kleiser (1986, private communication) reports that lower stability limits on the time-step arose in some of his time-dependent numerical experiments when the tau correction was ignored.

Collocation approximations have similar discrete effects. The basic discrete equations are, with D_N denoting the Chebyshev collocation derivative operator, and with $\hat{\mathbf{u}}$, $\hat{P}$, etc. now denoting fully discrete variables:

$$\nu D_N^2 \hat{u} - \lambda \hat{u} - ik\hat{P}|_{y=y_j} = -\hat{R}_x - \hat{b}_x|_{y=y_j} \qquad j = 0, \ldots, N \tag{7.3.46}$$

$$\hat{u}(\pm 1) = 0 \tag{7.3.47}$$

$$\nu D_N^2 \hat{v} - \lambda \hat{v} - D_N\hat{P}|_{y=y_j} = -\hat{R}_y - \hat{b}_y|_{y=y_j} \qquad j = 0, \ldots, N \tag{7.3.48}$$

$$\hat{v}(\pm 1) = 0 \tag{7.3.49}$$

$$ik\hat{u} + D_N\hat{v}|_{y=y_j} = 0 \qquad j = 0, \ldots, N, \tag{7.3.50}$$

where $\hat{b}_x = \hat{b}_y = 0$ for $j = 1, \ldots, N-1$. The discrete A-problem is

$$\begin{aligned} D_N^2\hat{P} - k^2\hat{P}|_{y=y_j} &= \hat{r} + D_N\hat{\sigma}|_{y=y_j} \qquad j = 1, \ldots, N-1 \\ D_N\hat{v}(\pm 1) &= 0 \\ \nu D_N^2\hat{v} - \lambda\hat{v} - D_N\hat{P}|_{y=y_j} &= -\hat{R}_y - \hat{\sigma}|_{y=y_j} \qquad j = 0, \ldots, N \\ \hat{v}(\pm 1) &= 0, \end{aligned} \tag{7.3.51}$$

and the discrete B-problem is

$$\begin{aligned} D_N^2\hat{P} - k^2\hat{P}|_{y=y_j} &= \hat{r} + D_N\hat{\sigma}|_{y=y_j} \qquad j = 1, \ldots, N-1 \\ \hat{P}(\pm 1) &= \hat{P}_\pm \\ \nu D_N^2\hat{v} - \lambda\hat{v} - D_N\hat{P}|_{y=y_j} &= -\hat{R}_y - \hat{\sigma}|_{y=y_j} \qquad j = 0, \ldots, N \\ \hat{v}(\pm 1) &= 0, \end{aligned} \tag{7.3.52}$$

where

$$\hat{r} = ik\hat{R}_x + D_N\hat{R}_y \qquad j = 0, \ldots, N, \tag{7.3.53}$$

and $\hat{\sigma} = \hat{b}_y$ denotes the boundary correction.

The B-problem may itself be solved by an influence-matrix method. Let

$$\hat{\sigma}_t = \begin{cases} 1 & y = +1 \\ 0 & \text{elsewhere} \end{cases}$$
$$\hat{\sigma}_b = \begin{cases} 1 & y = -1 \\ 0 & \text{elsewhere}' \end{cases} \tag{7.3.54}$$

and let $\hat{P}_0$, $\hat{v}_0$ be the solution to

$$\begin{aligned} D_N^2 \hat{P} - k^2 \hat{P}|_{y=y_j} &= \hat{r}|_{y=y_j} \qquad & j = 1, \ldots, N-1 \\ \hat{P}(\pm 1) &= \hat{P}_\pm \\ \nu D_N^2 \hat{v} - \lambda \hat{v} - D_N \hat{P}|_{y=y_j} &= -\hat{R}_y|_{y=y_j} \qquad & j = 1, \ldots, N-1 \\ \hat{v}(\pm 1) &= 0. \end{aligned} \tag{7.3.55}$$

Let $\hat{P}_t$, $\hat{v}_t$ solve

$$\begin{aligned} D_N^2 \hat{P} - k^2 \hat{P}|_{y=y_j} &= D_N \hat{\sigma}|_{y=y_j} \qquad & j = 1, \ldots, N-1 \\ \hat{P}(\pm 1) &= 0 \\ \nu D_N^2 \hat{v} - \lambda \hat{v} - D_N \hat{P}|_{y=y_j} &= 0 \qquad & j = 1, \ldots, N-1 \\ \hat{v}(\pm 1) &= 0, \end{aligned} \tag{7.3.56}$$

with $\hat{\sigma} = \hat{\sigma}_t$, and let $\hat{P}_b$, $\hat{v}_b$ solve (7.3.56) with $\hat{\sigma} = \hat{\sigma}_b$. The solution to (7.3.52) has the form

$$\begin{aligned} \hat{P} &= \hat{P}_0 + \alpha_t \hat{P}_t + \alpha_b \hat{P}_b \\ \hat{v} &= \hat{v}_0 + \alpha_t \hat{v}_t + \alpha_b \hat{v}. \end{aligned} \tag{7.3.57}$$

The constants α_t, α_b are the solutions to

$$\begin{aligned} &[1 - (\nu D_N^2 \hat{v}_t - \lambda \hat{v}_t - D_N \hat{P}_t)]\alpha_t - (\nu D_N^2 \hat{v}_b - \lambda \hat{v}_b - D_N \hat{P}_b)\alpha_b \\ &\quad = -(\nu D_N^2 \hat{v}_0 - \lambda \hat{v}_0 - D_N \hat{P}_0 + \hat{R}_y) \qquad \text{at } y = +1 \\ &-(\nu D_N^2 \hat{v}_t - \lambda \hat{v}_t - D_N \hat{P}_t)\alpha_t + [1 - (\nu D_N^2 \hat{v}_b - \lambda \hat{v}_b - D_N \hat{P}_b)]\alpha_b \\ &\quad = -(\nu D_N^2 \hat{v}_0 - \lambda \hat{v}_0 - D_N \hat{P}_0 + \hat{R}_y) \qquad \text{at } y = -1, \end{aligned} \tag{7.3.58}$$

which is a simple 2×2 system.

The discrete Helmholtz equations which arise in the *B*-problem may be solved directly by one-dimensional versions of the direct techniques described in Sec. 5.1.3 or else by the iterative methods discussed in Chap. 5.

The solution of the *A*-problem (with the collocation correction included) requires the equivalent of six complex Helmholtz solutions (and only four if $\hat{P}_t$, $\hat{v}_t$, $\hat{P}_b$ and $\hat{v}_b$ are stored).

An influence-matrix technique for two-dimensional channel flow (periodic in x) in the streamfunction-vorticity formulation has been described by Dennis and Quartapelle (1983).

7.3.2. Splitting Methods

An alternative approach to the coupled methods is the splitting technique (or fractional-step scheme) discussed in Yanenko (1971) and in Marchuk (1975). Chorin (1968) and Temam (1968) developed splitting methods for the incompressible Navier–Stokes equations, and more recently, Glowinski (1984) has described several such schemes combined with a finite-element spatial discretization. The splitting scheme proposed by Zang and Hussaini (1986) uses the staggered grid in y given by (7.3.6) and (7.3.7), and consists of an advection-diffusion step

$$\begin{aligned} \frac{\partial \mathbf{u}}{\partial t} &= \mathbf{u} \times \boldsymbol{\omega} + \nabla \cdot [\nu(\nabla \mathbf{u} + \nabla \mathbf{u}^T)] && \text{in } \Omega \\ \mathbf{u} &= \mathbf{g} && \text{at } y = \pm 1 \end{aligned} \tag{7.3.59}$$

followed by a pressure correction

$$\begin{aligned} \frac{\partial \mathbf{u}}{\partial t} &= -\nabla P && \text{in } \Omega \\ \nabla \cdot \mathbf{u} &= 0 && \text{in } \Omega \\ v &= 0 && \text{at } y = \pm 1 \\ \frac{\partial u}{\partial t} &= -\frac{\partial P}{\partial x} && \text{at } y = \pm 1, \end{aligned} \tag{7.3.60}$$

where (7.3.59) is integrated from t_n to the intermediate time $t_{n+1/2}$ with $\mathbf{g}$ denoting the boundary condition on this intermediate step; then (7.3.60) is integrated from $t_{n+1/2}$ to t_{n+1}. The pressure step may be viewed as a projection of the velocity field onto a divergence-free space. Note that in the velocity step all of the velocity components are specified at the boundary, whereas in the pressure step only the normal velocity component is specified on the boundary, and the differential equation itself is used for the tangential velocity component there.

The boundary conditions for each step are determined by the condition that the individual step be well-posed. At the end of a full step the flow will be divergence-free, but there will be a slip velocity at the boundary. If the boundary conditions on the intermediate, velocity step are

$$\mathbf{g}^{n+1/2} = 0, \tag{7.3.61}$$

then the slip velocity at the end of the full step will be $O(\Delta t)$. This can be reduced by resorting to intermediate boundary conditions of the type suggested by Fortin, Peyret and Temam (1971):

$$g_x^{n+1/2} = \Delta t \frac{\partial P^n}{\partial x} \tag{7.3.62a}$$

$$g_y^{n+1/2} = 0. \tag{7.3.62b}$$

This reduces the final slip velocity to $O(\Delta t^2)$. Higher order corrections are also possible, for example,

$$g_x^{n+1/2} = \Delta t \frac{\partial P^n}{\partial x} + (\Delta t)^2 \frac{\partial^2 P^n}{\partial t \partial x} \cong \Delta t \left(2 \frac{\partial P^n}{\partial x} - \frac{\partial P^{n-1}}{\partial x} \right), \tag{7.3.63}$$

which reduces the slip velocity to $O(\Delta t^3)$ when backward Euler is used for the pressure step. As noted by Zang and Hussaini (1986), replacing the right-hand side of (7.3.62b) with $\Delta t\ \partial P^n/\partial y$ leads to explosive numerical instability, whereas replacing it (in the constant viscosity case) with $\Delta t\ \nu\ \partial^2 v^n/\partial y^2$ leads to an accurate, well-behaved solution.

The most general solution technique for the implicit viscous terms in (7.3.59) is the spectral multigrid iterative scheme described in Sec. 5.5. This can handle viscosity which depends on all the spatial coordinates as well as time. If the viscosity is constant in space and time, then this term can be inverted efficiently by Fourier transforming in x and using the Chebyshev tau method in y.

The pressure correction must be performed in a manner consistent with the fully discrete equations. Apply a Fourier transform in x and use Chebyshev collocation in y. Let C_0 and C_+ be the matrices which represent the operations of computing Chebyshev coefficients from function values at the velocity nodes (7.3.6) and the pressure nodes (7.3.7), respectively. (The matrix C_0 represents the operations in (5.6.6) and C_+^{-1} corresponds to (5.6.7).) Let D represent the differentiation operator in terms of the Chebyshev coefficients.

The discrete divergence operator is

$$\mathbb{D}_N \cdot \hat{\mathbf{u}} = (C_+^{-1} C_0)[ik\hat{u} + C_0^{-1} D C_0 \hat{v}]. \tag{7.3.64}$$

(The operator $C_+^{-1}C_0$ represents interpolation from the cell edges to the cell centers.) The discrete gradient operator is

$$\mathbb{G}_N \hat{P} = (C_0^{-1} C_+)(ik\hat{P}, C_+^{-1} D C_+ \hat{P}). \tag{7.3.65}$$

Letting $\hat{Q} = \Delta t \hat{P}$, the fully discrete version of (7.3.60) is

$$\hat{\mathbf{u}}^{n+1} = \hat{\mathbf{u}}^{n+1/2} - \mathbb{G}_N \hat{Q} \qquad \text{at interior cell edges} \tag{7.3.66}$$

$$\left.\begin{aligned} \hat{u}^{n+1} &= \hat{u}^{n+1/2} - (\mathbb{G}_N \hat{Q})_x \\ \hat{v}^{n+1} &= 0 \end{aligned}\right\} \qquad \text{at } y = \pm 1 \tag{7.3.67}$$

$$\mathbb{D}_N \cdot \hat{\mathbf{u}}^{n+1} = 0 \qquad \text{at cell centers.} \tag{7.3.68}$$

Equations (7.3.66) and (7.3.67) combine to yield

$$\hat{\mathbf{u}}^{n+1} = Z_N(\hat{\mathbf{u}}^{n+1/2} - \mathbb{G}_N \hat{Q}), \tag{7.3.69}$$

where Z_N represents the operation of setting the boundary values of the y component to zero. Applying (7.3.68) to (7.3.69) we arrive at

$$\mathscr{L}\hat{Q} = \hat{F}, \tag{7.3.70}$$

with

$$\mathscr{L}\hat{Q} = \mathbb{D}_N \cdot Z_N \mathbb{G}_N \hat{Q} \tag{7.3.71}$$

$$\hat{F} = \mathbb{D}_N \cdot Z_N \hat{\mathbf{u}}^{n+1/2}. \tag{7.3.72}$$

The pressure is computed from (7.3.70) and the velocities are then adjusted via (7.3.66)–(7.3.67). The equation for the pressure is singular for $k = 0$, but this merely reflects the indeterminacy of the pressure to within an additive constant. Note that no pressure boundary condition is included in (7.3.70). The right-hand side, however, contains the desired boundary conditions on the normal velocity. (The variable P which appears in this splitting scheme differs from the true pressure by a quantity of order $\nu\,\Delta t$ (Kim and Moin (1985)).

Since there is no variable-viscosity term in the pressure equation, a simple iterative scheme such as minimum residual (see Sec. 5.3) can be effective, although multigrid methods can be employed here as well.

A different splitting method, developed by Orszag and Kells (1980), has been employed extensively. They used a three-step operator splitting technique for solving the incompressible Navier–Stokes equations for the plane Poiseuille flow problem with constant viscosity. In the subsequent discussion, we follow Marcus (1984a), who has given a clear and detailed discussion of this scheme.

The first fractional-step uses an Adams–Bashforth method to advance the non-linear advection terms

$$\mathbf{u}^{n+1/3} = \mathbf{u}^n + \Delta t(\tfrac{3}{2}\mathbf{u}^n \times \boldsymbol{\omega}^n - \tfrac{1}{2}\mathbf{u}^{n-1} \times \boldsymbol{\omega}^{n-1}). \tag{7.3.73}$$

The second fractional-step applies the pressure correction

$$\hat{\mathbf{u}}^{n+2/3} = \hat{\mathbf{u}}^{n+1/3} - \Delta t \hat{\nabla} \mathbf{P}^{n+1}, \tag{7.3.74}$$

enforcing the divergence-free condition on the intermediate field so that

$$\Delta t \hat{\Delta} P^{n+1} = \hat{\nabla} \cdot \hat{\mathbf{u}}^{n+1/3} \tag{7.3.75}$$

with the no-flux boundary condition

$$\hat{v}^{n+2/3} = 0, \tag{7.3.76}$$

or equivalently

$$\Delta t \frac{\partial \hat{P}^{n+1}}{\partial y} = \hat{v}^{n+2/3}. \tag{7.3.77}$$

The operator $\hat{\Delta}$ is the Laplacian operator in Fourier-physical space (in k and y) and is equivalent to the composition of the operators $\hat{\nabla}\cdot$ and $\hat{\nabla}$ defined by (7.3.14) and (7.3.13).

In the third and final step, the viscous correction is carried out via

$$\hat{\mathbf{u}}^{n+1} = \hat{\mathbf{u}}^{n+2/3} + \frac{\nu \Delta t}{2}(\hat{\Delta}\hat{\mathbf{u}}^{n+2/3} + \hat{\Delta}\hat{\mathbf{u}}^{n+1}) \tag{7.3.78}$$

subject to the no-slip boundary condition on the new velocity field.

The stability of the explicit non-linear step is governed by the Courant–Friedrich–Lewy (CFL) condition. It is sometimes possible to break the advection term into a mean advection term (which is linear with its coefficient dependent on the only inhomogeneous direction) and the remainder. Then the mean advection term can be treated implicitly, thereby relaxing the CFL condition. Note that the first and second steps have inviscid rather than no-slip boundary conditions. This choice is consistent with the character of the operator which occurs in these steps. The use of no-slip boundary conditions here can lead to numerical instability.

The operator-splitting error due to solving (7.3.74) is of $O(\Delta t)$ for the velocity and $O(\Delta t^{1/2})$ for the pressure in the interior of the domain. This error can be reduced using Richardson's extrapolation. The boundary error due to operator-splitting in the normal pressure gradient and normal diffusion terms is of $O(1)$ and cannot be reduced by Richardson's extrapolation. (An asymptotic analysis of the splitting errors has been provided by Orszag, Israeli and Deville (1986). They discuss many alternative intermediate boundary conditions.) These errors appear to cause no serious problem in plane Poiseuille flow calculations. However, Marcus determined the boundary error to cause serious inaccuracies in the Taylor–Couette flow problem. He ascribed the greater sensitivity of the rotating cylinder problem to the fact that its dynamics are driven by the motion of the boundary rather than by a mean pressure gradient.

An obvious way to remove the operator-splitting error is to solve for the pressure with the following correct boundary condition

$$\Delta t \frac{\partial \hat{P}^{n+1}}{\partial n} = \hat{v}^{n+1} + \frac{\nu \Delta t}{2} \hat{\Delta} \hat{v}^{n+1}. \tag{7.3.79}$$

This equation cannot be solved directly as v^{n+1} is an unknown quantity at this stage.

Marcus removed the splitting error by a Green's function (or influence-matrix) technique. He solves the pressure step (7.3.75) with homogeneous Dirichlet boundary conditions on $\hat{P}$ and then applies the viscous step to produce a preliminary velocity $\bar{\mathbf{u}}^{n+1}$. This is corrected via

$$\hat{\mathbf{u}}^{n+1} = \bar{\mathbf{u}}^{n+1} + \alpha \hat{\mathbf{G}}_1 + \beta \hat{\mathbf{G}}_2, \tag{7.3.80}$$

where $\hat{\mathbf{G}}_1$ and $\hat{\mathbf{G}}_2$ are pre-computed Green's functions for the pressure and viscous steps and the constants α and β are determined from the condition (7.3.22). Only two Helmholtz solutions are required per step, but there is a substantial storage requirement for the Green's functions. Marcus uses a collocation discretization in y. He does not produce a collocation solution to (7.3.46)—(7.3.50), since he makes use of the continuous differentiation operators in deriving his Green's function method. Thus, the final solution is not strictly divergence-free.

7.3.3. Galerkin Methods

A class of spectral Galerkin methods for wall-bounded flows based on Jacobi–Fourier polynomials was proposed by Leonard. He suggested the use of velocity trial functions which are divergence-free and also satisfy the viscous boundary conditions. This was first applied to pipe flow by Leonard and Wray (1982). Moser, Moin and Leonard (1983) developed the method for both straight and curved channels. In the spirit of this section we shall present a two-dimensional version of their straight channel algorithm.

First, the velocity field is expanded as

$$\mathbf{u}(x, y, t) = \sum_{j=0}^{N_y} \sum_{k=-N_x/2}^{N_x/2-1} \alpha_{jk} \hat{\mathbf{u}}_j(k, y) e^{2\pi ikx/L}. \tag{7.3.81}$$

The trial functions are vector functions satisfying the divergence-free constraint and the no-slip boundary conditions:

$$\nabla \cdot [\hat{\mathbf{u}}_j(\mathbf{k}, y) e^{2\pi ikx/L}] = 0 \tag{7.3.82}$$

$$\hat{\mathbf{u}}_j(k, y) = 0 \qquad \text{at } y = \pm 1. \tag{7.3.83}$$

The test functions $\boldsymbol{\phi}_{jk}$ are chosen as

$$\boldsymbol{\phi}_{jk}(x, y) = \boldsymbol{\xi}_j(k, y) e^{2\pi ikx/L} \tag{7.3.84}$$

to satisfy the divergence-free constraint and the no-slip condition at the boundaries:

$$\nabla \cdot \boldsymbol{\phi}_{jk} = 0 \tag{7.3.85}$$

$$\boldsymbol{\xi}_j(k, y) = 0 \qquad \text{at } y = \pm 1. \tag{7.3.86}$$

We define an inner product

$$(\boldsymbol{\phi}, \boldsymbol{\psi}) = \int_0^L dx \int_{-1}^1 dy\, \boldsymbol{\phi} \bar{\boldsymbol{\psi}}. \tag{7.3.87}$$

Substituting (7.3.81) into (7.3.1) with constant viscosity, and taking the inner product with the test functions (7.3.84), the following set of equations are obtained:

$$\left(\frac{\partial \mathbf{u}}{\partial t}, \boldsymbol{\phi}_{jk}\right) = -\nu(\Delta \mathbf{u}, \boldsymbol{\phi}_{jk}) + (\mathbf{u} \times \boldsymbol{\omega}, \boldsymbol{\phi}_{jk}). \tag{7.3.88}$$

Notice that the term $(\nabla P, \boldsymbol{\phi}_{jk}) = -(P, \nabla \cdot \boldsymbol{\phi}_{jk}) = 0$, and thus, the pressure is effectively eliminated from the problem. This relationship requires only that $\boldsymbol{\phi}_{jk} \cdot \hat{\mathbf{n}}$ vanish at $y = \pm 1$, and is guaranteed by (7.3.86). In their explanation of this method, Moser et al. state they require only this "inviscid" boundary condition on their test functions. However, the functions they actually do choose (see (7.3.92)–(7.3.95)) satisfy (7.3.86). From a mathematical point of view, the test functions ought to satisfy the full no-slip conditions (Pasquarelli, Quarteroni and Sacchi–Landriani (1987)). Notice that in this discretization the divergence operator is indeed the adjoint of the (negative) gradient operator.

The resulting equations are uncoupled in k. Each set can be written in the compact form

$$A\frac{d\boldsymbol{\alpha}}{dt} = \nu B\boldsymbol{\alpha} + \mathbf{F}, \tag{7.3.89}$$

where A and B are $(N_y + 1) \times (N_y + 1)$ matrices with elements

$$A_{ij} = (\hat{\mathbf{u}}_i, \boldsymbol{\xi}_j) \tag{7.3.90}$$

$$B_{ij} = \left(\frac{d^2\hat{\mathbf{u}}_i}{dy^2} - \left(\frac{2\pi k}{L}\right)^2 \hat{\mathbf{u}}_i, \boldsymbol{\xi}_j\right), \tag{7.3.91}$$

and $\mathbf{F}$ represents a similar contribution from the non-linear term.

The freedom in the choice of the vectors $\hat{\mathbf{u}}_j$ and $\boldsymbol{\xi}_j$ is exercised in favor of those which yield matrices A and B with small bandwidths. A convenient choice is

$$\hat{\mathbf{u}}_j = \begin{pmatrix} if_j' \\ \dfrac{2\pi k}{L} f_j \end{pmatrix} \tag{7.3.92}$$

$$f_j(\pm 1) = f_j'(\pm 1) = 0,$$

and

$$\boldsymbol{\xi}_j = \begin{pmatrix} ig_j' \\ \dfrac{2\pi k}{L} g_j \end{pmatrix} \tag{7.3.93}$$

$$g_j(\pm 1) = 0.$$

(The case $k = 0$ is treated separately.) Equation (7.3.89) is a set of ordinary differential equations which can be solved by any standard explicit or implicit numerical scheme. Note that explicit schemes for the viscous term have no computational advantage unless A is much sparser than B.

A simple choice for the quasi-orthogonal functions f_j and g_j is:

$$f_j(y) = (1 - y^2)^2 T_j(y) \tag{7.3.94}$$

$$g_j(y) = \left(\frac{T_{j+2}(y)}{j(j+1)} - \frac{2T_j(y)}{(j+1)(j-1)} + \frac{T_{j-2}(y)}{j(j-1)}\right) \Big/ 4(1 - y^2)^{1/2}. \tag{7.3.95}$$

Thus, the trial functions are polynomials, whereas the test functions are polynomials divided by $\sqrt{1 - y^2}$. Notice that $f_j'(\pm 1) = g_j'(\pm 1) = 0$. Although the trial and test functions satisfy the same conditions, they are chosen from different spaces of functions. Hence, this method is properly referred to as a Petrov–Galerkin method. The factor $(1 - y^2)^{-1/2}$ is included in the test functions, so that the inner products in (7.3.88) involve integrals of Chebyshev polynomials, low degree polynomials, and the Chebyshev weight. As a result, the test and trial functions are quasi-orthogonal—the inner products in (7.3.88) are non-zero only for a small separation between the order of the test and trial functions.

This Galerkin method may be interpreted in another way. The test functions $\boldsymbol{\phi}_{jk}$ are now taken to be given by (7.3.84), (7.3.93) and (7.3.95), *without* the $(1 - y^2)^{-1/2}$ factor in the last equation, and the inner product (7.3.87) is replaced by one which includes the Chebyshev weight. In the new inner product, $(\nabla P, \boldsymbol{\phi}_{jk})$ still vanishes, because the integration-by-parts operation yields

$$-\int_0^L dx \int_{-1}^1 dy\, P\nabla \cdot (\boldsymbol{\phi}_{jk} w),$$

where $w(y) = (1 - y^2)^{-1/2}$, and the divergence term is identically zero. From this point of view both the trial and test functions are polynomials which satisfy no-slip conditions. However, the trial functions are themselves divergence-free, whereas the test functions are not.

Spalart (1986) applied a related spectral Galerkin method within Leray's (1933) weak formulation of the incompressible Navier–Stokes equations. In this case, (7.3.88) is replaced by

$$\left(\frac{\partial \mathbf{u}}{\partial t}, \boldsymbol{\phi}_{jk}\right) = (\mathbf{u} \times \boldsymbol{\omega}, \boldsymbol{\phi}_{jk}) - \nu(\nabla \mathbf{u}, \nabla \boldsymbol{\phi}_{jk}), \qquad (7.3.96)$$

and the test and trial functions are identical. For channel flow one must sacrifice either the use of a fast transform (in y) or else the banded structure of the matrices A and B. The use of Chebyshev polynomials in the basis functions makes A and B full matrices, whereas Legendre polynomials do not admit a fast transform. Spalart's method was developed, not for the channel, but for the parallel boundary layer (see Sec. 7.3.5).

7.3.4. Other Confined Flows

The algorithms described above can all be extended in a straightforward fashion to the Rayleigh-Benard problem (see Chandrasekhar (1961)) in a channel. The conventional description of this convection problem uses the Boussinesq equations, which consist of the incompressible Navier–Stokes equations, with a term linear in the temperature added to the normal momentum equation, plus an additional equation for the temperature. McLaughlin and Orszag (1982) used the Orszag–Kells splitting method in their three-dimensional simulations of Rayleigh–Benard transition.

The algorithms described above can also be applied in cylindrical coordinates to flow in a curved channel and to flow between concentric, rotating cylinders (Taylor–Couette flow), provided that the flow is assumed to be periodic in the axial direction. The use of cylindrical coordinates introduces geometric factors (inverse powers of r) into the equations. They pose no difficulty for the explicitly treated terms in spectral algorithms. But they may affect the efficiency of the solution of the implicit terms. The coupled method of Zang and Hussaini (1985a) is essentially unaffected by such terms, since the solution scheme is iterative. However, the direct solution methods used in the Moin–Kim and Kleiser–Schumann algorithms become more expensive. In cylindrical geometry the tau method is still applicable, although the bandwidth of the matrix increases and the cost roughly triples. The matrix-diagonalization technique is an attractive alternative to the tau method for the direct solution of the implicit equations. However, it does increase the asymptotic operation count of the entire algorithm from $O(N_x N_y N_z \log_2 N)$ to $O(N_x N_y^2 N_z)$, where $N = \max(N_x, N_y, N_z)$. Likewise, for the operator splitting techniques, the iterative method of Zang and Hussaini (1986) is essentially unaffected by the geometric terms, whereas either a more expensive tau method or a matrix-diagonalization technique is required for the direct schemes. Marcus (1984a, 1984b) and King et al. (1984) used the latter in their simulations of Taylor–Couette flow by the Green's function method.

The practicality of the Galerkin methods depends critically upon finding trial and test functions which produce narrow bandwidths in the implicit equations. Suitable combinations of Chebyshev polynomials were devised for the curved channel problem by Moser, Moin and Leonard (1983). Nevertheless, due to the increased bandwidth, the implicit step of Galerkin curved channel algorithm is three times as expensive as the corresponding plane channel one. This algorithm has been used by Moser and Moin (1987) in an extensive study of turbulent curved channel flow.

Pipe flow algorithms have been devised by Leonard and Wray (1982) and Orszag and Patera (1983). The former used a Galerkin method with a special class of shifted Jacobi polynomials. They handle the coordinate singularity at the origin automatically. Although the implicit equations have small bandwidth, this method has an asymptotic operation count of $O(N_x N_y^2 N_z)$ due to the lack of a fast Jacobi transform for evaluating the non-linear terms. Orszag and Patera (1983) used a splitting method for their simulations of pipe flow transition. They dealt with the coordinate singularity by using expansions such as the one in (3.4.8), which incorporate the radial behavior of each mode. The implicit equations were solved by matrix diagonalization.

The Green's function method has been applied by Marcus and Tuckerman (1987a, 1987b) to transition problems for axisymmetric flow between concentric rotating spheres (spherical Taylor–Couette flow).

7.3.5. Unbounded Flows

Calculations for true boundary-layer flows (see Fig. 1.2) require a method that can handle two inhomogeneous directions. Methods for this problem are discussed in Sec. 7.4. A simpler and yet useful problem with but one inhomogeneous direction is the so-called parallel boundary layer (Fig. 6.8). Here the boundary-layer profile at some interesting streamwise location is singled out. It becomes an exact solution to the steady Navier–Stokes equations with an appropriate forcing term.

As noted above, a true Galerkin method was developed for the parallel boundary layer by Spalart (1986). He used both the even and odd polynomials on [0, 1] (see Sec. 2.5.3) in addition to a special basis function (for each Fourier wavenumber) which is essentially $e^{-|k|y}$. This represents the exponentially-decaying irrotational component. The remaining, rotational part of the flow also decays exponentially as y tends to infinity, but at a much faster rate. Spalart uses this decomposition into rotational and irrotational components to improve the resolution of the rotational component, which is much more confined to a thin boundary layer than the irrotational component. Thus, an exponential mapping yields finite, but high order accuracy for the rotational part of the flow. If the exponential mapping is combined with a special set of Jacobi polynomials, then a small bandwidth results for the implicit terms. However, the evaluation of the explicit terms requires $O(N_x N_y^2 N_z)$ operations since a fast transform is not available.

Both the Galerkin method for the boundary layer and the more conventional collocation algorithms of Orszag and Patera (1983) and Laurien (1986) apply, in effect, zero-perturbation boundary conditions as y tends to infinity. An alternative is to place the top boundary at a finite distance $y_{\max}$ and apply asymptotic boundary conditions there. Malik, Zang and Hussaini (1985) used the boundary conditions

$$\frac{d\mathbf{u}}{dy} = -|k|\mathbf{u}, \tag{7.3.97}$$

which follow from the behavior of the inviscid linearized problem.

For large $y_{\max}$, both asymptotic and zero-perturbation boundary conditions perform comparably, but the asymptotic boundary conditions are clearly better for smaller $y_{\max}$ (Malik et al.).

Zang and Hussaini (1985b) have performed extensive numerical simulations of the parallel boundary layer with the iterative coupled method described for channel flow in Sec. 7.3.1. They also simulated variable-viscosity flows (due to temperature fluctuations) using the flexibility provided by their general iterative solution scheme.

Free shear layer problems are posed on the infinite domain $y = (-\infty, \infty)$ with quiescent boundary conditions at $y = \pm\infty$. In such problems, splitting errors are inconsequential. The free mixing layer has been simulated numerically by Cain, Reynolds and Ferziger (1981) and by Metcalfe et al. (1987) under the assumption of periodicity in x (and z). Cain et al. (1981) used the mapping (2.5.16) in y together with Fourier series in ξ. (The mapping (2.5.15) is inappropriate because the mean flow has different values at $y = \pm\infty$.) The Poisson and Helmholtz equations can be solved in $O(N_x N_y N_z \log_2 N)$ operations by the direct method described in Sec. 5.1.1. Metcalfe et al. have used both algebraic and exponential mappings (see (2.5.17) and (2.5.18)) combined with a Chebyshev polynomial expansion in ξ. They used the matrix-diagonalization technique for the solution of the implicit equations.

A numerical method for a different model of a free mixing layer was used by Riley and Metcalfe (1980) and by Metcalfe et al. (1987). The simulation is conducted on a finite domain in y, say, $y \in (0, \pi)$, with no mapping. Free-slip boundary conditions are applied at $y = 0$ and $y = \pi$. This is achieved by using a cosine expansion in y for u (and w) and a sine expansion for v. Consequently, there is a infinite array of image mixing layers stacked in the y-direction. Curry et al. (1983) have used a similar expansion for a Rayleigh–Benard problem with free-slip boundary conditions.

7.3.6. Aliasing in Transition Calculations

The focus in this section has been on algorithms suitable for simulations of transition to turbulence. The physical problems of interest are characterized by a strong instability. As the flow evolves, vortical structures become increasingly complicated. Hence, the resolution requirements increase with time. In this regard, isotropic turbulence simulations behave in the opposite manner: the small-scale structures become less significant in the course of the calculation.

Krist and Zang (1987) have provided an assessment of the effects of aliasing on such calculations. Figures 7.3 and 7.4 compare the results of four simulations of the same transition problem in channel flow: (a) $32 \times 64 \times 32$ aliased, (b) $64 \times 64 \times 64$ aliased, (c) $96 \times 128 \times 162$ aliased and (d) $32 \times 64 \times 32$ horizontally (x and z) de-aliased. At $t = 15$, only cases (b) and (c) are entirely satisfactory. Non-physical oscillations are clearly visible in case (a) and barely so in case (d). At $t = 18.75$ only case (c) is reliable. The de-aliased run is less troubled by oscillations than the corresponding aliased one. It is, however, more expensive. Since adequate resolution can be achieved by refining the aliased calculation, and de-aliased calculations require refinements as well, there is no fundamental difficulty with a collocation method.

The Reynolds numbers used in typical wall-bounded transition calculations are much larger than those used in homogeneous turbulence studies. The effects of viscosity are felt primarily in the vicinity of the wall and the standard collocation grids are suitably fine there. The use of the standard form of the momentum equation in transition calculations leads to numerical instabilities

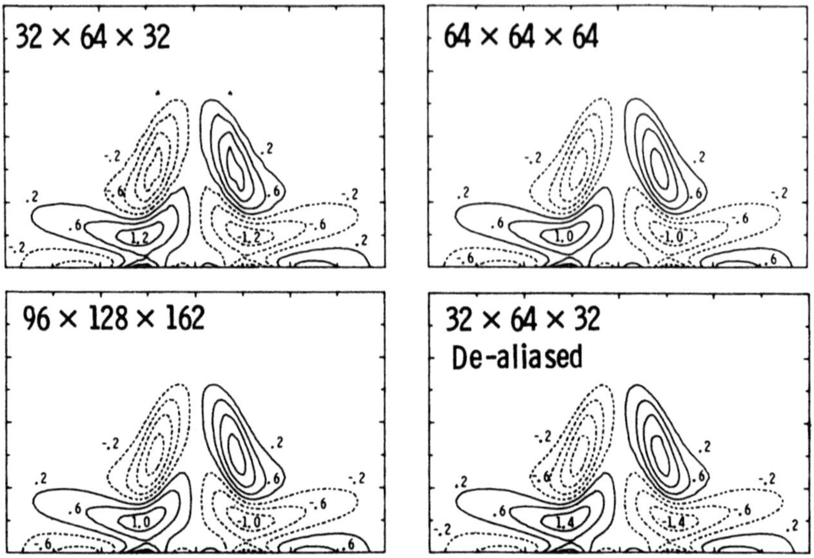

Figure 7.3. Streamwise vorticity in a crucial $y - z$ plane at $t = 15$ for several channel flow simulations.

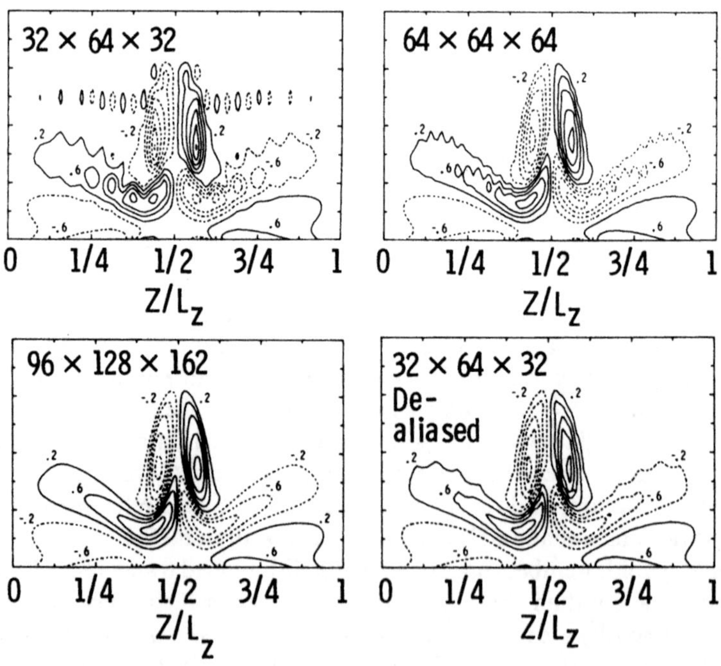

Figure 7.4. Streamwise vorticity in a crucial $y - z$ plane at $t = 18.75$ for several channel flow simulations.

at Reynolds numbers of interest. The use of the rotation form is even more crucial here than for homogeneous turbulence simulations.

7.4. Flows with Multiple Inhomogeneous Directions

Flows with but a single inhomogeneous direction, such as infinite channel flow, the parallel boundary layer, and the classical Taylor–Couette flow between infinite, rotating cylinders are only idealizations of the flows which occur in nature. The streamwise direction in channel and boundary-layer flow is, in fact, inhomogeneous (especially for the boundary layer for which even the mean flow depends on x) and rotating cylinders do have finite length. Spectral algorithms for these problems must have two Chebyshev directions. We focus here on algorithms for flow in a two-dimensional cavity (Fig. 7.5). The extension to three-dimensional problems is straightforward, using the techniques of the previous section if it may justifiably be taken to be periodic, and extending the methods discussed here if it must be treated as inhomogeneous.

The spatially-developing boundary layer (Fig. 7.6) is a problem of considerable physical interest. The semi-infinite domain in y can be handled by mapping, as discussed in Sec. 7.3.5. However, inflow/outflow boundary conditions are required at the two streamwise boundaries. The correct way to impose such boundary conditions has still not been resolved satisfactorily (but, see Fasel (1976) and Patera (1984)). Hence, to illustrate the essentials of the various algorithms we choose the (physically uninteresting) problem of flow in a two-dimensional cavity. The basic equations are again given by (7.3.1) and (7.3.2) on the domain $\Omega = (-1, 1)^2$ with the boundary conditions

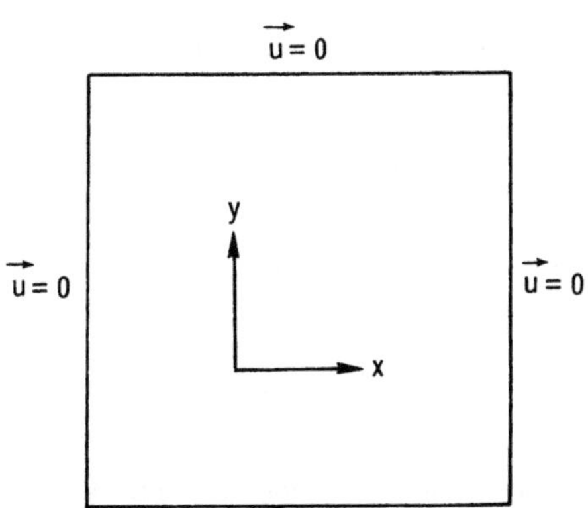

Figure 7.5. Two-dimensional cavity flow geometry.

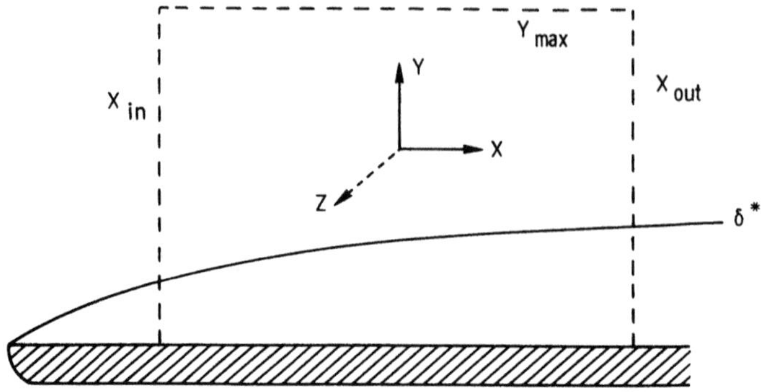

Figure 7.6. Computational domain for a spatially growing boundary layer. There is an inflow boundary at $x = x_{in}$, and an outflow boundary at $x = x_{out}$. Domain truncation in y is applied at $y = y_{max}$.

$$\begin{aligned} \mathbf{u} &= 0 \qquad x = \pm 1 \\ \mathbf{u} &= 0 \qquad y = \pm 1. \end{aligned} \tag{7.4.1}$$

7.4.1. Choice of Mesh

In the case of collocation methods, the first consideration is the precise grid to be employed. Three possibilities are illustrated in Fig. 7.7. The standard, or non-staggered, grid uses the Gauss–Lobatto points (in both x and y) as the nodes for all the variables. The half-staggered grid uses the Gauss–Lobatto points for the velocity and the Gauss points for the pressure. The fully-staggered grid uses different nodes for each primitive variable. The u component of velocity is defined at the Gauss–Lobatto points in x and the Gauss points (plus the boundary points) in y. The v component is handled in reverse fashion. The pressure is defined at the Gauss points in both directions.

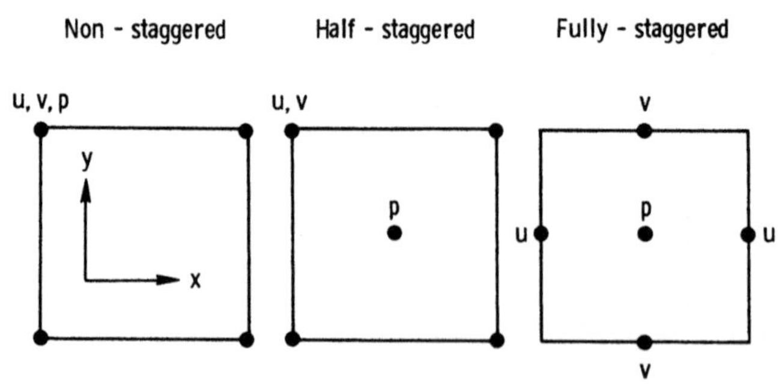

Figure 7.7. Alternative velocity and pressure nodes for flows with two inhomogeneous directions.

The standard grid is the most straightforward one to implement, but it suffers from the presence of seven spurious pressure modes. In a collocation method, spurious modes for the pressure are characterized as those non-constant pressures which have vanishing x-derivatives at all the interior u nodes and vanishing y-derivatives at all the interior v nodes. (See also the discussion at the beginning of Sec. 7.3.1.) Such pressures have no effect upon the velocity, since the interior velocity nodes are where the momentum equations are enforced. The specific spurious modes for the standard grid are the line mode $P = T_{N_x}(x)$, the column mode $P = T_{N_y}(y)$, the checkerboard mode $P = T_{N_x}(x)T_{N_y}(y)$, and four corner modes $P = T'_{N_y}(x)(1 \pm x) \cdot T'_{N_y}(y)(1 \pm y)$. A corner mode is one which vanishes at all the nodes (including those on the boundary) save at one of the corners. Its gradient vanishes at all the interior nodes and on the nodes on the two sides which do not form the corner. In a practical sense, since these modes do not affect the velocity, all that needs to be done with them is to filter them from the pressure before making use of the computed solution. Huberson and Morchoisne (1983) discuss some filtering techniques. Theoretical studies of these algorithms must also account properly for the spurious modes. See Sec. 11.3 for more details.

The implementation of the half-staggered grid is only slightly more involved. The same type of interpolation used in the Zang–Hussaini algorithm discussed in Sec. 7.3.2 applies here. This discretization contains a single spurious mode: $P = T'_{N_x}(x)T'_{N_y}(y)$. At the interior velocity nodes this has the form

$$P(x_j, y_k) = \frac{(-1)^{j+k}}{\sin\left(\dfrac{(2j+1)\pi}{2N_x}\right)\sin\left(\dfrac{(2k+1)\pi}{2N_y}\right)}, \tag{7.4.2}$$

which is close to a pure checkerboard pattern. In addition to possessing fewer spurious modes, it also voids the need for a pressure boundary condition. Montigny-Rannou and Morchoisne (1987) have recently described an algorithm which uses the half-staggered grid.

The fully-staggered grid is obviously the most cumbersome one to use. The u velocity, for example, is represented as

$$u(x, y) = \sum_{j=0}^{N_x} \sum_{k=0}^{N_y+1} \tilde{u}_{jk} T_j(x) T_k(y). \tag{7.4.3}$$

This has the standard number of nodes in the x-direction, but, in order to impose $u = 0$ at $y = \pm 1$, two nodes are needed in addition to the standard Gauss nodes. However, it possesses no spurious modes (Bernardi and Maday (1986a)). The calculation of the collocation derivatives of u in the x-direction is straightforward. The computation of collocation derivatives with respect to y should be accomplished by writing $u(x, y) = (1 - y^2)u_b(x, y)$ and using the Leibnitz rule with $u_b(x, y)$ differentiated via collocation. Since $u_b(x, y)$ is a polynomial in y of degree $\leqslant N_y - 1$, its Chebyshev coefficients can be computed exactly from its values at the Gauss points in y. As discussed in

Appendix B, this, as well as the inverse operation, can be performed via the FFT.

As of this writing, essentially all the spectral computations for flows with multiple inhomogeneous directions have used the non-staggered grid and we shall focus on it hereafter.

7.4.2. Coupled Methods

An early coupled method was developed by LeQuere and Alziary de Roquefort (1985). It is an extension to two Chebyshev directions of the Kleiser–Schumann influence-matrix method in the tau formulation. The influence-matrix, taking into account redundancies at the corners, is of size $(2N_x + 2N_y - 5) \times (2N_x + 2N_y - 5)$. (Since there is no Fourier direction, it does not decouple into small 2×2 systems.) The method described in their paper does not employ a tau correction, and so the velocity field in the interior is not strictly divergence-free.

Tuckerman (1988) has recently provided a general formulation of the influence matrix method for spectral discretizations. It encompasses collocation and tau methods in Cartesian, cylindrical and spherical geometries, and it includes the appropriate correction needed to achieve a truly divergence-free velocity field. In the case of the rectangle, the size of the influence matrix is $4(Nx + Ny - 1) \times 4(Nx + Ny - 1)$.

Haldenwang (1984) has developed an iterative scheme for solving the coupled equations without the need for storing an influence-matrix. The primitive variables are written as

$$\begin{aligned} \mathbf{u}(x, y, t) &= \sum_{m=0}^{N_x} \sum_{n=0}^{N_y} \tilde{\mathbf{u}}_{mn} T_m(x) T_n(y) \\ P(x, y, t) &= \sum_{m=0}^{N_x} \sum_{n=0}^{N_y} \tilde{P}_{mn} T_m(x) T_n(y) \end{aligned} \tag{7.4.4}$$

and the standard grid is used. Assuming, as before, Crank–Nicolson for the implicit terms, the basic collocation equations for a coupled method are

$$\nu \mathbb{L}_N \mathbf{u} - \lambda \mathbf{u} - \mathbb{G}_N P = -\mathbf{R} \qquad \begin{aligned} m &= 1, \ldots, N_x - 1 \\ n &= 1, \ldots, N_y - 1 \end{aligned} \tag{7.4.5}$$

$$\mathbb{D}_N \cdot \mathbf{u} = 0 \qquad \begin{aligned} m &= 0, \ldots, N_x \\ n &= 0, \ldots, N_y \end{aligned} \tag{7.4.6}$$

$$\mathbf{u} = 0 \qquad \text{boundary nodes,} \tag{7.4.7}$$

where $\mathbb{G}_N$, $\mathbb{D}_N$ and $\mathbb{L}_N$ represent the two-dimensional discrete gradient, divergence and Laplacian operators and $\lambda = 2/\Delta t$. Haldenwang uses the following iterative scheme to solve these equations

$$\begin{aligned} \mathbb{L}_N P^{(k+1)} &= \mathbb{D}_N \cdot \mathbf{R} - C^{(k)} && \text{in } \Omega \\ \frac{\partial P^{(k+1)}}{\partial n} &= a^{(k)} && \text{on } \partial\Omega \end{aligned} \tag{7.4.8}$$

$$\begin{aligned} \nu \mathbb{L}_N \mathbf{u}^{(k+1)} - \lambda \mathbf{u}^{(k+1)} &= -\mathbb{G}_N P^{(k+1)} - \mathbf{R} && \text{in } \Omega \\ \mathbf{u} \cdot \hat{\boldsymbol{\tau}} &= 0 && \text{on } \partial\Omega \\ \frac{\partial}{\partial n}(\mathbf{u} \cdot \hat{\mathbf{n}}) &= 0 && \text{on } \partial\Omega \end{aligned} \tag{7.4.9}$$

$$a^{(k+1)} = a^{(k)} - (\nu \mathbb{L}_N - \lambda I)|_{\partial\Omega} \mathbf{u}^{(k+1)}, \tag{7.4.10}$$

and $C(\mathbf{u}, P, \mathbf{R})$ is the error (or commutator) which arises from commuting the discrete divergence operator with (7.4.5) to arrive at (7.4.8)—see Haldenwang (1984) for details. The implicit equations in (7.4.8)—(7.4.10) can be solved by the techniques discussed in Chap. 5. The rate of convergence of this iteration clearly depends upon ν, Δt, N_x and N_y. Haldenwang and Labrosse (1986) report that between two and eight iterations are required for free-convection problems on 64^2 and 32^3 grids.

Most of the complexities of coupled methods disappear if the diffusion term is treated explicitly. It then becomes a relatively simple matter to solve for the pressure with zero-divergence boundary conditions. Ku, Taylor and Hirsch (1987) have described the implementation of such a method. But, of course, the explicit stability limit for the method can be quite severe on all but fairly coarse meshes.

7.4.3. Splitting Methods

Splitting methods are, if anything, even more attractive for problems with multiple inhomogeneous directions, because the relative cost of a true coupled method to a splitting method is higher than for a single inhomogeneous direction—the influence-matrix is much larger and several iterations are needed for the Haldenwang algorithm.

Streett and Hussaini (1987) have implemented the two-dimensional generalization of the split method discussed at the beginning of Sec. 7.3.2. They used a non-staggered grid, in part because of difficulties in preconditioning the consistent pressure equation on the half-staggered grid. In addition to the extra spurious modes that this entails, there is the issue of the appropriate pressure boundary condition. They impose $\partial P/\partial n = 0$ on the grounds that this condition is consistent with the zero-normal velocity boundary condition on the pressure step. The third-order intermediate boundary condition (7.3.63) is employed. At the end of a full step the velocity is divergence-free throughout Ω, and typical slip velocities are many orders of magnitude smaller than the interior velocities. The implicit equations can be solved by either matrix-diagonalization or iterative techniques.

7.4.4. Other Methods

Métivet (1987) has devised a splitting method in which the diffusion term is treated explicitly and an implicit finite-difference diffusion operator is added to increase the allowable time-step. This operator is further approximated by an ADI procedure. This finite-difference operator produces a substantial deterioration in the time-accuracy of the method as well as preventing the diffusion term from being unconditionally stable. She formulates the continuity step as a minimization problem and uses conjugate gradients for its solution. This method is useful in problems for which transient phenomena are not as delicate as they are in transition applications and in typical aerodynamic applications for which only the steady-state is of interest. Metivet has also generalized the algorithm to deal with curved domains.

Vanel, Peyret and Bontoux (1985) and Ehrenstein and Peyret (1986) have developed tau and collocation influence-matrix techniques for the streamfunction-vorticity formulation. The algorithmic subtlety here is, that like the pressure in primitive-variable formulations, there is no physical boundary condition for the vorticity.

Ouazzani and Peyret (1984) and Ouazzani, Peyret and Zakaria (1985) have developed spectral versions of the artificial-compressibility method in primitive variables. They also use an ADI scheme for the viscous term and have used both spectral and finite-difference treatments of this term.

Leonard (1984) has developed a set of basis functions suitable for a two-dimensional version of the Galerkin method. Efficient solutions schemes are, however, still lacking.

7.5. Mixed Spectral/Finite-Difference Methods

The focus of this book is on numerical methods which employ spectral discretizations in all coordinate directions. There have, of course, been numerous incompressible Navier–Stokes computations which used mixed spectral/finite-difference methods, i.e., algorithms with spectral discretizations in some directions and finite-differences in the others. The parallel boundary-layer transition calculations of Wray and Hussaini (1984) fall into this category. They used a Fourier spectral method in two periodic directions and second-order finite-differences in the normal direction. A slightly different spectral/finite-difference method was used by Moin and Kim (1982) in their large-eddy simulations of turbulent channel flow and by Biringen (1985) in a study of active control in channel flows. More recently, Eidson, Hussaini and Zang (1986) have used a similar algorithm in a high resolution direct simulation of a turbulent Rayleigh–Benard flow.

CHAPTER 8

Compressible Flow

8.1. Introduction

Spectral methods have been far less common for compressible flow calculations than for incompressible ones. A principal reason is that few compressible flows possess the high degree of regularity in the primitive variables that is ideal for spectral approximations. Incompressible flows possess singularities primarily for geometric reasons, such as sharp edges or corners. Compressible flows, however, are also subject to singularities arising from non-linear wave propagation. An examination of the Euler equations (see Sec. 1.3.2) is instructive. The one-dimensional version of (1.3.12)—(1.3.14) is

$$\begin{aligned}\frac{\partial \rho}{\partial t} + u\frac{\partial \rho}{\partial x} + \rho\frac{\partial u}{\partial x} &= 0\\ \frac{\partial u}{\partial t} + u\frac{\partial u}{\partial x} + \frac{1}{\rho}\frac{\partial p}{\partial x} &= 0\\ \frac{\partial p}{\partial t} + u\frac{\partial p}{\partial x} + \gamma p\frac{\partial u}{\partial x} &= 0,\end{aligned} \tag{8.1.1}$$

or the hyperbolic system

$$\frac{\partial Q}{\partial t} + A\frac{\partial Q}{\partial x} = 0, \tag{8.1.2}$$

where

$$Q = (\rho, u, p) \tag{8.1.3}$$

and

$$A = \begin{pmatrix} u & \rho & 0 \\ 0 & u & 1/\rho \\ 0 & \gamma p & u \end{pmatrix}. \tag{8.1.4}$$

The eigenvalues of A are u, $u + c$ and $u - c$, where the sound speed c is given by (1.3.2). The character of the flow depends upon the relative magnitudes of u and c. The Mach number is the ratio

$$M = \frac{|u|}{c}. \tag{8.1.5}$$

If $M > 1$, then all the eigenvalues have the same sign. The flow is called supersonic and all information propagates downstream. If $M < 1$, then the eigenvalue $u - c$ has the opposite sign from u and $u + c$ (assuming $u > 0$). The flow is called subsonic and information is able to propagate upstream.

Supersonic flow usually adjusts to downstream obstacles by undergoing a discontinuous change referred to as a shock. The Euler equations themselves do not hold right at a shock. The conditions there must be derived from the physical conservation laws which are given by the integral version of (8.1.1). These produce the Rankine–Hugoniot conditions

$$\begin{aligned} [\rho u] &= 0 \\ [\rho u^2 + p] &= 0 \\ [(e + p)u] &= 0. \end{aligned} \tag{8.1.6}$$

The brackets denote the jump across the shock. There is an additional constraint, which is the condition that entropy does not decrease across the shock. (This follows from the fact that the Euler equations are the limit of the compressible Navier–Stokes equations for vanishingly small dissipation.)

Of course, if the flow is subsonic everywhere, or if it is smooth even though supersonic, then there is no a priori reason to be skeptical of applying spectral methods. But spectral methods do appear to be ill-suited to problems with discontinuities in the solution or even in some of its low order derivatives. The standard integration-by-parts estimate of the size of the spectral coefficients of such non-smooth functions implies that they decay in a slow algebraic manner. Left as is, this behavior produces the familiar Gibbs oscillations in the solution together with slow, global convergence of the numerical results to the true solution. Nevertheless, techniques have been developed which enable useful and accurate information to be extracted from spectral solutions to some discontinuous problems.

Our intention in this chapter is to focus on those aspects of spectral methods which are unique to compressible flow and have not been discussed in the preceding chapters. The nature of the non-linearities in the compressible flow equations make collocation methods the only practical ones. In the case of Fourier methods we use the standard collocation points and for non-periodic problems, the Chebyshev Gauss–Lobatto points are chosen. The use of the alternative Gauss and Gauss–Radau points appears to be only practical for special scalar equations and linear systems.

8.2. Boundary Conditions for Hyperbolic Problems

A correct treatment of the boundary conditions is essential for an effective spectral calculation. Non-periodic, hyperbolic systems present the greatest difficulty. Not only does the correct number of boundary conditions depend upon the type of boundary and the structure of the equation, but so also does the permissible form of the boundary conditions. The two most important types of boundaries are artificial boundaries and wall boundaries. The former situation arises for problems which are posed in an unbounded domain. Since it is economical to keep the computational domain as small as possible, one encounters the task of specifying boundary conditions at an artificial boundary. The latter situation arises from directions in which the domain is bounded. Here one is faced with the task of updating variables at the boundaries in a way which is consistent with the boundary conditions as well as the differential equation.

If incorrect boundary conditions are imposed in the numerical scheme, the resulting errors will propagate into the computational domain. If these errors propagate and/or grow sufficiently rapidly, they will destroy the solution. In an explicit finite-difference scheme the errors have a finite rate of propagation. Moreover, if the scheme is dissipative the growth of the errors will be retarded or perhaps even suppressed. However, spectral methods have little dissipation to slow the growth of the errors, and because of their global character the errors immediately affect the entire domain.

Numerical experience has confirmed that spectral methods are far more sensitive than finite-difference methods to the boundary treatment. On the other hand, spectral methods require no special formulas for derivatives at the boundary, whereas finite-difference methods typically do.

An instructive example of sensitivity to boundary conditions was provided by Gottlieb, Gunzburger and Turkel (1982). Consider the system

$$\frac{\partial}{\partial t}\begin{pmatrix} u \\ v \end{pmatrix} = \begin{pmatrix} \frac{1}{2} & 1 \\ 1 & \frac{1}{2} \end{pmatrix} \frac{\partial}{\partial x}\begin{pmatrix} u \\ v \end{pmatrix}, \qquad \begin{matrix} -1 < x < 1 \\ t > 0 \end{matrix} \tag{8.2.1}$$

with boundary conditions

$$\begin{aligned} u(-1,t) &= \sin(-2+3t) + \cos(-2-t) \\ u(+1,t) &= \sin(2+3t) + \cos(2-t) \end{aligned} \tag{8.2.2}$$

and initial conditions

$$\begin{aligned} u(x,0) &= \sin(2x) + \cos(2x) \\ v(x,0) &= \sin(2x) - \cos(2x). \end{aligned} \tag{8.2.3}$$

This a well-posed problem. Let a Chebyshev collocation scheme which uses the Gauss–Lobatto points (2.4.14) be employed together with a modified

Euler time-discretization. The most straightforward boundary treatment is to update both u and v in the interior according to (8.2.1), to fix u at the boundaries via (8.2.2), and to update v at the boundaries according to (8.2.1). Note that the x-derivatives of u and v which are required in (8.2.1) are readily available at the boundaries from the usual differentiation formula. Computed solutions to this problem are strongly unstable. The reason is that the use of (8.2.1) for v at the boundaries is an incorrect extrapolation of the PDE to the boundary. Equation (8.2.1) can be transformed to read

$$\frac{\partial}{\partial t}\begin{pmatrix} u+v \\ u-v \end{pmatrix} = \begin{pmatrix} 3/2 & 0 \\ 0 & -1/2 \end{pmatrix} \frac{\partial}{\partial x}\begin{pmatrix} u+v \\ u-v \end{pmatrix}. \tag{8.2.4}$$

The characteristic variable $(u + v)$ is propagated to the left with speed 3/2 and $(u - v)$ moves to the right with speed 1/2. Hence, the Chebyshev collocation method for the scalar case (see (12.1.28)) suggests that the partial differential equation for $u + v$ be collocated at $x = -1$, and the partial differential equation for $u - v$ at $x = 1$. Since the time advancing scheme is explicit, this scheme is equivalently accomplished by advancing both the physical unknowns at each boundary using (8.2.1), and then retaining only the linear combination corresponding to outgoing characteristic variable. More precisely, let u_{prelim} denote the value of u at a boundary derived from the partial differential equation and use a similar notation for v_{prelim}. Then, the final values of u and v at the boundaries satisfy the conditions

$$\begin{aligned} u(-1) &= \sin(-2+3t) + \cos(-2-t) \\ u(+1) &= \sin(2+3t) + \cos(2-t) \\ u(-1) - v(-1) &= u_{\text{prelim}}(-1) - v_{\text{prelim}}(-1) \\ u(+1) + v(+1) &= u_{\text{prelim}}(+1) + v_{\text{prelim}}(+1). \end{aligned} \tag{8.2.5}$$

The first two equations came from (8.2.2) and the last two from (8.2.4). Computed solutions for these boundary conditions exhibit no problems, and indeed, spectral accuracy is achieved as the discretization is refined. The stability limit for the time advancing scheme is the same as for the scalar case (see Sec. 4.3).

The boundary treatment described above amounts to advancing the differential system at all the Gauss–Lobatto points ignoring the boundary conditions, and then applying the characteristic boundary conditions. However, this recipe may fail if an implicit time advancing scheme is used. For instance, if the differential system (8.2.1) is advanced by one step of the Crank–Nicolson method and then the characteristic boundary corrections are applied, the resulting stability limit is roughly three quarters of the stability limit for the modified Euler method (Canuto and Quarteroni (1987)).

The remedy to this consists, of course, in also treating the boundary conditions implicitly. At each boundary point two equations have to be satisfied: one physical boundary condition from (8.2.2) and the partial differential equation (8.2.4) for the outgoing characteristic variable. Both of these equations have to be included in the implicit linear system for each time-step. As observed by Canuto and Quarteroni, the matrix corresponding to the spatial part of the differential system has eigenvalues with negative real parts. It follows that if the time-derivative is discretized by an implicit A-stable method such as the Crank–Nicolson scheme, no stability restriction on the time-step will occur. This is particularly appealing in the simulation of slow transients or in the convergence to steady state.

The previous considerations extend in a straightforward manner to the case of a general hyperbolic system. Again, the correct strategy consists of collocating at the boundary points the linear combinations of the physical equations which correspond to the outgoing characteristic variables.

The importance of basing the boundary conditions of finite-difference schemes upon the characteristic variables has been stressed for many years by Moretti (1968). Gustafsson, Kreiss and Sundström (1972) and Osher (1969) have developed a rigorous mathematical framework for ascertaining the stability of interior and boundary difference schemes for initial-boundary value problems. Their discussions implicitly suggest the use of characteristic variables in the boundary conditions, since the stability criterion is given for the equations in characteristic form. This mathematical theory is rather difficult to understand. However, Trefethen (1983) has recently provided a useful physical interpretation of the technical stability criterion in terms of group velocity. The particular boundary condition formula given in (8.2.5) has been stressed by Gottlieb and Turkel (1985), who simplified the stability criterion for systems by showing that it could be reduced to the criterion for a scalar equation. A useful collection of papers on numerical boundary conditions is contained in NASA Conference Publication 2201. None of the theoretical results cited above, of course, have been proven for spectral methods. They have, however, been used as a guide for selecting boundary conditions in such calculations.

For non-linear problems, such as those which arise in gas dynamics, one often resorts to the use of linearized characteristic variables. The simplest point of linearization is the most recent time-level. In explicit spectral methods the major source of error is usually the time-discretization. The linearization error is often far smaller.

From the mathematical point of view, it is possible to give an explicit expression for the effect of the boundary conditions upon the overall accuracy of the scheme. Consider, for instance, a Chebyshev collocation-forward Euler approximation to the system of the conservation laws

$$\frac{\partial \mathbf{u}}{\partial t} + \frac{\partial \mathbf{f}}{\partial x}(\mathbf{u}) = 0, \tag{8.2.6}$$

where $\mathbf{u}$, $\mathbf{f}(\mathbf{u})$ are vectors in $\mathbb{R}^d$. Let $\tilde{\mathbf{u}}^{n+1}$ be the preliminary values of the numerical solution (which is a polynomial of degree N) obtained by applying in a straightforward way the forward Euler formula at all the points. Let $\mathbf{u}^{n+1}$ denote the values equal to $\tilde{\mathbf{u}}^{n+1}$ at the interior nodes and modified according to the previous discussion at the boundary nodes. The error equation for this approximation (see Sec. 10.6) reads as follows:

$$\mathbf{u}^{n+1} = \mathbf{u}^n - \Delta t \frac{\partial}{\partial x} \mathbf{f}_N(\mathbf{u}^n) + \frac{1}{2}\{\boldsymbol{\tau}_+^{n+1}(1+x) + \boldsymbol{\tau}_-^{n+1}(1-x)\} T_N'(x), \quad (8.2.7)$$

where $(\partial/\partial x)\mathbf{f}_N(\mathbf{u})$ denotes the collocation derivative of $\mathbf{f}(\mathbf{u})$ and

$$\begin{aligned} \boldsymbol{\tau}_+^{n+1} &= \frac{1}{N^2} \frac{\mathbf{u}_0^{n+1} - \tilde{\mathbf{u}}_0^{n+1}}{\Delta t} \\ \boldsymbol{\tau}_-^{n+1} &= \frac{1}{N^2} \frac{\mathbf{u}_N^{n+1} - \tilde{\mathbf{u}}_N^{n+1}}{\Delta t}. \end{aligned} \quad (8.2.8)$$

To examine the conservation properties of the scheme, we integrate (8.2.7) from -1 to 1, obtaining

$$\begin{aligned} \int_{-1}^{1} \mathbf{u}^{n+1}\,dx &= \int_{-1}^{1} \mathbf{u}^n\,dx - \Delta t[\mathbf{f}(\mathbf{u}^n)]_{-1}^{+1} \\ &\quad + \frac{\delta_N}{N^2}\left\{\frac{\mathbf{u}_0^{n+1} - \tilde{\mathbf{u}}_0^{n+1}}{\Delta t} - \frac{\mathbf{u}_N^{n+1} - \tilde{\mathbf{u}}_N^{n+1}}{\Delta t}\right\}, \end{aligned} \quad (8.2.9)$$

with $\delta_N = 2(1 + 1/(N^2 - 1))$. Thus, the scheme is globally conservative up to an error which decays as $N \to \infty$ and depends on the boundary conditions. Moreover, using (8.2.7) again, it is possible to prove that the consistency error of this method is first-order in time and infinite-order in space. Thus, the boundary conditions do not destroy the spectral accuracy of the Chebyshev method.

Implicit in the preceding discussion has been the condition that the differential initial-boundary value problem is well-posed. This places a certain restriction on the allowable boundary conditions. Numerical experience for linear problems indicates that if a spectral method is used with well-posed boundary conditions implemented by the characteristic correction method, then the boundary treatment will produce no instabilities. An elementary discussion of how to determine whether the initial-boundary value problem is well-posed is given by Oliger and Sundström (1978).

The numerical boundary conditions for multidimensional non-linear problems have often been implemented by applying a one-dimensional characteristic correction scheme in the direction normal to the boundary. For problems with strong non-linear effects at the boundary, compatibility conditions are more robust. Several examples of these are given in Sec. 8.5.

8.3. Basic Results for Scalar Nonsmooth Problems

The naive use of spectral methods on hyperbolic problems with discontinuous solutions produces oscillatory numerical results. The oscillations arising directly from the discontinuity have a Gibbs-like, high frequency character. These oscillations are not in themselves insurmountable, for according to a result of Lax (1978), they should contain sufficient information to permit the reconstruction of the correct physical solution from the visually disturbing numerical one.

Further theoretical support for the use of spectral methods on non-smooth (including non-linear) problems was furnished by Gottlieb, Lustman and Orszag (1981). Following an argument due to Lax and Wendroff (1960), they proved that if the sequence u^N ($N \geqslant 0$) of the solutions produced by a Fourier or Chebyshev collocation scheme for the equation

$$\frac{\partial u}{\partial t} + \frac{\partial f}{\partial x}(u) = 0 \tag{8.3.1}$$

is bounded and converges almost everywhere, as $N \to \infty$, then the limit is a weak solution of (8.3.1). This means that it satisfies the equation

$$\iint \left(u \frac{\partial \phi}{\partial t} - f(u) \frac{\partial \phi}{\partial x} \right) dx\, dt = \int u(x,0)\phi(x,0)\, dx \tag{8.3.2}$$

for all smooth functions ϕ which vanish for large t and on the boundary of the domain. The limit solution thus satisfies the jump conditions

$$[f(u)] = 0. \tag{8.3.3}$$

Hence, any shocks that are present are propagated with the correct speed.

Moreover, despite the presence of high frequency oscillations in the spectral representation of a shock, the transition between the pre-shock and the post-shock states always occurs within one mesh interval. This property was already evident in the Gibbs phenomenon (see Fig. 2.5) and has been heuristically justified by Gottlieb et al. on the basis of the accuracy of spectral methods. Thus, a very accurate shock position is inherent in a spectral solution.

Although reassuring, these results do not imply that a time-dependent calculation will yield a convergent solution. Variable-coefficient and non-linear terms provide a mechanism for spreading the discontinuity-induced high frequency oscillations over the entire frequency spectrum as the solution evolves. Moreover, some additional dissipative or filtering mechanism may be needed to stabilize the calculation.

As we noted in Sec. 2.1.4, the pronounced oscillations that arise in both the truncated Fourier series and the trigonometric interpolating polynomials of discontinuous functions may be suppressed, or smoothed, by a gradual

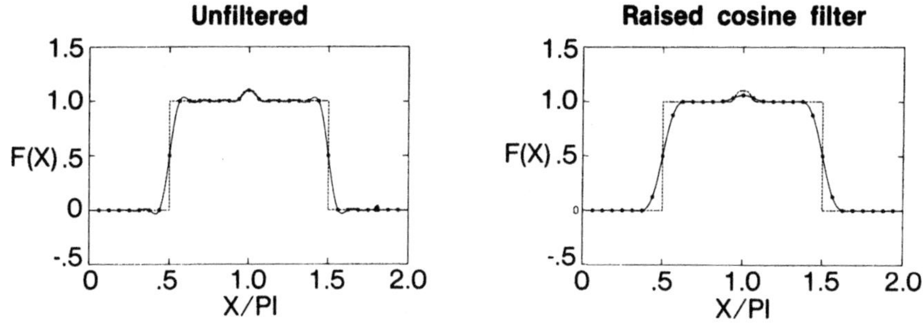

Figure 8.1. Effect of a raised cosine filter upon a periodic square wave plus a Gaussian bump. The dashed line indicates the exact function, the circles indicate the value of the post-filtered approximation at the collocation points, and the solid line indicates the interpolating polynomial.

tapering of the Fourier coefficients. The reduction in the oscillations is achieved at the cost of a deterioration in the representation of the high frequency information. One consequence is that the effective region of the discontinuity is broadened. Another is that small wavelength features away from the discontinuity are adversely affected. An illustration is provided in Fig. 8.1. This is a slight modification of the square wave example that appeared in Chap. 2 in Figs. 2.1 and 2.5. Superimposed upon the basic square wave is a narrow Gaussian bump. The top diagram displays an interpolating polynomial which has just enough points to resolve the bump. The bottom diagram shows the effect of the raised cosine tapering. The Gibbs oscillations are no longer in evidence. However, the bump is now poorly represented and the effective width of the discontinuity has been doubled. From the point of view of spectral methods for PDEs, the interesting theoretical question is the convergence rate of the original interpolating polynomial and the filtered polynomial to both the function itself and its derivative.

A detailed examination of the effect of filtering on the accuracy of spectral solutions of linear systems of hyperbolic equations was made by Majda, McDonough and Osher (1978). We will summarize their conclusion about a semi-discrete Fourier collocation method for

$$\frac{\partial u}{\partial t}+\frac{\partial u}{\partial x}=0, \qquad \begin{array}{l} t>0 \\ x\in(0,2\pi) \end{array} \tag{8.3.4}$$

with periodic boundary conditions and discontinuous initial condition

$$u(x_0)=u_0(x), \tag{8.3.5}$$

with a single jump discontinuity located at a collocation point.

If the Fourier collocation method is applied in the conventional manner to (8.3.4) and (8.3.5), then in a region which excludes the discontinuity, the maximum error, for any $t>0$, decays as N^{-2}. However, it is possible to achieve a convergence rate of infinite order by a proper filtering of the initial condition. This filtering is applied to the *continuous* Fourier coefficients of $u_0(x)$. The new, filtered initial condition is

$$u_f(x) = \sum_{k=-N/2+1}^{N/2-1} \sigma\left(\frac{2\pi k}{N}\right)\hat{u}_k, \tag{8.3.6}$$

where

$$\hat{u}_k = \frac{1}{2\pi}\int_0^{2\pi} u_0(x)e^{-ikx}\,dx,$$

and $\sigma(\theta)$ is a filtering function with the properties

$$\begin{aligned} &\sigma(\theta) \text{ is } C^\infty \text{ on } \mathbb{R} \\ &\sigma(\theta) \text{ is unity is a neighborhood of the origin} \\ &\sigma(\theta) = 0 \quad \text{for} \quad |\theta| \geqslant \pi. \end{aligned} \tag{8.3.7}$$

(The application of this filtering to the *discrete* Fourier coefficients of $u_0(x)$ still leads to second-order convergence.)

Some of the filtering functions that have been employed are

$$\sigma(\theta) = \frac{\sin\theta}{\theta} \qquad \text{(Lanczos)} \tag{8.3.8}$$

$$\sigma(\theta) = \tfrac{1}{2}(1 + \cos\theta) \qquad \text{(raised cosine)} \tag{8.3.9}$$

$$\sigma(\theta) = \sigma_0^4(35 - 84\sigma_0 + 70\sigma_0^2 - 20\sigma_0^3) \qquad \text{(sharpened raised cosine),} \tag{8.3.10}$$

where σ_0 is the raised cosine given by (8.3.9), and

$$\sigma(\theta) = \begin{cases} 1 & |\theta| \leqslant \theta_c \\ e^{-\alpha(|\theta|-\theta_c)^4} & \theta_c \leqslant |\theta| \leqslant \pi. \end{cases} \qquad \text{(exponential cut-off)} \tag{8.3.11}$$

These filters are listed above in order of decreasing strength and are illustrated in Fig. 8.2. The Lanczos and raised cosine filters are classical. Equation (8.3.10) represents one of a number of standard formulas for sharpening a basic filter (Hamming (1977)). The exponential cut-off was mentioned explicitly by Majda et al. These filters are reasonable compromises between the ideal conditions (8.3.7) and practical considerations.

Kopriva (1987) has illustrated the effect of several filters on (8.3.4) with the initial condition

$$u_0(x) = (x - \pi)\left[1 + \sin\frac{x}{2}\right]. \tag{8.3.12}$$

This function has a simple jump discontinuity at $x = 0$. Figure 8.3 compares

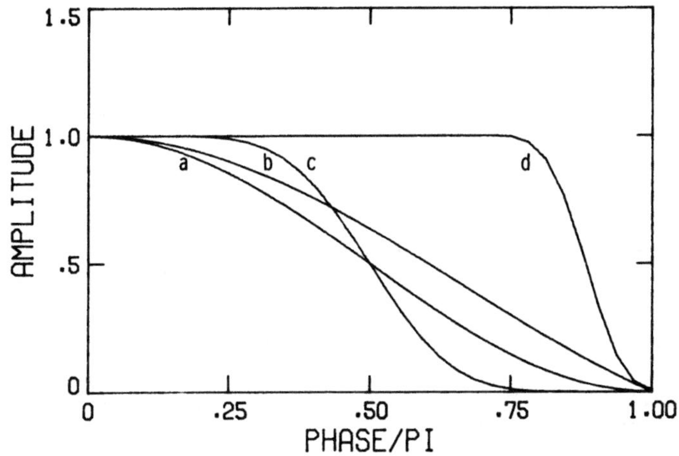

Figure 8.2. Filters used in the spectral calculations. (a) Raised cosine; (b) Lanczos; (c) sharpened raised cosine; (d) exponential cut-off.

various filtering functions for the semi-discrete Fourier collocation approximation at $t \cong 3\pi/2$. The top portion shows the solution itself for $N = 64$ and the bottom displays the pointwise errors for $N = 8$, 16, 32 and 64. The compromises are apparent. The raised cosine filter has the mildest oscillations near the discontinuity, but the largest errors elsewhere. Indeed, these errors decay algebraically. The sharpened raised cosine has very small errors away from the discontinuity, but pronounced oscillations nearby and a broad region of large error. Neither of these filters satisfies all of the conditions (8.3.7). The exponential filter is perhaps the closest approximation to these conditions and is, in fact, the filter recommended by Majda et al. Nevertheless, the results for it are not very impressive. The broad region of large error is especially disturbing.

Even the slight additional complexity of introducing a variable wave speed into the scalar wave equation required more modification to the Fourier collocation method than just filtering the initial condition. In the constant-coefficient case there is no mechanism to generate high frequency components in the solution during the evolution process. In the variable-coefficient case, however, the interaction of the coefficient with the solution generates high frequency components. These must be controlled to retain high accuracy. Majda et al. showed that applying a filtering during the derivative evaluations will suffice.

Further difficulties arise in two dimensions. Here, the region of influence of the initial discontinuity grows linearly with time rather than remaining a single point as it does in one dimension. Majda et al. proved that a scheme

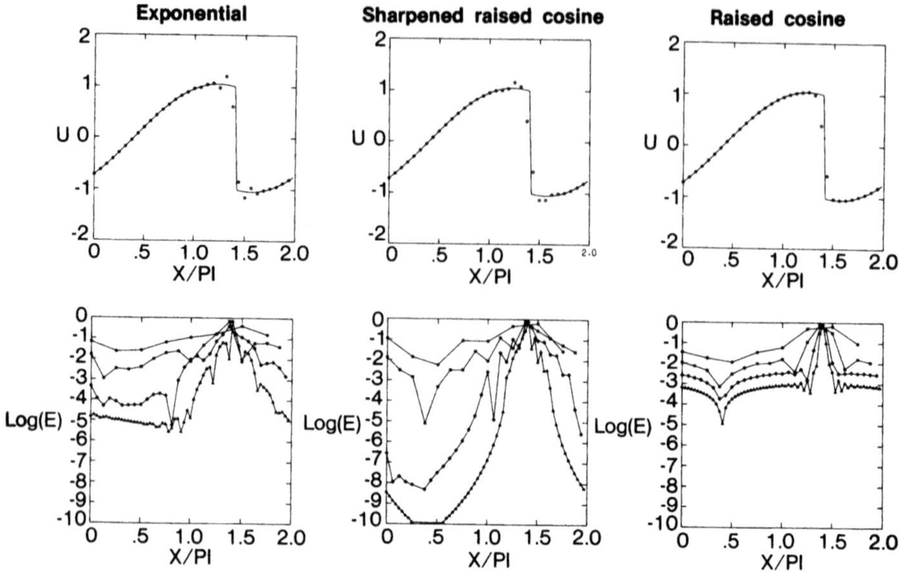

Figure 8.3. Filtered Fourier collocation solutions to a linear discontinuous problem (top) and pointwise error (bottom). (Courtesy of D. Kopriva.)

with proper filtering can produce infinite-order convergence in a domain which excludes the region of influence. Unfortunately, the contaminated region spreads linearly with time.

No theoretical results yet exist for the stability or convergence of spectral methods on non-linear hyperbolic problems, not even for those with smooth solutions. The linear results of Majda et al. are discouraging because they would appear to require that at every time-step a filtering procedure be applied to the continuous Fourier coefficients, and these are obviously unavailable.

A number of computations have been performed on non-linear, non-smooth problems. The types of filtering procedures are more varied than those used for linear problems. These filtering procedures may be classified as follows:

(1) *Pre-processing*. The initial condition is filtered in terms of its continuous Fourier coefficients as described by Majda, McDonough and Osher.
(2) *Derivative filtering*. In the computation of spatial derivatives the term ik is replaced with $ik\sigma(2\pi k/N)$, i.e.,

$$\left.\frac{du}{dx}\right|_j = \sum_{k=-(N/2)+1}^{(N/2)-1} ik\sigma(2\pi k/N)\tilde{u}_k e^{ikx_j}. \tag{8.3.13}$$

(3) *Solution smoothing.* At regular intervals in the course of advancing the solution in time, the current solution values are smoothed in Fourier space i.e., u_j is replaced with

$$\sum_{k=-N/2}^{N/2-1} \sigma(2\pi k/N)\tilde{u}_k e^{ikx_j}. \tag{8.3.14}$$

(4) *Cosmetic post-processing.* This is similar to solution smoothing except that the solution is smoothed for display purposes only. If the calculation is continued, it is done so with the unaltered solution.

The choice of filter will determine what Fourier frequencies will be modified. The $k = 0$ component of the Fourier decomposition is the only one which contributes to the average value. Thus, in order to conserve the value of the solution, $\sigma(0) = 1$ is required. The filters shown in Fig. 8.2 all have this property. The effect of the filter on higher frequencies, however, is usually more difficult to assess in non-linear problems.

The raised cosine (also known as the von Hann window) admits a simple physical interpretation. It is algebraically equivalent to

$$\begin{aligned} u_j &\leftarrow \frac{u_{j-1} + 2u_j + u_{j+1}}{4} \\ &= u_j + \frac{(\Delta x)^2}{4} \frac{u_{j-1} - 2u_j + u_{j+1}}{(\Delta x)^2}. \end{aligned} \tag{8.3.15}$$

This is clearly a second-order artificial viscosity term with the coefficient

$$\frac{(\Delta x)^2}{4(\Delta t/N_f)}, \tag{8.3.16}$$

where N_f is the number of time-steps between applications of the filter. Figure 8.2 suggests that the other types of filters amount to non-physical viscosities: they damp preferentially the high frequency components of the solution, but in a different manner than a physical viscosity.

The frequency of applying the filter is analogous to the selection of the size of the artificial viscosity for finite-difference methods. Applied too often, a strong filter like the Lanczos filter will unacceptably smear out a shock. Frequent applications of a weak filter such as the exponential cut-off may not be enough even to stabilize the calculation.

Cosmetic post-processing is the weakest filtering of all, for it makes no change in the solution itself. It may be viewed as a means of making presentable those solutions which exhibit a Gibbs phenomenon.

Gottlieb, Lustman and Orszag (1981) developed a post-processing technique that consists of subtracting simple discontinuous functions from the computed solution. They locate the discontinuity and determine its strength by comparing the spectrum of the numerical solution with the spectrum of a step function. They then subtract the step function jump from the numerical solution and apply a weak conventional filter to the difference. Further refinements were provided by Abarbanel, Gottlieb and Tadmor (1986). Smoothing procedures based on convolutions in physical space have recently been developed by Gottlieb (1985) and Abarbanel et al. A mathematical account of this scheme is furnished in Sec. 12.1.4.

8.4. Homogeneous Turbulence

The Fourier spectral algorithms for incompressible homogeneous turbulence discussed in Sec. 7.2 can be extended to the compressible case. Provided that the velocity fluctuations are sufficiently small that no shocks develop, the solution will be smooth. Of course, the flow field will contain features on scales as small as the dissipation scales, and these must be resolved.

The inviscid non-linearities of the compressible Navier–Stokes equations are not merely quadratic, as they are for compressible flow. (Moreover, for simulations of real fluids the viscous terms contain transcendental non-linearities arising from appropriate empirical laws for the transport coefficients.) A strict Fourier Galerkin approximation, therefore, is quite unwieldy and a collocation approach is preferable. An incompressible version was discussed in Sec. 7.2.4. As is typical of collocation methods, numerical stability may depend on the precise form of the equation. Feiereisen, Reynolds and Ferziger (1981) recommended the following version:

$$\frac{\partial \rho}{\partial t} + \nabla\cdot(\rho \mathbf{u}) = 0 \tag{8.4.1}$$

$$\frac{\partial}{\partial t}(\rho \mathbf{u}) + \frac{1}{2}\nabla\cdot(\rho \mathbf{u}\mathbf{u}) + \frac{1}{2}\rho \mathbf{u}\cdot\nabla\mathbf{u} + \frac{1}{2}\mathbf{u}\nabla\cdot(\rho\mathbf{u}) + \nabla p = \nabla\cdot\underline{\tau} \tag{8.4.2}$$

$$\frac{\partial p}{\partial t} + \mathbf{u}\cdot\nabla p + \gamma p \nabla\cdot\mathbf{u} = (\gamma - 1)\nabla\cdot(k\nabla T) + (\gamma - 1)\Phi, \tag{8.4.3}$$

where $\underline{\tau}$ and Φ are given by (1.3.5) and (1.3.6). Equation (8.4.2) differs from the standard form (1.3.4) in that the identity

$$\nabla\cdot(\rho\mathbf{u}\mathbf{u}) = \rho\mathbf{u}\cdot\nabla\mathbf{u} + \mathbf{u}\nabla\cdot(\rho\mathbf{u}) \tag{8.4.4}$$

has been employed. Feiereisen et al. demonstrated that a Fourier collocation method applied to the inviscid version of (8.4.1)—(8.4.3) conserves mass—$\sum \rho_{l,m,n}$, momentum—$\sum (\rho\mathbf{u})_{l,m,n}$, and energy—$\sum (1/(\gamma - 1)p + \frac{1}{2}\rho|\mathbf{u}|^2)_{l,m,n}$—in the absence of time-discretization errors. These conservation properties are desirable on physical grounds, but there is no firm evidence that they are essential. Moreover, they are not exactly the type of conservation properties that are most useful for ensuring temporal numerical stability—they are not quadratic in the dependent variables.

Feiereisen et al. used a purely explicit algorithm and performed several 64^3 simulations of isotropic turbulence and homogeneous turbulence in uniform shear. (Their algorithm incorporated the compressible version of Rogallo's transformation that was described in Sec. 7.2.3.) For such an algorithm the time-step limitation is imposed by the acoustic terms at low Mach numbers and by the advection terms otherwise.

Erlebacher, Hussaini, Speziale and Zang (1987) developed a semi-implicit version of this algorithm. In this splitting technique, the first step integrates the equations

$$\begin{gathered} \frac{\partial \rho}{\partial t} = 0 \\ \frac{\partial}{\partial t}(\rho \mathbf{u}) + \frac{1}{2}\nabla\cdot(\rho\mathbf{u}\mathbf{u}) + \frac{1}{2}\rho\mathbf{u}\cdot\nabla\mathbf{u} + \frac{1}{2}\mathbf{u}\cdot\nabla(\rho\mathbf{u}) = \nabla\cdot\underline{\tau} \\ \frac{\partial p}{\partial t} + \mathbf{u}\cdot\nabla p + \gamma p\nabla\cdot\mathbf{u} - c_0^2\nabla\cdot(\rho\mathbf{u}) = (\gamma-1)\nabla\cdot(k\nabla T) + (\gamma-1)\Phi \end{gathered} \tag{8.4.5}$$

from t_n to $t_{n+1/2}$, and then the equations

$$\begin{gathered} \frac{\partial \rho}{\partial t} + \nabla\cdot(\rho\mathbf{u}) = 0 \\ \frac{\partial}{\partial t}(\rho\mathbf{u}) + \nabla p = 0 \\ \frac{\partial p}{\partial t} + c_0^2\nabla\cdot(\rho\mathbf{u}) = 0 \end{gathered} \tag{8.4.6}$$

are integrated from $t_{n+1/2}$ to t_n in the second step. The constant c_0 is the average value of the sound speed c at $t = t_n$.

The second step may be integrated analytically. In Fourier space, (8.4.6) becomes

$$\begin{gathered} \frac{\partial \tilde{\rho}}{\partial t} + i\mathbf{k}\cdot\tilde{\mathbf{m}} = 0 \\ \frac{\partial \tilde{\mathbf{m}}}{\partial t} + i\mathbf{k}\tilde{p} = 0 \\ \frac{\partial \tilde{p}}{\partial t} + ic_0^2\mathbf{k}\cdot\tilde{\mathbf{m}} = 0, \end{gathered} \tag{8.4.7}$$

where $\mathbf{m} = \rho\mathbf{u}$ and Fourier transformed quantities are denoted by tildes. Let

$$\tilde{A} = \tilde{p}^{n+1/2}$$
$$\tilde{B} = i\frac{c_0}{k}\mathbf{k}\cdot\tilde{\mathbf{m}}^{n+1/2}. \tag{8.4.8}$$

The result of the second step is

$$\begin{aligned}
\tilde{\rho}^{n+1} &= \tilde{\rho}^{n+1/2} + \frac{1}{c_0^2}[\tilde{A}\cos(c_0 k\Delta t) + \tilde{B}\sin(c_0 k\Delta t) - \tilde{A}] \\
\tilde{\mathbf{m}}^{n+1} &= \tilde{\mathbf{m}}^{n+1/2} - \frac{i\mathbf{k}}{c_0 k}[\tilde{A}\sin(c_0 k\Delta t) - \tilde{B}\cos(c_0 k\Delta t) + \tilde{B}] \\
\tilde{p}^{n+1} &= \tilde{A}\cos(c_0 k\Delta t) + \tilde{B}\sin(c_0 k\Delta t),
\end{aligned} \tag{8.4.9}$$

where $k = |\mathbf{k}|$.

In the first step the principal terms responsible for the acoustic waves have been removed. Thus, one expects the time-step limitation to depend upon $\mathbf{u} + |c - c_0|$ rather than $\mathbf{u} + c$. This is clearly a substantial advantage at low Mach numbers. Since the second step is integrated analytically, no time-step limit arises from it for stability reasons. If one is truly interested in all the details arising from the sound waves, then the time-step must be small enough to resolve the temporal evolution of these waves. But, if only the largest-scale sound waves are of interest, then this splitting method is useful.

Erlebacher et al. have used this algorithm for numerous 96^3 and 128^3 simulations of isotropic turbulence. Figure 8.4 is a comparison between an incompressible case and a compressible one at a Mach number of 0.6. The energy spectra are similar, but compressibility effects are obvious in the density fluctuations. The splitting method allowed the compressible calcula-

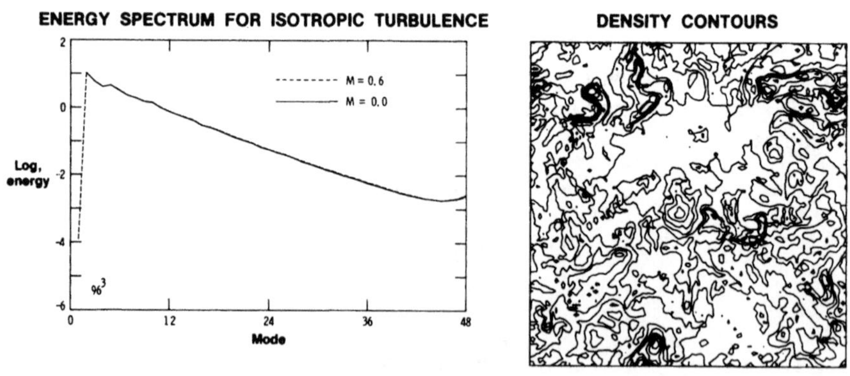

Figure 8.4. Energy spectra from incompressible and compressible 96^3 simulations of isotropic turbulence; density fluctuations in a two-dimensional plane for compressible turbulence.

tions to be run at a time-step three times larger than possible with the purely explicit algorithm.

Several investigations of two-dimensional compressible turbulence have been performed with Fourier spectral methods. Delorme (1984) developed a class of schemes employing a semi-implicit treatment of the advection and/or diffusion terms. The semi-implicit time-discretization of the advection terms was based on the Lerat schemes mentioned in Sec. 4.4.3. Leorat et al. (1985) have used a fairly standard spectral collocation algorithm for investigating two-dimensional turbulence in a supersonic stream.

8.5. Shock-Capturing

8.5.1. Potential Flow

An important simplification of the Euler equations results for flows which are steady and irrotational. In this case a velocity potential ϕ exists:

$$\mathbf{u} = \nabla\phi. \tag{8.5.1}$$

The continuity equation reduces to

$$\nabla\cdot(\rho\nabla\phi) = 0, \tag{8.5.2}$$

with the density ρ related to the velocity potential by Bernoulli's law

$$\frac{1}{2}|\mathbf{u}|^2 + \frac{\gamma}{\gamma - 1}\frac{p}{\rho} = f(s), \tag{8.5.3}$$

which is valid along streamlines. This differs from the incompressible case discussed in Sec. 6.2 by the inclusion of the non-linear density in (8.5.2). If the entropy s is uniform, then ρ follows easily from ϕ. For two-dimensional flow, the potential equation is

$$(u^2 - c^2)\frac{\partial^2\phi}{\partial x^2} + 2uv\frac{\partial^2\phi}{\partial x\partial y} + (v^2 - c^2)\frac{\partial^2\phi}{\partial y^2} = 0, \tag{8.5.4}$$

where $u = \partial\phi/\partial x$ and $v = \partial\phi/\partial y$. This equation is non-linear. Standard characteristic analysis reveals that this equation is elliptic for $u^2 + v^2 < c^2$ and hyperbolic for $u^2 + v^2 > c^2$. These are the subsonic and supersonic cases, respectively. The challenging case arises when the flow is of mixed type over the domain, i.e., it consists of both subsonic and supersonic flow. This type of flow is referred to as transonic. Generally, there will be a shock wave where the flow compresses from supersonic to subsonic. This is associated with a discontinuity in the gradient of the potential.

Fairly accurate predictions for a number of transonic flows of practical interest can be made on the basis of the compressible potential equation. The mixed elliptic-hyperbolic type precludes purely elliptic or purely hyperbolic solution procedures. The numerical solution of the potential equation became feasible first for finite-difference methods and only after the introduction of type-dependent differencing by Murman and Cole (1971). The review by Hall (1981) provides an exhaustive history of computational approaches to the potential equation.

Until the recent work of Streett (1983) the discretization procedures for the potential equation were invariably based on low order finite-volume, finite-difference, or finite-element methods. Streett solved the two-dimensional full potential equation (applying boundary conditions at the actual airfoil surface). In this work a numerical conformal mapping (also generated by Fourier techniques) was used to transform the airfoil onto the unit circle. The coordinate system is illustrated in Fig. 8.5. Moreover, the calculations were actually performed in terms of the reduced potential G, which is defined by

$$G = \phi - \left(R + \frac{1}{R}\right)\cos\Theta - E\tan^{-1}[\sqrt{1 - M_\infty^2}\tan\Theta], \qquad (8.5.5)$$

where ϕ is the potential, R and Θ are the computational polar coordinates, E is the circulation and M_∞ is the Mach number at infinity. The last term represents the singularity at infinity of ϕ. The reduced potential itself is regular except for a possible shock and therefore is a better choice of dependent variable. This substitution is used in (8.5.2) and the boundary conditions on G are

$$\begin{aligned} \frac{\partial G}{\partial R} &= 0 \qquad \text{at } R = 1, \\ G &\to 0 \qquad \text{as } R \to \infty, \end{aligned} \qquad (8.5.6)$$

and of regularity at the sharp trailing edge of the airfoil.

The spectral method employs a Fourier series representation in Θ. Constant grid spacing in Θ corresponds to a convenient dense spacing in the physical plane at the leading and trailing edges. The domain in R (with a large, but finite outer cutoff) is mapped onto the standard Chebyshev domain $[-1, 1]$ by an analytical stretching transformation that clusters the collocation points near the airfoil surface. The discrete spectral equations are solved with a multigrid procedure using the approximate factorization relaxation scheme discussed in Chap. 5.

The flow past an NACA 0012 airfoil at 4° angle of attack and a freestream Mach number of 0.5 is distinctly non-linear, but still entirely subsonic. Since the reduced potential is smooth, the spectral solution on a relatively coarse grid captures all the essential details of the flow. The pressure on the upper and lower airfoil surfaces as computed from a spectral calculation using

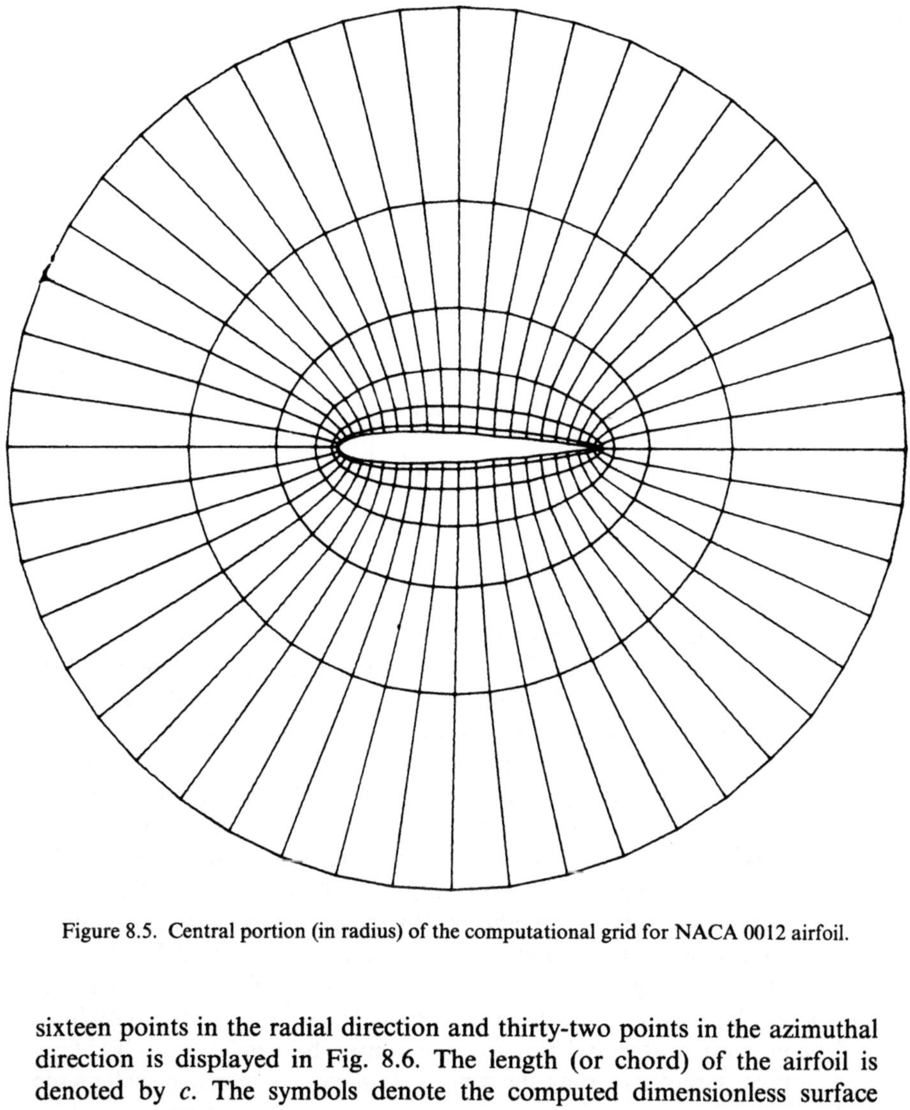

Figure 8.5. Central portion (in radius) of the computational grid for NACA 0012 airfoil.

sixteen points in the radial direction and thirty-two points in the azimuthal direction is displayed in Fig. 8.6. The length (or chord) of the airfoil is denoted by c. The symbols denote the computed dimensionless surface pressure coefficient

$$C_p = \frac{p - p_\infty}{\frac{1}{2}\rho_\infty u_\infty^2}$$

at the collocation points. For comparison, a benchmark result on a 64×384 grid from the finite-difference code FLO36 (Jameson (1979)) is shown as a solid line. The finite-difference and the spectral results are identical to plotting accuracy. Streett's grid refinement study suggests at least sixth-order accuracy for the spectral method.

Figure 8.6. Surface pressure for a 16×32 spectral solution to a subsonic airfoil.

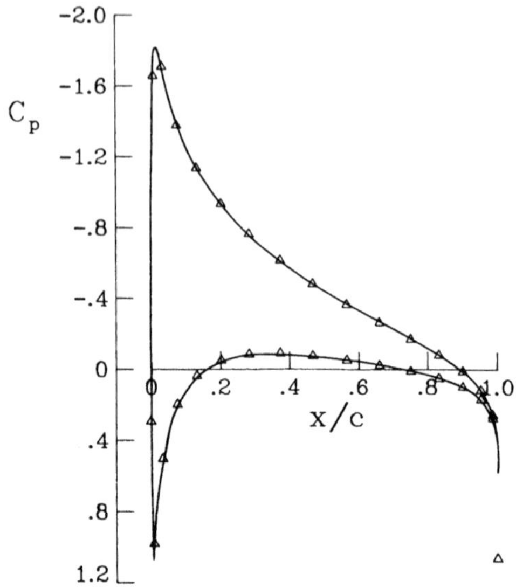

A dramatic demonstration of the efficiency of the spectral method for nonlinear, subsonic potential flow was furnished by Hussaini, Streett and Zang (1983) who determined the critical freestream Mach number at which the potential flow past a circular cylinder first develops a supersonic region. This spectral calculation represents an alternative to the asymptotic series method employed by Van Dyke and Guttmann (1983) to arrive at the estimate $M_{\text{crit}} = .39823780 \pm .00000001$. Combining subsonic calculations at subcritical Mach numbers with extrapolation, Hussaini et al. estimated the critical freestream Mach number to be $M_{\text{crit}} = .3982415 \pm .0000002$.

The spectral algorithm is so efficient that all of the requisite calculations consumed less than twenty minutes to CPU time on the CDC Cyber 175 and were performed on grids with no more than 2000 points. A comparable calculation by existing finite-difference codes would likely exhibit only first-order convergence. It would be far more expensive both in terms of CPU time and storage, surely exceeding the central memory of a machine such as the CDC Cyber 175.

A more challenging application of this method is to transonic potential flow. There is first the non-trivial matter of developing a convergent iterative scheme and then the question of accuracy in the converged solution. The most expedient technique for dealing with the mixed elliptic-hyperbolic nature of the transonic problem is to use the artificial density approach of Hafez, South and Murman (1979). In its simplest form the density ρ in (8.5.2) is replaced with $\tilde{\rho}$, where

$$\tilde{\rho} = \rho - \mu \overleftarrow{\delta} \rho, \tag{8.5.7}$$

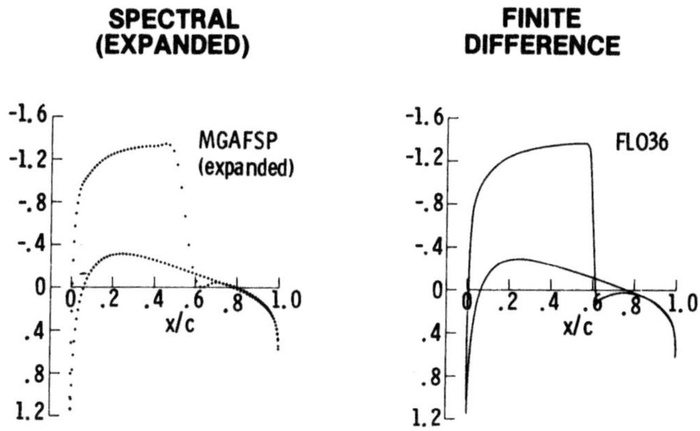

Figure 8.7. Surface pressures for expanded spectral (MGAFSP) and finite difference (FLO36) solutions to a transonic airfoil.

with

$$\mu = \max\left\{0, 1 - \frac{1}{M^2}\right\},$$

where M is the local Mach number and $\overleftarrow{\delta}\rho$ is an upwind first-order (undivided) difference. The spectral calculations of Streett, Zang and Hussaini (1985) employed a higher order artificial density formula. The spectral method also required a weak filtering technique to deal with some high frequency oscillations generated by the shock.

A lifting transonic case is provided by the NACA 0012 airfoil at $M_\infty = 0.75$ and $\alpha = 2°$. A shock appears on the upper surface for these conditions. Lifting transonic cases are especially difficult for spectral methods since the solution will always have significant content in the entire frequency spectrum: the shock populates the highest frequencies of the grid and the lift is predominantly on the scale of the entire domain. An iterative scheme must therefore be able to damp error components across the spectrum.

Surface pressure distributions from the spectral and finite-difference codes are shown in Fig. 8.7. The respective computational grids are 18×64 and 32×192. The latter is the default grid for the production finite-difference code. Spectral results obtained by trigonometrically interpolating the 18×64 grid results onto a much finer grid are used to reveal the wealth of detail that is provided by the relative coarse spectral grid.

8.5.2. Ringleb Flow

Spectral shock-capturing techniques for the Euler equations have not yet been as successful as they have been for the potential equation. The problem is not

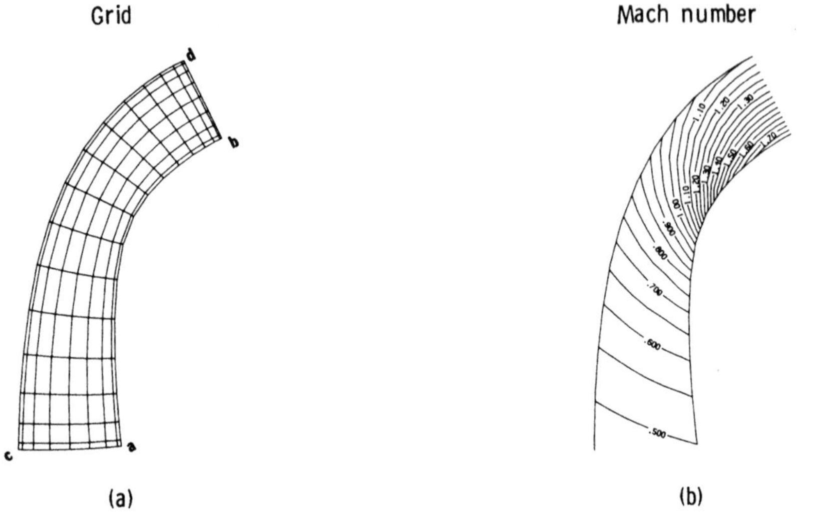

Figure 8.8. (a) The 16 × 8 Chebyshev–Chebyshev grid used for the computation of transonic Ringleb flow. (b) Computed Mach number contours of the transonic Ringleb flow for the 16 × 8 grid.

some intrinsic difficulty of compressible flow. The difficulties clearly arise only when a shock is present. Indeed, Kopriva et al. (1984) have exhibited Chebyshev spectral solutions that converge exponentially fast for special shock-free, two-dimensional transonic and supersonic flows.

These numerical solutions were produced for the Ringleb (1940) flow, a class of steady, isentropic flows between curved walls which admit an analytic solution. The computational domains for the transonic and supersonic cases are shown in Figs. 8.8(a) and 8.9(a) for the transonic and supersonic cases, respectively. The curved boundaries are readily handled by transforming to potential-streamfunction coordinates (ϕ, ψ) of the exact, steady solution. These new, orthogonal coordinates can be calculated from the exact solution. The computational domain is rectangular in the new coordinates: $(\phi_L, \phi_R) \times (\psi_B, \psi_T)$, where, in Fig. 8.8(a), **ab** corresponds to ψ_B, **cd** to ψ_T, **ca** to ϕ_L and **bd** to ϕ_R.

This flow has constant entropy since there are no wall boundary layers nor any shocks to produce irreversible losses. Thus, the energy equation is superfluous and the continuity equation may be written in terms of P, the natural logarithm of pressure. Furthermore, the momentum equation need not be used in the conservation form. Thus, (1.3.12)—(1.3.14) reduce to

$$\frac{\partial Q}{\partial t} + B\frac{\partial Q}{\partial \phi} + C\frac{\partial Q}{\partial \psi} = 0, \tag{8.5.8}$$

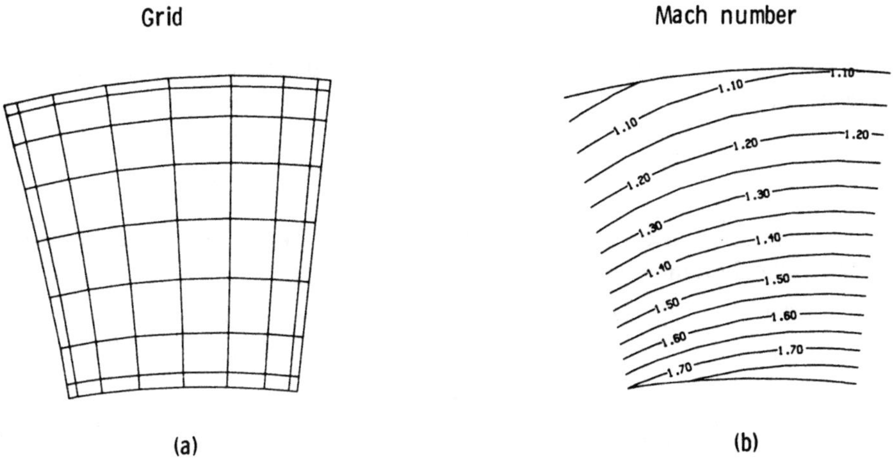

Figure 8.9. (a) The 8 × 8 Chebyshev–Chebyshev grid used for the computation of supersonic Ringleb flow. (b) Computed Mach number contours of the supersonic Ringleb flow for the 8 × 8 grid.

where

$$Q = (P, u, v), \tag{8.5.9}$$

and the coefficient matrices are

$$B = \begin{pmatrix} U & \gamma\phi_x & \gamma\phi_y \\ \dfrac{c^2}{\gamma}\phi_x & U & 0 \\ \dfrac{c^2}{\gamma}\phi_y & 0 & U \end{pmatrix}, \quad C = \begin{pmatrix} V & \gamma\psi_x & \gamma\psi_y \\ \dfrac{c^2}{\gamma}\psi_x & V & 0 \\ \dfrac{c^2}{\gamma}\psi_y & 0 & V \end{pmatrix}. \tag{8.5.10}$$

The contravariant velocity components are

$$\begin{aligned} U &= u\phi_x + v\phi_y \\ V &= u\psi_x + v\psi_y. \end{aligned} \tag{8.5.11}$$

Kopriva et al. discretized (8.5.8) in space by Chebyshev collocation and in time by a second-order Runge–Kutta method.

The Ringleb problem has both solid wall and inflow/outflow boundaries. The outflow boundary **bd** is chosen to be supersonic. The inflow boundary **ac** can be either a subsonic or a supersonic boundary.

The only physical boundary conditions are that the contravariant velocity

$$V = 0 \tag{8.5.12}$$

at the two walls. The tangential momentum equation at the walls can be written as

$$\frac{\partial U}{\partial t} + U\left(\frac{\partial u}{\partial \phi}\phi_x + \frac{\partial v}{\partial \phi}\phi_y\right) + \frac{c^2}{\gamma}|\nabla\phi|^2\frac{\partial P}{\partial \phi} = 0. \tag{8.5.13}$$

An equation for the wall pressure can be obtained by combining the pressure and normal momentum equations:

$$\begin{aligned}\frac{\partial P}{\partial t} = \pm c|\nabla\psi|\frac{\partial P}{\partial \psi} - \Bigg[U\frac{\partial P}{\partial \phi} + \gamma\left(\frac{\partial u}{\partial \psi}\psi_x + \frac{\partial u}{\partial \phi}\phi_x + \frac{\partial v}{\partial \psi}\psi_y + \frac{\partial v}{\partial \phi}\phi_y\right) \\ \mp \frac{\gamma U}{c|\nabla\psi|}\left(\frac{\partial u}{\partial \phi}\psi_x + \frac{\partial v}{\partial \phi}\psi_y\right)\Bigg],\end{aligned} \tag{8.5.14}$$

where the upper sign applies to the upper wall and the lower sign to the lower wall boundary. The spatial derivatives in (8.5.13) and (8.5.14) are computed from the Chebyshev interpolants and the Runge-Kutta method is used to find the new wall values of U and P. The new U together with (8.5.12) for V enable the updated Cartesian velocity components to be extracted. Kopriva et al. observed that the most crucial part of the wall boundary condition was the pressure equation. They comment on various alternatives to (8.5.14) which proved unsatisfactory. The success of (8.5.14) may be attributed to the fact that it expresses the compatibility relation for the characteristics intersecting the wall from the interior of the flow.

The supersonic outflow and inflow boundaries pose no difficulties. At the inflow all the quantities are specified. The outflow requires no boundary condition.

The two conditions specified for the subsonic inflow are the total enthalpy and the angle of the flow. A compatibility condition combining the normal momentum equation and the pressure equation is

$$\begin{aligned}\frac{\partial P}{\partial t} + (U - c|\nabla\phi|)\frac{\partial P}{\partial \phi} - \frac{\gamma}{|\nabla\phi|c}\left[\frac{\partial U}{\partial t} + U\left(\frac{\partial u}{\partial \phi}\phi_x + \frac{\partial v}{\partial \phi}\phi_y\right)\right] \\ = -\gamma\left(\frac{\partial u}{\partial \psi}\psi_x + \frac{\partial u}{\partial \phi}\phi_x + \frac{\partial v}{\partial \psi}\psi_y + \frac{\partial v}{\partial \phi}\phi_y\right).\end{aligned} \tag{8.5.15}$$

Since the total enthalpy is taken to be a constant along the inflow boundary, another relation between P and U can be obtained by differentiating the total enthalpy equation in time

$$\frac{\partial P}{\partial t} = -\frac{U}{|\nabla\phi|^2}e^{-P(\gamma-1)/\gamma}\frac{\partial U}{\partial t}. \tag{8.5.16}$$

Solving (8.5.15) and (8.5.16) allows both $\partial P/\partial t$ and $\partial U/\partial t$ to be computed.

Table 8.1. Maximum error in p for MacCormack and spectral computation of transonic Ringleb flow

Grid	MacCormack	Spectral
8×4	2.6×10^{-2}	2.2×10^{-2}
16×8	1.1×10^{-2}	1.9×10^{-3}
32×16	3.2×10^{-3}	5.0×10^{-5}

Table 8.2. Maximum error in p for MacCormack and spectral computation of supersonic Ringleb flow

Grid	MacCormack	Spectral
4×4	2.2×10^{-2}	7.5×10^{-4}
8×8	4.1×10^{-3}	1.1×10^{-6}
16×16	1.0×10^{-3}	6.6×10^{-11}

From the computed U and the fact that $V = 0$, the Cartesian velocities are calculated.

The initial condition for the calculations was the exact solution of the continuous problem. The solution was marched forward in time until the discrete equations were satisfied. No filtering was needed for either the supersonic or the transonic cases. The difference between the discrete solution and the exact solution measures the accuracy of the method. For comparison the same equations were solved by the MacCormack (1969) finite-difference method.

The computed Mach number contours are shown in Figs. 8.8(b) and 8.9(b). Convergence tables are provided in Tables 8.1 and 8.2. The results for supersonic flow are better than those for transonic flow (for a fixed number of grid points.) This occurs simply because the computational domain is smaller for the supersonic case. The sharpest gradients in the transonic flow occur near the sonic line, which is where the flow is accelerating from subsonic to supersonic. Once this feature has been resolved, the error decays rapidly.

Two lessons can be drawn from this example: (1) spectral methods can work well for non-linear systems of hyperbolic or mixed elliptic/hyperbolic equations with smooth solutions; and (2) effective numerical boundary conditions are based on the physical boundary conditions and the mathematical characterization of the problem. The availability of the exact solution makes this problem a good candidate for assessing the numerical boundary conditions. (A GAMM workshop was held on just this topic in the late 1970s (Förster (1978).)

8.5.3. Astrophysical Nozzle

Let us return now to problems containing discontinuities. Discussions of shock-capturing techniques are easiest for the Fourier spectral methods. The discrete operator is simple, the collocation points are uniformly distributed and the boundary conditions pose no difficulty. One non-trivial test case for spectral shock-capturing techniques uses an approximate set of equations derived by Woodward (1975) for studying the time development of shock waves in a spiral galaxy. The equations which describe an isothermal gaseous component, i.e., $p = c^2\rho$, in a very thin disk galaxy are

$$\frac{\partial u}{\partial t} + \frac{\partial(\rho u)}{\partial \phi} = 0$$

$$\frac{\partial(\rho u)}{\partial t} + \frac{\partial}{\partial \phi}[\rho(u^2 + c^2)] = 2\Omega(v - v_0)\rho + \rho\varepsilon \sin\phi \qquad (8.5.17)$$

$$\frac{\partial(\rho v)}{\partial t} + \frac{\partial(\rho u v)}{\partial \phi} = -\left(\frac{\kappa^2}{2\Omega}\right)(u - u_0)\rho,$$

where κ and Ω are rotational frequencies, and (u_0, v_0) is the mean flow. The boundary conditions are periodic in the spiral phase ϕ with period 2π. The last term in the middle equation represents the gravitational forcing of the gas from the spiral field of the much more massive stellar component. These equations have some significant differences from the Euler equations, notably the forcing term and the asymmetrical role of the velocity components. A more detailed explanation of this physical problem and the approximations used in deriving (8.5.17) is available in Woodward (1975).

The parameter ε is a dimensionless measure of the strength of the gravitational forcing. If this forcing is sufficiently strong, then the steady-state solution to this astrophysical problem contains a shock. Behind the shock there is a region of rapid decompression, and further downstream occur some features due to the second harmonic of the forcing. The steady-state solution, then, is more complex than some standard test problems (such as the one-dimensional shock tube) whose solutions are merely piecewise linear functions. The challenge for the spectral method is to capture the shock and to suppress its attendant oscillations without also destroying the remaining structure of the solution.

Fourier spectral calculations were performed for this problem by Hussaini et al. (1985a). They used the steady-state solution to (8.5.17) as the initial condition. The transients that would be generated by some other initial condition take a very long time to damp out, because spectral methods have very low inherent damping and there are no boundaries out of which transients can convect. The temporal discretization uses the second-order Adams–Bashforth predictor, third-order Adams–Moulton corrector discussed in Chap. 4. The calculations were run for 500 time-steps. The specific test

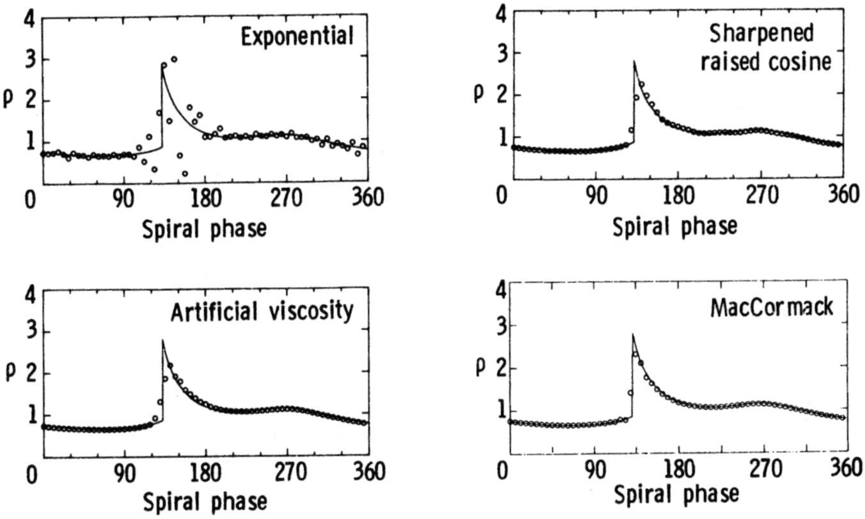

Figure 8.10. Computed density for the astrophysical model problem (circles). The solid line is the exact solution.

problem uses (in units which are not of interest here) $c = 8.56$, $\Omega = 21.37$, $u_0 = 13.5$, $\kappa = 26.75$, $v_0 = 115$, and $\varepsilon = 72.92$.

Figure 8.10 shows the effect of applying the weak exponential cut-off filter (with $\theta_c = 0.7\pi$ and $\alpha = 5$) every fifty time-steps. Both high and low frequency oscillations are quite evident. Even if a Lanczos cosmetic post-processing step is added, the results are not appreciably improved. The failure of the weak filtering strategy on this problem may appear puzzling in view of the success a similar strategy has achieved on a standard shock tube problem (Gottlieb, Lustman and Orszag (1981)). We attribute the difference to the strong expansion in the post-shock region in the present problem. Without appreciably stronger smoothing, something resembling an expansion shock will form in this region. The dip in the density plot is not smoothed by a weak filter and it grows until the density becomes negative and the calculation breaks down. This more difficult problem evidently requires more drastic filtering.

Such smoothing is provided by applying the sharpened raised cosine filter every fifty time-steps. As Fig. 8.10 indicates, the expansion is now adequately controlled and the shock is captured in two points. But notice that there are some low frequency errors in the vicinity of the second harmonic near a spiral phase of 270°. An alternative form of strong filtering is to add a non-linear artificial viscosity. In his second-order MacCormack's method calculations for the two-dimensional version of this problem, Liebovitch (1978) used a fourth-order viscosity. In the density equation this was proportional to

$$\nu_{i+1/2}(\rho_{i+1} - \rho_i) - \nu_{i-1/2}(\rho_i - \rho_{i-1}), \qquad (8.5.18)$$

where

$$\nu_{i+1/2} = |u_{i+1} - u_i|.$$

Similar terms were used for the two momentum equations with the appropriate momentum variable replacing the density variable in (8.4.18). Figure 8.10 shows the results of employing this non-linear artificial viscosity instead of any type of filtering. Compared with the previous case there is now one more point in the shock, but the solution is smoother, particularly near the second harmonic. Away from the shock, similar RMS errors are exhibited by these two solutions on eight, sixteen, thirty two and sixty four point grids. However, both error decay rates are only first-order. Nothing approaching "spectral accuracy", i.e., exponential convergence, has been obtained for this problem.

The last frame in Fig. 8.10 provides a comparison with a second-order MacCormack's finite-difference solution. Even this simple method yields better results than the best spectral method. Moreover, upwind finite-difference methods do better yet on this problem (Van Albada, Van Leer and Roberts (1982)). Clearly, a substantial improvement in filtering techniques for spectral shock-capturing methods will have to occur before this procedure can be recommended.

Spectral shock-capturing calculations for non-periodic problems have been reported by several workers. Gottlieb et al. applied a Chebyshev collocation method together with an elaborate filtering technique—based on fitting the final solution to simple discontinuous functions—to a standard one-dimensional shock-tube problem. They required no extra dissipation to stabilize the calculation, perhaps because a Chebyshev method contains some intrinsic dissipation. Taylor, Myers and Albert (1981) used some conventional artificial viscosity methods in a spectral shock-capturing solution of a similar problem. Gottlieb, Lustman and Streett (1984) and Sakell (1984) have computed two-dimensional shock reflection problems. All of these model problems have solutions which are piecewise linear. The flows contain no real small-scale structure which serves as a challenge to the filtering procedure. To date, no spectral shock-capturing calculation has been performed for a structured solution which retains spectral accuracy.

8.6. Shock-Fitting

The shock-induced oscillations that plague shock-capturing methods can be avoided by resorting to shock-fitting techniques (Moretti (1968, 1972)). Here, the shock front is a computational boundary whose shape and motion are generated during the calculation. Since the flow within the computational

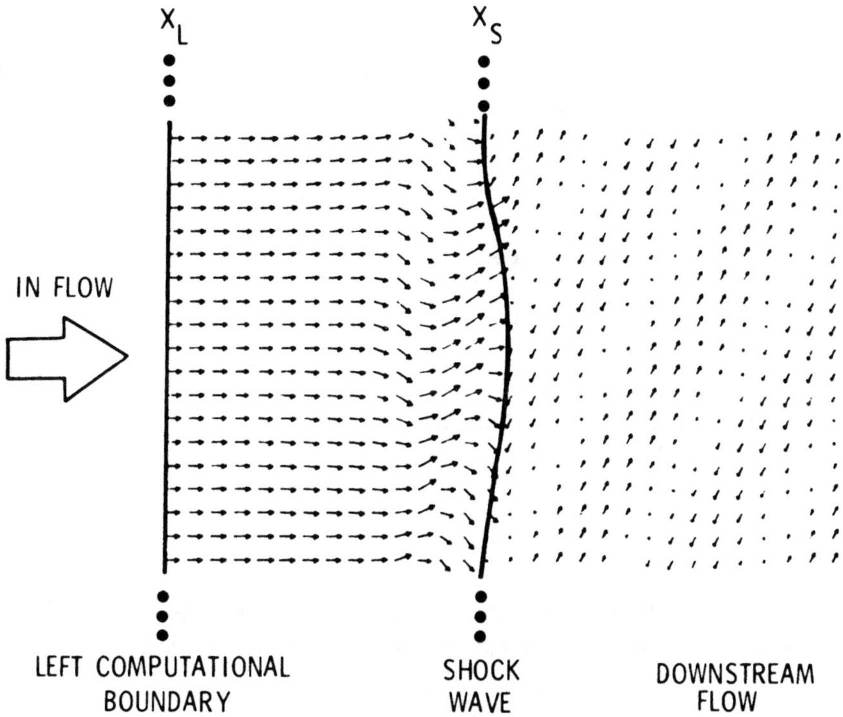

Figure 8.11. Typical shock-fitted time-dependent flow model in the physical plane. The downstream flow consists of a plane shear wave inclined at an angle of 30° to the shock front.

domain is smooth, there is reason to expect a shock-fitted solution to be highly accurate. Shock-fitted spectral solutions to the Euler equations were first presented by Salas, Zang and Hussaini (1982) and were described in more detail by Hussaini et al. (1985b).

A model problem which has been used to study the interaction of a shock wave with a vortex or with idealized turbulence is illustrated in Fig. 8.11. At time $t = 0$ an infinite, normal shock at $x = 0$ separates a rapidly moving, uniform fluid on the left from the fluid on the right, which is in a quiescent state except for some specified fluctuation. The initial conditions are chosen so that in the absence of any fluctuation the shock moves uniformly in the positive x-direction with a Mach number (relative to the fluid on the right) denoted by M_s. In the presence of fluctuations the shock front will develop ripples. The shape of the shock is described by the function $x_s(y, t)$. The numerical calculations are used to determine the state of the fluid in the region between the shock front and some suitable left boundary $x_L(t)$, and also to determine the motion and shape of the shock front itself.

The physical domain in which the fluid motion is computed is given by

$$\begin{aligned} x_L(t) \leqslant x \leqslant x_s(y,t) \\ 0 \leqslant y \leqslant y_l \\ t \geqslant 0. \end{aligned} \tag{8.6.1}$$

For the shock/turbulence interaction problem illustrated in Fig. 8.11, periodic boundary conditions in y are appropriate. The change of variables

$$\begin{aligned} X &= \frac{x - x_L(t)}{x_s(y,t) - x_L(t)} \\ Y &= y/y_l \\ T &= t, \end{aligned} \tag{8.6.2}$$

produces the computational domain

$$\begin{aligned} 0 \leqslant X \leqslant 1 \\ 0 \leqslant Y \leqslant 1 \\ T \geqslant 0. \end{aligned} \tag{8.6.3}$$

The fluid motion is modeled by the non-isentropic, two-dimensional Euler equations. In terms of the computational coordinates these are

$$\frac{\partial Q}{\partial T} + B\frac{\partial Q}{\partial X} + C\frac{\partial Q}{\partial Y} = 0, \tag{8.6.4}$$

where

$$Q = [P, u, v, s], \tag{8.6.5}$$

$$B = \begin{pmatrix} U & \gamma X_x & \gamma X_y & 0 \\ \frac{c^2}{\gamma} X_x & U & 0 & 0 \\ \frac{c^2}{\gamma} X_y & 0 & U & 0 \\ 0 & 0 & 0 & U \end{pmatrix}, \tag{8.6.6}$$

and

$$C = \begin{pmatrix} V & \gamma Y_x & \gamma Y_y & 0 \\ \frac{c^2}{\gamma} Y_x & V & 0 & 0 \\ \frac{c^2}{\gamma} Y_y & 0 & V & 0 \\ 0 & 0 & 0 & V \end{pmatrix}. \tag{8.6.7}$$

The contravariant velocity components are given by

$$U = X_t + uX_x + vX_y$$

and

$$V = Y_t + uY_x + vY_y. \tag{8.6.8}$$

The appropriate discretization is Chebyshev–Fourier collocation in space combined with a standard time-discretization.

The Rankine–Hugoniot conditions are used both to determine the flow variables (P, u, v and s) immediately upstream of the shock and to determine the shock position. Let the subscripts 1 and 2 denote the variables on the downstream (right) and upstream (left) sides of the shock. Since all the quantities on the downstream side are prescribed, the flow variables on the upstream side follow routinely from the Rankine–Hugoniot relations. Of course, these relations must be employed in a manner which accounts for the shock velocity and curvature.

A few preliminary definitions are needed for the equation which determines the shock position as a function of the computational time T. Let $\hat{\mathbf{N}}$ be the unit normal to the shock front. Its components in the physical plane are

$$\hat{\mathbf{N}} = \frac{(1, \partial x_s/\partial y)}{\sqrt{1 + (\partial x_s/\partial y)^2}}. \tag{8.6.9}$$

Let $\mathbf{u}_s$ denote the normal velocity of a point on the shock. Then

$$\mathbf{u}_s = u_s \hat{\mathbf{N}} \tag{8.6.10}$$

and $x_s(Y, T)$ can be obtained by integrating with respect to T the projection of u_s onto the X-direction. If the incoming normal velocity relative to the shock is denoted by u_{rel}, then

$$u_{\text{rel}} = \mathbf{u}_1 \cdot \hat{\mathbf{N}} - u_s, \tag{8.6.11}$$

and the relative Mach number is

$$M_{\text{rel}} = u_{\text{rel}}/c_1. \tag{8.6.12}$$

This numerical method presumes that M_{rel} is always greater than 1.

The Rankine–Hugoniot relations imply that

$$P_2 = P_1 + \ln\left[\gamma M_{\text{rel}}^2 - \frac{\gamma - 1}{2}\right] + \ln\left[\frac{2}{\gamma + 1}\right]. \tag{8.6.13}$$

The equation for the shock acceleration is obtained by differentiating (8.6.12) and (8.6.13) and then combining the results to obtain

$$\frac{\partial u_s}{\partial T} = A - \frac{c_1}{2\gamma M_{\text{rel}}}\left(\frac{\partial P_2}{\partial T} - \frac{\partial P_1}{\partial T}\right)\left(\gamma M_{\text{rel}}^2 - \frac{\gamma - 1}{2}\right) - M_{\text{rel}}\frac{\partial c_1}{\partial T}, \tag{8.6.14}$$

where

$$A = \frac{\partial \mathbf{u}_1}{\partial T} \cdot \hat{\mathbf{N}} + \mathbf{u}_1 \cdot \frac{\partial \hat{\mathbf{N}}}{\partial T}. \tag{8.6.15}$$

The time derivatives on the right-hand side of (8.6.14) are obtained from (8.6.4) using spectral approximations to the spatial derivatives. The shock velocity is obtained by integrating (8.6.14) with respect to T.

The collocation grid in the computational plane is fixed and uniform. Since the shock front moves to the right in the course of the calculation, the corresponding discrete grid in the physical plane is expanding. Thus, the effective resolution in the x-direction continually decreases during the evolution. Eventually, the resolution of any calculation will become inadequate and the results will no longer be reliable. Fortunately, in many situations the important information can be extracted before this occurs, especially if the initial grid is taken to be a fine one.

The correct boundary conditions at both the left and right boundaries depend upon the relative shock Mach number. If $\gamma = 1.4$ and $M_s \geqslant 2.08$, then the flow behind the shock is supersonic. In this case both boundaries are supersonic inflow boundaries and it is appropriate to prescribe all variables there. If $M_s < 2.08$, then these boundaries are subsonic inflow ones. The advisable procedure here is to base the numerical boundary conditions on the linearized characteristics of the Euler equations. At the left (subsonic) boundary the (linearized) characteristic variable corresponding to the outgoing characteristic direction is

$$R^- = P - \frac{\gamma}{c} u. \tag{8.6.16}$$

Similarly,

$$R^+ = P + \frac{\gamma}{c} u \tag{8.6.17}$$

corresponds to the outgoing characteristic direction at the right (subsonic) boundary.

A set of successful numerical boundary conditions on the left is obtained by first calculating preliminary values of all quantities at the left boundary and then incorporating the given values of s, v, and R^+ as

$$\begin{aligned} s &= s_{\text{given}} \\ v &= v_{\text{given}} \\ P + \frac{\gamma}{c} u &= R^+_{\text{given}} \\ P - \frac{\gamma}{c} u &= P_{\text{prelim}} - \frac{\gamma}{c} u_{\text{prelim}}. \end{aligned} \tag{8.6.18}$$

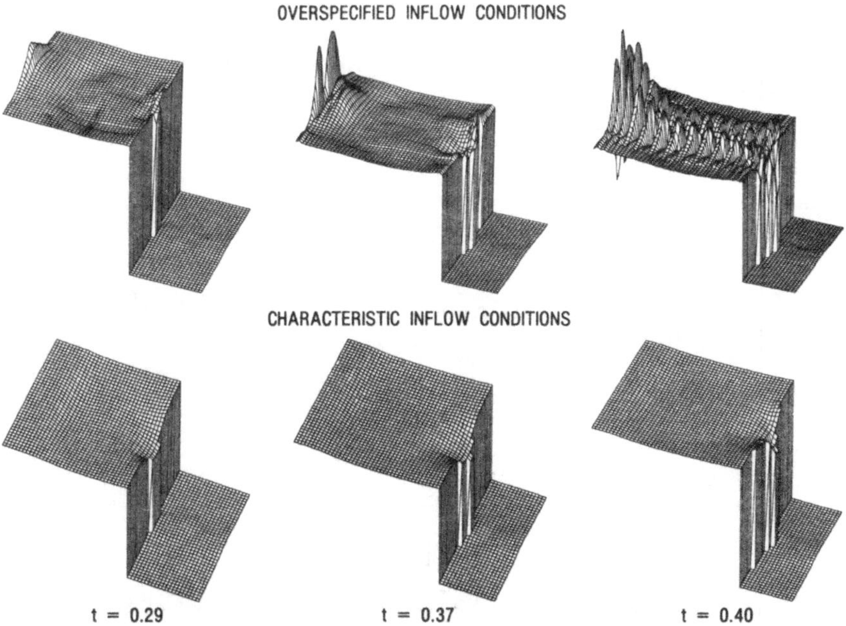

Figure 8.12. Computed pressure distribution for a Mach 1.3 Karman vortex street.

Thus, the PDE is used to update the appropriate characteristic combination of variables at the boundary. For the right boundary a similar characteristic correction procedure can be incorporated into the evaluation of the $\partial P_2/\partial T$ term in (8.6.14). This characteristic affects the shock velocity. The periodic boundary conditions in y are satisfied automatically.

The effectiveness of these characteristic boundary conditions is illustrated in Fig. 8.12. The shock Mach number $M_s = 1.3$. Thus, the inflow boundary is subsonic. If all the inflow quantities are specified, the solution rapidly degenerates, as shown in the top row of the figure. Even at $t = 0.40$ no physical signals have had a chance to propagate from the shock front to the left boundary. However, numerical signals reach there immediately in a spectral method. This boundary instability is cured by the use of the characteristic boundary conditions (8.6.18).

The non-linear interaction of plane waves with shocks was examined at length by Zang, Hussaini and Bushnell (1984). An important issue is the dependence of the amplification coefficients, i.e., the ratios of upstream to downstream wave amplitudes, upon Mach number and the angle of incidence of the plane wave (see Fig. 8.11). The numerical method used there was similar to the one described above, but employed second-order finite differences in place of the present Chebyshev–Fourier spectral discretization. Detailed comparisons were made with the predictions of linear theory. The linear results turned out to be surprisingly robust, remaining valid at very low (but

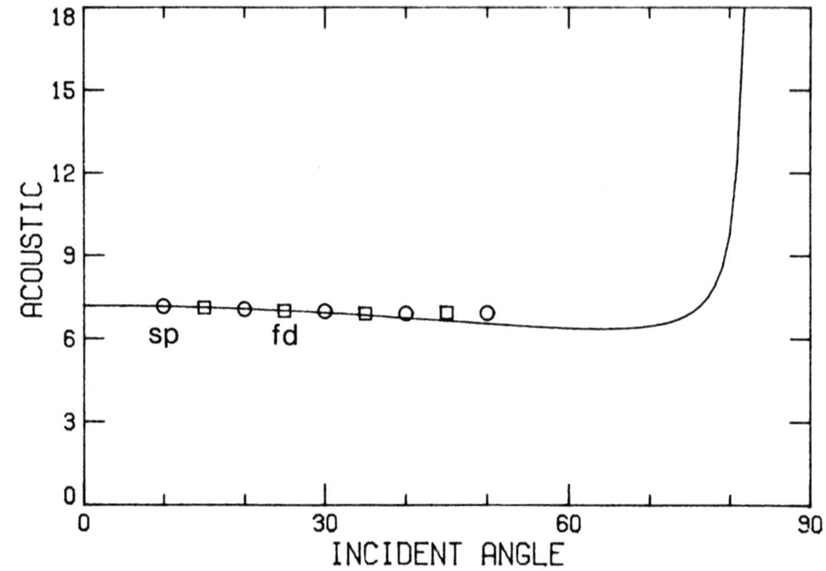

Figure 8.13. Angular dependence of the acoustic amplification coefficient for a Mach 3 shock. Spectral (circles) and finite-difference (squares) solutions are compared with the linear result (solid line).

still supersonic) Mach numbers and at very high incident wave amplitudes. The only substantial disagreement occurred for incident waves whose wave fronts were nearly perpendicular to the shock front. This type of shock/turbulence interaction is a useful test of the spectral technique, because the method can be calibrated in the regions for which linear theory has been shown to be valid.

Figure 8.13 presents a comparison of the acoustic wave amplification predicted by spectral methods, finite-difference methods and linear theory. As is discussed by Zang et al., linear theory is quite reliable at angles below, say, 45°. The finite-difference results were obtained with the second-order MacCormack's method. The finite-difference grid was 64×16 and the spectral grid was 32×8. Figure 8.13 shows that both methods produce the same results. A head-to-head comparison of both methods for the $\theta_1 = 10°$ case is provided in Table 8.3. The "exact" value is taken from linear theory. Since the amplitude of the incident acoustic wave is so small, it should come as no surprise that four points in the y-direction suffice for the spectral calculation. Note that the standard deviations are substantially smaller for the spectral method. These results suggest that the spectral method requires only half as many grid points in each coordinate direction.

Spectral shock-fitting has been applied to several other problems. Salas et al. computed the interaction of shocks with an isolated vortex, a Karman

Table 8.3. Grid dependence of acoustic transmission coefficient

Grid	Finite-difference	Chebyshev–Fourier spectral
16 × 4	6.403 ± 2.652	7.257 ± 0.587
16 × 8	6.427 ± 2.626	7.257 ± 0.587
32 × 4	7.105 ± 0.453	7.158 ± 0.022
32 × 8	7.134 ± 0.471	7.158 ± 0.022
32 × 16	7.139 ± 0.497	7.158 ± 0.022
64 × 16	7.163 ± 0.078	7.157 ± 0.017
128 × 16	7.152 ± 0.022	
"exact"	7.156	7.156

vortex street and an isolated hot spot. In these applications, the flow is not periodic in y. A hyperbolic tangent mapping in y was combined with a Chebyshev discretization in the mapping variable. Hussaini et al. (1985b) computed the steady flow past a circular cylinder in a uniform stream.

A possible alternative to the shock-fitting method, albeit one which has not yet been attempted for spectral discretizations, is the method of integral relations, which has been developed for finite-difference schemes (Holt (1977)).

8.7. Reacting Flows

An emerging application field for spectral methods is reacting flows. These flows are especially challenging because they contain sharp gradients in both space and time, and because most real flows involve dozens or even hundreds of species. Flame fronts and shock waves are an additional complication. Some of the important features are mixing rates, ignition, and flame holding.

There are a number of simplifying assumptions which lead to more tractable, but less realistic, models of reacting flows. The most drastic of these is that the reactions proceed without heat release and that the Mach number is so low that the flow may be treated as incompressible. Riley, Metcalfe and Orszag (1986) have performed some three-dimensional simulations of a 2 species, time-developing mixing layer. They used the same type of free-slip boundary conditions that were discussed in Sec. 7.3.5 and obtained good agreement with both similarity theory and experimental data.

McMurtry, et al. (1986) employed a low Mach number approximation which includes some mild heat release effects but neglects the acoustic modes. They peformed some two-dimensional calculations which indicate that the first-order effect of heat release is to reduce the rate of mixing.

Drummond, Hussaini and Zang (1986) applied a Chebyshev spectral method to a supersonic quasi-one-dimensional diverging nozzle flow with a simple but quite stiff 2 species hydrogen-air reaction. The spectral method proved to be quite economical compared with a benchmark finite-difference result. The Chebyshev grid point distribution was quite well-adapted to the sharp gradients at the nozzle inflow, but less well suited to the fairly uniform outflow region.

Bayliss and Matkowsky (1987) applied an adaptive Chebyshev method to a one-dimensional reaction-diffusion problem. This problem exhibits relaxation oscillations. The position of the flame front changes with time and the adaptive grid strategy was applied to provide resolution in this region of steep gradients.

CHAPTER 9

Global Approximation Results

In this chápter we present error estimates for the approximation of functions by orthogonal polynomials. The results will cover the following topics:

(i) inverse inequalities for polynomials concerning summability and differentiability;
(ii) error estimates for the truncation error $u - P_N u$, where $P_N u$ denotes the truncated "Fourier" series of u;
(iii) existence, uniqueness and error estimates for the polynomials of best approximation in L^p or Sobolev norms;
(iv) error estimates for the interpolation error $u - I_N u$, where $I_N u$ denotes the polynomial interpolating u at a selected set of points in the domain.

Many of the results we present are taken from the general theory of approximation by polynomials. Their interest extends beyond the boundaries of approximation theory, since they are applied to the convergence analysis of spectral methods (see Chap. 10). We include proofs of those results which are most significant for the analysis of spectral methods.

In all the estimates contained in this chapter, C will denote a positive constant which depends upon the type of norm involved in the estimate, but which is independent of both the function u and the integer N.

9.1. Fourier Approximation

In this section, as well as throughout the remaining chapters, we will deal with trigonometric polynomials of degree up to N, rather than $N/2$ as in the previous chapters. This change is motivated by the desire for simplicity in the mathematical notation. Thus, we denote here by S_N the space of the trigonometric polynomials of degree up to N:

$$S_N = \text{span}\{e^{ikx} | -N \leqslant k < N\}. \tag{9.1.1}$$

9.1.1. Inverse Inequalities for Trigonometric Polynomials

We consider the problem of the equivalence of the L^p norms for trigonometric polynomials. We recall that the L^p norm of a function u over $(0, 2\pi)$ is defined as follows:

$$\|u\|_{L^p(0,2\pi)} = \left(\int_0^{2\pi} |u(x)|^p \, dx\right)^{1/p} \qquad 1 \leqslant p < \infty \tag{9.1.2}$$

and

$$\|u\|_{L^\infty(0,2\pi)} = \sup_{0 \leqslant x \leqslant 2\pi} |u(x)| \qquad p = \infty. \tag{9.1.3}$$

The set of functions for which each particular norm is finite forms a Banach space denoted by $L^p(0, 2\pi)$ (see (A.9.f)). The following several inequalities enable one to relate the norms of a given polynomial in different L^p spaces.

If p, q are any real numbers such that $1 \leqslant p \leqslant q \leqslant \infty$, and if $u \in L^q(0, 2\pi)$, then $u \in L^p(0, 2\pi)$, and $\|u\|_{L^p(0,2\pi)} \leqslant C\|u\|_{L^q(0,2\pi)}$, where C depends on p and q. If u is a periodic function with a finite expansion this inequality can be inverted. Indeed, the following *Nikolski's inequality* holds:

$$\|\phi\|_{L^q(0,2\pi)} \leqslant CN^{1/p-1/q}\|\phi\|_{L^p(0,2\pi)} \qquad \text{for all } \phi \in S_N. \tag{9.1.4}$$

A different kind of inverse inequality, the *Bernstein inequality*, relates the norm of a function $u \in S_N$ to that of its derivatives. For all real p, $1 \leqslant p \leqslant \infty$, and for all integers $r \geqslant 1$

$$\|\phi^{(r)}\|_{L^p(0,2\pi)} \leqslant N^r \|\phi\|_{L^p(0,2\pi)} \qquad \text{for all } \phi \in S_N, \tag{9.1.5}$$

where $\phi^{(r)}$ denotes the derivative of order r of ϕ.

9.1.2. Estimates for the Truncation and Best Approximation Errors

Let P_N: $L^2(0, 2\pi) \to S_N$ be the orthogonal projection upon S_N in the inner product of $L^2(0, 2\pi)$ (see (2.1.10)):

$$(u - P_N u, \phi) = 0 \qquad \text{for all } \phi \in S_N.$$

With the present definition of S_N (see (9.1.1)), $P_N u$ is the truncated Fourier series of u, i.e.,

$$P_N\left(\sum_{k=-\infty}^{\infty} \hat{u}_k \phi_k\right) = \sum_{k=-N}^{N-1} \hat{u}_k \phi_k,$$

where $\phi_k(x) = e^{ikx}$.

A natural family of norms for the modern numerical analysis of differential equations is comprised of the Sobolev norms. Hence, we present approximation results with respect to these norms. We recall that the Sobolev norm of integer order $m \geqslant 0$ is given by

$$\|u\|_{H^m(0,2\pi)} = \left\{\sum_{k=0}^{m} \int_0^{2\pi} |u^{(k)}(x)|^2 \, dx\right\}^{1/2}. \tag{9.1.6}$$

The reader unfamiliar with Sobolev spaces can think of $u^{(k)}$ as the classical (continuous) derivative of u of order k. However, this norm can be defined for a wider class of functions. These form a Hilbert space, called $H^m(0, 2\pi)$, which is introduced in (A.11.a). We are concerned here with functions periodic on $(0, 2\pi)$. We consider the subspace $H_p^m(0, 2\pi)$ of $H^m(0, 2\pi)$, which consists of functions whose first $m - 1$ derivatives are periodic (see (A.11.d)). Since $(e^{ikx})' = ike^{ikx}$, it follows that for any $u = \sum_{k=-\infty}^{\infty} \hat{u}_k \phi_k \in H_p^m(0, 2\pi)$, the norm $\|u\|_{H^m(0,2\pi)}$ is equivalent to

$$\|u\|_m = \left\{ \sum_{k=-\infty}^{\infty} (1 + |k|^{2m}) |\hat{u}_k|^2 \right\}^{1/2}, \tag{9.1.7}$$

i.e., for some positive constants C_1 and C_2 which are independent of u

$$C_1 \|u\|_{H^m(0,2\pi)} \leqslant \|u\|_m \leqslant C_2 \|u\|_{H^m(0,2\pi)}.$$

The spaces $H_p^m(0, 2\pi)$ consist of functions for which it is permissible to differentiate termwise the Fourier series m times, provided the convergence is in the square mean. For instance, $H_p^1(0, 2\pi)$ is the space of all functions u for which

$$u' = \sum_{k=-\infty}^{\infty} ik\hat{u}_k \phi_k \qquad \text{in } L^2(0, 2\pi). \tag{9.1.8}$$

This means that the Fourier series of u' converges in the squared mean to the derivative of u. Result (9.1.8) is a direct consequence of the commutability of the operators d/dx and P_N on $H_p^1(0, 2\pi)$, i.e.,

$$(P_N u)' = P_N u' \qquad \text{for all } u \in H_p^1(0, 2\pi).$$

This, in turn follows from the identity

$$2\pi \widehat{(u')}_k \equiv (u', \phi_k) = -(u, \phi_k') = ik(u, \phi_k) = ik\hat{u}_k \qquad \text{for all } k.$$

Since u is in $H_p^1(0, 2\pi)$, the first inner product is well-defined. By the same arguments, a similar characterization can be given also for $H_p^m(0, 2\pi)$. It is enough to replace the first derivative with the m-th order derivative in (9.1.8).

The first error estimate we present concerns the truncation error in the L^2 norm. We recall that, by definition, $P_N u$ is the best approximation of u in the L^2-norm among all the functions in S_N. One has

$$\|u - P_N u\|_{L^2(0,2\pi)} \leqslant CN^{-m} \|u^{(m)}\|_{L^2(0,2\pi)} \qquad \text{for any } u \in H_p^m(0, 2\pi),\ m \geqslant 0. \tag{9.1.9}$$

This follows from the Parseval identity (2.1.14). Indeed,

$$\frac{1}{\sqrt{2\pi}} \|u - P_N u\|_{L^2(0,2\pi)} = \left(\sum_{|k| \gtrsim N} |\hat{u}_k|^2 \right)^{1/2} = \left(\sum_{|k| \gtrsim N} \frac{1}{|k|^{2m}} |k|^{2m} |\hat{u}_k|^2 \right)^{1/2}$$
$$\leqslant N^{-m} \left(\sum_{|k| \gtrsim N} |k|^{2m} |\hat{u}_k|^2 \right)^{1/2},$$

where the symbol $\sum_{k \gtrsim N}$ has been introduced in (2.1.16). The last bracket can be bounded by the L^2-norm of $u^{(m)}$; hence, (9.1.9) follows.

Moreover, we can estimate the truncation error in higher Sobolev norms as follows:

$$\|u - P_N u\|_{H^l(0,2\pi)} \leqslant CN^{l-m} \|u^{(m)}\|_{L^2(0,2\pi)} \tag{9.1.10}$$

for any $m \geqslant 0$ and any $0 \leqslant l \leqslant m$. The proof of (9.1.10) is very similar to the one of (9.1.9). Indeed,

$$\|u - P_N u\|_{H^l(0,2\pi)} = \left\{ \sum_{|k| \gtrsim N} (1 + |k|^{2l}) |\hat{u}_k|^2 \right\}^{1/2}$$

$$\leqslant 2 \left\{ \sum_{|k| \gtrsim N} |k|^{2m-2(m-l)} |\hat{u}_k|^2 \right\}^{1/2} \leqslant CN^{l-m} \|u^{(m)}\|_{L^2(0,2\pi)}.$$

We have seen that truncation and differentiation commute. Hence, $P_N u$ is the best approximation of u in S_N for any Sobolev norm. However, it is not so if we consider the L^p norms, $1 \leqslant p \leqslant \infty$. An estimate of $u - P_N u$ in these norms can be given as a consequence of a preliminary investigation of the best approximation error. Results of this kind are known as *Jackson's theorems*. We shall recall here those applied to the forthcoming convergence analysis.

The first result is concerned with the best approximation in S_N relative to the maximum norm; it states that for any $m \geqslant 0$

$$\inf_{\phi \in S_N} \|u - \phi\|_{L^\infty(0,2\pi)} \leqslant \frac{\pi}{2} N^{-m} \|u^{(m)}\|_{L^\infty(0,2\pi)}. \tag{9.1.11}$$

The next result concerns best approximation errors in L^p for the whole range $1 \leqslant p < \infty$:

$$\inf_{\phi \in S_N} \|u - \phi\|_{L^p(0,2\pi)} \leqslant CN^{-m} \|u^{(m)}\|_{L^p(0,2\pi)}. \tag{9.1.12}$$

In the two previous estimates we have assumed that the m-th order derivative of u (in the sense of periodic distributions, see (A.10.c)) belongs to the space $L^p(0, 2\pi)$ for which the norm on the right-hand side is finite.

We deal now with the evaluation of the truncation error $u - P_N u$ in the L^p-norms, $1 \leqslant p \leqslant \infty$. We recall first that if $u \in L^p(0, 2\pi)$, with $1 < p < \infty$, then its Fourier series converges, i.e.,

$$\|u - P_N u\|_{L^p(0,2\pi)} \longrightarrow 0 \qquad \text{as } N \to \infty. \tag{9.1.13}$$

This result includes and generalizes the property (2.1.9) which corresponds to the case $p = 2$. Furthermore, if $1 < p < \infty$,

$$\|u - P_N u\|_{L^p(0,2\pi)} \leqslant C \inf_{\phi \in S_N} \|u - \phi\|_{L^p(0,2\pi)}. \tag{9.1.14}$$

Hence, $P_N u$ approximates u in the L^p-norms with the same order as the best approximation. If $p = 1$ or $p = \infty$, inequality (9.1.14) still holds provided the constant C is replaced by $C(1 + \log N)$.

9.1.3. Estimates for the Interpolation Error

Let $I_N u \in S_N$ denote the trigonometric interpolant of the function u at the nodes $x_j = \pi j/N, j = 0, \ldots, 2N-1$ (see (2.1.25), where on the right-hand side $N/2$ must be replaced by N). We shall give some approximation results for the interpolation error $u - I_N u$. For the estimate in the L^2-norm we have

$$\|u - I_N u\|_{L^2(0,2\pi)} \leqslant CN^{-m}\|u^{(m)}\|_{L^2(0,2\pi)} \quad \text{for all } u \in H_p^m(0,2\pi) \text{ with } m \geqslant 1. \tag{9.1.15}$$

A comparison of (9.1.9) and (9.1.15) reveals that the interpolation error behaves asymptotically like the truncation error. A proof of this estimate will be presented at the end of the section.

The following result provides an estimate of the interpolation error in the maximum norm:

$$\|u - I_N u\|_{L^\infty(0,2\pi)} \leqslant C(\log N)N^{-m}\|u^{(m)}\|_{L^\infty(0,2\pi)}. \tag{9.1.16}$$

The result (9.1.15) allows one to estimate the *aliasing error* $R_N u = I_N u - P_N u$ (see (2.1.30)). Indeed, since by (2.1.32), $\|R_N u\|_{L^2(0,2\pi)} \leqslant \|u - I_N u\|_{L^2(0,2\pi)}$, one gets

$$\|R_N u\|_{L^2(0,2\pi)} \leqslant CN^{-m}\|u^{(m)}\|_{L^2(0,2\pi)} \tag{9.1.17}$$

under the same hypotheses as (9.1.15). *The important implication of this estimate is that the aliasing error is asymptotically no worse than the interpolation error in the L^2-norm.*

An evaluation of the interpolation error in all Sobolev norms is now possible, and it is given by the estimate

$$\|u - I_N u\|_{H^l(0,2\pi)} \leqslant CN^{l-m}\|u^{(m)}\|_{L^2(0,2\pi)} \quad \text{for } 0 \leqslant l \leqslant m \text{ and } u \in H_p^m(0,2\pi) \text{ with } m \geqslant 1. \tag{9.1.18}$$

This inequality follows directly from the preceding results. It is a consequence of (9.1.17), (9.1.10) and the Bernstein inequality (9.1.5) used with $p = 2$ and $r = l$. Indeed we get

$$\begin{aligned}\|u - I_N u\|_{H^l(0,2\pi)} &\leqslant \|u - P_N u\|_{H^l(0,2\pi)} + \|R_N u\|_{H^l(0,2\pi)} \\ &\leqslant CN^{l-m}\|u^{(m)}\|_{L^2(0,2\pi)} + CN^l\|R_N u\|_{L^2(0,2\pi)} \\ &\leqslant CN^{l-m}\|u^{(m)}\|_{L^2(0,2\pi)}.\end{aligned}$$

As a particular relevant case of (9.1.18), one can estimate the error produced in evaluating the pseudospectral derivative of a function (see Sec. 2.1.3):

$$\|u' - (I_N u)'\|_{L^2(0,2\pi)} \leqslant CN^{1-m}\|u^{(m)}\|_{L^2(0,2\pi)} \tag{9.1.19}$$

for all $u \in H_p^m(0, 2\pi)$, $m \geqslant 1$. Equivalently, recalling the identity (2.1.27) and noting that $u'(x_j) = I_N u'(x_j)$ for $j = 0, \ldots, 2N-1$, one has under the same hypotheses

$$\left(\frac{\pi}{N}\sum_{j=0}^{2N-1}[u'(x_j) - (I_N u)'(x_j)]^2\right)^{1/2} \leqslant CN^{1-m}\|u^{(m)}\|_{L^2(0,2\pi)}. \tag{9.1.20}$$

When the function u is analytic, the error $u' - I_N u'$ decays exponentially in N. Precisely, if u is a 2π-periodic analytic function in the strip $|\text{Im}\, z| < \eta_0$, then

$$\|u' - (I_N u)'\|_{L^2(0,2\pi)} \leqslant C(\eta)Ne^{-N\eta/2} \tag{9.1.21}$$

for all η, $0 < \eta < \eta_0$.

Proof of (9.1.15). For each function u: $(0, 2\pi) \to \mathbb{C}$ we consider the function $\mathscr{F}u$: $(0, 2\pi N) \to \mathbb{C}$ such that $\mathscr{F}u(x) = u(x/N)$ for all $x \in (0, 2\pi N)$. Then we define

$$S_N^* = \{\mathscr{F}\phi |\ \phi \in S_N\}.$$

Let $x_j = j\pi/N$, $j = 0, \ldots, 2N-1$, be the interpolation points, and set $\theta_j = Nx_j$ for $j = 0, \ldots, 2N-1$. We denote by I_N^* the interpolation operator with respect to these points, i.e., for all $u \in C^0([0, 2\pi N])$,

$$I_N^* u \in S_N^*, \quad I_N^* u(\theta_j) = u(\theta_j) \quad \text{for } j = 0, \ldots, 2N-1. \tag{9.1.22}$$

The following three relations can be easily proved:

$$\mathscr{F}(I_N u) = I_N^*(\mathscr{F}u) \qquad \text{for all } u \in C^0([0, 2\pi]) \tag{9.1.23}$$

$$I_N^* u = u \qquad \text{for all } u \in S_N^* \tag{9.1.24}$$

$$\|u^{(l)}\|_{L^2(0,2\pi)} = N^{l-1/2}\|(\mathscr{F}u)^{(l)}\|_{L^2(0,2\pi N)} \qquad l \geqslant 0. \tag{9.1.25}$$

Then, if we denote by E the identity operator (i.e., $E(u) = u$ for all u), it follows that:

$$\begin{aligned}\|u - I_N u\|_{L^2(0,2\pi)} &= N^{-1/2}\|\mathscr{F}u - I_N^*(\mathscr{F}u)\|_{L^2(0,2\pi N)} \\ &= N^{-1/2}\|(E - I_N^*)(\mathscr{F}u - \mathscr{F}(P_N u))\|_{L^2(0,2\pi N)} \\ &\leqslant N^{-1/2}\|E - I_N^*\|_{\mathscr{L}_m}\|\mathscr{F}(u - P_N u)\|_{H^m(0,2\pi N)}.\end{aligned} \tag{9.1.26}$$

We have denoted by $\mathscr{L}_m = \mathscr{L}(H_p^m(0, 2\pi N), L^2(0, 2\pi N))$ the space of all linear and continuous applications from $H_p^m(0, 2\pi N)$ into $L^2(0, 2\pi N)$ (see (A.3)). Using (9.1.25) and (9.1.10) gives

$$\begin{aligned}\|\mathscr{F}(u - P_N u)\|^2_{H^m(0,2\pi N)} &= \sum_{l=0}^{m} N^{1-2l}\|(u - P_N u)^{(l)}\|_{L^2(0,2\pi)} \\ &\leqslant CN^{1-2m}\|u^{(m)}\|_{L^2(0,2\pi)}.\end{aligned}$$

Then, from (9.1.26) we obtain

$$\|u - I_N u\|_{L^2(0,2\pi)} \leqslant CN^{-m}\|u^{(m)}\|_{L^2(0,2\pi)}\|E - I_N^*\|_{\mathscr{L}_m}. \tag{9.1.27}$$

Since $\|E\|_{\mathscr{L}_m} = 1$, it remains to prove that there is a constant C independent of N

such that:

$$\|I_N^*\|_{\mathscr{L}_m} \leqslant C. \tag{9.1.28}$$

We note that (see (A.3))

$$\|I_N^*\|_{\mathscr{L}_m} = \sup\{\|I^*v\|_{L^2(0,2\pi N)} | v \in H_p^m(0, 2\pi N), \|v\|_{H^m(0,2\pi N)} = 1\}. \tag{9.1.29}$$

Using (9.1.23) and (9.1.25) it follows that

$$\begin{aligned} \|I_N^* v\|_{L^2(0,2\pi N)} &= N^{1/2}\|I_N(\mathscr{F}^{-1}v)\|_{L^2(0,2\pi)} = N^{1/2}\left\{\int_0^{2\pi} |I_N(\mathscr{F}^{-1}v)|^2\,dx\right\}^{1/2} \\ &= N^{1/2}\left\{\frac{\pi}{N}\sum_{j=0}^{2N-1} |(\mathscr{F}^{-1}v)(x_j)|^2\right\}^{1/2} \\ &= \sqrt{\pi}\left\{\sum_{j=0}^{2N-1} |v(\theta_j)|^2\right\}^{1/2}. \end{aligned} \tag{9.1.30}$$

We can write $[0, 2\pi N] = \bigcup_{j=0}^{2N-1} [\theta_j, \theta_{j+1}]$, and, by the Sobolev inequality (see (A.12)) we get for each $m \geqslant 1$

$$|v(\theta_j)| \leqslant C\|v\|_{H^m(\theta_j,\theta_{j+1})} \qquad \text{for } j = 0, \ldots, 2N-1.$$

Thus,

$$\sum_{j=0}^{2N-1} |v(\theta_j)|^2 \leqslant C\|v\|^2_{H^m(0,2\pi N)},$$

and (9.1.28) follows now from (9.1.29) and (9.1.30). □

Bibliograpical Notes

Nikolskii's inequality has been proven in Nikolskii (1951). In Butzer and Nessel (1971) one can find proofs of the Bernstein inequality (Theorem 2.3.1 and Corollary 2.3.2), estimate (9.1.12) (Theorem 2.2.3), estimate (9.1.14) (Proposition 9.3.8), and the convergence result (9.1.13) (Theorem 9.3.6). Estimate (9.1.11) is proved in Cheney (1966, p. 145). Estimate (9.1.15) was first proved by Kreiss and Oliger (1979). The proof given here is due to Pasciak (1980), who actually proved (9.1.18). Estimate (9.1.16) was proven by Jackson (1930, p. 123). Finally, inequality (9.1.20) has been established by Tadmor (1986).

9.2. Sturm–Liouville Expansions

In this section we consider expansions with respect to eigenfunctions of Sturm–Liouville problems. We refer for notation to Sec. 2.2.1. We analyze the decay properties of the coefficients of a function with respect to such a basis, distinguishing between regular and singular Sturm–Liouville problems.

We assume that the coefficients p, q and w satisfy the assumptions made in Sec. 2.2.1. Moreover, we suppose that $\int_{-1}^{1} w(x)^{-1}\,dx < +\infty$.

9.2.1. Regular Sturm–Liouville Problems

If the function p is bounded from below by a positive constant, say $p(x) \geqslant p_0 > 0$, then the two boundary conditions, to be specified in (2.2.1) assume the form

$$\begin{aligned} \alpha_1 u(-1) + \beta_1 u'(-1) &= 0 \qquad \alpha_1^2 + \beta_1^2 \neq 0 \\ \alpha_2 u(1) + \beta_2 u'(1) &= 0 \qquad \alpha_2^2 + \beta_2^2 \neq 0, \end{aligned} \tag{9.2.1}$$

for suitable $\alpha_1, \beta_1, \alpha_2, \beta_2$. In this case we are speaking of a *regular* Sturm–Liouville boundary value problem.

Under the assumption that $\alpha_1\beta_1 \leqslant 0$ and $\alpha_2\beta_2 \geqslant 0$, it is known (see, e.g., Courant and Hilbert (1953, vol. I)), that the eigenvalues of the regular Sturm–Liouville problem (2.2.1), (9.2.1) form an infinite, unbounded sequence of non-negative numbers $0 \leqslant \lambda_0 \leqslant \cdots \leqslant \lambda_k \leqslant \lambda_{k+1} < \cdots$ and have multiplicity 1. The corresponding eigenfunctions ϕ_k, determined up to a constant, have exactly k zeroes in the open interval $(-1, 1)$. The asymptotic behavior of the eigenvalues as $k \to \infty$ is given by the formula

$$\lim_{k\to\infty} \frac{k^2}{\lambda_k} = \frac{\pi^2}{4} \int_{-1}^{1} \sqrt{w/p}\,dx. \tag{9.2.2}$$

The asymptotic behavior of the eigenfunctions depends on the type of boundary conditions. For instance, for the Neumann boundary conditions $u'(-1) = u'(1) = 0$, one has

$$\phi_k(x) = A_k \cos\frac{\pi}{2}k(x+1) + \frac{O(1)}{k} \qquad k \to \infty.$$

Eigenfunctions are mutually orthogonal with respect to the weighted inner product

$$(u, v)_w = \int_{-1}^{1} u(x)v(x)w(x)\,dx \tag{9.2.3}$$

namely

$$(\phi_k, \phi_m)_w = 0 \qquad \text{if } k \neq m. \tag{9.2.4}$$

Moreover, the system $\{\phi_k, k = 0, 1, \ldots,\}$ is complete in the weighted $L_w^2(-1, 1)$ space (see (A.9.g)). This means that if we define the sequence of the "Fourier" coefficients of a function $u \in L_w^2(-1, 1)$ as

$$\hat{u}_k = (u, \phi_k)_w \qquad k = 0, 1, \ldots$$

(ϕ_k ia assumed to be normalized by $\|\phi_k\|_{L_w^2(-1,1)} = 1$), and we set

$$P_N u = \sum_{k=0}^{N} \hat{u}_k \phi_k \qquad \text{for } N \text{ integer } > 0,$$

then

$$\|u - P_N u\|_{L^2_w(-1,1)} \to 0 \qquad \text{as } N \to +\infty.$$

In other words, the "Fourier" series $\sum_{k=0}^{\infty} \hat{u}_k \phi_k$ of u is convergent to u in the weighted squared mean for any $u \in L^2_w(-1,1)$.

Local convergence properties require more regularity on u. For instance, as in the case of the Fourier expansion, if u is of bounded variation on $[-1,+1]$ (see (A.8)), $P_N u(x)$ converges pointwise to $[u(x^+) + u(x^-)]/2$ for any $x \in [-1,1]$, (see, e.g., Titchmarsh (1962)).

The rate of decay of the coefficients of a function $u \in L^2_w(-1,1)$ depends on its regularity but also on the fulfillment of a suitable set of boundary conditions. This can be seen as follows. Equation (2.2.1) and integration by parts yield

$$\begin{aligned}
\hat{u}_k = (u, \phi_k)_w &= \frac{1}{\lambda_k} \int_{-1}^{1} u[-(p\phi_k')' + q\phi_k]\,dx \\
&= \frac{1}{\lambda_k} \int_{-1}^{1} [-(pu')' + qu]\phi_k\,dx - \frac{1}{\lambda_k}[p(\phi_k' u - \phi_k u')]_{-1}^{1} \qquad (9.2.5)\\
&= \frac{1}{\lambda_k}\left(\frac{1}{w} Lu, \phi_k\right)_w - \frac{1}{\lambda_k}[p(\phi_k' u - \phi_k u')]_{-1}^{1}.
\end{aligned}$$

This deduction is rigorous under the assumption that

$$u_{(1)} \equiv \frac{1}{w} Lu \in L^2_w(-1,1), \qquad (9.2.6)$$

which means—due to the regularity of the elliptic operator L—that the second derivative of u must be square integrable with respect to the weight $1/w$. Under this hypothesis, u and u' are continuous up to the boundary.

Now, *if u satisfies the boundary conditions* (9.2.1), the boundary term in (9.2.5) vanishes, so that

$$\hat{u}_k = \frac{1}{\lambda_k}(u_{(1)}, \phi_k)_w.$$

The iteration of this argument yields $\hat{u}_k = 1/(\lambda_k)^m (u_{(m)}, \phi_k)_w$, for $m \geqslant 2$, provided $u_{(m)} \equiv (1/w) L u_{(m-1)} \in L^2_w(-1,1)$ *and* $u_{(m-1)}$ satisfies the boundary conditions (9.2.1). We deduce the asymptotic decay estimate

$$|\hat{u}_k| \leqslant \frac{C}{k^{2m}} \|u_{(m)}\|_{L^2_w(-1,1)}.$$

If for some m, $u_{(m)}$ does not satisfy (9.2.1), then $\hat{u}_k$ decays no faster than $1/k^{2m}$, even if u is infinitely smooth. In this case u cannot be approximated with spectral accuracy by the system of the ϕ_k's.

9.2.2 Singular Sturm–Liouville Problems

A singular Sturm–Liouville problem occurs when p vanishes for at least one point on the boundary. We will consider here only the case $p(-1) = p(1) = 0$. The boundary conditions (9.2.1) are replaced by conditions on the type of singularities allowed on the boundary. Precisely, one requires the solution u to satisfy

$$p(x)u'(x) \to 0 \qquad \text{as } x \to \pm 1. \tag{9.2.7}$$

Let us assume that u is square integrable with respect to both the weights q and w, and that u' is square integrable with respect to the weight p, i.e.,

$$u \in X = \{v \in L^2_w(-1,1) \cap L^2_q(-1,1) | v' \in L^2_p(-1,1)\}.$$

(X is a Hilbert space for the norm $\|v\|^2 = \int_{-1}^{1} v^2 w\,dx + \int_{-1}^{1} v^2 q\,dx + \int_{-1}^{1} (v')^2 p\,dx$.) Then it is possible to give the following variational formulation of (2.2.1):

$$\int_{-1}^{1} (pu'v' + quv)\,dx = \lambda \int_{-1}^{1} uvw\,dx \qquad \text{for all } v \in X. \tag{9.2.8}$$

This takes into account the new boundary conditions in a natural way. As for the regular Sturm–Liouville problem, the eigenvalues of (9.2.8) form an unbounded sequence of non-negative real numbers $0 \leqslant \lambda_0 \leqslant \cdots \lambda_k \leqslant \cdots$; each of them has finite multiplicity. The system of corresponding eigenfunctions ϕ_k is orthogonal and complete in $L^2_w(-1,1)$. In order to prove these results, let us consider the following problem:

$$\begin{cases} u \in X \\ \displaystyle\int_{-1}^{1} (pu'v' + quv + uvw)\,dx = \int_{-1}^{1} fvw\,dx \qquad \text{for all } v \in X. \end{cases} \tag{9.2.9}$$

For each $f \in L^2_w(-1,1)$, there exists a unique solution to this problem. This follows from the Riesz representation theorem (see (A.1.d)), since the left-hand side of (9.2.9) is precisely the inner product in X. Let T: $L^2_w(-1,1) \to L^2_w(-1,1)$ be the linear operator which maps f into u. The eigenvalues λ of (9.2.8) are obtained from the eigenvalues μ of T by the relation $\lambda + 1 = \mu^{-1}$. The eigenfunctions are the same. It is immediate that T is a symmetric positive operator in the inner product of $L^2_w(-1,1)$, and that each eigenvalue of T is $\leqslant 1$. Moreover, one can prove that T is compact (see (A.3)). The proof of this property is based on the observation that if u is the solution of (9.2.9), then $(pu')' \in L^1(-1,1)$ and pu' is continuous on $[-1,1]$; thus, one can apply

Ascoli's Theorem (see, e.g., Taylor (1958, Sec. 5.5)). At this point we can invoke a fundamental result of spectral analysis in Hilbert spaces (see, e.g., Taylor (1958, Theorem 6.4-D)), which states that the eigenvalues of T form an infinite sequence of positive numbers which converges to 0. The corresponding eigenfunctions form a complete orthogonal basis in $L^2_w(-1,1)$. This yields the desired properties for the eigenvalues of (9.2.8).

In order to investigate the behavior of the expansion coefficients $\hat{u}_k = (u, \phi_k)_w$ of a function $u \in L^2_w(-1,1)$ with respect to the system of eigenfunctions of a singular Sturm–Liouville problem, we proceed as in (9.2.5):

$$\begin{aligned} \hat{u}_k &= \frac{1}{\lambda_k}\int_{-1}^{1} (p\phi_k' u' + q\phi_k u)\,dx \qquad \text{(by (9.2.8))} \\ &= \frac{1}{\lambda_k}\int_{-1}^{1} [-(pu')' + qu]\phi_k\,dx + \frac{1}{\lambda_k}[pu'\phi_k]_{-1}^{1} \qquad (9.2.10) \\ &= \frac{1}{\lambda_k}\left(\frac{1}{w}Lu, \phi_k\right)_w + \frac{1}{\lambda_k}[pu'\phi_k]_{-1}^{1}. \end{aligned}$$

Again, this holds provided (9.2.6) is satisfied. Note that under this assumption, pu' is continuous up to the boundary, since

$$\begin{aligned} |(pu')(x_1) - (pu')(x_2)| &= \left|\int_{x_1}^{x_2} (pu')'\,dx\right| \\ &\leqslant \left(\int_{x_1}^{x_2} \frac{1}{w}[(pu')']^2\right)^{1/2}\left(\int_{x_1}^{x_2} w\right)^{1/2} \end{aligned}$$

Thus, condition (9.2.7) makes sense, and it implies that the boundary term in (9.2.8) vanishes. We stress that, unlike the case of regular Sturm-Liouville boundary value problems, (9.2.7) is just a *regularity* assumption on u over the closed interval $[-1,1]$, i.e., u is not required to satisfy specific boundary conditions. One can easily check that (9.2.7) is satisfied if, for instance, $(p/w)u'' \in L^2_w(-1,1)$. Again, one can iterate the argument and get the representation $\hat{u}_k = 1/(\lambda_k)^m (u_{(m)}, \phi_k)_w$ provided $u_{(m)} \equiv (1/w)Lu_{(m-1)} \in L^2_w(-1,1)$ and $u_{(m-1)}$ satisfies (9.2.7) for $m \geqslant 2$. In the cases of interest (see Secs. 2.3.1 and 2.4.1), $\lambda_k = O(k^2)$ as $k \to \infty$. Hence, the expansion coefficients of u decay faster than algebraically under the sole assumption that u be infinitely differentiable.

This result does not necessarily hold if q is unbounded in $[-1,1]$. For instance, let us consider the singular Sturm–Liouville boundary value problem (Bessel equation) after changing the interval to $[0,2]$:

$$\begin{cases} -(xu')' + \dfrac{n^2}{x}u = \lambda x u \qquad 0 < x < 2, \\ \text{with } u(2) = 0 \text{ and } u \text{ bounded near } 0. \end{cases}$$

For $n \neq 0$, the condition $u_{(m)} \in L^2_w(-1, 1)$ forces $u_{(m)}$ to vanish at $x = 0$, since q^2/w is not integrable. In order to achieve spectral accuracy in this case, an infinite number of boundary conditions must be satisfied even though the operator is singular.

We conclude this section by showing that the only polynomial eigenfunctions of a singular Sturm–Liouville problem are the Jacobi polynomials. Actually, if $\phi_k = (1/(\lambda_k w_k))L\phi_k$ is a polynomial of degree k for $k = 0, 1, 2, \ldots$, it is readily seen that q/w is a polynomial of degree zero (i.e., $q(x) = q_0 w(x)$), and that p/w and p'/w are, respectively, polynomials of degree two and one. Since p must vanish at the boundary, necessarily one has $w(x) = c_1(1-x)^\alpha(1+x)^\beta$ and $p(x) = c_2(1-x)^{\alpha+1}(1+x)^{\beta+1}$ with $-1/2 \leqslant \alpha, \beta \leqslant 1/2$.

9.3. Discrete Norms

Before stating the approximation results for the Legendre and the Chebyshev polynomials, we give here some general theoretical results concerning the discrete inner product $(u, v)_N$ defined in (2.2.24). This bilinear form is a high precision approximation of the inner product $(u, v)_w$ for which the polynomials p_k are orthogonal. The function

$$\|v\|_N = (v, v)_N^{1/2}, \tag{9.3.1}$$

is the associated norm for the polynomials of $\mathbb{P}_N$. If the quadrature points $\{x_j\}$ are of Gauss or Gauss-Radau type, then $\|v\|_N = \|v\|_{L^2_w(-1,1)}$ for all $v \in \mathbb{P}_N$. If the $\{x_j\}$ are of Gauss-Lobatto type, this equality holds for $v \in \mathbb{P}_{N-1}$, but in general $\|p_N\|_N \neq \|p_N\|_{L^2_w(-1,1)}$. However, the discrete norm $\|v\|_N$ is uniformly equivalent to the norm $\|v\|_{L^2_w(-1,1)}$ in the more important cases, such as Legendre, Chebyshev or other Jacobi polynomials. This means that there exist positive constants C_1 and C_2, independent of N, such that

$$C_1 \|\phi\|_{L^2_w(-1,1)} \leqslant \|\phi\|_N \leqslant C_2 \|\phi\|_{L^2_w(-1,1)} \qquad \text{for all } \phi \in \mathbb{P}_N. \tag{9.3.2}$$

This result has been established by Canuto and Quarteroni (1982a). For the Legendre and Chebyshev polynomials one has

$$1 \leqslant \frac{\|p_N\|_N}{\|p_N\|_{L^2_w(-1,1)}} = \begin{cases} \sqrt{2} & \text{(Chebyshev)} \\ \sqrt{2 + \dfrac{1}{N}} & \text{(Legendre)} \end{cases}$$

as a consequence of (2.2.23), (2.3.13) and (2.4.18). Thus, (9.3.2) holds with $C_1 = 1$ and $C_2 = \sqrt{3}$, thanks to the orthogonality of the polynomials p_k.

The uniform equivalence of the discrete and continuous norms is used in a variety of ways in the analysis of stability and convergence, as will be seen in Chap. 10. For instance, at each stage of the analysis one may use whichever of the two norms is more convenient, and, if desired, convert to the other norm by the uniform equivalence property. Moreover, error estimates obtained for the continuous norm can be readily converted to error estimates in the discrete norm, and conversely.

A trivial application of (9.3.2) is the estimate

$$\|v\|_N \leqslant C_2 \|I_N v\|_{L^2_w(-1,1)}, \tag{9.3.3}$$

which holds for all the continuous functions on $[-1, 1]$.

The difference between the L^2_w-inner product $(u, v)_w$ and the discrete inner product $(u, v)_N$ can be bounded in terms of truncation and interpolation errors. Such estimates will be used in the convergence analysis of the subsequent chapters. Hereafter we denote by u any continuous function on $[-1, 1]$, and by ϕ any polynomial of $\mathbb{P}_N$.

For the *Gauss and Gauss-Radau integration*, we have

$$|(u, \phi)_w - (u, \phi)_N| \leqslant \|u - I_N u\|_{L^2_w(-1,1)} \|\phi\|_{L^2_w(-1,1)}. \tag{9.3.4}$$

Indeed, from (2.2.25) and (2.2.26) we get

$$(u, \phi)_w - (u, \phi)_N = (u, \phi)_w - (I_N u, \phi)_w;$$

hence (9.3.4) follows from the Cauchy-Schwarz inequality.

For the *Gauss-Lobatto integration*, if (9.3.2) holds, then there exists a positive constant C independent of N such that

$$\begin{aligned} |(u, \phi)_w - (u, \phi)_N| \leqslant C(&\|u - P_{N-1} u\|_{L^2_w(-1,1)} \\ &+ \|u - I_N u\|_{L^2_w(-1,1)}) \|\phi\|_{L^2_w(-1,1)}. \end{aligned} \tag{9.3.5}$$

Actually we have

$$\begin{aligned} &|(u, \phi)_w - (u, \phi)_N| \\ &\quad = |(u, \phi)_w - (P_{N-1} u, \phi)_w + (P_{N-1} u, \phi)_w - (I_N u, \phi)_N| \\ &\quad \leqslant |(u - P_{N-1} u, \phi)_w| + |(P_{N-1} u - I_N u, \phi)_N| \qquad \text{(by (2.2.25))} \\ &\quad \leqslant C(\|u - P_{N-1} u\|_{L^2_w(-1,1)} + \|P_{N-1} u - I_N u\|_N) \|\phi\|_{L^2_w(-1,1)} \\ &\qquad\qquad \text{(by the Cauchy–Schwarz inequality and (9.3.2))} \\ &\quad \leqslant C(2\|u - P_{N-1} u\|_{L^2_w(-1,1)} + \|u - I_N u\|_{L^2_w(-1,1)}) \|\phi\|_{L^2_w(-1,1)} \qquad \text{(by (9.3.2))}; \end{aligned}$$

whence (9.3.5) follows.

9.4. Legendre Approximations

We present in this section some results concerning polynomial approximations by Legendre polynomials.

9.4.1. Inverse Inequalities for Algebraic Polynomials

We recall here the inverse inequalities concerning summability and differentiability for algebraic polynomials on the interval $(-1, 1)$. These results are expressed in terms of L^p-norms, which are defined as follows:

$$\|u\|_{L^p(-1,1)} = \left(\int_{-1}^{1} |u(x)|^p \, dx\right)^{1/p} \qquad 1 \leqslant p < +\infty \tag{9.4.1}$$

and

$$\|u\|_{L^\infty(-1,1)} = \sup_{-1 \leqslant x \leqslant 1} |u(x)| \qquad \text{for } p = \infty. \tag{9.4.2}$$

These are the norms of the Banach spaces $L^p(-1, 1)$ defined in (A.9.f).

The inverse inequality concerning summability states that for any real p and q with $1 \leqslant p \leqslant q \leqslant \infty$, there exists a positive constant C such that

$$\|\phi\|_{L^q(-1,1)} \leqslant CN^{2(1/p-1/q)} \|\phi\|_{L^p(-1,1)} \qquad \text{for all } \phi \in \mathbb{P}_N. \tag{9.4.3}$$

On the other hand, the inverse inequality concerning differentiation states that for any p with $2 \leqslant p \leqslant +\infty$, and for all integers $r \geqslant 1$, there exists a positive constant C such that

$$\|\phi^{(r)}\|_{L^p(-1,1)} \leqslant CN^{2r} \|\phi\|_{L^p(-1,1)}. \tag{9.4.4}$$

The exponent of N in both (9.4.3) and (9.4.4) is the smallest possible. However, it is exactly twice the exponent in the Fourier inverse inequalities (9.1.4) and (9.1.5) or in the corresponding uniform grid finite-difference and finite-element inequalities. This has some important consequences for the stability and convergence analysis of orthogonal polynomial spectral methods. Result (9.4.4) is used in Sec. 11.4 to discuss the growth with N of the eigenvalues of the first and second derivative operators. In Sec. 4.3 we show that explicit spectral methods have a more restrictive time-step limitation than, say, finite-difference methods.

9.4.2. Estimates for the Truncation and Best Approximation Errors

As for the Fourier system, we will measure several approximation errors for the Legendre system in terms of Sobolev norms. The most commonly used Sobolev norm of order $m \geqslant 0$ is given by

$$\|u\|_{H^m(-1,1)} = \left(\sum_{k=0}^{m} \int_{-1}^{1} |u^{(k)}(x)|^2 \, dx\right)^{1/2} \tag{9.4.5}$$

Again, one can consider $u^{(k)}$ to be the classical continuous derivative of u of order k. These norms can actually be defined for less regular functions, which form a Hilbert space called $H^m(-1, 1)$. This space is introduced in (A.11.a).

The truncation error $u - P_N u$, where $P_N u = \sum_{k=0}^{N} \hat{u}_k L_k$ is the truncated Legendre series of u, can be estimated as follows. For all $u \in H^m(-1,1)$, $m \geqslant 0$, one has

$$\|u - P_N u\|_{L^2(-1,1)} \leqslant C N^{-m} \|u\|_{H^m(-1,1)}. \tag{9.4.6}$$

The truncated Legendre series $P_N u$ is the polynomial of best approximation of u in the L^2-norm. One can consider the problem of the best polynomial approximation of u with respect to a general norm. For any normed linear space X, and any $u \in X$, it is known that there exists a polynomial $\phi^* \in \mathbb{P}_N$ such that

$$\|u - \phi^*\|_X = \inf_{\phi \in \mathbb{P}_N} \|u - \phi\|_X; \tag{9.4.7}$$

ϕ^* is called a best approximation polynomial of u in the norm of X. We are interested in the case where $X = L^p(-1,1)$ for $1 \leqslant p \leqslant \infty$. For these norms ϕ^* is unique.

The best approximation error in any L^p-norm with $2 < p \leqslant +\infty$, decays as the truncation error in the L^2-norm

$$\inf_{\phi \in \mathbb{P}_N} \|u - \phi\|_{L^p(-1,1)} \leqslant C N^{-m} \sum_{k=0}^{m} \|u^{(k)}\|_{L^p(-1,1)}. \tag{9.4.8}$$

This estimate holds for all the functions u whose (distributional) derivatives of order up to m belong to $L^p(-1,1)$.

The rate of convergence of the truncation error in L^p norms, $p > 2$, is not as fast as the rate of convergence of the best approximation. For instance, for any function u with an m-th derivative of bounded variation (see (A.8)), one has

$$\|u - P_N u\|_{L^\infty(-1,1)} \leqslant C N^{1/2-m} V(u^{(m)}), \tag{9.4.9}$$

where $V(u^{(m)})$ is the total variation of $u^{(m)}$. Comparing this result with (9.4.8) for $p = \infty$, and noting that a function of bounded variation is certainly bounded, we see that the rate of convergence of the truncation error is slower by at least a factor of $\sqrt{N}$.

In those cases for which the truncation error of the derivatives is relevant, the following estimate extends (9.4.6) to higher-order Sobolev norms:

$$\|u - P_N u\|_{H^l(-1,1)} \leqslant C N^{-1/2} N^{2l-m} \|u\|_{H^m(-1,1)}, \tag{9.4.10}$$

for $u \in H^m(-1,1)$ with $m \geqslant 1$ and for any l such that $1 \leqslant l \leqslant m$. Note that in the important case $l = m = 1$, this inequality does not imply convergence of the derivative of the truncated series. Indeed, it is possible to construct a function u such that the truncated Legendre series converges in $L^2(-1,1)$ but not in $H^1(-1,1)$. Thus, the derivative of the series does not converge.

A simple manifestation of this phenomena is provided by considering a sequence of functions rather than a series. In particular, let

$$u^{(N)} = \frac{1}{N+1} L_{N+1} - \frac{1}{N-1} L_{N-1}.$$

The norm $\|u^{(N)}\|_{H^1(-1,1)}$ is bounded, as can be verified by using the Parseval equality to evaluate the norm and then using (2.3.15), which expresses the coefficients of the derivative in terms of the coefficients of the function. Nevertheless, in a similar fashion one obtains

$$\|u^{(N)} - P_N u^{(N)}\|_{H^1(-1,1)} \sim \sqrt{N}.$$

Fourier series are better behaved in this regard. If u itself is in $H^1(0, 2\pi)$, then the derivative of the truncated series of u is at least bounded. The analogous example is

$$u^{(N)}(x) = \frac{1}{N+1} e^{i(N+1)x} - \frac{1}{N-1} e^{i(N-1)x}.$$

Clearly,

$$\|u^{(N)} - P_N u^{(N)}\|_{H^1(0,2\pi)} = \sqrt{2\pi\left(1 + \frac{1}{(N+1)^2}\right)}.$$

The difference between the two types of expansions can be attributed to the loss of two powers of N in (9.4.4) for every derivative as opposed to only one power of N in the Fourier case.

The function $u(x) = |x|^{3/2}$ displayed in Fig. 9.1 is almost in $H^2(-1, 1)$, i.e., for all real $p < 2$

$$\int_{-1}^{1} |u''(x)|^p \, dx < \infty.$$

Result (9.4.10) then implies that $(P_N u)'$ converges to u' in the L^2-norm. But it does not imply convergence in the L^∞-norm, as is evident from the figure. Indeed, a sharp upper bound in the maximum norm over all functions in

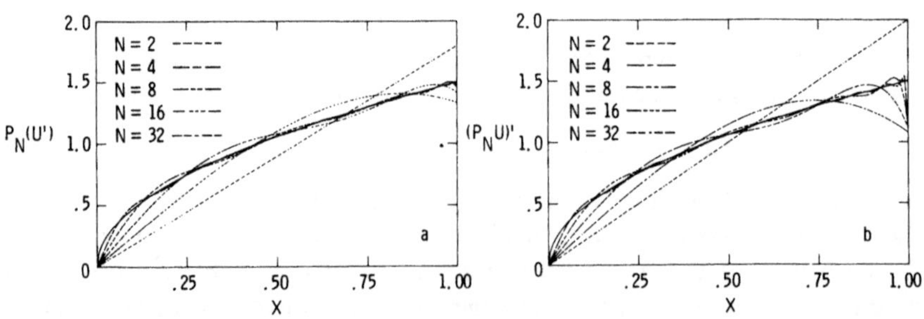

Figure 9.1. Several versions of Legendre differentiation for $u(x) = |x|^{3/2}$ on $[-1, 1]$. The exact result is indicated by the solid curves and the approximate results for $N = 2, 4, 8, 16$ and 32 are indicated by the dashed curves. Only the right half of the interval is shown. (a) $P_N u'$; (b) $(P_N u)'$.

$H^2(-1,1)$ can be obtained from the Sobolev inequality (A.12) and the estimate (9.4.10):

$$\|u' - (P_N u)'\|_{L^\infty(-1,1)} \leqslant C\|u - P_N u\|^{1/2}_{H^1(-1,1)} \|u - P_N u\|^{1/2}_{H^2(-1,1)}$$
$$\leqslant C N^{1/2} \|u\|_{H^2(-1,1)}.$$

On the other hand, Fig. 9.1 suggests that $P_N u'$ does converge to u' in the L^∞-norm. This is true for all functions in $H^2(-1,1)$, as follows from the estimate (9.4.9) applied with u' replacing u and with $m = 1$.

The rate of decay in (9.4.10) is not optimal in the sense that the best approximation error has a faster rate of convergence in the same norms. We will confine the discussion here to the $H^1(-1,1)$ norm. Since $H^1(-1,1)$ is a Hilbert space, the best approximation polynomial for u is the orthogonal projection of u upon $\mathbb{P}_N$ in the scalar product which induces the norm of $H^1(-1,1)$. This is defined as

$$((u,v)) = \int_{-1}^{1} (u'v' + uv)\,dx \qquad \text{for all } u, v \in H^1(-1,1). \tag{9.4.11}$$

Then, the polynomial $P_N^1 u \in \mathbb{P}_N$ such that

$$((P_N^1 u, \phi)) = ((u, \phi)) \qquad \text{for all } \phi \in \mathbb{P}_N \tag{9.4.12}$$

satisfies the identity

$$\|u - P_N^1 u\|_{H^1(-1,1)} = \inf_{\phi \in \mathbb{P}_N} \|u - \phi\|_{H^1(-1,1)}. \tag{9.4.13}$$

The approximation error (9.4.13) satisfies, for all $u \in H^m(-1,1)$, with $m \geqslant 1$, the following estimate

$$\|u - P_N^1 u\|_{H^1(-1,1)} \leqslant C N^{1-m} \|u\|_{H^m(-1,1)}. \tag{9.4.14}$$

On the other hand, the error $u - P_N^1 u$ in the L^2-norm satisfies

$$\|u - P_N^1 u\|_{L^2(-1,1)} \leqslant C N^{-m} \|u\|_{H^m(-1,1)}. \tag{9.4.15}$$

The exponent of N is the same here as it is for the best approximation error in the L^2-norm. The proofs given in the next section for similar results concerning Chebyshev approximations can be easily adapted to prove the two previous estimates.

An illustration of both the $L^2(-1,1)$ and $H^1(-1,1)$-projections is provided in Fig. 9.2 again for the function $u(x) = |x|^{3/2}$. The maximum pointwise error for the H^1-projection appears to decay slightly faster than the corresponding error for the L^2-projection (see Fig. 9.2(a) and (c)). In fact, for all functions $u \in H^m(-1,1)$, $m \geqslant 1$, one has

$$\|u - P_N u\|_{L^\infty(-1,1)} \leqslant C N^{(3/4)-m} \|u\|_{H^m(-1,1)} \tag{9.4.16}$$

and

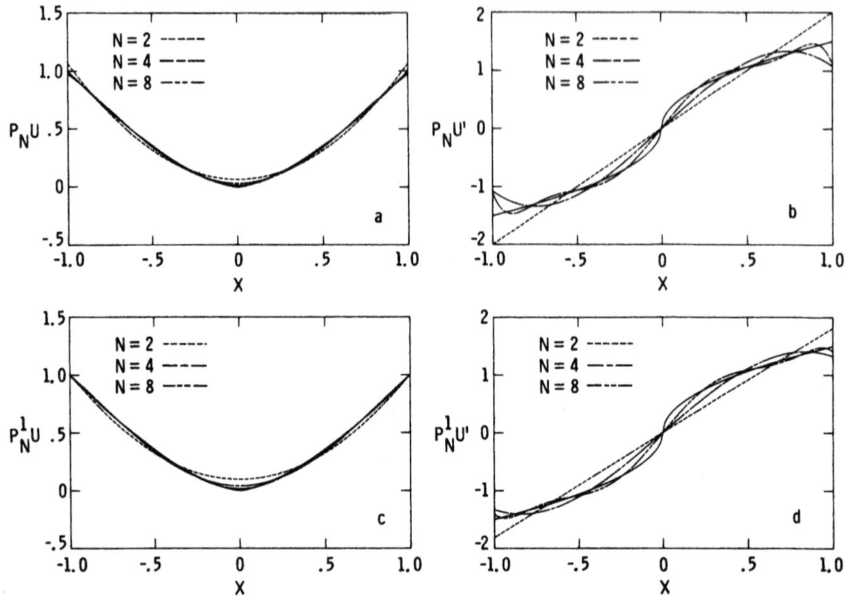

Figure 9.2. $L^2(-1,1)$ and $H^1(-1,1)$ Legendre projections for $u(x) = |x|^{3/2}$. The exact result is indicated by the solid curves and the approximate results for $N = 2, 4$ and 8 by the dashed curves.
(a) u under the $L^2(-1,1)$ projection for u;
(b) u' under the $L^2(-1,1)$ projection for u;
(c) u under the $H^1(-1,1)$ projection for u;
(d) u' under the $H^1(-1,1)$ projection for u.

$$\|u - P_N^1 u\|_{L^\infty(-1,1)} \leqslant CN^{(1/2)-m} \|u\|_{H^m(-1,1)}. \tag{9.4.17}$$

These estimates follow from the Sobolev inequality (A.12) together with the previous estimates in the Sobolev norms: (9.4.16) is obtained using (9.4.6) and (9.4.10) with $l = 1$; (9.4.17) is a consequence of (9.4.14) and (9.4.15). On the other hand, it is evident in Fig. 9.2(b) and (d) that the H^1-projection is definitely superior to the L^2-projection in the approximation of the first derivative of u.

The approximation results in the Sobolev norms are of importance for the analysis of spectral approximations of boundary value problems. In this case it may be more appropriate to project not just onto the space of polynomials, but onto the space of polynomials satisfying the boundary data. The result (9.4.14) holds for this projection as well (provided, of course, that u satisfies the same boundary data). Let us consider, for instance, homogeneous Dirichlet conditions at both endpoints of the interval $(-1,1)$. The functions of $H^1(-1,1)$ which satisfy such conditions form a subspace which is usually denoted by $H_0^1(-1,1)$, (see (A.11.c)), i.e.,

$$H_0^1(-1,1) = \{u \in H^1(-1,1) | u(-1) = u(1) = 0\}. \tag{9.4.18}$$

Similarly, the polynomials of degree N which vanish at the endpoints form a subspace $\mathbb{P}_N^0$ of $\mathbb{P}_N$,

$$\mathbb{P}_N^0 = \{v \in \mathbb{P}_N | v(-1) = v(1) = 0\}. \tag{9.4.19}$$

The inner product which is most commonly used for functions in $H_0^1(-1,1)$ is defined by

$$[u, v] = \int_{-1}^{1} u'(x)v'(x)\,dx \qquad \text{for } u, v \in H_0^1(-1,1). \tag{9.4.20}$$

It induces a norm on $H_0^1(-1,1)$ which is equivalent to the H^1-norm, due to the Poincaré inequality (A.13) (see also (A.11.c)). The H_0^1-projection of a function $u \in H_0^1(-1,1)$ upon $\mathbb{P}_N^0$ is the polynomial $P_N^{1,0}u \in \mathbb{P}_N^0$ such that

$$[P_N^{1,0}u, \phi] = [u, \phi] \qquad \text{for all } \phi \in \mathbb{P}_N^0. \tag{9.4.21}$$

We have the error estimate

$$\|u - P_N^{1,0}u\|_{H^k(-1,1)} \leqslant CN^{k-m}\|u\|_{H^m(-1,1)} \tag{9.4.22}$$

for all $u \in H^m(-1,1)$ vanishing at the boundary, with $m \geqslant 1$ and $k = 0, 1$.

The error bound (9.4.14) extends to higher order Sobolev norms as follows. Let $P_N^l u$ be the orthogonal projection of u onto $\mathbb{P}_N$ under the inner product of $H^l(-1,1)$ which induces the norm (9.4.5) (with $m = l$). Then

$$\|u - P_N^l u\|_{H^k(-1,1)} \leqslant CN^{k-m}\|u\|_{H^m(-1,1)}, \tag{9.4.23}$$

for $m \geqslant l$, $0 \leqslant k \leqslant l$. The same estimate holds if we replace P_N^l by $P_N^{l,\lambda}$ $(0 \leqslant \lambda \leqslant l-1)$, the orthogonal projection operator from the subspace of $H^l(-1,1)$ of the functions vanishing at the boundary with their derivatives of order up to λ, upon the subspace of $\mathbb{P}_N$ of the polynomials satisfying the same boundary conditions.

Finally, if $k > l$, i.e., if the norm in which the error is measured is stronger than the norm for which the error is minimal, then the exponent of N in all the previous estimates is $2k - l - \frac{1}{2} - m$.

9.4.3. Estimates for the Interpolation Error

We consider now the interpolation error. Let x_j, $0 \leqslant j \leqslant N$, be the Gauss, or the Gauss-Radau, or the Gauss-Lobatto points relative to the Legendre weight $w(x) \equiv 1$, considered in Sec. 2.3.1. Let $I_N u$ denote the polynomial of degree N which interpolates u at one of these sets of points. We give some estimates for the interpolation error $u - I_N u$ in the norms of the Sobolev spaces $H^l(-1,1)$. In the familiar $L^2(-1,1)$ norm, whenever $u \in H^m(-1,1)$ with $m \geqslant 1$,

$$\|u - I_N u\|_{L^2(-1,1)} \leqslant CN^{1/2}N^{-m}\|u\|_{H^m(-1,1)}. \tag{9.4.24}$$

The generalization of this formula for $1 \leqslant l \leqslant m$ is

$$\|u - I_N u\|_{H^l(-1,1)} \leqslant C N^{1/2} N^{2l-m} \|u\|_{H^m(-1,1)}. \tag{9.4.25}$$

A particular case of the last inequality is the following estimate of the error between exact and Legendre collocation differentiation

$$\|u' - (I_N u)'\|_{L^2(-1,1)} \leqslant C N^{5/2-m} \|u\|_{H^m(-1,1)}. \tag{9.4.26}$$

According to (9.3.3) and (9.4.24), the same estimate holds if the continuous L^2-norm of the error is replaced by the discrete L^2-norm at the interpolation points.

We conclude this section by providing an estimate for the integration error, arising from the use of Gauss quadrature formulae relative to the Legendre weight. Assume that a $N+1$-point Gauss, or Gauss-Radau, or Gauss-Lobatto quadrature formula relative to the Legendre weight is used to integrate the product $u\phi$, where $u \in H^m(-1,1)$ for some $m \geqslant 1$ and $\phi \in \mathbb{P}_N$. Then combining (9.3.4) or (9.3.5) with (9.4.24) and (9.4.6), one can show that

$$\left| \int_{-1}^{1} u(x)\phi(x)\,dx - (u,\phi)_N \right| \leqslant C N^{1/2-m} \|u\|_{H^m(-1,1)} \|\phi\|_{L^2(-1,1)}. \tag{9.4.27}$$

Bibliographical Notes

The inverse inequality (9.4.3) is proven, e.g., in Timan (1963, p. 236). The inequality (9.4.4) for $p = \infty$ is the classical Markov inequality (see, e.g., Timan (1963, p. 218)); for $p = 2$ we refer to Babuska, Szabo, and Katz (1980) or Canuto and Quarteroni (1982a), where different proofs are given; for $2 < p < \infty$, it can be obtained by interpolation of spaces (see Quarteroni (1984)). Estimates (9.4.6) and (9.4.10) have been obtained by Canuto and Quarteroni (1982a). The discussion on the optimality of the truncation error in higher Sobolev norms is also based on results from this paper. For the existence and uniqueness of the polynomials of best approximation in the L^p-norms we refer to Nikolskii (1975, Theorem 1.3.6), and Timan (1963, pp. 35–40). Estimate (9.4.8) is proven in Quarteroni (1984), while estimate (9.4.9) is due to Jackson (1930, Theorem XV). The estimates (9.4.14), (9.4.15) and (9.4.22) for the H^1- and H_0^1-projection operators are due to Maday and Quarteroni (1981), while their extension to higher order projections (9.4.23) has been carried out by Maday (1987a). The results of Sec. 9.4.3 have been proven by Canuto and Quarteroni (1982a).

9.5. Chebyshev Approximations

This section will be dedicated to the Chebyshev approximation, and will be similar in spirit to the section on Legendre approximation. Since the Chebyshev polynomials are orthogonal with respect to the non-constant weight $w(x) = (1 - x^2)^{-1/2}$, it is natural to frame the results in terms of weighted L^p and Sobolev spaces.

9.5.1 Inverse Inequalities for Polynomials

We define weighted L^p-norms as follows:

$$\|u\|_{L^p_w(-1,1)} = \left(\int_{-1}^{1} |u(x)|^p w(x)\, dx\right)^{1/p} \qquad \text{for } 1 \leqslant p < \infty, \tag{9.5.1}$$

and we set again

$$\|u\|_{L^\infty_w(-1,1)} = \sup_{-1 \leqslant x \leqslant 1} |u(x)| = \|u\|_{L^\infty(-1,1)}. \tag{9.5.2}$$

The space of functions for which a particular norm is finite forms a Banach space, indicated by $L^p_w(-1,1)$ (see (A.9.g)).

The inverse inequality concerning the summability in the Chebyshev L^p-norm for polynomials states that for any p and q, $1 \leqslant p \leqslant q \leqslant +\infty$, there exists a positive constant C such that for each $\phi \in \mathbb{P}_N$

$$\|\phi\|_{L^q_w(-1,1)} \leqslant (2N)^{(1/p-1/q)} \|\phi\|_{L^p_w(-1,1)}. \tag{9.5.3}$$

Note that the power of N is half the corresponding power in the Legendre estimate (9.4.3).

The inverse inequality concerning differentiation states that for any p, $2 \leqslant p \leqslant \infty$ and any integer $r \geqslant 1$, there exists a positive constant C such that for any $\phi \in \mathbb{P}_N$

$$\|\phi^{(r)}\|_{L^p_w(-1,1)} \leqslant CN^{2r} \|\phi\|_{L^p_w(-1,1)}. \tag{9.5.4}$$

Note that this estimate shares with the Legendre estimate (9.4.4) the double power of N on the right-hand side.

9.5.2 Estimates for the Truncation and Best Approximation Errors

The natural Sobolev norms in which to measure approximation errors for the Chebyshev system involve the Chebyshev weight in the quadratic averages of the error and its derivatives over the interval $(-1, 1)$. Thus, we set

$$\|u\|_{H^m_w(-1,1)} = \left(\sum_{k=0}^{m} \int_{-1}^{1} |u^{(k)}(x)|^2 w(x)\, dx\right)^{1/2}. \tag{9.5.5}$$

The Hilbert space associated to this norm is denoted by $H^m_w(-1,1)$, and is introduced in (A.11.b).

The truncation error $u - P_N u$, where now $P_N u = \sum_{k=0}^{N} \hat{u}_k T_k$ is the truncated Chebyshev series of u, satisfies the inequality

$$\|u - P_N u\|_{L^2_w(-1,1)} \leqslant CN^{-m} \|u\|_{H^m_w(-1,1)}, \tag{9.5.6}$$

for all $u \in H_w^m(-1, 1)$, with $m \geqslant 0$. This is a particular case of the estimate for the truncation error in the weighted L^p-norms, which reads as follows:

$$\|u - P_N u\|_{L_w^p(-1,1)} \leqslant C\sigma_N(p) N^{-m} \sum_{k=0}^{m} \|u^{(k)}\|_{L_w^p(-1,1)}, \tag{9.5.7}$$

for all functions u whose distributional derivatives of order up to m belong to $L_w^p(-1, 1)$. The constant $\sigma_N(p)$ equals 1 for $1 < p < \infty$, and $1 + \log N$ for $p = 1$ or $p = \infty$. As a consequence of this result, one gets an optimal estimate for the error of best approximation in the L_w^p-norms for $1 < p < +\infty$. (Note that this error in the norm of $L_w^\infty(-1, 1) = L^\infty(-1, 1)$ is estimated in (9.4.8).)

The truncation error in higher order Sobolev norms is estimated by the inequality

$$\|u - P_N u\|_{H_w^l(-1,1)} \leqslant C N^{-1/2} N^{2l-m} \|u\|_{H_w^m(-1,1)}, \tag{9.5.8}$$

for $u \in H_w^m(-1, 1)$, with $m \geqslant 1$ and $1 \leqslant l \leqslant m$. Thus, the asymptotic behavior of the Chebyshev truncation error is the same as for Legendre polynomials; hence, it is non-optimal with respect to the exponent of N.

In order to define the polynomial of best approximation in $H_w^1(-1, 1)$ we introduce the inner product

$$((u, v))_w = \int_{-1}^{1} (u'v' + uv) w \, dx \qquad \text{for all } u, v \in H_w^1(-1, 1), \tag{9.5.9}$$

and define the related orthogonal projection on $\mathbb{P}_N$ to be the polynomial $P_N^1 u \in \mathbb{P}_N$ such that

$$((P_N^1 u, \phi))_w = ((u, \phi))_w \qquad \text{for all } \phi \in \mathbb{P}_N. \tag{9.5.10}$$

The corresponding general estimate is

$$\|u - P_N^1 u\|_{H_w^k(-1,1)} \leqslant C N^{k-m} \|u\|_{H_w^m(-1,1)} \tag{9.5.11}$$

for all $u \in H_w^m(-1, 1)$ with $m \geqslant 1$, and $k = 0, 1$. Fig. 9.3 provides an example of the different behavior of the $L_w^2(-1, 1)$ and $H_w^1(-1, 1)$ projections. In higher order Sobolev norms one can prove the following result. For all integer l such that $0 \leqslant l \leqslant m$, and for every function $u \in H_w^m(-1, 1)$, there exists a polynomial $u^N \in \mathbb{P}_N$ such that

$$\|u - u^N\|_{H_w^k(-1,1)} \leqslant C N^{k-m} \|u\|_{H_w^m(-1,1)}, \tag{9.5.12}$$

for $0 \leqslant k \leqslant l$. The polynomial u^N can be defined as the orthogonal projection of u upon $\mathbb{P}_N$ in an inner product on $H_w^l(-1, 1)$ which induces a norm equivalent to $\|u\|_{H_w^l(-1,1)}$.

These estimates extend to functions satisfying prescribed boundary data in the same way that the Legendre estimates did. For instance, assume that u is a function in $H_w^1(-1, 1)$ which vanishes at $x = \pm 1$, i.e., u belongs to the subspace of $H_w^1(-1, 1)$

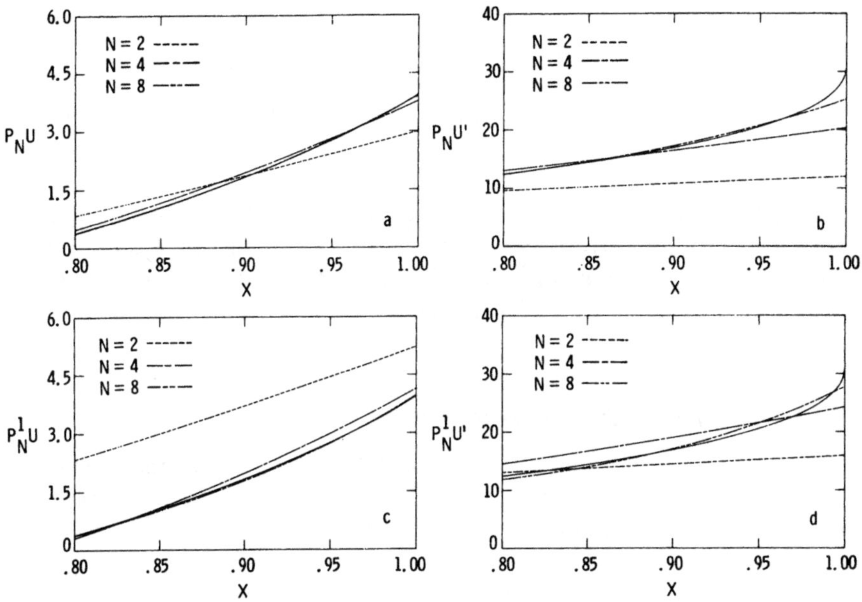

Figure 9.3. $L_w^2(-1,1)$ and $H_w^1(-1,1)$ Chebyshev projections for $u(x) = \frac{1}{48}[2\pi^2(\theta-\pi)^2 - (\theta-\pi)^4] - x$ where $\theta = \cos^{-1} x$. The exact result is indicated by the solid curves and the approximate results for $N = 2$, 4 and 8 by the dashed curves.
(a) u under the $L_w^2(-1,1)$ projection for u;
(b) u' under the $L_w^2(-1,1)$ projection for u;
(c) u under the $H_w^1(-1,1)$ projection for u;
(d) u' under the $H_w^1(-1,1)$ projection for u.

$$H_{w,0}^1(-1,1) = \{v \in H_w^1(-1,1) | v(-1) = v(1) = 0\} \qquad (9.5.13)$$

(see (A.11.c)). The projection of u upon $\mathbb{P}_N^0$ (see (9.4.19)) in the norm of this space is the polynomial $P_N^{1,0}u \in \mathbb{P}_N^0$ such that

$$[P_N^{1,0}u, \phi]_w = [u, \phi]_w \qquad \text{for all } \phi \in \mathbb{P}_N^0. \qquad (9.5.14)$$

Here, we use the natural inner product in $H_{w,0}^1(-1,1)$

$$[u, v]_w = \int_{-1}^{1} u'v'w\,dx \qquad \text{for } u, v \in H_{w,0}^1(-1,1) \qquad (9.5.15)$$

(see (A.11.c)). For the projector $P_N^{1,0}$ we have the following estimate

$$\|u - P_N^{1,0}u\|_{H_w^1(-1,1)} \leqslant CN^{1-m}\|u\|_{H_w^m(-1,1)}, \qquad (9.5.16)$$

for all $u \in H_w^m(-1,1) \cap H_{w,0}^1(-1,1)$, $m \geqslant 1$.

Furthermore, one can find a polynomial $u^N \in \mathbb{P}_N^0$ whose distance from u decays in an optimal way both in the H_w^1-norm and in the L_w^2-norm, i.e.,

$$\|u - u^N\|_{H_w^k(-1,1)} \leqslant CN^{k-m}\|u\|_{H_w^m(-1,1)}, \qquad (9.5.17)$$

for $k = 0$ and $k = 1$. For instance, u^N can be defined as the solution of the following Galerkin problem

$$\int_{-1}^{1} (u - u^N)'(\phi w)'\,dx = 0 \qquad \text{for all } \phi \in \mathbb{P}_N^0, \tag{9.5.18}$$

(see Sec. 11.1).

Finally, we mention that if u belongs to $H_w^l(-1,1)$ and vanishes at the boundary with the derivatives of order up to λ for an integer $\lambda \leqslant l - 1$, then one can find a polynomial u^N satisfying the same boundary conditions as u such that an estimate like (9.5.12) holds.

9.5.3. Estimates for the Interpolation Error

We consider now the interpolation error. Let $I_N u \in \mathbb{P}_N$ denote the interpolant of u at any of the three families of Chebyshev-Gauss points (2.4.12) or (2.4.13) or (2.4.14). Then the following estimate holds:

$$\|u - I_N u\|_{L_w^2(-1,1)} \leqslant CN^{-m} \|u\|_{H_w^m(-1,1)}, \tag{9.5.19}$$

if $u \in H_w^m(-1,1)$ for some $m \geqslant 1$. In higher order Sobolev norms one has

$$\|u - I_N u\|_{H_w^l(-1,1)} \leqslant CN^{2l-m} \|u\|_{H_w^m(-1,1)}, \tag{9.5.20}$$

for $0 \leqslant l \leqslant m$. As a consequence, we have

$$\|u' - (I_N u)'\|_{L_w^2(-1,1)} \leqslant CN^{2-m} \|u\|_{H_w^m(-1,1)}. \tag{9.5.21}$$

The same estimate holds in the discrete L_w^2-norms at the interpolation points. When the function u is analytic, the error $u' - (I_N u)'$ decays exponentially in N. Precisely, if u is analytic in $[-1, 1]$ and has a regularity ellipse whose sum of semi-axes equals $e^{\eta_0} > 1$, then

$$\|u' - (I_N u)'\|_{L_w^2(-1,1)} \leqslant C(\eta) N^2 e^{-N\eta}, \tag{9.5.22}$$

for all η, $0 < \eta < \eta_0$.

The interpolation error in the maximum norm is also of interest. An estimate of it is given by

$$\|u - I_N u\|_{L^\infty(-1,1)} \leqslant CN^{1/2-m} \|u\|_{H_w^m(-1,1)} \tag{9.5.23}$$

under the same assumptions as for (9.5.19).

By (9.5.6) and (9.5.19) we can obtain an estimate for the integration error produced by a Gauss-type quadrature formula relative to the Chebyshev weight. If $u \in H_w^m(-1,1)$ for some $m \geqslant 1$ and $\phi \in \mathbb{P}_N$, then using (9.3.4) and (9.3.5) we get the following result:

$$\left| \int_{-1}^{1} u(x)\phi(x)w(x)\,dx - (u,\phi)_N \right| \leqslant CN^{-m} \|u\|_{H_w^m(-1,1)} \|\phi\|_{L_w^2(-1,1)}. \tag{9.5.24}$$

9.5.4. Proofs of Some Approximation Results

We present in this section the proofs of some of the most relevant approximation error estimates given in the previous subsection. We confine ourselves to estimates in Hilbert norms of the truncation, interpolation and projection operators. Indeed, these are precisely the error estimates that most frequently occur in this book for the convergence analysis of spectral methods.

PROOF OF (9.5.4). We consider only the case $p = 2$ and $r = 1$. Let $\phi = \sum_{k=0}^{N} \hat{\phi}_k T_k$. By (2.4.22) we obtain $\phi' = \sum_{k=0}^{N-1} \hat{\phi}_k^{(1)} T_k$, with

$$c_k \hat{\phi}_k^{(1)} = 2 \sum_{\substack{p=k+1 \\ p+k \text{ odd}}}^{N} p\hat{\phi}_p,$$

where the coefficients c_k are defined in (2.4.10). The Cauchy-Schwarz inequality and the identity $\sum_{m=1}^{N} m^2 = N(N+1)(2N+1)/6$ give

$$(c_k \hat{\phi}_k^{(1)})^2 \leqslant 4 \left(\sum_{\substack{p=k+1 \\ p+k \text{ odd}}}^{N} p^2 \right) \left(\sum_{\substack{p=k+1 \\ p+k \text{ odd}}}^{N} (\hat{\phi}_p)^2 \right) \leqslant 4 \frac{N(N+1)(2N+1)}{6} \sum_{p=0}^{N} (\hat{\phi}_p)^2.$$

On the other hand, from (2.4.9) we have

$$\|\phi'\|^2_{L^2_w(-1,1)} = \sum_{k=0}^{N-1} \frac{\pi c_k}{2} (\hat{\phi}_k^{(1)})^2$$

$$\leqslant \frac{\pi}{3} N(N+1)(2N+1) \sum_{k=0}^{N-1} \frac{1}{c_k} \sum_{p=0}^{N} (\hat{\phi}_p)^2 \leqslant CN^4 \|\phi\|^2_{L^2_w(-1,1)}.$$

Although the proof of (9.5.4) may seem very crude, the exponent of N in (9.5.4) cannot be reduced. To convince oneself, it is sufficient to consider the function $\phi = \sum_{\substack{k=0 \\ k+N \text{ even}}}^{N} T_k$, for which one has $\|\phi'\|_{L^2_w(-1,1)} \simeq N^2 \|\phi\|_{L^2_w(-1,1)}$. □

PROOF OF (9.5.6). We shall make use of the transformation

$$x \in (-1, 1), \quad u(x) \to u^*(\theta) = u(\cos\theta), \quad \theta \in (0, 2\pi). \tag{9.5.25}$$

Since $\theta = \arccos x$, we have $d\theta/dx = -w(x)$ (the Chebyshev weight); thus,

$$\|u\|^2_{L^2_w(-1,1)} = \frac{1}{2} \|u^*\|^2_{L^2(0,2\pi)}. \tag{9.5.26}$$

It follows that the map $u \to u^*$ is an isomorphism between $L^2_w(-1,1)$ and the subspace of $L^2(0, 2\pi)$ of the even real functions. Moreover, it maps $H^m_w(-1,1)$ into the space of periodic functions $H^m_p(0, 2\pi)$ (see (A.11.d)). Indeed, since $u \in C^{m-1}([-1,1])$, then $u^* \in C^{m-1}(-\infty, +\infty)$ and it is 2π-periodic with all the derivatives of order up to $m-1$, whence $u^* \in H^m_p(0, 2\pi)$. Finally, since $|dx/d\theta| = |-\sin\theta| \leqslant 1$, we also have

$$\|u^*\|_{H^m(0,2\pi)} \leqslant C\|u\|_{H^m_w(-1,1)} \qquad \text{for } m \geqslant 1. \tag{9.5.27}$$

Let P_N^* denote the symmetric truncation of the Fourier series up to degree N, i.e.,

$$P_N^* \left(\sum_{k=-\infty}^{\infty} \hat{v}_k e^{ik\theta} \right) = \sum_{k=-N}^{N} \hat{v}_k e^{ik\theta}.$$

It is easily seen that

$$(P_N u)^* = P_N^* u^* \qquad \text{for all } u \in L^2_w(-1,1). \tag{9.5.28}$$

Indeed, since $u(x) = \sum_{k=0}^{\infty} \hat{u}_k T_k(x)$, $u^*(\theta) = \sum_{k=0}^{\infty} \hat{u}_k \cos k\theta = \sum_{k=0}^{\infty} \hat{u}_k (e^{ik\theta} + e^{-ik\theta})/2$, whence (9.5.28). Now, from (9.5.26) and (9.1.9) one gets

$$\|u - P_N u\|_{L^2_w(-1,1)} = \frac{1}{\sqrt{2}} \|u^* - P_N^* u^*\|_{L^2(0,2\pi)} \leqslant CN^{-m} \|u^{*(m)}\|_{L^2(0,2\pi)}.$$

Now (9.5.6) follows by (9.5.27). □

Proof of (9.5.8). We consider the case $l = 1$ only. The result corresponding to $l > 1$ follows by an inductive procedure. Using the triangle inequality and the estimate (9.5.6), we obtain

$$\begin{aligned} \|u - P_N u\|_{H^1_w(-1,1)} &\leqslant \|u - P_N u\|_{L^2_w(-1,1)} + \|u' - P_N u'\|_{L^2_w(-1,1)} \\ &\quad + \|P_N u' - (P_N u)'\|_{L^2_w(-1,1)} \\ &\leqslant CN^{1-m} \|u\|_{H^m_w(-1,1)} + \|P_N u' - (P_N u)'\|_{L^2_w(-1,1)}. \end{aligned} \tag{9.5.29}$$

In order to bound the last term let us expand u and u' in Chebyshev polynomials as

$$u = \sum_{k=0}^{\infty} \hat{u}_k T_k \qquad u' = \sum_{k=0}^{\infty} \hat{u}_k^{(1)} T_k.$$

Let us show that the polynomial $q_N = P_N u' - (P_N u)'$ has the form

$$q_N = \begin{cases} \hat{u}_N^{(1)} \phi_0^N + \hat{u}_{N+1}^{(1)} \phi_1^N & \text{if } N \text{ is even} \\ \hat{u}_{N+1}^{(1)} \phi_0^N + \hat{u}_N^{(1)} \phi_1^N & \text{if } N \text{ is odd,} \end{cases} \tag{9.5.30}$$

where $\phi_0^N = \sum_{\substack{k=0 \\ k \text{ even}}}^{N} (1/c_k) T_k$ and $\phi_1^N = \sum_{\substack{k=1 \\ k \text{ odd}}}^{N} T_k$. We can assume first that u is continuous with all its derivatives in $[-1, 1]$, so that (see (2.4.22))

$$c_k \hat{u}_k^{(1)} = 2 \sum_{\substack{p=k+1 \\ p+k \text{ odd}}}^{\infty} p \hat{u}_p \qquad k = 0, 1, 2, \ldots.$$

The series is absolutely convergent, since each $\hat{u}_p$ decays faster than any power of $1/p$ (this follows from (9.5.6)). Still using (2.4.22) we get

$$(P_N u)' = \sum_{k=0}^{N-1} \hat{v}_k T_k \qquad \text{with } c_k \hat{v}_k = 2 \sum_{\substack{p=k+1 \\ p+k \text{ odd}}}^{N} p \hat{u}_p,$$

thus

$$c_k(\hat{u}_k^{(1)} - \hat{v}_k) = \begin{cases} 2 \displaystyle\sum_{\substack{p=N+2 \\ p+N \text{ even}}}^{\infty} p \hat{u}_p = \hat{u}_{N+1}^{(1)} & \text{if } k + N \text{ is odd} \\ 2 \displaystyle\sum_{\substack{p=N+1 \\ p+N \text{ odd}}}^{\infty} p \hat{u}_p = \hat{u}_N^{(1)} & \text{if } k + N \text{ is even,} \end{cases}$$

whence, the result (9.5.30) if u is smooth. Next, we remove this assumption. If u is just in $H_w^1(-1,1)$, it can be approximated by a sequence of infinitely differentiable functions u_n (see (A.11.b)), for which (9.5.30) holds. Then we can pass to the limit as $n \to \infty$, since both sides of (9.5.30) are continuous in the norm of $H_w^1(-1,1)$.

From estimate (9.5.6) it follows that

$$|\hat{u}_{N+1}^{(1)}| \leqslant \|u' - P_N u'\|_{L_w^2(-1,1)} \leqslant CN^{1-m}\|u\|_{H_w^m(-1,1)},$$

and similarly for $\hat{u}_N^{(1)}$. On the other hand,

$$\|\phi_0^N\|_{L_w^2(-1,1)}^2 = \sum_{\substack{k=0 \\ k \text{ even}}}^{N} \frac{2}{c_k \pi} \simeq N \qquad \|\phi_1^N\|_{L_w^2(-1,1)}^2 = \sum_{\substack{k=1 \\ k \text{ odd}}}^{N} \frac{2}{\pi c_k} \simeq N.$$

Thus, noting that ϕ_0^N and ϕ_1^N are orthogonal, we have

$$\|P_N u' - (P_N u)'\|_{L_w^2(-1,1)} \leqslant CN^{(3/2)-m}\|u\|_{H_w^m(-1,1)}, \tag{9.5.31}$$

whence (9.5.8) by (9.5.29). □

As for the Legendre expansion, the exponent of N in (9.5.8) is optimal, in the sense that one cannot expect a faster decay of the error for all $u \in H_w^m(-1,1)$.

PROOF OF (9.5.11) IN THE CASE $k = 1$. Let us set

$$V = \left\{ v \in H_w^1(-1,1) | \hat{v}_0 = \frac{1}{\pi}\int_{-1}^{1} vT_o w\, dx = 0 \right\}. \tag{9.5.32}$$

V is a Hilbert space for the inner product $[u, v]_w$ defined in (9.5.15). Actually, if $v \in V$, there exists at least one point $\xi \in (-1,1)$ where $v(\xi) = 0$. Hence, the Poincaré inequality (A.13) holds, and $\|v\|_V = [v, v]_w^{1/2} = \|v'\|_{L_w^2(-1,1)}$ is a norm equivalent to the standard norm of $\|v\|_{H_w^1(-1,1)}$. For each $u \in H_w^1(-1,1)$, let us define the polynomial

$$u^N(x) = \alpha + \int_{-1}^{x} (P_{N-1}u')(s)\, ds. \tag{9.5.33}$$

As usual, $P_{N-1}v$ is the truncation of degree $N-1$ of the Chebyshev series of v. The constant α is chosen in such a way that $\widehat{(u^N)}_0 = \hat{u}_0$. Then, by (9.5.6) it follows that

$$\|u - u^N\|_V = \|u' - P_{N-1}u'\|_{L_w^2(-1,1)} \leqslant CN^{1-m}\|u\|_{H_w^m(-1,1)}. \tag{9.5.34}$$

The result (9.5.11) for $k = 1$ follows, noting that

$$\|u - P_N^1 u\|_{H_w^1(-1,1)} \leqslant \|u - v\|_{H_w^1(-1,1)} \qquad \text{for all } v \in \mathbb{P}_N.$$ □

In order to prove (9.5.11) for $k = 0$, we need the following regularity result.

Lemma 9.1. *For each $g \in L_w^2(-1,1)$, there exists a unique $\psi \in H_w^1(-1,1)$ such that*

$$\int_{-1}^{1} (\psi' v' + \psi v) w\, dx = \int_{-1}^{1} g v w\, dx, \qquad \text{for all } v \in H_w^1(-1,1). \tag{9.5.35}$$

Moreover, $\psi \in H_w^2(-1,1)$ and there is a constant $C > 0$ such that

$$\|\psi\|_{H_w^2(-1,1)} \leqslant C\|g\|_{L_w^2(-1,1)}. \tag{9.5.36}$$

Proof. Since the left-hand side of (9.5.35) is the inner product of $H_w^1(-1,1)$, the existence and uniqueness of ψ follows from the Riesz representation theorem (see (A.1.d)). Choosing $v = \psi$ in (9.5.35), we get

$$\|\psi\|_{H_w^1(-1,1)} \leqslant \|g\|_{L_w^2(-1,1)}. \tag{9.5.37}$$

Letting v vary in $\mathscr{D}(-1,1)$ (this space is defined in (A.10)), we obtain from (9.5.35)

$$-(\psi' w)' = (g - \psi)w \qquad \text{in the sense of distributions} \tag{9.5.38}$$

(see (A.10.a)). Next, we show that $\psi' w$ is continuous in $[-1,1]$. Indeed, for any x_1, $x_2 \in (-1,1)$ it follows by (9.5.38) and the Cauchy-Schwarz inequality (see (A.2)) that

$$|(\psi' w)(x_1) - (\psi' w)(x_2)| = \left|\int_{x_1}^{x_2} (g - \psi) w \, dx\right|$$
$$\leqslant \|g - \psi\|_{L_w^2(-1,1)} |\arccos x_2 - \arccos x_1|^{1/2}.$$

Hence, $(\psi' w)(\pm 1)$ makes sense. Multiplying (9.5.38) by $v \in H_w^1(-1,1)$ and integrating by parts yields

$$[\psi' w v]_{-1}^{1} = \int_{-1}^{1} \psi' v' w \, dx - \int_{-1}^{1} (g - \psi) v w \, dx \qquad \text{for all } v \in H_w^1(-1,1).$$

Hence, $\psi' w(-1) = \psi' w(1) = 0$ by (9.5.35). By (9.5.38), $-\psi'' = (g - \psi) - \psi'(w'/w)$. Thus, it remains to prove that $\psi'(w'/w) \in L_w^2(-1,1)$. Since $w'/w = xw^2$, we have

$$\int_{-1}^{1} (\psi' w'/w)^2 w \, dx \leqslant \int_{-1}^{1} (\psi')^2 w^5 \, dx.$$

Moreover,

$$\int_{-1}^{0} (\psi')^2 w^5 \, dx = \int_{-1}^{0} \left[\int_{-1}^{x} (\psi' w)' \, d\xi\right]^2 w^3 \, dx$$
$$= \int_{-1}^{0} \left[w^2 \int_{-1}^{x} (\psi - g) w \, d\xi\right]^2 w^{-1} \, dx$$
$$\leqslant C \int_{-1}^{0} \left[\frac{1}{1+x} \int_{-1}^{x} (\psi - g) w \, d\xi\right]^2 \sqrt{1+x} \, dx.$$

Using the Hardy inequality (A.14) with $\alpha = 1/2$, $a = -1$ and $b = 0$ we obtain

$$\int_{-1}^{0} (\psi')^2 w^5 \, dx \leqslant C \int_{-1}^{0} (\psi - g)^2 w \, dx.$$

Similarly, we can prove that $\int_0^1 (\psi')^2 w^5 \, dx \leqslant C \int_0^1 (\psi - g)^2 w \, dx$. Therefore, we conclude that $\psi'' \in L_w^2(-1,1)$, with

$$\|\psi''\|_{L_w^2(-1,1)} \leqslant C\left(\|\psi\|_{L_w^2(-1,1)} + \|g\|_{L_w^2(-1,1)}\right).$$

This gives (9.5.36), using (9.5.37). □

PROOF OF (9.5.11) IN THE CASE $k = 0$. We use a well-known duality argument, based on the identity

$$\|u - P_N^1 u\|_{L^2_w(-1,1)} = \sup_{\substack{g \in L^2_w(-1,1) \\ g \neq 0}} \frac{\displaystyle\int_{-1}^{1} (u - P_N^1 u) g w \, dx}{\|g\|_{L^2_w(-1,1)}}.$$

Let ψ be the solution of (9.5.35) corresponding to a given g. Then, choosing $v = u - P_N^1 u$ in (9.5.35) and recalling the definition of P_N^1 we get

$$\begin{aligned}\int_{-1}^{1} (u - P_N^1 u) g w \, dx &= \int_{-1}^{1} [\psi'(u - P_N^1 u)' + \psi(u - P_N^1 u)] w \, dx \\ &= \int_{-1}^{1} [(\psi - P_N^1 \psi)'(u - P_N^1 u)' + (\psi - P_N^1 \psi)(u - P_N^1 u)] w \, dx.\end{aligned}$$

The Cauchy-Schwarz inequality, estimate (9.5.11) with $k = 1$, and (9.5.36) yield

$$\begin{aligned}\left| \int_{-1}^{1} (u - P_N^1 u) g w \, dx \right| &\leqslant \|\psi - P_N^1 \psi\|_{H^1_w(-1,1)} \|u - P_N^1 u\|_{H^1_w(-1,1)} \\ &\leqslant C N^{-1} \|\psi\|_{H^2_w(-1,1)} \|u - P_N^1 u\|_{H^1_w(-1,1)} \\ &\leqslant C N^{-1} \|g\|_{L^2_w(-1,1)} \|u - P_N^1 u\|_{H^1_w(-1,1)}.\end{aligned}$$

Then,

$$\|u - P_N^1 u\|_{L^2_w(-1,1)} \leqslant C N^{-1} \|u - P_N^1 u\|_{H^1_w(-1,1)}.$$

Hence, the desired result follows again using (9.5.11) with $k = 1$. □

PROOF OF (9.5.16). Let us define u^N as in (9.5.33), now with $\alpha = 0$. Next, define

$$R_N u(\xi) = \int_{-1}^{\xi} \left(P_{N-1} u' - \frac{1}{2} u^N(1) \right) dx,$$

so that $R_N u \in \mathbb{P}_N^0$. We have by the triangle inequality

$$\|u' - (R_N u)'\|_{L^2_w(-1,1)} \leqslant \|u' - P_{N-1} u'\|_{L^2_w(-1,1)} + \frac{1}{2} \left(\int_{-1}^{1} w \, dx \right)^{1/2} |u^N(1)|.$$

On the other hand, by the Cauchy-Schwarz inequality one has

$$\begin{aligned}|u^N(1)| = |u(1) - u^N(1)| &= \left| \int_{-1}^{1} (u' - P_{N-1} u') \, dx \right| \\ &\leqslant \left(\int_{-1}^{1} w^{-1} \, dx \right)^{1/2} \|u' - P_{N-1} u'\|_{L^2_w(-1,1)}.\end{aligned}$$

Using (9.5.6) and the two previous inequalities, we obtain

$$\|u' - (R_N u)'\|_{L^2_w(-1,1)} \leqslant C N^{1-m} \|u\|_{H^m_w(-1,1)}.$$

Finally, (9.5.16) follows since $P_N^{1,0} u$ is the polynomial of best approximation of u in the norm associated to the $H^1_{w,0}$-inner product (9.5.15). □

Proof of (9.5.17). We define $u^N \in \mathbb{P}_N^0$ to be the solution of the problem

$$a(u - u^N, v) = 0 \qquad \text{for all } v \in \mathbb{P}_N^0, \tag{9.5.39}$$

where $a(\phi, \psi) = \int_{-1}^{1} \phi'(\psi w)' \, dx$ (see (11.1.11)). This is precisely the polynomial defined in (9.5.18). It is shown in Sec. 11.1 that the bilinear form $a(\phi, \psi)$ defined on $H_{w,0}^1(-1,1) \times H_{w,0}^1(-1,1)$ satisfies the hypotheses of the Lax-Milgram Theorem (A.5) (see (11.1.9) and (11.1.10)). Then the existence and uniqueness of u^N is assured. Moreover, by the coercivity and the continuity of a we get

$$\begin{aligned}\|u - u^N\|_{H_w^1(-1,1)}^2 &\leqslant C_1 a(u - u^N, u - u^N) = C_1 a(u - u^N, u - v) \qquad \text{(by (9.5.39))} \\ &\leqslant C_2 \|u - u^N\|_{H_w^1(-1,1)} \|u - v\|_{H_w^1(-1,1)},\end{aligned}$$

for all $v \in \mathbb{P}_N^0$. Thus,

$$\|u - u^N\|_{H_w^1(-1,1)} \leqslant C_2 \inf_{v \in \mathbb{P}_N} \|u - v\|_{H_w^1(-1,1)}. \tag{9.5.40}$$

Estimate (9.5.17) for $k = 1$ follows now from (9.5.16). In order to prove (9.5.17) for $k = 0$, we use a duality argument similar to the one we have used to prove (9.5.11). We have

$$\|u - u^N\|_{L_w^2(-1,1)} = \sup_{\substack{g \in L_w^2(-1,1) \\ g \neq 0}} \frac{\int_{-1}^{1} (u - u^N) g w \, dx}{\|g\|_{L_w^2(-1,1)}}. \tag{9.5.41}$$

For each fixed $g \in L_w^2(-1,1)$, $g \neq 0$, let $\psi = \psi(g) \in H_{w,0}^1(-1,1)$ be the solution of the problem

$$a(v, \psi) = \int_{-1}^{1} g v w \, dx \qquad \text{for all } v \in H_{w,0}^1(-1,1), \tag{9.5.42}$$

which is uniquely defined since the transpose form $a^T(u, v) = a(v, u)$ satisfies again the hypotheses of the Lax-Milgram theorem. A very technical argument allows us to prove that $\psi \in H_w^2(-1,1)$ and

$$\|\psi\|_{H_w^2(-1,1)} \leqslant C \|g\|_{L_w^2(-1,1)}. \tag{9.5.43}$$

Then, using (9.5.42) and (9.5.18), we obtain for each $\psi^N \in \mathbb{P}_N^0$

$$\begin{aligned}\left| \int_{-1}^{1} (u - u^N) g w \, dx \right| &= |a(u - u^N, \psi)| = |a(u - u^N, \psi - \psi^N)| \\ &\leqslant C \|u - u^N\|_{H_w^1(-1,1)} \|\psi - \psi^N\|_{H_w^1(-1,1)}.\end{aligned}$$

Using (9.5.17) with $k = 1$ for both u and ψ yields

$$\left| \int_{-1}^{1} (u - u_N) g w \, dx \right| \leqslant C N^{-m} \|\psi\|_{H_w^2(-1,1)} \|u\|_{H_w^m(-1,1)}.$$

Now estimate (9.5.17) with $k = 0$ follows using (9.5.41) and (9.5.43). □

Proof of (9.5.19) and (9.5.20). We consider the Gauss-Lobatto interpolation points $x_j = \cos(\pi j / N)$, for $j = 0, \ldots, N$. The proof for the other two sets of points

(Gauss and Gauss-Radau one) is similar. We still make use of the mapping (9.5.25). We define

$$\tilde{S}_N = \left\{ v: (0, 2\pi) \to \mathbb{C} \,\middle|\, v(\theta) = \sum_{k=-N}^{N} \hat{v}_k e^{ik\theta}, \hat{v}_N = \hat{v}_{-N} \right\},$$

and, for every $v \in C^0([0, 2\pi])$, we denote by $I_N^* v$ the unique function of $\tilde{S}_N$ that interpolates v at the points $\theta_j = \pi j/N$, for $j = 0, \ldots, 2N$. Note that these points are symmetrically distributed around the point $\theta = \pi$. Moreover, for each continuous function $u: [-1, 1] \to \mathbb{R}$, both u^* and $(I_N u)^*$ are even functions with respect to the point $\theta = \pi$. Therefore,

$$(I_N u)^* = I_N^* u^* \in \tilde{S}_N. \tag{9.5.44}$$

Now we use the error estimate (9.1.15) for the Fourier interpolation and we obtain, by (9.5.26) and (9.5.27)

$$\|u - I_N u\|_{L^2_w(-1,1)} = \frac{1}{\sqrt{2}} \|u^* - I_N^* u^*\|_{L^2(0,2\pi)} \leqslant CN^{-m} \|u\|_{H^m_w(-1,1)}, \tag{9.5.45}$$

i.e., (9.5.19). For $m \geqslant 1$, the inverse inequality (9.5.4) yields

$$\|u - I_N u\|_{H^l_w(-1,1)} \leqslant \|u - P_N u\|_{H^l_w(-1,1)} + CN^{2l} \|P_N u - I_N u\|_{L^2_w(-1,1)}.$$

Now (9.5.20) follows using (9.5.6) and (9.5.45). □

Bibliographical Notes

A proof of the inverse inequality (9.5.3) is given in Quarteroni (1984). The inequality (9.5.4) has been established by Canuto and Quarteroni (1982a) for $p = 2$ and extended to arbitrary p by Quarteroni (1984). Estimates (9.5.6) and (9.5.8) are proven in Canuto and Quarteroni (1982a), while estimate (9.5.7) has been obtained by Quarteroni (1984). Inequalities (9.5.11), (9.5.16) and (9.5.17) are due to Maday and Quarteroni (1981); here, we give a different proof of (9.5.11). The extension of these results to higher order norms has been carried out by Maday (1987a). Finally, the results given in Sec. 9.5.3 are due to Canuto and Quarteroni (1982), except for (9.5.22), which is due to Tadmor (1986).

9.6. Other Polynomial Approximations

The orthogonal systems described so far have been the ones most commonly used in building up spectral approximations to partial differential equations. Other relevant sets of orthogonal polynomials guarantee spectral accuracy as well, but only recently have some of these been used in actual computations.

9.6.1. Jacobi Polynomials

The Jacobi polynomials $\{p_k^{(\alpha,\beta)}(x), k = 0, 1, 2, \ldots\}$ are the eigenfunctions of the singular Sturm-Liouville problem (2.2.1) where $p(x) = (1-x)^{\alpha+1}(1+x)^{\beta+1}$, ($\alpha$ and $\beta > -1$), $q(x) \equiv 0$ and $w(x) = (1-x)^{\alpha}(1+x)^{\beta}$. The eigenvalue corresponding to p_k is $\lambda_k = k(k+\alpha+\beta+1)$. Note that the Legendre polynomials correspond to the choice $\alpha = \beta = 0$, while the Chebyshev polynomials of the second kind correspond to $\alpha = \beta = -1/2$.

Jacobi polynomials for other choices of α and β have no better asymptotic approximation properties than do Chebyshev polynomials. Although a fast transform is not available for them, they can lead to small matrix bandwidths in Galerkin methods (Sec. 7.3.3). The mathematical difficulties in the analysis of Chebyshev methods which arise from the singularity of the Chebyshev weight are shared by the Jacobi methods.

We will not provide here a general approximation theory for the Jacobi polynomials. However, some Chebyshev approximations to hyperbolic problems (see Sec. 12.1.2) are shown to be stable in some weighted norms corresponding to a Jacobi weight. In such cases, the following results of best approximation are of interest.

For each Jacobi weight $w(x) = (1-x)^{\alpha}(1+x)^{\beta}$ with $-\frac{1}{2} \leqslant \alpha, \beta \leqslant \frac{1}{2}$ and for all functions $u \in H_w^m(-1,1)$, one has

$$\inf_{\phi \in \mathbb{P}_N} \|u - \phi\|_{L_w^2(-1,1)} \leqslant CN^{-m} \|u\|_{H_w^m(-1,1)} \qquad m \geqslant 0 \tag{9.6.1}$$

and

$$\inf_{\phi \in \mathbb{P}_N} \|u - \phi\|_{H_w^1(-1,1)} \leqslant CN^{1-m} \|u\|_{H_w^m(-1,1)} \qquad m \geqslant 1. \tag{9.6.2}$$

Here, the norms of $L_w^2(-1,1)$ and $H_w^m(-1,1)$ are defined as in (9.5.1) and (9.5.5) respectively. (For a proof, see Canuto and Quarteroni (1982b).)

9.6.2. Laguerre and Hermite Polynomials

The Laguerre polynomials $\{l_k(x), k = 0, 1, \ldots\}$ are the eigenfunctions of the singular Sturm-Liouville problem (2.2.1) on the semi-infinite interval $(0, +\infty)$, with $p(x) = xe^{-x}$, $q(x) = 0$ and $w(x) = e^{-x}$. The eigenvalue corresponding to l_k is $\lambda_k = k$. The Hermite polynomials $\{H_k(x), k = 0, 1, \ldots\}$ are the eigenfunctions of (2.2.1) on the real axis with $p(x) = e^{-x^2}$, $q(x) = 0$ and $w(x) = e^{-x^2}$. The eigenvalue corresponding to H_k is $\lambda_k = 2k$.

By adapting the arguments of Sec. 9.2 to the case of an unbounded interval, one can prove that the coefficients of the Laguerre (or Hermite) expansion of a smooth function defined over $(0, +\infty)$ (respectively, over $(-\infty, +\infty)$) decay faster than algebraically.

An expansion based on Laguerre polynomials has been recently investigated by Maday, Pernaud-Thomas, and Vandeven (1985). In this paper, approximation results for truncation and interpolation operators are given. We present them hereafter. For any positive real number α define

$$L^2_{e^{-\alpha x}} = \left\{ f\colon [0, +\infty) \to \mathbb{R} \text{ measurable} \,\middle|\, \|f\|_{0,\alpha} = \left\{ \int_0^\infty f^2(x) e^{-\alpha x}\, dx \right\}^{1/2} < \infty \right\},$$

and for each $m \geqslant 0$, define

$$H^m_{e^{-\alpha x}} = \left\{ f \in L^2_{e^{-\alpha x}} \,\middle|\, \|f\|_{m,\alpha} = \left\{ \sum_{k=0}^{m} \|f^{(k)}\|^2_{0,\alpha} \right\}^{1/2} < \infty \right\},$$

For each $u \in L^2_{e^{-\alpha x}}$, let $P_N u$ be its projection upon $\mathbb{P}_N$ with respect to the inner product of $L^2_{e^{-\alpha x}}$, i.e., $P_N u \in \mathbb{P}_N$ or,

$$\int_0^\infty (u - P_N u)\phi e^{-\alpha x}\, dx = 0 \qquad \text{for all } \phi \in \mathbb{P}_N.$$

The following error estimate holds for all $\varepsilon > 0$:

$$\|u - P_N u\|_{k,1} \leqslant C N^{k-(m/2)} \|u\|_{m,1-\varepsilon} \qquad 0 \leqslant k \leqslant m. \tag{9.6.3}$$

Concerning interpolation, let us consider the $N + 1$ points $\{x_j, j = 0, \ldots, N\}$, where $x_0 = 0$, and x_j, for $j = 1, \ldots, N$, are the zeroes of $l'_{N+1}(x)$, the derivative of the $(N + 1)$-th Laguerre polynomials. For each continuous function u on $\mathbb{R}^+$, let $I_N u \in \mathbb{P}_N$ be the interpolant of u at the points $\{x_j\}$. Then, for all $\varepsilon > 0$

$$\|u - I_N u\|_{k,1} \leqslant C N^{k-(m-1/2)} \|u\|_{m,1-\varepsilon} \qquad 0 \leqslant k \leqslant m,\ m \geqslant 1. \tag{9.6.4}$$

9.7. Approximation Results in Several Dimensions

We shall now extend to several space dimensions some of the approximation results we presented in the previous sections for a single-space variable. The three expansions of Fourier, Legendre, and Chebyshev will be considered. However, we will only be concerned with those Sobolev-type norms which are most frequently applied to the convergence analysis of spectral methods.

9.7.1. Fourier Approximations

Let us consider the domain $\Omega = (0, 2\pi)^d$ in $\mathbb{R}^d$, for $d = 2$ or 3, and denote an element of $\mathbb{R}^d$ by $\mathbf{x} = (x_1, \ldots, x_d)$. The space $L^2(\Omega)$, as well as the Sobolev spaces $H^m_p(\Omega)$ of periodic functions, are defined in Appendix A (see (A.9.h) and (A.11.d)). Since Ω is the Cartesian product of d copies of the interval $(0, 2\pi)$, it is natural to use as an orthogonal system in $L^2(\Omega)$, the tensor product of the trigonometric system in $L^2(0, 2\pi)$. Thus, we set

$$\phi_{\mathbf{k}}(\mathbf{x}) = e^{i\mathbf{k}\cdot\mathbf{x}} \qquad \text{with} \quad \mathbf{k}\cdot\mathbf{x} = k_1 x_1 + \cdots + k_d x_d \tag{9.7.1}$$

by analogy with (2.1.1), and

$$S_N = \text{span}\{\phi_{\mathbf{k}}(\mathbf{x}) \mid -N \leqslant k_j \leqslant N-1 \quad \text{for } j=1,\ldots,d\}. \tag{9.7.2}$$

Moreover, we still denote by P_N the orthogonal projection operator from $L^2(\Omega)$ upon S_N. Then, for any $u \in L^2(\Omega)$ we have

$$P_N u = \sum_{\|\mathbf{k}\| \lesssim N} \hat{u}_{\mathbf{k}} \phi_{\mathbf{k}} \qquad \hat{u}_{\mathbf{k}} = \left(\frac{1}{2\pi}\right)^d \int_\Omega u(\mathbf{x}) \overline{\phi_{\mathbf{k}}(\mathbf{x})}\, d\mathbf{x}, \tag{9.7.3}$$

where the above summation is extended to all $\mathbf{k} \in \mathbb{Z}^d$ such that $-N \leqslant k_j \leqslant N-1$, for $j = 1, \ldots, d$. The following result provides an estimate in all Sobolev norms for the remainder of the Fourier series of u:

$$\|u - P_N u\|_{H^l(\Omega)} \leqslant C N^{l-m} \|u\|_{H^m(\Omega)} \qquad \text{for } 0 \leqslant l \leqslant m. \tag{9.7.4}$$

It can be obtained for all $u \in H_p^m(\Omega)$ by a proof that mimics the one of (9.1.10).

Concerning interpolation, let us introduce the $(2N)^d$ points:

$$\begin{aligned} &\mathbf{x}_j = (x_{j_1}, \ldots, x_{j_d}) \qquad \text{where } x_l = \frac{\pi}{N} l, \\ &\text{and } 0 \leqslant j_m \leqslant 2N-1 \qquad \text{for } m = 1, \ldots, d. \end{aligned} \tag{9.7.5}$$

For every function u continuous in the closure of Ω, we denote by $I_N u$ the function of S_N interpolating u at the points (9.7.5). By analogy with the one-dimensional case (cf. (2.1.25) and (2.1.26)) one has

$$I_N u = \sum_{\|\mathbf{k}\| \lesssim N} \tilde{u}_{\mathbf{k}} \phi_{\mathbf{k}} \qquad \tilde{u}_{\mathbf{k}} = \left(\frac{1}{2N}\right)^d \sum_{\mathbf{j}} u(x_{\mathbf{j}}) \overline{\phi_{\mathbf{k}}(x_{\mathbf{j}})}, \tag{9.7.6}$$

where $\tilde{u}_{\mathbf{k}}$ is the k-th discrete Fourier coefficient of u. The error estimate for this interpolation is the following:

$$\|u - I_N u\|_{H^l(\Omega)} \leqslant C N^{l-m} \|u\|_{H^m(\Omega)} \qquad \text{for } 0 \leqslant l \leqslant m. \tag{9.7.7}$$

It holds for all $u \in H_p^m(\Omega)$ with $m > d/2$. For $l = 0$, the proof can be done as for (9.1.15) by mapping Ω into the reference domain $\Omega_N = (0, 2\pi N)^d$. For $l > 0$, the estimate (9.7.7) is obtained using the corresponding one for $l = 0$, the estimate (9.7.4), and the following inverse inequality

$$\|\phi\|_{H^m(\Omega)} \leqslant C N^{m-k} \|\phi\|_{H^k(\Omega)} \qquad \begin{array}{l} \text{for } 0 \leqslant k \leqslant m, \\ \text{for all } \phi \in S_N, \end{array} \tag{9.7.8}$$

which extends (9.1.5) for $p = 2$.

9.7.2. Legendre Approximations

We consider now the domain $\Omega = (-1, 1)^d$ in $\mathbb{R}^d$ with $d = 2$ or 3, and we still denote an element of $\mathbb{R}^d$ by $\mathbf{x} = (x_1, \ldots, x_d)$. We denote by $L^2(\Omega)$ the space of square integrable functions in Ω and by $H^m(\Omega)$ the corresponding Sobolev space of order m, (see (A.9.h) and (A.11.a)). We denote by $\mathbb{P}_N$ the space of algebraic polynomials of degree up to N in each variable x_i, for $i = 1, \ldots, d$, and we consider the tensor product of the Legendre polynomials,

$$\phi_{\mathbf{k}}(\mathbf{x}) = L_{k_1}(x_1)\ldots L_{k_d}(x_d) \qquad \text{for } \mathbf{k} = (k_1, \ldots, k_d) \in \mathbb{N}^d \tag{9.7.9}$$

as a basis for $L^2(\Omega)$. Also, we denote by P_N the orthogonal projection operator from $L^2(\Omega)$ upon $\mathbb{P}_N$, so that (see (2.3.9))

$$P_N u = \sum_{\|\mathbf{k}\| \leqslant N} \hat{u}_{\mathbf{k}} \phi_{\mathbf{k}}, \qquad \hat{u}_{\mathbf{k}} = \prod_{i=1}^{d} \left(k_i + \frac{1}{2}\right) \cdot \int_\Omega u(\mathbf{x}) \phi_{\mathbf{k}}(\mathbf{x})\, d\mathbf{x} \tag{9.7.10}$$

for all $u \in L^2(\Omega)$. The summation in (9.7.10) is extended to all multi-integers $\mathbf{k} \in \mathbb{N}^d$ such that $0 \leqslant k_i \leqslant N$, for $i = 1, \ldots, d$.

Concerning the truncation error, the following estimate holds for all $u \in H^m(\Omega)$, $m \geqslant 0$:

$$\|u - P_N u\|_{H^l(\Omega)} \leqslant C N^{\sigma(l) - m} \|u\|_{H^m(\Omega)} \qquad 0 \leqslant l \leqslant m, \tag{9.7.11}$$

where $\sigma(l) = 0$ if $l = 0$ and $\sigma(l) = 2l - \frac{1}{2}$ for $l > 0$.

We consider now the operators of orthogonal projection for the inner product of the Sobolev spaces $H^1(\Omega)$ and $H_0^1(\Omega)$ (the latter space being defined in (A.11.c)). By analogy with (9.4.12), we set

$$\begin{aligned} &P_N^1 \colon H^1(\Omega) \to \mathbb{P}_N \qquad \text{such that} \\ &((P_N^1 u, \phi)) = ((u, \phi)) \qquad \text{for all } \phi \in \mathbb{P}_N, \end{aligned} \tag{9.7.12}$$

where $((u, v)) = \int_\Omega (\nabla u \cdot \nabla v + uv)\, d\mathbf{x}$ is the inner product of $H^1(\Omega)$. Moreover, if we denote by $\mathbb{P}_N^0$ the subspace of $\mathbb{P}_N$ of those polynomials vanishing at the boundary of Ω, we set by analogy with (9.4.21):

$$\begin{aligned} &P_N^{1,0} \colon H_0^1(\Omega) \to \mathbb{P}_N^0 \qquad \text{such that} \\ &[P_N^{1,0} u, \phi] = [u, \phi] \qquad \text{for all } \phi \in \mathbb{P}_N^0, \end{aligned} \tag{9.7.13}$$

where $[u, v] = \int_\Omega \nabla u \cdot \nabla v\, d\mathbf{x}$ denotes the inner product of $H_0^1(\Omega)$ (see (A.11.c)). For all $u \in H^m(\Omega)$ (respectively, $H^m(\Omega) \cap H_0^1(\Omega)$) with $m \geqslant 1$, set $u^N = P_N^1 u$ (respectively $P_N^{1,0} u$). Then the following estimates hold:

$$\|u - u^N\|_{H^k(\Omega)} \leqslant C N^{k-m} \|u\|_{H^m(\Omega)} \qquad 0 \leqslant k \leqslant 1. \tag{9.7.14}$$

These estimates are optimal, and generalize to more dimensions those given in Sec. 9.4.2 for a single-space variable. They can be extended to higher order Sobolev norms, and to cover different kinds of boundary behaviour of u. This very general result reads as follows:

Let l and m be two integers such that $0 \leqslant l \leqslant m$, and let λ be another integer so that $0 \leqslant \lambda \leqslant l$. Let u be a function of $H^m(\Omega)$ such that, if $\lambda \geqslant 1$, u vanishes at the boundary together with its derivatives of order up to $\lambda - 1$. Then there exists a polynomial $u^N \in \mathbb{P}_N$ having the same boundary behavior as u, such that

$$\|u - u^N\|_{H^k(\Omega)} \leqslant CN^{k-m}\|u\|_{H^m(\Omega)} \qquad \text{for } 0 \leqslant k \leqslant l. \tag{9.7.15}$$

Finally, we consider multidimensional Legendre interpolation. We set

$$\mathbf{x_j} = (x_{j_1}, \ldots, x_{j_d}), \quad \text{for } \mathbf{j} = (j_1, \ldots, j_d) \in \mathbb{N}^d \quad \|\mathbf{j}\| \leqslant N, \tag{9.7.16}$$

where $\{x_m, 0 \leqslant m \leqslant N\}$ is one of the Gauss-Legendre quadrature families (2.3.10), (2.3.11), or (2.3.12). Then we denote by I_N the interpolation operator at the points (9.7.16), i.e., for each continuous function u, $I_N u \in \mathbb{P}_N$ satisfies

$$(I_N u)(\mathbf{x_j}) = u(\mathbf{x_j}) \qquad \text{for all } \mathbf{j} \in \mathbb{N}^d,\ \|\mathbf{j}\| \leqslant N. \tag{9.7.17}$$

We can represent $I_N u$ as follows:

$$I_N u = \sum_{\|k\| \leqslant N} \tilde{u}_{\mathbf{k}} \phi_{\mathbf{k}},$$

with (9.7.18)

$$\tilde{u}_{\mathbf{k}} = (\gamma_{k_1} \ldots \gamma_{k_d})^{-1} \sum_{\mathbf{j}} u(\mathbf{x_j}) \phi_{\mathbf{k}}(\mathbf{x_j}) w_{j_1} \ldots w_{j_d},$$

where γ_k's are defined in (2.3.13) and the w_m's are one of the weights (2.3.10)—(2.3.12). The interpolation error estimate is

$$\|u - I_N u\|_{H^l(\Omega)} \leqslant CN^{2l+d/2-m}\|u\|_{H^m(\Omega)} \qquad \text{for } 0 \leqslant l \leqslant m, \tag{9.7.19}$$

for all $u \in H^m(\Omega)$, with $m > d/2$.

9.7.3. Chebyshev Approximations

Unless otherwise specified, we keep the notation of the previous section. Instead of (9.7.9) we set now

$$\phi_{\mathbf{k}}(\mathbf{x}) = T_{k_1}(x_1) \ldots T_{k_d}(x_d) \qquad \text{for } \mathbf{k} = (k_1, \ldots k_d) \in \mathbb{N}^d.$$

This is a basis of $L^2_w(\Omega)$, the space of the measurable functions on Ω which are square integrable for the multidimensional Chebyshev weight $w(\mathbf{x}) = \{\prod_{i=1}^d (1 - x_i^2)\}^{-1/2}$ (see (A.9.h)). For each $u \in L^2_w(\Omega)$, the truncation of its Chebyshev series is given by

$$P_N u = \sum_{\|\mathbf{k}\| \leqslant N} \hat{u}_{\mathbf{k}} \phi_{\mathbf{k}},$$

with (9.7.20)

$$\hat{u}_{\mathbf{k}} = \left(\frac{2}{\pi}\right)^d \left(\prod_{i=1}^d \frac{1}{c_{k_i}}\right) \int_\Omega u(\mathbf{x}) \phi_{\mathbf{k}}(\mathbf{x}) w(\mathbf{x})\, d\mathbf{x},$$

(see (2.4.10) and (2.4.11)). Denoting by $H^m_w(\Omega)$ the weighted Sobolev spaces relative to the Chebyshev weight (see (A.11.b)) the remainder of the Chebyshev series of a function $u \in H^m_w(-1, 1)$, $m \geqslant 0$, can be bounded as follows:

$$\|u - P_N u\|_{H^l_w(\Omega)} \leqslant C N^{\sigma(l)-m} \|u\|_{H^m_w(\Omega)} \qquad 0 \leqslant l \leqslant m, \tag{9.7.21}$$

where $\sigma(l) = 0$ if $l = 0$, and $\sigma(l) = 2l - \frac{1}{2}$ if $l > 0$.

Concerning the projection errors in the higher order Sobolev norms, we have essentially the same kind of results as for the Legendre expansion, except that now the estimates are not quite so general. Precisely, let us define the operator

$$\begin{aligned} &P^1_N \colon H^1_w(\Omega) \to \mathbb{P}_N \qquad \text{such that} \\ &((P^1_N u, \phi))_w = ((u, \phi)_w \qquad \text{for all } \phi \in \mathbb{P}_N \end{aligned} \tag{9.7.22}$$

where $((u, v))_w = \int_\Omega (\nabla u \cdot \nabla v + uv) w \, dx$ is the inner product of $H^1_w(\Omega)$. Moreover, we define the operator

$$\begin{aligned} &P^{1,0}_N \colon H^1_{w,0}(\Omega) \to \mathbb{P}^0_N \qquad \text{such that} \\ &[P^{1,0}_N u, \phi]_w = [u, \phi]_w \qquad \text{for all } \phi \in \mathbb{P}^0_N. \end{aligned} \tag{9.7.23}$$

Here $[u, v]_w = \int_\Omega (\nabla u) \cdot (\nabla v) w \, dx$ is the inner product of $H^1_{w,0}(\Omega)$ (see (A.11.c)).

For each $u \in H^m_w(\Omega)$ (respectively, $H^1_{w,0}(\Omega) \cap H^m_w(\Omega)$), with $m \geqslant 1$, we set $u^N = P^1_N u$ (respectively $P^{1,0}_N u$). Then we have the estimate

$$\|u - u^N\|_{H^1_w(\Omega)} \leqslant C N^{1-m} \|u\|_{H^m_w(\Omega)}. \tag{9.7.24}$$

A similar estimate can also be obtained for the orthogonal projection operator $P^2_N \colon H^2_w(\Omega) \to \mathbb{P}_N$, and for the projector $\tilde{P}^2_N \colon H^2_w(\Omega) \cap H^1_{w,0}(\Omega) \to \mathbb{P}^0_N$. In both cases, setting $u^N = P^2_N u$ (or $u^N = \tilde{P}^2_N u$) we obtain

$$\|u - u^N\|_{H^2_w(\Omega)} \leqslant C N^{2-m} \|u\|_{H^m_w(\Omega)} \qquad m \geqslant 2. \tag{9.7.25}$$

Concerning Chebyshev interpolation in Ω, let the interpolation points be defined as in (9.7.16), where now the x_m's are any of the nodes (2.4.12), (2.4.13), or (2.4.14). The Chebyshev interpolation at these points is defined as in (9.7.17) or (9.7.18), where the γ_k's are defined in (2.4.18) and the w_m's are defined in (2.4.12), (2.4.13), or (2.4.14). The interpolation error estimate is

$$\|u - I_N u\|_{H^l_w(\Omega)} \leqslant C N^{2l-m} \|u\|_{H^m_w(\Omega)} \qquad 0 \leqslant l \leqslant m, \tag{9.7.26}$$

for all $u \in H^m_w(\Omega)$ with $m > d/2$.

9.7.4. Blended Fourier and Chebyshev Approximations

Several spectral approximations provide a numerical solution which is a finite expansion in terms of trigonometric (Fourier) polynomials in some Cartesian directions, and of algebraic (Chebyshev) polynomials in the others.

This is typically the case of those problems set in Cartesian geometry, whose physical solution is periodic with respect to one (or more variables), and submitted to Dirichlet or Neumann boundary conditions in the direction of the remaining variables.

We consider here for the sake of simplicity a two-dimensional domain, say $\Omega = (-1, 1) \times (0, 2\pi)$, but what we are going to present is extendable in an obvious manner to a domain of the form $\Omega = (-1, 1)^{d_1} \times (0, 2\pi)^{d_2}$ for d_1, $d_2 \geqslant 1$. We introduce first some notation. For each integer M we denote by $\mathbb{P}_M$ the space of algebraic polynomials in the variable x of degree up to M. Moreover, for each integer N we denote by S_N the space

$$S_N = \operatorname{span}\{e^{iky} \mid -N \leqslant k \leqslant N-1\}.$$

Then we define the space $V_{M,N}$ as the tensor product of $\mathbb{P}_M$ and S_N, i.e.,

$$V_{M,N} = \left\{\phi(x, y) = \sum_{m=0}^{M} \sum_{n=-N}^{N-1} a_{mn} T_m(x) e^{iny}, a_{mn} \in \mathbb{C}\right\}.$$

Let us denote by $L_y^2(H_{w,x}^k)$ the space of the measurable functions u: $(0, 2\pi) \to H_w^k(-1, 1)$ such that

$$\|u\|_{k,0} = \left\{\int_0^{2\pi} \|u(\cdot, y)\|^2_{H_w^k(-1,1)}\, dy\right\}^{1/2} < \infty. \tag{9.7.27}$$

For $k = 0$, this norm will be denoted briefly by

$$\|u\|_0 = \left\{\int_0^{2\pi} dy \int_{-1}^{1} |u(x, y)|^2 w(x)\, dx\right\}^{1/2}. \tag{9.7.28}$$

Moreover, for any positive integer h we define

$$H_y^h(L_{w,x}^2) = \left\{u \in L^2(\Omega) \,\middle|\, \frac{\partial^j u}{\partial y^j} \in L_y^2(L_{w,x}^2),\ 0 \leqslant j \leqslant h\right\};$$

the norm is given by

$$\|u\|_{0,h} = \left\{\sum_{j=0}^{h} \left\|\frac{\partial^j u}{\partial y^j}\right\|_0^2\right\}^{1/2}. \tag{9.7.29}$$

Next, for any positive integers h, k we define

$$H^{k,h}(\Omega) = L_y^2(H_{w,x}^k) \cap H_y^h(L_{w,x}^2). \tag{9.7.30}$$

The norm of this space will be

$$\|u\|_{k,h} = \{\|u\|_{k,0}^2 + \|u\|_{0,h}^2\}^{1/2}. \tag{9.7.31}$$

Finally, in order to take into account the periodicity in the y-variable, let $C_p^\infty(\Omega)$ denote the set of those functions continuous with all their derivatives up to the boundary of Ω, and 2π-periodic with all their derivatives with respect to the y-direction. Then we define $H_p^{k,h}(\Omega)$ as the Hilbert space of the functions which are limit of Cauchy sequences in $C_p^\infty(\bar{\Omega})$ with respect to the norm $\|u\|_{k,h}$. The Hilbert space $H_p^m(\Omega)$ is defined similarly, with respect to the norm

$$\|u\|_m = \left\{\sum_{\mathbf{j}} \int_0^{2\pi} dy \int_{-1}^{1} \left|\frac{\partial^{j_1+j_2}u}{\partial x^{j_1}\partial y^{j_2}}\right|^2 w(x)\,dx\right\}^{1/2}, \tag{9.7.32}$$

where $\mathbf{j} = (j_1, j_2)$ is such that $0 \leqslant j_1 + j_2 \leqslant m$.

For any function $u \in L_y^2(L_{w,x}^2)$, let $P_{M,N}u$ denote the projection of u upon $V_{M,N}$, i.e.,

$$P_{M,N}u = \sum_{m=0}^{M} \sum_{n=-N}^{N-1} \hat{u}_{mn} T_m(x) e^{iny}, \tag{9.7.33}$$

where

$$\hat{u}_{mn} = \frac{1}{c_k \pi^2} \int_0^{2\pi} \int_{-1}^{1} u(x,y) T_m(x) e^{-iny} w(x)\,dx\,dy.$$

The c_k's are given in (2.4.10).

If $u \in H_p^{k,h}(\Omega)$ for some $k, h \geqslant 0$, then

$$\|u - P_{M,N}u\|_0 \leqslant C(M^{-k} + N^{-h})\|u\|_{k,h}. \tag{9.7.34}$$

The proof of this result can be done as follows. Denote by P_M^C and P_N^F the L^2-orthogonal projection operators upon $\mathbb{P}_M$ and S_N in the Chebyshev and Fourier expansions, respectively. Then,

$$\begin{aligned} u - P_N^F P_M^C u = {} & -(u - P_M^C u) + P_N^F(u - P_M^C u) \\ & + (u - P_N^F u) + (u - P_M^C u). \end{aligned} \tag{9.7.35}$$

Now (9.7.34) follows, noting that $\|u - P_{M,N}u\|_0 \leqslant \|u - P_N^F P_M^C u\|_0$ and using (9.1.9) and (9.5.6).

In higher order norms, the best approximation error can be estimated by a splitting technique similar to the one used in (9.7.35). For instance, using instead of P_M^C the $H_w^1(-1,1)$-orthogonal projector $(P_M^1)^C$ defined in (9.5.10), it follows that

$$\|u - P_N^F (P_M^1)^C u\|_1 \leqslant C\{M^{1-k}\|u\|_{k-1,1} + N^{1-h}\|u\|_{1,h-1}\},$$

where we used (9.1.10) and (9.5.11)). Thus,

$$\inf_{v \in V_{M,N}} \|u - v\|_1 \leqslant C(M^{1-k} + N^{1-h})\|u\|_{k,h} \qquad \text{if } k, h \geqslant 1. \tag{9.7.36}$$

Obviously, a similar estimate holds if u and v are assumed to vanish on the sides $x = -1$ and $x = 1$ of the boundary of Ω. It is enough to take the operator defined in (9.5.14) and to use (9.5.16) instead of (9.5.11). Best approximation error estimates in higher norms can be proven similarly.

Concerning interpolation, let us consider the points

$$\xi_{ij} = \left(\cos\frac{\pi i}{M}, \frac{\pi j}{N}\right), \qquad 0 \leqslant i \leqslant M, \quad 0 \leqslant j \leqslant 2N - 1. \tag{9.7.37}$$

Then, denote by I_M^C the usual Chebyshev interpolation operator with respect to the points $\{\cos(\pi i/M)\}$ and by I_N^F the Fourier interpolant relative to the points $\{\pi j/N\}$. Of course, $I_{M,N} = I_M^C I_N^F (= I_N^F I_M^C)$ is the interpolation operator relative to the points $\{\xi_{ij}\}$, i.e., for all $u \in C^0(\bar{\Omega})$,

$$I_{M,N}u \in V_{M,N}: I_{M,N}u(\xi_{ij}) = u(\xi_{ij}), \tag{9.7.38}$$

for $0 \leqslant i \leqslant M$ and $0 \leqslant j \leqslant 2N-1$.

By (9.1.18) and (9.5.20) we obtain if $k^{-1} + h^{-1} < 2$:

$$\begin{aligned}\|u - I_{M,N}u\|_0 &\leqslant \|u - I_M^C u\|_0 + \|I_M^C(u - I_N^F u)\|_0 \\ &\leqslant \|u - I_M^C u\|_0 + \|(u - I_N^F u) - I_M^C(u - I_N^F u)\|_0 + \|u - I_N^F u\|_0 \\ &\leqslant C\{M^{-k}\|u\|_{k,0} + M^{-1}\|u - I_N^F u\|_{1,0} + N^{-h}\|u\|_{0,h}\}.\end{aligned}$$

Moreover,

$$\|u - I_N^F u\|_{1,0} \leqslant C(M^{-1}N)N^{-h}\left\{\sum_{j=0}^{h-1}\left\|\frac{\partial^j u}{\partial y^j}\right\|_{1,0}^2\right\}^{1/2}.$$

If h and k are sufficiently large, then the last norm on the right-hand side can be bounded by the norm in (9.7.31) (see Lions and Magenes (1972)). Finally, if we assume that there exist two constants α and β for which $\alpha \leqslant M^{-1}N \leqslant \beta$, we conclude that

$$\|u - I_{M,N}u\|_0 \leqslant C(M^{-k} + N^{-h})\|u\|_{k,h}. \tag{9.7.39}$$

Bibliographical Notes

The estimate (9.7.7) is due to Pasciak (1980). The results on the truncation and interpolation operators for both the Legendre and the Chebyshev systems can be found in Canuto and Quarteroni (1982a); the exponent in (9.7.19) has been improved over the original estimate using the polynomial satisfying (9.7.15) (for $\lambda = 0$) in the proof of Theorem 3.2 in Canuto and Quarteroni (1982a). Maday (1981) proved the estimates on the higher order projection operators. Sharper estimates concerning interpolation of blended Fourier and Chebyshev expansions are given in Quarteroni (1987a).

A blending of Fourier and Legendre expansion can be considered as well, and the analysis can be carried out in a similar way (see Bernardi, Maday and Métivet (1987a)).

In some cases, a coupling of Fourier and finite-element (or finite-difference) approximations may be more appropriate. This is typically the case of problems in complex geometries, for instance in a domain of the form $\Omega \times (0, 2\pi)$, where Ω is a polygonal region of $\mathbb{R}^d$ ($d \geqslant 2$) and the solution is periodic with respect to the last variable. In such a situation, the greater flexibility of finite elements over spectral expansions can be exploited in order to resolve the non-Cartesian geometry of Ω. An analysis of this kind of approximation can be found in Canuto, Maday and Quarteroni (1982), and in Mercier and Raugel (1982).

CHAPTER 10

Theory of Stability and Convergence for Spectral Methods

In this chapter we present a fairly general approach to the stability and convergence analysis of spectral methods. We confine ourselves to linear problems. Analysis of several non-linear problems is presented in Chaps. 11 and 12. For time-dependent problems, only the discretizations in space are considered. Stability for fully discretized time-dependent problems is discussed in Chap. 4 by a classical eigenvalue analysis, and in Chap. 12 by variational methods.

It may be worthwhile to specify precisely what is meant here by stability of a spatial approximation based on a spectral method. A scheme will be called *stable* if it is possible to control the discrete solution by the data in a way independent of the discretization parameter N (the degree of the polynomials used). This means that a suitable norm of the solution is bounded by a constant multiple of a suitable norm of the data, and all the norms involved, as well as the constant, do not depend on N. In other words, for a fixed data, all the discrete solutions produced by the spectral scheme, as N tends to infinity, lie in a bounded subset of a normed linear space.

The three most representative methods of spectral type, i.e., Galerkin, tau, and collocation, are considered. We begin with a reexamination of the three examples of Chap. 1. The aim here is to introduce the salient aspects of the different methods of analysis. We then proceed to the general theory with the objective of achieving a unified methodology. Time-independent problems are considered first, and then both parabolic and hyperbolic equations are analyzed. All spectral schemes are interpreted as projection methods over a finite-dimensional space of polynomials with respect to a certain inner product. The stability is proved either by the energy method or by a generalized variational principle. The convergence analysis uses stability results and the results of approximation theory given in Chap. 9 for several projection operators. Applications of these general results to the analysis of many pertinent examples are given.

10.1. The Three Examples Revisited

Some basic aspects of the analysis of stability and convergence for spectral methods can be illustrated by considering the three examples already discussed in Chap. 1. The nature of the theory presented in this section is deliberately pedestrian, since the purpose is to introduce the reader to the more sophisticated and abstract mathematics in the remaining sections of this chapter.

10.1.1. A Fourier Galerkin Method for the Wave Equation

The linear hyperbolic problem

$$\begin{cases} \dfrac{\partial u}{\partial t} - \dfrac{\partial u}{\partial x} = 0 & 0 < x < 2\pi, \quad t > 0 \\ u(x,t) & 2\pi\text{-periodic in } x \\ u(x,0) = u_0(x) & 0 < x < 2\pi, \end{cases}$$

was approximated in Sec. 1.2.1 by the Galerkin scheme (1.2.3). For any $t \geqslant 0$, $u^N(x,t)$ is a trigonometric polynomial of degree N in x, i.e., $u^N(t) \in S_N$ where

$$S_N = \text{span}\{e^{ikx} \mid -N \leqslant k \leqslant N-1\},$$

(see (9.1.1)). (Note that in this part of the book we are following the convention that Fourier series are truncated at degree N rather than degree $N/2$.) The solution u^N satisfies the variational relation

$$\int_0^{2\pi} \left[\frac{\partial u^N}{\partial t} - \frac{\partial u^N}{\partial x}\right] \overline{v(x)}\, dx = 0 \qquad \text{for all } v \in S_N, \tag{10.1.1}$$

which is equivalent to (1.2.3) since the ψ_k's are a basis in S_N, and by the initial condition (1.2.11)

$$u^N(0) = P_N u_0 = \sum_{k=-N}^{N-1} \hat{u}_{0,k} e^{ikx}.$$

For any $t > 0$, let us set $v(x) = u^N(x,t)$ in (10.1.1). An integration-by-parts yields

$$\mathbb{R}e \int_0^{2\pi} \frac{\partial u^N}{\partial x} \bar{u}^N\, dx = \frac{1}{2}\{|u^N(2\pi,t)|^2 - |u^N(0,t)|^2\} = 0$$

by the periodicity condition. It follows that

$$\frac{1}{2}\frac{d}{dt}\int_0^{2\pi} |u^N(x,t)|^2\, dx = \mathbb{R}e \int_0^{2\pi} \frac{\partial u^N}{\partial t} \bar{u}^N\, dx = 0,$$

i.e., the L^2-norm (in space) of the spectral solution is constant in time. Therefore, for any $t > 0$

$$\int_0^{2\pi} |u^N(x,t)|^2\, dx = \int_0^{2\pi} |P_N u_0(x)|^2\, dx \leqslant \int_0^{2\pi} |u_0(x)|^2\, dx.$$

Since the right-hand side is a constant, the Galerkin scheme (1.2.3) is stable in the L^2-norm.

On the other hand, projecting the equation $(\partial u/\partial t) - (\partial u/\partial x) = 0$ on S_N yields the result that the truncated Fourier series $P_N u$ of the exact solution u satisfies at any $t > 0$

$$\int_0^{2\pi} \left(\frac{\partial}{\partial t} P_N u - \frac{\partial}{\partial x} P_N u\right) \overline{v(x)}\, dx = 0 \qquad \text{for all } v \in S_N.$$

This is the same variational relation which defines u^N. Since $u^N = P_N u$ at time $t = 0$, it follows that

$$u^N = P_N u \qquad \text{for all } t \geqslant 0.$$

Since $P_N u$ converges to u as N tends to infinity, the approximation is convergent. Moreover, (9.1.9) provides an estimate of the error between the exact and the spectral solution. For all $t > 0$, we have:

$$\int_0^{2\pi} |u(x,t) - u^N(x,t)|^2\, dx \leqslant CN^{-2m} \int_0^{2\pi} \left|\frac{\partial^m u}{\partial x^m}(x,t)\right|^2 dx.$$

10.1.2. A Chebyshev Collocation Method for the Heat Equation

Consider now the linear heat equation

$$\frac{\partial u}{\partial t} - \frac{\partial^2 u}{\partial x^2} = 0 \qquad -1 < x < 1, \quad t > 0,$$

with homogeneous Dirichlet conditions

$$u(-1,t) = u(1,t) = 0 \qquad t > 0$$

and initial condition

$$u(x,0) = u_0(x) \qquad -1 < x < 1.$$

A Chebyshev collocation scheme was considered for this problem in Sec. 1.2.2 (see (1.2.26)—(1.2.28)). For any $t > 0$, the spectral solution u^N is an algebraic polynomial of degree N on the interval $(-1, 1)$, vanishing at the endpoints. It is defined through the collocation equations

$$\frac{\partial u^N}{\partial t}(x_k,t) - \frac{\partial^2 u^N}{\partial x^2}(x_k,t) = 0 \qquad k = 1, \ldots, N-1 \qquad (10.1.2)$$

and the initial condition

$$u^N(x_k, 0) = u_0(x_k) \qquad k = 0, \ldots, N.$$

The collocation points are given by $x_k = \cos(k\pi/N)$ (see (1.2.29) or (2.4.14)). They are the nodes of the Gauss–Lobatto quadrature formula relative to the

Chebyshev weight $w(x) = 1/\sqrt{1-x^2}$, whose weights are given by $w_0 = w_N = \pi/2N$, and $w_k = \pi/N$ if $k = 1, \ldots, N-1$ (see (2.4.14)). This property will be constantly used in the subsequent analysis of Chebyshev collocation methods. Its relevance in the theory of spectral methods was first pointed out by Gottlieb (1981).

Let us multiply the k-th equation of (10.1.2) by $u^N(x_k, t)w_k$ and sum over k. We get

$$\frac{1}{2}\frac{d}{dt}\sum_{k=0}^{N}[u^N(x_k,t)]^2 w_k - \sum_{k=0}^{N}\frac{\partial^2 u^N}{\partial x^2}(x_k,t)u^N(x_k,t)w_k = 0, \quad (10.1.3)$$

where we are allowed to include the boundary points in the sum since u^N vanishes there. The product $(\partial^2 u^N/\partial x^2)u^N$ is a polynomial of degree $2N-2$; hence, by the exactness of the quadrature formula (see (2.2.17)),

$$-\sum_{k=0}^{N}\frac{\partial^2 u^N}{\partial x^2}(x_k,t)u^N(x_k,t)w_k = -\int_{-1}^{1}\frac{\partial^2 u^N}{\partial x^2}(x,t)u^N(x,t)w(x)\,dx.$$

In Sec. 11.1.2 it is proved, as a part of a general result, that the right-hand side is positive, and actually dominates a weighted "energy" of the solution, i.e.,

$$-\int_{-1}^{1}\frac{\partial^2 u^N}{\partial x^2}(x,t)u^N(x,t)w(x)\,dx \geqslant \frac{1}{4}\int_{-1}^{1}\left[\frac{\partial u^N}{\partial x}(x,t)\right]^2 w(x)\,dx.$$

Then from (10.1.3) it follows that

$$\frac{1}{2}\frac{d}{dt}\sum_{k=0}^{N}[u^N(x_k,t)]^2 w_k + \frac{1}{4}\int_{-1}^{1}\left[\frac{\partial u^N}{\partial x}(x,t)\right]^2 w(x)\,dx \leqslant 0,$$

whence

$$\sum_{k=0}^{N}[u^N(x_k,t)]^2 w_k + \frac{1}{2}\int_0^t\int_{-1}^{1}\left[\frac{\partial u^N}{\partial x}(x,s)\right]^2 w(x)\,dx\,ds \leqslant \sum_{k=0}^{N}[u_0(x_k)]^2 w_k.$$

The sum on the left-hand side represents the *discrete* L^2-norm of the solution with respect to the Chebyshev weight. It does not coincide with the continuous L^2-norm $\int_{-1}^{1}[u^N(x,t)]^2 w(x)\,dx$, since $(u^N)^2$ is a polynomial of degree $2N$. However, as pointed out in Sec. 9.3, (see (9.3.2)), it is uniformly equivalent to this norm, i.e.,

$$\int_{-1}^{1}[u^N(x,t)]^2 w(x)\,dx \leqslant \sum_{k=0}^{N}[u^N(x_k,t)]^2 w_k \leqslant 2\int_{-1}^{1}[u^N(x,t)]^2 w(x)\,dx.$$

On the other hand, the sum on the right-hand side can be bounded, for instance, by twice the square of the maximum of the data on the interval $[-1, 1]$. We conclude that for any $t > 0$

$$\int_{-1}^{1} [u^N(x,t)]^2 w(x)\,dx + \frac{1}{2}\int_0^t \int_{-1}^{1} \left[\frac{\partial u^N}{\partial x}(x,s)\right]^2 w(x)\,dx\,ds$$
$$\leqslant 2 \max_{-1 \leqslant x \leqslant 1} |u_0(x)|^2.$$

This proves that the Chebyshev collocation scheme is stable. Note that this stability estimate provides a bound for both the weighted L^2-norm at any given time and also the weighted "energy" norm integrated over the time interval $(0, t)$.

The convergence of the approximation can be proved by a simple argument. Assume the exact solution u to be smooth enough. Its interpolant $\tilde{u} = I_N u$, defined in Sec. 2.2.3, satisfies the collocation equations:

$$\frac{\partial \tilde{u}}{\partial t}(x_k, t) - \frac{\partial^2 \tilde{u}}{\partial x^2}(x_k, t) = r(x_k, t) \qquad t > 0, \quad k = 1, \ldots, N-1,$$

with the truncation error $r = (\partial^2/\partial x^2)(u - \tilde{u})$. Hence, the difference $e = \tilde{u} - u^N$, which is a polynomial of degree N vanishing at the boundary points, satisfies the equations

$$\frac{\partial e}{\partial t}(x_k, t) - \frac{\partial^2 e}{\partial x^2}(x_k, t) = r(x_k, t) \qquad t > 0, \quad k = 1, \ldots, N-1.$$

and the initial condition $e(x_k, 0) \equiv 0$.

The same analysis previously used yields

$$\frac{1}{2}\frac{d}{dt}\sum_{k=0}^{N} [e(x_k,t)]^2 w_k + \frac{1}{4}\int_{-1}^{1} \left[\frac{\partial e}{\partial x}(x,t)\right]^2 w(x)\,dx$$
$$\leqslant \sum_{k=0}^{N} r(x_k,t) e(x_k,t) w_k$$
$$\leqslant \frac{1}{2}\sum_{k=0}^{N} [r(x_k,t)]^2 w_k + \frac{1}{2}\sum_{k=0}^{N} [e(x_k,t)]^2 w_k.$$

Here we have used by Cauchy–Schwarz inequality (see (A.2)). By the Gronwall lemma (see (A.15)) we get

$$\sum_{k=0}^{N} [e(x_k,t)]^2 w_k + \frac{1}{2}\int_0^t \int_{-1}^{1} \left[\frac{\partial e}{\partial x}(x,s)\right]^2 w(x)\,dx\,ds$$
$$\leqslant \exp(t) \int_0^t \sum_{k=0}^{N} [r(x_k,s)]^2 w_k\,ds. \tag{10.1.4}$$

If we drop the second term on the left-hand side, we get an estimate on the discrete L^2-norm of the error $u - u^N$ at the collocation points:

$$\sum_{k=0}^{N} [u(x_k,t) - u^N(x_k,t)]^2 w_k \leqslant \exp(t) \int_0^t \sum_{k=0}^{N} [r(x_k,s)]^2 w_k\,ds.$$

Hence, the scheme is convergent provided the truncation error vanishes as N tends to infinity. Now we have

$$
\begin{aligned}
\sum_{k=0}^{N} [r(x_k, s)]^2 w_k &= \sum_{k=0}^{N} [I_N r(x_k, s)]^2 w_k \\
&\leqslant 2 \int_{-1}^{1} [I_N r(x, s)]^2 w(x)\, dx \\
&= 2 \int_{-1}^{1} \left[\left(I_N \frac{\partial^2 u}{\partial x^2} - \frac{\partial^2}{\partial x^2}(I_N u) \right)(x, s) \right]^2 w(x)\, dx \\
&\leqslant 4 \int_{-1}^{1} \left[\left(\frac{\partial^2 u}{\partial x^2} - I_N \frac{\partial^2 u}{\partial x^2} \right)(x, s) \right]^2 w(x)\, dx \\
&\quad + 4 \int_{-1}^{1} \left[\frac{\partial^2}{\partial x^2}(u - I_N u)(x, s) \right]^2 w(x)\, dx,
\end{aligned}
$$

where we have used the equivalence (9.3.2) between discrete and continuous L^2-norms. Applying estimate (9.5.20) in evaluating the right-hand side, we obtain the estimate

$$
\begin{aligned}
&\left(\sum_{k=0}^{N} [u(x_k, t) - u^N(x_k, t)]^2 w_k \right)^{1/2} \\
&\qquad \leqslant C N^{4-m} \exp\left(\frac{t}{2}\right) \left(\int_0^t \|u(s)\|^2_{H_w^m(-1,1)}\, ds \right)^{1/2},
\end{aligned}
\tag{10.1.5}
$$

where the norm on the right-hand side is defined in (A.11.b), and C is a constant independent of N and u.

Using (10.1.4) once more, one can derive an estimate for the spatial derivative of the error, i.e.,

$$
\begin{aligned}
&\left(\int_0^t \int_{-1}^{1} \left[\left(\frac{\partial u}{\partial x} - \frac{\partial u^N}{\partial x} \right)(x, s) \right]^2 w(x)\, dx\, ds \right)^{1/2} \\
&\qquad \leqslant C N^{4-m} \exp\left(\frac{t}{2}\right) \left(\int_0^t \|u(s)\|^2_{H_w^m(-1,1)}\, ds \right)^{1/2}.
\end{aligned}
$$

This inequality proves that the approximation is convergent and the error decays faster than algebraically when the solution is infinitely smooth.

The previous analysis allows us to prove the convergence of the method in square mean norms by a transparent argument, namely, the comparison between the spectral solution and the Chebyshev interpolant of the exact solution at the collocation nodes. However, the rate of decay of the error predicted by this theory is not optimal, that is, it is slower than the one corresponding to the best approximation. According to the previous estimate, the energy norm of the error decays at least like N^{4-m}, while the error of best approximation in the same norm decays like N^{1-m} (see Sec. 9.5.2).

A more careful analysis allows us to state that the error for the collocation approximation considered here is actually asymptotic with the best approximation error, i.e., the following estimate can be obtained:

$$\left(\int_{-1}^{1} [(u-u^N)(x,t)]^2 w(x)\,dx\right)^{1/2} + \left(\int_0^t \int_{-1}^{1} \left[\left(\frac{\partial u}{\partial x} - \frac{\partial u^N}{\partial x}\right)(x,s)\right]^2 w(x)\,dx\right)^{1/2}$$

$$\leqslant CN^{1-m}\left\{\int_0^t \left(\left\|\frac{\partial u}{\partial t}(s)\right\|^2_{H_w^{m-2}(-1,1)} + \|u(s)\|^2_{H_w^m(-1,1)}\right) ds\right\}^{1/2}. \qquad (10.1.6)$$

In deriving this estimate, it is crucial that the spectral solution be compared with a projection $\tilde{u} = R_N u$ of the exact solution onto the space of polynomials of degree N, which behaves asymptotically as the best approximation of u in the energy norm. The details of this analysis are given in Example 3 of Sec. 10.5.1.

10.1.3. A Legendre Tau Method for the Poisson Equation

In Sec. 1.2.3 we considered the homogeneous Dirichlet problem for the Poisson equation in the square $\Omega = (-1, 1) \times (-1, 1)$:

$$\begin{cases} \Delta u = f & -1 < x, y < 1, \\ u = 0 & \text{if } x = \pm 1 \text{ or } y = \pm 1. \end{cases}$$

This problem was approximated by the following Legendre tau method. Let $\mathbb{P}_N$ denote the space of polynomials in two variables x, y of degree at most N in each variable. The spectral solution u^N belongs to $\mathbb{P}_N$ and is defined by

$$\int_\Omega \Delta u^N \phi\,dx\,dy = \int_\Omega f\phi\,dx\,dy \qquad \text{for all } \phi \in \mathbb{P}_{N-2}, \qquad (10.1.7)$$

and by the boundary condition

$$u^N(x, y) = 0 \qquad \text{if } x = \pm 1 \text{ or } y = \pm 1. \qquad (10.1.8)$$

The last condition was imposed in (1.2.52) in a variational way, i.e., it was translated into a set of linear relations among the Legendre coefficients. Since the problem is intrinsically formulated in a variational way, it is natural to try to derive the stability of the method from an appropriate choice of the test function ϕ in (10.1.7). Both the choices $\phi = u^N$ and $\phi = \Delta u^N$—which would immediately give stability—are not allowed, since these functions are polynomials of degree higher than $N-2$. They could be projected onto the space $\mathbb{P}_{N-2}$ of the admissible functions for (10.1.7). Instead, we adopt a different strategy. Since u^N vanishes at the boundary of the square $(-1, 1) \times (-1, 1)$, it can be factored as

$$u^N(x, y) = (1 - x^2)(1 - y^2)q(x, y) \qquad \text{for a } q \in \mathbb{P}_{N-2}.$$

We choose $\phi = -q$ in (10.1.7). Denote by $b(x, y)$ the bubble function $(1 - x^2)(1 - y^2)$. Applying Green's formula twice (in which $\partial/\partial n$ is the outward normal derivative on the boundary $\partial\Omega$ of the square), we have:

$$\begin{aligned} -\int_\Omega \Delta u^N q\,dx\,dy &= \int_\Omega \nabla(bq)\cdot\nabla q\,dx\,dy - \int_{\partial\Omega} \frac{\partial(bq)}{\partial n} q\,d\sigma \\ &= \int_\Omega b|\nabla q|^2\,dx\,dy + \frac{1}{2}\int_\Omega \nabla b\cdot\nabla(q^2)\,dx\,dy - \int_{\partial\Omega}\frac{\partial b}{\partial n} q^2\,d\sigma \\ &= \int_\Omega b|\nabla q|^2\,dx\,dy - \frac{1}{2}\int_\Omega \Delta b q^2\,dx\,dy - \frac{1}{2}\int_{\partial\Omega}\frac{\partial b}{\partial n} q^2\,d\sigma. \end{aligned} \tag{10.1.9}$$

Each term on the right-hand side is positive. On the other hand, the right-hand side of (10.1.7) can be bounded by the Cauchy–Schwarz inequality as follows:

$$\begin{aligned} &\left|\int_\Omega fq\,dx\,dy\right| \\ &\quad = \left|\int_\Omega \frac{f}{\sqrt{|\Delta b|}}\sqrt{|\Delta b|}\,q\,dx\,dy\right| \leqslant \left(\int_\Omega \frac{f^2}{|\Delta b|}\,dx\,dy\right)^{1/2}\left(\int_\Omega |\Delta b|\,q^2\,dx\,dy\right)^{1/2} \\ &\quad \leqslant \int_\Omega \frac{f^2}{|\Delta b|}\,dx\,dy + \frac{1}{4}\int_\Omega |\Delta b|\,q^2\,dx\,dy. \end{aligned}$$

By (10.1.9) and this inequality one gets

$$\int_\Omega b|\nabla q|^2\,dx\,dy + \frac{1}{4}\int_\Omega |\Delta b|\,q^2\,dx\,dy \leqslant \int_\Omega \frac{f^2}{|\Delta b|}\,dx\,dy. \tag{10.1.10}$$

The integral on the right-hand side is certainly finite if f is bounded in Ω. Finally, using the identity $\nabla u^N = b\nabla q + q\nabla b$, and noting that $b \leqslant 1$ and $|\nabla b|^2 \leqslant 2|\Delta b|$ we have:

$$\begin{aligned} \int_\Omega |\nabla u^N|^2\,dx\,dy &\leqslant 2\int_\Omega b^2|\nabla q|^2\,dx\,dy + 2\int_\Omega |\nabla b|^2 q^2\,dx\,dy \\ &\leqslant 2\int_\Omega b|\nabla q|^2\,dx\,dy + 4\int_\Omega |\Delta b|\,q^2\,dx\,dy, \end{aligned}$$

whence, by (10.1.10)

$$\int_\Omega |\nabla u^N|^2\,dx\,dy \leqslant 16\int_\Omega \frac{f^2}{|\Delta b|}\,dx\,dy. \tag{10.1.11}$$

This proves the stability of the Legendre tau method in the energy norm. Indeed, $(\int_\Omega |\nabla u^N|^2\,dx\,dy)^{1/2}$ is a norm for u^N, since u^N is zero on $\partial\Omega$.

In order to derive the convergence of the scheme, let $\tilde{u}$ denote a polynomial of degree N vanishing on $\partial\Omega$, to be chosen later as a suitable approximation of the exact solution u. Then $e = \tilde{u} - u^N$ satisfies

$$\int_\Omega \Delta e\phi \, dx\, dy = \int_\Omega \Delta(\tilde{u} - u)\phi \, dx\, dy \qquad \text{for all } \phi \in \mathbb{P}_{N-2}.$$

By the previous argument we get

$$\int_\Omega |\nabla e|^2 \, dx\, dy \leqslant 16 \int_\Omega \frac{|\Delta(u - \tilde{u})|^2}{|\Delta b|} dx\, dy,$$

whence, by the triangle inequality,

$$\int_\Omega |\nabla(u - u^N)|^2 \, dx\, dy \leqslant 2 \int_\Omega |\nabla(u - \tilde{u})|^2 \, dx\, dy + C_1 \max_{(x,y)\in\Omega} |\Delta(u - \tilde{u})|^2, \tag{10.1.12}$$

where $C_1 = 32C_0$, and C_0 is the value of the integral of $1/|\Delta b|$ over Ω (C_0 is less than 3).

There are various choices of $\tilde{u}$ which ensure that the right-hand side vanishes as N tends to infinity, thus implying the convergence of the method in the energy norm. For instance, if one chooses as $\tilde{u}$ the interpolant of u at the Gauss–Lobatto points (x_i, y_j), $0 \leqslant i, j \leqslant N$, on the square $[-1, 1] \times [-1, 1]$, then using (9.7.19) the square root of the right-hand side can be bounded by

$$CN^{7-m}\left(\sum_{k=0}^{m} \int_\Omega |D^k u|^2 \, dx\, dy\right)^{1/2}.$$

This, however, is not the best rate of convergence. A more clever choice of $\tilde{u}$, involving orthogonal projections in Sobolev spaces of high order, yields the estimate for all real $p > 4$

$$\left(\int_\Omega |\nabla(u - u^N)|^2 \, dx\, dy\right)^{1/2} \leqslant CN^{2-m}\left(\int_\Omega \sum_{k=0}^{m} |D^k u|^p \, dx\, dy\right)^{1/p}. \tag{10.1.13}$$

The details are given in Sacchi–Landriani (1986).

10.2. Towards a General Theory

In the previous section a mathematical analysis was sketched for the stability and convergence properties of three representative spectral methods. This analysis relied in a fundamental way upon interpreting the schemes as projection methods over suitable subspaces with respect to the appropriate inner products. The projection analysis is certainly natural for the Galerkin and tau methods. It appears, however, to be unnatural for the collocation method, which is invariably implemented in a pointwise manner. Unfortunately, in all but the simplest cases, the pointwise analysis of collocation methods is not only far more difficult than their projection analysis, it is also less precise, i.e., the error estimates suggest a lower rate of convergence than is achieved in practice. (The mathematical reasons for this are similar to those that make optimal error estimates easier to obtain for finite-element methods than for finite-difference methods.) An additional reason for preferring the projection analysis of collocation methods is that it enables all spectral methods to be discussed in terms of the same general theory.

As we noted in the introduction of Chap. 1, the finite-dimensional space on which the equation is projected is not necessarily the same finite-dimensional space in which the spectral solution lies. Galerkin methods invariably use the same space for both purposes. The Legendre tau approximation discussed in Sec. 10.1.3 is an example of a situation in which the two spaces differ. Many familiar collocation methods also use two different spaces. It follows that a unified approach to the theory must necessarily involve two families of finite-dimensional spaces, one for the trial functions and the other for the test functions.

The most straightforward technique for establishing the stability of the spectral schemes—the so-called energy method—is based on choosing the solution itself as the test function. This approach is successful if the spaces of the trial and test functions coincide, and if the spectral operator is positive with respect to a suitable inner product (as occurred in the first two examples of the previous section). If either of these hypotheses is not satisfied, then the energy method cannot be used. In an alternative strategy which is often invoked, stability is proven by building up a suitable test function which depends in some way on the spectral solution. This was the strategy employed in the last example of Sec. 10.1. Generally speaking, the inequality that is associated with the energy method and which ensures stability must be replaced by a more general inequality. Mathematically, this inequality amounts to the requirement that the spectral operator be an isomorphism (i.e., a continuous invertible map) between the spaces of trial and test functions, and that a suitable norm of its inverse be bounded independently of the discretization parameter.

The convergence analysis given for the introductory examples of this chapter used the standard technique of systematically comparing the spectral solution with a projection of the exact solution onto the space of the trial functions. This strategy is essentially the same as that used in the proof of the Lax–Richtmyer equivalence theorem (which states that for consistent approximations, stability is equivalent to convergence).

The last two examples in Sec. 10.1 show that the error estimate (i.e., the rate of decay of the error) predicted by this approach is extremely sensitive to the approximation properties of the particular projection of the exact solution which one chooses in this analysis. Both the truncated series and the interpolant of the exact solution appear to be viable candidates for the projection. However, the rates of decay predicted by choosing these functions are asymptotically worse than the errors of best approximation in the same norms. (This point has already been emphasized in Chap. 9.) Typically, one chooses a projection of the exact solution which yields the same approximation properties as the best approximation. Such projection operators were introduced in Secs. 9.4.2, 9.5.2 and 9.7 and will play a key role in the subsequent convergence analysis.

10.3. General Formulation of Spectral Approximations to Linear Steady Problems

Let Ω be an open bounded domain in $\mathbb{R}^d$, with piecewise smooth boundary $\partial\Omega$. We assume we want to approximate the boundary value problem

$$\begin{cases} Lu = f & \text{in } \Omega, \qquad (10.3.1\\ Bu = 0 & \text{on } \partial\Omega_b, \qquad (10.3.2) \end{cases}$$

where L is a linear differential operator in Ω and B is a set of linear boundary differential operators on a part (or the whole) of $\partial\Omega$ that we call $\partial\Omega_b$.

We assume that there exists a Hilbert space X such that L is an unbounded operator in X (see (A.1) and (A.3)). We will denote by (u, v) the inner product in X and by $\|u\| = (u, u)^{1/2}$ the associated norm. Typically, X will be a space of real or complex functions defined in Ω, which are square integrable with respect to a suitable *weight* function. Hereafter, by weight function we shall mean a continuous and strictly positive function in Ω, which is properly or improperly integrable.

The domain of definition of L, i.e., the subset $D(L)$ of those functions u of X for which Lu is still an element of X, is supposed to be a dense subspace of X (see (A.6)). Thus, L is a linear operator from $D(L)$ to X. For instance, let us consider the second-derivative operator $L = -d^2/dx^2$ on the interval $\Omega = (-1, 1)$. If $w(x) = 1/\sqrt{1-x^2}$ is the Chebyshev weight function, we set $X = L^2_w(-1, 1) = \{v \,|\, \int_{-1}^{1} v^2(x)w(x)\,dx < \infty\}$ with $(u, v) = \int_{-1}^{1} u(x)v(x)w(x)\,dx$. Then L is an unbounded operator in X, whose domain is

$$D(L) = \left\{ v \in C^1(-1, 1) \,\middle|\, \frac{d^2 v}{dx^2} \in L^2_w(-1, 1) \right\},$$

where the derivative is taken in the sense of distributions (see (A.10)).

We assume that the boundary operators make sense when applied to all the functions of the domain $D(L)$. Prescribing the boundary conditions (10.3.2) amounts to restricting the domain of L to the subspace $D_B(L)$ of $D(L)$

defined by

$$D_B(L) = \{v \in D(L) | Bv = 0 \text{ on } \partial\Omega_b\},$$

which again we assume to be dense in X. Hence, we consider L as acting between $D_B(L)$ and X

$$L\colon D_B(L) \subset X \to X,$$

and problem (10.3.1)—(10.3.2) can be written as

$$\begin{cases} u \in D_B(L) \\ Lu = f \end{cases} \tag{10.3.3}$$

for $f \in X$ (the equality is between two functions in X). In the previous example, the operator L can be supplemented, for instance, either by Dirichlet boundary conditions $Bu(\pm 1) \equiv u(\pm 1) = 0$, or by Neumann boundary conditions $Bu(\pm 1) \equiv u_x(\pm 1) = 0$. Notice that in both cases the boundary conditions make sense for functions of $D_B(L)$, which are continuous with their first derivative. The density of $D_B(L)$ into $L^2_w(-1, 1)$ is a consequence of the density of $\mathscr{D}(-1, 1)$ into $L^2_w(-1, 1)$ (see (A.9)).

Before introducing the spectral approximations to (10.3.3) we recall some conditions which guarantee the well-posedness of the problem.

The simplest case occurs when the operator L satisfies a *coercivity condition*. Let us assume that there is a Hilbert space $E \subseteq X$ with norm $\|u\|_E$, for which there exists a positive constant C such that $\|u\| \leqslant C\|u\|_E$ for all $u \in E$. E is the subspace of the functions $u \in X$ with "finite" energy, the energy being precisely given by $\|u\|_E^2$. We assume that $D_B(L) \subseteq E$ and that there exist constants $\alpha > 0$ and $\beta > 0$ such that

$$\alpha \|u\|_E^2 \leqslant (Lu, u) \qquad \text{for all } u \in D_B(L) \tag{10.3.4}$$

$$|(Lu, v)| \leqslant \beta \|u\|_E \|v\|_E \qquad \text{for all } u \in D_B(L) \text{ and } v \in E. \tag{10.3.5}$$

Inequality (10.3.4) states that L is a positive operator, which is coercive over E, while (10.3.5) is a continuity condition for L (in the sense that (Lu, v) depends continuously on u and v). If $D_B(L)$ is densely contained in E, then hypotheses (10.3.4) and (10.3.5) are generally sufficient to ensure the existence of a unique solution u of (10.3.3).[1] The solution u depends continuously on f, namely:

$$\|u\|_E \leqslant C\|f\|.$$

---

[1] In fact, the bilinear form (Lu, v) can be extended in a unique way to be defined for all functions $u, v \in E$, so as still to satisfy the hypotheses (10.3.4), (10.3.5). Then, the Lax–Milgram theorem (see (A.5)) assures us that there exists a unique $u \in E$ which is a weak solution of (10.3.1) and (10.3.2), i.e., which satisfies

$$(Lu, v) = (f, v), \qquad \text{for all } v \in E.$$

The final step consists of proving that the weak solution is indeed a strong solution, i.e., it satisfies (10.3.3).

Going back to the example considered above, let us assume that Dirichlet boundary conditions are prescribed for the operator $L = -d^2/dx^2$. Then conditions (10.3.4) and (10.3.5) are satisfied with $E = H^1_{w,0}(-1, 1)$ (see (A.11.c)), which is a Hilbert space for the norm

$$\|u\|_E = \left(\int_{-1}^{1} |u_x|^2 w \, dx\right)^{1/2}.$$

This result will be proven in Chap. 11 (see Theorem 11.1).

The positivity condition (10.3.4) is the most immediate condition which guarantees the well-posedness of problem (10.3.3). However, there are situations for which it is not fulfilled. In such cases, one can resort to a more general condition, known as *inf-sup condition*, which we now present.

Let $W \subseteq X$ and $V \subseteq X$ be Hilbert spaces, whose norms will be denoted by $\|u\|_W$ and $\|u\|_V$ respectively. We assume that the inclusion of V into X is continuous, in the sense that $\|v\|_V \leqslant C\|v\|$, for a suitable constant C. We suppose that $D_B(L)$ is contained in W as a dense subspace, and that there exist constants $\alpha > 0$ and $\beta > 0$ such that

$$0 < \sup_{u \in D_B(L)} (Lu, v) \qquad \text{for all } v \in V, \tag{10.3.6}$$

$$\alpha \|u\|_W \leqslant \sup_{\substack{v \in V \\ v \neq 0}} \frac{(Lu, v)}{\|v\|_V} \qquad \text{for all } u \in D_B(L), \tag{10.3.7}$$

$$|(Lu, v)| \leqslant \beta \|u\|_W \|v\|_V \qquad \text{for all } u \in D_B(L) \text{ and } v \in V. \tag{10.3.8}$$

Conditions (10.3.6)—(10.3.8) generally assure that the problem (10.3.3) has a unique solution which depends continuously on the data[2], i.e.,

$$\|u\|_W \leqslant C\|f\|.$$

Note that conditions (10.3.6) and (10.3.7) are implied by the coercivity condition (10.3.4) by choosing $V = W = E$.

As an example, consider a second-order operator of the form $Lu = -(a(x)u_x)_x$ in the interval $\Omega = (-1, 1)$, where $a(x)$ is a smooth, strictly positive function. It will be supplemented by homogeneous Dirichlet boundary conditions. The operator L can still be defined on $X = L^2_w(-1, 1)$, its domain of definition once again being $D(L) = \{v \in C^1(-1, 1) | v_{xx} \in L^2_w(-1, 1)\}$. The coercivity condition (10.3.4) may not be satisified with $E = H^1_{w,0}(-1, 1)$. However, conditions (10.3.6)—(10.3.8) are fulfilled if we take $W = V = H^1_{w,0}(-1, 1)$ (see the discussion in the subsequent Example 6).

Another example is given by the operator $Lu = -u_{xx} + u$ supplemented by homogeneous Neumann boundary conditions. For this problem, condi-

[2] Again, existence and uniqueness of a weak solution follow from (10.3.6)—(10.3.8) using an extended form of the Lax–Milgram theorem (see Necas (1962)). Next, existence and uniqueness of a strong solution can be recovered.

tions (10.3.6)—(10.3.8) are fulfilled with the choice $V = L^2_w(-1,1)$ and $W = \{u \in H^2_w(-1,1) | u_x(\pm 1) = 0\}$ (see Example 7).

We will describe in general terms the process which leads to the definition of a spectral approximation of problem (10.3.3). The discussion on methods of Galerkin, tau, and collocation type, given in Sec. 10.4, will be based on the framework we are going to state.

The operator L is approximated, in a suitable sense, by a family of linear operators L_N depending on an integer $N > 0$. For instance, in a collocation scheme L_N is obtained from L by replacing exact derivatives by collocation ones (see Secs. 2.1.3, 2.3.2, and 2.4.2). Each L_N is defined on a finite-dimensional subspace $X_N \subset X$, in which we look for the approximate solution to the problem (10.3.3), and it maps X_N into X. Usually X_N is contained in $D_B(L)$, i.e., each function of X_N satisfies exactly the prescribed boundary conditions. However, there are cases (see Sec. 10.5.1), where the spectral solution satisfies the boundary conditions in an approximate way only.

The spectral approximation $u^N \in X_N$ is defined by requiring that a suitable projection of $L_N u^N - f$ over a finite-dimensional subspace Y_N of X be zero. More precisely, let Q_N be a linear projection operator, from a subspace Z of X onto Y_N. We assume that Y_N is contained in Z, that L_N maps X_N into Z, and finally that the data f belongs to Z (see Fig. 10.1). For Galerkin and tau approximations one chooses $Z = X$ (recall that X is a space of square integrable functions). In collocation approximations, which involve pointwise values of the operator and the data, the space Z may be chosen as the space of the functions continuous up to the boundary of Ω. Then, the spectral approximation is defined by

$$\begin{cases} u^N \in X_N \\ Q_N(L_N u^N - f) = 0. \end{cases} \tag{10.3.9}$$

Obviously, in order that (10.3.9) define a unique solution u^N, it is necessary that X_N and Y_N have the same dimension. We shall see that Y_N will be

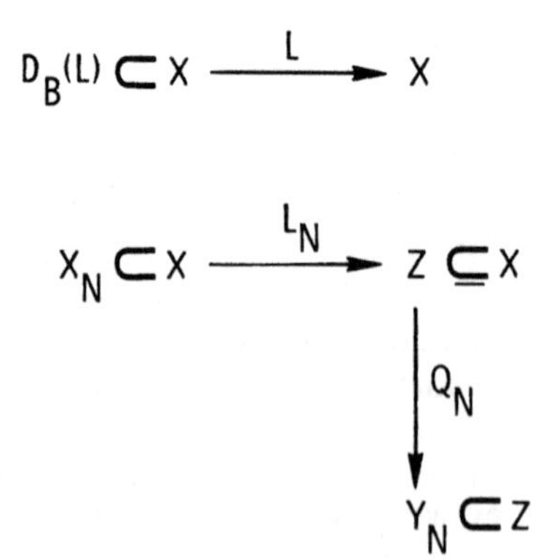

Figure 10.1. The spaces and the projection operators involved in spectral approximation.

chosen equal to X_N in some cases (e.g., Galerkin and collocation methods for problems with Dirichlet boundary conditions). In other situations (like tau methods) Y_N is actually different from X_N. We will always assume that Q_N is an orthogonal projection, that is, Q_N is defined by the relations

$$\begin{cases} Q_N: Z \to Y_N \\ (z - Q_N z, v)_N = 0 \qquad \text{for all } v \in Y_N. \end{cases} \tag{10.3.10}$$

Here, $(u, v)_N$ denotes a bilinear form on Z which is an inner product on Y_N. For Galerkin and tau methods, where $Z = X$, $(u, v)_N$ is the inner product (u, v) of X. For collocation methods $(u, v)_N$ is defined through the values of u and v at the collocation points only. This explains why $(u, v)_N$ is *not* defined on the whole space X, but on a proper subspace Z of functions for which point values are meaningful. Under this hypothesis, (10.3.9) can be equivalently written in a variational form as

$$\begin{cases} u^N \in X_N \\ (L_N u^N - f, v)_N = 0 \qquad \text{for all } v \in Y_N. \end{cases} \tag{10.3.11}$$

This formulation attests to the fact that any spectral scheme is actually a method of *weighted residuals*. The choice of Y_N and Q_N defines the way the residual $L_N u - f$ is minimized.

10.4. Galerkin, Collocation and Tau Methods

In this section, we will provide a general formulation of the three fundamental types of spectral methods. The formulation will be given in a way that fits into the general framework introduced above and at the same time permits the construction of an algorithm for the solution. The essential elements for each method are the space of the trial and of the test functions, the projection operator, and the inner product. Several examples of approximations to steady boundary value problems will be discussed for each method. General theorems will be given which guarantee stability and convergence results for each method. Some of the cumbersome details will be omitted.

Galerkin, collocation and tau methods are not the only schemes of spectral type which can be conceived and which are actually used in applications. Indeed, for some problems a method which combines the three schemes may be the most flexible and efficient. An important example is provided by algorithms for the Navier–Stokes equations, which often couple a tau discretization of the diffusive term with a pseudospectral method for the convective term. Such combined schemes can be often analyzed using elements of the theory presented separately here for the three fundamental schemes.

The Space $\mathrm{Pol}_N(\Omega)$

In what follows, we maintain the same notation used in Sec. 10.3. However, we now specify that the domain Ω, in which the problem (10.3.1) has to be solved, is the product of the intervals $(0, 2\pi)$ or $(-1, 1)$ according to the type of prescribed boundary conditions. Precisely, we set

$$\Omega = \prod_{k=0}^{d} I_k,$$

where $I_k = (0, 2\pi)$ if periodicity is required in the x_k-direction, and $I_k = (-1, 1)$ otherwise. Thus, Ω may be either the physical domain or the computational domain on which the original problem has been mapped, as is done in many applications (see Secs. 2.5 and 3.5).

For each integer N, the spectral approximation involves functions which in each variable are either trigonometric or algebraic polynomials of degree N. We shall denote by $\mathrm{Pol}_N(\Omega)$ the set of these functions. Precisely, $\mathrm{Pol}_N(\Omega)$ *is the space of the continuous functions* $u\colon \Omega \to \mathbb{C}$ *such that* u *is a trigonometric polynomial of degree* $\leqslant N$ *in the variables* x_k *for which* $I_k = (0, 2\pi)$, *and an algebraic polynomial of degree* $\leqslant N$ *in the remaining variables*. If there are no directions of periodicity, the functions of $\mathrm{Pol}_N(\Omega)$ will be real-valued.

It will always be assumed that for all N, $\mathrm{Pol}_N(\Omega)$ is contained in the domain of definition $D(L)$ of the operator L.

10.4.1. Galerkin Methods

Let X_N be the subspace of $\mathrm{Pol}_N(\Omega)$ of the functions which satisfy the boundary conditions, so that $X_N \subset D_B(L)$. Choose a basis $\{\phi_k, k \in J\}$ in X_N, where J is a set of indices. The ϕ_k's need not be orthogonal in the inner product of X. A Galerkin method is defined by the equations

$$\begin{cases} u^N \in X_N \\ (Lu^N, \phi_k) = (f, \phi_k) \qquad \text{for all } k \in J. \end{cases} \tag{10.4.1}$$

The unknowns are the coefficients α_k in the expansion $u^N = \sum_{k \in J} \alpha_k \phi_k$. Equation (10.4.1) is equivalent to

$$\begin{cases} u^N \in X_N \\ (Lu^N, v) = (f, v) \qquad \text{for all } v \in X_N. \end{cases} \tag{10.4.2}$$

It follows that with respect to the general formulation (10.3.11), a Galerkin method is defined by the choice: $Y_N = X_N$ and $(u, v)_N = (u, v)$, the inner product of X. We note that Q_N is the orthogonal projection from X into X_N in the inner product of X. Moreover, we have assumed that $L_N = L$, as occurs in most applications. A generalization of the Galerkin method is the so-called Petrov–Galerkin method. With this method, test functions differ from trial functions, though they individually satisfy the boundary conditions. In this case, we have $X_N \neq Y_N$, and (10.4.2) is replaced by

$$\begin{cases} u^N \in X_N \\ (Lu^N, v) = (f, v), \qquad \text{for all } v \in Y_N. \end{cases}$$

An example is given by Leonard's method for the incompressible Navier–Stokes equations (see Sec. 7.3.3).

EXAMPLE 1. THE HELMHOLTZ EQUATION IN THE SQUARE WITH PERIODIC BOUNDARY CONDITIONS. Let us consider the boundary value problem

$$\begin{cases} -\Delta u + \lambda u = f \qquad \text{in } \Omega = (0, 2\pi) \times (0, 2\pi), \\ u \text{ periodic in } \Omega \end{cases}$$

with $\lambda > 0$ and $f \in L^2(\Omega)$. The Galerkin solution u^N belongs to $X_N = \text{span}\{e^{i(kx+my)} | -N \leqslant k, m \leqslant N-1\}$ and satisfies

$$\int_\Omega (-\Delta u^N + \lambda u^N) e^{-i(kx+my)}\, dx\, dy$$

$$= \int_\Omega f e^{-i(kx+my)}\, dx\, dy \qquad -N \leqslant k, m \leqslant N-1.$$

Equivalently, the Fourier coefficients $\hat{u}_{km}^N$ of u^N are defined in terms of the Fourier coefficients $\hat{f}_{km}$ of f by the set of linear relations

$$(k^2 + m^2 + \lambda)\hat{u}_{km}^N = \hat{f}_{km} \qquad -N \leqslant k, m \leqslant N-1.$$

Thus, $X = L^2(\Omega)$ and $(u, v) = \int_\Omega u(x, y)\overline{v(x, y)}\, dx\, dy$.

EXAMPLE 2. THE POISSON EQUATION IN THE SQUARE WITH DIRICHLET BOUNDARY CONDITIONS. Let us consider the problem

$$\begin{cases} -\Delta u = f \qquad \text{in } \Omega = (-1, 1) \times (-1, 1), \\ \quad u = 0 \qquad \text{on the boundary } \partial\Omega. \end{cases}$$

Denote by $X_N = \{v \in \mathbb{P}_N | v = 0 \text{ on } \partial\Omega\}$ the space of algebraic polynomials of degree at most N in each variable, vanishing on the boundary of the square. A basis for X_N is given by

$$\phi_{km}(x, y) = \phi_k(x)\phi_m(y) \qquad 2 \leqslant k, m \leqslant N,$$

where

$$\phi_k(x) = \begin{cases} T_k(x) - T_0(x) & k \text{ even} \\ T_k(x) - T_1(x) & k \text{ odd} \end{cases} \tag{10.4.3}$$

if the Chebyshev polynomials introduced in Sec. 2.4 are used, or

$$\phi_k(x) = \begin{cases} L_k(x) - L_0(x) & k \text{ even} \\ L_k(x) - L_1(x) & k \text{ odd,} \end{cases} \tag{10.4.4}$$

if the Legendre polynomials introduced in Sec. 2.3 are used. The Galerkin equations to be satisfied by $u^N \in X_N$ are

$$-\int_\Omega \Delta u^N \phi_{km} w(x,y)\,dx\,dy = \int_\Omega f\phi_{km} w(x,y)\,dx\,dy,$$

where $w(x,y) = w(x)w(y)$ and $w(x)$ is either the Chebyshev or the Legendre weight according to whether (10.4.3) or (10.4.4) is used for the basis. In the present example we choose $X = L^2_w(\Omega)$ and $(u,v) = \int_\Omega uvw\,dx\,dy$. (see (A.9.h)).

We are now concerned with the *stability and convergence properties* of Galerkin approximations. The simplest case occurs when the operator L satisfies the coercivity condition (10.3.4) and the continuity condition (10.3.5). Then we have

$$\alpha \|u\|_E^2 \leqslant (Lu, u) \qquad \text{for all } u \in X_N \tag{10.4.5}$$

and

$$|(Lu, v)| \leqslant \beta \|u\|_E \|v\|_E \qquad \text{for all } u, v \in X_N. \tag{10.4.6}$$

If (10.4.5) *holds, then the Galerkin approximation* (10.4.2) *is stable, in the sense that the estimate*

$$\|u^N\|_E \leqslant C\|f\| \tag{10.4.7}$$

holds with a constant $C > 0$ *independent of* N.

Actually, choosing as test function in (10.4.2) the solution itself, and using the coercivity condition (10.4.5) on the left-hand side, and the Cauchy–Schwarz inequality on the right-hand side, one has

$$\alpha \|u^N\|_E^2 \leqslant (Lu^N, u^N) = (f, u^N) \leqslant \|f\|\,\|u^N\|.$$

Recalling that $\|u^N\| \leqslant C\|u^N\|_E$, we have (10.4.7). Inequality (10.4.7) also proves that (10.4.2) has a unique solution.

When (10.4.5) is satisfied, the stability of the approximation (10.4.2) is achieved by the *energy method* (and (10.4.7) is referred to as an *energy inequality*).

If stability is assured, convergence is a consequence of a consistency hypothesis, according to the Lax–Richtmyer theorem.

The consistency hypothesis assures that X is well-approximated by the family of the X_N's. More precisely, assume that there exists a dense subspace $\mathscr{W} \subseteq D_B(L)$, ($\mathscr{W}$ will be a space of sufficiently smooth functions), and for all $N > 0$, a projection operator

$$R_N\colon \mathscr{W} \to X_N \tag{10.4.8}$$

such that for $N \to \infty$

$$\|u - R_N u\|_E \to 0 \qquad \text{for all } u \in \mathscr{W}. \tag{10.4.9}$$

Under this consistency hypothesis, the approximation (10.4.2) is convergent.

Actually, $e = u^N - R_N u$ satisfies, by (10.4.2)

$$(Le, v) = (L(u - R_N u), v) \qquad \text{for all } v \in X_N.$$

Then by (10.4.5) and (10.4.6), it follows that

$$\|e\|_E \leqslant \frac{\beta}{\alpha} \|u - R_N u\|_E.$$

Since $u - u^N = u - R_N u - e$ we deduce the error estimate

$$\|u - u^N\|_E \leqslant \left(1 + \frac{\beta}{\alpha}\right) \|u - R_N u\|_E. \tag{10.4.10}$$

This inequality implies convergence for all $u \in \mathscr{W}$ due to the assumption (10.4.9). (Note that convergence occurs even if u is just a function in E, provided $\mathscr{W}$ is dense in E.) The above equality states the well-known fact that the error of the Galerkin approximation behaves like the error of best approximation in the norm for which the stability is proven (*Cea's lemma*).

In order to check the consistency hypothesis, one could choose as R_N the orthogonal projection operator onto X_N with respect to the inner product (u, v) of X. However, the orthogonal projection of u with respect to the inner product of X is generally *not* the best approximation of u in the energy norm among the elements in X_N. (This has been noticed throughout Chap. 9.) Thus, this choice (which nevertheless allows us to prove convergence) is not the best possible one from the point of view of the analysis of convergence: *the rate of decay of the error predicted by estimate* (10.4.10) *is generally slower than the real one.*

An optimal error estimate can be derived by a more sophisticated choice of the projection operator R_N. $R_N u$ is usually chosen as the best approximation of u in X_N with respect to the E-norm, or as an element in X_N which asymptotically behaves like the best approximation in the E-norm, namely:

$$\|u - R_N u\|_E \leqslant C \inf_{v \in X_N} \|u - v\|_E$$

for a constant C independent of N. This error can be bounded according to the estimates presented in Chap. 9.

We go back to Examples 1 and 2 with the purpose of proving stability and convergence for the approximations considered therein.

EXAMPLE 1 (continued from page 331). Using integration-by-parts and the periodicity condition we have

$$\int_\Omega (-\Delta u + \lambda u)\bar{u}\, dx\, dy = \int_\Omega (|\nabla u|^2 + \lambda |u|^2)\, dx\, dy$$

$$\geqslant \min(\lambda, 1) \int_\Omega (|\nabla u|^2 + |u|^2)\, dx\, dy,$$

for all $u \in X_N$. The integral on the right-hand side is precisely the square of the norm $\|u\|_1$ in the Hilbert space $H_p^1(\Omega)$ defined in (A.11.d). Hence the stability condition (10.4.5) is verified with $E = H_p^1(\Omega)$ and $\alpha = \min(\lambda, 1)$, and the approximation is stable according to (10.4.7). Condition (10.4.6) follows easily by integrating by parts and using the Cauchy–Schwarz inequality. As regards the convergence analysis, the truncation operator P_N defined in (9.7.3) gives the best approximation errors in the norm of any $H_p^m(\Omega)$, $m \geqslant 0$. Therefore, we can choose this operator as R_N in (10.4.8). Using the estimate (9.7.4), we get the optimal error bound

$$\|u - u^N\|_1 \leqslant CN^{1-m} \|u\|_m \qquad m \geqslant 1,$$

where

$$\|u\|_m^2 = \sum_{j=0}^{m} \sum_{k=0}^{j} \int_\Omega |D_x^k D_y^{j-k} u|^2 \, dx \, dy.$$

EXAMPLE 2 (continued from page 331). In the Legendre case, all $u \in X_N$ satisfy

$$-\int_\Omega \Delta u u \, dx \, dy = \int_\Omega |\nabla u|^2 \, dx \, dy.$$

Since u is zero on $\partial\Omega$, the L^2-norm of its gradient controls the norm $\|u\|_1 = \{\int_\Omega (|u|^2 + |\nabla u|^2) \, dx \, dy\}^{1/2}$ of $H^1(\Omega)$, according to the Poincaré inequality (A.13). We choose E to be the subspace $H_0^1(\Omega)$ of the functions in $H^1(\Omega)$ which vanish on $\partial\Omega$ (see (A.11.c)). E is a Hilbert space under the same norm as $H^1(\Omega)$. Thus, (10.4.5) is verified and the scheme is stable.

In order to prove the convergence, $R_N u$ is chosen to be the best approximation of u among the functions in X_N in the norm of E. By (9.7.14) and (10.4.10) we conclude that the optimal error estimate

$$\|u - u^N\|_1 \leqslant CN^{1-m} \|u\|_m \qquad m \geqslant 1$$

holds, where

$$\|u\|_m^2 = \sum_{0 \leqslant k+j \leqslant m} \int_\Omega |D_x^k D_y^j u|^2 \, dx \, dy.$$

In the Chebyshev case, it is not immediate that the quantity

$$-\int_\Omega \Delta u u w \, dx \, dy = \int_\Omega |\nabla u|^2 w \, dx \, dy + \int_\Omega u \nabla u \nabla w \, dx \, dy$$

is positive, due to the presence of the Chebyshev weight. However, (see Sec. 11.1) the right-hand side actually controls the norm

$$\|u\|_{1,w} = \left\{ \int_\Omega (u^2 + |\nabla u|^2) w \, dx \, dy \right\}^{1/2}$$

of the weighted Sobolev space $H_w^1(\Omega)$ (defined in (A.11.b)). Thus, we have the same stability and convergence results as above, provided the Chebyshev weight is inserted in all the norms.

So far we have assumed that the Galerkin approximation (10.4.2) satisfies the discrete coercivity condition (10.4.5). There are cases in which this condition is not fulfilled (cf. Example 6). Another way of getting stability and convergence results is to check a discrete form of the "inf-sup" condition (10.3.6) and (10.3.7). This condition is also suitable for the analysis of Petrov–Galerkin methods. We refer to the next subsection on tau methods for the detailed description of this approach.

10.4.2. Tau Methods

Tau methods are used for non-periodic problems. The definition of these methods is particularly simple for problems in one space dimension. We begin with this case, and then we consider the general situation.

We assume that the differential problem (10.3.1) is defined in the interval $\Omega = (-1, 1)$ and we recall that $\partial\Omega_b$ is the set of the endpoints where the boundary conditions (10.3.2) are imposed.

Let $\{\phi_k, k = 0, 1, \ldots\}$ be a system of algebraic polynomials, orthogonal with respect to the inner product $\int_{-1}^{1} u(x)v(x)w(x)\,dx$, where $w > 0$ is a weight function on $(-1, 1)$. We assume that each ϕ_k is a polynomial of effective degree k. The tau solution is a polynomial of degree N, $u^N = \sum_{k=0}^{N} \alpha_k \phi_k$, whose coefficients in the expansion according to this basis are the unknowns of the problem. They are determined in the following way. Denote by β the number of boundary conditions prescribed at the endpoints of the interval. The differential equation (10.3.1) is projected onto the space of polynomials of degree $N - \beta$:

$$\int_{-1}^{1} Lu^N \phi_k w\,dx = \int_{-1}^{1} f\phi_k w\,dx \qquad k = 0, 1, \ldots, N - \beta, \tag{10.4.11}$$

and the boundary conditions (10.3.2) are imposed exactly on $\partial\Omega_b$:

$$\sum_{k=0}^{N} \alpha_k B\phi_k = 0 \qquad \text{at the points of } \partial\Omega_b. \tag{10.4.12}$$

Conditions (10.4.12) are necessary, since the basis functions do not automatically satisfy the boundary conditions, unlike the basis used in a Galerkin method.

In order to cast a tau method in the framework of Sec. 10.3, we set $X = L_w^2(-1, 1)$,

$$X_N = \{v \in \mathbb{P}_N \,|\, Bv = 0 \text{ at the points of } \partial\Omega_b\} \tag{10.4.13}$$

and

$$Y_N = \mathbb{P}_{N-\beta}. \tag{10.4.14}$$

Then the tau method is equivalent to

$$\begin{cases} u^N \in X_N \\ (Lu^N, v) = (f, v) \qquad \text{for all } v \in Y_N. \end{cases} \tag{10.4.15}$$

With respect to the general setting (10.3.10), in a tau method the projector Q_N is the orthogonal projection operator from X upon Y_N relative to the inner product (u, v) of X.

EXAMPLE 3. THE DIRICHLET PROBLEM FOR A SECOND-ORDER ELLIPTIC OPERATOR IN THE INTERVAL $(-1, 1)$. Consider the problem

$$\begin{cases} Lu \equiv -u_{xx} + \lambda^2 u = f \qquad -1 < x < 1, \quad \lambda \in \mathbb{R} \\ u(-1) = u(1) = 0. \end{cases}$$

We look for a tau solution u^N expanded in Chebyshev polynomials. Thus, we assume that $f \in L^2_w(-1, 1)$ (w being the Chebyshev weight), and we determine the solution $u^N(x) = \sum_{k=0}^N \alpha_k T_k(x)$ by the conditions

$$\begin{cases} \displaystyle\int_{-1}^1 (-u^N_{xx} + \lambda^2 u^N)(x) T_k(x) w(x)\, dx \\ \qquad = \displaystyle\int_{-1}^1 f(x) T_k(x) w(x)\, dx \qquad \text{for } k = 0, 1, \ldots, N-2, \\ \displaystyle\sum_{k=0}^N \alpha_k (-1)^k = \sum_{k=0}^N \alpha_k = 0. \end{cases}$$

In the present case, $X_N = \{v \in \mathbb{P}_N | v(-1) = v(1) = 0\}$ and $Y_N = \mathbb{P}_{N-2}$.

EXAMPLE 4. THE NEUMANN PROBLEM FOR A SECOND-ORDER ELLIPTIC OPERATOR IN THE INTERVAL $(-1, 1)$. Consider the problem

$$\begin{cases} Lu \equiv -u_{xx} + u = f \qquad -1 < x < 1, \\ u_x(-1) = u_x(1) = 0. \end{cases}$$

Again, we look for a tau solution u^N expanded in Chebyshev polynomials. Thus, $u^N(x) = \sum_{k=0}^N \alpha_k T_k(x)$ is determined by the conditions

$$\begin{cases} \displaystyle\int_{-1}^1 (-u^N_{xx} + u^N)(x) T_k(x) w(x)\, dx \\ \qquad = \displaystyle\int_{-1}^1 f(x) T_k(x) w(x)\, dx \qquad \text{for } k = 0, 1, \ldots, N-2 \\ \displaystyle\sum_{k=0}^{N-1} \beta_k (-1)^k = \sum_{k=0}^{N-1} \beta_k = 0, \end{cases} \tag{10.4.16}$$

where the β_k's are the coefficients of the Chebyshev expansion of the derivative u_x^N (see (2.4.22)). We now have $X_N = \{v \in \mathbb{P}_N | v_x(-1) = v_x(1) = 0\}$ and $Y_N = \mathbb{P}_{N-2}$.

We consider now the d-dimensional case. The domain Ω is the product of d copies of the interval $(-1, 1)$, and the functions of $\text{Pol}_N(\Omega)$ are algebraic polynomials in each variable. We assume that on a given side of the boundary the same kind of boundary conditions are given. We exclude, for example, the use of Dirichlet boundary conditions on part of a side and Neumann boundary conditions on the rest of the side.

A basis in $\text{Pol}_N(\Omega)$ can be built as a product of the basis functions $\{\phi_k\}$ in each variable. Define the lattice

$$J = \{\mathbf{k} = (k_1, \ldots, k_n) | k_i \text{ is an integer, with } 0 \leqslant k_i \leqslant N \text{ for } i = 1, \ldots, d\},$$

and let

$$\phi_{\mathbf{k}}(\mathbf{x}) = \phi_{k_1}(x_1) \ldots \phi_{k_d}(x_d).$$

Then $\{\phi_{\mathbf{k}}, \mathbf{k} \in J\}$ is a basis in $\text{Pol}_N(\Omega)$, which is orthogonal for the inner product

$$(u, v) = \int_{-1}^{1} w(x_1)\, dx_1 \cdots \int_{-1}^{1} u(\mathbf{x}) v(\mathbf{x}) w(x_d)\, dx_d.$$

The solution of a spectral tau scheme is a polynomial in $\text{Pol}_N(\Omega)$ expanded in this basis. Its coefficients in this expansion are determined by two sets of linear equations. The first set is obtained by requiring that the residual $L_N u^N - f$ be orthogonal to a family of basis functions of reduced degree. The $\phi_{\mathbf{k}}$'s which are retained as test functions are the ones whose degree in each direction is at most N minus the number of boundary conditions prescribed on the sides orthogonal to that direction. More precisely, for each $i = 1, \ldots, d$ denote by β_i the total number of boundary conditions prescribed on the sides $x_i = \pm 1$. Define the sublattice

$$J_e = \{\mathbf{k} = (k_1, \ldots, k_d) \in J | 0 \leqslant k_i \leqslant N - \beta_i \text{ for } i = 1, \ldots, d\}.$$

where the subscript e stands for equation. (See Fig. 10.2 for an example.) The differential equation is enforced by requiring that the tau solution $u^N \in \text{Pol}_N(\Omega)$ satisfies the set of equations

$$(Lu^N, \phi_{\mathbf{k}}) = (f, \phi_{\mathbf{k}}) \qquad \text{for all } \mathbf{k} \in J_e. \tag{10.4.17}$$

The remaining equations are obtained by imposing the boundary conditions. These give a set of algebraic relations involving the coefficients of u^N with respect to the orthogonal basis $\{\phi_{\mathbf{k}} | \mathbf{k} \in J\}$.

The most direct way of taking into account the boundary conditions in a tau method consists, for each side, of projecting separately, upon the space of polynomials of degree N, the equation to be satisfied at the boundary

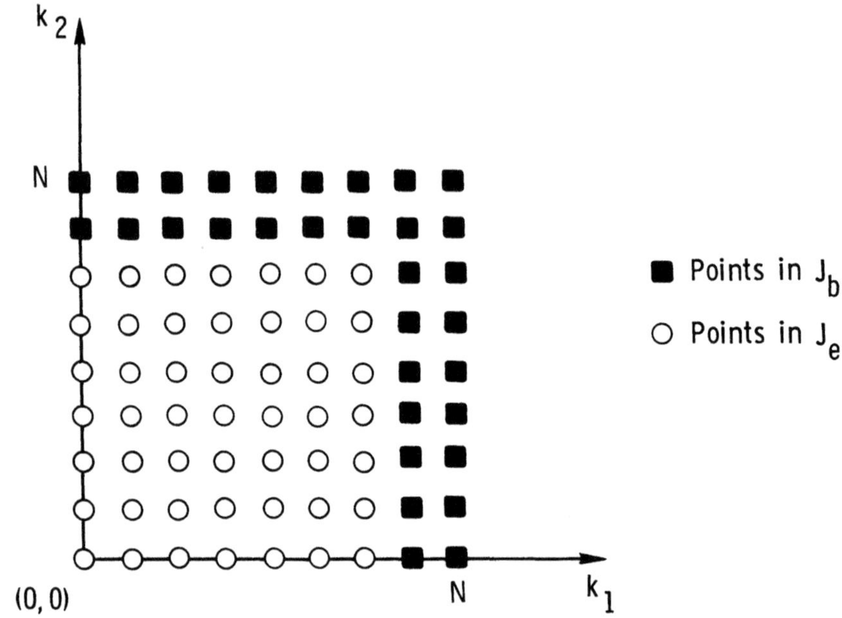

Figure 10.2. The set J in frequency space for the tau approximation to the Dirichlet boundary value problem for the Laplace equation in the square (example 5).

(see, for instance, Example 1.2.3 in Chap. 1). This method may lead to an overdetermined set of boundary equations due to possible continuity conditions at the corners (in two dimensions) or edges (in three dimensions). In the quoted example, the number of equations represented by (1.2.54) is $4N + 4$, while only $4N$ independent equations have to be added to (1.2.53) in order to determine u^N. The rank of the system is only $4N$.

We describe hereafter a mathematically rigorous procedure of boundary projection, which leads to the correct number of linearly independent boundary equations. To this end, define the inner product $(u, v)_{\partial\Omega_b}$ between two functions u, v on $\partial\Omega_b$ as follows. If S is a side of $\partial\Omega_b$ orthogonal to the direction x_i, let σ be the independent variable on S and let $\tilde{w}(\sigma) = \prod_{\substack{j=1 \\ j\neq i}}^{d} w(x_j)$. Then we set

$$(u, v)_{\partial\Omega_b} = \sum_{\text{sides of } \partial\Omega_b} \int_S u(\sigma)v(\sigma)\tilde{w}(\sigma)\,d\sigma. \tag{10.4.18}$$

Next, we consider the set of indices $J_b = J - J_e$ and take into account the boundary conditions (10.3.2) by requiring that the tau solution u^N satisfy the set of equations

$$(Bu^N, \phi_{\mathbf{k}})_{\partial\Omega_b} = 0 \qquad \text{for all } \mathbf{k} \in J_b. \tag{10.4.19a}$$

Condition (10.4.9a) involves the traces of the $\phi_{\mathbf{k}}$'s on $\partial\Omega_b$ only, with $\mathbf{k} \in J_b$. These traces are linearly independent on $\partial\Omega_b$ and actually they generate the space $C^0(\partial\Omega_b; N)$ of all the continuous functions on $\partial\Omega_b$ which are polynomials of degree up to N on each side of $\partial\Omega_b$. The proof of this property is not hard, but it is rather technical, and it will be left to the reader. Thus, (10.4.19a) is equivalent to

$$(Bu^N, \psi)_{\partial\Omega_b} = 0 \qquad \text{for all } \psi \in C^0(\partial\Omega_b; N). \tag{10.4.19b}$$

Any convenient basis in $C^0(\partial\Omega_b; N)$ can be used to enforce (10.4.19a), such as a basis whose functions are non-zero on at most n contiguous sides of $\partial\Omega_b$.

We conclude that a multidimensional tau method is represented again by (10.4.15), where now $X = L^2_w(\Omega)$ (see (A.9.h)) and

$$X_N = \{v \in \mathrm{Pol}_N(\Omega) | (Bv, \phi_{\mathbf{k}})_{\partial\Omega_b} = 0 \text{ for all } \mathbf{k} \in J_b\}; \tag{10.4.20}$$

$$Y_N = \mathrm{span}\{\phi_{\mathbf{k}} | \mathbf{k} \in J_e\}. \tag{10.4.21}$$

EXAMPLE 5. A LEGENDRE TAU METHOD FOR THE POISSON EQUATION. We consider again the tau approximation introduced in Sec. 1.2.3 and analyzed in Sec. 10.1.3. The aim here is to incorporate this scheme in the previous general framework.

The tau solution is expanded into Legendre polynomials $\phi_{\mathbf{k}}(x,y) = L_{k_1}(x)L_{k_2}(y)$, namely, $u^N(x,y) = \sum_{k=0}^{N}\sum_{m=0}^{N} \hat{u}_{km} L_k(x) L_m(y)$. Thus the natural choice for the Hilbert space X is the space $L^2(\Omega)$, (Ω being the square $(-1,1) \times (-1,1)$), with inner product

$$(u, v) = \int_{-1}^{1}\int_{-1}^{1} u(x,y)v(x,y)\,dx\,dy.$$

The boundary conditions are prescribed over the whole boundary of Ω, hence $\partial\Omega_b = \partial\Omega$ and the boundary inner product takes the form

$$(u, v)_{\partial\Omega} = \int_{-1}^{1} u(x,-1)v(x,-1)\,dx + \int_{-1}^{1} u(x,1)v(x,1)\,dx$$
$$+ \int_{-1}^{1} u(-1,y)v(-1,y)\,dy + \int_{-1}^{1} u(1,y)v(1,y)\,dy.$$

Exactly one boundary condition is prescribed on each side of Ω; hence we have

$$J_e = \{(k_1, k_2) | 0 \leqslant k_1, k_2 \leqslant N-2\}$$

and

$$J_b = \{(k_1, k_2) | N-1 \leqslant k_i \leqslant N, \text{ for at least one index } i = 1, 2\}.$$

Thus, equations (1.2.51) are nothing but (10.4.17), while equations (1.2.54) clearly imply (10.4.19b). We look now for a basis of $C^0(\partial\Omega; N)$, the space of polynomials of degree N on each side of Ω, which are continuous at the corners. Define for $k \geqslant 2$, $l_k(x) = L_k(x) - L_{\bar{k}}(x)$, where $\bar{k} = k(\text{mod } 2)$, (i.e., $\bar{k} = 0$, if k is even, $\bar{k} = 1$, if k is odd). Thus, $l_k(+1) = l_k(-1) = 0$. Furthermore, set $l_{\pm}(x) = L_N(x) \pm L_{N-1}(x)$, so that $l_{\pm}(\pm 1) \neq 0$ and $l_{\pm}(\mp 1) = 0$. Each of the functions

$$\Psi_{(k,+)}(x, y) = l_k(x) l_+(y) \qquad k \geqslant 2$$

is a linear combination of basis functions $\phi_{\mathbf{k}}(x, y)$ with $\mathbf{k} \in J_b$; hence (10.4.19) yields

$$\begin{aligned}(u^N, \Psi_{(k,+)})_{\partial\Omega} &= \int_{-1}^{1} u^N(x, 1) l_k(x)\, dx \\ &= \sum_{m=0}^{N} [\hat{u}_{km} + \hat{u}_{\bar{k},m}] = 0 \qquad 2 \leqslant k \leqslant N.\end{aligned} \tag{10.4.22a}$$

In the same way, the test functions $\Psi_{(k,-)}(x, y) = l_k(x) l_-(y)$ and $\Psi_{(\pm,k)}(x, y) = l_{\pm}(x) l_k(y)$ yield, respectively, the relations

$$\sum_{m=0}^{N} (-1)^m [\hat{u}_{km} - \hat{u}_{\bar{k},m}] = 0 \qquad 2 \leqslant k \leqslant N \tag{10.4.22b}$$

$$\sum_{k=0}^{N} [\hat{u}_{km} + \hat{u}_{k,\bar{m}}] = 0 \qquad 2 \leqslant m \leqslant N \tag{10.4.22c}$$

$$\sum_{k=0}^{N} (-1)^k [\hat{u}_{km} - \hat{u}_{k,\bar{m}}] = 0 \qquad 2 \leqslant m \leqslant N \tag{10.4.22d}$$

Finally, the test functions $\Psi_{(\pm,\pm)}(x, y) = l_{\pm}(x) l_{\pm}(y)$ give the remaining relations

$$\left\{\begin{aligned} \sum_{m=0}^{N} [\hat{u}_{Nm} + \hat{u}_{N-1,m}] + \sum_{k=0}^{N} [\hat{u}_{kN} + \hat{u}_{k,N-1}] = 0 \\ \sum_{m=0}^{N} (-1)^m [\hat{u}_{Nm} + \hat{u}_{N-1,m}] + \sum_{k=0}^{N} [\hat{u}_{kN} - \hat{u}_{k,N-1}] = 0 \\ \sum_{m=0}^{N} [\hat{u}_{Nm} - \hat{u}_{N-1,m}] + \sum_{k=0}^{N} (-1)^k [\hat{u}_{kN} + \hat{u}_{k,N-1}] = 0 \\ \sum_{m=0}^{N} (-1)^m [\hat{u}_{Nm} - \hat{u}_{N-1,m}] + \sum_{k=0}^{N} (-1)^k [\hat{u}_{kN} - \hat{u}_{k,N-1}] = 0. \end{aligned}\right. \tag{10.4.23}$$

Note that the functions $\Psi_{(k,\pm)}$, and $\Psi_{(\pm,k)}$ are nonzero on one side of Ω, while $\Psi_{(\pm,\pm)}$ are non-zero on two contiguous sides of Ω. We conclude that (10.4.22) and (10.4.23) are equivalent to (10.4.19b).

For the present scheme, one has $X_N = \{v \in \mathbb{P}_N | v \equiv 0 \text{ on } \partial\Omega\}$ and $Y_N = \mathbb{P}_{N-2}$. Here, $\mathbb{P}_N$ is the space of algebraic polynomials of degree $\leqslant N$ in each variable.

□

We are now concerned with the problem of *stability and convergence* for the tau approximation (10.4.15). Since the space X_N of basis functions is different from the space Y_N of test functions, the natural approach now is the discrete form of the "inf-sup" condition given in Sec. 10.3. We assume, therefore, that the operator L satisfies (10.3.6)—(10.3.8). Moreover, we assume here that $X_N \subseteq W$ and $Y_N \subseteq V$ for all $N > 0$. Then we have the following stability condition, which is due to Babuška (see, e.g., Babuška and Aziz (1972)).

If there exists a constant $\bar{\alpha} > 0$ independent of N such that

$$\bar{\alpha}\|u\|_W \leqslant \sup_{\substack{v \in Y_N \\ v \neq 0}} \frac{(Lu, v)}{\|v\|_V} \qquad \textit{for all } u \in X_N, \tag{10.4.24}$$

then

$$\|u^N\|_W \leqslant C\|f\| \tag{10.4.25}$$

for a constant C independent of N.

The inequality (10.4.25) implies that (10.4.15) has a unique solution (since X_N and Y_N have the same dimension), and the approximation is stable. Inequality (10.4.25) is obtained by dividing each term in (10.4.15) by $\|v\|_V$, then taking the supremum over all $v \in Y_N$ and using (10.4.24) together with the continuity of the inclusion of V into X.

Concerning the *convergence* of the method, as for the Galerkin approximation, let R_N be a linear operator from a dense subspace $\mathscr{W} \subseteq D_B(L)$ into X_N, such that for $N \to \infty$

$$\|u - R_N u\|_W \to 0 \qquad \text{for all } u \in \mathscr{W}. \tag{10.4.26}$$

By an argument similar to that used for proving (10.4.10), the following error estimate between the solution of (10.3.3) and the tau solution of (10.4.15) can be established:

$$\|u - u^N\|_W \leqslant \left(1 + \frac{\beta}{\bar{\alpha}}\right)\|u - R_N u\|_W. \tag{10.4.27}$$

Thus, the tau method is convergent.

A stability condition of "inf-sup" type can also be given for Galerkin approximations. Obviously, it is obtainable from (10.4.24) by replacing Y_N with X_N. The coercivity condition (10.4.5) is nothing but a particular form of this condition, in which $W = V = E$. Actually, (10.4.24) can be written as

$$\alpha\|u\|_E \leqslant \frac{(Lu, u)}{\|u\|_E} \qquad \text{for all } u \in X_N, \quad u \neq 0, \tag{10.4.28}$$

which is clearly implied by (10.4.5).

We go back to Examples 3, 4, and 5 in order to show that the tau approximations considered there are stable and convergent.

EXAMPLE 3 (continued from page 336). Throughout Examples 3 and 4 we will use the simplified notation $\|u\|_{m,w}$ instead of $\|u\|_{H^m_w(-1,1)}$ for $m \geqslant 0$ (see (9.5.5)). If u is any polynomial of degree N which vanishes on the boundary, then $v = -u_{xx}$ is a polynomial of degree $N-2$, and

$$(Lu, v) = \int_{-1}^{1} (u_{xx})^2 w\, dx + \lambda^2 \int_{-1}^{1} u_x(uw)_x\, dx$$

$$\geqslant \|u_{xx}\|_{0,w}^2 + \frac{\lambda^2}{4}\|u_x\|_{0,w}^2 \geqslant C\|u\|_{2,w}^2.$$

We have used (11.1.14) and the Poincaré inequality (A.13) (if $\lambda = 0$ this inequality must be used twice). Therefore, the "inf-sup" condition (10.4.24) is satisfied if we choose $W = H^2_w(-1,1)$ and $V = L^2_w(-1,1)$, and we have the estimate

$$\|u^N\|_{2,w} \leqslant C\|f\|_{0,w} \tag{10.4.29}$$

for a constant independent of N and λ. The convergence of the method can be established as a consequence of (10.4.27) by defining the projection operator R_N as follows. Let $P^2_N u$ be an algebraic polynomial of degree $\leqslant N$ which satisfies (9.5.12) for $l = 2$. It need not vanish at the boundary; hence, we subtract from it the linear polynomial $p_1(x)$ which coincides with $P^2_N u$ at the boundary. It is easily seen that using the inclusion $H^1_w(-1,1) \subset C^0([-1,1])$ (see (A.11.a)) one has

$$\|p_1\|_{2,w} \leqslant \|u - P^2_N u\|_{2,w} \leqslant CN^{2-m}\|u\|_{m,w} \qquad m \geqslant 2.$$

We set $R_N u = P^2_N u - p_1$ and we obtain the optimal convergence estimate

$$\|u - u^N\|_{2,w} \leqslant CN^{2-m}\|u\|_{m,w} \qquad m \geqslant 2. \tag{10.4.30}$$

We note here that $v = -u_{xx}$ is not the only test function which allows us to obtain a stability estimate for the scheme under consideration. Actually, if we set $v = P_{N-2}u$, we have

$$(Lu, v) = -\int_{-1}^{1} u_{xx} P_{N-2} u w\, dx + \lambda^2 \int_{-1}^{1} u P_{N-2} u w\, dx$$

$$= \int_{-1}^{1} u_x(uw)_x\, dx + \lambda^2 \int_{-1}^{1} (P_{N-2}u)^2 w\, dx.$$

This yields the estimate

$$\frac{1}{2}\|u^N_x\|_{0,w} + \lambda\|P_{N-2}u\|_{0,w} \leqslant C\|f\|_{0,w}. \tag{10.4.31}$$

If $\lambda \gg 1$, (10.4.31) contains the new information that the L^2_w-norm of P_{N-2} is $O(1/\lambda)$. This kind of result has been used by Canuto and Sacchi–Landriani

(1986) in the analysis of the Kleiser–Schumann method for the Navier–Stokes equations (see Sec. 11.3.3).

EXAMPLE 4 (continued from page 336). Let us begin with the stability analysis. Note that for all $u \in X_N$, $P_{N-2}Lu = Lu - (u - P_{N-2}u)$, where P_{N-2} is the orthogonal projection operator on $\mathbb{P}_{N-2}$. Hence,

$$\begin{aligned}(Lu, P_{N-2}Lu) &= \|Lu\|_{0,w}^2 - (Lu, u - P_{N-2}u) \\ &\geqslant \|Lu\|_{0,w}^2 - \|Lu\|_{0,w}\|u - P_{N-2}u\|_{0,w}.\end{aligned} \tag{10.4.32}$$

Now, by (9.5.6), we have $\|u - P_{N-2}u\|_{0,w} \leqslant C_0 N^{-2}\|u\|_{2,w}$. Moreover, it is possible to prove the "a priori" estimate

$$\|u\|_{2,w} \leqslant C_1 \|Lu\|_{0,w},$$

for a suitable constant $C_1 > 0$. By (10.4.32) we get

$$(Lu, P_{N-2}Lu) \geqslant (1 - C_0C_1N^{-2})\|Lu\|_{0,w}^2 \geqslant (2C_1^2)^{-1}\|u\|_{2,w}^2,$$

provided N is so large that $1 - C_0C_1N^{-2} \geqslant 1/2$. Since $\|P_{N-2}Lu\|_{0,w} \leqslant C_3\|u\|_{2,w}$, we conclude that the estimate

$$\frac{(Lu, P_{N-2}Lu)}{\|P_{N-2}Lu\|_{0,w}} \geqslant \frac{1}{2C_1^2C_3}\|u\|_{2,w} \tag{10.4.33}$$

holds.

This proves that the scheme (10.4.16) satisfies the stability condition (10.4.24), if we define $W = \{v \in H_w^2(-1,1) | v_x(-1) = v_x(1) = 0\}$ and $V = L_w^2(-1,1)$.

The convergence analysis is straightforward, in view of (10.4.27). Define the projector R_N onto X_N as

$$(R_N u)(x) = u(-1) + \int_{-1}^{x} (P_{N-1}^{1,0}u_x)(\xi)\,d\xi,$$

where $P_{N-1}^{1,0}$ is introduced in (9.5.14). Then it is easy to prove that $\|u - R_N u\|_{2,w} \leqslant CN^{2-m}\|u\|_{m,w}$; whence, by (10.4.27) we get the optimal error estimate

$$\|u - u^N\|_{2,w} \leqslant CN^{2-m}\|u\|_{m,w}.$$

EXAMPLE 5 (continued from page 340). In Sec. 10.1.3 the test function designed to prove stability was $q(x,y) = u^N(x,y)/[(1-x^2)(1-y)^2]$. This appears to be a natural choice for tau approximations to homogeneous Dirichlet boundary value problems. Actually, any $u \in X_N$ can be split into the product $u = bq$, where q is a polynomial of the space Y_N and b is a polynomial of minimal degree which vanishes on $\partial\Omega_b$.

If for a suitable choice of Hilbert spaces W and V, there exist positive constants α_1 and α_2 independent of N such that

$$\alpha_1 \|u\|_W^2 \leqslant (Lu, q) \qquad \text{for all } u \in X_N, \tag{10.4.34}$$

$$\|q\|_V \leqslant \alpha_2 \|u\|_W \qquad \text{for all } u \in X_N, \tag{10.4.35}$$

then (10.4.24) is satisfied with $\bar{\alpha} = \alpha_1/\alpha_2$.

In the current example we set $b(x, y) = (1 - x^2)(1 - y^2)$ and we define the norms

$$\|u\|_W = \left(\int_\Omega b|\nabla q|^2\,dx\,dy + \frac{1}{2}\int_\Omega |\Delta b|\,q^2\,dx\,dy\right)^{1/2} \quad \text{with } q = u/b,$$

and

$$\|v\|_V = \left(\int_\Omega |\Delta b|\,v^2\,dx\,dy\right)^{1/2}$$

(W and V being defined as the weighted Sobolev spaces of the functions for which these norms, respectively, are finite). In the present example, however, the continuity condition (10.3.8) is not verified. Rather we have by the Cauchy–Schwarz inequality

$$\left|\int_\Omega \Delta uv\,dx\,dy\right| \leqslant \left(\int_\Omega \frac{|\Delta u|^2}{|\Delta b|^2}\,dx\,dy\right)^{1/2} \left(\int_\Omega \Delta b v^2\,dx\,dy\right)^{1/2}.$$

Hence, the operator L turns out to be continuous with respect to a stronger norm than the norm of W. More precisely, if we define $\|u\|_{\tilde{W}}$ to be the maximum of the values attained in Ω by any derivative of u of order up to 2, (mathematically, the Sobolev space $\tilde{W}$ for which such a norm is finite is denoted by $W^{2,\infty}(\Omega)$), we have

$$|(Lu, v)| \leqslant \beta \|u\|_{\tilde{W}} \|v\|_V \qquad \text{for all } u \in \tilde{W} \text{ and } v \in V,$$

with $\beta = (\int_\Omega (1/|\Delta b|^2)\,dx\,dy)^{1/2}$. The convergence estimate (10.4.27) has to be modified to

$$\|u - u^N\|_W \leqslant C\|u - R_N u\|_{\tilde{W}}.$$

Using this inequality and a suitable projection operator, one gets the estimate (10.1.13) given in Sec. 10.1.

10.4.3. Collocation Methods

To define a collocation method, one gives as many distinct points

$$x_k \qquad k \in J \text{ (a set of indices)} \tag{10.4.36}$$

in the domain Ω or on its boundary $\partial\Omega$, as the dimension of the space $\mathrm{Pol}_N(\Omega)$

in which the spectral solution is sought. At a number of these points, located on $\partial\Omega$, the boundary conditions are imposed. The remaining points are used to enforce the differential equation.

We assume that for any $k \in J$, there exists a polynomial $\phi_k \in \mathrm{Pol}_N(\Omega)$, necessarily unique, such that

$$\phi_k(x_m) = \begin{cases} 1 & \text{if } k = m, \\ 0 & \text{if } k \neq m. \end{cases} \tag{10.4.37}$$

This is certainly true in all the applications, where the points (10.4.36) are products of distinct points in each space variable. The ϕ_k's form a basis for the polynomials of degree N, since $v(x) = \sum_{k \in J} v(x_k)\phi_k(x)$ for all $v \in \mathrm{Pol}_N(\Omega)$. A collocation method is obtained by requiring that the differential equation be satisfied at a number of points $\{x_k\}$ (those in the interior of the domain, and possibly some on the boundary), and that the boundary conditions (or, possibly, some of them) be satisfied at the remaining x_k's. To be precise, let J be divided into two disjoint subsets J_e and J_b, such that if $k \in J_b$, the x_k's are on the part $\partial\Omega_b$ of the boundary where the boundary conditions (10.3.2) are prescribed. Moreover, let L_N be an approximation to the operator L in which derivatives are taken via collocation at the points x_k's (see Secs. 2.1.3, 2.3.2, and 2.4.2). The collocation solution is a polynomial $u^N \in \mathrm{Pol}_N(\Omega)$ which satisfies the equations

$$\begin{cases} L_N u^N(x_k) = f(x_k) & \text{for all } k \in J_e, \quad (10.4.38) \\ Bu^N(x_k) = 0 & \text{for all } k \in J_b. \quad (10.4.39) \end{cases}$$

The unknowns in a collocation method are the values of u^N at the points (10.4.36), i.e., the coefficients of u^N with respect to the Lagrange basis (10.4.37). The set J_b is empty in Fourier approximations for periodic problems since the trigonometric polynomials are themselves periodic. However, J_b may be empty even in approximations to non-periodic problems. In these cases, the boundary conditions are taken into account in the definition of the operator L_N (see Example 4 of Sec. 10.5.1; see also Sec. 3.3).

We will now set the collocation method (10.4.38)—(10.4.39) into the framework given in Sec. 10.3. To this end, we introduce a bilinear form $(u, v)_N$ on the space $Z = C^0(\bar{\Omega})$ of the functions continuous up to the boundary of Ω by fixing a family of weights $w_k > 0$ and setting

$$(u, v)_N = \sum_{k \in J} u(x_k)\overline{v(x_k)}w_k. \tag{10.4.40}$$

The existence of the Lagrange basis (10.4.37) ensures that (10.4.40) is an inner product on $\mathrm{Pol}_N(\Omega)$. Consequently, we define a *discrete norm* on $\mathrm{Pol}_N(\Omega)$ as

$$\|u\|_N = \{(u, u)_N\}^{1/2} \qquad \text{for } u \in \mathrm{Pol}_N(\Omega). \tag{10.4.41}$$

The basis of the ϕ_k's is orthogonal under the *discrete* inner product (10.4.40).

We make the assumption that the nodes $\{x_k\}$ and the weights $\{w_k\}$ are such that

$$(u,v)_N = (u,v) \qquad \text{for all } u,\, v \text{ such that } uv \in \mathrm{Pol}_{2N-1}(\Omega). \tag{10.4.42}$$

This means that the discrete inner product (10.4.40) must approximate with enough precision the inner product of X. Condition (10.4.42) introduces a constraint in the choice of the collocation points. In all the applications, this assumption is fulfilled since the x_k's are the knots of quadrature formulas of Gaussian type.

Let X_N be the space of the polynomials of degree $\leqslant N$ which satisfy the boundary conditions (10.4.39), i.e.,

$$X_N = \{v \in \mathrm{Pol}_N(\Omega) \,|\, Bv(x_k) = 0 \text{ for all } k \in J_b\}. \tag{10.4.43}$$

Then the collocation method is equivalently written as

$$\begin{cases} u^N \in X_N \\ (L_N u^N, \phi_k)_N = (f, \phi_k)_N \qquad \text{for all } k \in J_e. \end{cases} \tag{10.4.44}$$

If Y_N is the space spanned by the ϕ_k's with $k \in J_e$, i.e.,

$$Y_N = \{v \in \mathrm{Pol}_N(\Omega) \,|\, v(x_k) = 0 \text{ for all } k \in J_b\}, \tag{10.4.45}$$

then (10.4.44) can be written as

$$\begin{cases} u^N \in X_N \\ (L_N u^N, v)_N = (f, v)_N \qquad \text{for all } v \in Y_N. \end{cases} \tag{10.4.46}$$

This is precisely (10.3.11). Equivalently, (10.4.44) can be written in the form

$$Q_N(L_N u^N - f) = 0 \tag{10.4.47}$$

(see (10.3.9)). For a collocation approximation, $Q_N v$ is the polynomial of degree N matching v at the interior points $\{x_k, k \in J_e\}$ and vanishing at the boundary points $\{x_k, k \in J_b\}$.

Note that in the special case where all the boundary conditions are of Dirichlet type, i.e., if $Bv \equiv v$, one has $X_N = Y_N$. In this case the collocation method can be viewed as a Galerkin method in which the *continuous* inner product (u, v) is replaced by the *discrete* inner product $(u, v)_N$ (compare (10.4.46) with (10.4.2)).

Another similarity can be established between collocation and tau methods. Indeed, from the tau equations (10.4.17) and (10.4.19) one can obtain the collocation equations (10.4.38) and (10.4.39) formally by replacing the continuous inner product with the discrete one, and taking as ϕ_k the Lagrange basis functions (10.4.37). In both methods, the basis in which the solution is expanded is orthogonal with respect to the inner product involved in the scheme.

EXAMPLE 6. THE DIRICHLET PROBLEM FOR A VARIABLE-COEFFICIENT SECOND-ORDER OPERATOR IN THE INTERVAL $(-1,1)$. We consider the problem

$$\begin{cases} -(au_x)_x = f & -1 < x < 1 \\ u(-1) = u(1) = 0, \end{cases}$$

where $a(x)$ is continuously differentiable and satisfies $a(x) \geqslant \alpha_0 > 0$ in $[-1,1]$, and f is continuous.

For a fixed integer $N > 0$, set $J = \{0, 1, \ldots, N\}$ and choose as points (10.4.36) the nodes $\{x_k, k \in J\}$ of the $(N+1)$-points Gauss–Lobatto quadrature formula with respect to the Chebyshev or the Legendre weight. If $\{w_k, k \in J\}$ are the corresponding weights, assumption (10.4.42) is satisfied. Denote by $I_N v$ the polynomial of degree N which interpolates a continuous function v at the points $\{x_k, k \in J\}$. The collocation approximation to u is a polynomial u^N of degree N which satisfies the equations

$$\begin{cases} -[I_N(au_x^N)]_x(x_k) = f(x_k) & k = 1, \ldots, N-1 \\ u^N(x_0) = u^N(x_N) = 0. \end{cases} \tag{10.4.48}$$

Thus, the operator $Lu = -(au_x)_x$ has been approximated by the spectral operator $L_N u = -[I_N(au_x)]_x$. Problem (10.4.48) corresponds to (10.4.38)–(10.4.39), with $J_e = \{1, \ldots, N-1\}$ and $J_b = \{0, N\}$. The space X can be chosen here as $L_w^2(-1,1)$, where w is either the Chebyshev or the Legendre weight function. The spaces X_N and Y_N coincide in this case, and one has

$$X_N = Y_N = \{v \in \mathbb{P}_N | v(-1) = v(1) = 0\},$$

where, as usual, $\mathbb{P}_N$ denotes the space of algebraic polynomials of degree $\leqslant N$ in the variable x.

EXAMPLE 7. THE NEUMANN PROBLEM FOR A CONSTANT-COEFFICIENT ELLIPTIC OPERATOR IN THE INTERVAL $(-1,1)$. The problem

$$\begin{cases} -u_{xx} + u = f & -1 < x < 1, \\ u_x(-1) = u_x(1) = 0 \end{cases} \tag{10.4.49}$$

can be approximated by the following collocation method

$$\begin{cases} (-u_{xx}^N - u^N)(x_k) = f(x_k) & 1 \leqslant k \leqslant N-1, \\ u_x^N(-1) = u_x^N(1) = 0, \end{cases} \tag{10.4.50}$$

where u^N is an algebraic polynomial of degree N and $\{x_k | k \in J\}$ are the points introduced in the previous example. Again, we set $X = L_w^2(-1,1)$, whereas now

$$X_N = \{v \in \mathbb{P}_N | v_x(-1) = v_x(1) = 0\}$$

and

$$Y_N = \{v \in \mathbb{P}_N | v(-1) = v(1) = 0\}.$$

We consider now the *stability and convergence properties* of the collocation approximation (10.4.38)—(10.4.39). As for the Galerkin approximation, the simplest situation occurs when the operator L satisfies the coercivity condition (10.3.4) and the continuity condition (10.3.5) with respect to a suitable energy space E.

Again we assume that $X_N \subseteq E$ for all $N > 0$. Moreover, we assume that for all $u \in X_N$, $\|u\|_N \leqslant C\|u\|_E$ with $C > 0$ independent of N (see (10.4.41)). A stability condition for the approximation (10.4.46), by analogy with condition (10.3.4), is as follows:

If there exists a constant $\bar{\alpha} > 0$ (independent of N) such that

$$\bar{\alpha}\|u\|_E^2 \leqslant (Q_N L_N u, u)_N \qquad \text{for all } u \in X_N, \tag{10.4.51}$$

then the approximation is stable, in the sense that the estimate

$$\|u^N\|_E \leqslant C\|f\|_N \tag{10.4.52}$$

holds with a constant $C > 0$ independent of N.

Actually, one has

$$\bar{\alpha}\|u^N\|_E^2 \leqslant (Q_N L_N u^N, u^N)_N = (Q_N f, u^N)_N \leqslant \|Q_N f\|_N \|u^N\|_N \leqslant C\|f\|_N \|u^N\|_E.$$

We use here the fact that Q_N is the projection operator upon Y_N with respect to the discrete inner product $(u, v)_N$.

We go now to the convergence analysis. Let R_N be a projection operator from a dense subspace $\mathscr{W}$ of $D_B(L)$ upon X_N. For each $u \in \mathscr{W}$, we further require $R_N u$ to satisfy the exact boundary conditions, i.e.,

$$R_N \colon \mathscr{W} \to X_N \cap D_B(L). \tag{10.4.53}$$

The following error bound between the exact and the collocation solutions holds:

$$\|u - u^N\|_E \leqslant \left(1 + \frac{\beta}{\bar{\alpha}}\right)\|u - R_N u\|_E + \frac{|(LR_N u, e) - (Q_N L_N R_N u, e)_N|}{\|e\|_E}$$
$$+ \frac{|(f, e) - (Q_N f, e)_N|}{\|e\|_E}, \tag{10.4.54}$$

with $e = u^N - R_N u$.

Assume for the moment that (10.4.54) is proven. It follows that *convergence is assured if the following three conditions are fulfilled*:

$$\|u - R_N u\|_E \to 0 \tag{10.4.55.1}$$

as $N \to \infty$, *for all* $u \in \mathscr{W}$;

$$\sup_{\substack{v \in X_N \\ v \neq 0}} \frac{(LR_N u, v) - (Q_N L_N R_N u, v)_N}{\|v\|_E} \to 0 \tag{10.4.55.2}$$

as $N \to \infty$, *for all* $u \in \mathscr{W}$;

$$\sup_{\substack{v \in X_N \\ v \neq 0}} \frac{(f, v) - (Q_N f, v)_N}{\|v\|_E} \to 0 \tag{10.4.55.3}$$

as $N \to \infty$, *for all* $f \in Z$ *smooth enough.*

Proof of (10.4.54). From (10.3.1) and (10.4.47) it follows that for any $v \in X_N$

$$(Lu, v) = (f, v) \tag{10.4.56}$$

and

$$(Q_N L_N u^N, v)_N = (Q_N f, v)_N. \tag{10.4.57}$$

On the other hand:

$$(Q_N L_N e, v)_N = (Q_N L_N u^N, v)_N - (Q_N L_N R_N u, v)_N.$$

Adding and subtracting (Lu, v) and using (10.4.56) and (10.4.57) we obtain:

$$(Q_N L_N e, v)_N = (Q_N f, v)_N - (f, v) + (L(R_N u - u), v) + (LR_N u, v) - (Q_N L_N R_N u, v)_N.$$

Taking $v = e$ and using the hypotheses (10.3.5) and (10.4.51) it follows:

$$\bar{\alpha}\|e\|_E^2 \leqslant |(Q_N f, e)_N - (f, e)| + \beta\|R_N u - u\|_E\|e\|_E + |(LR_N u, e) - (Q_N L_N R_N u, e)_N|.$$

Now (10.4.54) follows using the triangle inequality $\|u - u^N\|_E \leqslant \|u - R_N u\|_E + \|e\|_E$. □

The positivity condition (10.4.51) is the most immediate condition which guarantees the well-posedness of problem (10.4.46). However, there are situations where (10.4.51) is not fulfilled. This occurs for instance when the norms involved in the stability and convergence analysis depend on weight functions like the Chebyshev norms presented in Chap. 9. In these cases, the discrete analog of the "inf-sup" condition provides a more general criterion for checking the stability of the scheme.

Let us assume that the operator L satisfies conditions (10.3.6) to (10.3.8). Assume that for all $N > 0$, $X_N \subset W$ and $Y_N \subset V$. Moreover, assume that $\|v\|_N \leqslant C\|v\|_V$ for all v in Y_N, with $C > 0$ independent of N. Then we have the following stability condition for problem (10.4.46).

If there exists a constant $\bar{\alpha} > 0$ *independent of* N *such that*

$$\bar{\alpha}\|u\|_W \leqslant \sup_{\substack{v \in Y_N \\ v \neq 0}} \frac{(L_N u, v)_N}{\|v\|_V} \qquad \textit{for all } u \in X_N \tag{10.4.58}$$

then

$$\|u^N\|_W \leqslant C\|f\|_N \tag{10.4.59}$$

for a constant C independent of N.

The proof of (10.4.59) is a slight modification of the one of (10.4.25) pertaining to tau approximations.

Concerning the *convergence* of the method, one can estimate the error $u - u^N$ according to the following formula:

$$\|u - u^N\|_W \leqslant \left(1 + \frac{\beta}{\alpha}\right)\|u - R_N u\|_W + \sup_{\substack{v \in Y_N \\ v \neq 0}} \frac{|(LR_N u, v) - (L_N R_N u, v)_N|}{\|v\|_V} + \sup_{\substack{v \in Y_N \\ v \neq 0}} \frac{|(f, v) - (f, v)_N|}{\|v\|_V}. \tag{10.4.60}$$

As in the previous case, R_N is a projection operator from a dense subspace $\mathscr{W} \subseteq D_B(L)$ into $X_N \cap D_B(L)$. According to (10.4.60), *the approximation is convergent if the three following conditions hold true*:

$$\|u - R_N u\|_W \to 0 \tag{10.4.61.1}$$

as $N \to \infty$ for all $u \in \mathscr{W}$;

$$\sup_{\substack{v \in Y_N \\ v \neq 0}} \frac{(LR_N u, v) - (L_N R_N u, v)_N}{\|v\|_V} \to 0 \tag{10.4.61.2}$$

as $N \to \infty$, for all $u \in \mathscr{W}$;

$$\sup_{\substack{v \in Y_N \\ v \neq 0}} \frac{(f, v) - (f, v)_N}{\|v\|_V} \to 0 \tag{10.4.61.3}$$

as $N \to \infty$, for all $f \in Z$ smooth enough.

These are precisely the conditions to be checked in any specific situation in order to prove the convergence and to establish the rate of decay of the error.

Proof. The proof of (10.4.60) mimics the proof of (10.4.54). The error $e = u^N - R_N u$ satisfies:

$$(L_N e, v)_N = (L(u - R_N u), v) + (LR_N u, v) - (LR_N u, v)_N + (f, v)_N - (f, v),$$

for any $v \in Y_N$. We divide both sides by $\|v\|_V$ and take the supremum over all the functions in Y_N. Then, (10.4.60) follows from (10.4.58) and (10.3.8). □

We emphasize that the stability and convergence estimates given for the collocation problem include as special cases the ones for the Galerkin and tau approximations, provided that the discrete inner product is replaced by the continuous one. The last two terms appearing in the right-hand side of the convergence estimate (10.4.60) for collocation are precisely due to the use of quadrature formulas in the collocation scheme. Therefore, the conditions

(10.4.58) and (10.4.61) are the most general ones which assure stability and convergence for the general spectral approximation (10.3.9).

We want to bring the attention of the reader to the concept of *algebraic stability* introduced by Gottlieb and Orszag (1977) for approximations by spectral methods.

In both the stability criteria (10.4.51) and (10.4.58) we require that the constant $\bar{\alpha}$ be independent of N. This is not necessary for the convergence of the method. The constant $\bar{\alpha}$ may depend on N in an algebraic way, i.e., it may be of the form $\bar{\alpha} = O(N^{-r})$ for a suitable $r > 0$. In this case, convergence is still assured, according to the estimates (10.4.54) and (10.4.60), provided that the exact solution u is so smooth that the deviation $u - R_N u$ vanishes fast enough. Precisely, convergence occurs if $\|u - R_N u\|_E$ (or $\|u - R_N u\|_W$) decays as $O(N^{-r'})$ for an $r' > r$. This is a slightly different form of the concept of algebraic stability presented in Gottlieb and Orszag (1977, Sec. 5).

The previous abstract results will be now applied to prove the stability and convergence of the collocation approximations considered in Examples 6 and 7.

EXAMPLE 6 (continued from page 347). The stability and convergence analysis is easy if the Legendre points are used. In this case, the scheme satisfies a stability condition of the type (10.4.51). To check this result, let us start by observing that

$$(Q_N L_N u, u)_N = (L_N u, u)_N \qquad \text{for all } u \in X_N,$$

since Q_N is now the orthogonal projection onto X_N for the discrete inner product $(u, v)_N$. Furthermore, for all $u \in X_N$

$$\begin{aligned}(L_N u, u)_N &= -\int_{-1}^{1} [I_N(au_x)]_x u \, dx = \int_{-1}^{1} I_N(au_x) u_x \, dx \\ &= \sum_{k=0}^{N} a(x_k)[u_x(x_k)]^2 w_k \geqslant \alpha_0 \sum_{k=0}^{N} [u_x(x_k)]^2 w_k \\ &= \alpha_0 \int_{-1}^{1} [u_x(x)]^2 \, dx.\end{aligned}$$

Each change between integral and sum is allowed since the integrands are polynomials of degree at most $2N - 1$. Thus, (10.4.51) holds with $E = H_0^1(-1, 1)$ due to the Poincaré inequality (see (A.13) and (A.11.c)).

Let us consider now the Chebyshev collocation points. If the coefficient a in (10.4.48) is constant, say $a \equiv 1$, the scheme still fulfills the stability condition (10.4.51), with $E = H_{w,0}^1(-1, 1)$ defined in (A.11.c). Actually,

$$(L_N u, u)_N = -\int_{-1}^{1} u_{xx} u w \, dx,$$

which dominates the norm of $H^1_{w,0}(-1,1)$ as shown in Sec. 11.1 (see (11.1.14)). If $a(x)$ is not constant in the interval $(-1,1)$, the operator L_N may be indefinite in the inner product $(u,v)_N$. This can be seen by the following heuristic argument (which, however, can be made mathematically rigorous). For N large enough, $(L_N u, u)_N$ approaches $(Lu, u) = \int_{-1}^{1} a u_x (uw)_x \, dx$. Now $u_x(uw)_x$ may be strictly negative in a region excluding the endpoints and the origin, though its average on $(-1,1)$ is positive according to (11.1.14). Thus, if a is large in this region and small elsewhere, (Lu, u) and consequently $(L_N u, u)_N$ are strictly negative. The argument in turn shows that the stability condition (10.4.51) may not be satisfied in this case. However, it is possible to prove that the collocation scheme (10.4.48) is stable according to the more general stability condition (10.4.58), where $W = V = H^1_{w,0}(-1,1)$. More precisely, for any polynomial $u \in X_N$, it is possible to construct a polynomial $v \in X_N$, which depends on u but is different from it, such that $\|v\|_{1,w} \leqslant C\|u\|_{1,w}$, and $(L_N u, v)_N \geqslant \tilde{\alpha}\|u\|^2_{1,w}$ for two positive constants C and $\tilde{\alpha}$ independent of N. This clearly implies (10.4.48). The proof is rather technical, and can be found in Canuto and Quarteroni (1984).

The convergence of the approximation can be proved by checking the conditions (10.4.55) for the Legendre points, and the conditions (10.4.61) for the Chebyshev points. In both cases, an optimal error estimate is obtained by choosing as $R_N u$ the best polynomial approximation of u in the norm of $H^1_{w,0}$, as defined in (9.4.21) or (9.5.14). The precise result is

$$\|u - u^N\|_{H^1_w(-1,1)} \leqslant C_1 N^{1-m}(\|u\|_{H^m_w(-1,1)} + \|f\|_{H^{m-1}_w(-1,1)}),$$

where the norms are defined in (9.4.5) or (9.5.5).

EXAMPLE 7 (continued from page 348). Each $v \in Y_N$ can be written as $v(x) = z(x)(1-x^2)$, with $z \in \mathbb{P}_{N-2}$. Thus, Y_N can be identified with $\mathbb{P}_{N-2}$, in the sense that $Y_N = (1-x^2)\mathbb{P}_{N-2}$. In this example, the variational formulation (10.4.46) reads as follows:

$$\sum_{k=0}^{N} [-u^N_{xx} + u^N](x_k) z(x_k)(1-x_k^2) w_k = \sum_{k=0}^{N} f(x_k) z(x_k)(1-x_k^2) w_k \quad \text{for all } z \in \mathbb{P}_{N-2}. \tag{10.4.62}$$

Due to the relation (2.2.17), the higher order term on the left-hand side can be integrated exactly, namely,

$$-\sum_{k=0}^{N} u^N_{xx}(x_k) z(x_k)(1-x_k^2) w_k = -\int_{-1}^{1} u^N_{xx}(x) z(x) \eta(x)\, dx, \tag{10.4.63}$$

where $\eta(x) = \sqrt{1-x^2}$ is a Jacobi weight on the interval $(-1,1)$. So one is naturally led to establish the stability of (10.4.50) in a norm depending on the weight η. Actually, if we choose $z = -u_{xx}^N$ in (10.4.62), then (10.4.63) is precisely the square of the norm of u_{xx}^N in $L_\eta^2(-1, 1)$, i.e., $\int_{-1}^{1} [u_{xx}^N(x)]^2 \eta(x)\, dx$. In view of the stability condition (10.4.58), this observation suggests the choice of space W as

$$W = \{v \in L_\eta^2(-1, 1) | v_{xx} \in L_\eta^2(-1, 1)\},$$

with norm

$$\|v\|_W^2 = \int_{-1}^{1} [v^2(x) + v_{xx}^2(x)]\eta(x)\, dx.$$

The natural norm for the test functions z is the norm of $L_\eta^2(-1, 1)$. In terms of the original test functions $v = (1 - x^2)z$, this norm reads as

$$\int_{-1}^{1} z^2(x)\eta(x)\, dx = \int_{-1}^{1} \left(\frac{v(x)}{1 - x^2}\right)^2 \eta(x)\, dx = \|v\|_V^2. \qquad (10.4.64)$$

Thus, V will be the space of those functions for which the right-hand side of (10.4.64) is finite.

Within this framework it can be shown that the stability and convergence conditions (10.4.58) and (10.4.61) hold. The following error estimate can be proven:

$$\|u - u^N\|_W \leqslant CN^{2-m}(\|u\|_{H_w^m(-1,1)} + \|f\|_{H_w^{m-1}(-1,1)}) \quad m \geqslant 2.$$

Details can be found in Canuto and Quarteroni (1984).

10.5. General Formulation of Spectral Approximations to Linear Evolution Equations

Our attention now turns to an abstract formulation of spectral approximations to time-dependent problems. It is based on the same mathematical setting introduced in Sec. 10.3. We will retain the same notation here without referring repeatedly to Sec. 10.3.

We will analyze *semi-discrete* approximations only, namely the time variable will not be discretized. Time-marching methods to be used in combination with spectral approximations are discussed in detail in Chap. 4 and analyzed in Chap. 12.

Consider the initial-boundary value problem

$$\begin{cases} u_t + Lu = f & \text{in } \Omega \times (0, +\infty) \quad (10.5.1) \\ Bu = 0 & \text{on } \partial\Omega_b \times (0, +\infty) \quad (10.5.2) \\ u = u_0 & \text{in } \Omega \text{ for } t = 0. \quad (10.5.3) \end{cases}$$

The initial value u_0 is a function belonging to the space X, and the right-hand side f is a function of the variable t with values in X, i.e., $f(t) \in X$ for each $t > 0$. A solution for this problem is an X-valued function $u(t)$ such that u is continuous for all $t \geqslant 0$, du/dt exists and is continuous for all $t > 0$, $u(0) = u_0$, $u(t) \in D_B(L)$ for all $t > 0$, and (10.5.1) holds for all $t > 0$. In compact notation:

$$\begin{cases} u \in C^1(0, +\infty; X), \qquad u(t) \in D_B(La) & \text{for } t > 0 \\ \dfrac{du}{dt}(t) + Lu(t) = f(t) & \text{for } t > 0 \\ u(0) = u_0. \end{cases} \tag{10.5.4}$$

We assume that problem (10.5.4) is well-posed. For a rigorous difinition of well-posedness, and for conditions assuring the well-posedness, we refer, e.g., to Hille and Philips (1957) or to Richtmyer (1978, Chap. 16).

Any spectral approximation to the time-independent problem (10.3.3), as defined in Sec. 10.3, yields in a natural way a semi-discrete spectral approximation to the evolution problem (10.5.4). Precisely, assume that $f(t) \in Z$ for all $t \geqslant 0$, and for all $N > 0$ let $u_0^N \in X_N$ be a suitable approximation to u_0. Then, the spectral approximation $u^N(t)$ is for all $t \geqslant 0$ a continuously differentiable function with values in X_N which satisfies

$$\begin{cases} Q_N\left[\dfrac{du^N}{dt}(t) + L_N u^N(t) - f(t)\right] = 0 & \text{for all } t > 0 \\ u^N(0) = u_0^N. \end{cases} \tag{10.5.5}$$

Again, the problem can be written variationally. The spectral solution u^N satisfies the system of ordinary differential equations:

$$\begin{cases} u^N \in C^1(0, +\infty; X_N) \\ \left(\dfrac{du^N}{dt}(t) + L_N u^N(t) - f(t), v\right)_N = 0 \qquad \text{for all } v \in Y_N, \quad t > 0 \\ u^N(0) = u_0^N. \end{cases} \tag{10.5.6}$$

Galerkin, Tau and Collocation Approximations

The formulation (10.5.6) summarizes various spectral approximations to the evolution problem (10.5.4). In particular, the Galerkin, tau and collocation schemes, which have been defined in Sec. 10.4 for steady problems, apply in the present situation also. The time-derivative term du^N/dt is treated formally in the same way as the right-hand side f. Each of these procedures transforms (10.5.5) into a system of ordinary differential equations whose unknowns are the "Fourier" coefficients of the solution if a Galerkin or tau method is used, or the grid values of the solution if a collocation method is preferred. From a mathematical point of view, each of these methods is defined by the same choice of the spaces X_N, Y_N and the inner product $(u, v)_N$ made in Sec. 10.4. It is therefore straightforward to extend the material of that section to the case of time-dependent problems.

10.5.1. Conditions for Stability and Convergence: The Parabolic Case

In order to discuss questions of stability (in space) and convergence for spectral approximations to time-dependent problems, we distinguish between equations of parabolic and hyperbolic type. Roughly speaking, the parabolic case is characterized by the fact that the operator L is coercive with respect to a norm which is stronger than the one of X. Instead, in the hyperbolic case L is just non-negative, or semi-bounded, with respect to the inner product of X. We make these concepts rigorous starting with the parabolic case.

As for time-independent problems, the simplest stability condition arises from an *energy* inequality. We will assume henceforth that all the hypotheses made in Sec. 10.3.1 hold true; in particular *we assume that the spatial operator* L *satisfies the continuity condition* (10.3.5) *and the coercivity condition* (10.3.4).

We consider first a *Galerkin approximation* for which (10.5.6) takes the form

$$\left(\frac{du^N}{dt}(t) + Lu^N(t) - f(t), v\right) = 0 \qquad \text{for all } v \in X_N, \quad t > 0. \tag{10.5.7}$$

We assume that the hypotheses (10.4.5) and (10.4.6) hold. Then, taking $v = u^N(t)$ in (10.5.7) we get for each $t > 0$:

$$\frac{1}{2}\frac{d}{dt}\|u^N(t)\|^2 + \alpha\|u^N(t)\|_E^2 \leqslant (f(t), u^N(t)).$$

Now, applying the algebraic inequality $ab \leqslant (1/4\varepsilon)a^2 + \varepsilon b^2$ with $\varepsilon = \alpha/2$ to the right-hand side, we can find a constant C depending on α but independent of N such that we have for all $t > 0$

$$\|u^N(t)\|^2 + \alpha \int_0^t \|u^N(s)\|_E^2\, ds \leqslant \|u_0^N\|^2 + C\int_0^t \|f(s)\|^2\, ds. \tag{10.5.8}$$

This proves the *stability* of the Galerkin approximation.

Concerning its *convergence*, let us set $e(t) = R_N u(t) - u^N(t)$, where R_N is the projection operator introduced in (10.4.8). Then, the error function $e(t)$ satisfies the inequality

$$\frac{1}{2}\frac{d}{dt}\|e\|^2 + \alpha\|e\|_E^2 \leqslant |(u_t - R_N u_t, e) + (L(u - R_N u), e)|. \tag{10.5.9}$$

For any function $g \in X$, we can define the new norm

$$\|g\|_{E^*} = \sup_{\substack{v \in E \\ v \neq 0}} \frac{(g, v)}{\|v\|_E}. \tag{10.5.10}$$

This is the norm of g in the dual space E^* of E (see (A.1.c)). Note that $\|g\|_{E^*} \leqslant C\|g\|_X$, since $\|v\|_X \leqslant C\|v\|_E$ for all $v \in E$. Then, using the above definition and the continuity of the operator L (see (10.4.6)), it follows that

$$|(u_t - R_N u_t, e) + (L(u - R_N u), e)| \leqslant C\{\|u_t - R_N u_t\|_{E^*} + \|u - R_N u\|_E\}\,\|e\|_E.$$

Therefore, for all $t > 0$ the following error estimate can be inferred from (10.5.9):

$$\|e(t)\|^2 + \alpha \int_0^t \|e(s)\|_E^2\,ds \leqslant \|e(0)\|^2 + C\left\{\int_0^t \|(u_t - R_N u_t)(s)\|_{E^*}^2\,ds + \int_0^t \|(u - R_N u)(s)\|_E^2\,ds\right\}, \tag{10.5.11}$$

where C is a constant independent of N.

We conclude that *the approximation is convergent if each term on the right-hand side tends to 0 as $N \to \infty$ for u, u_t and u_0 regular enough.* In particular, this is true if the hypothesis (10.4.9) holds uniformly in t for time-dependent functions $u = u(t)$ and $u_t = u_t(t)$ in a suitable class. The approximation results given in Chap. 9 guarantee this property.

EXAMPLE 1. A FOURIER GALERKIN METHOD FOR THE HEAT EQUATION. Consider the one-dimensional heat equation

$$\begin{cases} u_t - u_{xx} = f & 0 < x < 2\pi, \quad t > 0 \\ u(x,0) = u_0(x) & 0 < x < 2\pi \\ u(x,t) & 2\pi\text{-periodic in } x \text{ for all } t \geqslant 0. \end{cases} \tag{10.5.12}$$

Its Galerkin approximation consists of looking for a function $u^N(t) \in S_N$, where S_N is the space of trigonometric polynomials defined in (9.1.1), which satisfies

$$(u_t^N - u_{xx}^N - f, v) = 0 \qquad \text{for all } v \in S_N, \quad t > 0 \tag{10.5.13}$$

and $u^N(0) = P_N u_0$ (see (2.1.7)). In this case, the operator $L = -\partial^2/\partial x^2$ satisfies the following energy identity

$$(Lu, u) = \int_0^{2\pi} |u_x|^2\,dx.$$

The square root of the right-hand side is just a semi-norm for the space $E = H_p^1(0, 2\pi)$ (see (A.11.d)). However, using the change of variable $u^N(t) \to w^N(t) = e^{-t}u^N(t)$, (10.5.13) becomes

$$(w_t^N - w_{xx}^N + w^N - e^t f, v) = 0 \qquad \text{for all } v \in S_N, \quad t > 0. \tag{10.5.14}$$

The "new" operator $L = -(\partial^2/\partial x^2) + 1$ satisfies the coercivity estimate (10.3.4); hence, stability and convergence follows by the previous general results. We choose $R_N = P_N$ in (10.5.9) and we observe that for all $v \in H_p^1(0, 2\pi)$ we have by (9.1.9)

$$|(u_t - P_N u_t, v)| = |(u_t - P_N u_t, v - P_N v)| \leqslant CN^{1-m} \|u_t\|_{H^{m-2}(0,2\pi)} \|v\|_{H^1(0,2\pi)}.$$

Hence, $\|u_t - R_N u_t\|_{[H^1_p(0,2\pi)]^*} \leqslant CN^{1-m}\|u_t\|_{H^{m-2}(0,2\pi)}$. Thus, we obtain the following error estimate which holds for all $t > 0$ and $m \geqslant 1$:

$$\|u(t) - u^N(t)\|_{L^2(0,2\pi)} + \left(\int_0^t \|(u-u^N)(s)\|^2_{H^1(0,2\pi)}\,ds\right)^{1/2}$$
$$\leqslant CN^{1-m}\left(\int_0^t \|u_t(s)\|^2_{H^{m-2}(0,2\pi)}\,ds + \int_0^t \|u(s)\|^2_{H^m(0,2\pi)}\,ds\right)^{1/2}. \qquad \square$$

We consider now *tau approximations* of problem (10.5.4). The tau method has been introduced for steady problems in Sec. 10.4.2. When applied to the evolution equation (10.5.4), it yields the scheme

$$\left(\frac{du^N}{dt}(t) + Lu^N(t) - f(t), v\right) = 0 \qquad \text{for all } v \in Y_N, \quad t > 0. \quad (10.5.15)$$

Thus, *stability* can be obtained provided the following inequality holds:

$$(Lu, Q_N u) \geqslant \bar{\alpha}\|u\|_E^2 \qquad \text{for all } u \in X_N, \qquad (10.5.16)$$

where $\bar{\alpha}$ is a positive constant and Q_N is the orthogonal projection upon Y_N in the inner product of X. Indeed, choosing $v = Q_N u^N(t)$ as test function, we obtain the following stability result

$$\|Q_N u^N(t)\|^2 + \bar{\alpha}\int_0^t \|u^N(s)\|_E^2\,ds$$
$$\leqslant \|u_0^N\|^2 + C\int_0^t \|f(s)\|^2\,ds, \quad t > 0. \qquad (10.5.17)$$

Proceeding as done for the Galerkin approximation the convergence inequality takes now the form

$$\|Q_N e(t)\|^2 + \bar{\alpha}\int_0^t \|e(s)\|_E^2\,ds \leqslant \|e(0)\|^2$$
$$+ C\int_0^t \|Q_N(u_t - R_N u_t)(s)\|^2_{E^*}\,ds + \int_0^t \|Q_N L(u - R_N u)(s)\|^2_{E^*}\,ds. \quad (10.5.18)$$

This inequality, together with the approximation results of Chap. 9, allows one to prove the *convergence* of the scheme.

EXAMPLE 2. A LEGENDRE TAU METHOD FOR THE HEAT EQUATION. We consider the initial-boundary value problem

$$\begin{cases} u_t - u_{xx} = f & -1 < x < 1, \quad t > 0 \\ u(-1,t) = u(1,t) = 0, & t > 0 \\ u(x,0) = u_0(x), & -1 < x < 1. \end{cases}$$

The solution $u^N(x,t)$ of the Legendre tau approximation of this problem is for all $t \geqslant 0$ a polynomial of degree N in x, which is zero at $x = \pm 1$ and satisfies the equations

$$\begin{cases} \displaystyle\int_{-1}^{1} [u_t^N(x,t) - u_{xx}^N(x,t)]v(x)\,dx = \int_{-1}^{1} f(x,t)v(x)\,dx, \quad t > 0 \\ \qquad\qquad\qquad\qquad\qquad\qquad\qquad\qquad \text{for all } v \in \mathbb{P}_{N-2} \\ \displaystyle\int_{-1}^{1} [u^N(x,0) - u_0(x)]v(x)\,dx = 0. \end{cases} \tag{10.5.19}$$

It follows that this scheme conforms to the abstract form (10.5.5) if we set $X_N = \{u \in \mathbb{P}_N | u(-1) = u(1) = 0\}$, $Y_N = \mathbb{P}_{N-2}$, and $(u, v) = \int_{-1}^{1} u(x)v(x)\,dx$. Hence, X is the space $L^2(-1,1)$ and the projection Q_N: $L^2(-1,1) \to \mathbb{P}_{N-2}$ is the truncation P_{N-2} of the Legendre series.

For all $u \in X_N$ we have

$$-\int_{-1}^{1} u_{xx} P_{N-2} u\,dx = -\int_{-1}^{1} u_{xx}^N u\,dx = \int_{-1}^{1} (u_x)^2\,dx.$$

It follows that the stability condition (10.5.16) is verified with $E = H_0^1(-1,1)$ (see (A.11.c)), since $\|u\|_E = (\int_{-1}^{1} (u_x)^2\,dx)^{1/2}$ is a norm for this space (see (A.13)). Hence, the Legendre tau approximation (10.5.19) is stable, and (10.5.17) gives for all $t > 0$ the estimate

$$\|P_{N-2}u^N(t)\|_{L^2(-1,1)}^2 + \int_0^t \|u_x(s)\|_{L^2(-1,1)}^2\,ds$$
$$\leqslant \|u_0\|_{L^2(-1,1)}^2 + C \int_0^t \|f(s)\|_{L^2(-1,1)}^2\,ds.$$

A bound for the error $u - u^N$ can be derived from the estimate (10.5.18). The operator R_N is chosen as the orthogonal projection on X_N in the norm of $H_0^1(-1,1)$, as defined in (9.4.21). We bound each term on the right-hand side of (10.5.18). The first term is bounded by $C\|u_0 - R_N u_0\|_{L^2(-1,1)}$. Concerning the second term we have for each $v \in H_0^1(-1,1)$:

$$\begin{aligned} (P_{N-2}(u_t - R_N u_t), v) &= (u_t - R_N u_t, v) - (u_t - R_N u_t, v - P_{N-2} v) \\ &= ((u_t - R_N u_t)_x, (\phi - R_N \phi)_x) - (u_t - R_N u_t, v - P_{N-2} v), \end{aligned}$$

where ϕ is the only function in $H_0^1(-1,1)$ satisfying $-\phi_{xx} = v$. Then using the approximation results for the operators P_{N-2} and R_N given in (9.4.6) and (9.4.22), respectively, and recalling (10.5.10) we obtain

$$\|P_{N-2}(u_t - R_N u_t)\|_{E^*} \leqslant C N^{1-m} \|u_t\|_{H^{m-2}(-1,1)}. \tag{10.5.20}$$

For the last term of (10.5.18) we have for all $v \in H_0^1(-1,1)$:

$$\begin{aligned} (P_{N-2}(u - R_N u)_{xx}, v) &= -((u - R_N u)_x, v_x) - ((u - R_N u)_{xx}, v - P_{N-2} v) \\ &= -((u - R_N u)_x, v_x) - (u_{xx} - P_{N-2} u_{xx}, v - P_{N-2} v). \end{aligned}$$

Here we have used the fact that both $P_{N-2}u_{xx}$ and $(R_N u)_{xx}$ are orthogonal to $v - P_{N-2}v$. Using the same approximation results as before, we deduce

$$\|P_{N-2}(u - R_N u)_{xx}\|_{E^*} \leqslant CN^{1-m}\|u\|_{H^m(-1,1)}. \qquad (10.5.21)$$

Combining the previous results we obtain the final error estimate, for all $t > 0$ and all $m \geqslant 2$:

$$\|u(t) - P_{N-2}u^N(t)\|_{L^2(-1,1)} + \left(\int_0^t \|(u_x - u_x^N)(s)\|^2_{L^2(-1,1)}\, ds\right)^{1/2}$$
$$\leqslant CN^{1-m}\left(\int_0^t (\|u_t(s)\|^2_{H^{m-2}(-1,1)} + \|u(s)\|^2_{H^m(-1,1)})\, ds\right)^{1/2}. \qquad (10.5.22)$$

□

Finally, let us consider *collocation approximations* to (10.5.4). We recall that collocation methods for steady problems have been introduced in Sec. 10.3.3. For simplicity, we assume in (10.5.6) that $Y_N = X_N$, which is the case when the boundary conditions are of Dirichlet type. Moreover, we assume that the discrete operator L_N satisfies the coercivity inequality

$$(L_N u, u)_N \geqslant \bar{\alpha}\|u\|_E^2 \qquad \text{for all } u \in X_N. \qquad (10.5.23)$$

The technique already applied to the other spectral schemes yields for each $t > 0$ the *stability* inequality:

$$\|u^N(t)\|_N^2 + \bar{\alpha}\int_0^t \|u^N(s)\|_E^2\, ds \leqslant \|u_0^N\|_N^2 + C\int_0^t \|f(s)\|_N^2\, ds. \qquad (10.5.24)$$

We recall here that the discrete norm $\|u\|_N = \sqrt{(u,u)_N}$ can be controlled by $C\|u\|_X$ for all $u \in \mathrm{Pol}_N(\Omega)$, with C independent of N (see Sec. 9.3).

Concerning the *convergence* of the approximation, the following estimate, which is the counterpart of estimate (10.4.54) for evolution equations, holds for all $t > 0$:

$$\|e(t)\|_N^2 + 2\bar{\alpha}\int_0^t \|e(s)\|_E^2\, ds$$
$$\leqslant \|e(0)\|_N^2 + C\int_0^t \|u_t - R_N u_t\|_{E^*}^2\, ds + \int_0^t \|u - R_N u\|_E^2\, ds$$
$$+ C\int_0^t \left(\frac{(R_N u_t, e) - (R_N u_t, e)_N}{\|e\|_E}\right)^2 ds \qquad (10.5.25)$$
$$+ \int_0^t \left(\frac{(LR_N u, e) - (L_N R_N u, e)_N}{\|e\|_E}\right)^2 ds$$
$$+ \int_0^t \left(\frac{(f,e) - (f,e)_N}{\|e\|_E}\right)^2 ds.$$

This estimate can be obtained by adapting to the present situation the proof of estimate (10.5.18), taking into account the extra errors due to the discrete inner product.

EXAMPLE 3. A CHEBYSHEV COLLOCATION METHOD FOR THE HEAT EQUATION WITH DIRICHLET BOUNDARY CONDITIONS. We consider again the scheme presented in Sec. 1.2.2. This scheme is analyzed in Sec. 10.1.2, where it is actually proven that the stability condition (10.5.23) holds in this case with $E = H^1_{w,0}(-1,1)$ defined in (A.11.b) and $\|u\|_E = (\int_{-1}^{1} |\partial u/\partial x|^2 w(x)\,dx)^{1/2}$. Moreover, it is claimed there that the optimal error bound (10.1.6) holds. Indeed, this estimate is an immediate consequence of the general estimate (10.5.25).

We choose as $R_N u$ the orthogonal projection of u upon $\mathbb{P}_{N-1}$ rather than $\mathbb{P}_N$ with respect to the inner product of $H^1_{w,0}(-1,1)$ (see (9.5.14)). Then the three last terms of (10.5.25) are zero in the current situation, while the two remaining ones can be handled as in Example 2.

EXAMPLE 4. A CHEBYSHEV COLLOCATION METHOD FOR THE HEAT EQUATION WITH NEUMANN BOUNDARY CONDITIONS. Let us define the following Chebyshev collocation approximation of the initial-boundary value problem

$$\begin{cases} u_t - u_{xx} = 0 & \text{in } -1 < x < 1, \quad t > 0, \\ u_x(-1,t) = u_x(1,t) = 0 & t > 0 \\ u(x,0) = u_0(x) & -1 < x < 1. \end{cases}$$

For each $N > 0$, consider the Gauss–Lobatto Chebyshev points $x_k = \cos(k\pi/N)$, $k = 0, \ldots, N$. Given any polynomial v of degree N, denote by $B_0 v$ the unique polynomial of degree N such that $(B_0 v)(x_k) = v(x_k)$ if $k = 1, \ldots, N-1$, and $(B_0 v)(x_0) = (B_0 v)(x_N) = 0$. Thus, B_0 is a linear operator which maps $\mathbb{P}_N(-1,1)$ into the subset of $\mathbb{P}_N(-1,1)$ consisting of polynomials that vanish on the boundary.

The collocation approximation $u^N = u^N(x,t)$ is for all $t \geqslant 0$, a polynomial of degree N in x, which satisfies the set of equations

$$\begin{cases} u_t^N(x_k,t) - (B_0 u_x^N)_x(x_k,t) = 0 & \text{for all } t > 0 \text{ and } k = 0, \ldots, N \\ u^N(x_k,0) = u_0(x_k) & k = 0, \ldots, N. \end{cases} \tag{10.5.26}$$

Note that u^N is not required to satisfy the boundary conditions. They are rather taken into account within the differential equation, via the operator B_0. Since no boundary conditions are imposed directly, the differential equation is collocated at the boundary points also. This indirect way of imposing the boundary conditions is discussed in Sec. 3.3.

The approximation (10.5.26) can be written as (10.5.6) if we set $X_N = Y_N = \mathbb{P}_N$, $L_N = -(\partial/\partial x)(B_0(\partial/\partial x))$ and

$$(u, v)_N = \sum_{k=0}^{N} u(x_k)v(x_k)w_k,$$

the w_k being the weights of the Gauss–Lobatto quadrature formula for the Chebyshev weight w. Thus it is natural to set $X = L_w^2(-1, 1)$, with inner product $(u, v) = \int_{-1}^{1} u(x)v(x)(1 - x^2)^{-1/2}\,dx$.

The stability and convergence analysis for this approximation can be based on the results obtained in the previous example in the following way. Since $(B_0 u_x)_x$ is a polynomial of degree less than N, (10.5.26) is actually equivalent to the equation

$$u_t^N - (B_0 u_x^N)_x = 0 \qquad \text{for all } x \in (-1, 1) \text{ and } t > 0, \tag{10.5.27}$$

with initial condition

$$u^N(x, 0) = (I_N u_0)(x) \qquad \text{for all } x \in (-1, 1). \tag{10.5.28}$$

$I_N u_0$ denotes the polynomial of degree N which interpolates u_0 at the nodes x_k, $k = 0, \ldots, N$.

If we set $U^N(x, t) = (B_0 u_x^N)(x, t)$, then U^N is for all $t \geqslant 0$, a polynomial of degree N in x, which vanishes at the boundary and satisfies the collocation equations

$$\begin{cases} U_t^N(x_k, t) - U_{xx}^N(x_k, t) = 0 & \text{for all } t > 0 \text{ and } k = 1, \ldots, N-1 \\ U^N(x_k, 0) = U_0(x_k) & k = 0, \ldots, N \end{cases}$$

with $U_0(x) = B_0[I_N u_0]_x(x)$. This can be seen by differentiating (10.5.27) and (10.5.28) with respect to x. It follows from the results of Sec. 10.1.2 that the approximation is stable, in the sense that

$$\|B_0 u_x^N(t)\|_{L_w^2(-1,1)}^2 + \frac{1}{2}\int_0^t \|(B_0 u_x^N)_x(s)\|_{L_w^2(-1,1)}^2\,ds \leqslant \|B_0[(I_N u_0)_x]\|_{L_w^2(-1,1)}^2,$$

for all $t > 0$. Moreover, the convergence estimate (10.1.6) reads as follows:

$$\|u_x(t) - B_0 u_x^N(t)\|_{L_w^2(-1,1)} + \left(\int_0^t \|u_{xx}(s) - (B_0 u_x^N)_x(s)\|_{L_w^2(-1,1)}^2\,ds\right)^{1/2}$$
$$\leqslant CN^{2-m}\left\{\|u_0\|_{H_w^m(-1,1)} + \left(\int_0^t \|u_t(s)\|_{H_w^{m-2}(-1,1)}^2\,ds\right)^{1/2}\right. \tag{10.5.29}$$
$$\left. + \left(\int_0^t \|u(s)\|_{H_w^m(-1,1)}^2\,ds\right)^{1/2}\right\}$$

for all $t \geqslant 0$ and $m \geqslant 2$.

Finally, an estimate for the norm of the error in $L_w^2(-1, 1)$ can be easily obtained by observing that, due to (10.5.27), the error $u - u^N$ satisfies the differential equation

$$(u - u^N)_t - [u_{xx} - (B_0 u_x)_x] = 0 \qquad \text{in } (-1, 1) \times (0, +\infty).$$

Multiplying by $u - u^N$, integrating in space, and using the Cauchy–Schwarz inequality, we get for all $t > 0$

$$\frac{1}{2}\frac{d}{dt}\|u - u^N\|^2_{L^2_w(-1,1)} \leqslant \|u_{xx} - (B_0 u_x)_x\|_{L^2_w(-1,1)}\|u - u^N\|_{L^2_w(-1,1)}.$$

The Gronwall lemma (see (A.15)) and (10.5.29) yield the estimate

$$\|u(t) - u^N(t)\|_{L^2_w(-1,1)}$$
$$\leqslant CN^{2-m}\exp\left(\frac{t}{2}\right)\left[\|u_0\|_{H^m_w(-1,1)} + \left(\int_0^t \|u_t(s)\|^2_{H^{m-2}_w(-1,1)}\,ds\right)^{1/2}\right.$$
$$\left. + \left(\int_0^t \|u(s)\|^2_{H^m_w(-1,1)}\,ds\right)^{1/2}\right],$$

for all $t > 0$ and all $m \geqslant 2$.

We conclude this section with the following remark. The "inf-sup" condition (10.4.24) or (10.4.58) discussed in Sec. 10.4 for steady problems, does not guarantee by itself the stability of discrete approximations to evolution problems. Actually, this condition assures that the spatial operator is non-singular, but it does not provide any information about the distribution of its spectrum. Hence, it cannot imply the well-posedness of the time-dependent problem.

An analysis for evolution equations could be based on the application of an "inf-sup" condition for the operator $L + \tau I$, (τ varying in a region of the complex plane) which is obtained by Laplace transforming the equation in time.

This method, which has been successfully proposed in Babuška and Aziz (1972) for finite-element analysis, seems promising, but it has yet to be applied to the analysis of spectral methods.

10.5.2. Conditions for Stability and Convergence: The Hyperbolic Case

The *energy* approach for equations of hyperbolic type takes the following general form. It is assumed that there exists a Hilbert space $E \subset X$ with norm $\|u\|_E$, such that $D_B(L) \subset E$ and $\|u\| \leqslant C\|u\|_E$ for all $u \in E$. Moreover, it is assumed that

$$\|Lu\| \leqslant C\|u\|_E \qquad \text{for all } u \in D_B(L), \tag{10.5.30}$$

with a constant $C > 0$, and that the operator L satisfies the non-negativity property

$$0 \leqslant (Lu, u) \qquad \text{for all } u \in D_B(L). \tag{10.5.31}$$

The discrete counterpart of this property is

$$0 \leqslant (L_N u, Q_N u)_N \qquad \text{for all } u \in X_N. \tag{10.5.32}$$

If this assumption is fulfilled for all $N > 0$, the approximation scheme (10.5.6) *is stable, namely, the following estimate holds*:

$$\|Q_N u^N(t)\|_N^2 \leqslant \|u_0^N\|_N^2 + \exp(t) \int_0^t \|f(s)\|_N^2 \, ds \qquad \textit{for all } t > 0. \tag{10.5.33}$$

The result is easily obtained by taking $v = Q_N u^N(t)$ in (10.5.6) and using the Gronwall lemma (see (A.15)). We recall that for Galerkin and tau approximations, the symbol $\|u\|_N$ in (10.5.33) denotes the norm $\|u\|$ of the space X, whereas for collocation methods it takes the usual meaning of the discrete norm at the collocation nodes. Also note that whenever the space of trial functions X_N is equal to the one of test functions Y_N (this happens for all Galerkin methods, and for collocation approximations to scalar hyperbolic problems), we have $Q_N u^N = u^N$ in (10.5.33).

To state the *convergence* estimates, we set as usual $e(t) = R_N u(t) - u^N(t)$, where R_N is a suitable projection operator defined as in (10.4.8). The equation verified by the error function $Q_N e(t)$ is obtainable from (10.5.4) and (10.5.6). The assumption (10.5.32), together with the Gronwall lemma, allows us to get a bound for $\|Q_N e(t)\|$. Then, by the triangle inequality $\|u - Q_N u\| \leqslant \|u - Q_N R_N u\| + \|Q_N e\|$, we obtain the desired convergence estimate. For clarity, we specify the error estimate for the three cases of Galerkin, tau and collocation methods.

For *Galerkin approximations* the following inequality can be obtained for all $t > 0$:

$$\begin{aligned} &\frac{1}{2}\|u(t) - u^N(t)\|^2 \\ &\quad \leqslant \|u(t) - R_N u(t)\|^2 + \|u_0^N - R_N u_0\|^2 \\ &\quad\quad + C\exp(t)\left\{\int_0^t (\|(u_t - R_N u_t)(s)\|^2 + \|(u - R_N u)(s)\|_E^2)\, ds\right\}. \end{aligned} \tag{10.5.34}$$

For *tau approximations* we have for all $t > 0$:

$$\begin{aligned} &\frac{1}{2}\|u(t) - Q_N u^N(t)\|^2 \\ &\quad \leqslant \|u(t) - Q_N R_N u(t)\|^2 + \|Q_N(u_0^N - R_N u_0)\|^2 \\ &\quad\quad + C\exp(t) \int_0^t (\|(u_t - R_N u_t)(s)\|^2 + \|(u - R_N u)(s)\|_E^2)\, ds. \end{aligned} \tag{10.5.35}$$

Finally, we consider *collocation approximations*. For simplicity, we assume that $X_N = Y_N$. Moreover, we suppose that the discrete and continuous norms are uniformly equivalent on X_N, i.e.,

$$C_1 \|u\| \leqslant \|u\|_N \leqslant C_2 \|u\| \qquad \text{for all } u \in X_N$$

with two constants C_1 and C_2 independent of N. This condition is always fulfilled in the cases of interest, as has been shown in Chap. 9 (see Sec. 9.3). In this situation, for all $t > 0$ the convergence estimate takes the form

$$\begin{aligned}
&\|u(t) - u_N(t)\|^2 \\
&\leqslant C\Bigg\{\|u(t) - R_N u(t)\|^2 + \|u_0^N - R_N u_0\|^2 \\
&\quad + \exp(t)\Bigg[\int_0^t (\|u_t - R_N u_t\|^2 + \|u - R_N u\|_E^2)\, ds \\
&\quad + \int_0^t \left(\frac{(R_N u_t, e) - (R_N u_t, e)_N}{\|e\|}\right)^2 ds \\
&\quad + \int_0^t \left(\frac{(LR_N u, e) - (L_N R_N u, e)_N}{\|e\|}\right)^2 ds + \int_0^t \left(\frac{(f, e) - (f, e)_N}{\|e\|}\right)^2 ds\Bigg]\Bigg\}.
\end{aligned} \tag{10.5.36}$$

The three last terms in the right-hand side originate from the aliasing error. Again, the convergence of the methods is guaranteed if each term on the right-hand sides of (10.5.34), (10.5.35), or (10.5.36) vanishes as $N \to \infty$. This can be proven for regular solutions using the approximation results given in Chap. 9. Now we present some examples.

EXAMPLE 5. A LEGENDRE TAU METHOD FOR THE EQUATION $u_t + u_x = f$. We consider the initial-boundary value problem:

$$\begin{cases} u_t + u_x = f & -1 < x < 1, \quad t > 0 \\ u(-1, t) = 0 & t > 0 \\ u(x, 0) = u_0(x) & -1 < x < 1. \end{cases} \tag{10.5.37}$$

As usual, let $L_k(x)$ denote the k-th Legendre polynomial. The Legendre tau approximate solution $u^N(x, t) = \sum_{k=0}^N \alpha_k(t) L_k(x)$ to this problem is defined by the set of equations

$$\begin{cases} \displaystyle\int_{-1}^{1} [u_t^N + u_x^N](x, t) L_k(x)\, dx = \int_{-1}^{1} f(x, t) L_k(x)\, dx \\ \qquad\qquad\qquad\qquad \text{for } k = 0, \ldots, N-1, \quad t > 0 \\ \displaystyle\sum_{k=0}^{N} (-1)^k \alpha_k(t) = 0 \qquad\qquad t > 0 \\ \displaystyle\alpha_k(0) = (k + \tfrac{1}{2}) \int_{-1}^{1} u_0(x) L_k(x)\, dx \qquad k = 0, \ldots, N-1. \end{cases} \tag{10.5.38}$$

This scheme fits into the general formulation (10.5.6), provided one sets $X = L^2(-1, 1)$, $X_N = \{u \in \mathbb{P}_N | u(-1) = 0\}$, $Y_N = \mathbb{P}_{N-1}$, $L_N = L = \partial/\partial x$ and $(u, v)_N = (u, v) = \int_{-1}^{1} u(x)v(x)\,dx$. The projection Q_N is the orthogonal projection P_{N-1} over the space of polynomials of degree up to $N-1$ with respect to this inner product (see (2.2.6)). The continuity condition (10.5.30) holds with $E = H^1(-1, 1)$, (this space is defined in (A.11.a)). Moreover, since u_x is a polynomial of degree $N-1$, one has

$$\int_{-1}^{1} u_x P_{N-1} u\,dx = \int_{-1}^{1} u_x u\,dx = \frac{1}{2}u^2(1),$$

which proves that both (10.5.31) and (10.5.32) are satisfied. It follows that the scheme is stable, namely, for all $t > 0$ (10.5.33) yields the estimate

$$\|P_{N-1}u^N(t)\|^2_{L^2(-1,1)} \leqslant \|u_0\|^2_{L^2(-1,1)} + \exp(t)\int_0^t \|f(s)\|^2_{L^2(-1,1)}\,ds. \quad (10.5.39)$$

We apply now the general convergence estimate (10.5.35) to the present situation. It is convenient to choose $R_N u$ as the best approximation of u in $X_{N-1} \subset X_N$ with respect to the norm of $E = H^1(-1, 1)$. In this case $Q_N R_N u = R_N u$. It is possible to prove an error estimate for R_N similar to (9.4.22), namely

$$\|u - R_N u\|_{H^k(-1,1)} \leqslant CN^{k-m}\|u\|_{H^m(-1,1)} \qquad k = 0 \text{ or } 1 \text{ and } m \geqslant 1. \quad (10.5.40)$$

Noting that $Q_N(u_0^N - R_N u_0) = P_{N-1}u_0 - R_N u_0$, using (10.5.40) and (9.4.6) we obtain from (10.5.35):

$$\|u(t) - P_{N-1}u^N(t)\|_{L^2(-1,1)} \leqslant CN^{1-m}\exp\left(\frac{t}{2}\right) \times \left[\int_0^t (\|u_t(s)\|^2_{H^{m-1}(-1,1)} + \|u(s)\|^2_{H^m(-1,1)})\,ds\right]^{1/2}, \quad (10.5.41)$$

which holds for all $t > 0$ and $m \geqslant 1$. We have bounded $\|u_0\|_{H^{m-1}(-1,1)}$ and $\|u(t)\|_{H^{m-1}(-1,1)}$ by the last integral on the right-hand side of the previous inequality. This is allowed by classical results of functional analysis (see, e.g., Lions and Magenes (1972)).

The stability and convergence analysis for the scheme (10.5.38) can be also carried out using a test function different from $Q_N u^N$ (or $Q_N e$). Indeed, take $v(t) = (u^N(t))/b$ as test function in (10.5.6) with $b(x) = 1 + x$, and define a new inner product $[u, v] = \int_{-1}^{1} u(x)v(x)(dx/b(x))$. Then, setting $|||u||| = [u, u]^{1/2}$, we have

$$\frac{1}{2}\frac{d}{dt}|||u^N(t)|||^2 + [u_x^N(t), u^N(t)] = [f(t), u^N(t)], \qquad t > 0.$$

Integrating by parts, we have

$$[u_x^N, u^N] = \frac{1}{2}\int_{-1}^{1} v^2\,dx + v^2(1).$$

Moreover,

$$[f, u^N] = \int_{-1}^{1} fv\,dx \leqslant \|f\|_{L^2(-1,1)}\|v\|_{L^2(-1,1)} \leqslant \frac{1}{2}\|f\|_{L^2(-1,1)}^2 + \frac{1}{2}\|v\|_{L^2(-1,1)}^2.$$

On the other hand, it is evident that $|||u^N(t)|||^2 \geqslant \frac{1}{2}\|u^N(t)\|_{L^2(-1,1)}^2$. Therefore, integrating in time we obtain

$$\|u^N(t)\|_{L^2(-1,1)}^2 \leqslant \|u_0^N\|_{L^2(-1,1)}^2 + \int_0^t \|f(s)\|_{L^2(-1,1)}^2\,ds, \quad t > 0. \qquad (10.5.42)$$

We stress that with this new stability estimate all frequencies of the solution u^N are controlled. Moreover, the bound on the right-hand side of (10.5.42) does not blow up in time, unlike the one in (10.5.39). Concerning convergence, by the usual argument, one can obtain the following error estimate

$$\|u(t) - u_N(t)\|_{L^2(-1,1)} \leqslant CN^{1-m}\left\{\int_0^t \left(\|u_t(s)\|_{H^{m-1}(-1,1)}^2 + \|u(s)\|_{H^m(-1,1)}^2\right) ds\right\}^{1/2}, \qquad (10.5.43)$$

which improves (10.5.41).

EXAMPLE 6. A CHEBYSHEV TAU METHOD FOR THE EQUATION $u_t - xu_x = f$. We consider the initial-boundary value problem:

$$\begin{cases} u_t - xu_x = f & -1 < x < 1, \quad t > 0 \\ u(-1, t) = u(1, t) = 0 & t > 0 \\ u(x, 0) = u_0(x) & -1 < x < 1. \end{cases} \qquad (10.5.44)$$

The Chebyshev tau solution $u^N(x, t) = \sum_{k=0}^N \alpha_k(t) T_k(x)$ of this problem is defined by the conditions

$$\begin{cases} \displaystyle\int_{-1}^{1} [u_t^N(x,t) - xu_x^N(x,t)]\,T_k(x)w(x)\,dx = \int_{-1}^{1} f(x,t)\,T_k(x)w(x)\,dx \\ \qquad\qquad k = 0, \ldots, N-2, \text{ and } t > 0 \\ \displaystyle\sum_{k=0}^{N} (-1)^k \alpha_k(t) = \sum_{k=0}^{N} \alpha_k(t) = 0 \qquad t > 0 \\ \displaystyle\alpha_k(0) = \frac{2}{c_k \pi}\int_{-1}^{1} u_0(x)\,T_k(x)w(x)\,dx \qquad k = 0, \ldots, N-2. \end{cases} \qquad (10.5.45)$$

Here $T_k(x)$ is the k-th Chebyshev polynomial, $w(x) = (1 - x^2)^{-1/2}$ is the Chebyshev weight and the c_k's are defined in (2.4.10).

Problem (10.5.45) can be expressed in the form (10.5.6) by setting $X = L^2_w(-1, 1)$, $X_N = \{u \in \mathbb{P}_N | u(-1) = u(1) = 0\}$, $Y_N = \mathbb{P}_{N-2}$, $L_N = L = -x(\partial/\partial x)$ and $(u, v)_N = (u, v)_w = \int_{-1}^{1} u(x)v(x)w(x)\,dx$. The projection operator Q_N is the orthogonal projection operator P_{N-2} over $\mathbb{P}_{N-2}$ with respect to the Chebyshev inner product $(u, v)_w$.

The positivity condition (10.5.32) takes the form

$$-\int_{-1}^{1} xu_x P_{N-2}uw\,dx \geqslant 0 \qquad \text{for all } u \in X_N.$$

It is satisfied in the current example since one has

$$-\int_{-1}^{1} xu_x P_{N-2}uw\,dx = \int_{-1}^{1} xu_x(u - P_{N-2}u)w\,dx + \frac{1}{2}\int_{-1}^{1} u^2(xw)_x\,dx.$$

The last term is positive since $xw(x)$ is an increasing function. The other term, using (2.4.4) and (2.4.22), equals $\frac{1}{2}N\hat{u}_N^2 + \frac{1}{2}(N-1)\hat{u}_{N-1}^2$, (where $\hat{u}_N$ and $\hat{u}_{N-1}$ denote the two last Chebyshev coefficients of u); hence, it is positive. The convergence analysis follows along the guidelines of the previous example.

A different approach consists of choosing $v = u/b$, where $b(x) = 1 - x^2$, as a test function. A straightforward calculation reveals that:

$$(Lu, v)_w = -\int xu_x vw\,dx = \frac{1}{2}\int_{-1}^{1} v^2 \frac{1}{w}\,dx + \frac{3}{2}\int_{-1}^{1} v^2x^2w\,dx$$
$$\geqslant \frac{1}{2}\int_{-1}^{1} u^2w\,dx + \frac{3}{2}\int_{-1}^{1} v^2x^2w\,dx. \tag{10.5.46}$$

Then, proceeding as in the previous example, stability and convergence inequalities like (10.5.42) and (10.5.43) can be proven, relative to the weighted Chebyshev norms.

EXAMPLE 7. FOURIER GALERKIN AND COLLOCATION APPROXIMATIONS TO A TWO-DIMENSIONAL ADVECTION EQUATION. We consider the advection equation:

$$\begin{cases} u_t + \mathbf{b}\cdot\nabla u + \nabla\cdot(\mathbf{b}u) = 0 & \mathbf{x} \in \Omega = (0, 2\pi)^2, \quad t > 0 \\ u(\mathbf{x}, 0) = u_0(\mathbf{x}) & \mathbf{x} \in \Omega \\ u(\mathbf{x}, t) \text{ periodic in } \mathbf{x} & t > 0. \end{cases} \tag{10.5.47}$$

We have set $\mathbf{x} = (x_1, x_2)$ and we assume that $\mathbf{b} = (b_1(\mathbf{x}), b_2(\mathbf{x}))$ and u_0 are given regular and periodic functions. Denote by $\mathbf{k} = (k_1, k_2)$ any couple of integers (positive or negative). Then $\mathbf{k}\cdot\mathbf{x} = k_1x_1 + k_2x_2$ denotes the Euclidean inner product of $\mathbb{R}^2$. Finally we denote by J the set of multi-indexes $\mathbf{k} = (k_1, k_2)$ such that $-N \leqslant k_i \leqslant N - 1$ for $i = 1, 2$.

The Fourier Galerkin approximation to u is the function $u^N(\mathbf{x}, t) = \sum_{\mathbf{k}\in J} \alpha_{\mathbf{k}}(t)e^{i\mathbf{k}\cdot\mathbf{x}}$ which satisfies the equations

$$\begin{cases} \displaystyle\int_\Omega [u_t^N + Lu^N](\mathbf{x}, t) e^{-i\mathbf{k}\cdot\mathbf{x}}\, d\mathbf{x} = 0 & \text{for } \mathbf{k} \in J,\ t > 0 \\ \alpha_{\mathbf{k}}(0) = \dfrac{1}{2\pi} \displaystyle\int_\Omega u_0(\mathbf{x}) e^{-i\mathbf{k}\cdot\mathbf{x}}\, d\mathbf{x} = 0 & \text{for } \mathbf{k} \in J. \end{cases} \tag{10.5.48}$$

Here $Lu = \mathbf{b}\cdot\nabla u + \nabla\cdot(\mathbf{b}u)$ is the linear operator associated to the problem (10.5.47).

Problem (10.5.48) is a particular case of (10.5.6) corresponding to the choice: $X_N = Y_N = \mathrm{span}\{e^{i\mathbf{k}\cdot\mathbf{x}}, \mathbf{k} \in J\}$, $(u, v)_N = (u, v) = \int_\Omega u(\mathbf{x})\overline{v(\mathbf{x})}\, d\mathbf{x}$, $L_N = L$, and $Q_N = P_N$, the orthogonal projection from $X = L^2(\Omega)$ onto X_N.

The continuity property (10.5.30) holds, taking as E the space $H_p^1(\Omega)$, defined in (A.11.d). Furthermore, integrating by parts and using periodicity yields

$$(Lu, u) = 0 \qquad \text{for all } u \in H_p^1(\Omega); \tag{10.5.49}$$

hence, (10.5.32) holds. From (10.5.33), it follows that (10.5.48) is a stable approximation to (10.5.47), namely

$$\|u^N(t)\|_{L^2(\Omega)} \leqslant \|P_N u_0\|_{L^2(\Omega)} \leqslant \|u_0\|_{L^2(\Omega)}. \tag{10.5.50}$$

Moreover, taking $R_N u = P_N u$, the convergence estimate (10.5.34) gives in the present situation the following inequality for all $t > 0$ and $m \geqslant 1$

$$\|u(t) - u^N(t)\|_{L^2(\Omega)} \leqslant CN^{1-m} \exp\left(\frac{t}{2}\right)\left(\int_0^t (\|u_t(s)\|^2_{H^{m-1}(\Omega)} + \|u(s)\|^2_{H^m(\Omega)})\, dt\right)^{1/2}. \tag{10.5.51}$$

Let us now introduce the $4N^2$ collocation points $\mathbf{x}_{jk} = (x_j, x_k)$, $0 \leqslant j, k \leqslant 2N - 1$, with $x_j = \pi j/N$, and denote by $I_N u \in X_N$ the interpolant of u at these points. The Fourier collocation approximation to u is the function $u^N(\mathbf{x}, t) = \sum_{\mathbf{k}\in J} \beta_{\mathbf{k}}(t) e^{i\mathbf{k}\cdot\mathbf{x}}$ satisfying the equations

$$\begin{cases} [u_t^N + L_N u^N](\mathbf{x}_{jk}, t) = 0 & \text{for } t > 0 \text{ and } 0 \leqslant j, k \leqslant 2N - 1 \\ u^N(\mathbf{x}_{jk}, 0) = u_0(\mathbf{x}_{jk}) & \text{for } 0 \leqslant j, k \leqslant 2N - 1. \end{cases} \tag{10.5.52}$$

Here $L_N u = \mathbf{b}\cdot\nabla u + \nabla\cdot I_N(\mathbf{b}u)$ for all $u \in X_N$. This scheme can be written in the general form (10.5.6) by setting

$$(u, v)_N = \left(\frac{\pi}{N}\right)^2 \sum_{0 \leqslant j,k \leqslant 2N-1} u(\mathbf{x}_{jk})\overline{v(\mathbf{x}_{jk})}.$$

Due to (2.1.27) we note that $(u, v)_N = (u, v)$ for all $u, v \in X_N$. Then, an easy computation shows that $(L_N u, u)_N = 0$ for all $u \in X_N$. This proves that the collocation scheme is quadratically conservative, as discussed in Sec. 4.5. Moreover, since $\|u\|_N^2 \equiv (u, u)_N = \|u\|^2_{L^2(\Omega)}$ for all $u \in X_N$, the stability estimate (10.5.33) gives

$$\|u^N(t)\|_{L^2(\Omega)} \leqslant \|I_N u_0\|_{L^2(\Omega)} \leqslant \max_{x \in \overline{\Omega}} |u_0(x)|.$$

Furthermore, the same convergence estimate as (10.5.51) can be proven for the Fourier collocation solution, taking now $R_N u = I_N u$ in (10.5.36) and using the approximation properties of this operator (see Sec. 9.1.3).

The stability and convergence analysis for the approximation schemes (10.5.48) and (10.5.52) has been given first by Pasciak (1980).

EXAMPLE 8. A CHEBYSHEV COLLOCATION APPROXIMATION TO A ONE-DIMENSIONAL ADVECTION EQUATION IN THE INTERVAL. We consider the one-dimensional advection equation with variable coefficients:

$$\begin{cases} u_t + (bu)_x + b_0 u = f & -1 < x < 1, \quad t > 0 \\ u(x,0) = u_0(x) & -1 < x < 1, \end{cases} \tag{10.5.53}$$

where $b(x)$ and $b_0(x)$ are two smooth functions. The boundary conditions for this problem must be prescribed at those points of the boundary where flux is entering. For instance, assuming that both $b(-1)$ and $b(1)$ are positive, (10.5.53) must be supplemented with the condition:

$$u(-1,t) = 0 \qquad t > 0. \tag{10.5.54}$$

We consider this situation first. A Chebyshev collocation approximation can be defined as follows. Let

$$\begin{aligned} x_j &= \cos\left(-\pi + \frac{2\pi j}{2N+1}\right) & 0 \leqslant j \leqslant N \\ w_0 &= \frac{\pi}{2N+1} \qquad w_j = 2w_0 & 1 \leqslant j \leqslant N, \end{aligned} \tag{10.5.55}$$

be, respectively, the nodes and the weights of the Chebyshev–Gauss–Radau quadrature formula having as prescribed boundary node $x_0 = -1$ (in (2.4.13) the prescribed node is $x = 1$). For all $t \geqslant 0$, the collocation approximation to u is the polynomial $u^N(t) \in \mathbb{P}_N$ satisfying

$$\begin{cases} [u_t^N + L_N u^N](x_j,t) = f(x_j,t) & 1 \leqslant j \leqslant N, \quad t > 0 \\ u^N(x_j,0) = u_0(x_j) & 0 \leqslant j \leqslant N, \\ u^N(x_0,t) = 0 & t > 0. \end{cases} \tag{10.5.56}$$

Here L_N is the *skew-symmetric* decomposition of the operator $Lu = (bu)_x + b_0 u$, namely

$$L_N u = \tfrac{1}{2}\{bu_x^N + [I_N(bu)]_x\} + [\tfrac{1}{2}(I_N b)_x + b_0]u$$

and I_N denotes the interpolation operator with respect to the nodes $\{x_j\}$. We set $X_N = \{u \in \mathbb{P}_N | u(-1) = 0\}$ and $Y_N = X_N$. Moreover, we define a discrete inner product as follows:

$$(u,v)_N = \sum_{j=0}^{N} u(x_j)v(x_j)\tilde{w}_j \qquad \tilde{w}_j = (1-x)w_j. \tag{10.5.57}$$

Then (10.5.56) can be equivalently written in the form (10.5.6) taking $u_0^N = I_N u_0$. The stability and convergence analysis can be carried out according to the theory of this section, setting

$$X = L^2_{\tilde{w}}(-1,1) \quad \text{where } \tilde{w}(x) = (1-x)\frac{1}{\sqrt{1-x^2}} = \left(\frac{1-x}{1+x}\right)^{1/2}. \quad (10.5.58)$$

Hereafter, we sketch the main steps of this analysis. The details can be found in Canuto and Quarteroni (1982b). The continuity hypothesis (10.5.30) is valid if we choose $E = H^1_{\tilde{w}}(-1,1)$, the weighted Sobolev space of order 1 with respect to the weight $\tilde{w}$, which is defined by analogy with $H^1_w(-1,1)$ (see (A.11.b)). Moreover, we have

$$\begin{aligned} &(u,v)_N = (u,v) \quad \text{if } uv \in \mathbb{P}_{2N-1} \text{ and} \\ &\|u\|_N = (u,u)_N^{1/2} \quad \text{is uniformly equivalent to } \|u\| \text{ for all } u \in \mathbb{P}_N. \end{aligned} \quad (10.5.59)$$

Here (u, v) and $\|u\|$ denote as usual the inner product and the norm of X. Using these properties, one has

$$(L_N u, u)_N \geqslant 0 \qquad \text{for all } u \in X_N.$$

Then (10.5.32) holds, and the method is stable in the norm of X. From the estimate (10.5.36), taking $R_N u = I_N u$, using (10.5.59) and the approximation results of Sec. 9.5.3 it can be proven that for all $t > 0$ and $m \geqslant 2$

$$\|u(t) - u^N(t)\| \leqslant CN^{2-m} \exp\left(\frac{t}{2}\right) \left\{ \int_0^t \|u_t(s)\|^2_{H^{m-1}_w(-1,1)} + \|u(s)\|^2_{H^m_w(-1,1)}) \, ds \right\}^{1/2}, \quad (10.5.60)$$

where w is the Chebyshev weight.

The same analysis can be carried out for the boundary conditions required by different boundary values of $b(x)$. Precisely, instead of (10.5.54) we demand that either

$$u(1,t) = 0 \quad t > 0 \quad \text{if } b(-1) < 0, b(1) < 0, \quad (10.5.61)$$

or

$$u(\pm 1, t) = 0 \quad t > 0 \quad \text{if } b(-1) > 0, b(1) < 0. \quad (10.5.62)$$

Finally, if

$$b(-1) < 0 \qquad b(1) > 0, \quad (10.5.63)$$

no boundary conditions should be prescribed. In these three different situations, the collocation scheme is defined as in (10.5.49). The collocation points are the nodes of the Gauss–Chebyshev quadrature formula including those boundary points where boundary conditions are given. In the analysis, the weight $\tilde{w}(x)$ becomes $\tilde{w}(x) = \varepsilon(x)(1/\sqrt{1-x^2})$, and $\varepsilon(x)$ is $(1 + x)$ or 1 or

$(1 - x^2)$ corresponding to (10.5.61) or to (10.5.62) or to (10.5.63). The same kind of stability and convergence results can be proven.

10.6. The Error Equation

It has been shown in Secs. 10.3 and 10.5 that any spectral method is defined through a projection of the differential equation onto a finite-dimensional space of polynomials. Thus, the spectral solution is characterized by a set of Weighted Residual Equations (see (10.3.11) and (10.5.6)).

It is also useful to characterize the spectral solution as the exact solution of a suitable differential problem. This problem is of the same type as the original problem to be discretized. It only differs in the forcing term, which takes into account the error committed by the spectral projection. The new differential equation is called the *error equation* of the method.

The error equation can be exploited in deriving the stability and convergence properties of spectral schemes. It was first used for this purpose by Dubiner (1977) and Gottlieb and Orszag (1977). Since the spectral solution satisfies the error equation pointwise over the whole domain, it is also possible to deduce from it local information on the qualitative behavior of the solution, as opposed to the global information produced by variational methods. On the other hand, the analysis based on the error equation is usually confined to simple model problems.

For brevity, our discussion of the error equation will be limited to evolution problems only. However, a similar discussion could be carried out for steady or eigenvalue problems as well. In what follows, we refer for both notation and hypotheses to the abstract formulation of spectral approximations for evolution problems given in Sec. 10.5.

In particular, we recall that for all $t > 0$, the spectral solution $u^N(t)$ belongs to a finite-dimensional space X_N, and that the spectral operator L_N maps X_N into a space Z, which is either a space of square-integrable functions or a space of continuous functions on the domain Ω. We have assumed that $X_N \subseteq Z$ and the data $f(t) \in Z$ for all $t > 0$. Hence, $u_t^N + L_N u^N - f$ is an element of Z for all $t \geqslant 0$. By definition $Q_N[u_t^N + L_N u^N - f] = 0$ (see (10.5.5)), where Q_N is a projection upon a finite-dimensional space Y_N.

The error equation arises from the trivial decomposition

$$w = Q_N w + Q_N^* w \qquad \text{for all } w \in Z,$$

where

$$Q_N^* w = w - Q_N w.$$

Taking into account (10.5.5), one has

$$u_t^N + L_N u^N - f = Q_N^*[u_t^N + L_N u^N - f], \tag{10.6.1}$$

or equivalently

$$u_t^N + L_N u^N = Q_N^*[u_t^N + L_N u^N] + Q_N f. \qquad (10.6.2)$$

This is precisely the error equation. The right-hand side of (10.6.1) represents the error generated pointwise by the spectral approximation scheme. It is precisely from the analysis of this error that one can infer information about the spectral solution. In all the relevant schemes, the space Z contains the space $\mathrm{Pol}_N(\Omega)$ of the polynomials of degree N, introduced at the beginning of Sec. 10.4. Thus, we make here the assumptions that X_N and Y_N are contained in $\mathrm{Pol}_N(\Omega)$ and that the spectral operator L_N actually maps X_N into $\mathrm{Pol}_N(\Omega) \subset Z$. The last assumption is certainly true if L_N has constant coefficients.

Under these hypotheses, $Q_N^*[u_t^N + L_N u^N]$ is a polynomial in $\mathrm{Pol}_N(\Omega)$ for all $t > 0$. Hence, it can be expanded according to any basis $\{\phi_k | k \in J\}$ in $\mathrm{Pol}_N(\Omega)$, as

$$Q_N^*[u_t^N + L_N u^N] = \sum_{k \in J} \tau_k(t)\phi_k \qquad t \geqslant 0. \qquad (10.6.3)$$

This expression takes a simplified form in some relevant cases.

Full Fourier Approximations

If the boundary conditions are all periodic, $\mathrm{Pol}_N(\Omega)$ is a space of trigonometric polynomials, and $X_N = Y_N = \mathrm{Pol}_N(\Omega)$. Thus, $Q_N^* v = 0$ for all $v \in \mathrm{Pol}_N(\Omega)$, and the error equation becomes

$$u_t^N + L_N u^N = Q_N f. \qquad (10.6.4)$$

As a simple example, let us consider the Fourier Galerkin approximation to the heat equation which has been presented in Example 1 of Sec. 10.5. In this case the spectral solution u^N satisfies the following error equation

$$u_t^N - u_{xx}^N = P_N f \qquad 0 < x < 2\pi, \quad t > 0,$$

where $P_N f$ is the truncation of order N of the Fourier series of f (see (2.1.7)). For a collocation approximation to the same heat problem (10.5.12), the error equation satisfied by the spectral solution u^N would be

$$u_t^N - u_{xx}^N = I_N f \qquad 0 < x < 2\pi, \quad t > 0,$$

where now $I_N f$ is the interpolant of f at the collocation points (see (2.1.25)).

Collocation and Tau Methods for Non-periodic Boundary Conditions

For collocation methods, the natural basis in $\mathrm{Pol}_N(\Omega)$ is the Lagrange basis associated to the collocation points, which has been introduced in (10.4.37). This basis is orthogonal with respect to the inner product $(u, v)_N$ defined in (10.4.40).

On the other hand, in tau methods, $\mathrm{Pol}_N(\Omega)$ is represented in terms of the orthogonal basis with respect to the inner product (u, v) of X.

Note that for all $v \in \mathrm{Pol}_N(\Omega)$, $Q_N^* v$ is orthogonal to any polynomial in Y_N in the inner product $(u, v)_N$. This follows from the definition of $Q_N^* v$. Hence $Q_N^* v$ has no components along the elements in Y_N. In particular, (10.6.3) becomes

$$Q_N^*[u_t^N + L_N u^N] = \sum_{k \in J_b} \tau_k(t)\phi_k \qquad t \geqslant 0. \tag{10.6.5}$$

This expansion, recalling the definition of the set J_b, shows that the error on the left-hand side of (10.6.5) arises from the process by which the boundary conditions are taken into account in the spectral scheme.

An explicit representation of the coefficients $\tau_k(t)$ can be derived from (10.6.2), using the orthogonality of the basis functions in $\mathrm{Pol}_N(\Omega)$. One immediately has for all $t > 0$:

$$\tau_k(t) = \frac{1}{(\phi_k, \phi_k)_N}\left(\frac{d}{dt}(u^N, \phi_k)_N + (L_N u^N, \phi_k)_N\right) \qquad \text{for all } k \in J_b.$$

As an example, consider Chebyshev approximations to the heat equation

$$\begin{cases} u_t - u_{xx} = f & -1 < x < 1, \quad t > 0 \\ u(-1, t) = u(1, t) = 0 & t > 0 \\ u(x, 0) = u_0(x) & \text{for } -1 < x < 1. \end{cases} \tag{10.6.6}$$

The error equation pertaining to the Chebyshev tau approximation of (10.6.6) is

$$u_t^N - u_{xx}^N = \tau_N(t) T_N + \tau_{N-1}(t) T_{N-1} + P_{N-2} f. \tag{10.6.7}$$

Here $\tau_k(t) = da_k/dt$ for $k = N$, $N - 1$, where $a_k(t)$ are the Chebyshev coefficients of the expansion of u^N, and $P_{N-2} f$ is the truncation of order $N - 2$ of the Chebyshev series of f (see (2.2.16)).

The collocation approximation to (10.6.6) has the form

$$u_t^N - u_{xx}^N = \{\tau_0(t)(1 + x) + \tau_N(t)(1 - x)\} T_N' + I_N f, \tag{10.6.8}$$

where

$$\tau_0(t) = \frac{2}{N^2} u_t^N(1, t) - u_{xx}^N(1, t),$$

$$\tau_N(t) = (-1)^N \frac{2}{N^2} u_t^N(-1, t) - u_{xx}^N(-1, t),$$

and $I_N f$ is the interpolant of f at the Chebyshev collocation points.

The error equation has been extensively used to derive stability estimates for constant-coefficient equations in the 1977 monograph by Gottlieb and Orszag (see Secs. 7 and 8) and in several papers by Gottlieb and coworkers which have appeared in the last few years. In this book, an example of analysis based on the error equation is reported in Sec. 12.1.2, where the tau method for the equation $u_t + u_x = 0$ is considered.

The error equation method applies successfully only to constant-coefficient problems. On the other hand, the method makes possible a local investigation of the behavior of the solution in physical as well as in transform space. Among the results which have been obtained by means of the error equation are: information on the asymptotic behavior of tau solutions to hyperbolic problems (see Dubiner's thesis and the review of his results given in the appendix of Gottlieb, Hussaini and Orszag (1984)); properties of the spectra of several Chebyshev and Legendre differencing operators (see Gottlieb and Lustman (1983b), Lustman (1986), Funaro (1987a)); stability of spectral approximation for hyperbolic systems (Tal–Ezer (1986b)), Gottlieb, Lustman and Tadmor (1987b)); and stability in the maximum norm for solutions of singular perturbation problems (Canuto (1987)).

CHAPTER 11

Steady, Smooth Problems

A fairly advanced theory exists for spectral approximations to steady, smooth problems. The purpose of this chapter is to review the status of this theory. Computational aspects are discussed in Chaps. 5 to 8.

Problem of Poisson type will be considered first. We shall present the proof of several of the fundamental results which have been invoked repeatedly in Chap. 10. Then results will be given for a fairly general advection-diffusion equation. The linear case can be placed in the general setting of Chap. 10, but for the non-linear problem we need to introduce a new and more generally applicable approximation theorem. Such a theorem also plays a fundamental role in the numerical analysis of spectral approximations to the steady incompressible Navier–Stokes equations. This topic is treated next. We discuss, as well, compatibility conditions which have to be satisfied in choosing discrete velocities and pressure, and we detail the analysis in the case of the Kleiser and Schumann method.

We end the chapter by presenting an analysis of the behavior of the eigenvalues of several spectral approximations to model differential operators.

11.1. The Poisson Equation

Numerous spectral algorithms for the numerical simulation of physical phenomena require the approximate solution of one or more Poisson equations of the type

$$-\Delta u = f \tag{11.1.1}$$

in a bounded domain $\Omega \subset \mathbb{R}^d$ ($d = 1, 2, 3$). Here $\Delta = \sum_{i=1}^{d} \partial^2/\partial x_i^2$ denotes the Laplace operator in d space variables, u is the unknown function, and f is a given data.

Among the boundary conditions which are more commonly associated to the Poisson equation (11.1.1) are homogeneous Dirichlet conditions:

$$u = 0 \qquad \text{on } \partial\Omega. \tag{11.1.2}$$

Thus, as usual in spectral methods, we assume that the computational domain is the Cartesian product of d copies of the interval $(-1, 1)$, i.e., $\Omega = (-1, 1)^d$.

In Chap. 10 we discussed from a general point of view conditions which guarantee the convergence of spectral approximations to boundary value problems. These conditions concern from one side the properties of approximation of the space of polynomials chosen to represent the discrete solution, and from the other side the fulfillment of suitable properties of coercivity by the differential operator and by its spectral approximation.

Several examples have been given in Chap. 10 to illustrate the application of the theory to specific problems. Some of them pertained to the Poisson equation with Dirichlet boundary conditions, in one or more space dimensions.

Hereafter, we collect the most relevant theoretical facts about the Laplace operator submitted to homogeneous Dirichlet boundary conditions, and about its approximations of spectral type. We show that the coercivity conditions of Chap. 10 are fulfilled with a natural choice of the norms.

Non-periodic boundary value problems are usually approximated by Legendre or Chebyshev methods. From a theoretical point of view, the analysis of Chebyshev methods is more involved, due to the presence of the singular weight. Thus, it is convenient to treat separately Legendre and Chebyshev methods.

11.1.1. Legendre Methods

The natural norms in which to set the analysis of these methods are the norms of the standard (non-weighted) Sobolev spaces $H^m(\Omega)$ (see (A.11.a)). A central role is played by the Hilbert space $H_0^1(\Omega)$, defined in (A.11.c).

The operator $L = -\Delta$ is a linear unbounded operator in $L^2(\Omega)$ (see (A.3)). Supplemented with homogeneous Dirichlet boundary conditions, its domain of definition is the dense subspace $D_B(L) = \{v \in H^2(\Omega): v|_{\partial\Omega} = 0\}$. If $u \in D_B(L)$ and $v \in H_0^1(\Omega)$, integration-by-parts yields

$$-\int_\Omega \Delta u v \, dx = \int_\Omega \nabla u \cdot \nabla v \, dx. \tag{11.1.3}$$

(In this section dx denotes $dx_1 \ldots dx_d$.) The right-hand side is precisely the inner product of the Hilbert space $H_0^1(\Omega)$ (see (A.11.c)). It follows that the coercivity and continuity assumptions (10.3.4) and (10.3.5) are satisfied with the choice $E = H_0^1(\Omega)$.

Using (11.1.3) the following weak (or variational) formulation of the boundary value problem (11.1.1)—(11.1.2) is obtained: One looks for a function $u \in H_0^1(\Omega)$ such that

$$\int_\Omega \nabla u \cdot \nabla v \, dx = \int_\Omega f v \, dx \qquad \text{for all } v \in H_0^1(\Omega). \tag{11.1.4}$$

Here, we have assumed $f \in L^2(\Omega)$. A more general data $f \in H^{-1}(\Omega)$ (the dual space of $H_0^1(\Omega)$, (see (A.11.c)) is allowed, in which case the right-hand side has to be replaced by the duality pairing $\langle f, v \rangle$ between $H^{-1}(\Omega)$ and $H_0^1(\Omega)$. By the Riesz representation theorem (see (A.1.d)) there exists a unique solution of problem (11.1.4). If $f \in L^2(\Omega)$, then one can prove that the second derivatives of u are square integrable in Ω. Hence, we conclude that $u \in D_B(L)$.

Now we turn to the numerical approximations. Since the coercivity assumption (10.4.5) is fulfilled, it follows that the Legendre Galerkin method for (11.1.1)—(11.1.2) is stable (hence, convergent) in the $H_0^1(\Omega)$-norm, or equivalently, in the $H^1(\Omega)$-norm. The same conclusion holds for the Legendre collocation method, which uses the Gauss–Lobatto points (2.3.12) in each space direction. Actually, consider the discrete inner product

$$(u, v)_N = \sum_{j \in J} u(x_j) v(x_j) w_j, \tag{11.1.5}$$

where $\{x_j | j \in J\}$ are the Cartesian products of these points and $\{w_j | j \in J\}$ are the corresponding weights. Then if $u \in \mathbb{P}_N(\Omega)$ and $v \in \mathbb{P}_N^0(\Omega)$, the space of polynomials of degree N in each space variable vanishing on $\partial\Omega$, one has

$$(-\Delta u, v)_N = (\nabla u, \nabla v)_N. \tag{11.1.6}$$

This follows by integration-by-parts, since in each direction of differentiation the quadrature rule can be replaced by the exact integral, the integrand being a polynomial of degree at most $2N - 1$ in that direction. On the other hand, by (9.3.2), the right-hand side of (11.1.6) is an inner product on $\mathbb{P}_N^0(\Omega)$, which induces a norm equivalent to the $H_0^1(\Omega)$-norm. Thus, (10.4.51) is fulfilled with $E = H_0^1(\Omega)$.

For the analysis of the tau approximation to (11.1.1)—(11.1.2) we have to resort to the generalized "inf-sup" condition (10.4.24). In the one-dimensional case we endow $X_N = \mathbb{P}_N^0$ with the norm of $H^2(-1, 1)$ and $Y_N = \mathbb{P}_{N-2}$ with the norm of $L^2(-1, 1)$ and we choose as test function $v = -u_{xx}$. This yields the stability of the method in the norm of $H^2(-1.1)$. The two-dimensional case has been discussed in Example 5 of Sec. 10.4.2.

11.1.2. Chebyshev Methods

Let $w(x) = \sum_{i=1}^d (1 - x_i^2)^{-1/2}$ be the Chebyshev weight in dimension d. Chebyshev methods are naturally studied in the norms of the weighted Sobolev spaces $H_w^m(\Omega)$ (see (A.11.b)). Here we consider the operator $L = -\Delta$ as a linear unbounded operator in $L_w^2(\Omega)$. The domain of definition of L with Dirichlet boundary conditions is the dense subspace $D_B(L) = \{v \in H_w^2(\Omega): v|_{\partial\Omega} = 0\}$. This fact is immediate in one space dimension, whereas in more space dimensions it requires a complex proof due to Bernardi and Maday (1986b).

Let $u \in D_B(L)$ and $v \in H_{w,0}^1(\Omega)$ (see (A.11.c) for the definition of this space). Integrating by parts in a formal manner we get

$$-\int_\Omega \Delta u v w \, dx = \int_\Omega \nabla u \cdot \nabla(vw) \, dx. \tag{11.1.7}$$

The right-hand side is non-symmetric in its arguments u and v, due to the presence of the weight w. Let us set

$$a(u, v) = \int_\Omega \nabla u \cdot \nabla(vw) \, dx. \tag{11.1.8}$$

The bilinear form $a(u, v)$ is defined, continuous and coercive on the product space $H^1_{w,0}(\Omega) \times H^1_{w,0}(\Omega)$, as stated precisely in the following theorem.

Theorem 11.1.

(i) *There exists a constant* $\beta > 0$ *such that for all* $u, v \in H^1_{w,0}(\Omega)$

$$|a(u, v)| \leqslant \beta \|u_x\|_{L^2_w(\Omega)} \|v_x\|_{L^2_w(\Omega)}; \tag{11.1.9}$$

(ii) *there exists a constant* $\alpha > 0$ *such that for all* $u \in H^1_{w,0}(\Omega)$

$$\alpha \|u\|^2_{H^1_w(\Omega)} \leqslant a(u, u). \tag{11.1.10}$$

This result was proved by Canuto and Quarteroni (1981a) in dimension 1, and was extended to higher space dimensions by Funaro (1981).

Hereafter, we give the proof for the one-dimensional case, since it already contains all the essential elements of the analysis. The bilinear form (11.1.8) becomes

$$a(u, v) = \int_{-1}^{1} u_x (vw)_x \, dx, \tag{11.1.11}$$

where $w(x) = (1 - x^2)^{-1/2}$ is the Chebyshev weight. Let us start with the following inequality.

Lemma 11.1. *For all* $u \in H^1_{w,0}(-1, 1)$

$$\int_{-1}^{1} u^2(x) w^5(x) \, dx \leqslant \frac{8}{3} \int_{-1}^{1} u_x^2(x) w(x) \, dx. \tag{11.1.12}$$

PROOF. Let us split the left-hand side as

$$\int_{-1}^{1} u^2(x) w^5(x) \, dx = \int_{-1}^{0} u^2(x) w^5(x) \, dx + \int_{0}^{1} u^2(x) w^5(x) \, dx.$$

Since $w(x) \leqslant (1 - x)^{-1/2}$ if $0 \leqslant x \leqslant 1$,

$$\int_0^1 u^2(x) w^5(x) \, dx \leqslant \int_0^1 u^2(x)(1 - x)^{-5/2} \, dx$$

$$= \int_0^1 \left[\frac{1}{1 - x} \int_x^1 u_x(s) \, ds \right]^2 (1 - x)^{-1/2} \, dx.$$

Now we apply Hardy's inequality (A.14) with $\alpha = -1/2$ and we get

$$\int_0^1 u^2(x)w^5(x)\,dx \leqslant \frac{8}{3}\int_0^1 u_x^2(x)w(x)\,dx.$$

The same inequality holds over the interval $(-1, 0)$, whence the result. □

Let us prove part (i) of Theorem 11.1. Precisely, we will prove that *for all u and $v \in H^1_{w,0}(-1,1)$, the following inequality holds*:

$$\left|\int_{-1}^1 u_x(vw)_x\,dx\right| \leqslant \left(1+\sqrt{\frac{8}{3}}\right)\|u_x\|_{L^2_w(-1,1)}\|v_x\|_{L^2_w(-1,1)}. \tag{11.1.13}$$

Proof of (11.1.13). By the identity

$$\int_{-1}^1 u_x(vw)_x\,dx = \int_{-1}^1 u_x v_x w\,dx + \int_{-1}^1 u_x(vw_x w^{-1})w\,dx,$$

and the application of the Cauchy-Schwarz inequality (A.2) to both terms on the right-hand side one gets

$$|a(u,v)| \leqslant \|u_x\|_{L^2_w(-1,1)}\left\{\|v_x\|_{L^2_w(-1,1)} + \left(\int_{-1}^1 v^2 w_x^2 w^{-1}\,dx\right)^{1/2}\right\}.$$

Noting that $w_x = xw^3$, it follows using (11.1.12) that

$$\int_{-1}^1 v^2 w_x^2 w^{-1}\,dx \leqslant \int_{-1}^1 v^2 w^5\,dx \leqslant \frac{8}{3}\|v_x\|^2_{L^2_w(-1,1)},$$

whence the result. □

Finally, we prove part (ii) of Theorem 11.1. Precisely, we shall prove that *for all $u \in H^1_{w,0}(-1,1)$ the following inequality holds*:

$$\frac{1}{4}\|u_x\|^2_{L^2_w(-1,1)} \leqslant \int_{-1}^1 u_x(uw)_x\,dx. \tag{11.1.14}$$

Then, (11.1.10) will follow from the Poincaré inequality (A.13). (Note that the Poincaré inequality is implied by the inequality (11.1.10), since $w(x) \geqslant 1$.)

Proof of (11.1.14). By partial integration (which is allowed by (11.1.12)) one gets

$$\begin{aligned} a(u,u) &= \int_{-1}^1 (u_x)^2 w\,dx + \int_{-1}^1 uu_x w_x\,dx \\ &= \int_{-1}^1 (u_x)^2 w\,dx - \frac{1}{2}\int_{-1}^1 u^2 w_{xx}\,dx. \end{aligned} \tag{11.1.15}$$

In order to estimate the last integral on the right-hand side, let us use another expression for $a(u,u)$, namely

$$
\begin{aligned}
a(u,u) &= \int_{-1}^{1} [(u_x)^2 w^2 + u u_x w_x w] w^{-1}\, dx \\
&= \int_{-1}^{1} [(u_x w)^2 + 2u_x w u w_x + (u w_x)^2] w^{-1}\, dx \\
&\quad - \int_{-1}^{1} (u u_x w_x + u^2 w_x^2 w^{-1})\, dx \\
&= \int_{-1}^{1} [(uw)_x]^2 w^{-1}\, dx + \int_{-1}^{1} u^2 \left(\frac{w_{xx}}{2} - w_x^2 w^{-1}\right) dx.
\end{aligned}
\tag{11.1.16}
$$

By the identity $w_{xx} - 2w_x^2 w^{-1} = w^5$ we obtain

$$
\frac{1}{2}\int_{-1}^{1} u^2 w^5\, dx \leqslant a(u,u). \tag{11.1.17}
$$

On the other hand, since $w_{xx} = (1 + 2x^2)w^5$,

$$
\int_{-1}^{1} u^2 w_{xx} \leqslant 3 \int_{-1}^{1} u^2 w^5\, dx \leqslant 6a(u,u).
$$

Thus, recalling (11.1.15)

$$
a(u,u) \geqslant \int_{-1}^{1} (u_x)^2 w\, dx - 3a(u,u),
$$

or, equivalently,

$$
4a(u,u) \geqslant \int_{-1}^{1} (u_x)^2 w\, dx,
$$

whence the result. □

Let us now turn to the general d-dimensional case. Theorem 11.1. essentially states that the Laplace operator with homogeneous Dirichlet boundary conditions fulfills the coercivity and the continuity conditions (10.3.4) and (10.3.5) with respect to the Hilbert space $E = H^1_{w,0}(\Omega)$. In Sec. 10.3 we made the general claim that whenever these conditions apply to a boundary value problem, its well-posedness can be established. Let us check this statement in the present situation. Problem (11.1.1)—(11.1.2) can be formulated in a weak (or variational) form which involves the Chebyshev weight as follows: One looks for a function $u \in H^1_{w,0}(\Omega)$ such that

$$
\int_{\Omega} \nabla u \cdot \nabla(vw)\, dx = \int_{\Omega} fvw\, dx \qquad \text{for all } v \in H^1_{w,0}(\Omega). \tag{11.1.18}
$$

(The data f is assumed to belong to $L^2_w(\Omega)$). By Theorem 11.1, we can apply the Lax-Milgram Theorem (see (A.5)) to this problem; this assures the existence of a unique solution. Finally, one can prove (this is technical) that the solution not only belongs to $H^1_{w,0}(\Omega)$, but indeed it is more regular, i.e.,

$u \in H^2_w(\Omega)$ (Bernardi and Maday (1986b)). Thus, given an arbitrary data $f \in L^2_w(\Omega)$, there exists a unique solution in $D_B(L)$ to the problem (11.1.1)—(11.1.2).

Let us now consider the numerical approximation of this problem by Chebyshev methods. The Galerkin method is proven to be stable in the norm of $H^1_{w,0}(\Omega)$ as a direct consequence of Theorem 11.1. Here, we apply it to functions u and v, which are polynomials of degree N in each space variable and which vanish on the boundary (i.e., $u, v \in \mathbb{P}^0_N(\Omega)$). Theorem 11.1 ensures that the assumptions (10.4.5) and (10.4.6) are satisfied. The corresponding convergence estimate, based on (10.4.10) and the approximation estimates (9.5.16) or (9.7.24), reads as follows:

$$\|u - u^N\|_{H^1_w(\Omega)} \leqslant CN^{1-m}\|u\|_{H^m_w(\Omega)} \qquad \text{for } m \geqslant 1. \tag{11.1.19}$$

The two-dimensional case has been considered in Example 2 of Sec. 10.4.1.

The stability of the collocation method which uses the Gauss-Lobatto points (2.4.14) for the Chebyshev weight in each space direction, follows from a specific version of Theorem 11.1. In dimension one, the stability is actually a direct consequence of Theorem 11.1, since

$$\sum_{j=0}^{N} u_{xx}(x_j)u(x_j)w_j = \int_{-1}^{1} u_{xx}uw\,dx \qquad \text{for all } u \in \mathbb{P}_N.$$

Thus, condition (10.4.51) is fulfilled. This result has been applied in the second example of Sec. 10.1. Let us now detail the stability analysis in dimension two. The collocation solution of (11.1.1)—(11.1.2) is a polynomial $u^N \in \mathbb{P}^0_N(\Omega)$ (Ω is the square $(-1,1) \times (-1,1)$) satisfying

$$-\Delta u^N = f \qquad \text{at } \mathbf{x}_{ij} \qquad \text{for } 1 \leqslant i,j \leqslant N \tag{11.1.20}$$

where

$$\mathbf{x}_{ij} = \left(\cos\frac{i\pi}{N}, \cos\frac{j\pi}{N}\right) \qquad 0 \leqslant i,j \leqslant N. \tag{11.1.21}$$

Setting

$$(u,v)_N = \sum_{i,j=0}^{N} u(\mathbf{x}_{ij})v(\mathbf{x}_{ij})w_iw_j \tag{11.1.22}$$

let us define the bilinear form on $\mathbb{P}^0_N(\Omega) \times \mathbb{P}^0_N(\Omega)$

$$a_N(u,v) = -(\Delta u, v)_N. \tag{11.1.23}$$

Then, (11.1.20) is equivalent to the variational equations

$$a_N(u^N, v) = (f, v)_N \qquad \text{for all } v \in \mathbb{P}^0_N(\Omega). \tag{11.1.24}$$

Using the exactness of the Gauss–Lobatto formula and integration-by-parts one gets the identity

$$a_N(u,v) = (\nabla u, \nabla(vw)w^{-1})_N \tag{11.1.25}$$

for all u and $v \in \mathbb{P}_N^0(\Omega)$. Note that $\nabla(vw)w^{-1} \in \mathbb{P}_N(\Omega)^2$. Although $a_N(u,v)$ does not equal $a(u,v)$ (the form defined in (11.1.8)) for all u, $v \in \mathbb{P}_N^0(\Omega)$, it nonetheless retains the same continuity and coercivity properties of the form $a(u,v)$. Precisely, the following result has been proved by Funaro (1981).

Theorem 11.2.

(i) *There exists a constant* $\tilde{\beta} > 0$ *independent of* N *such that for all* $u, v \in \mathbb{P}_N^0(\Omega)$

$$|a_N(u,v)| \leqslant \tilde{\beta} \|u\|_{H^1_{w,0}(\Omega)} \|v\|_{H^1_{w,0}(\Omega)}; \tag{11.1.26}$$

(ii) *there exists a constant* $\tilde{\alpha} > 0$ *independent of* N *such that for all* $u \in \mathbb{P}_N^0(\Omega)$

$$\tilde{\alpha} \|u\|^2_{H^1_{w,0}(\Omega)} \leqslant a_N(u,u). \tag{11.1.27}$$

It follows that the stability condition (10.4.51) is satisfied with $E = H^1_{w,0}(\Omega)$. Thus, the energy method of Sec. 10.4.3 can be applied to obtain the stability and the convergence of the scheme (11.1.20) (see (10.4.52) and (10.4.54)). Moreover, the approximation results of Sec. 9.5 and 9.7 yield the following error estimate:

$$\|u - u^N\|_{H^1_w(\Omega)} \leqslant CN^{1-m}\{\|u\|_{H^m_w(\Omega)} + \|f\|_{H^{m-1}_w(\Omega)}\} \tag{11.1.28}$$

provided $m > 2$.

11.1.3. Other Boundary Value Problems

So far we have discussed the Dirichlet boundary value problem for the Poisson equation. The analysis can be extended to cover other boundary conditions (such as Neumann or Robin conditions) as well as more general second-order elliptic operators.

In most cases, the "energy method" corresponding to the stability condition (10.4.5) or (10.4.51) turns out to be inadequate, and one has to resort to the more general coercivity condition of the type (10.4.24) or (10.4.58).

Examples of analysis for different elliptic boundary value problems have been given throughout Chap. 10. Example 6 of Sec. 10.4.3 contains a discussion of the Dirichlet boundary value problem for a second-order elliptic operator in dimension one, with variable coefficients in the higher order term. The Neumann problem with the direct treatment of the boundary conditions (see Sec. 3.2) is considered in the subsequent Example 7, whereas the indirect treatment of Neumann or Robin conditions is analyzed in Example 4 of Sec. 10.5.1. All the previous examples concern collocation methods. A Chebyshev tau approximation to the one-dimensional Neumann problem is analyzed in Example 4 of Sec. 10.4.2.

11.2. Advection-Diffusion Equation

In this section we discuss the approximation by spectral methods of the one-dimensional, advection-diffusion problem:

$$\begin{cases} -\nu u_{xx} + pu_x + qu = f & -1 < x < 1, \quad \nu > 0 \\ u(-1) = u(1) = 0. \end{cases} \tag{11.2.1}$$

We shall consider both the linear case, in which $p = p(x)$ and $q = q(x)$ are two given functions, and the non-linear case in which $p = u$ and $q = 0$ (steady, viscous Burgers equation).

We shall confine ourselves to the analysis of Chebyshev approximations to (11.2.1), for the approximations based on other expansions (Fourier, Legendre) are easier to analyze.

11.2.1. Linear Advection-Diffusion Equation

We assume that p and q are some given functions, which depend smoothly upon x in the interval $[-1, 1]$.

In some cases, the analysis of Chebyshev spectral approximations to (11.2.1) can be carried out by directly applying the general theory developed in Sec. 10.3. This occurs whenever the differential operator

$$Lu = -\nu u_{xx} + pu_x + qu \tag{11.2.2}$$

satisfies the coercivity assumption (10.3.4) and its spectral approximations satisfy the corresponding conditions (10.4.5) or (10.4.51).

Let us introduce the bilinear form

$$b(u, v) = \nu \int_{-1}^{1} u_x(vw)_x \, dx + \int_{-1}^{1} pu_x vw \, dx + \int_{-1}^{1} quvw \, dx, \tag{11.2.3}$$

which is defined on the product $V \times V$, where $V = H^1_{w,0}(-1, 1)$. Note that the right-hand side of (11.2.3) has been obtained by taking the L^2_w-inner product of (11.2.2) with a test function $v \in V$ and using integration-by-parts. Condition (10.3.4) for L is satisfied if the form $b(u, v)$ is coercive on V, i.e., if there exists a constant $\alpha > 0$ such that

$$b(u, u) \geqslant \alpha \|u\|^2_{H^1_{w,0}(-1,1)} \qquad \text{for all } u \in H^1_{w,0}(-1, 1). \tag{11.2.4}$$

This occurs, for instance, in either of the two following cases:

(i) $\quad p^2(x) \leqslant \nu/\beta$ *and* $q(x) \geqslant 1/2$, *for a* $\beta > 2$ *and* $-1 < x < 1$.

Indeed, by the Cauchy-Schwarz inequality (A.2) it follows:

$$\int_{-1}^{1} pu_x uw\,dx \leqslant \|pu_x\|_{L^2_w(-1,1)}\|u\|_{L^2_w(-1,1)} \leqslant \frac{\nu}{2\beta}\|u_x\|^2_{L^2_w(-1,1)} + \frac{1}{2}\|u\|^2_{L^2_w(-1,1)}.$$

Then, using (11.1.14) gives

$$b(u,u) \geqslant \nu\left(\frac{1}{4} - \frac{1}{2\beta}\right)\|u_x\|^2_{L^2_w(-1,1)} + \int_{-1}^{1}\left(q(x) - \frac{1}{2}\right)u^2(x)w(x)\,dx$$

$$\geqslant \eta\nu\gamma^2\|u\|^2_{H^1_w(-1,1)},$$

where $\eta = (1/4) - (1/2\beta)$ is a positive constant, and γ is the constant in the Poincaré inequality (A.13.2).

(ii) $\qquad (pw)_x \leqslant 2qw \qquad$ *for* $-1 < x < 1$.

Indeed, noting that, by partial integration

$$\int_{-1}^{1} pu_x uw\,dx = -\frac{1}{2}\int_{-1}^{1}\frac{(pw)_x}{w}u^2 w\,dx,$$

we have, using again (11.1.14)

$$b(u,u) \geqslant \frac{\nu}{4}\|u_x\|^2_{L^2_w(-1,1)} + \int_{-1}^{1}\left(q - \frac{1}{2}\frac{(pw)_x}{w}\right)u^2 w\,dx.$$

Therefore, (11.2.4) holds with $\alpha = \nu(\gamma/2)^2$ if (ii) is satisfied.

Whenever inequality (11.2.4) holds, the Chebyshev Galerkin approximation to (11.2.1) is stable and convergent according to the energy method discussed in Sec. 10.4.1. The same result can be established for the Chebyshev collocation method, provided that the bilinear form $b_N(u,v)$ obtained from $b(u,v)$ by replacing exact sums by discrete sums at the collocation points, satisfies an estimate like (11.2.4). This occurs, for instance, if p and q are constant functions satisfying (i).

It is noteworthy that, even if condition (11.2.4) is not fulfilled, a transformation of problem (11.2.1) into an equivalent one whose associated bilinear form is positive is always possible. Assume for simplicity that $q \equiv 0$. Then, let $P(x)$ be a primitive of $p(x)$, and set $h(x) = \exp(-(1/\nu)P(x))$. From (11.2.1) we deduce that

$$-\nu h u_{xx} + hpu_x = hf.$$

Since $hp = -\nu h_x$, it follows that (11.2.1) can be formulated equivalently as follows:

$$\begin{cases} -\nu(hu_x)_x = g & -1 < x < 1 \\ u(-1) = u(1) = 0. \end{cases} \tag{11.2.5}$$

Here, we have set $g(x) = h(x)f(x)$. Since $h(x)$ is a positive function, problem (11.2.5) is of the same type as the one considered in Example 6 of Sec. 10.4.3. Therefore, spectral approximations to (11.2.5) can be analyzed as discussed there.

The reformulation of (11.2.1) as (11.2.5) is of more than theoretical interest. We noted in Chap. 5 that effective preconditionings are unavailable for problems with large first-derivative terms. On the other hand, iterative solutions of self-adjoint equations such as (11.2.5) are fairly straightforward.

The most general approach in the analysis of spectral approximations to problem (11.2.1) is based on a compactness argument, which exploits the fact that from a mathematical point of view, the leading term in (11.2.1) is the second-order, diffusion term. Thus, for a fixed $\nu > 0$ and for $N \to \infty$, the error produced by the spectral approximation to the first- and zero-order terms is negligible with respect to the error arising from the second-order term. This latter error can be estimated as shown in the previous section. A very simple example of the application of this technique is provided by Example 4 of Sec. 10.4.2. The detailed analysis for the Chebyshev approximation to (11.2.1), even in the presence of variable viscosity, has been given by Canuto and Quarteroni (1984).

The compactness argument alluded to here is essentially equivalent to applying to the linear problem (11.2.1) the implicit function theorem method described in the next subsection for the Burgers equation.

The methods of analysis described so far require that the diffusion parameter ν be large with respect to either the lower order coefficients or the inverse of the discretization parameter N. An analysis which avoids such a restriction has been provided by Canuto (1987) for the model boundary-layer problem

$$\begin{cases} -\nu u_{xx} + \mathscr{L}u = 0 \qquad -1 < x < 1 \quad \nu > 0 \\ u(-1) = 0,\ u(1) = 1 \end{cases} \tag{11.2.6}$$

where $\mathscr{L}u = u$ or $\mathscr{L}u = u_x$. We recall that the solution of problems like (11.2.6) with $\mathscr{L}u = u$ is part of the Kleiser–Schumann method for the spectral simulation of an incompressible flow in a channel (see Sec. 7.3.1). Since the differential operator in (11.2.6) has constant coefficients, the Chebyshev methods of Galerkin, tau or collocation type for problem (11.2.6) can be investigated by the error equation technique described in Sec. 10.6. If $u^N = \sum_{k=0}^{N} \hat{u}_k T_k(x)$ denotes any such spectral solution to (11.2.6), the analysis shows that for all $\nu > 0$ and all $N > 0$ the Chebyshev coefficients of u^N satisfy the bounds

$$0 < \hat{u}_k < \tfrac{1}{2} \qquad 0 \leqslant k \leqslant N \tag{11.2.7}$$

($k = 0$ may be an exception if $\mathscr{L}u = u_x$, N is even and ν is sufficiently small). Inequality (11.2.7) can be viewed as a sort of "maximum principle" in transform space, in the sense that all the Chebyshev coefficients of u^N are strictly positive. Note that the usual maximum principle in physical space does not hold for the spectral solutions to (11.2.6), as shown by the onset of a Gibbs phenomenon near $x = 1$ when ν becomes small compared with N^{-1}.

An important implication of (11.2.7) is that u^N is uniformly bounded in the interval $[-1, 1]$, *independently of N and ν*. In fact,

$$|u^N(x)| \leqslant \sum_{k=0}^{N} \hat{u}_k |T_k(x)| \leqslant \sum_{k=0}^{N} \hat{u}_k T_k(1) = u^N(1) = 1.$$

Thus, the spectral solutions, although possibly highly oscillatory, are stable in the maximum norm. Another consequence of (11.2.7) concerns the limit behavior of u^N as $\nu \to 0$ and $N \to \infty$. It can be shown that the maximum error $\|u - u^N\|_{L^\infty(-1,1)}$ between the exact solution u of (11.2.6) and any spectral approximation u^N satisfies an estimate of the form

$$\|u - u^N\|_{L^\infty(-1,1)} \leqslant \frac{C}{\nu N^4} \tag{11.2.8}$$

for the Helmholtz equation ($\mathscr{L}u = u$), and of the form

$$\|u - u^N\|_{L^\infty(-1,1)} \leqslant \frac{C}{\nu N^2} \tag{11.2.9}$$

for the advection-diffusion equation ($\mathscr{L}u = u_x$). Here $C > 0$ is a constant independent of ν and N. This proves that any spectral Chebyshev method is capable of accurately resolving a boundary layer with a number of modes which is inversely proportional to the square root of the boundary-layer width.

11.2.2. Steady Burgers Equation

We consider here the non-linear problem

$$\begin{cases} -\nu u_{xx} + u u_x = f & -1 < x < 1. \\ u(-1) = u(1) = 0 \end{cases} \tag{11.2.10}$$

We intend to show that Chebyshev (Galerkin and collocation) approximations to this problem are stable and convergent for all positive values of ν. This is the simplest example of the rigorous results that have been obtained for non-linear problems. We choose to outline the analysis in the general framework that has been used for more difficult non-linear problems such as the Navier–Stokes equations.

We assume that $f \in L^2_w(-1,1)$, where w is, as usual, the Chebyshev weight. Let $a(u,v)$ denote again the bilinear form (11.1.11) defined on the product space $H^1_{w,0}(-1,1) \times H^1_{w,0}(-1,1)$ and associated with the second-derivative operator with Dirichlet boundary conditions. Moreover, let us set

$$\lambda = \nu^{-1} \qquad G(\lambda, u) = \lambda(u u_x - f). \tag{11.2.11}$$

Each $u \in H^1_{w,0}(-1,1)$ is bounded in $[-1,1]$ (see (A.11.a)); hence, $G(\lambda,u) \in L^2_w(-1,1)$. Thus, we can consider the following weak formulation of problem (11.2.10):

$$\begin{cases} u \in H^1_{w,0}(-1,1) & \text{such that} \\ a(u,v) + (G(\lambda,u),v)_w = 0 & \text{for all } v \in H^1_{w,0}(-1,1). \end{cases} \tag{11.2.12}$$

Here $(z, v)_w$ denotes the inner product in $L^2_w(-1, 1)$. For each positive λ and each $u \in H^1_{w,0}(-1, 1)$, the linear form $v \to (G(\lambda, u), v)_w$ is continuous on $H^1_{w,0}(-1, 1)$. Hence, we can think of $G(\lambda, u)$ as an element of the dual space $H^{-1}_w(-1, 1)$ of $H^1_{w,0}(-1, 1)$ (see (A.1.c)), so that $(G(\lambda, u), v)_w = \langle G(\lambda, u), v \rangle$ for all $v \in H^1_{w,0}(-1, 1)$. (The symbol $\langle \cdot, \cdot \rangle$ denotes the duality pairing between $H^{-1}_w(-1, 1)$ and $H^1_{w,0}(-1, 1)$.)

Let T: $H^{-1}_w(-1, 1) \to H^1_{w,0}(-1, 1)$ be the linear operator which associates to an element $g \in H^{-1}_w(-1, 1)$ the solution $Tg \in H^1_{w,0}(-1, 1)$ of the problem

$$a(Tg, v) = \langle g, v \rangle \qquad \text{for all } v \in H^1_{w,0}(-1, 1). \tag{11.2.13}$$

This problem has indeed a unique solution since the bilinear form $a(u, v)$ satisfies the assumptions of the Lax–Milgram Theorem (A.5), as shown in Sec. 11.1. It follows that problem (11.2.12) can be written equivalently in the form

$$\begin{cases} u \in H^1_{w,0}(-1, 1) & \text{such that} \\ F(\lambda, u) \equiv u + TG(\lambda, u) = 0. \end{cases} \tag{11.2.14}$$

Many non-linear problems depending upon a parameter can be formulated in a manner similar to (11.2.14). In general, the linear operator T acts between the dual space V' of a Banach space V (see (A.1)) and the Banach space V itself, i.e.,

$$T: V' \to V. \tag{11.2.15}$$

It represents the inverse of the linear part of the differential problem (for instance, the inverse of the Stokes operator in the steady incompressible Navier–Stokes equations). The operator G maps $\mathbb{R} \times V$ into the dual space V' in a continuously differentiable way,

$$G: \mathbb{R} \times V \to V', \tag{11.2.16}$$

and represents the non-linear part of the problem. The full problem can be written as a non-linear equation in V, in the form

$$\begin{cases} u(\lambda) \in V & \text{solution of} \\ F(\lambda, u(\lambda)) \equiv u(\lambda) + TG(\lambda, u(\lambda)) = 0. \end{cases} \tag{11.2.17}$$

Here we have stressed the dependence of the solution upon the parameter λ, which is usually restricted to vary in a closed, bounded interval Λ of the real line.

Let us make the technical assumption that there exists a Banach space $W \subset V'$ such that

$$G(\lambda, u) \textit{ is a continuous mapping from } \mathbb{R}^+ \times V \textit{ into } W, \tag{11.2.18}$$

and

$$T \textit{ is a compact operator } \text{(see (A.3))} \textit{ from } W \textit{ into } V. \tag{11.2.19}$$

For the Burgers problem (11.2.10), these hypotheses are fulfilled, for instance, with the choice $W = L^2_w(-1, 1)$. In fact, if $g \in L^2_w(-1, 1)$, the solution $\xi = Tg$ of (11.2.13) (i.e., the solution of the boundary value problem $-\xi_{xx} = g$ in $-1 < x < 1$, with $\xi(-1) = \xi(1) = 0$) belongs to $H^2_w(-1, 1)$, which is compactly imbedded in $H^1_w(-1, 1)$.

We shall confine our analysis to the case of a *non-singular branch* of solutions $\{(\lambda, u(\lambda)): \lambda \in \Lambda\}$ of (11.2.17), i.e., to a branch of solutions along which the Fréchet derivative (see (A.4)) $D_uF(\lambda, u)$ of the map F with respect to the variable u is invertible. More precisely, we assume that there exists a positive constant $\alpha > 0$ such that

$$\|v + TD_uG(\lambda, u(\lambda))v\|_V \geqslant \alpha\|v\|_V \text{ for all } v \in V \text{ and all } \lambda \in \Lambda. \quad (11.2.20)$$

Here the symbol $D_uG(\lambda_0, u_0)$ denotes the Fréchet derivative of $G(\lambda, u)$ with respect to the variable u, computed at the point (λ_0, u_0). For problem (11.2.10), condition (11.2.20) amounts to the requirement that for all $g \in H^{-1}_w(-1, 1)$ and all $\lambda \in \Lambda$, the problem

$$\begin{cases} v \in H^1_{w,0}(-1, 1) \\ -v_{xx} + \lambda(u(\lambda)v_x + u_x(\lambda)v) = g \end{cases} \quad (11.2.21)$$

has a unique solution satisfying the inequality

$$\|v\|_{H^1_{w,0}(-1,1)} \leqslant C\|g\|_{H^{-1}_w(-1,1)}.$$

We are going now to introduce a general approximation to any problem which can be written in the form (11.2.17), provided the assumptions (11.2.18)—(11.2.20) are satisfied. Further, we shall state a general theorem to be used for the analysis of stability and convergence of such approximations. As a particular case, this theorem will be used to infer stability and convergence of both Galerkin and collocation Chebyshev approximations to the Burgers problem (11.2.10), which was previously written in the form (11.2.14).

For any integer N, let V_N be a finite-dimensional subspace of V, and let $G_N: \mathbb{R}^+ \times V_N \to V'$ be a suitable approximation to G. Further, let $T_N: V' \to V_N$ be a linear operator which approximates T. The following is a finite-dimensional approximation to problem (11.2.17):

$$\begin{cases} u^N(\lambda) \in V_N \qquad \text{solution of} \\ F_N(\lambda, u^N(\lambda)) \equiv u^N(\lambda) + T_NG_N(\lambda, u^N(\lambda)) = 0. \end{cases} \quad (11.2.22)$$

The next theorem, due to Maday and Quarteroni (1982a), is concerned with the convergence of the discrete solutions $\{(\lambda, u^N(\lambda)), \lambda \in \Lambda\}$ (problem (11.2.22)) to the non-singular branch of the exact solutions $\{(\lambda, u(\lambda)), \lambda \in \Lambda\}$ (problem (11.2.17)).

Theorem 11.3. *Assume that* (11.2.18)—(11.2.20) *hold. Moreover, assume that for some integer* $m \geqslant 2$, $G: \Lambda \times V \to W$ *is a* C^m *mapping, and* D^mG *is bounded* *over any bounded subset of* $\Lambda \times V$. *Concerning the discrete problem, we assume that*

$$\lim_{N\to\infty} \|T - T_N\|_{\mathscr{L}(W,V)} = 0. \tag{11.2.23}$$

(See (A.3) for the definition of the norm of a linear operator.) *About* G_N, *we assume that it is a* C^m *mapping from* $\Lambda \times V_N \to V'$, *and that there exists a positive function* K: $\mathbb{R}^+ \to \mathbb{R}^+$ *such that*:

$$\|D^l G_N[\lambda, v]\|_{\mathscr{L}_l(\Lambda\times V_N, W)} \leqslant K(|\lambda| + \|v\|_V) \qquad l = 1, \ldots, m. \tag{11.2.24}$$

(See again (A.3) for the definition of the norm of a multilinear operator.) *Further, we assume that there exists a projection operator* Π_N: $V \to V_N$ *satisfying*

$$\lim_{N\to\infty} \|v - \Pi_N v\|_V = 0 \qquad \text{for all } v \in V, \tag{11.2.25}$$

and such that

$$\lim_{N\to\infty} \sup_{\lambda\in\Lambda} \|D_u(G - G_N)[\lambda, \Pi_N u(\lambda)]\|_{\mathscr{L}(V_N, V')} = 0. \tag{11.2.26}$$

Then there exists a neighborhood Θ *of the origin in* V, *and for* N *large enough, a unique* C^m *mapping* $\lambda \in \Lambda \to u^N(\lambda) \in V_N$ *such that for all* $\lambda \in \Lambda$

$$F_N(\lambda, u^N(\lambda)) = 0 \qquad u^N(\lambda) - u(\lambda) \in \Theta, \tag{11.2.27}$$

and the following estimate holds

$$\begin{aligned}\|u(\lambda) - u^N(\lambda)\|_V \leqslant C(&\|u(\lambda) - \Pi_N u(\lambda)\|_V + \|(T - T_N)G(\lambda, u(\lambda)\|_V \\ &+ \|T_N(G - G_N)(\lambda, \Pi_N u(\lambda))\|_V),\end{aligned} \tag{11.2.28}$$

with a positive constant C *independent of* λ *and* N.

A qualitative interpretation of this theorem is in order. There are several assumptions on the approximations to the linear and non-linear components of the problem. Assumption (11.2.25) means that V is well-approximated by the sequence of subspaces V_N, and (11.2.23) means that the linear operator T is well-approximated by the sequence of operators T_N. Naturally enough, stricter requirements are placed on the approximation to the non-linear operator G. Assumption (11.2.24) means that the derivatives of G_N up to order m are locally Lipschitz continuous, and (11.2.26) states that the Fréchét derivative of G_N approximates that of G as $N \to \infty$.

The first conclusion, (11.2.27), is that, for fixed N, there is a unique branch of non-singular solutions and that these solutions are bounded independent of N. Finally, inequality (11.2.28) exhibits the dependence of the error on the approximation properties of Π_N, T_N and G_N.

Chebyshev Galerkin Approximation

We return now to problem (11.2.10), and its equivalent formulation (11.2.14), with G and T defined in (11.2.11), (11.2.13). For any $\lambda \in \mathbb{R}^+$, we look for a polynomial $u^N(\lambda) \in V_N = \{v \in \mathbb{P}_N(-1,1) | v(\pm 1) = 0\}$ which satisfies

$$a(u^N(\lambda), v) + (G(\lambda, u^N(\lambda)), v)_w = 0 \qquad \text{for all } v \in V_N. \tag{11.2.29}$$

This is a Chebyshev Galerkin approximation.

We define the operator T_N: $V' \to V_N$ by

$$a(T_N g, v) = \langle g, v \rangle \qquad \text{for all } v \in V_N. \tag{11.2.30}$$

Then it follows that $T_N = \Pi_N T$, where Π_N: $V \to V_N$ is the operator defined in (9.5.18), namely:

$$a(\Pi_N u - u, v) = 0 \qquad \text{for all } v \in V_N. \tag{11.2.31}$$

Owing to (11.2.30), the Chebyshev Galerkin approximation to (11.2.10) can be restated as follows:

$$\begin{cases} u^N(\lambda) \in V_N \qquad \text{solution of} \\ F_N(\lambda, u^N(\lambda)) \equiv u^N(\lambda) + T_N G(\lambda, u^N(\lambda)) = 0. \end{cases} \tag{11.2.32}$$

This is precisely the form (11.2.22); in the current situation, however, $G_N \equiv G$. To apply Theorem 11.3, we need to check that the assumptions (11.2.23)—(11.2.26) are fulfilled.

Property (11.2.25) follows from the fact that each function $v \in H^1_{w,0}(-1,1)$ can be approximated in the norm of $H^1_w(-1,1)$ by a sequence of more regular functions $\tilde{v}_n \in H^m_w(-1,1) \cap H^1_{w,0}(-1,1)$, with $m > 1$. Then one applies to each such $\tilde{v}_n$ the convergence estimate (11.1.19) for the Chebyshev Galerkin approximation (where $\tilde{v}_n^N$ is indeed $\Pi_N \tilde{v}_n$). In order to check (11.2.23), let us choose $W = L^2_w(-1,1)$. Recalling that $T_N = \Pi_N T$ we have

$$\|T - T_N\|_{\mathscr{L}(W,V)} = \sup_{g \in L^2_w(-1,1)} \frac{\|Tg - \Pi_N Tg\|_{H^1_w(-1,1)}}{\|g\|_{L^2_w(-1,1)}}.$$

Using again (11.1.19) and the definition of the operator T, we have

$$\|Tg - \Pi_N Tg\|_{H^1_w(-1,1)} \leqslant CN^{-1} \|Tg\|_{H^2_w(-1,1)} \leqslant C' N^{-1} \|g\|_{L^2_w(-1,1)},$$

whence (11.2.23) follows. Moreover, both (11.2.24) and (11.2.26) are trivially verified for all integers $m \geqslant 0$.

By (11.2.27) and (11.2.28) we conclude that for any branch $\{(\lambda, u(\lambda)), \lambda \in \Lambda\}$, $\Lambda \subset \mathbb{R}^+$, of non-singular solutions of (11.2.10), there exists a C^∞ mapping: $\lambda \in \Lambda \to u^N(\lambda) \in V_N$, such that $u^N(\lambda)$ is the only solution of the Chebyshev Galerkin approximation (11.2.29) in a neighborhood of $u(\lambda)$. Moreover, one has the estimate

$$\|u(\lambda) - u^N(\lambda)\|_{H^1_w(-1,1)} \leqslant C \|u(\lambda) - \Pi_N u(\lambda)\|_{H^1_w(-1,1)} \\ + \|TG(\lambda, u(\lambda)) - \Pi_N TG(\lambda, u(\lambda))\|_{H^1_w(-1,1)}.$$

Noting that from (11.2.14), $TG(\lambda, u(\lambda)) = -u(\lambda)$, and using again (11.1.19) we get the convergence estimate:

$$\|u(\lambda) - u^N(\lambda)\|_{H^1_w(-1,1)} \leqslant CN^{1-m}\|u(\lambda)\|_{H^m_w(-1,1)} \qquad m \geqslant 1, \quad (11.2.33)$$

for a constant C which depends only upon the parameter interval Λ.

Chebyshev Collocation Approximation

Let $x_j = \cos(\pi j/N)$, $j = 0, \ldots, N$ be the Chebyshev–Gauss–Lobatto points (see (2.4.14)) and let $I_N v$ be the interpolant of v at these points (see Sec. 2.2.3). We look now for a polynomial $u^N = u^N(\lambda)$ of degree N which satisfies

$$\begin{cases} -u^N_{xx} + \lambda(\tfrac{1}{2}(I_N(u^N)^2)_x - f) = 0 & \text{at } x = x_j, \quad 1 \leqslant j \leqslant N-1 \\ u^N(x_0) = u^N(x_N) = 0. \end{cases} \qquad (11.2.34)$$

Introducing the discrete inner product $(u, v)_N$ associated with the Chebyshev points $\{x_j\}$ (see (2.2.24)), we can restate this collocation problem as follows:

$$\begin{cases} u^N \in V_N & \text{solution of} \\ a(u^N, v) + \lambda(\tfrac{1}{2}(I_N(u^N)^2)_x - f, v)_N = 0 & \text{for all } v \in V_N. \end{cases} \qquad (11.2.35)$$

We have used (2.2.25) to replace $-(u^N_{xx}, v)_N$ by $a(u^N, v)$. We define the operator G_N: $\mathbb{R}^+ \times V_N \to V'$ by setting

$$\langle G_N(\lambda, v), \phi\rangle = \lambda(\tfrac{1}{2}(I_N(v^2))_x - f, \phi)_N \qquad \text{for all } \phi \in V.$$

Note that again by (2.2.25) we have

$$\langle G_N(\lambda, v), \phi\rangle = \lambda\{(\tfrac{1}{2}(I_N(u^N)^2)_x, \phi)_w - (f, \phi)_N\} \qquad \text{for all } \phi \in V_N.$$

If we define T_N: $V' \to V_N$ as in (11.2.30), then problem (11.2.34) fits into the general form (11.2.22).

The assumptions of Theorem 11.3 can be checked by very technical arguments, which will not be reported here. The interested reader can refer to the paper by Maday and Quarteroni (1982a). The conclusion of the analysis is that there exists a C^∞-mapping $\lambda \in \Lambda \to u^N(\lambda) \in V_N$ such that $u^N(\lambda)$ is the only solution of the Chebyshev collocation approximation (11.2.34) in a neighborhood of $u(\lambda)$, and such that the error estimate (11.2.28) holds.

Let us briefly work out this estimate in our particular case. The first two terms on the right-hand side can be handled as they were for the Chebyshev Galerkin approximation; we concentrate on the last term. For the sake of simplicity, we drop the dependence of u on λ. Moreover, let us set $\phi = \Pi_N u$ and $\psi = T_N(G - G_N)(\lambda, \Pi_N u)$. By (11.2.30) and (11.1.14) we have

$$\begin{aligned} \gamma\|\psi\|^2_{H^1_w(-1,1)} &\leqslant a(\psi, \psi) = \langle (G - G_N)(\lambda, \phi), \psi\rangle \\ &= \frac{\lambda}{2}([\phi^2 - I_N(\phi^2)]_x, \psi)_w + \lambda[(f, \psi) - (f, \psi)_N]. \end{aligned} \qquad (11.2.36)$$

Integrating by parts and using the Cauchy–Schwarz inequality together with inequality (11.1.12) yields

$$|([\phi^2 - I_N(\phi^2)]_x, \psi)_w| = \left| \int_{-1}^{1} [\phi^2 - I_N(\phi^2)](\psi w)_x \, dx \right|$$
$$\leqslant C \|\phi^2 - I_N(\phi^2)\|_{L^2_w(-1,1)} \|\psi\|_{H^1_w(-1,1)}.$$

Now, by the triangle inequality

$$\|(I - I_N)(\phi^2)\|_{L^2_w(-1,1)} \leqslant \|(I - I_N)(u^2)\|_{L^2_w(-1,1)} + \|(I - I_N)(u^2 - \phi^2)\|_{L^2_w(-1,1)}.$$

Assuming that $f \in H^{m-1}_w(-1,1)$ for some $m \geqslant 2$, it is easily seen using equation (11.2.10) that $u \in H^m_w(-1,1)$. Thus, by (9.5.20) we have

$$\|(I - I_N)(u^2)\|_{L^2_w(-1,1)} \leqslant CN^{-m} \|u\|^2_{H^m_w(-1,1)},$$

while again by (9.5.20) and the estimate (11.1.19) for $u^N = \Pi_N u = \phi$ it follows

$$\|(I - I_N)(u^2 - \phi^2)\|_{L^2_w(-1,1)} \leqslant C_1 N^{-1} \|u^2 - \phi^2\|_{H^1_w(-1,1)}$$
$$\leqslant C_1 N^{-1} \|u + \Pi_N u\|_{H^1_w(-1,1)} \|u - \Pi_N u\|_{H^1_w(-1,1)}$$
$$\leqslant C_2 N^{-m} \|u\|_{H^1_w(-1,1)} \|u\|_{H^m_w(-1,1)}.$$

Finally, the error on the forcing term in (11.2.36) can be handled as shown in Sec. 9.3 (see formula (9.3.5)), to give

$$|(f, \psi)_w - (f, \psi)_N| \leqslant CN^{1-m} \|f\|_{H^{m-1}_w(-1,1)} \|\psi\|_{L^2_w(-1,1)}.$$

The final result of the convergence analysis, here briefly sketched, is the following error estimate for the Chebyshev collocation approximation (11.2.34):

$$\|u(\lambda) - u^N(\lambda)\|_{H^1_w(-1,1)} \leqslant CN^{1-m} \{ \|u(\lambda)\|^2_{H^m_w(-1,1)} + \|f\|_{H^{m-1}_w(-1,1)} \}, \quad (11.2.37)$$

for a constant C which depends only upon the parameter interval Λ.

11.3. Navier–Stokes Equations

The mathematical analysis of spectral methods for the incompressible Navier–Stokes equations commenced relatively recently. So far, the analysis has been confined to time-independent problems. The main purpose of the mathematical investigation is to prove the convergence of the approximate polynomial solution to the exact solution of the Navier–Stokes equations, as the degree of the polynomials tends to infinity. An isolated (or non-singular) solution is considered in most cases, but the analysis can be extended to cover turning points and bifurcations as well. Error estimates in Sobolev norms are obtained, and they show that the error decays spectrally for infinitely smooth solutions.

Most of the existing algorithms of spectral type deal with the Navier–Stokes equations in the primitive-variable formulation. In this case, the velocity and the pressure cannot be approximated independently: a compatibility condition must be satisfied by the finite-dimensional spaces in which they are sought in order to have a solvable system. This is a well-known issue in the finite-element method (see, e.g., Girault and Raviart (1986)), and was discussed briefly in Chap. 7. In those spectral methods for which the continuity equation is discretized directly, the pressure can be affected by spurious components (parasitic modes) which deteriorate the accuracy of the method (see the following Sec. 11.3.1). Parasitic modes can be completely characterized mathematically, and the theoretical analysis can even indicate a way to filter them out (see Sec. 11.3.2). Other spectral methods, like the Kleiser–Schumann method described in Sec. 7.3.1 or the Zang–Hussaini method described in Sec. 7.3.2, replace the continuity equation by an equation for the pressure. In these cases, parasitic modes may be implicitly filtered out by the solution process.

In this section we shall present the essential details of the mathematical theory of spectral methods for the incompressible Navier–Stokes equations, and shall review the existing literature on the subject.

We consider the steady Navier–Stokes equations in primitive variables in a square domain $\Omega \subset \mathbb{R}^2$ of the plane under periodic and/or homogeneous Dirichlet boundary conditions. Thus, Ω is either $(0, 2\pi)^2$ (if fully periodic boundary conditions are imposed) or $(0, 2\pi) \times (-1, 1)$ (periodic/Dirichlet boundary conditions) or $(-1, 1)^2$ (fully Dirichlet boundary conditions). There is no difficulty in extending our discussion to the three-dimensional case. The results in the literature are actually given for three-dimensional problems, but we choose here the two-dimensional geometry for ease of exposition.

The velocity $\mathbf{u}$ belongs to the subspace V of $(H^1(\Omega))^2$ of the vector fields which satisfy the prescribed boundary conditions. The space $H^1(\Omega)$ is defined in (A.11.a). V is a closed subspace under the norm of $(H^1(\Omega))^2$, which we denote by $\|\mathbf{v}\|_V$. The pressure p belongs to the closed subspace Q of $L^2(\Omega)$ of the functions with zero average in Ω. The norm $\|q\|_Q$ is the $L^2(\Omega)$ norm.

We have in mind to approximate the following problem:

$$\begin{cases} \mathbf{u} \in V \qquad p \in Q \\ -\nu\Delta\mathbf{u} + \nabla p + \mathbf{u}\cdot\nabla\mathbf{u} = \mathbf{f} & \text{in } \Omega \\ \nabla\cdot\mathbf{u} = 0 & \text{in } \Omega. \end{cases} \tag{11.3.1}$$

The external force $\mathbf{f}$ is supposed to be square integrable in Ω, or even continuous in $\bar{\Omega}$ (more generally $\mathbf{f}$ can be an element in the dual space V' of V (see (A.1.c)).

First, we observe that the analysis of an approximation to the full Navier–Stokes problem can be reduced to the analysis of the corresponding approximation to the Stokes problem, by resorting to an implicit function theorem or to more sophisticated functional analysis results. More precisely, consider the Stokes problem

$$\begin{cases} \mathbf{u} \in V \qquad p \in Q \\ -\nu\Delta\mathbf{u} + \nabla p = \mathbf{g} \qquad \text{in } \Omega \\ \nabla \cdot \mathbf{u} = 0 \qquad \text{in } \Omega, \end{cases} \tag{11.3.2}$$

and define the operator T: $V' \to V \times Q$ as $T\mathbf{g} = (\mathbf{u}, p)$. Then define the non-linear operator G: $V \times Q \to V'$ as $G(\mathbf{u}, p) = \mathbf{f} - \mathbf{u} \cdot \nabla\mathbf{u}$. Setting $\mathbf{w} = (\mathbf{u}, p)$ and $F(\mathbf{w}) = \mathbf{w} - TG(\mathbf{w})$, the solution $\mathbf{w}$ of (11.3.1) can be characterized by the equation

$$F(\mathbf{w}) = 0. \tag{11.3.3}$$

Consider now a spectral approximation of (11.3.1), in which the velocity $\mathbf{u}^N$ is sought in a subspace V_N of V, the pressure p^N in a subspace Q_N of Q, the Stokes operator T is approximated by an operator T_N: $V' \to V_N \times Q_N$ and the non-linear operator G by an operator G_N: $V_N \times Q_N \to V'$. Set $\mathbf{w}^N = (\mathbf{u}^N, p^N)$ and $F_N(\mathbf{w}^N) = \mathbf{w}^N - T_N G_N(\mathbf{w}^N)$. As before, the approximate solution is defined by the equation

$$F_N(\mathbf{w}^N) = 0. \tag{11.3.4}$$

Abstract theorems concerning the approximation of non-linear equations like (11.3.3) by a family of finite-dimensional equations like (11.3.4) can be invoked at this point (see, e.g., Brezzi, Rappaz and Raviart (1980, 1981a, 1981b), Descloux and Rappaz (1982), Crouzeix and Rappaz (1987)). One can even take into account in the formulation the dependence of the solution upon the Reynolds number $\lambda = \nu^{-1}$, i.e., replace $F(\mathbf{w})$ by $F(\lambda, \mathbf{w})$ in (11.3.3) and $F_N(\mathbf{w}^N)$ by $F_N(\lambda, \mathbf{w}^N)$ in (11.3.4). An example of such abstract results, in the case of isolated solutions, is Theorem 11.3 given in Sec. 11.2. There it is shown how to use it in the analysis of spectral approximations to the Burgers equation. The analysis for the Navier–Stokes equations follows the same guidelines, although it is technically more involved. In order to check the assumptions, one exploits the approximation properties of T_N to T, using the stability results sketched hereafter and the consistency results given in Chap. 9. Since G_N is typically the collocation realization of G, estimates on the error between G_N and G can be obtained easily from the approximation results for the Fourier or Chebyshev interpolation (see Chap. 9, Secs. 9.1.3 or 9.5.3). Thus, the crucial point is the analysis of the error in the Stokes approximation. Hence, from now on we will deal with the Stokes problem (11.3.2) only.

11.3.1. Compatibility Conditions Between Velocity and Pressure

In most of the spectral algorithms in current use, the space V_N of the discrete velocities is the subspace of V of the polynomials of degree $\leqslant N$ in each variable. Here by polynomial we mean trigonometric polynomial in the direction(s) of periodicity, and algebraic polynomial in the remaining direction(s). From now on, we assume that V_N is defined in this way. An exception in the literature, is represented by the method by Moin, Moser and Leonard (1983), where the discrete velocities individually satisfy the continuity equation; hence they span a proper subspace of our V_N.

With our choice of the subspace V_N for the velocities, the most natural candidate for the space Q_N of the pressures is the subspace $\mathscr{P}_N$ of Q of all the polynomials of degree $\leqslant N$ in each variable. However, such a space may be "too large" compared to the space V_N, in the sense that the approximation scheme may fail to define a unique pressure p_N in $\mathscr{P}_N$ (the corresponding algebraic system being underspecified). When this occurs, Q_N has to be restricted to a proper subspace of $\mathscr{P}_N$, or in other words, one has to satisfy a *compatibility condition* between the spaces of velocity and pressure.

In order to investigate the phenomenon, let us assume that the momentum equation in (11.3.2) is discretized by a projection method of Galerkin, tau or collocation type. All the existing algorithms treat the momentum equation by one of these methods. They differ primarily in the approximation of the continuity equation. For the sake of simplicity, we consider a Galerkin projection method with respect to the $L^2(\Omega)$ inner product $(\mathbf{u}, \mathbf{v})$. (Thus, we couple a Fourier Galerkin method in the direction(s) of periodicity with a Legendre Galerkin method in the other direction(s).) The discretization reads as follows:

$$\begin{cases} \mathbf{u}^N \in V_N \qquad p^N \in \mathscr{P}_N \\ -\nu(\Delta \mathbf{u}^N, \mathbf{v}) + (\nabla p^N, \mathbf{v}) = (\mathbf{g}, \mathbf{v}) \qquad \text{for all } \mathbf{v} \in V_N. \end{cases} \tag{11.3.5}$$

Now assume that there exist non-zero $p \in \mathscr{P}_N$ such that

$$(\nabla p, \mathbf{v}) = 0 \qquad \text{for all } \mathbf{v} \in V_N. \tag{11.3.6}$$

Then each couple $(\mathbf{u}^N, p^N + p)$ is also a solution of (11.3.5). Thus, the pressure is not uniquely determined by (11.3.5). The pressures $p \in \mathscr{P}_N$ which satisfy (11.3.6) are called "spurious modes" or "parasitic modes." They form a linear subspace $X_N \subset \mathscr{P}_N$. In order to characterize X_N, we have to distinguish among different combinations of boundary conditions.

(a) *Fully Periodic Problem*

The space of velocities V_N is $[S_N \cap Q]^2$, where S_N denotes the space of the trigonometric polynomials of degree $\leqslant N$ in each variable and Q is, as before, the space of the square integrable functions with zero average in Ω. (We choose this normalization for the velocities since their zero mode is not affected by equation (11.3.5).) The space of pressures $\mathscr{P}_N$ is $S_N \cap Q$. If $p \in \mathscr{P}_N$, then $\nabla p \in V_N$; hence choosing $\mathbf{v} = \nabla p$ in (11.3.6), one gets $\nabla p \equiv 0$; thus, $p = 0$. It follows that for pure Fourier methods there are no spurious modes (other than the physical one $p =$ constant), i.e., X_N is empty. The spaces V_N, $\mathscr{P}_N$ are precisely those used in the Orszag–Patterson method.

(b) *Mixed Periodic—Non-periodic Problem*

V_N is the space $[S_N \otimes \mathbb{P}_N^0]^2$, where S_N is now the space of the trigonometric polynomials of degree $\leqslant N$ in one variable, and $\mathbb{P}_N^0$ is the subspace of $\mathbb{P}_N$ of the algebraic polynomials which vanish at $y = \pm 1$. The space $\mathscr{P}_N$ is $[S_N \otimes \mathbb{P}_N] \cap Q$.

Let us represent $p \in \mathscr{P}_N$ as $p(x,y) = \sum_{|k|,n \leqslant N} \hat{p}_{kn} e^{ikx} L_n(y)$, where $L_n(y)$ denotes the n-th Legendre polynomial. Equation (11.3.6) is equivalent to

$$\left(p, \frac{\partial v}{\partial y}\right) = 0 \qquad \text{for all } v \in S_N \otimes \mathbb{P}_N^0, \tag{11.3.7}$$

$$\left(p, \frac{\partial v}{\partial x}\right) = 0 \qquad \text{for all } v \in S_N \otimes \mathbb{P}_N^0. \tag{11.3.8}$$

A basis in $S_N \otimes \mathbb{P}_N^0$ is given by

$$\{e^{ikx}(1-y^2)L'_m(y) \,|\, |k| \leqslant N,\ 1 \leqslant m \leqslant N-1\}. \tag{11.3.9}$$

Using this basis in (11.3.7) and taking into account the differential equation (2.3.1) satisfied by the Legendre polynomials, one gets the set of equations

$$\hat{p}_{km} = 0 \qquad \text{for } |k| \leqslant N, \quad 1 \leqslant m \leqslant N-1, \tag{11.3.10}$$

which are equivalent to (11.3.7). Another basis for $S_N \otimes \mathbb{P}_N^0$ is given by

$$\{e^{ikx}[L_m(y) - L_\alpha(y)] \,|\, |k| \leqslant N, \quad 2 \leqslant m \leqslant N\}, \tag{11.3.11}$$

where $\alpha = m \pmod 2$. Using this basis in (11.3.8) yields the set of equations

$$(L_m, L_m)\hat{p}_{km} = (L_\alpha, L_\alpha)\hat{p}_{k\alpha} \qquad 0 < |k| \leqslant N, \quad 2 \leqslant m \leqslant N. \tag{11.3.12}$$

By (11.3.10) and (11.3.12), it follows that

$$\hat{p}_{km} = 0 \qquad \text{for all } k, m, \text{ except } \hat{p}_{00} \text{ and } \hat{p}_{0N}. \tag{11.3.13}$$

We conclude that there exists one spurious mode, other than the physical one. Thus, $\dim X_N = 1$ and $X_N = \operatorname{span}\{L_N(y)\}$.

(c) *Fully Non-periodic Problem*

V_N is the space $[\mathbb{P}_N^0 \otimes \mathbb{P}_N^0]^2$, and $\mathscr{P}_N = [\mathbb{P}_N \otimes \mathbb{P}_N] \cap Q$. We proceed as in the previous case, using now as test function v the product of a basis function $(1-s^2)L'_m(s)$ in the direction of differentiation, and a basis function $L_m(s) - L_\alpha(s)$ in the other direction. It is easily seen that the non-trivial solutions of (11.3.6) are spanned by the four modes

$$L_i(x)L_j(y), \qquad i,j = 0 \text{ or } N, \tag{11.3.14}$$

and by four other functions $\bar{p}_{ij} (i,j = 0 \text{ or } 1)$, which are suitable combinations of the remaining modes with the same parity:

$$\bar{p}_{ij} = {}' \sum_{\substack{k=i \,(\mathrm{mod}\, 2) \\ m=j \,(\mathrm{mod}\, 2)}} \bar{c}_{ijkm} L_k(x) L_m(y). \tag{11.3.15}$$

Thus, we have seven spurious modes other than the physical one, i.e., $\dim X_N = 7$. Spurious pressure modes were characterized mathematically by Bernardi, Maday and Métivet (1986).

If a Galerkin Chebyshev method is used instead, the previous results hold after replacing each L_l by T_l, the l-th Chebyshev polynomial. If a Chebyshev collocation method is used with collocation points the Cartesian product of the Gauss–Lobatto knots $x_j = \cos(j\pi/N)$, $(j = 0, \ldots, N)$, then the spurious modes are spanned by the eight functions

$$\begin{cases} T_i(x)T_j(y), & i, j = 0 \text{ or } N; \text{ and} \\ T_N'(x)T_N'(y)(1 \pm x)(1 \pm y). \end{cases} \tag{11.3.16}$$

(See Bernardi, Canuto and Maday (1986).)

11.3.2. Direct Discretization of the Continuity Equation: The "inf-sup" Condition

The most direct way of enforcing the divergence-free condition on the numerical flow consists of approximating the continuity equation by a projection method of Galerkin, tau or collocation type. Again, we choose the Galerkin method in the following discussion for its formal simplicity. Similar results apply for collocation methods with only slightly changed notation.

We consider the following approximation of the Stokes problem (11.3.2)

$$\begin{cases} \mathbf{u}^N \in V_N, \quad p^N \in Q_N, \\ -\nu(\Delta \mathbf{u}^N, \mathbf{v}) + (\mathbf{v}, \nabla p^N) = (\mathbf{g}, \mathbf{v}), & \text{for all } \mathbf{v} \in V_N, \\ (\nabla \cdot \mathbf{u}^N, q) = 0 & \text{for all } q \in Q_N. \end{cases} \tag{11.3.17}$$

We recall that V_N is the space of the polynomial fields of degree $\leqslant N$ in each variable which satisfy the boundary conditions; Q_N denotes a suitable subspace of the space $\mathscr{P}_N$ of the polynomials of degree $\leqslant N$ in each variable, which have zero average over the domain Ω; finally, $(\mathbf{u}, \mathbf{v})$ denotes the $L^2(\Omega)$ inner product. The Galerkin method used in practice for fully-periodic problems (the Orszag–Patterson method described in Sec. 7.2) can actually be written as (11.3.17). For non-periodic problems, it is preferable to resort to the tau or the collocation method. The tau method can be written in the form (11.3.17), except that the test functions $\mathbf{v}$ for the momentum equation are polynomials of degree $N-2$ in the non-periodic directions, and need not satisfy the boundary conditions. The method by Moin and Kim (1980), but with a conventional tau discretization of the momentum equation (see Sec. 7.3.1) is precisely of this type. A collocation method can also be cast in the framework (11.3.17), provided that the $L^2(\Omega)$ inner product is replaced by a discrete inner product at the collocation points (see, e.g., Bernardi, Maday and Métivet (1987a, 1987b)). One can even use different grids for the two sets of equations (see Métivet (1987), Bernardi and Maday (1986a)): this results in two different discrete inner products to be inserted in (11.3.17).

In order to discuss the stability and convergence properties of the approximation (11.3.17), we make use of a general result on the approximation of saddle-point problems such as (11.3.17), known as the "inf-sup" condition (see Brezzi (1974)). To this end, let us define the bilinear forms

$$a\colon V \times V \to \mathbb{R} \qquad a(\mathbf{u}, \mathbf{v}) = \nu(\nabla \mathbf{u}, \nabla \mathbf{v}) \tag{11.3.18}$$

$$b\colon V \times Q \to \mathbb{R} \qquad b(\mathbf{v}, q) = -(\nabla \cdot \mathbf{v}, q). \tag{11.3.19}$$

These forms are continuous, in the sense that there exist two constants $\gamma > 0$ and $\delta > 0$ such that

$$|a(\mathbf{u}, \mathbf{v})| \leqslant \gamma \|\mathbf{u}\|_V \|\mathbf{v}\|_V \qquad \text{for all } \mathbf{u}, \mathbf{v} \in V \tag{11.3.20}$$

$$|b(\mathbf{v}, q)| \leqslant \delta \|\mathbf{v}\|_V \|q\|_Q \qquad \text{for all } \mathbf{v} \in V, \text{ for all } q \in Q. \tag{11.3.21}$$

With these definitions, problem (11.3.17) fits into the abstract form

$$\begin{cases} \mathbf{u}^N \in V_N \qquad p^N \in Q_N \\ a(\mathbf{u}^N, \mathbf{v}) + b(\mathbf{v}, p^N) = (\mathbf{g}, \mathbf{v}) \qquad \text{for all } \mathbf{v} \in V_N \\ b(\mathbf{u}^N, q) = 0 \qquad \text{for all } q \in Q_N. \end{cases} \tag{11.3.22}$$

According to Brezzi's theorem, this problem has a unique solution if the following two conditions are satisfied:

(i) *Setting*

$$Z_N = \{\mathbf{v} \in V_N \mid b(\mathbf{v}, q) = 0 \quad \text{for all } q \in Q_N\}, \tag{11.3.23}$$

there exists a constant $\alpha_N > 0$ *such that*

$$a(\mathbf{v}, \mathbf{v}) \geqslant \alpha_N \|\mathbf{v}\|_V^2 \qquad \text{for all } \mathbf{v} \in Z_N; \tag{11.3.24}$$

(ii) *there exists a constant* $\beta_N > 0$ *such that*

$$\sup_{\mathbf{v} \in V_N} \frac{b(\mathbf{v}, q)}{\|\mathbf{v}\|_V} \geqslant \beta_N \|q\|_Q \qquad \text{for all } q \in Q_N. \tag{11.3.25}$$

Note that if we represent the system of equations (11.3.22) in matrix form with respect to a basis in V_N and Q_N as

$$\left[\begin{array}{c|c} A & B \\ \hline B^T & 0 \end{array}\right] \left[\begin{array}{c} \tilde{u} \\ \hline \tilde{q} \end{array}\right] = \left[\begin{array}{c} \tilde{g} \\ \hline 0 \end{array}\right],$$

then condition (11.3.24) says that A is positive-definite, whereas condition (11.3.25) is equivalent to the fact that B has maximal rank.

Condition (11.3.24) can be weakened to require that A be just non-singular (see Brezzi (1974), or the subsequent condition (11.3.42)). The weaker condition together with (11.3.25) are not only sufficient but also necessary for the well-posedness of problem (11.3.22). Under the assumptions (11.3.24) and

(11.3.25) the following estimate on the solution of (11.3.22) holds:

$$\|\mathbf{u}^N\|_V \leqslant \frac{\gamma}{\alpha_N}\|\mathbf{g}\|_{V'}, \tag{11.3.26}$$

$$\|p^N\|_Q \leqslant \frac{1}{\beta_N}\left(1+\frac{\gamma^2}{\alpha_N}\right)\|\mathbf{g}\|_{V'}, \tag{11.3.27}$$

where $\|\mathbf{g}\|_{V'}$ is the norm of $\mathbf{g}$ in the dual space V' (see (A.1.c)). It follows that the approximation (11.3.22) is stable if the constants α_N and β_N introduced in (11.3.24)—(11.3.25) are independent of N.

Concerning the convergence, the following abstract error estimates hold (see Girault and Raviart (1986)):

$$\|\mathbf{u}-\mathbf{u}^N\|_V \leqslant \left(1+\frac{\gamma}{\alpha_N}\right)\inf_{\mathbf{w}\in Z_N}\|\mathbf{u}-\mathbf{w}\|_V + \frac{\delta}{\alpha_N}\inf_{q\in Q_N}\|p-q\|_Q \tag{11.3.28}$$

(where $(\mathbf{u}, p)$ is the solution of (11.3.2) and Z_N is defined in (11.3.23)), and

$$\begin{aligned}\|p-p^N\|_Q \leqslant \frac{\gamma}{\beta_N}\left(1+\frac{\gamma}{\alpha_N}\right)&\inf_{\mathbf{w}\in Z_N}\|\mathbf{u}-\mathbf{w}\|_V \\ &+\left(1+\frac{\gamma\delta}{\alpha_N\beta_N}\right)\inf_{q\in Q_N}\|p-q\|_Q.\end{aligned} \tag{11.3.29}$$

We now apply the previous general results to the spectral approximation (11.3.17) of the Stokes problem. It is clear from (11.3.26)—(11.3.29) that the discrete spaces for the velocities V_N and for the pressure Q_N have to be chosen in such a way that the constants α_N and β_N, defined in (11.3.24)—(11.3.25) are independent of N, or at least decay with N at the lowest possible rate. In the latter case, estimates (11.3.28) and (11.3.29) will still imply convergence, provided that the infima on the right-hand side decay to zero sufficiently fast to compensate for the growth of the constants.

Let us consider condition (11.3.24) first. For all $\mathbf{v}\in V$, $a(\mathbf{v},\mathbf{v}) = \nu\int_\Omega |\nabla\mathbf{v}|^2\,dx\,dy$ is the square of a norm on V, equivalent to the standard $H^1(\Omega)$ norm of V. This is a consequence of the Poincaré inequality (A.13). It follows that inequality (11.3.24) holds for a constant $\alpha_N = \alpha > 0$ *independent* of N, no matter how the space Z_N is chosen.

Now we deal with condition (11.3.25) which represents the true compatibility condition between V_N and Q_N. We recall (see above) that we have chosen as V_N the subspace of V of the polynomial vectors of degree $\leqslant N$ in each variable. Then, it is clear that Q_N cannot contain any element in the space of spurious modes X_N, for otherwise the left-hand side of (11.3.25) would be zero on these elements, while the right-hand side would not. Therefore, one is led to choose as Q_N any supplementary space of X_N in $\mathscr{P}_N$, i.e., any subspace of $\mathscr{P}_N$ such that

$$\dim Q_N + \dim X_N = \dim \mathscr{P}_N; \qquad Q_N \cap X_N = \{0\}. \tag{11.3.30}$$

Most algorithms (see Sec. 7.4) actually use the whole of $\mathscr{P}_N$ to represent the pressure. The linear systems that arise in these algorithms for the Stokes problem are singular because of the presence of spurious modes. The solution procedures, whether direct or iterative, select the correct pressure (in Q_N) plus some specific linear combination of the spurious modes (in X_N). A filtering procedure is then applied to remove the spurious modes.

Note that any Q_N satisfying (11.3.30) will also satisfy (11.3.25) for a suitable constant β_N. Actually, if the left-hand side of (11.3.25) is zero, then $q \in X_N$; hence, by (11.3.30), $q = 0$. Thus, the left-hand side of (11.3.25) is a norm on Q_N, and therefore it is equivalent to any other norm on it, due to the finite dimension of Q_N.

We conclude that for each Q_N satisfying (11.3.30), the Stokes approximation (11.3.17) has a unique solution $(\mathbf{u}^N, p^N)$. Among the spaces satisfying this condition, one looks for those spaces which lead to the "best" (the largest) constant β_N. A natural candidate for Q_N is the orthogonal space of X_N in $\mathscr{P}_N$, with respect to the inner product of $L^2(\Omega)$.

Also note that one can choose as Q_N any proper subspace of a supplementary space of X_N in $\mathscr{P}_N$. It is clear, however, that the smaller is Q_N, the worse are the approximation properties of the scheme.

Let us now consider in some detail the three boundary conditions for the Stokes problem.

(a) *Fully Periodic Problem*

Since X_N is empty, we can choose $Q_N = \mathscr{P}_N$. Then (11.3.25) is fulfilled with $\beta_N = \beta > 0$ independent of N, (see Bernardi, Maday, and Métivet (1987b)). It is easily seen that the space Z_N defined in (11.3.23) is the subspace of V_N of the divergence-free velocities over all Ω. If P_N: $L^2(\Omega) \to S_N$ denotes the truncation of the Fourier series (see (2.1.8)), then $\nabla \cdot (P_N \mathbf{u}) = P_N \nabla \cdot \mathbf{u} = 0$ if $\mathbf{u}$ is divergence-free. Thus, we can choose $\mathbf{w} = P_N \mathbf{u} \in Z_N$ and $q = P_N p$ in (11.3.28)—(11.3.29). Using the approximation properties for the operator P_N (see Sec. 9.1), one gets the following error estimate for $m \geqslant 1$

$$\|\mathbf{u} - \mathbf{u}^N\|_{H^1(\Omega)} + \|p - p^N\|_{L^2(\Omega)} \leqslant CN^{1-m}(\|u\|_{H_p^m(\Omega)} + \|p\|_{H_p^{m-1}(\Omega)}), \quad (11.3.31)$$

where $C > 0$ is a constant independent of N and $H_p^s(\Omega)$ are Sobolev spaces of the periodic functions as defined in (A.11.d).

This kind of result was first obtained by Maday and Quarteroni (1982b) in their analysis of the Orszag–Patterson method, without using explicitly an "inf-sup" condition. They consider a Fourier Galerkin method, where the non-linear convective term is de-aliased, and a collocation method in which it is not. The analysis shows that in terms of rate of convergence the collocation method is just as good as the Galerkin method. We have here a rigorous demonstration for a complicated non-linear system that the aliasing errors are of the same order as the truncation error. We recall that this feature was already observed in Chap. 2, and numerical examples were furnished in Secs. 4.6 and 7.2.4. The same conclusions have been obtained by Bernardi, Maday, and Métivet (1987b) using the approach described above. An earlier convergence analysis of the Fourier approximation to the Navier–Stokes equations was given by Hald (1981). The convergence of the Fourier method for the Euler equations in the sphere has been proven by Del Prete (1983).

(b) *Periodic-Non-periodic Problem*

Bernardi, Maday and Métivet (1987a) have shown that with the choice $Q_N = S_N \otimes \mathbb{P}_{N-1}^0$ the constant β_N in (11.3.25) is bounded from below by CN^{-1}. On the other hand, for all divergence-free vectors $\mathbf{u}$, one can build up an element $\mathbf{w} \in V_N$ such that $\nabla \cdot \mathbf{w} \equiv 0$ in Ω and the estimate $\|\mathbf{u} - \mathbf{w}\|_V \leqslant CN^{1-m}\|\mathbf{u}\|_{H_p^m(\Omega)}$ holds (here $H_p^m(\Omega)$ is the Sobolev space of order m of the periodic functions in the x variable). This is accomplished by taking the H^1-projection upon $\mathbb{P}_{N-1}^0(-1,1)$ of each Fourier coefficient of order $\leqslant N$ of the second component of $\mathbf{u}$, then modifying it to have zero average on $(-1,1)$ without changing the boundary values, and finally defining the Fourier coefficient of the first component in order to satisfy the divergence-free condition. Obviously, $\mathbf{w}$ belongs to the space Z_N defined in (11.3.23). By the previous estimate and the error estimate for the best approximation of p in Q_N (see Sec. 9.7.4) we obtain from (11.3.28)—(11.3.29) the following convergence results:

$$\|\mathbf{u} - \mathbf{u}^N\|_{H^1(\Omega)} \leqslant CN^{1-m}\{\|u\|_{H_p^m(\Omega)} + \|p\|_{H_p^{m-1}(\Omega)}\}, \tag{11.3.32}$$

$$\|p - p^N\|_{L^2(\Omega)} \leqslant CN^{2-m}\{\|\mathbf{u}\|_{H_p^m(\Omega)} + \|p\|_{H_p^{m-1}(\Omega)}\}, \tag{11.3.33}$$

for a constant $C > 0$ independent of N and $m \geqslant 1$.

The same conclusions hold if one considers a collocation method. For instance, Bernardi, Maday, and Métivet (1987a, 1987b) have studied a collocation method in which a staggered grid in the direction of non-periodicity is used; the families of Gauss–Lobatto points and of Gauss points with respect to the Legendre weight (see (2.3.12) and (2.3.10)) serve, respectively, as the collocation grids for the momentum and continuity equations. Recall that a similar approach is followed in many of the algorithms for inhomogeneous flows described in Sec. 7.3.

Spectral (Fourier) methods can also be coupled with finite-element methods in the direction of non-periodicity. Canuto, Fujii and Quarteroni (1983) have studied an approximation of this type for a two-dimensional model of the Rayleigh convection problem (flow between two infinite slabs at different temperature, with periodic boundary conditions in the direction parallel to the slabs). The analysis is based on the fulfillment of an "inf-sup" condition as described above. The main result is that the numerical solution exhibits the same symmetry-breaking properties in the periodicity direction as the physical solution. This highly desirable feature is a consequence of having a Fourier rather than a finite-element method in this direction.

The effect of numerical integration in a coupled Fourier/finite-element method was analyzed by Canuto, Maday and Quarteroni (1984).

(c) *Fully Non-periodic Problem*

Bernardi, Maday and Métivet (1987b) exhibit a family of supplementary spaces of X_N in $\mathscr{P}_N$ for which the constant β_N in (11.3.25) is bounded from below by CN^{-2}. Again, for each divergence-free vector $\mathbf{u}$ it is possible to find a divergence-free vector $\mathbf{w} \in V_N$ such that $\|\mathbf{u} - \mathbf{w}\|_{H^1(\Omega)} \leqslant CN^{1-m} \|\mathbf{u}\|_{H^m(\Omega)}$. Actually, $\mathbf{u} = \nabla \times \phi$ with $\phi \in H_0^2(\Omega)$ (see (A.11.c)) and we define $\mathbf{w} = \nabla \times \phi_N$, where ϕ_N is the projection of ϕ defined in (9.4.23). (General results on the approximation of divergence-free vector fields by divergence-free polynomial fields have been given by Sacchi–Landriani and Vandeven (1987).) On the other hand, the space Q_N can be chosen so as to contain all the polynomials of degree $\leqslant \lambda N$ for a positive $\lambda < 1$. We apply the approximation results for the Legendre system (see Sec. 9.4) and we obtain the following error estimates:

$$\|\mathbf{u} - \mathbf{u}^N\|_{H^1(\Omega)} \leqslant CN^{1-m}\{\|\mathbf{u}\|_{H^m(\Omega)} + \|p\|_{H^{m-1}(\Omega)}\}; \tag{11.3.34}$$

$$\|p - p^N\|_{L^2(\Omega)} \leqslant CN^{3-m}\{\|\mathbf{u}\|_{H^m(\Omega)} + \|p\|_{H^{m-1}(\Omega)}\}. \tag{11.3.35}$$

Again, similar estimates can be obtained for various collocation methods. Bernardi, Maday and Métivet (1987b) consider one grid based on the Gauss–Lobatto points, while Bernardi and Maday (1986) show that the use of staggered grids can reduce the number of parasitic modes for the pressure.

So far, we have considered spectral approximations to the Stokes problem (11.3.2) which use Legendre polynomials in the directions of non-periodicity. In most applications, Chebyshev polynomials are preferred for their close relationship with Fourier polynomials. Using Chebyshev polynomials in a Galerkin projection method amounts to replacing in (11.3.17) the L^2-inner product (u, v) by the weighted L^2-inner product $(u, v)_w$ (w denotes, as usual, the Chebyshev weight in each direction of non-periodicity). Thus, we consider now the problem:

$$\begin{cases} \mathbf{u}^N \in V_N \qquad p^N \in Q_N \\ \nu(-\Delta \mathbf{u}^N, \mathbf{v})_w + (\mathbf{v}, \nabla p^N)_w = (\mathbf{g}, \mathbf{v})_w & \text{for all } \mathbf{v} \in V_N \\ (\nabla \cdot \mathbf{u}^N, q)_w = 0 & \text{for all } q \in Q_N. \end{cases} \tag{11.3.36}$$

Unlike (11.3.17), the new problem cannot be cast into the abstract form

(11.3.22). Actually, due to the Chebyshev weight, we have

$$(\nabla \cdot \mathbf{v}, p)_w \neq -(\mathbf{v}, \nabla p)_w,$$

i.e., the negative gradient and the divergence operators are not adjoint to each other in the weighted Chebyshev inner product. Thus, we are led to look for a more general abstract formulation of problem (11.3.36) which involves two different bilinear forms $b_1(\mathbf{v}, q)$ and $b_2(\mathbf{v}, q)$.

More precisely, let us introduce the following bilinear forms:

$$\begin{cases} a\colon V_w \times V_w \to \mathbb{R} \\ a(\mathbf{u}, \mathbf{v}) = \nu(\nabla \mathbf{u}, \nabla(\mathbf{v}w)); \end{cases} \tag{11.3.37}$$

$$\begin{cases} b_1\colon V_w \times Q_w \to \mathbb{R} \\ b_1(\mathbf{v}, q) = -(\nabla \cdot (\mathbf{v}w), q); \end{cases} \tag{11.3.38}$$

$$\begin{cases} b_2\colon V_w \times Q_w \to \mathbb{R}, \\ b_2(\mathbf{v}, q) = (\nabla \cdot \mathbf{v}, q)_w. \end{cases} \tag{11.3.39}$$

Here, the Sobolev spaces V_w and Q_w are defined as V and Q, except that integrals are weighted by the Chebyshev weight w in the direction(s) of non-periodicity. Using the results of Sec. 11.1, one can prove that these forms are well-defined and continuous. Then problem (11.3.36) can be written in the form:

$$\begin{cases} \mathbf{u}^N \in V_N \qquad p^N \in Q_N \\ a(\mathbf{u}^N, \mathbf{v}) + b_1(\mathbf{v}, p^N) = (\mathbf{g}, \mathbf{v})_w \qquad \text{for all } \mathbf{v} \in V_N \\ b_2(\mathbf{u}^N, q) = 0 \qquad \text{for all } q \in Q_N. \end{cases} \tag{11.3.40}$$

The solvability of (11.3.40) can be completely characterized in terms of "inf-sup" conditions on the forms a and b_i, $i = 1, 2$. These conditions generalize those given in Brezzi (1974) for problem (11.3.22). In order to state them, let us define

$$Z_N^{(i)} = \{\mathbf{v} \in V_N \mid b_i(\mathbf{v}, q) = 0, \text{ for all } q \in Q_N\} \quad i = 1, 2. \tag{11.3.41}$$

In Bernardi, Canuto and Maday (1986), it is proven that problem (11.3.40) has a unique solution if the following assumptions are fulfilled:

(i) *there exists a constant* $\alpha_N > 0$ *such that*

$$\sup_{\mathbf{v} \in Z_N^{(1)}} \frac{a(\mathbf{u}, \mathbf{v})}{\|\mathbf{v}\|_{V_w}} \geqslant \alpha_N \|\mathbf{u}\|_{V_w} \qquad \textit{for all } \mathbf{u} \in Z_N^{(2)}; \tag{11.3.42}$$

(ii) $$\dim Z_N^{(1)} = \dim Z_N^{(2)};$$

(iii) *there exist two constants* $\beta_N^{(i)} > 0$, $(i = 1, 2)$ *such that*

$$\sup_{\mathbf{v} \in V_N} \frac{b_i(\mathbf{v}, q)}{\|\mathbf{v}\|_{V_w}} \geqslant \beta_N^{(i)} \|q\|_{Q_w} \qquad \textit{for all } q \in Q_N. \tag{11.3.43}$$

These conditions are also necessary. Under the previous assumptions, one can get stability estimates similar to (11.3.26)—(11.3.27) and convergence estimates similar to (11.3.28)—(11.3.29). The application of this abstract tool of analysis to the Chebyshev case is quite technical (the details are given in Bernardi, Canuto and Maday (1986)). The conclusions are essentially comparable with the results for the Legendre case discussed above.

11.3.3. Discretizations of the Continuity Equation by an Influence-Matrix Technique: The Kleiser–Schumann Method

An alternative approach to the direct discretization of the continuity equation consists of discretizing a Poisson equation for the pressure. This equation is obtained by taking the divergence of the momentum equation in (11.3.2) and using the continuity equation to get

$$\Delta p = \nabla \cdot \mathbf{g} \qquad \text{in } \Omega. \tag{11.3.44}$$

A proper boundary condition to be associated with this equation is

$$\nabla \cdot \mathbf{u} = 0 \qquad \text{on } \partial\Omega. \tag{11.3.45}$$

As a matter of fact, if $(\mathbf{u}, p)$ is a (classical) solution of the Stokes system (11.3.2), then (11.3.44) and (11.3.45) hold. Conversely, if $(\mathbf{u}, p)$ is a (classical) solution of the momentum equation and of (11.3.44) and (11.3.45), then it is easily seen that the function $d = \nabla \cdot \mathbf{u}$ satisfies

$$\begin{cases} \Delta d = 0 & \text{in } \Omega \\ d = 0 & \text{on } \partial\Omega. \end{cases} \tag{11.3.46}$$

Thus, $\mathbf{u}$ is divergence-free in Ω, i.e., $(\mathbf{u}, p)$ is also a solution of (11.3.2).

The boundary condition (11.3.45) is of non-standard type, since a boundary condition for the pressure is usually associated with (11.3.44). However, Kleiser and Schumann (1980) suggested an efficient method for correctly imposing the boundary condition on the pressure in order to satisfy the continuity equation at the walls. Their method is described in Sec. 7.3.1.

The Kleiser–Schumann method has been theoretically investigated by Canuto and Sacchi–Landriani (1986), who considered a Fourier–Legendre tau method for the Stokes problem, and by Sacchi–Landriani (1987) who extended the analysis to the Fourier–Chebyshev case for the full Navier–Stokes equations. Here is a brief account of their analysis.

We maintain in this section the same notation of Sec. 7.3.1. Thus, the basic set of equations to be approximated is (7.3.21)—(7.3.23). Since the time-discretization does not affect the treatment of the continuity equation, we consider a time-independent version of the scheme, i.e., we set $\Delta t = \infty$ in the definition of λ. Furthermore, we assume for simplicity that $\nu = 1$, so that $\lambda = k^2$.

Canuto and Sacchi-Landriani first prove that the influence-matrix M_N which appears in (7.3.28) and which corresponds to the tau discretization of the "B-problem," is non-singular for all wavenumbers $k \neq 0$ and for all degree N of the polynomials.

Next, some estimates on the solutions of the approximate "B problems" are obtained. Hereafter, we denote by $\|v\|$ the norm of the space $L^2(-1, 1)$. For the solution $(\hat{v}_p^N, \hat{P}_p^N)$ of the tau approximation to the "B problem" with homogeneous Dirichlet boundary conditions for the pressure, the estimates

$$\|(\hat{v}_p^N)''\|^2 + \lambda\|(\hat{v}_p^N)'\|^2 + \lambda^2\|\hat{v}_p^N\|^2 \leqslant C\|\hat{\mathbf{R}}\|^2, \tag{11.3.47}$$

$$\|(\hat{P}_p^N)'\|^2 + \lambda\|\hat{P}_p^N\|^2 \leqslant C\|\hat{\mathbf{R}}\|^2 \tag{11.3.48}$$

hold with a constant C independent of N. These estimates are proven by a suitable choice of test functions in the tau equations for $\hat{v}_p^N$ and $\hat{P}_p^N$. As a by-product of these estimates we get an estimate for the right-hand side of the algebraic system (7.3.28). More precisely,

$$|(\hat{v}_p^N)'(\pm 1)| \leqslant C\lambda^{-1/4}\|\hat{\mathbf{R}}\|. \tag{11.3.49}$$

On the other hand, let $(\hat{v}_\pm^N, \hat{P}_\pm^N)$ be the tau solutions of the homogeneous B-problems with boundary conditions $\hat{P}_-^N(-1) = 1$, $\hat{P}_-^N(1) = 0$ and $\hat{P}_+^N(-1) = 0$, $\hat{P}_+^N(1) = 1$ respectively. Then we have the following estimates

$$\|(\hat{v}_\pm^N)''\|^2 + \lambda\|(\hat{v}_\pm^N)'\|^2 + \lambda^2\|\hat{v}_\pm^N\|^2 \leqslant C\lambda^{1/2}, \tag{11.3.50}$$

$$\|(\hat{P}_\pm^N)'\|^2 + \lambda\|\hat{P}_\pm^N\|^2 \leqslant C\lambda^{1/2}. \tag{11.3.51}$$

The estimates are a consequence of an analysis of the behavior of the Legendre coefficients of $(\hat{v}_\pm^N, \hat{P}_\pm^N)$, using a sort of maximum principle in frequency space (see Canuto (1987) for more details). As a by-product of (11.3.50), one proves that the Euclidean norm of the inverse of the influence-matrix M_N is bounded independently of λ and N, i.e.,

$$\|M_N^{-1}\|_2 \leqslant C. \tag{11.3.52}$$

From this estimate and (11.3.49) one gets an estimate on the solution of the algebraic system (7.3.28)

$$|\delta_\pm| \leqslant C\lambda^{-1/4}\|\mathbf{R}\|. \tag{11.3.53}$$

Recalling the decomposition (7.3.27), we obtain from (11.3.47)—(11.3.48), (11.3.50)—(11.3.51) an estimate for the approximate solution $(\hat{v}^N, \hat{P}^N)$ of the "A-problem," i.e.,

$$\|(\hat{v}^N)''\|^2 + \lambda\|(\hat{v}^N)'\|^2 + \lambda^2\|\hat{v}^N\|^2 \leqslant C\|\mathbf{R}\|^2, \tag{11.3.54}$$

$$\|(\hat{P}^N)'\|^2 + \lambda\|\hat{P}^N\|^2 \leqslant C\|\mathbf{R}\|^2. \tag{11.3.55}$$

The previous estimates concern one fixed mode (of wavenumber k) in the Fourier expansion. Since the dependence on k is explicit, it is possible to get from them a stability estimate on the complete solution produced by the Kleiser–Schumann method, which we denote by $(\mathbf{u}^N, p^N)$. Precisely, we have

$$\|\mathbf{u}^N\|_{H^2(\Omega)} + \|p^N\|_{H^1(\Omega)/\mathbb{R}} \leqslant C\|\mathbf{R}\|_{L^2(\Omega)}, \tag{11.3.56}$$

for a constant C independent of N. Here $\Omega = (0, 2\pi) \times (-1, 1)$.

The convergence analysis is carried out by comparing the approximate solution to a suitable orthogonal projection of the exact solution and by using the approximation results of Secs. 9.1 and 9.4 for checking consistency. Denoting here the exact solution by $(\mathbf{u}, p)$, one can prove the following convergence estimate

$$\begin{aligned} &\|\mathbf{u} - \mathbf{u}^N\|_{H^2(\Omega)} + \|p - p^N\|_{H^1(\Omega)/\mathbb{R}} \\ &\quad \leqslant CN^{2-m}\{\|u\|_{H^m_p(\Omega)} + \|p\|_{H^{m-1}_p(\Omega)}\}, \end{aligned} \tag{11.3.57}$$

for all $m \geqslant 2$ for which the right-hand side is finite. Here $H^m_p(\Omega)$ denotes the Sobolev space of order m of the functions periodic in the first variable.

The same estimate holds if we take into account the tau correction proposed by Kleiser and Schumann in order to satisfy exactly the divergence-free condition all over the domain. Thus, spectral accuracy is guaranteed for the velocity field in the Kleiser–Schumann method for the periodic channel whether or not the tau correction is applied. Furthermore, the method selects one pressure among all the approximate solutions of the momentum equation (see (11.3.5) and (11.3.6)). Since by (11.3.57) the exact pressure is approximated with spectral accuracy, we conclude that the computed pressure is not affected by the spurious mode exhibited in Sec. 11.3.1.

11.3.4. Navier–Stokes Equations in Streamfunction Formulation

The stream function formulation of the steady Navier–Stokes equations in the square $\Omega = (-1, 1)^2$ reads as follows:

$$\nu\Delta^2\psi - \frac{\partial}{\partial x}\left(\Delta\psi\frac{\partial\psi}{\partial y}\right) + \frac{\partial}{\partial y}\left(\Delta\psi\frac{\partial\psi}{\partial x}\right) = g. \tag{11.3.58}$$

If we submit the velocity field to homogeneous Dirichlet boundary conditions, then the streamfunction ψ satisfies the boundary conditions $\psi = \partial\psi/\partial n = 0$.

Maday and Métivet (1986) have studied a Chebyshev spectral and a Chebyshev pseudospectral approximation of this problem. They prove the convergence of the schemes and derive error estimates in weighted Sobolev spaces. In this framework, the natural space for the streamfunction ψ is the subspace $H^2_{w,0}(\Omega)$ of $H^2_w(\Omega)$ consisting of the functions which vanish with their normal derivative on the boundary of Ω (see (A.11.c)). The analysis relies on the mathematical result that the biharmonic operator Δ^2 with homogeneous Dirichlet boundary conditions is continuous and coercive in $H^2_{w,0}(\Omega)$. More precisely, Maday and Métivet prove that there exist two constants $\delta > 0$ and

$\gamma > 0$ such that

$$\int_\Omega \Delta\psi\Delta(\phi w)\,dx\,dy \leqslant \delta\|\psi\|_{H_w^2(\Omega)}\|\phi\|_{H_w^2(\Omega)} \quad \text{for all } \phi, \psi \in H_{w,0}^2(\Omega), \tag{11.3.59}$$

and

$$\gamma\|\psi\|^2_{H_w^2(\Omega)} \leqslant \int_\Omega \Delta\psi\Delta(\psi w)\,dx\,dy \qquad \text{for all } \psi \in H_{w,0}^2(\Omega). \tag{11.3.60}$$

Note that these inequalities extend those given in Sec. 11.1 for the Laplace operator Δ. Similar estimates hold with constants $\tilde{\delta}$ and $\tilde{\gamma}$ independent of N if ϕ and ψ are polynomials of degree N which vanish with the normal derivative at the boundary, and the biharmonic operator is collocated at the Gauss–Lobatto points for the Chebyshev weight.

11.4. The Eigenvalues of Some Spectral Operators

We shall give a brief theoretical discussion on the qualitative behavior of the eigenvalues of some relevant spectral approximations to the following differential operators: the pure second-derivative operator $Lu = -u_{xx}$; the advection-diffusion operator $Lu = -\nu u_{xx} + bu_x$; the first-order hyperbolic operator $Lu = u_x$. All the operators are associated with non-periodic boundary conditions.

11.4.1. The Discrete Eigenvalues for $Lu = -u_{xx}$

The boundary conditions we impose here are of Dirichlet type

$$u(+1) = u(-1) = 0 \tag{11.4.1a}$$

or of Neumann type

$$u_x(-1) = u_x(1) = 0. \tag{11.4.1b}$$

We will first consider the *collocation* method which uses the Gauss–Lobatto points $\{x_j\}_{j=0}^N$ (see Sec. 2.2.3) with respect to the Chebyshev or the Legendre weight $w(x)$. The corresponding eigenvalues λ are defined by the relations

$$-u_{xx}(x_j) = \lambda u(x_j) \qquad j = 1, \ldots, N-1, \tag{11.4.2}$$

where u is a non-trivial polynomial of degree N which satisfies the boundary conditions (11.4.1a) or (11.4.1b).

It has been proved by Gottlieb and Lustman (1983) that for the Chebyshev points $x_j = \cos(j\pi/N)$, the eigenvalues of (11.4.1)—(11.4.2) are all real, non-negative, and distinct. Gottlieb and Lustman actually prove their result for a wider class of boundary conditions than (11.4.1), namely for

$$\alpha u(1) + \beta u_x(1) = 0 \qquad \gamma u(-1) + \delta u_x(-1) = 0, \tag{11.4.3}$$

with α, β, $\gamma > 0$ and $\delta < 0$ (these conditions can be relaxed to allow α and γ, or β and δ to be zero). Starting from the error equation associated with (11.4.2) (see Sec. 10.6), their method consists of finding an explicit expression for the characteristic polynomial of the collocation matrix. Next, they prove that this polynomial satisfies an algebraic condition which implies that its roots are real, non-negative and simple. The method can be used to prove the same kind of result when the collocation points are the Legendre points defined in (2.3.12).

For the Dirichlet boundary conditions, it is easy to derive an upper- and a lower-bound for the eigenvalues of the collocation operator. Multiplying each equation (11.4.2) by $u(x_j)w_j$—where w_j is the j-th weight of the Gauss–Lobatto formula (2.2.17)—and summing up we get

$$-\sum_{j=0}^{N} u_{xx}(x_j)u(x_j)w_j = \lambda \sum_{j=0}^{N} u^2(x_j)w_j.$$

By the exactness of the quadrature rule over $\mathbb{P}_{2N-1}$, we get

$$\lambda = \frac{-\int_{-1}^{1} u_{xx}uw\,dx}{\|u\|_N^2}. \tag{11.4.4}$$

Here $\|u\|_N$ denotes the discrete L^2_w-norm of u (see (2.2.24)), which is uniformly equivalent to the standard norm $\|u\|_{L^2_w(-1,1)}$ (see (9.3.2)). Integrating by parts the numerator of (11.4.4) if w is the Legendre weight, or using inequalities (11.1.9) and (11.1.10) if w is the Chebyshev weight, we obtain the bounds

$$\bar{c}_1 \frac{\|u_x\|^2_{L^2_w(-1,1)}}{\|u\|^2_{L^2_w(-1,1)}} \leqslant \lambda \leqslant \bar{c}_2 \frac{\|u_x\|^2_{L^2_w(-1,1)}}{\|u\|^2_{L^2_w(-1,1)}}$$

for two constants $\bar{c}_1$ and $\bar{c}_2$ independent of N. Using the Poincaré inequality (A.13.2) on the left-hand side, and the inverse inequality (9.4.4) or (9.5.4) (with $p = 2$) on the right-hand side, we conclude that there exist two positive constants c_1, c_2 independent of N such that

$$0 < c_1 \leqslant \lambda \leqslant c_2 N^4. \tag{11.4.5}$$

When the Neumann boundary conditions (11.4.1b) are enforced indirectly as described in Sec. 3.3, the corresponding eigenvalue problem is

$$-(\tilde{u}_x)_x(x_j) = \lambda u(x_j) \qquad j = 0, \ldots, N, \tag{11.4.6}$$

where $\tilde{u}_x$ is the polynomial of degree N which vanishes at $x = \pm 1$ and is equal to u_x at the interior collocation points. Since both sides in (11.4.6) are polynomials of degree N, we have

$$-(\tilde{u}_x)_x = \lambda u \qquad -1 < x < 1. \tag{11.4.7}$$

If we set $\tilde{u}_x = v$ and differentiate (11.4.7), we get

$$-v_{xx}(x_j) = \lambda v(x_j) \qquad j = 1, \ldots, N-1,$$

which proves that any $\lambda \neq 0$ is also an eigenvalue of the Dirichlet problem; hence, in particular, it is positive, simple, and bounded by CN^4 as $N \to \infty$.

Let us consider now the *tau* approximation for the second-derivative operator. The corresponding eigenvalues are defined by

$$-\hat{u}_k^{(2)} = \lambda \hat{u}_k \qquad k = 0, \ldots, N-2, \tag{11.4.8}$$

where $\hat{u}_k$ and $\hat{u}_k^{(2)}$ denote respectively the k-th coefficient of u and of u_{xx} in the expansion according to the Chebyshev or the Legendre basis. As usual, the two highest coefficients of u are determined by the boundary conditions (11.4.1). An equivalent formulation of (11.4.8) is

$$-\int_{-1}^{1} u_{xx} v w \, dx = \lambda \int_{-1}^{1} u v w \, dx \qquad \text{for all } v \in \mathbb{P}_{N-2}. \tag{11.4.9}$$

For the Chebyshev method, the technique of Gottlieb and Lustman (1983) can be adapted to prove that the eigenvalues of (11.4.8)and (11.4.1) are real, non-negative, and distinct (D. Gottlieb (1986, private communication)). For the Dirichlet boundary conditions, the positivity of the eigenvalues is an easy consequence of their being real, since one can choose $v = -u_{xx}$ in (11.4.9) and use inequalities (11.1.9)—(11.1.10) to get the estimate

$$\bar{c}_1 \frac{\int_{-1}^{1} |u_{xx}|^2 w \, dx}{\int_{-1}^{1} |u_x|^2 w \, dx} \leqslant \lambda \leqslant \bar{c}_2 \frac{\int_{-1}^{1} |u_{xx}|^2 w \, dx}{\int_{-1}^{1} |u_x|^2 w \, dx}.$$

Since u is a polynomial vanishing at $x = \pm 1$, its first derivative u_x vanishes for at least one point in the interval $(-1, 1)$. Thus, we can apply to the function u_x the Poincaré inequality (A.13.2) and the inverse inequality (2.5.4) (with $p = 2$) to get an estimate of the type (11.4.5). For both Dirichlet and Neumann boundary conditions, the largest computed eigenvalue exhibits a growth asymptotical with N^4. The particular constants are given in Table 4.1.

The theory is instead very easy for the Legendre method. By choosing $v = -\bar{u}_{xx}$ in (11.4.9) and integrating by parts, one proves that λ has to be real and positive. The inverse inequality (9.4.4) ensures that λ can grow at most as $O(N^4)$. For Dirichlet boundary conditions, λ is uniformly bounded away from 0.

11.4.2. The Discrete Eigenvalues for $Lu = -\nu u_{xx} + b u_x$

We assume that ν is a strictly positive constant, while b is a smooth real function of x. Hereafter, we shall submit u to the Dirichlet boundary conditions (11.4.1a).

The exact eigenvalues of this operator are, in general, complex due to the presence of the first-order advective term. Moreover, multiplying the equation $Lu = \lambda u$ by $\bar{u}$ and using a standard integration-by-parts argument one gets

$$\mathbb{R}e\,\lambda = \frac{\nu \int_{-1}^{1} |u_x|^2\,dx - \frac{1}{2}\int_{-1}^{1} b_x |u|^2\,dx}{\int_{-1}^{1} |u|^2\,dx}.$$

This shows that the real part of the eigenvalues need not be positive, whenever ν is small and b_x is strictly positive. However, only a finite number of eigenvalues have negative real parts, and $\mathbb{R}e\,\lambda \geqslant -\beta_1/2$, where $\beta_1 = \max\{|b_x(x)|, -1 \leqslant x \leqslant 1\}$.

Let us discuss first the behavior of the eigenvalues of the spectral *tau* operator. They are defined by the existence of a non-trivial polynomial u of degree N vanishing at $x = \pm 1$ such that

$$\int_{-1}^{1} (-\nu u_{xx} + b u_x) v w\,dx = \lambda \int_{-1}^{1} u v w\,dx \qquad \text{for all } v \in \mathbb{P}_{N-2}. \quad (11.4.10)$$

An estimate on $\mathbb{R}e\,\lambda$ can be obtained by choosing $v = \mathbb{P}_{N-2}\bar{u}$, if we assume that u is normalized to have $\int_{-1}^{1} |P_{N-2}u|^2 w\,dx = 1$. By (11.1.14)

$$\mathbb{R}e \int_{-1}^{1} u_{xx}\bar{u} w\,dx \geqslant \frac{1}{4}\int_{-1}^{1} |u_x|^2 w\,dx,$$

while by the Cauchy–Schwarz inequality

$$\left|\mathbb{R}e \int_{-1}^{1} b u_x \mathbb{P}_{N-2}\bar{u} w\,dx\right| \leqslant \beta_0 \left(\int_{-1}^{1} |u_x|^2 w\,dx\right)^{1/2},$$

where $\beta_0 = \max\{|b(x)|, -1 \leqslant x \leqslant 1\}$.

Hence, $\mathbb{R}e\,\lambda \geqslant (\nu/4)\|u_x\|^2_{L^2_w(-1,1)} - \beta_0 \|u_x\|_{L^2_w(-1,1)}$ which implies

$$\mathbb{R}e\,\lambda \geqslant -\frac{\beta_0}{\nu}. \quad (11.4.11)$$

This proves that the real parts of the eigenvalues of the tau method are uniformly bounded from below.

For the Legendre tau method, it is possible to refine this estimate, showing that $\mathbb{R}e\,\lambda \geqslant -\beta_1/2$ as for the eigenvalues of the exact problem, provided that N is large enough. Actually,

$$\begin{aligned}
\mathbb{R}e \int_{-1}^{1} b u_x P_{N-2}\bar{u}\,dx &= \mathbb{R}e \int_{-1}^{1} b u_x \bar{u}\,dx - \mathbb{R}e \int_{-1}^{1} b u_x(\bar{u} - P_{N-2}\bar{u})\,dx \\
&= -\frac{1}{2}\int_{-1}^{1} b_x |u|^2\,dx - \mathbb{R}e \int_{-1}^{1} b u_x(\bar{u} - P_{N-2}\bar{u})\,dx \\
&= -\frac{1}{2}\int_{-1}^{1} b_x |P_{N-2}u|^2\,dx - \frac{1}{2}\int_{-1}^{1} b_x[|u|^2 - |P_{N-2}u|^2]\,dx \\
&\quad - \mathbb{R}e \int_{-1}^{1} b u_x(\bar{u} - P_{N-2}\bar{u})\,dx.
\end{aligned}$$

The last two integrals on the right-hand side are easily shown to be bounded by $CN^{-1}\|u_x\|^2_{L^2(-1,1)}$, according to the estimate (9.4.6). Here C depends on β_0 and β_1. Hence,

$$\mathbb{R}e\,\lambda \geqslant \left(\nu - \frac{C}{N}\right)\|u_x\|^2_{L^2(-1,1)} - \beta_1/2 \geqslant -\beta_1/2$$

if N is large enough.

For both the Chebyshev and the Legendre method, a bound for $|\lambda|$ is obtained by choosing $v = -\bar{u}_{xx}$ in (11.4.10) and taking the modulus of both sides. One gets

$$|\lambda| \leqslant 4\frac{\nu\|u_{xx}\|^2_{L^2_w(-1,1)} + \beta_0\|u_x\|_{L^2_w(-1,1)}\|u_{xx}\|_{L^2_w(-1,1)}}{\|u_x\|_{L^2_w(-1,1)}},$$

whence by the inverse inequality (9.4.4) or (9.5.4)

$$|\lambda| \leqslant \nu O(N^4) + \beta_0 O(N^2). \tag{11.4.12}$$

Numerical experiments show that the largest modulus of the eigenvalues grows according to this estimate.

The eigenvalues of the *collocation* operator for the advection-diffusion problem are defined by the relation

$$-\nu u_{xx}(x_j) + b(x_j)u_x(x_j) = \lambda u(x_j) \qquad j = 1, \ldots, N-1, \tag{11.4.13}$$

where again u is a non-trivial polynomial of degree N, zero at $x = \pm 1$. Equivalently, we have

$$-\nu \int_{-1}^{1} u_{xx} v w\, dx + (bu_x, v)_N = \lambda(u, v)_N \tag{11.4.14}$$

for all $v \in \mathbb{P}_N$ such that $v(\pm 1) = 0$, where $(u, v)_N$ is defined in (2.2.24).

The theoretical estimates derived above also hold for the eigenvalues of (11.4.13). It is enough to adapt the arguments previously used, taking into account the exactness of the quadrature formula related to the collocation nodes (see (2.2.25)), and the uniform equivalence of the continuous and discrete norms over $\mathbb{P}_N$ (see Sec. 9.3).

Numerical experiments for collocation approximations to the operators

$$Lu = -\nu u_{xx} + u_x$$

and

$$Lu = -\nu u_{xx} + x u_x$$

support the estimates (11.4.11) and (11.4.12). In the former case for Legendre approximations, all the eigenvalues have non-negative real parts, whereas, for Chebyshev approximations there are some eigenvalues with negative real parts when ν and N are small. In the latter case for Legendre approximations, the real parts of the eigenvalues are bounded from below by $-\frac{1}{2}$, whereas, for Chebyshev approximations the real parts of the eigenvalues can have quite large negative values when ν and N are small.

11.4.3. The Discrete Eigenvalues for $Lu = u_x$

We associate to this operator the boundary conditions

$$u(1) = 0. \qquad (11.4.15)$$

The eigenvalues arising from the *tau* approximation of this operator are defined by the existence of a non-trivial polynomial u of degree N vanishing at $x = 1$ and such that

$$\hat{u}_k^{(1)} = \lambda \hat{u}_k \qquad k = 0, \ldots, N-1, \qquad (11.4.16)$$

where $\hat{u}_k$ and $\hat{u}_k^{(1)}$ denote respectively the k-th coefficient of u and of u_x in the expansion according to the Chebyshev or the Legendre basis. Equation (11.4.16) is equivalent to the variational form

$$(u_x, v)_w = \lambda (u, v)_w \qquad \text{for all } v \in \mathbb{P}_{N-1}. \qquad (11.4.17)$$

The eigenvalues of (11.4.16) are complex numbers. Their real parts are all strictly negative. In order to show this result, let us consider first the Chebyshev method. Equation (11.4.17) yields the error equation (see Sec. 10.6)

$$u_x = \lambda u + \alpha T_N \qquad -1 < x < 1; \qquad (11.4.18)$$

by equating the coefficients of T_N on both sides we get

$$\alpha = -\lambda \hat{u}_N \frac{\pi}{2}.$$

Let us multiply equation (11.4.18) by $(1 + x)\bar{u}_x(x) w(x)$ and integrate over $(-1, 1)$. It is easily checked using (2.4.22) that the N-th Chebyshev coefficient of the function $(1 + x)\bar{u}_x$ is $N\bar{\hat{u}}_N$. Thus, setting $\tilde{w}(x) = (1 + x)w(x)$ we have

$$\int_{-1}^{1} |u_x|^2 \tilde{w}(x)\, dx = \lambda \left[\int_{-1}^{1} u\bar{u}_x \tilde{w}(x)\, dx - N\frac{\pi}{2} |\hat{u}_N|^2 \right].$$

Note that $\mathbb{R}e \int_{-1}^{1} u\bar{u}_x \tilde{w}\, dx = -\frac{1}{2}\int_{-1}^{1} |u|^2 \tilde{w}_x\, dx < 0$, whence it follows that $\mathbb{R}e\, \lambda < 0$.

A bound for the modulus of λ can be obtained by setting $v = P_{N-1}\bar{u}$ in (11.4.17) and using the Cauchy–Schwarz inequality to get

$$|\lambda| \leqslant \frac{\|u_x\|_{L^2_w(-1,1)}}{\|P_{N-1} u\|_{L^2_w(-1,1)}}.$$

One can prove (following the argument used in Canuto and Quarteroni (1982a) to obtain the inverse inequality (9.5.4)) that there exists a constant

$C > 0$ independent of N such that

$$\|u_x\|_{L^2_w(-1,1)} \leqslant CN^2 \|P_{N-1}u\|_{L^2_w(-1,1)} \qquad \text{for all } u \in \mathbb{P}_N \text{ such that } u(1) = 0.$$

Thus, one obtains the estimate

$$|\lambda| \leqslant O(N^2) \qquad \text{as } N \to \infty. \tag{11.4.19}$$

The eigenvalues of the Chebyshev tau method (11.4.16) are plotted in Fig. 4.1.

For the eigenvalues of the Legendre tau method, the negativity of the real parts follows immediately setting $v = \bar{u}_x$ in (11.4.17), since $\mathbb{R}e \int_{-1}^{1} u\bar{u}_x \, dx = -\frac{1}{2}|u(-1)|^2 < 0$. On the other hand, the eigenvalues of the Legendre tau method differ qualitatively from those of the Chebyshev tau method, in that their largest modulus satisfies an estimate of the form

$$|\lambda| \leqslant O(N) \qquad \text{as } N \to \infty, \tag{11.4.20}$$

instead of (11.4.19) (see Fig. 4.2). This rather surprising fact has been proven by Dubiner (1977) in an unpublished paper, using an asymptotic analysis. Tal-Ezer (1986) takes advantage from this property in defining a fully discrete approximation to the hyperbolic equation $u_t = u_x$ in which the stability condition on the time step is $\Delta t \leqslant O(N^{-1})$ rather than $\Delta t \leqslant O(N^{-2})$. On the other hand, when the Legendre tau method is applied to a system of hyperbolic equations, the corresponding eigenvalues grow again at the rate of $O(N^2)$, as predicted by the inverse inequality (9.4.4).

We consider now *collocation* methods. We choose here the collocation points to be the Gauss–Lobatto points $\{x_j\}_{j=0}^{N}$ for the Chebyshev or the Legendre weight, as defined in Sec. 2.2.3. Other choices of collocation points are possible.

The eigenvalues of the collocation operator are defined by the set of equations

$$\begin{aligned} u_x(x_j) &= \lambda u(x_j) \qquad j = 1, \ldots, N \\ u(x_0) &= 0 \end{aligned} \tag{11.4.21}$$

provided u is a non-trivial polynomial of degree N.

For the Chebyshev points, the sign property of the real parts of λ follows from a stability result, due to Gottlieb and Turkel (1985), for the associated time-dependent problem

$$\begin{cases} u_t(x_j, t) = u_x(x_j, t) & j = 1, \ldots, N, \quad t > 0 \\ u(x_0, t) = 0 & t > 0 \\ u(x_j, 0) = u_0(x_j) & j = 0, \ldots, N. \end{cases} \tag{11.4.22}$$

As discussed in Sec. 12.1.2 (see (12.1.28) and following), for each $N > 0$, there exists a spatial norm of u which remains bounded for all times $t > 0$. This clearly implies that the eigenvalues of the spatial operator in (11.4.22) have non-positive real parts. Moreover, an estimate of the form

$$|\lambda| \leqslant O(N^2) \tag{11.4.23}$$

for each eigenvalue follows easily from the identity

$$\int_{-1}^{1} u_x \bar{u} w \, dx = \lambda \sum_{j=0}^{N} |u(x_j)|^2 w_j, \tag{11.4.24}$$

taking into account (9.5.4) and (9.3.2). This identity is obtained in the usual way by multiplying the j-th equation in (11.4.21) by $\bar{u}(x_j)w_j$, summing up over $j = 0, \ldots, N$ and using (2.2.25). The estimate (11.4.23) is sharp, as confirmed by numerical experiments (see Fig. 4.3).

The analysis for the Legendre collocation operator is easier. Equation (11.4.24) (where now $w \equiv 1$) implies the positivity of $\mathbb{R}e\,\lambda$ (since $\mathbb{R}e \int_{-1}^{1} u_x \bar{u}\, dx = -\frac{1}{2}|u(-1)|^2 < 0$) as well as the growth estimate (11.4.23).

CHAPTER 12

Transient, Smooth Problems

In this chapter, we present the numerical analysis of spectral methods for linear hyperbolic and parabolic problems.

12.1. Linear Hyperbolic Equations

The purpose of this section is to review the theoretical results currently available for spectral approximations to linear hyperbolic problems.

The analysis will be concerned primarily with one-dimensional problems. The model problem we consider is

$$\begin{cases} u_t + a(x)u_x = 0 & \text{for } t > 0, \\ u(x,0) = u_0(x), \end{cases} \tag{12.1.1}$$

supplemented with proper boundary conditions. The real functions a and u_0 are assumed to be smooth. Throughout this chapter we denote, for each t, by $u(t)$ the function v such that $v(x) = u(x, t)$. Since both periodic and non-periodic boundary conditions are relevant in applications, but require different techniques in the analysis, they will be considered separately hereafter.

12.1.1. Periodic Boundary Conditions

In (12.1.1), u, u_0 and a are supposed to be 2π-periodic functions. Let us first recall that the solution u is defined by the formula

$$u(x,t) = u_0(x_0(x,t)), \tag{12.1.2}$$

where $x_0(x, t)$ denotes the value at $\tau = 0$ of the solution of the backward initial value problem

$$\begin{cases} \dfrac{dx}{d\tau} = a(x) & 0 \leqslant \tau \leqslant t \\ x(t) = x. \end{cases}$$

According to (12.1.2), the maximum norm of u on the interval $(0, 2\pi)$ (see

(A.9.f)) is constant in time, i.e.,

$$\|u(t)\|_{L^\infty(0,2\pi)} = \|u_0\|_{L^\infty(0,2\pi)} \qquad t > 0. \tag{12.1.3}$$

On the other hand, the L^2-norm of u, although finite for all $t > 0$, may grow exponentially in time. Indeed, multiplying (12.1.1) by u and integrating by parts over $(0, 2\pi)$, we get

$$\frac{d}{dt}\int_0^{2\pi} u^2\,dx - \int_0^{2\pi} a_x u^2\,dx = 0;$$

whence

$$\|u(t)\|^2_{L^2(0,2\pi)} \leqslant e^{\alpha t}\|u_0\|^2_{L^2(0,2\pi)} \qquad t > 0, \tag{12.1.4}$$

where $\alpha = \max_{0\leqslant x\leqslant 2\pi} a_x(x)$. This estimate actually describes the behavior of the L^2-norm of the solution on a finite time interval. Take for instance the case $a(x) = \sin x$; choose the initial data u_0 such that $u_0(x) = 1$ if $|x| \leqslant \varepsilon$, $u_0(x) = 0$ if $\varepsilon < |x| \leqslant \pi$ and let $\varepsilon \to 0$. (Although this example is not for a smooth function, one can easily regularize it).

However, the L^2-norm of u is bounded independently of t when a is of one sign. In fact, in this case (12.1.1) is equivalent to

$$\frac{1}{a}u_t + u_x = 0,$$

which, by multiplication by u and integration-by-parts, yields

$$\frac{d}{dt}\int_0^{2\pi} \frac{1}{a(x)} u^2(x,t)\,dx = 0,$$

or

$$\|u(t)\|^2_{L^2(0,2\pi)} \leqslant \frac{\max\limits_{0\leqslant x\leqslant 2\pi} |a(x)|}{\min\limits_{0\leqslant x\leqslant 2\pi} |a(x)|}\|u_0\|^2_{L^2(0,2\pi)}. \tag{12.1.5}$$

Finally, we recall that if the functions a and u_0 are globally smooth, then so is u; this follows from (12.1.2)—(12.1.3). Nevertheless, u develops gradients (in space) which grow exponentially in time at each point ξ where a changes sign with strictly negative derivative. Indeed, by differentiating (12.1.2) at $x = \xi$ and recalling that $u(\xi, t) = u_0(\xi)$, we obtain

$$u_x(\xi, t) = e^{-a_x(\xi)t} u_{0,x}(\xi). \tag{12.1.6}$$

This exponential steepening of the solution near these special points poses a difficulty for any numerical approximation of (12.1.1).

Let us now consider spectral methods for this problem. A semi-discrete Fourier approximation is $u^N(t)$, where $u^N(x, t)$ is a trigonometric polynomial of degree N in x, i.e., $u_N(t) \in S_N$ (where S_N is defined in (9.1.1)). It can be defined by a *Galerkin method*:

$$\begin{cases} \hat{u}_{k,t} + \widehat{(au_x^N)}_k = 0 & -N \leqslant k \leqslant N-1, \quad t > 0 \\ \hat{u}_k(0) = \hat{u}_{0,k} & -N \leqslant k \leqslant N-1. \end{cases} \tag{12.1.7}$$

Here $\hat{u}_k$ denotes the k-th Fourier coefficient of u^N. Another way of defining u^N is by a *collocation method*:

$$\begin{cases} u_t^N(x_j, t) + a(x_j)u_x^N(x_j, t) = 0 & j = 0, \ldots, 2N-1, \quad t > 0 \\ u^N(x_j, 0) = u_0(x_j) & j = 0, \ldots, 2N-1 \end{cases} \tag{12.1.8}$$

where $x_j = j\pi/N$.

We discuss now the stability and convergence properties of these methods. The Galerkin solution satisfies, by (12.1.7),

$$\begin{cases} (u_t^N + au_x^N, v) = 0 & \text{for all } v \in S_N, \ t > 0 \\ u^N(0) = P_N u_0, \end{cases} \tag{12.1.9}$$

where $(u, v) = \int_0^{2\pi} u\bar{v}\,dx$ and P_N is the L^2-projection operator upon S_N. Setting $v = u^N$ we obtain

$$\frac{d}{dt}\int_0^{2\pi} |u^N|^2\,dx - \int_0^{2\pi} a_x|u^N|^2\,dx = 0;$$

whence

$$\|u^N(t)\|_{L^2(0,2\pi)}^2 \leqslant e^{\alpha t}\|u_0\|_{L^2(0,2\pi)}^2 \qquad t > 0. \tag{12.1.10}$$

This estimate is the same as the one for the exact solution of (12.1.1) (see (12.1.4)). Thus, the L^2-norm of the Fourier Galerkin solution is bounded independently of N on every finite time interval $[0, T]$. On the other hand, for each fixed N the L^2-norm of u^N is allowed to grow exponentially as $t \to \infty$, precisely as may the L^2-norm of the exact solution, according to (12.1.4).

There are examples in which $\|u^N(t)\|_{L^2(0,2\pi)}$ does grow exponentially in time as $t \to \infty$. This happens, for instance, for the equation $u_t + \sin(\delta x - \gamma)u_x = 0$, as reported in Gottlieb (1981, Sec. 3). Such a phenomenon is attributed (see Gottlieb (1981), Gottlieb, Orszag and Turkel (1981)) to the eventual insufficient resolution of the numerical scheme (for a fixed N), which surfaces as soon as excessively steep gradients are developed in the solution. According to the mechanism described by (12.1.6), oscillations which grow in time are produced in the numerical solution. However, if resolution is improved, i.e., if N is increased, then the growth with time of $\|u^N(t)\|_{L^2(0,2\pi)}$ is retarded.

The fact that oscillations are bounded independently of N on every fixed time interval can also be established by investigating the behavior of higher order Sobolev norms of the spectral solution. Setting $v = -u_{xx}^N$ in (12.1.9) we get

$$\frac{1}{2}\frac{d}{dt}\int_0^{2\pi} |u_x^N|^2\,dx - \int_0^{2\pi} au_x^N\bar{u}_{xx}^N\,dx = 0;$$

whence

$$\|u_x^N(t)\|^2_{L^2(0,2\pi)} \leqslant e^{\alpha t}\|u_{0,x}\|^2_{L^2(0,2\pi)}. \tag{12.1.11}$$

This estimate together with (12.1.10) proves that $u^N(x,t)$ is bounded independently of N for all fixed intervals $0 \leqslant t \leqslant T$.

Finally, the convergence theory established in Sec. 10.5.2 and the approximation estimate (9.1.10) allow us to derive the following error estimate from the stability bound (12.1.10):

$$\|u(t) - u^N(t)\|_{L^2(0,2\pi)} \leqslant Ce^{\alpha t}N^{1-m} \max_{0\leqslant\tau\leqslant t} \|u(\tau)\|_{H_p^m(0,2\pi)}, \quad m \geqslant 1 \tag{12.1.12}$$

(the Sobolev space $H_p^m(0,2\pi)$ is defined in (A.11.d)).

We turn now to the Fourier collocation method (12.1.8). If $a(x)$ does not vanish in $[0,2\pi]$, then (12.1.8) can be written as

$$\frac{1}{a(x_j)}u_t^N(x_j,t) + u_x^N(x_j,t) = 0 \qquad j = 0,\ldots,2N-1.$$

Let us multiply each equation by $\bar{u}^N(x_j,t)(\pi/N)$, and sum up over j. By the exactness of the quadrature rule for polynomials of degree $\leqslant N$ (see Sec. 2.1.2), and by the skew-symmetry of the spatial operator, we get

$$\frac{d}{dt}\sum_{j=0}^{2N-1}\frac{1}{a(x_j)}|u^N(x_j,t)|^2\frac{\pi}{N} = 0;$$

whence

$$\|u^N(t)\|^2_{L^2(0,2\pi)} \leqslant \frac{\max_{0\leqslant x\leqslant 2\pi}|a(x)|}{\min_{0\leqslant x\leqslant 2\pi}|a(x)|}\|I_N u_0\|^2_{L^2(0,2\pi)}, \tag{12.1.13}$$

where $I_N u_0$ is the trigonometric interpolant of u_0 at the collocation nodes. This proves the stability of the method, provided that the initial data is continuous or of bounded variation. Such a result was first established by Gottlieb (1981). Again, the convergence of the method can be inferred using the technique described in Sec. 10.5.2.

For an arbitrary coefficient $a(x)$ which changes sign in the domain, stability results are as yet unproven. On the other hand, no proof of instability of the method for $N \to \infty$ on a fixed time interval has been given. Numerical experiments (see Kreiss and Oliger (1979), Gottlieb, Orszag and Turkel (1981)) show that the collocation solution behaves in a manner qualitatively similar to the Galerkin solution, i.e., its L^2-norm appears to be bounded independently of N on every finite time interval $0 \leqslant t \leqslant T$, even though its L^2-norm may diverge as $t \to \infty$. Gottlieb, Orszag and Turkel (1981) were able to prove a similar result when $a(x)$ is a trigonometric polynomial of degree 1. Precisely, they proved an estimate of the form

$$\|u^N(t)\|_{L^2(0,2\pi)} \leqslant e^{\beta t}(\|u_0\|_{L^2(0,2\pi)} + N|\tilde{u}_{0,N}|),$$

where $\beta > 0$ depends upon the coefficients of a, and $\tilde{u}_{0,N}$ is the N-th discrete Fourier coefficient of the initial data u_0. The right-hand side is bounded independently of N if u_0 is of bounded variation in $[0, 2\pi]$ (according to Zygmund (1968, Chap. X, Theorem 4.8)). The case $a(x) = \sin 2x$ has been discussed by Tal–Ezer (1987a).

A stability result can be proven for two variants of the collocation method—the skew-symmetric version and the filtered version. We begin by considering a Fourier collocation approximation of (12.1.1) in which the spatial term is discretized in a skew-symmetric way (see Gottlieb and Orszag (1977), Kreiss and Oliger (1979), Pasciak (1980)). Since au_x can be decomposed as

$$au_x = \tfrac{1}{2}[au_x + (au)_x] - \tfrac{1}{2}a_x u,$$

one considers the scheme

$$u_t^N(x_j, t) + \tfrac{1}{2}[au_x^N + D_N(au^N)](x_j, t) - \tfrac{1}{2}a_x(x_j)u^N(x_j, t) = 0 \qquad j = 0, \ldots, 2N-1, \tag{12.1.14}$$

where D_N represents the collocation derivative operator at the collocation points (see Sec. 2.1.3). Since, by (2.1.27),

$$\mathbb{R}e(D_N(au^N), u^N)_N = -\mathbb{R}e(au^N, u_x^N)_N = -\mathbb{R}e(au_x^N, u^N)_N$$

we obtain, by multiplying (12.1.13) by $\bar{u}^N(x_j, t)(\pi/N)$ and summing over j,

$$\frac{d}{dt}\|u^N(t)\|_{L^2(0,2\pi)}^2 \leqslant \alpha \|u^N(t)\|_{L^2(0,2\pi)}^2, \qquad t > 0,$$

where again $\alpha = \max_{0 \leqslant x \leqslant 2\pi} a_x(x)$. Thus,

$$\|u^N(t)\|_{L^2(0,2\pi)}^2 \leqslant e^{\alpha t}\|I_N u_0\|_{L^2(0,2\pi)}^2, \tag{12.1.15}$$

which proves stability. Again by the methods of Sec. 10.5.2 one can prove the following convergence estimate (Pasciak (1980):

$$\|u(t) - u^N(t)\|_{L^2(0,2\pi)} \leqslant CN^{1-m}\|u_0\|_{H_p^m(0,2\pi)}, \quad m \geqslant 1. \tag{12.1.16}$$

(The two-dimensional version of this scheme is discussed in Sec. 10.5.2, Example 7.)

Although the skew-symmetric decomposition offers the theoretical warranty of producing stable computations, its use has not become very popular. This is mainly due to the extra cost involved in evaluating the skew-symmetrically decomposed form rather than the simpler form (12.1.8). No computational examples have yet appeared which indicate that this extra cost is sufficiently rewarded by an improvement in the accuracy of the results. Gottlieb, Orszag and Turkel (1981) report their negative experience with the skew-symmetric form. Moreover, the skew-symmetric decomposition does not prevent the onset of oscillations near the points where sharp gradients are developed.

Since oscillations, as well as the possible instability of the numerical solution, are due to the growth of the higher order modes, the preferred alternative to the skew-symmetric decomposition consists of inserting in the scheme (12.1.8) a filtering or smoothing mechanism. This can be accomplished by using the scheme

$$u_t^N(x_j, t) + a(x_j)(\mathscr{S}_N u_x^N)(x_j, t) = 0 \quad j = 0, \ldots, 2N-1, \quad (12.1.17)$$

where $\mathscr{S}_N\colon S_N \to S_N$ is a smoothing operator acting in transform space (see Sec. 2.1.4). The computational effort required by this process is generally relatively modest.

The class of filters proposed by Kreiss and Oliger (1979) offers the theoretical advantage of making easier the derivation of a stability estimate in the L^2-norm. Here is a short description of their method. Fix three real, strictly positive constants m, s, and j. Let M denote the largest integer $\leqslant (1 - (1/m))N$. For each $u = \sum_{k=-N}^{N} \hat{u}_k e^{ikx} \in S_N$, define $u_M \in S_M$ to be the truncation of u of order M, i.e., $u_M = \sum_{|k| \leqslant M} \hat{u}_k e^{ikx}$. Then, the smoothing operator $\mathscr{S}_N$ is defined as $\mathscr{S}_N u = \sum_{k=-N}^{N} \sigma_k \hat{u}_k e^{ikx}$, where

$$\sigma_k = \begin{cases} 1 \quad \text{if } |k| \leqslant M \text{ or } |\hat{u}_k| \leqslant \dfrac{\gamma \|u_M\|_{L^2(0,2\pi)}}{|2\pi k|^s} \\[2ex] \dfrac{\gamma \|u_M\|_{L^2(0,2\pi)}}{|2\pi k|^s |\hat{u}_k|} \qquad \text{otherwise.} \end{cases} \quad (12.1.18)$$

Note that $\mathscr{S}_N$ is stable in the L^2-norm, i.e., $\|\mathscr{S}_N u\|_{L^2(0,2\pi)} \leqslant \|u\|_{L^2(0,2\pi)}$, and leaves unchanged the lower portion of the spectrum, i.e., $\mathscr{S}_N u_M = u_M$. Moreover, $\mathscr{S}_N$ leaves unchanged the polynomials which are "sufficiently smooth", in the sense that

$$\left\| \frac{d^s u}{dx^s} \right\|_{L^2(0,2\pi)} \leqslant \delta \|u\|_{L^2(0,2\pi)} \qquad \text{for a suitable constant } \delta > 0,$$

provided that m and j are properly chosen as functions of δ (see Kreiss and Oliger (1979, Lemma 4.2)).

The operator $\mathscr{S}_N$ prescribes a minimal rate of decay of the higher order coefficients, since $|\sigma_k \hat{u}_k| \leqslant O(|k|^{-s})$. Thus, according to (9.1.7), $\mathscr{S}_N$ enforces a minimal regularity for this part of the spectrum. This suggests that the choice of the actual value of the parameter s should be based upon a priori information on the regularity of the exact solution of (12.1.1).

Kreiss and Oliger prove that with their filter the solution of (12.1.17) satisfies the estimate

$$\frac{d}{dt}\|u^N(t)\|^2_{L^2(0,2\pi)} \leqslant \left\{ \max_{0 \leqslant x \leqslant 2\pi} |(I_N a)_x| + O(N^{2-s}) \right\} \|u^N(t)\|^2_{L^2(0,2\pi)}, \quad (12.1.19)$$

provided that the k-th Fourier coefficient of a decays at least as fast as $|k|^{-s}$. Thus, if $s > 2$, (12.1.19) implies that the L^2-norm of $u^N(t)$ is bounded independently of N on every finite time interval.

Smoothing operators other than Kreiss and Oliger's can be used in (12.1.17) in order to stabilize the computation: for instance, the Lanczos and the raised cosine smoothing (see Sec. 2.1.4), and the exponential filter (8.3.11) considered by Majda, McDonough and Osher (1978). Again, there are no computational examples which indicate that the use of these filtering methods produces for linear problems more stable results than the straightforward collocation method. Although there is no proof that the collocation method is stable, computational experience suggests that it is so.

12.1.2. Non-Periodic Boundary Conditions

We assume that (12.1.1) holds in the interval $-1 < x < 1$, and that the value of u is prescribed for $t > 0$ at the inflow boundary points. This means that u is required to satisfy the conditions

$$\begin{aligned} u(-1,t) &= g_-(t) \qquad \text{if } a(-1) > 0 \\ u(1,t) &= g_+(t) \qquad \text{if } a(1) < 0 \end{aligned} \qquad t > 0, \tag{12.1.20}$$

where $g_\pm$ are smooth data. Under these boundary conditions, problem (12.1.1), (12.1.20) is well-posed in the L^2-norm, since by multiplication of (12.1.1) by u and partial integration we have

$$\frac{d}{dt}\int_{-1}^{1} u^2\,dx - \int_{-1}^{1} a_x u^2\,dx + \sigma_+ a(1) g_+^2 - \sigma_- a(-1) g_-^2 \leqslant 0,$$

where

$$\sigma_\pm = \begin{cases} 0 & \text{if } (\pm 1)a(\pm 1) \leqslant 0 \\ 1 & \text{otherwise.} \end{cases}$$

It follows that

$$\begin{aligned} \|u(t)\|^2_{L^2(0,2\pi)} &\leqslant e^{\alpha t}\|u_0\|^2_{L^2(0,2\pi)} \\ &\quad + \int_0^t e^{\alpha(t-s)}\{-\sigma_+ a(1) g_+^2(s) + \sigma_- a(-1) g_-^2(s)\}\,ds, \end{aligned} \tag{12.1.21}$$

with $\alpha = \max_{-1\leqslant x\leqslant 1} a_x(x)$.

In the analysis of Chebyshev methods, the most natural norms in which to seek the stability of the approximation involve the Chebyshev weight $w(x) = (1 - x^2)^{-1/2}$. However, as pointed out by Gottlieb and Orszag (1977) and Gottlieb and Turkel (1985), the initial-boundary value problem (12.1.1), (12.1.20) need not be well-posed in such a norm. A simple counterexample (Gottlieb and Orszag (1977)) is provided by the constant-coefficient problem

$$u_t + u_x = 0 \qquad u(-1,t) = 0, \tag{12.1.22}$$

with the initial condition

$$u(x,0) = u_0^\varepsilon(x) = \begin{cases} 1 - \dfrac{|x|}{\varepsilon} & \text{if } |x| \leqslant \varepsilon \\ 0 & \text{if } |x| > \varepsilon. \end{cases} \tag{12.1.23}$$

It is easily seen that the L^2_w-norm of the solution, i.e.,

$$\|u(t)\|^2_{L^2_w(-1,1)} = \int_{-1}^{1} u^2(x,t) \frac{dx}{\sqrt{1-x^2}}$$

satisfies the relations

$$\|u_0^\varepsilon\|_{L^2_w(-1,1)} = O(\varepsilon^{1/2}) \qquad \text{but } \|u(1)\|_{L^2_w(-1,1)} = O(\varepsilon^{1/4}).$$

Since ε is arbitrarily small, the problem is not stable in the L^2_w-norm. Clearly, the same negative result holds for hyperbolic systems, as an effect of the propagation and reflection of waves at the boundary points, where the weight w becomes unbounded.

Greater freedom in the choice of the weighted norm in which to seek stability is obtained by allowing the weight function $\tilde{w}$ to be a product of the Chebyshev weight $w(x)$ times a rational function $r(x)$, i.e.,

$$\tilde{w}(x) = r(x)w(x). \tag{12.1.24}$$

For the scalar equation (12.1.1) a good choice consists of taking r of the form $r(x) = (1-x)^\alpha(1+x)^\beta$, where α and β equal 0 or 1 in such a way that $r(x)$ vanishes at the inflow boundary points for (12.1.1) (see Gottlieb and Orszag (1977), Gottlieb (1981), Canuto and Quarteroni (1982b)). When the boundary conditions are homogeneous, the stability in the $L^2_{\tilde{w}}$-norm follows from the identity

$$\frac{d}{dt}\int_{-1}^{1} u^2 \tilde{w}\, dx - \int_{-1}^{1} \{a_x + \tilde{w}^{-1}(a\tilde{w}_x)\} u^2 \tilde{w}\, dx = 0,$$

by observing that the coefficient in braces is bounded from above by a finite constant. Note that now waves always propagate toward boundary points where the weight vanishes. In the case of non-homogeneous boundary conditions, stability can be inferred from the homogeneous case, provided that $g_\pm(t)$ are differentiable functions.

Unfortunately, such a choice of weights does not work for systems of hyperbolic equations, due to the coupling of the unknowns at the boundary. Gottlieb and Turkel (1985) consider the system

$$\begin{cases} u_t - u_x = 0 \\ v_t + v_x = 0 \end{cases}$$

with boundary conditions

$$\begin{aligned} u(1,t) &= \alpha v(1,t) \\ v(-1,t) &= \beta u(-1,t) \end{aligned} \qquad \alpha\beta \neq 0$$

and initial conditions $v(x,0) \equiv 0$, $u(x,0) = u_0^\varepsilon(x)$ given in (12.1.23). It is easily seen that at time $t = 1 + \varepsilon$ one has

$$\int_{-1}^{1} v^2(x, 1+\varepsilon)\tilde{w}(x)\,dx = O(\varepsilon^{1/2}),$$

where $\tilde{w}$ is either the Chebyshev weight or the modified weight $\tilde{w}(x) = (1-x)w(x)$.

Now let us go back to the scalar equation (12.1.1) in order to discuss stability results for spectral approximations. Most of the available results assume a very special or simple form of the function $a(x)$. Hereafter, we treat the three cases of boundary condition separately.

(a) *One inflow boundary point*:
Let us consider the problem

$$\begin{cases} u_t + u_x = 0 & -1 < x < 1, \quad t > 0 \\ u(-1,t) = 0 & t > 0 \\ u(x,0) = u_0(x) & -1 < x < 1. \end{cases} \tag{12.1.25}$$

The *Chebyshev tau* method for this problem defines the solution $u^N(t) \in \mathbb{P}_N$ by the conditions

$$\begin{cases} \displaystyle\int_{-1}^{1} (u_t^N + u_x^N) T_k(x) w(x)\,dx = 0 & k = 0, \ldots, N-1 \\ u^N(-1,t) = 0, \qquad u^N(x,0) = u_0^N(x) & -1 < x < 1 \end{cases} \tag{12.1.26}$$

where $u_0^N \in \mathbb{P}_N$ is a suitable approximation of the initial data. The stability for this method was proved by Gottlieb and Orszag (1977); later, Mercier (1982) considered the same problem from a more abstract point of view. The error equation (see Sec. 10.6) for (12.1.26) is

$$u_t^N + u_x^N = \hat{u}_{N,t} T_N, \qquad -1 < x < 1, \quad t > 0,$$

where $\hat{u}_N$ is the N-th Chebyshev coefficient of $u^N(t)$. Let us take the Chebyshev inner product of both sides with the function $(1-x)u_{xt}^N$. Recalling that the N-th Chebyshev coefficient of this function is $-N\hat{u}_{N,t}$ according to (2.4.22), we have after integration-by-parts

$$-\int_{-1}^{1} (u_t^N)^2 [(1-x)w(x)]_x\,dx + \frac{1}{2}\frac{d}{dt}\int_{-1}^{1} (u_x^N)^2 (1-x)w(x)\,dx + N(\hat{u}_{N,t})^2 = 0.$$

Since the weight $\tilde{w}(x) = (1 - x)w(x)$ is non-increasing in $(-1, 1)$, we obtain

$$\int_{-1}^{1} (u_x^N(x,t))^2 \tilde{w}(x)\,dx \leqslant \int_{-1}^{1} (u_{0,x}^N)^2 \tilde{w}(x)\,dx. \tag{12.1.27}$$

This proves that the spatial gradient of u^N is stable in the $L^2_{\tilde{w}}$-norm provided that the initial data is smooth enough to guarantee that the right-hand side of (12.1.27) is bounded independently of N. If u_0^N is chosen to be the Chebyshev truncation $P_N u_0$ of u_0, this requires $u_0 \in H^2_w(-1, 1)$ according to (9.5.8) (note that $\tilde{w}(x) \leqslant w(x)$, $-1 < x < 1$). If u_0^N is defined as the H^1_w-projection of u_0 upon $\mathbb{P}_N$ (see (9.5.10)), then it is enough that $u_0 \in H^1_w(-1, 1)$. Finally, we remark that by the Poincaré inequality (A.13),

$$|||u||| = \left(\int_{-1}^{1} (u_x)^2 \tilde{w}\,dx\right)^{1/2}$$

is a norm over the space of the sufficiently smooth functions which vanish at $x = -1$. This norm is stronger than the Chebyshev norm; actually, if $u(-1) = 0$, we have by the Cauchy–Schwarz inequality

$$u^2(x) = \left(\int_{-1}^{x} u_x(\xi)\,d\xi\right)^2 \leqslant \left(\int_{-1}^{1} (u_x)^2\,\tilde{w}\,dx\right)\cdot\left(\int_{-1}^{x} \frac{d\xi}{\tilde{w}(\xi)}\right).$$

Integrating over $(-1, 1)$ with respect to the measure $w(x)\,dx$ yields

$$\|u\|_{L^2_w(-1,1)} \leqslant |||u|||.$$

Thus, estimate (12.1.27) implies the boundedness of the Chebyshev norm of $u^N(t)$ for all $t > 0$.

We consider now the *collocation* method for problem (12.1.25). Given $N + 1$ distinct points $-1 = x_N < x_{N-1} < \cdots x_0 \leqslant 1$ in the interval $[-1, 1]$, the collocation solution $u^N(t) \in \mathbb{P}_N$ is defined by the equations

$$\begin{cases} [u_t^N + u_x^N](x_j, t) = 0 & j = 0, \ldots, N-1, \quad t > 0 \\ u^N(-1, t) = 0 & \\ u^N(x_j, 0) = u_0(x_j) & j = 0, \ldots, N. \end{cases} \tag{12.1.28}$$

The collocation points most commonly used are the Gauss–Lobatto points

$$x_j = \cos\frac{\pi j}{N} \qquad j = 0, \ldots, N, \tag{12.1.29}$$

(see (2.4.14)). Numerical evidence shows that the scheme (12.1.28) with these points is stable in the weighted L^2-norm of Chebyshev type

$$\|u\|_{L^2_{\tilde{w}}(-1,1)} = \left(\int_{-1}^{1} u^2(x)\tilde{w}(x)\,dx\right)^{1/2} \qquad \tilde{w}(x) = (1 - x)w(x)$$

(for which the exact problem is well-posed), or rather in the equivalent discrete norm

$$|||u|||_N^2 = \sum_{j=0}^{N} u^2(x_j)(1 - x_j)w_j,$$

where w_j are the Gauss–Lobatto weights (2.4.14). Unfortunately, no proof is available so far to support rigorously this result. Gottlieb and Turkel (1985) have instead introduced a different norm (which depends upon N), namely

$$\|u\|_*^2 = \sum_{j=0}^{N} u^2(x_j)(1 + x_j)\left(1 - \frac{x_j}{2}\right)w_j + \frac{\pi}{4N}[2\hat{u}_N - \hat{u}_{N-1}]^2,$$

and have shown that $\|u^N(t)\|_*$ can be bounded in terms of the initial data u_0 for all $t > 0$. The proof is given in Solomonoff and Turkel (1986).

The lack of a stability proof in the more usual Chebyshev norm appears related to the insufficient precision of the quadrature rule which uses these points, with respect to the weight $\tilde{w}(x) = (1 - x)w(x)$ for which problem (12.1.25) is well-posed. Actually, the formula

$$\sum_{j=0}^{N} v(x_j)(1 - x_j)w_j = \int_{-1}^{1} v(x)\tilde{w}(x)\,dx \tag{12.1.30}$$

is exact only for $v \in \mathbb{P}_{2N-2}$, whereas the polynomial $u^N u_x^N$ has degree $2N - 1$. Therefore, other sets of collocation points have been considered, for which (12.1.30) holds whenever $v \in \mathbb{P}_{2N-1}$. Gottlieb (1981) proposed to retain all but one of the Gauss–Lobatto points relative to the degree $N + 1$, namely

$$x_j = \cos\frac{(j + 1)\pi}{N + 1} \qquad j = 0, \ldots, N. \tag{12.1.31}$$

In this case, (12.1.31) is actually exact for polynomials of degree $2N$. From (12.1.28) we obtain

$$\frac{d}{dt}\sum_{j=0}^{N} [u^N(x_j, t)]^2(1 - x_j)w_j + \sum_{j=0}^{N} [(u^N)^2]_x(x_j, t)(1 - x_j)w_j = 0;$$

whence

$$\frac{d}{dt}\int_{-1}^{1} (u^N)^2\tilde{w}\,dx = -\int_{-1}^{1} (u^N)^2\tilde{w}_x\,dx \leqslant 0. \tag{12.1.32}$$

This implies the stability of the method. Of course, such a result does not by itself imply stability for the Gauss–Lobatto points (12.1.29). Computationally, it is still possible to use the FFT algorithm to compute the Chebyshev coefficients $\hat{u}_k$, $(k = 0, \ldots, N)$ of the polynomial u^N, starting from its values at the points (12.1.31). In fact, one can compute by the FFT the Chebyshev coefficients $\hat{u}_k$, $(k = 0, \ldots, N)$ of the polynomial v of degree $N + 1$ which equals u^N at the points (12.1.31) and is zero at $x = +1$, and then use the relation (Gottlieb (1981)),

$$\hat{u}_k = \hat{v}_k + (-1)^{N+k}\frac{2}{c_k}\hat{v}_{N+1} \qquad k = 0, \ldots, N-1. \tag{12.1.33}$$

Another choice of collocation points for the scheme (12.1.28), which leads to the stability estimate (12.1.32), is the set of the Gauss–Radau points (see (2.4.13))

$$x_j = \cos\left(\frac{2j+1}{2N+1}\pi\right) \qquad j = 0, \ldots, N, \tag{12.1.34}$$

which have been proposed by Mercier (1982). More details on the stability analysis concerning these collocation points is provided in Sec. 10.5.2, Example 8. Again, a Fast Fourier Transform can be used in computing the Chebyshev coefficients of a polynomial from its values at the points (12.1.34) and conversely. Since such points are the projection on the real-axis of the roots of unity of order $2N + 1$, an FFT for an odd number of frequencies, rather that the more common FFT for power of two frequencies, is needed.

(b) *Two inflow boundary points*:
Now, we consider the problem

$$\begin{cases} u_t - xu_x = 0 & -1 < x < 1, \quad t > 0 \\ u(-1,t) = u(1,t) = 0 & t > 0 \\ u(x,0) = u_0(x) & -1 < x < 1. \end{cases} \tag{12.1.35}$$

Let us denote by $Lu = -xu_x$ the spatial operator which appears in this initial-boundary value problem. We assume that u belongs to the space $H^1_{w,0}(-1,1)$ (see (A.11.c)). Using the Cauchy–Schwarz inequality and Lemma 11.1, it is easily seen that the bilinear form

$$[Lu, v] = -\int_{-1}^{1} xu_x v \frac{w(x)}{1-x^2}\,dx$$

is finite for all u and v in $H^1_{w,0}(-1,1)$. Moreover, by partial integration

$$[Lu, u] = \frac{1}{2}\int_{-1}^{1} u^2(x)\left(\frac{x}{(1-x^2)^{3/2}}\right)_x dx \geqslant 0.$$

This implies that the operator L with homogeneous boundary conditions is semi-definite in the norm

$$\|u\|^2 = \int_{-1}^{1} \frac{u^2(x)}{1-x^2} w(x)\,dx. \tag{12.1.36}$$

Thus, problem (12.1.35) is well-posed in this norm.

Let $u^N(t)$ denote the solution of the *Chebyshev collocation* method

$$u_t^N(x_j, t) - x_j u_x^N(x_j, t) = 0 \qquad j = 1, \ldots, N-1, \quad t > 0 \quad (12.1.37)$$

plus boundary and initial conditions, for the Gauss–Lobatto points (12.1.29). Multiplying the j-th equation of (12.1.37) by $(u^N(x_j, t)/(1 - x_j^2))w_j$ and summing up over j, one gets

$$\begin{aligned}\frac{1}{2}\frac{d}{dt}\sum_{j=1}^{N-1}\frac{[u^N(x_j,t)]^2}{1-x_j^2}w_j &= \sum_{j=1}^{N-1} u_x^N(x_j,t)\frac{u^N(x_j,t)}{1-x_j^2}x_j w_j \\ &\leqslant \sum_{j=0}^{N} u_x^N(x_j,t)\frac{u^N(x_j,t)}{1-x_j^2}x_j w_j \\ &= \int_{-1}^{1} u_x^N(x,t)\frac{u^N(x,t)}{1-x^2}xw(x)\,dx \leqslant 0.\end{aligned}$$

We have used the fact that $u_x^N(x,t)[u^N(x,t)/(1-x^2)]x$ tends to $[(u^N)^2]_x(\pm 1, t) \geqslant 0$, as $x \to \pm 1$. Thus, the Chebyshev collocation method is stable in the norm (12.1.36).

The *Chebyshev tau* method for problem (12.1.35) is analyzed in Example 6 of Sec. 10.5.2.

(c) *No inflow boundary points*:
Finally, we consider the problem

$$\begin{cases} u_t + xu_x = 0 & -1 < x < 1, \quad t > 0 \\ u(x,0) = u_0(x) & -1 < x < 1. \end{cases} \qquad (12.1.38)$$

If u^N is either the tau or the collocation solution, then $u_t^N + xu_x^N$ is a polynomial of degree N, which is orthogonal to all the polynomials of degree N in the tau scheme, or is zero at $N + 1$ distinct points in the collocation scheme. Hence,

$$u_t^N + xu_x^N \equiv 0 \qquad -1 \leqslant x \leqslant 1, \quad t > 0.$$

Therefore, the stability of both schemes follows from the well-posedness of problem (12.1.38) in the L^2-norm.

A Legendre tau approximation for a control problem involving hyperbolic equations has been discussed by Ito and Teglas (1986).

12.1.3. Hyperbolic Systems

We end this section with a brief account of the existing theoretical results on spectral approximations to *hyperbolic systems*. Consider a linear system of the form

$$\mathbf{u}_t + A\mathbf{u}_x = 0 \qquad \mathbf{u}(x,0) = \mathbf{u}_0(x), \qquad (12.1.39)$$

where $\mathbf{u}(x, t)$ is an n-dimensional vector and $A = A(x, t)$ is a smooth non-singular symmetric matrix. At the boundary, one has to prescribe the incoming variables in terms of the outgoing variables: assuming $A(\pm 1, t)$ to be non-singular, there exist orthogonal matrices $T(\pm 1, t)$ such that

$$T^{-1}(\pm 1,t)A(\pm 1, t)T(\pm 1, t) = \mathrm{diag}(\Lambda^{I}(\pm 1, t), \Lambda^{II}(\pm 1, t)) = \Lambda(\pm 1, t), \tag{12.1.40}$$

where $\Lambda^{I}(\pm 1, t)$, (respectively $\Lambda^{II}(\pm 1, t)$) are positive-definite (respectively negative-definite) diagonal matrices. Setting $\mathbf{v}(\pm 1, t) = T^{-1}(\pm 1, t)\mathbf{u}(\pm 1, t)$ and splitting $\mathbf{v}$ according to (12.1.40), i.e., $\mathbf{v} = (\mathbf{v}^{I}, \mathbf{v}^{II})^{T}$, the boundary conditions to be prescribed are

$$\mathbf{v}^{I}(\pm 1, t) = S(\pm 1, t)\mathbf{v}^{II}(\pm 1, t) + G(\pm 1, t), \tag{12.1.41}$$

where G is a smooth vector and $S(\pm 1, t)$ are rectangular reflection matrices. If one assumes that the boundary conditions satisfy the dissipativity condition

$$(\pm 1)\{S^{T}(\pm 1, t)\Lambda^{I}(\pm 1, t)S(\pm 1, t) + \Lambda^{II}(\pm 1, t)\} < 0, \tag{12.1.42}$$

then the spatial L^2-norm of the solution is stable in time, i.e., it can be bounded in terms of the initial and boundary data. On the other hand, as previously seen, a hyperbolic system may not be stable in any of the weighted Chebyshev L^2-norms.

These two facts led Reyna (1982) to propose a Chebyshev collocation method with a smoothing consisting of a high-mode cut-off in Legendre transform space rather than in Chebyshev transform space. The extra cost compared with a more conventional Chebyshev smoothing is the multiplication by a matrix that transforms between Chebyshev expansions and Legendre expansions. The size of this full matrix depends upon the amount of smoothing. The resulting numerical scheme is stable and convergent in the L^2-norm and the stability estimate corresponds to the estimate for the continuous problem.

A stability and convergence analysis based on the use of the Fourier–Laplace transform has been carried out by Gottlieb, Lustman, and Tadmor (1987a, 1987b). The study was initiated by Lustman (1984), who investigated conditions under which the spectral solution of a constant-coefficient hyperbolic system decays in time whenever the exact solution does so. Gottlieb, Lustman and Tadmor assume that the coefficient matrix A in (12.1.39) is constant, and that the reflection matrices $S(\pm 1)$ defined in (12.1.41) satisfy the dissipativity requirement

$$\|S(+1)\| \cdot \|S(-1)\| \leqslant 1 - \delta \qquad \delta > 0, \tag{12.1.43}$$

(here $\|S\|$ denotes the Euclidean norm of the matrix S). According to (12.1.43), waves originating at one of the boundaries are not amplified when reflected at the other one. Thus, the solution does not grow in time, in the sense that the following estimate holds for all $\eta > 0$:

$$\eta \int_0^\infty e^{-2\eta t} \|\mathbf{u}(t)\|^2 \, dt \leqslant C \int_0^\infty e^{-2\eta t} |\mathbf{G}(t)|^2 \, dt, \qquad (12.1.44)$$

provided that the initial data u_0 is zero. In the previous estimate $\|\mathbf{u}(t)\|$ denotes a suitable spatial weighted norm of $\mathbf{u}$, while $|\mathbf{G}(t)|$ is the Euclidean norm of the vector $(G(-1,t), G(+1,t))^T$. Any spectral approximation to the system (12.1.39) in diagonal form can be represented as

$$\mathbf{v}_t^N + \Lambda \mathbf{v}_x^N = Q(x)\boldsymbol{\tau}(t). \qquad (12.1.45)$$

This is the error equation corresponding to the method (see Sec. 10.6). Each component of (12.1.45) has the form

$$v_t^N + \lambda v_x^N = \tau(t) q(x), \qquad (12.1.46)$$

where v^N is a N-th degree polynomial in the x variable, λ is a non-zero constant, $q(x)$ is the N-th degree polynomial which characterizes the specific spectral method, and the coefficients $\tau(t)$ are determined by the set of boundary conditions. In order to carry out their analysis, Gottlieb et al. introduce the new variable $e^{-\eta t} v^N(x,t)$ and Fourier transform the resulting equation with respect to time (this method was introduced by Kreiss (1970)). In the transformed variables (12.1.46) becomes

$$s\hat{v}^N + \lambda \hat{v}_x^N = q(x)\hat{\tau} \qquad s = \eta + i\xi, \qquad (12.1.47)$$

with ξ denoting the real dual variable corresponding to time. One remarkable aspect of the analysis is that (algebraic) stability for the system can be proven as a consequence of a sufficiency criterion which deals exclusively with the properties of the scalar equation. Thus, the difficulty inherent in the coupling of the scalar equations through the boundary conditions is avoided. More precisely, it is assumed that for the scalar problem (12.1.47) with $\lambda > 0$, an (algebraic) stability estimate of the form

$$(\eta - \eta_0) \|\hat{v}^N(s)\|^2 \leqslant C N^{2\alpha} |\hat{v}^N(-1,s)|^2 \qquad (12.1.48)$$

holds for all s with $\mathbb{R}e\, s = \eta > \eta_0$, with suitable constants η_0 and α. The norm on the left-hand side is a suitable weighted L^2-norm. Moreover, the solution of (12.1.47), again with $\lambda > 0$, has to satisfy the inequality

$$|\hat{v}^N(1,s)| \leqslant |\hat{v}^N(-1,s)| \qquad \text{for } \mathbb{R}e\, s > \eta_0. \qquad (12.1.49)$$

Obviously, estimates similar to (12.1.48)—(12.1.49) must hold when $\lambda < 0$. Under these assumptions one can prove that for the spectral solution of (12.1.45) the following stability estimate holds for all $\eta > \eta_0$:

$$(\eta - \eta_0) \int_0^\infty e^{-2(\eta-\eta_0)t} \|\mathbf{v}^N(t)\|^2 \, dt \leqslant C N^{2\alpha} \int_0^\infty e^{-2(\eta-\eta_0)t} |\mathbf{G}(t)|^2 \, dt. \qquad (12.1.50)$$

The convergence of the method is then proven by comparing the numerical solution to a projection of the exact solution.

The stability assumptions on the scalar problem are fulfilled by a Chebyshev or Legendre collocation method, which uses as collocation points the zeros of T'_{N+1} or L'_{N+1}. For these methods, the weight function in the spatial L^2-norm is given by $(1 \pm x)w(x)$, where $w(x)$ is either the Chebyshev or the Legendre weight and the sign is chosen in order to annihilate the weight at the outflow boundary. The exponent α in (12.1.50) is 1/2 for the Chebyshev case and 1 for the Legendre case. The previous analysis does not apply to the more popular collocation method which uses as interior collocation knots the points (12.1.29), i.e., the zeros of T'_N, and which collocates at the boundaries those linear combinations of the equations that correspond to outgoing characteristic variables (see Sec. 8.2 for more details on this method).

A collocation Legendre method which uses the zeros of the Legendre polynomials has been proposed and analyzed by Tal–Ezer (1986b).

12.1.4. Spectral Accuracy for Non-Smooth Solutions

The convergence results presented in the previous sections are meaningful under the assumption that the exact solution be smooth enough, in the sense that it belongs to a Sobolev space of sufficiently high order. In hyperbolic problems, however, discontinuities in the data are propagated toward the interior of the domain, and if the operator is non-linear discontinuities can even develop in a finite time starting from smooth data.

If global convergence at a spectral rate is unattainable in such cases, at least one can hope to achieve spectral accuracy away from discontinuities, i.e., in those regions where the solution is smooth. The results of the analysis by Majda, McDonough and Osher (1978), briefly summarized in Sec. 8.3, indicate that it is not realistic to expect spectral accuracy directly in the numerical solution obtained by a standard collocation scheme. The use of a proper filter in transform space (see Sec. 2.1.4) may dramatically improve the global accuracy of the solution, as documented in Sec. 8.3. However, only a finite-order decay of the error is usually observed in practical applications.

As opposed to global smoothing, one can post-process the collocation solution by a local smoothing, in order to recover spectral accuracy. The idea is based on the observation that while the pointwise convergence of a high order polynomial approximation to a discontinuous solution is very slow, the convergence in a weighted mean—the weight being a smooth function—is very fast, because oscillations kill each other on the average. Local smoothing can be carried out by a convolution in physical space with a localized function, and hence by a weighted mean which approximates exceedingly well the exact value of the solution.

From a rigorous mathematical point of view, the convergence in the mean can be measured in terms of a *Sobolev norm of negative order*. For simplicity, let us confine ourselves to the case of periodic functions. Each function

$f \in L^2(0, 2\pi)$ defines a continuous linear form on the space $H_p^s(0, 2\pi)$ $(s \geqslant 0)$ (introduced in (A.11.d)), given by the mapping $\phi \to (f, \phi) = \int_0^{2\pi} f(x)\overline{\phi(x)}\,dx$. Thus, f can be identified with an element in the dual space of $H_p^s(0, 2\pi)$, here denoted by $H_p^{-s}(0, 2\pi)$ (see (A.1.c)). Its norm in this space is given by

$$\|f\|_{-s} = \sup_{\phi \in H_p^s(0,2\pi)} \frac{|(f, \phi)|}{\|\phi\|_s}. \tag{12.1.51}$$

For the remainder of this section $\|\phi\|_s$ denotes the norm of ϕ in $H_p^s(0, 2\pi)$. As usual, let $P_N f \in S_N$ be the truncation of the Fourier series of f to N modes. We want to estimate the error $f - P_N f$ in a negative Sobolev norm. If (u, v) denotes the $L^2(0, 2\pi)$-inner product, by definition of P_N we have for all $\phi \in H_p^s(0, 2\pi)$

$$(f - P_N f, \phi) = (f - P_N f, \phi - P_N \phi).$$

Hence,

$$\begin{aligned} |(f - P_N f, \phi)| &\leqslant \|f - P_N f\|_0 \|\phi - P_N \phi\|_0 \\ &\leqslant C N^{-s} \|\phi\|_s \|f\|_0. \end{aligned}$$

Here we have used (9.1.9). Thus, we obtain the estimate

$$\|f - P_N f\|_{-s} \leqslant C N^{-s} \|f\|_0, \qquad s \geqslant 0. \tag{12.1.52}$$

Note that even though f is merely square integrable, the truncation error in a negative Sobolev norm decays at a rate which depends solely upon the order of the norm.

As first pointed out by Mercier (1981), the previous argument can be extended to get an estimate in negative norms for the error between the exact and the spectral solutions to a linear hyperbolic problem. Let L be a linear, first-order hyperbolic operator with smooth periodic coefficients, such that $(Lu, u) \geqslant 0$ for all $u \in H_p^1(0, 2\pi)$. Denote by $u = u(t)$ the solution of the following initial-boundary value problem:

$$\begin{cases} u_t + Lu = 0 & 0 < x < 2\pi, \quad t > 0 \\ u \quad 2\pi\text{-periodic in } x & \\ u(0) = u_0 \in L^2(0, 2\pi). & \end{cases} \tag{12.1.53}$$

Let $u^N = u^N(t) \in S_N$ be the solution of the following Galerkin approximation of (12.1.53):

$$\begin{cases} (u_t^N + Lu^N, v) = 0 & \text{for all } v \in S_N, \quad t > 0 \\ (u^N(0) - u_0, v) = 0 & \text{for all } v \in S_N. \end{cases} \tag{12.1.54}$$

We want to estimate the quantity $(u(t) - u^N(t), \phi)$, where $\phi \in H_p^s(0, 2\pi)$. To this end, let L^* be the adjoint of L, i.e., $(L^* w, v) = (w, Lv)$ for all v and $w \in H_p^1(0, 2\pi)$. Define $w = w(t)$ to be the solution of the hyperbolic problem

$$\begin{cases} w_t + L^*w = 0 & 0 < x < 2\pi, \quad t > 0 \\ w \quad 2\pi\text{-periodic in } x & \\ w(0) = \phi. & \end{cases} \tag{12.1.55}$$

Next, consider the corresponding Galerkin approximation $w^N = w^N(t) \in S_N$ which satisfies

$$\begin{cases} (w_t^N + L^*w^N, v) = 0 & \text{for all } v \in S_N,\ t > 0 \\ (w^N(0) - \phi, v) = 0 & \text{for all } v \in S_N. \end{cases} \tag{12.1.56}$$

For a fixed $t > 0$ we have

$$\begin{aligned} (u(t) - u^N(t), \phi) &= (u(t) - u^N(t), w(0)) \\ &= (u(t), w(0)) - (u^N(t), w^N(0)). \end{aligned}$$

Set $\tilde{w}(s) = w(t - s)$. Then, for $0 < s < t$

$$\frac{d}{ds}(u(s), \tilde{w}(s)) = (u_s, \tilde{w}) + (u, \tilde{w}_s) = -(Lu, \tilde{w}) + (u, L^*\tilde{w}) = 0.$$

Thus,

$$(u(t), w(0)) = (u_0, w(t)). \tag{12.1.57}$$

Similarly,

$$(u^N(t), w^N(0)) = (u^N(0), w^N(t)) = (u_0, w^N(t)). \tag{12.1.58}$$

It follows from (12.1.57) that

$$(u(t) - u^N(t), \phi) = (u_0, w(t) - w^N(t)). \tag{12.1.59}$$

Under the assumptions on L, if ϕ belongs to $H_p^s(0, 2\pi)$, then the solution to (12.1.55) belongs to $H_p^s(0, 2\pi)$ for all times and $\|w(t)\|_s \leqslant C\|\phi\|_s$, (see, e.g., Taylor (1981)). Moreover, the theory of Sec. 10.5.2 yields the error estimate

$$\|w(t) - w^N(t)\|_0 \leqslant CN^{-s}\|\phi\|_s.$$

Thus, we obtain the error estimate in negative Sobolev norm

$$\|u(t) - u^N(t)\|_{-s} \leqslant CN^{-s}\|u_0\|_0 \qquad s \geqslant 0. \tag{12.1.60}$$

The previous proof can be suitably adapted to cover the case of a Fourier collocation approximation.

Finally, we are going to use (12.1.60) in order to show that it is possible to use the information contained in $u^N(t)$ to approximate $u(t)$ with spectral accuracy at each point where u is smooth. The idea, already sketched in Mercier (1981), has been developed independently by Gottlieb and coworkers, both theoretically and computationally (see Gottlieb (1985), Gottlieb and Tadmor (1986), Abarbanel, Gottlieb and Tadmor (1986)).

Let us drop the dependence upon time in all the functions which appear hereafter. Assume that at time $t > 0$ the solution u of (12.1.53) is infinitely smooth in an open neighborhood J of a point $x_0 \in [0, 2\pi]$. Let us choose an infinitely differentiable function $\rho = \rho(x)$ such that ρ is identically zero outside J, ρ is non-negative everywhere, and $\rho(x_0) = 1$. Thus, the function ρu is everywhere smooth and $(\rho u)(x_0) = u(x_0)$. For each fixed $M > 0$, the maximum error between ρu and its Fourier truncation $P_M(\rho u)$ can be estimated according to (9.1.12)—(9.1.14):

$$\|\rho u - P_M(\rho u)\|_{L^\infty(0,2\pi)} \leqslant C(1 + \log M)M^{-s}\|\rho u\|_{s,\infty}, \quad (s \geqslant 0).$$

The norm $\|\rho u\|_{s,\infty}$ is the maximum modulus over $(0, 2\pi)$ of all the derivatives of ρu of order up to s. Such a quantity can be bounded by a constant, depending upon ρ, times the maximum modulus over J of the derivatives of u of order up to s. This latter quantity is finite by assumption and will be denoted by $\|u\|_{s,\infty,J}$. Thus,

$$|u(x_0) - P_M(\rho u)(x_0)| \leqslant C(1 + \log M)M^{-s}\|u\|_{s,\infty,J}. \tag{12.1.61}$$

On the other hand, we have the following representation of $P_M(\rho u)$ as a convolution integral (see (2.1.44))

$$P_M(\rho u)(x_0) = \frac{1}{2\pi}\int_0^{2\pi} D_M(x_0 - y)\rho(y)u(y)\,dy, \tag{12.1.62}$$

where D_M is the Dirichlet kernel, used here with the classical notation ((2.1.45) with N replaced by $2M$). For a fixed M, the function $\phi(y) = D_M(x_0 - y)\rho(y)$ is an infinitely smooth, periodic function. Thus, we can apply (12.1.60) and get

$$\begin{aligned}\left|\int_0^{2\pi} D_M(x_0 - y)\rho(y)u(y)\,dy - \int_0^{2\pi} D_M(x_0 - y)\rho(y)u^N(y)\,dy\right| \\ \leqslant CN^{-s}\|u_0\|_0\|\phi\|_s.\end{aligned} \tag{12.1.63}$$

The norm $\|\phi\|_s$ can be bounded by $C(1 + M)^{s+1}\|\rho\|_s$. Finally, we choose $M = N^\beta$ with $0 < \beta < 1$ and we denote by

$$Ru^N(x_0) = \frac{1}{2\pi}\int_0^{2\pi} D_{N^\beta}(x_0, y)\rho(y)u^N(y)\,dy, \tag{12.1.64}$$

the regularized value of u^N at the point x_0. Note that this value can only be evaluated exactly once the Fourier coefficients of ϕ are known; in practice, in order to evaluate the integral in (12.1.64) one can use a trapezoidal rule with sufficiently many points. By (12.1.61) and (12.1.63) we obtain the following error estimate

$$|u(x_0) - Ru^N(x_0)| \leqslant C_1(1 + \log N)N^{-s\beta} + C_2 N^{-s+\beta(1+s)}, \tag{12.1.65}$$

where the constants depend upon Sobolev norms of ρ and u over the interval J. We conclude that $u(x_0)$ can be approximated with spectral accuracy starting from the knowledge of the Galerkin approximation u^N. A balance of the errors in (12.1.65) is achieved by choosing $\beta = 1/2$.

A number of generalizations of the previous results are possible. First, one can consider a collocation approximation, in which case the integral in (12.1.64) is replaced by the trapezoidal rule. An extra error term due to aliasing is added, but the asymptotic behavior of the error is the same. Next one can consider Legendre or Chebyshev methods for non-periodic problems. An integral representation of the truncation operator, similar to (12.1.62), is still available. The Dirichlet kernel has to be replaced by the kernel

$$K_M(\xi) = \sum_{k=0}^{M} (k + \tfrac{1}{2}) L_k(\xi) L_k(0) \tag{12.1.66}$$

in a Legendre method, and by the kernel

$$K_M(\xi) = \frac{2}{\pi} \sum_{k=0}^{M} T_k(\xi) T_k(0) \tag{12.1.67}$$

in a Chebyshev method. For the details we refer to Gottlieb and Tadmor (1985) and Abarbanel, Gottlieb and Tadmor (1986).

From a computational point of view, one has to choose a proper cut-off function ρ whose support is in the region of smoothness of the solution, and also choose a value for β. As usual, the method may require a fine-tuning of the parameters for the problem at hand. Gottlieb and Tadmor (1985) consider the piecewise C^∞ function

$$u(x) = \begin{cases} \sin\dfrac{x}{2} & 0 \leqslant x < \pi \\[2ex] -\sin\dfrac{x}{2} & \pi \leqslant x < 2\pi \end{cases} \tag{12.1.68}$$

and use an exponential cut-off function. Denoting by u^N the truncation of the Fourier series, the results listed in Tables 12.1 and 12.2 have been reported.

Table 12.1. Results of smoothing of the spectral approximation of $u(x)$, $N = 64$

$x_\nu = \dfrac{\pi\nu}{8}$ ν equals	$\lvert u(x_\nu) - u^N(x_\nu)\rvert$	$\lvert u - Ru^N\rvert$ at $x = x_\nu$
2	6.4 (−3)	4.8 (−6)
3	1.0 (−2)	5.9 (−6)
4	1.5 (−2)	7.7 (−6)
5	2.3 (−2)	12.9 (−6)

Table 12.2. Results of smoothing of the spectral approximation of $u(x)$, $N = 128$

$x_\nu = \frac{\pi\nu}{8}$ ν equals	$\lvert u(x_\nu) - u^N(x_\nu)\rvert$	$\lvert u - Ru^N\rvert$ at $x = x_\nu$
2	3.2 (−3)	5.8 (−10)
3	5.2 (−3)	7.9 (−10)
4	7.8 (−3)	6.3 (−10)
5	1.1 (−2)	1.1 (−10)

The unsmoothed error decays linearly in N^{-1}, whereas spectral accuracy is clearly documented for the regularized approximation.

12.2. Heat Equation

In this section, several spectral approximations to the heat equation will be reviewed. Semi-discrete (continuous-in-time) approximations will be considered first. Then, the fully-discrete approximation based on the "θ-method" will be analyzed.

As usual, we focus our discussion on Chebyshev approximations. For the analysis of Fourier approximations see, for example, Bernardi (1982).

12.2.1. Semi-Discrete Approximation

We consider the one-dimensional heat equation with Dirichlet boundary conditions:

$$\begin{cases} u_t - u_{xx} = 0 & -1 < x < 1, \quad t > 0 \\ u(-1, t) = u(1, t) = 0 & t > 0 \\ u(x, 0) = u_0(x) & -1 < x < 1. \end{cases} \tag{12.2.1}$$

A semi-discrete approximation to this problem, based on the Chebyshev collocation method, has been discussed in Secs. 10.1.2 and 10.5.1 (see Example 3). Using the general theory developed in Sec. 10.5.1, it was possible to prove the error estimate (10.1.6).

We recall that a semi-discrete approximation to (12.2.1) based on the Fourier Galerkin method has been analyzed in Example 1 of Sec. 10.5.1, while a Legendre tau method has been discussed in Example 2 of Sec. 10.5.1. In Example 4 of the same section, a Chebyshev collocation approximation to the heat equation with Neumann boundary conditions has been analyzed.

Consider now the two-dimensional heat equation:

$$\begin{cases} u_t - \Delta u = 0 & x \in \Omega = (-1,1)^2, \quad t > 0 \\ u(x,t) = 0 & x \in \partial\Omega, \quad t > 0 \\ u(x,0) = u_0(x) & x \in \Omega. \end{cases} \tag{12.2.2}$$

The bilinear form $a(u, v)$ associated with the Laplacian operator with Dirichlet boundary conditions was introduced in (11.1.8). Semi-discrete Galerkin approximations to (12.2.2) can be analyzed using the method of Sec. 10.5.1, for $a(u, v)$ is a continuous and coercive form (see Theorem 11.1). The convergence estimate is still of the same form as (10.1.6).

The analysis of semi-discrete Chebyshev collocation approximations to (12.2.2) requires more attention. The semi-discrete solution is a function u^N which is, for each $t > 0$, a polynomial of degree N in each variable satisfying

$$u_t^N - \Delta u^N = 0 \qquad \text{at } x_{ij} \qquad \text{for } 1 \leqslant i, j \leqslant N-1. \tag{12.2.3}$$

The points $\{x_{ij}\}$ are the Chebyshev–Gauss–Lobatto nodes on Ω, defined in (11.1.21). Moreover, u^N vanishes at those points x_{ij} belonging to the boundary of Ω (i.e., if at least one of the indices i or j takes the value 0 or N). Let w_{ij} be the weights of the Gauss–Lobatto formula relative to the $(N+1)^2$ points x_{ij}. Then, using the definitions (11.1.22) and (11.1.23), it is easily seen that u^N satisfies

$$(u_t^N, v)_N + a_N(u^N, v) = 0 \qquad \text{for all } v \in \mathbb{P}_N^0. \tag{12.2.4}$$

Here, $\mathbb{P}_N^0$ denotes the space of algebraic polynomials of degree N in either variable, vanishing on the boundary of Ω. As recalled in Sec. 11.1, the bilinear form $a_N(u, v)$ is continuous and coercive on $\mathbb{P}_N^0$ (see Theorem 11.2). Thus, the energy method of Sec. 10.5.1 can be used to get the stability and convergence estimates (10.5.24) and (10.5.25) for the semi-discrete collocation approximation to the heat equation (12.2.2). Using the approximation results of Sec. 9.7, the inequality (10.5.25) yields the following error estimate:

$$\begin{aligned} &\|u(t) - u^N(t)\|_{L_w^2(\Omega)} + \left\{ \int_0^t \|\nabla(u - u^N)(\tau)\|_{L_w^2(\Omega)}^2 \, d\tau \right\}^{1/2} \\ &\leqslant CN^{1-m} \left\{ \int_0^t (\|u(\tau)\|_{H_w^m(\Omega)}^2 + \|u_t(\tau)\|_{H_w^{m-2}(\Omega)}^2 \right. \\ &\quad \left. + \|f(\tau)\|_{H_w^{m-1}(\Omega)}^2) \, d\tau \right\}^{1/2} \end{aligned} \tag{12.2.5}$$

for all $t > 0$.

Similar results can be proven for semi-discrete collocation methods based on Legendre expansions.

12.2.2. Fully Discrete Approximation

Semi-discrete approximations to the heat equation yield systems of ordinary differential equations. Finite-difference methods can be used for their solution. Unfortunately, as seen in Chap. 4, explicit methods have a severe stability restriction on the time-step Δt of the form $\Delta t \leqslant C/N^4$, where C is a constant and N is the number of Chebyshev points to be used in each space dimension. To avoid such a restriction, implicit time-discretization methods are customary.

In this section, the properties of stability and convergence of the "θ-method" will be investigated. As it is well known, this method includes, among others, both the forward and backward Euler methods (for $\theta = 0$ and $\theta = 1$, respectively), as well as the Crank–Nicolson method (for $\theta = \frac{1}{2}$).

Let $\Delta t > 0$ be the time-step, let $t^k = k\Delta t$, and let ϕ_j^k denote the value of the function ϕ for $x = x_j$ and $t = t^k$, where $x_j = \cos \pi j/N$. The fully discrete approximation to (12.2.1) reads as follows:

For any $k \geqslant 0$, U^k is a polynomial of degree N which satisfies:

$$\begin{cases} U_j^{k+1} - U_j^k - \Delta t\{\theta(U_{xx})_j^{k+1} + (1-\theta)(U_{xx})_j^k\} = 0 & 1 \leqslant j \leqslant N-1 \\ U_0^{k+1} = U_N^{k+1} = 0 & \\ U_j^0 = u_0(x_j) & 0 \leqslant j \leqslant N. \end{cases} \tag{12.2.6}$$

If $\theta = 0$, this is an explicit method. We assume, hereafter, that $\frac{1}{2} \leqslant \theta \leqslant 1$.

By standard arguments, (12.2.6) can be restated as follows: $U^k \in \mathbb{P}_N^0$ for all $k \geqslant 0$, and satisfies:

$$(U^{k+1} - U^k, v)_N + \Delta t a(\theta U^{k+1} + (1-\theta)U^k, v) = 0 \quad \text{for all } v \in \mathbb{P}_N^0. \tag{12.2.7}$$

Furthermore, $U^0 = I_N u_0$.

For convenience of notation we denote here by $\|v\|_r$ $(r \geqslant 0)$ the norm $\|v\|_{H_w^r(-1,1)}$ defined in (A.11.b) (in particular, $\|v\|_0$ is the norm of v in $L_w^2(-1,1)$). To prove stability, let us take $v = \theta U^{k+1} + (1-\theta)U^k$ in (12.2.7). By (11.1.14), we get:

$$\theta \|U^{k+1}\|_N^2 + (1-2\theta)(U^{k+1}, U^k)_N - (1-\theta)\|U^k\|_N^2 + \frac{\Delta t}{4}\|\theta U_x^{k+1} + (1-\theta)U_x^k\|_0^2 \leqslant 0. \tag{12.2.8}$$

Since $1 - 2\theta \leqslant 0$, the Cauchy–Schwarz inequality gives:

$$(1-2\theta)(U^{k+1}, U^k)_N \geqslant (\tfrac{1}{2} - \theta)(\|U^{k+1}\|_N^2 + \|U^k\|_N^2).$$

Then from (12.2.8), it follows that

$$\|U^{k+1}\|_N^2 \leqslant \|U^k\|_N^2 \qquad \text{for all } k \geqslant 0,$$

and thus

$$\|U^k\|_N \leqslant \|U^0\|_N \qquad \text{for all } k \geqslant 0. \tag{12.2.9}$$

This shows that the scheme (12.2.6) is *unconditionally stable* if $\theta \in [1/2, 1]$.

We prove now that the L^2_w-norm of $u^k - U^k$ tends to zero as both Δt and $1/N$ tend to zero. Using the function $\tilde{u}(t) = \Pi_N u(t)$ defined through (11.2.31), it follows that:

$$(\theta \tilde{u}_t^{k+1} + (1-\theta)\tilde{u}_t^k, v)_w + a(\theta \tilde{u}^{k+1} + (1-\theta)\tilde{u}^k, v) = (\delta^k, v)_w,$$

for all $v \in \mathbb{P}_N^0$ and $k \geqslant 0$, where $\delta^k = \theta(\tilde{u} - u)_t^{k+1} + (1-\theta)(\tilde{u} - u)_t^k$. Then, setting $e^k = U^k - \tilde{u}^k$ and using (12.2.7), we obtain

$$\begin{aligned}\frac{1}{\Delta t}(U^{k+1} - U^k, v)_N - (\theta \tilde{u}_t^{k+1} + (1-\theta)\tilde{u}_t^k, v)_N + a(\theta e^{k+1} + (1-\theta)e^k, v)\\ = -(\delta^k, v)_w - E(\theta \tilde{u}_t^{k+1} + (1-\theta)\tilde{u}_t^k, v),\end{aligned} \tag{12.2.10}$$

where $E(\theta, \psi) \equiv (\phi, \psi)_w - (\phi, \psi)_N$.

Using the standard approximation results (11.1.19) and (9.3.5), we obtain

$$|(\delta^k, v)_w + E(\theta \tilde{u}_t^{k+1} + (1-\theta)\tilde{u}_t^k, v)| \leqslant C_1 \|\gamma^k(u)\|_0 \|v\|_N, \tag{12.2.11}$$

where the definition of $\gamma^k(u)$ is obvious and

$$\|\gamma^k(u)\|_0 \leqslant C_2 N^{-r}(\|u_t^k\|_r + \|u_t^{k+1}\|_r), \qquad r \geqslant 1. \tag{12.2.12}$$

Now let $z = z(t)$ be any continuously differentiable function in the semi-infinite inverval $(0, +\infty)$, and define

$$\varepsilon^k(z) = \frac{1}{\Delta t}(z^{k+1} - z^k) - (\theta z_t^{k+1} + (1-\theta)z_t^k).$$

If $z \in C^2(0, +\infty)$, then using the Taylor formula with the integral form of the remainder gives

$$\varepsilon^k(z) = \frac{1}{\Delta t}\int_{t^k}^{t^{k+1}} (s - (1-\theta)t^{k+1} - \theta t^k) z_{tt}(s)\, ds;$$

whence

$$|\varepsilon^k(z)| \leqslant \max(\theta, 1-\theta) \int_{t^k}^{t^{k+1}} |z_{tt}(s)|\, ds \leqslant \int_{t^k}^{t^{k+1}} |z_{tt}(s)|\, ds. \tag{12.2.13}$$

If $\theta = 1/2$, and $z \in C^3(0, +\infty)$, then a better estimate is obtained from a higher-order Taylor formula, namely

$$\varepsilon^k(z) = \frac{1}{2\Delta t}\int_{t^k}^{t^{k+1}} (t^k - s)(t^{k+1} - s) z_{ttt}(s)\, ds;$$

whence

$$|\varepsilon^k(z)| \leqslant \frac{\Delta t}{8}\int_{t^k}^{t^{k+1}} |z_{ttt}(s)|\, ds. \tag{12.2.14}$$

From (12.2.10) we obtain, using the above definition of ε^k,

$$\frac{1}{\Delta t}(e^{k+1} - e^k, v)_N + a(\theta e^{k+1} + (1-\theta)e^k, v)$$
$$= -\{(\varepsilon^k(\tilde{u}), v)_N + (\delta^k, v)_w + E(\theta\tilde{u}_t^{k+1} + (1-\theta)\tilde{u}_t^k, v)\}.$$

Taking $v = \theta e^{k+1} + (1-\theta)e^k$, proceeding in a manner similar to the stability proof, using (12.2.11) and the Cauchy–Schwarz inequality, we obtain

$$\frac{1}{2\Delta t}(\|e^{k+1}\|_N^2 - \|e^k\|_N^2) \leqslant (\|\varepsilon^k(\tilde{u})\|_N + C_1\|\gamma^k(u)\|_0)\|\theta e^{k+1} + (1-\theta)e^k\|_N$$
$$\leqslant 2\|\varepsilon^k(\tilde{u})\|_N^2 + 2C_1^2\|\gamma^k(u)\|_0^2 + \frac{1}{2}\|e^k\|_N^2 + \frac{1}{2}\|e^{k+1}\|_N^2.$$

The last inequality has been obtained using twice the inequality $ab \leqslant a^2 + \frac{1}{4}b^2$ for $a, b \in \mathbb{R}$. Therefore,

$$\|e^{k+1}\|_N^2 \leqslant \frac{1+\Delta t}{1-\Delta t}\|e^k\|_N^2 + \frac{4\Delta t}{1-\Delta t}(\|\varepsilon^k(\tilde{u})\|_N^2 + C_1^2\|\gamma^k(u)\|_0^2).$$

Applying the above estimate recursively yields

$$\|e^k\|_N^2 \leqslant \left(\frac{1+\Delta t}{1-\Delta t}\right)^k \|e^0\|_N^2 + \frac{4\Delta t}{1-\Delta t}\sum_{j=0}^{k-1}\left(\frac{1+\Delta t}{1-\Delta t}\right)^{k-j-1}$$
$$\cdot(\|\varepsilon^j(\tilde{u})\|_N^2 + C_1^2\|\gamma^j(u)\|_0^2) \tag{12.2.15}$$
$$\leqslant \left(\frac{1+\Delta t}{1-\Delta t}\right)^k \|e^0\|_N^2 + 8\sum_{j=0}^{k-1}\Delta t(\|\varepsilon^j(\tilde{u})\|_N^2 + C_1^2\|\gamma^j(u)\|_0^2.$$

The last inequality holds if Δt is sufficiently small (e.g., it is enough to assume that $\Delta t \leqslant 1/2$). Since $e^0 = U^0 - \tilde{u}(0) = I_N u_0 - \Pi_N u_0$ from (11.1.19) and (9.5.19), it follows using (9.3.2) that

$$\|e^0\|_N \leqslant 2\|e^0\|_0 \leqslant C_2 N^{-r}\|u_0\|_r \qquad r \geqslant 1. \tag{12.2.16}$$

We are going now to estimate the term $\|\varepsilon^j(\tilde{u})\|_N$. Since $\varepsilon^j(\tilde{u}) \in \mathbb{P}_N^0$, using (9.3.2), (12.2.14) (if $\theta = 1/2$), and the Cauchy–Schwarz inequality yields

$$\|\varepsilon^j(\tilde{u})\|_N^2 \leqslant 4\|\varepsilon^j(u)\|_0^2 \leqslant \frac{1}{16}\Delta t^2 \int_{-1}^{1}\left(\int_{t^j}^{t^{j+1}} |\tilde{u}_{ttt}(s)|\, ds\right)^2 w(x)\, ds$$
$$\leqslant \frac{1}{2}\left(\frac{\Delta t}{2}\right)^3 \int_{t^j}^{t^{j+1}} \|\tilde{u}_{ttt}(s)\|_0^2\, ds;$$

therefore,

$$8\sum_{j=0}^{k-1}\Delta t\|e^j(\tilde{u})\|_N^2 \leqslant \frac{1}{2}\Delta t^4 \int_0^{t^k} \|\tilde{u}_{ttt}(s)\|_0^2\, ds. \tag{12.2.17}$$

Finally, from (12.2.15)—(12.2.17), and (12.2.12) it follows that

$$\|e^k\|_N^2 \leqslant C_3\left(\frac{1+\Delta t}{1-\Delta t}\right)^k N^{-2r}\|u_0\|_r^2 + \sum_{j=0}^{k-1}(\|u_t^j\|_r + \|u_t^{j+1}\|_r)^2$$

$$+ \Delta t^4 C_4\left(\frac{1+\Delta t}{1-\Delta t}\right)^k \int_0^{t^k} \|\tilde{u}_{ttt}(s)\|_0^2\, ds.$$

If we use now the triangle inequality:

$$\|u^k - U^k\|_0 \leqslant \|u^k - \Pi_N u^k\|_0 + \|e^k\|_0,$$

then from (11.1.19) and the above estimate of e^k, we obtain

$$\|u^k - U^k\|_0 \leqslant M(u)N^{-r} + C_5\left(\frac{1+\Delta t}{1-\Delta t}\right)^{k/2}\left(\int_0^{t^k} \|\tilde{u}_{ttt}(s)\|_0^2\, ds\right)^{1/2}\Delta t^2, \quad k \geqslant 0, \tag{12.2.18}$$

$$M(u) = C_6\left\{\|u^k\|_r + \left(\frac{1+\Delta t}{1-\Delta t}\right)^{k/2}\|u_0\|_r^2 + \sum_{j=0}^{k-1}(\|u_t^j\|_r + \|u_t^{j+1}\|_r)^2\right\}^{1/2}$$

The above convergence estimate has been obtained for $\theta = 1/2$. If $\theta \in (\frac{1}{2}, 1]$, then using (12.2.13) instead of (12.2.14), it is easily seen that the estimate (12.2.18) still holds provided Δt^2 is replaced by Δt, and $\tilde{u}_{ttt}$ by $\tilde{u}_{tt}$.

We finally note that since $\tilde{u}_t = \Pi_N u_t$, the time derivatives of $\tilde{u}$ can be replaced with those of u, using (12.2.2) in a straightforward way.

Fully discrete approximations to the heat equation, using the θ-method, and based on either Fourier or Legendre expansions of the discrete solution U^k, possess the same kind of stability and convergence properties. (Notation should be changed in the obvious way.) The analysis for the two-dimensional heat equation (12.2.2) is presented in Bressan and Quarteroni (1986b).

12.3. Advection-Diffusion Equation

In this section, we shall analyze a Chebyshev collocation approximation to the time-dependent advection-diffusion equation. Continuous-in-time approximations and fully discrete approximations will be considered.

For case of exposition we focus on one-dimensional problems. The extension of the forthcoming discussion to multidimensional problems is achievable following the usual guidelines.

12.3.1. Semi-Discrete Approximation

Let us consider the problem:

$$\begin{cases} u_t - \nu u_{xx} + p u_x + q u = 0 & -1 < x < 1, \quad t > 0 \\ u(-1,t) = u(1,t) = 0 & t > 0 \\ u(x,0) = u_0(x) & -1 < x < 1. \end{cases} \tag{12.3.1}$$

The viscosity parameter is $\nu > 0$, and p and q are some given functions of x and t. The semi-discrete Chebyshev collocation approximation reads as follows:

$$\begin{cases} u_t^N - \nu u_{xx}^N + p u_x^N + q u^N = 0 & \text{at } x = x_j, \text{ for } j = 1, \dots, N-1 \\ u^N = 0 & \text{at } x = x_0 \text{ and } x = x_N, \end{cases} \tag{12.3.2}$$

for all $t \geqslant 0$. Here u^N is a polynomial of degree N, for any time t, and $x_j = \cos(\pi j/N)$ are the Chebyshev–Gauss–Lobatto nodes. We note that the transformation $u(t) \to e^{-\lambda t}u(t)$, $\lambda > 0$, will transform problem (12.3.1) into a problem of the same type, where q is replaced by $q + \lambda$. Provided λ is a sufficiently large number, the bilinear form $b(u, v)$ (see (11.2.3)) associated with the transformed problem is coercive on the Sobolev space $H^1_{w,0}(-1, 1)$. This result can be proven using the argument of Sec. 11.2.1 (see ii).

For all time t, problem (12.3.2) can now be written as follows:

$$(u_t^N, v)_N + \nu a(u^N, v) + (p u_x^N + q u^N, v)_N = 0 \tag{12.3.3}$$

for all $v \in \mathbb{P}_N^0$. Here, $a(u, v)$ denotes the bilinear form defined in (11.1.11), and $(u, v)_N$ denotes the usual discrete Chebyshev inner product.

The coercivity of the form $b(u, v)$ implies that of the form

$$b_N(u, v) = \nu a(u, v) + (p u_x + q u, v)_N \qquad u, v \in \mathbb{P}_N^0.$$

Thus, problem (12.3.3) can be analyzed following the guidelines of Sec. 10.5.1. The achievable error estimate has a form similar to (12.2.5). In the present case, however, the constant C will also depend on p and q.

The details of this analysis can be found, for example, in Canuto and Quarteroni (1981a).

12.3.2. Fully Discrete Approximation

We are going to show that using the θ-method for the diffusive part of (12.3.2) and an explicit method for the advective term yields an unconditionally stable method on any finite time interval. The above is the most common way to advance in time equation (12.3.2) or its counterpart having non-linear advection. Indeed, at any time-step, the resulting linear system is associated with the same matrix, which can therefore be preconditioned and/or factored once for all.

In order to simplify the exposition, we assume here that $q \equiv 0$ and p is a constant. Moreover, we analyze the backward Euler method only ($\theta = 1$). With the usual notation, the fully discrete problem takes the form:

$$\begin{cases} \dfrac{U^{k+1} - U^k}{\Delta t} - \nu U_{xx}^{k+1} + p U_x^k = 0 & \text{at } x = x_j, \text{ for } j = 1, \dots, N-1, \\ U^{k+1} = 0 & \text{at } x = x_0 \text{ and } x = x_N. \end{cases} \tag{12.3.4}$$

For all $k \geqslant 0$, U^k is a polynomial of degree N. The above problem can be written as follows:

$$\left(\frac{U^{k+1}-U^k}{\Delta t}, v\right)_N + \nu a(U^{k+1}, v) + p(U_x^k, v)_N = 0 \qquad \text{for all } v \in \mathbb{P}_N^0. \quad (12.3.5)$$

Take $v = U^{k+1}$ in (12.3.5). By (11.1.14) we get

$$\nu a(U^{k+1}, U^{k+1}) \geqslant \frac{\nu}{4} \| U_x^{k+1} \|_0^2.$$

Moreover, by the Cauchy–Schwarz inequality it follows that

$$\left(\frac{U^{k+1}-U^k}{\Delta t}, U^{k+1}\right)_N \geqslant \frac{1}{2\Delta t}(\| U^{k+1} \|_N^2 - \| U^k \|_N^2).$$

Finally, integrating by parts we get

$$\begin{aligned} |p(U_x^k, U^{k+1})_N| = |p(U_x^k, U^{k+1})_w| &= \left| p \int_{-1}^{1} U^k (U^{k+1} w)_x \, dx \right| \\ &\leqslant C |p| \, \| U^k \|_0 \| U_x^{k+1} \|_0 \\ &\leqslant \frac{\nu}{4} \| U_x^{k+1} \|_0^2 + \frac{C^2 |p|^2}{\nu} \| U^k \|_0^2. \end{aligned}$$

From (12.3.5) we then obtain

$$\frac{1}{2\Delta t}(\| U^{k+1} \|_N^2 - \| U^k \|_N^2) \leqslant \frac{C^2 |p|^2}{\nu} \| U^k \|_0^2;$$

whence

$$\| U^{k+1} \|_N^2 \leqslant \left(1 + \frac{2C^2 |p|^2}{\nu} \Delta t\right) \| U^k \|_0^2.$$

A recursive application of this inequality yields

$$\| U^k \|_N^2 \leqslant \exp\left(\frac{2C^2 |p|^2}{\nu} k \Delta t\right) \| U^0 \|_0^2.$$

We conclude that for any k such that $k\Delta t \leqslant T$, the following bound holds

$$\| U^k \|_N \leqslant \exp\left(\frac{C^2 |p|^2}{\nu} T\right) \| U^0 \|_0.$$

This yields unconditional stability. Indeed, at any time-level $t^k \leqslant T$, and for any choice of N and Δt, the L_w^2-norm of the discrete solution is bounded by the norm of the initial data times a constant which is independent of both N and Δt.

The convergence analysis for (12.3.4) can now be worked out along the same guidelines of Sec. 12.2.2.

The generalization of these results to the case in which p and q are functions of x and t is given in Quarteroni (1983) for one space variable and in Bressan and Quarteroni (1986b) for two space variables. The unsteady Burgers equation

$$u_t - \nu u_{xx} + u u_x = 0 \qquad -1 < x < 1, \quad t > 0, \quad \nu > 0, \qquad (12.3.6)$$

with homogeneous boundary conditions is analyzed in Bressan and Quarteroni (1986a). In all these cases, the Chebyshev collocation approximation for the spatial discretization is coupled with a semi-implicit time-advancing scheme of the same form as the one used in (12.3.4). For any finite time interval $(0, T)$, unconditional stability is proven, as well as an error estimate exhibiting spectral convergence in N and an algebraic rate of decay with respect to Δt.

Other Fourier approximations of parabolic type problems can be found in the literature. We mention here the works by Guo and Manozanjan (1985), Pasciak (1982) and Quarteroni (1986b, 1987b).

CHAPTER 13

Domain Decomposition Methods

13.1. Introduction

In the previous chapters we have concentrated on the application and theory of spectral methods for problems in simple domains. There have been a number of recent developments on the use of spectral techniques in more general geometries. The basic idea has been to partition the complete domain of the problem into several subdomains. One situation in which this approach is useful is illustrated in Fig. 13.1. The approximation is spectral if increased accuracy is obtained by increasing the order of approximation in a fixed number of subdomains, rather than by resorting to a further partitioning. In this particular example it is clear that at least two subdomains are required in order to use spectral methods at all. Additional advantages may arise by using separate subdomains Ω_2, Ω_3, and Ω_4 instead of a single subdomain which is their union. The partitioning illustrated in Fig. 13.1 leads to a distribution of Chebyshev collocation points which improves the resolution of the approximation. Moreover, one would expect the corresponding algebraic problem to be better-conditioned because there is a less extreme ratio of the largest to smallest grid spacings. Finally, the use of subdomains facilitates the implementation of spectral methods on parallel computers, especially those with local memory.

In the case illustrated in Fig. 13.1 the partitions intersect only at the boundaries. It is also possible to use subdomains which overlap. This is illustrated in Fig. 13.2.

Greater generality can be achieved by combining mapping techniques with partitioning. Each of the subdomains in Fig. 13.3 can be mapped into the corresponding rectangular domain in Fig. 13.1. Even in problems where a single map can be used, the partitioning techniques may be useful both for easing the task of determining the map and for producing a grid with minimal distortion.

The partitioning technique has been employed for a number of years in finite-difference and finite-element methods. In the context of spectral methods, it dates from the late 1970s. Delves and Hall (1979) introduced a method which they called the global element method. Orszag (1980) described

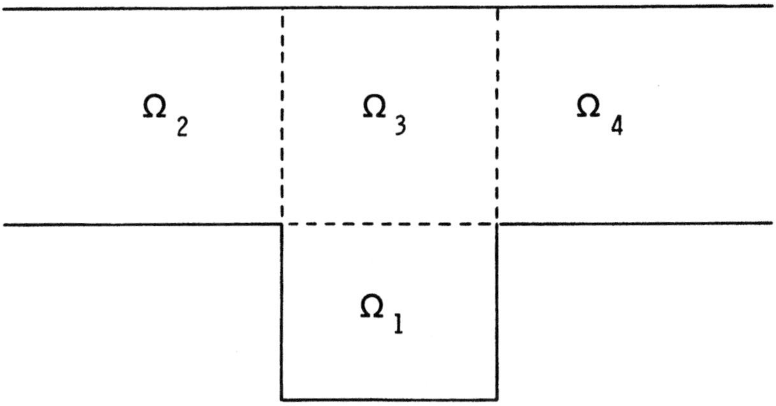

Figure 13.1. A non-overlapping domain decomposition for a grooved channel.

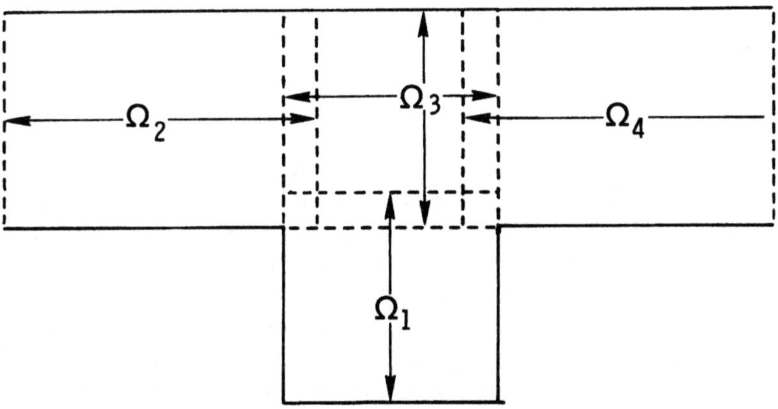

Figure 13.2. An overlapping domain decomposition for a grooved channel.

a technique for patching at interfaces. Morchoisne (1984) developed a method based on overlapping multiple domains. Patera (1984) used a variational formulation for what he termed the spectral-element method. Discretization and solution techniques will be discussed in detail in the next two sections. These will be followed by a theoretical discussion of the methods. We stress that spectral domain decomposition methods are a recent and rapidly evolving subject. The interested reader is advised to keep abreast of the literature.

A crucial aspect of any domain decomposition method is the manner in which solutions on contiguous domains are matched. Patching methods take a classical (pointwise) view of the differential equation. If the equation has

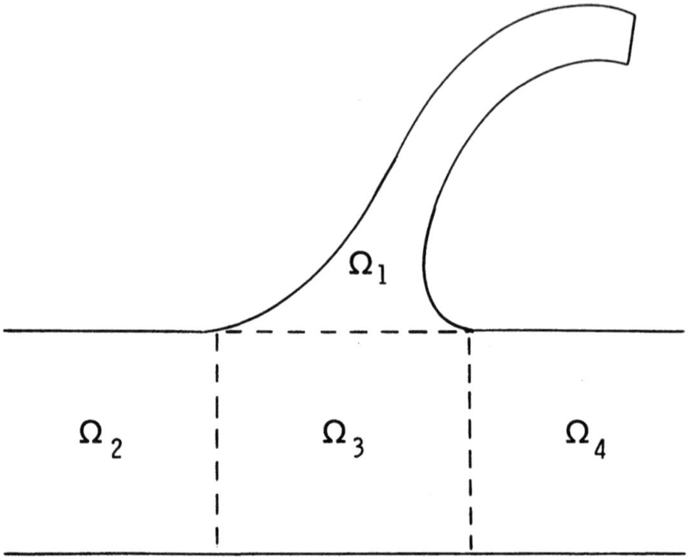

Figure 13.3. A domain which can be mapped into rectangular subdomains.

order d, then at the interface of contiguous domains the solution and all its derivatives of order up to $d - 1$ must be continuous. For second-order problems this is typically enforced by requiring that the solution u be continuous and that the normal derivatives $\partial u/\partial n$ match at the interface, in the sense that they are equal and opposite. The condition on the normal derivative may be replaced with a continuity condition on any other directional (but not tangential) derivative. These continuity conditions are discretized by enforcing them at selected points, and thus are satisfied exactly by any approximation.

Alternatively, the differential equation may be posed variationally. The standard weak formulation is based on an integration-by-parts procedure over the entire physical domain. This implicitly assumes continuity of u and its first $d - 1$ derivatives at all points, and, in particular, at the interfaces between subdomains. In most cases continuity of u is built into the choice of trial functions. Continuity of the derivatives of u occurs as a natural interface condition. By reversing the integration-by-parts procedure on the decomposed domain, one can deduce directly from the variational principle the continuity of the derivatives of u at the interfaces. In a variational domain decomposition method, the natural interface conditions are satisfied with increasing precision as the order of approximation is increased.

In an overlapping domain method only the continuity of the function is imposed, but this matching occurs on the boundary of the intersection between two adjacent domains. In the discrete case this is enforced at selected points on the boundary.

13.2. Patching Methods

Most collocation versions of spectral domain decomposition methods use variations or extensions of the patching technique originally suggested by Orszag (1980). In this section, we first establish the notation we shall use to explain this method. Then we present a detailed description of the discretization procedure. Some algorithmic details are covered next. Finally, some representative applications are presented.

13.2.1. Notation

The notation for a domain decomposition method is necessarily cumbersome. In our exposition of patching methods we shall confine ourselves to three geometries.

A one-dimensional domain decomposition for the interval $\Omega = (a, b)$ is illustrated in Fig. 13.4. It is broken into two subdomains, $\Omega_1 = (a_1, a_2)$ and $\Omega_2 = (a_2, a_3)$, where $a_1 = a$ and $a_3 = b$. The solution to a given differential equation on Ω is denoted by u. Its restrictions to $\bar{\Omega}_1$ and $\bar{\Omega}_2$ are denoted by u_1 and u_2, respectively. The integers N_1 and N_2 are used to indicate the degrees of approximation on these intervals. The approximate solution on Ω is denoted by u^N and its restrictions to $\bar{\Omega}_1$ and $\bar{\Omega}_2$ by u_1^N and u_2^N, respectively. The collocation points in $\bar{\Omega}_1$ are denoted by $x_j^{(1)}, j = 0, \dots, N_1$ and those in $\bar{\Omega}_2$ by $x_j^{(2)}, j = 0, \dots, N_2$.

A domain decomposition for a rectangle is sketched in Fig. 13.5. The interface between the subdomains Ω_1 and Ω_2 is denoted by $\Gamma_{1,2}$. In general, the collocation points along the interface of two adjoining domains will not coincide. We denote by J_1 the set of interface collocation points for Ω_1 and by J_2 those for Ω_2.

The third model decomposition includes the complication of corner points. It is illustrated in Fig. 13.6. The interface points of $\bar{\Omega}_s$ along the interface $\Gamma_{s,t}$ are denoted by $J_{s,t}$, and those of $\bar{\Omega}_t$ along this interface are denoted by $J_{t,s}$.

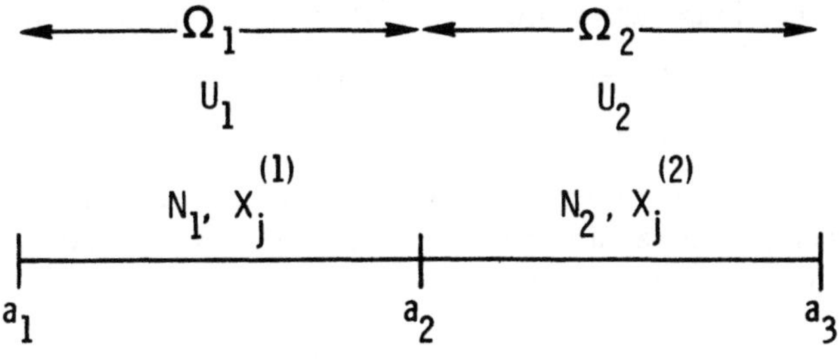

Figure 13.4. A one-dimensional domain decomposition into two subdomains.

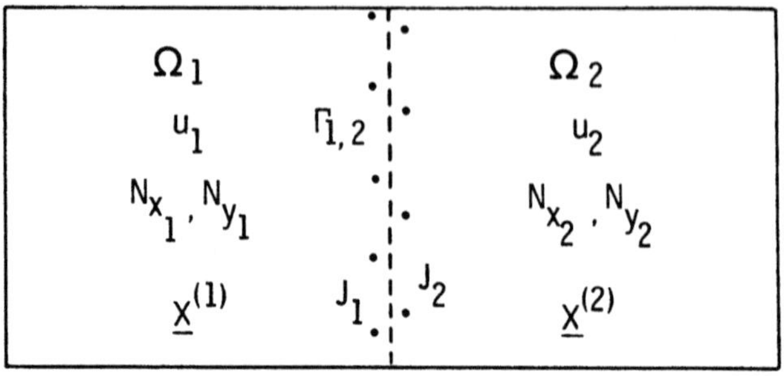

Figure 13.5. A two-dimensional domain decomposition into two subdomains.

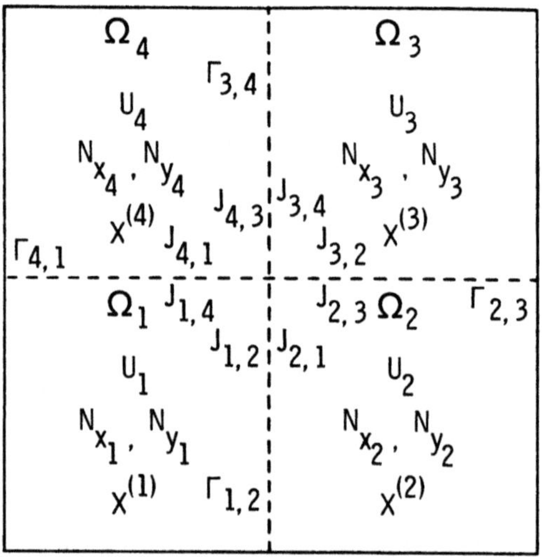

Figure 13.6. A two-dimensional domain decomposition into four subdomains.

13.2.2. Discretization

One-Dimensional Elliptic Problems

Consider a patching method on $\Omega = (a, b)$ for the linear problem

$$\begin{cases} Lu \equiv -\nu u_{xx} + \alpha u_x + \lambda u = f & \text{in } \Omega \\ u(a) = u(b) = 0 \end{cases} \tag{13.2.1}$$

where ν, α and λ are constants and $\nu > 0$. Suppose that the domain Ω is decomposed as illustrated in Fig. 13.4. Setting $u_1 = u|_{\Omega_1}$ and $u_2 = u|_{\Omega_2}$, the PDE problem may be formulated as

$$\begin{aligned} Lu_1 &= f \quad \text{in } \Omega_1 \\ u_1(a_1) &= 0 \end{aligned} \tag{13.2.2}$$

$$\begin{aligned} Lu_2 &= f \qquad \text{in } \Omega_2 \\ u_2(a_3) &= 0 \end{aligned} \tag{13.2.3}$$

$$u_1(a_2) = u_2(a_2) \tag{13.2.4}$$

$$\frac{du_1}{dx}(a_2) = \frac{du_2}{dx}(a_2). \tag{13.2.5}$$

A demonstration that this formulation is equivalent to (13.2.1) is given in Sec. 13.5.1.

The approximate solution u^N is represented on Ω_s, $s = 1, 2$ as a Chebyshev series:

$$u_s^N(x) = \sum_{k=0}^{N_s} \tilde{u}_{s,k} T_k(\xi), \tag{13.2.6}$$

where the computational coordinate ξ is in the standard domain $[-1, 1]$ and is related to x by

$$x = \frac{a_{s+1} - a_s}{2}\xi + \frac{a_{s+1} + a_s}{2}. \tag{13.2.7}$$

The collocation points in x are related to the Chebyshev–Gauss–Lobatto points by

$$x_j^{(s)} = \frac{a_{s+1} - a_s}{2}\cos\frac{\pi j}{N_s} + \frac{a_{s+1} + a_s}{2} \qquad j = 0, \ldots, N_s. \tag{13.2.8}$$

The fundamental unknowns are, of course, the nodal values

$$u_{s,j} \equiv u_s^N(x_j^{(s)}) \qquad \begin{aligned} s &= 1, 2 \\ j &= 0, \ldots, N_s. \end{aligned} \tag{13.2.9}$$

The series representation (13.2.6) is used for the approximation of derivatives via Chebyshev collocation.

The collocation equations for the PDE (13.2.1) are

$$\left\{ \begin{aligned} & Lu_1^N - f|_{x=x_j^{(1)}} = 0 \qquad j = 1, \ldots, N_1 - 1 && (13.2.10) \\ & u_1^N(a_1) = 0 && \\ & Lu_2^N - f|_{x=x_j^{(2)}} = 0 \qquad j = 1, \ldots, N_2 - 1 && (13.2.11) \\ & u_2^N(a_3) = 0, && \end{aligned} \right.$$

and the interface continuity conditions are

$$u_1^N(a_2) = u_2^N(a_2) \tag{13.2.12}$$

$$\frac{du_1^N}{dx}(a_2) = \frac{du_2^N}{dx}(a_2). \tag{13.2.13}$$

Equations (13.2.10)—(13.2.13) constitute $N_1 + N_2 + 2$ independent conditions for the $N_1 + N_2 + 2$ unknown nodal values. Solution techniques are discussed in the next section. This formulation of the patching interface conditions for an elliptic problem was proposed by Orszag (1980).

Let us consider now a non-linear generalization of (13.2.1), namely,

$$L(u) \equiv [G(u)]_x = f, \qquad (13.2.14)$$

where the flux $G(u)$ is given by

$$G(u) = g(u) - \nu u_x \qquad (13.2.15)$$

and $g(u)$ is a non-linear function of u. The appropriate interface conditions are continuity of u and G. The discretization is thus (13.2.10)—(13.2.12) plus

$$G(u_1^N(a_2)) = G(u_2^N(a_2)). \qquad (13.2.16)$$

Equations (13.2.13) and (13.2.16) are mathematically equivalent when $g(u)$ is continuous. Algebraically, however, (13.2.13) is linear and (13.2.16) is non-linear.

Macaraeg and Streett (1986) devised a patching method discretization in which the pointwise flux balance condition (13.2.16) is replaced by an integral version over the two subdomains adjacent to the interface point:

$$G(u_1^N(a_1)) + \int_{a_1}^{a_2} f(u_1^N)\,dx = G(u_2^N(a_3)) - \int_{a_2}^{a_3} f(u_2^N)\,dx. \qquad (13.2.17)$$

The integrals in (13.2.17) may be evaluated by the Clenshaw–Curtis formula (Davis and Rabinowitz (1984)), which is, for N even

$$\int_{-1}^{1} h(\xi)\,d\xi = \sum_{j=0}^{N} w_j h(\xi_j)$$

$$w_j = \begin{cases} \dfrac{1}{N^2 - 1} & j = 0 \text{ or } N \\[2ex] \dfrac{4}{N} \displaystyle\sum_{k=0}^{N/2} \bar{c}_k^{-1} \frac{\cos \dfrac{2\pi jk}{N}}{1 - 4k^2} & j = 1, \dots, N-1 \end{cases} \qquad (13.2.18)$$

where $\bar{c}_k = 2$ for $k = 0, N/2$ and $\bar{c}_k = 1$ for $k = 1, \dots, N/2 - 1$, and ξ_j are the Chebyshev–Gauss–Lobatto points.

One-Dimensional Hyperbolic Problems

Hyperbolic problems require different interface conditions than elliptic ones. Consider the one-dimensional wave equation

$$\begin{cases} u_t + u_x = 0 & x \in \Omega = (a, b), \quad t > 0 \\ u(a, t) = u_L(t) & t > 0 \\ u(x, 0) = u_0(x) & x \in \Omega \end{cases} \qquad (13.2.19)$$

with Ω decomposed as shown in Fig. 13.4 and u^N approximated as described by (13.2.6). A semi-discrete patching method for this problem reads

$$\frac{\partial u_1^N}{\partial t} + \frac{\partial u_1^N}{\partial x}\bigg|_{x=x_j^{(1)}} = 0 \qquad j = 1, \ldots, N_1 - 1$$
$$u_1^N(a, t) = u_L(t) \tag{13.2.20}$$

$$\frac{\partial u_2^N}{\partial t} + \frac{\partial u_2^N}{\partial x}\bigg|_{x=x_j^{(2)}} = 0 \qquad j = 0, \ldots, N_2 - 1$$
$$u_1^N(a_2, t) = u_2^N(a_2, t) \tag{13.2.21}$$

and a general interface condition of the form

$$\frac{\partial u^N}{\partial t} + \eta \frac{\partial u_1^N}{\partial x}\bigg|_{x=a_2} + (1-\eta)\frac{\partial u_2^N}{\partial x}\bigg|_{x=a_2} = 0. \tag{13.2.22}$$

Only the last equation requires comment. The constant η determines the weighting of the contributions from the two subdomains in the update equation for the common interface point. The most obvious choices for the weighting factor are $\eta = 1$ and $\eta = 1/2$. The former choice produces a pure upwind scheme and the latter corresponds to a simple average of the two subdomains' contributions.

The upwind choice means that the Ω_1-problem is solved independently and that the Ω_2-problem takes as a boundary condition for $u_2(a_2, t)$, the value $u_1(a_2, t)$ from the Ω_1-problem. This approach conforms to the mathematical properties of the scalar hyperbolic equation (13.2.19). The results in Sec. 12.1 can be extended in a straightforward fashion to assure stability (in time) of this patching method, regardless of the values of N_1 and N_2.

The temporal stability of the simple averaging procedure, however, is dependent upon N_1 and N_2. Kopriva (1986a) presented numerical evidence that it is unstable if $N_1 > N_2$ and $a_2 = \frac{1}{2}(a + b)$. A precise characterization of the weighting factors η, as a function of $a_2/(a_1 + a_3)$, N_1 and N_2, which produce a temporally stable scheme is not available at present.

The upwind interface condition extends in a natural way to quasilinear hyperbolic systems, such as

$$\mathbf{u}_t + A(\mathbf{u})\mathbf{u}_x = 0 \qquad \text{in } \Omega, \tag{13.2.23}$$

by a diagonalization procedure. Let P be a non-singular matrix which reduces the $n \times n$ matrix $A(\mathbf{u}^N(a_2))$ to the diagonal matrix

$$\Lambda = P^{-1}AP. \tag{13.2.24}$$

(The columns of P are the eigenvectors of A and the diagonal entries of Λ are the corresponding eigenvalues.) Suppose that the unknowns are ordered so that

$$\Lambda(u) = \begin{pmatrix} \Lambda^I & 0 \\ 0 & \Lambda^{II} \end{pmatrix},$$

where Λ^I and Λ^{II} are diagonal matrices such that $\Lambda^I > 0$ and $\Lambda^{II} < 0$ for $x = a_2$. Let q be the dimension of Λ^I. Canuto and Quarteroni (1987) proposed taking appropriate characteristic combinations of the PDE at the interface. This yields, with $\mathbf{u}_1^N = \mathbf{u}_2^N = \mathbf{u}^N$ at the interface,

$$\begin{aligned} &\sum_{k=1}^{n} P_{jk}^{-1}(\mathbf{u}^N)\left\{\frac{\partial \mathbf{u}_1^N}{\partial t} + A(\mathbf{u}^N)\frac{\partial \mathbf{u}_1^N}{\partial x}\right\}_k = 0 \qquad 1 \leqslant j \leqslant q \\ &\sum_{k=1}^{n} P_{jk}^{-1}(\mathbf{u}^N)\left\{\frac{\partial \mathbf{u}_2^N}{\partial t} + A(\mathbf{u}^N)\frac{\partial \mathbf{u}_2^N}{\partial x}\right\}_k = 0 \qquad q < j \leqslant n, \end{aligned} \tag{13.2.25}$$

where the subscript k on the braces denotes the k-th component. This method has been shown numerically to be stable for a non-linear gas dynamic problem by Canuto and Quarteroni (1987).

Alternatively, the upwind interface condition can be written as

$$\left.\frac{\partial \mathbf{u}^N}{\partial t} + \frac{1}{2}[A + |A|]\frac{\partial \mathbf{u}_1^N}{\partial x} + \frac{1}{2}[A - |A|]\frac{\partial \mathbf{u}_2^N}{\partial x}\right|_{x=a_2} = 0, \tag{13.2.26}$$

where

$$|A| = P|\Lambda|P^{-1}. \tag{13.2.27}$$

(This formulation is equivalent to (13.2.25) for a constant-coefficient problem.)

Kopriva (1986a) proposed an upwind dominant interface condition which avoids the computation of the matrix absolute value in (13.2.27) and requires knowledge solely of the eigenvalues of A. This scheme consists of approximating the $|A|$ terms in (13.2.26) via

$$|A| \cong \lambda^* I. \tag{13.2.28}$$

The choice of λ^* depends upon the eigenvalues of A and on whether one subdomain is more coarsely approximated than the other. (The approximation in Ω_2 is coarser than the one in Ω_1 if $(a_3 - a_2)/N_2 < (a_2 - a_1)/N_1$.) Kopriva (1986a) recommends that λ^* be chosen so that it is greater than any eigenvalue of A corresponding to a characteristic variable propagating from a coarse domain to a finer one.

Two-Dimensional Problems

Consider the two-dimensional problem

$$\begin{cases} Lu \equiv -\nu\Delta u + \boldsymbol{\alpha}\cdot\nabla u + \lambda u = f & \text{in } \Omega \\ u = 0 & \text{on } \partial\Omega \end{cases} \tag{13.2.29}$$

on the domain Ω illustrated in Fig. 13.5. Like the one-dimensional problem (13.2.1), this can be decomposed into problems on each subdomain together with conditions at interfaces prescribing the continuity of u and that the sum of the outward normal derivatives vanishes. We enforce

$$\begin{aligned} u_1 &= u_2 \qquad &\text{on } \Gamma_{1,2} \\ \frac{\partial u_1}{\partial n_1} &= -\frac{\partial u_2}{\partial n_2} \qquad &\text{on } \Gamma_{1,2}. \end{aligned} \tag{13.2.30}$$

Here, n_s denotes the unit outward normal to Ω_s, $s = 1, 2$. The solution is represented as a double Chebyshev series in an obvious generalization of (13.2.6)—(13.2.8). The collocation conditions on $\partial\Omega$ and the interior points of Ω_1 and Ω_2 are straightforward. If $N_{y_1} = N_{y_2}$, then the continuity condition (13.2.12) is enforced at all interior points of $\Gamma_{1,2}$. Likewise, the derivative matching condition (13.2.13) is enforced at these points.

Consider now the case $N_{y_1} < N_{y_2}$. Continuity of u must take precedence over continuity of the normal derivative of u at the interface. Hence, on the Ω_2 side of $\Gamma_{1,2}$ we require continuity at the right collocation points,

$$u_1^N(\mathbf{x}_{N_{x_2},j}^{(2)}) = u_2^N(\mathbf{x}_{N_{x_2},j}^{(2)}) \qquad j = 1, \ldots, N_{y_2} - 1, \tag{13.2.31}$$

and normal derivative matching is enforced at the left collocation points,

$$\frac{\partial u_1^N}{\partial x}(\mathbf{x}_{0,j}^{(1)}) = \frac{\partial u_2^N}{\partial x}(\mathbf{x}_{0,j}^{(1)}) \qquad j = 1, \ldots, N_{y_1} - 1. \tag{13.2.32}$$

Interpolation is required to evaluate u_1^N at the points $\mathbf{x}_{N_{x_2},j}$. This should be accomplished via the Chebyshev series itself. An explicit expression for the prolongation interpolation operator P is available in (5.5.17), where $N_c = N_{y_1}$ and $N_f = N_{y_2}$. Equation (13.2.31) can be written

$$\sum_{k=0}^{N_{y_1}} P_{jk} u_1^N(\mathbf{x}_{0,k}^{(1)}) = u_2^N(\mathbf{x}_{N_{x_2},j}^{(2)}) \qquad j = 1, \ldots, N_{y_2} - 1, \tag{13.2.33}$$

and (13.2.32) is equivalent to

$$\left.\frac{\partial u_1^N}{\partial x}\right|_{\mathbf{x}=\mathbf{x}_{0,j}^{(1)}} = \sum_{k=0}^{N_{y_2}} R_{jk} \left.\frac{\partial u_2^N}{\partial x}\right|_{\mathbf{x}=\mathbf{x}_{N_{y_2},k}^{(2)}} \qquad j = 1, \ldots, N_{y_1} - 1, \tag{13.2.34}$$

where R is the restriction interpolation operator defined by (5.5.19).

The integral flux balance technique of Macaraeg and Streett (1986) applies to a two-dimensional equation of the form

$$L(u) \equiv [G(u)]_x + [H(u)]_y = f. \tag{13.2.35}$$

The generalization of the flux balance condition (13.2.17) is again straightforward if $N_{y_1} = N_{y_2}$. If $N_{y_1} < N_{y_2}$, then it takes the form

$$
\begin{aligned}
G(u_1^N(a_1, y_j^{(1)})) &- \int_{a_1}^{a_2} \{f(u_1^N) - [H(u_1^N)]_y\}|_{y=y_j^{(1)}}\, dx \\
&= G(u_2^N(a_3, y_j^{(1)})) + \int_{a_2}^{a_3} \{f(u_2^N) - [H(u_2^N)]_y\}|_{y=y_j^{(1)}}\, dx.
\end{aligned}
\tag{13.2.36}
$$

The right-hand side is evaluated by first computing it at the points $y_j^{(2)}$ and then restricting it to the points $y_j^{(1)}$.

The only extra complication presented by patching methods for the domain illustrated in Fig. 13.6 is the presence of an interior "corner" point that belongs to more than two subdomains. There are three conditions at the corner stemming from the continuity requirements of the solution in the four domains. It is sufficient to impose continuity of either normal derivative at the corner. The jump in the other normal derivative is spectrally small. In the integral flux balance method, the fourth condition is obtained by integrating the equation in both variables over all subdomains adjoining the corner point.

The original patching method, in which continuity of the normal derivatives is required, extends to an arbitrary number of subdomains without difficulty. The extension of the integral flux balance method is achieved by performing the integrals in the flux balance condition only over adjacent subdomains. The corner point condition is likewise obtained by a two-dimensional integral over all subdomains containing the corner point.

Patching methods for two-dimensional hyperbolic systems are obtained as similar generalizations of the one-dimensional algorithms.

13.2.3. Solution Techniques

Patching methods for the spatial part of hyperbolic problems have generally been coupled with explicit time-discretizations. Elliptic problems, of course, require implicit methods as do implicit time-discretizations of parabolic problems. Equations (13.2.1) and (13.2.29) are typical of the problems for which implicit patching methods are employed. Direct Gauss elimination is one option for solving the implicit equations, albeit expensive in terms of storage and time. The matrices which represent the global algebraic system have a block structure due to the domain decomposition, and only adjacent subdomains (or blocks) are coupled. Block Gauss elimination, then, should be considered. So, too, should static condensation (see Sec. 13.3.2). These methods are most attractive in the context of time-dependent problems in which the factorization of the global matrices need only be performed once.

An influence-matrix technique is a useful alternative. Sec. 7.3.1 contains an extensive discussion of an influence-matrix method for satisfying the Stokes equations which arise in algorithms for the unsteady Navier-Stokes equations. Macaraeg and Streett (1986) have adapted this to spectral patching methods. This enables the solution to the global system to be obtained at the price of two local solutions to Dirichlet problems for each subdomain. The influence-matrix is constructed by solving on each subdomain Dirichlet problems that have boundary data which is 1 at a particular interface point and 0 elsewhere, and then computing the resultant jumps in the normal derivatives at all the interface points. In the first step, arbitrary values are assigned to the interior interface nodes. Then the residuals for the interface conditions are evaluated. These will be non-zero, but the influence-matrix is used to infer the correct interface values. The local solutions are then recomputed with the correct interface values and the desired solution is obtained. The local solutions themselves can be obtained by the methods discussed in Chap. 5.

Several types of iterative methods have been used in applications. McCrory and Orszag (1980) used a global iterative method, with finite-difference preconditioning employed to accelerate a truncated conjugate gradient scheme. At each step of this scheme a solution to a finite-difference (or finite-element) approximation to the problem on all of Ω is obtained. This requires the inversion of a large sparse matrix.

The spectral operator which must be inverted enforces the PDE at the interior of the subdomains, the boundary conditions on $\partial\Omega$ and a Neumann-type condition on the interfaces. In Sec. 5.2.3 we note that finite-difference preconditioning for elliptic operators with Neumann boundary conditions has potential pitfalls. Deville and Mund (1985) resorted to finite-element preconditioning, which implemented globally accounts for the interface naturally. The spectral residual at the interface (for the one-dimensional problem) is given by

$$\frac{1}{\varepsilon}\left[\frac{du_1^N}{dx} - \frac{du_2^N}{dx}\right]\bigg|_{x=a_2}, \tag{13.2.37}$$

where ε is a small constant on the order of the mesh spacing adjacent to the interface. For finite N, the interface condition (13.2.13) is not satisfied exactly. It is only in the limit $N \to \infty$ that derivative matching is achieved.

Various local iterative strategies have been devised. An arbitrary initial guess is made for the unknown solution (or its normal derivative) at the interface points. The iteration proceeds by the separate solution of boundary value problems on each subdomain and the subsequent relaxation of the interface values for the next iteration.

Metivet and Morchoisne (1982) based the relaxation on a penalty method. For the simple one-dimensional problem (13.2.1) with the domain Ω decomposed as shown in Fig. 13.4, their scheme uses an initial guess of $u_1^{(0)}(a_2)$ (where for convenience, the superscript N has been dropped) and the iteration, for $n \geqslant 1$, is

$$\begin{cases} Lu_1^{(n)} = f|_{x=x_j^{(1)}} & j = 1, \ldots, N_1 - 1 \\ u_1^{(n)}(a_1) = 0 & \\ u_1^{(n)}(a_2) = u^{(n-1)}(a_2) & \end{cases} \tag{13.2.38}$$

and

$$\begin{cases} Lu_2^{(n)} = f|_{x=x_j^{(2)}} & j = 1, \ldots, N_1 - 1 \\ u_2^{(n)}(a_2) = u^{(n-1)}(a_2) \\ u_2^{(n)}(a_3) = 0 \end{cases} \tag{13.2.39}$$

coupled with the relaxation

$$u^{(n)}(a_2) = \frac{1}{2}\left[u_1^{(n)}(a_2) + u_2^{(n)}(a_2) + \frac{1}{\varepsilon}\left(\frac{du_1^{(n)}}{dx}(a_2) - \frac{du_2^{(n)}}{dx}(a_2)\right)\right], \tag{13.2.40}$$

where ε is a small positive constant. This penalty term forces the converged solution to satisfy the interface condition on the normal derivative.

Zanolli (1987) has proposed and analyzed an alternative scheme for problem (13.2.1) on two subdomains with $\alpha = 0$. It consists of iterations between the two problems on Ω_1 and Ω_2 as follows:

$$\begin{cases} Lu_1^{(n)} = f|_{x=x_j^{(1)}} & j = 1, \ldots, N_1 - 1 \\ u_1^{(n)}(a_1) = 0 \\ u_1^{(n)}(a_2) = \lambda_n \end{cases} \tag{13.2.41}$$

and

$$\begin{cases} Lu_2^{(n)} = f|_{x=x_j^{(2)}} & j = 1, \ldots, N_2 - 1 \\ u_2^{(n)}(a_3) = 0, \\ \dfrac{du_2^{(n)}}{dx}(a_2) = \dfrac{du_1^{(n)}}{dx}(a_2). \end{cases} \tag{13.2.42}$$

The initial values λ_0 and λ_1 are arbitrary, and for $n \geqslant 2$ the values of λ_n are calculated recursively according to

$$\lambda_n = \theta u_2^{(n-1)}(a_2) + (1 - \theta)\lambda_{n-1} \qquad n \geqslant 2. \tag{13.2.43}$$

where θ is the relaxation parameter. A rigorous convergence analysis for the above procedure is given in Funaro, Quarteroni and Zanolli (1987). They also discuss the application of this iteration strategy to two-dimensional problems with two subdomains. They prove that in some circumstances (one-dimensional problems or two-dimensional problems with Ω_1 and Ω_2 having the same measure) there is an optimal value of θ which yields exact convergence after only two iterations. Moreover, they indicate an effective dynamical choice of the relaxation parameter.

13.2.4. Examples

Numerous examples of patching method solutions to one- and two-dimensional problems can be found in the references cited above. Here we furnish some examples for two-dimensional problems. The first is the Poisson equation

$$\begin{cases} \Delta u = \cos(\pi x/4)\cos(\pi y/4) & \text{in } \Omega \\ u = 0 & \text{on } \partial\Omega \end{cases} \tag{13.2.44}$$

where $\Omega = (-2, 2) \times (-2, 2)$. Macaraeg and Streett (1986) solved this by the integral flux balance method. The domain decomposition and the numerical solution are illustrated in Fig. 13.7. The discretization used $N_{x_s} = N_{y_s} = 8$ in each of four subdomains. The computed solution is evidently well-behaved at the interfaces as well as at the corner point.

Figure 13.8 shows the decomposed domain (in part (a)) and the computed

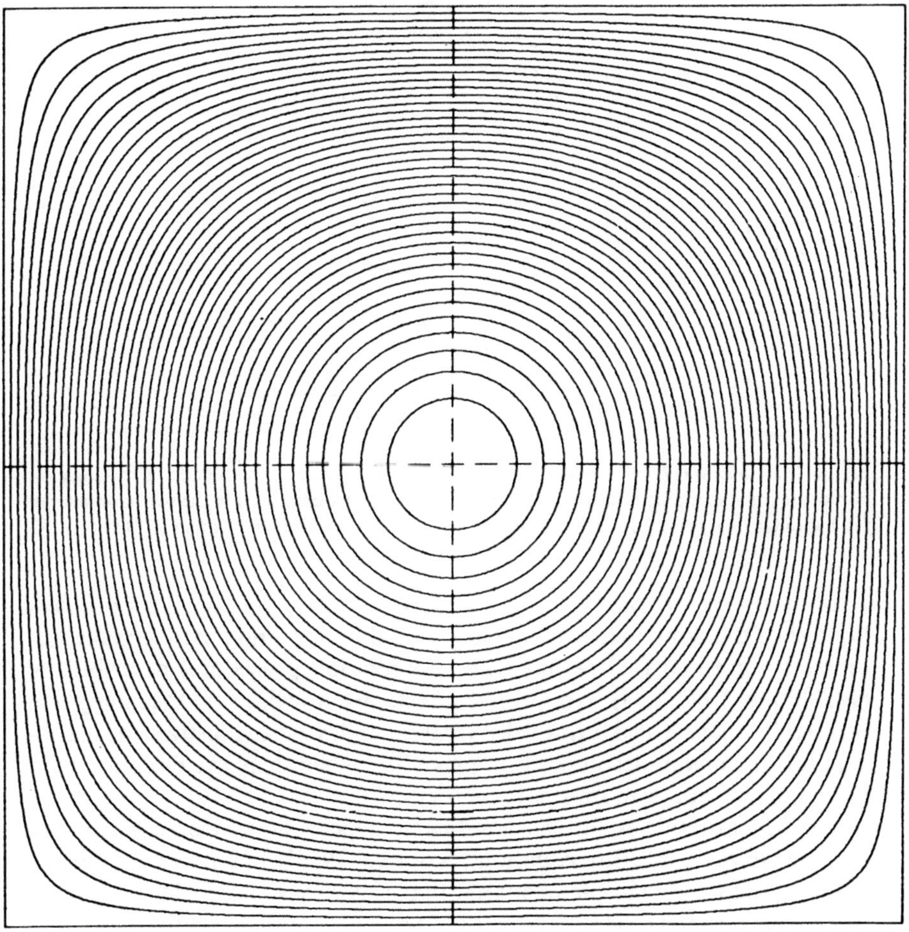

Figure 13.7. A spectral solution to Poisson's equation with four subdomains. The subdomain interfaces are indicated by the dashed lines and the numerical solution is indicated by the solid contour lines. (Courtesy of M. Macaraeg and C. Streett.)

(a)

(b)

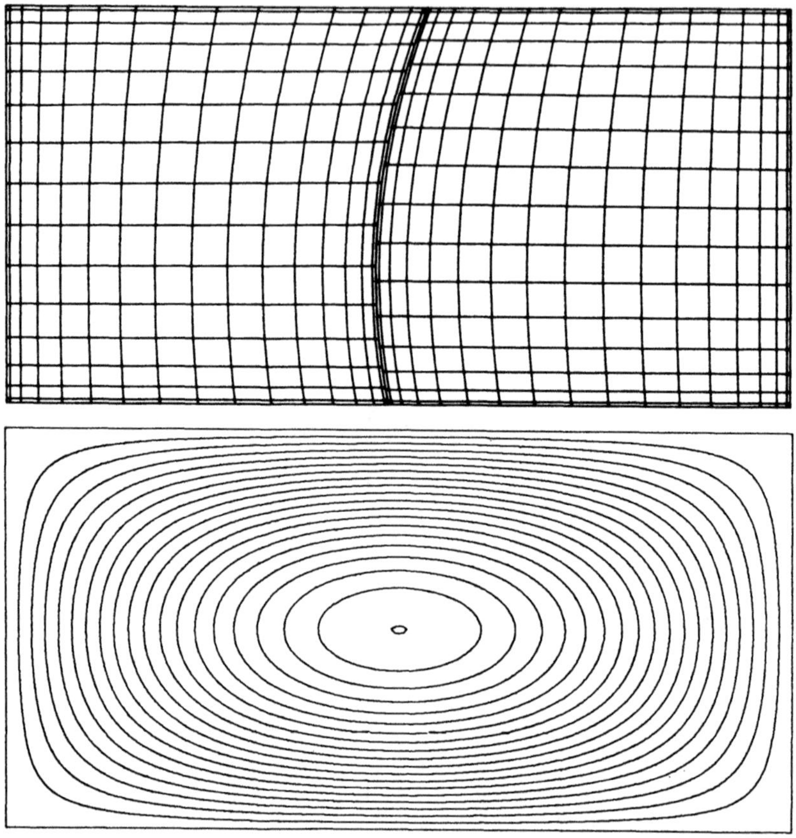

Figure 13.8. Spectral subdomain patching solution to Poisson's equation on a rectangle. A non-orthogonal mapping is used and different resolutions are employed in the two subdomains: (a) computational grid, (b) computed solution. (Courtesy of M. Macaraeg and C. Streett.)

solution (in part (b)) to (13.2.44) on $\Omega = (-2, 2) \times (-1, 1)$. This problem is a test of the domain decomposition procedure for a non-orthogonal coordinate system and for the interface interpolation procedures described by (13.2.33) and (13.2.34). No adverse effects are in evidence. Macaraeg and Streett (1986) have also furnished several examples which illustrate the usefulness of the domain decomposition procedure when there are discontinuities in boundary data or in transport coefficients.

The last example is for the non-linear, hyperbolic system given by the inviscid compressible flow equations. Kopriva (1986b) has used a domain decomposition method in conjunction with the spectral shock-fitting techniques discussed in Sec. 8.6. He used the fully upwind interface condition given in (13.2.26). The particular problem he studied was the acoustic response to the interaction of a Mach 1.25 shock with a isolated vortex. Part (a) of Fig. 13.9 shows the grid for a single domain solution and part (a) of Fig. 13.10 gives

(a)

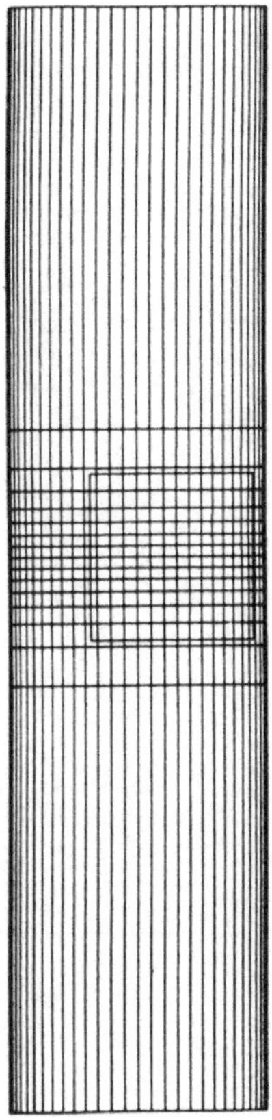

(b)

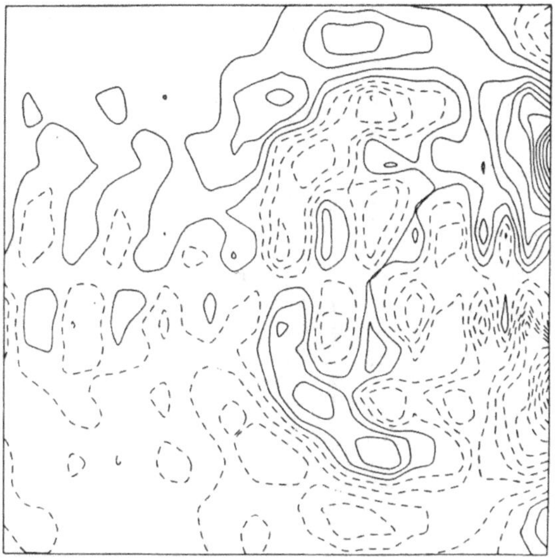

Figure 13.9. Single domain calculation of shock/vortex interaction: (a) computational grid, (b) pressure field in insert of part (a). (Courtesy of D. Kopriva.)

the domain decomposition for a calculation with three subdomains. Both grids contain the same total number of points. Note the much better horizontal resolution achieved by the domain decomposition method in the region of interest. Parts (b) of Figs. 13.9 and 13.10 show the pressure field predicted by both calculations in the region denoted by the inserts in parts (a). The single domain solution is afflicted by appreciable oscillations due to inadequate resolution. The domain decomposition solution, however, is quite regular.

13.3. Variational Methods

13.3.1. Formulation

The essential aspects of the variational spectral domain decomposition methods can be gleaned from an examination of their application to the

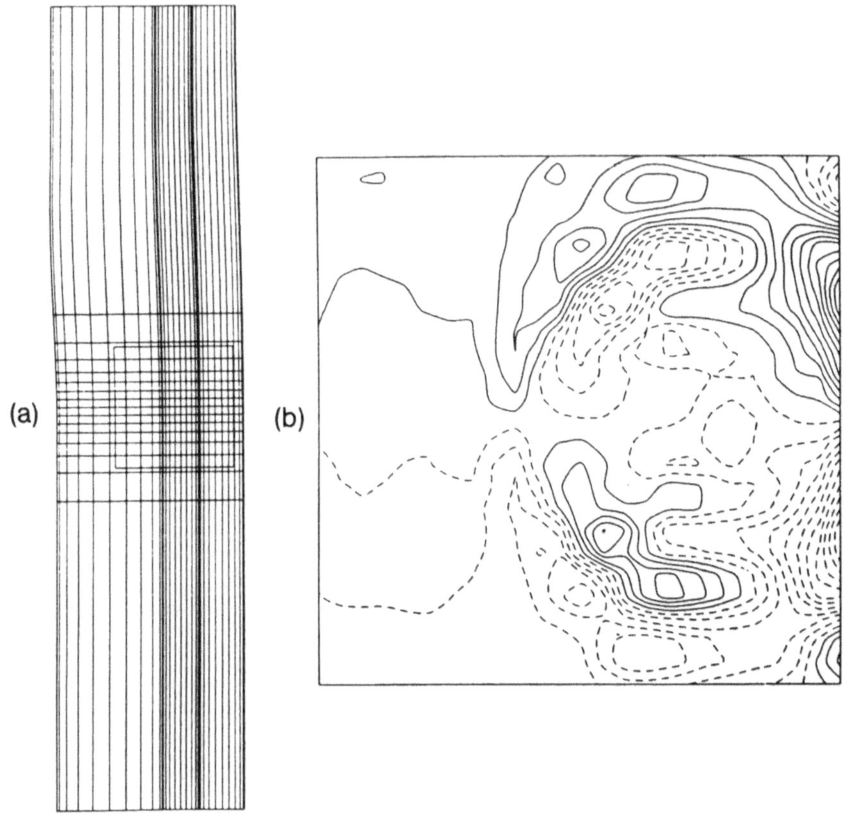

Figure 13.10. Triple domain patching calculation of shock/vortex interaction: (a) computational grid, (b) pressure field in insert of part (a). (Courtesy of D. Kopriva.)

two-dimensional Helmholtz equation

$$\begin{cases} -\Delta u + \lambda u = f & \text{in } \Omega \\ u = 0, & \text{on } \partial\Omega \end{cases} \tag{13.3.1}$$

where λ is a non-negative constant. An equivalent, variational formulation of this problem is that u is the solution to

$$\int_\Omega (\nabla u \cdot \nabla v + \lambda uv)\, dx\, dy = \int_\Omega fv\, dx\, dy \qquad \text{for all } v \in E, \tag{13.3.2}$$

where E is the space of all functions which vanish on $\partial\Omega$ and which, together with their first derivatives, are square integrable over Ω. This equivalence is demonstrated in Sec. 13.5.1.

Now suppose that Ω is decomposed into non-overlapping, rectangular subdomains Ω_s, for $s = 1, \ldots, S$. On each subdomain the approximation is presumed to be an element of $\mathbb{P}_{N_{x_s}} \otimes \mathbb{P}_{N_{y_s}}$, i.e., a polynomial of degree $\leqslant N_{x_s}$ in x and of degree $\leqslant N_{y_s}$ in y. Let N_s denote the pair (N_{x_s}, N_{y_s}) and define $\mathbb{P}_{N_s}$ to be $\mathbb{P}_{N_{x_s}} \otimes \mathbb{P}_{N_{y_s}}$. Finally, let N refer to $\{N_1, \ldots, N_S\}$.

In a variational domain decomposition method the space of trial (and test) functions X_N is given by

$$X_N = \{v \in C^0(\bar{\Omega}) | v_s \in \mathbb{P}_{N_s} \text{ for } s = 1, \ldots, S \text{ and } v = 0 \text{ on } \partial\Omega\}, \quad (13.3.3)$$

where v_s denotes the restriction of v to Ω_s. The space X_N is therefore composed of continuous functions which are piecewise polynomials defined on the decomposition of the physical domain Ω.

The solution of problem (13.3.2) is approximated by a function $u^N \in X_N$ satisfying

$$\sum_{s=1}^{S} \int_{\Omega_s} (\nabla u_s^N \cdot \nabla v + \lambda u_s^N v)\, dx\, dy = \sum_{s=1}^{S} \langle f, v \rangle_s \qquad \text{for all } v \in X_N. \quad (13.3.4)$$

For each s, $\langle f, v \rangle_s$ is a convenient approximation of the integral $\int_{\Omega_s} fv\, dx\, dy$. Equation (13.3.4) is a variational formulation with trial and test functions which are continuous across element (or subdomain) boundaries. Flux continuity, i.e., continuity of the normal derivative, at element interfaces, is not satisfied for fixed N, but only as part of the convergence process, i.e., when N tends to infinity (we mean here that all N_{x_s}, N_{y_s} tend to ∞).

The formulation (13.3.4) is fairly general, and we shall briefly refer to it as a variational method for problem (13.3.1). It includes, among others, the finite-element approach in its more standard form, the h-version (see, e.g., Strang and Fix (1973), and Ciarlet (1978)). The less conventional p-version of the finite-element method can be also cast in the form (13.3.4). In the p-version approach the decomposition is kept fixed, while convergence is achieved by increasing each N_s, i.e., by using local polynomial expansions of increasing order (we refer to the study by Babuška, Szabo and Katz (1981), and, for a more complete analysis of the p-version method, to Dorr (1984) and Guo and Babuška (1985)).

13.3.2. The Spectral-Element Method

The spectral-element method, proposed by Patera (1984), has the same viewpoint as the p-version of the finite-element method. Therefore, the discrete approximation to (13.3.1) satisfies (13.3.4), and, on each subdomain (here called element), u_s^N has a polynomial expansion of high degree. The essential difference between the p-version of the finite-element method and the spectral-element method lies in the choice of the basis for the trial and test functions in (13.3.4). Thus, the algebraic realizations of the variational formulation (13.3.4) differ. In the p-version case, Legendre-type polynomials are used as basis functions on each element Ω_s. In the spectral-element method, on the other hand, the basis in Ω_s is formed by the $M_s = (N_{x_s} + 1) \times (N_{y_s} + 1)$ Lagrange interpolants at the M_s nodes of the Gauss-Lobatto quadrature formula relative to the Chebyshev weight function.

We use the one-dimensional Helmholtz problem to illustrate some algebraic details of the spectral-element method:

$$\begin{cases} -u_{xx} + \lambda u = f & \text{in } \Omega = (a, b), \\ u(a) = u(b) = 0. \end{cases} \tag{13.3.5}$$

The domain Ω is decomposed into S subintervals (or elements, or subdomains) (a_s, a_{s+1}), with $a_1 = a$ and $a_{S+1} = b$. For each $s = 1, \ldots, S$, the local element coordinate system is denoted by

$$\xi^{(s)} \equiv \Phi^{(s)}(x) = \frac{2}{l_s}(x - a_s) - 1 \qquad l_s = a_{s+1} - a_s \tag{13.3.6}$$

(note that $\xi^{(s)} \in [-1, 1]$ if $x \in [a_s, a_{s+1}]$). Moreover, we denote by $x_j^{(s)}$ the images through $(\Phi^{(s)})^{-1}$ of the Chebyshev points

$$\xi_j^{(s)} = \cos\frac{\pi j}{N_s} \qquad \text{for } j = 0, \ldots, N_s. \tag{13.3.7}$$

The spectral-element approximation to (13.3.5) takes the following form. We introduce the finite-dimensional space

$$X_N = \{v \in C^0([a, b]) | v_s \in \mathbb{P}_{N_s}, \text{ for } s = 1, \ldots, S, v(a) = v(b) = 0\}, \tag{13.3.8}$$

and we look for a function $u^N \in X_N$ such that

$$\sum_{s=1}^{S} \int_{a_s}^{a_{s+1}} (u_{s,x}^N v_x + \lambda u_s^N v)\, dx = \sum_{s=1}^{S} \langle f, v \rangle_s \qquad \text{for all } v \in X_N. \tag{13.3.9}$$

In (13.3.9) the right-hand side is defined as follows:

$$\langle f, v \rangle_s = \int_{a_s}^{a_{s+1}} (I_N^{(s)} f) v\, dx \tag{13.3.10}$$

and $I_N^{(s)} f \in \mathbb{P}_N$ is the interpolant of f at the Chebyshev nodes $\{x_j^{(s)}\}$. The trial functions on each subinterval (a_s, a_{s+1}) are the images $H_i^{(s)} \in \mathbb{P}_{N_s}$ of the Lagrange interpolants at the points (13.3.7) according to the mapping $(\Phi^{(s)})^{-1}$. Therefore, $H_i^{(s)}(\xi_j^{(s)}) = \delta_{ij}$ for $i, j = 0, \ldots, N_s$, and

$$H_i^{(s)}(\xi^{(s)}) = \frac{2}{N_s} \sum_{n=0}^{N_s} \frac{1}{\bar{c}_i \bar{c}_n} T_n(\xi_i^{(s)}) T_n(\xi^{(s)}), \tag{13.3.11}$$

where, as usual, $\bar{c}_0 = \bar{c}_{N_s} = 2$ and $\bar{c}_k = 1$ for $1 \leqslant k \leqslant N_s - 1$. With respect to the local coordinate system, the solution u_s^N has the following expansion

$$u_s^N(\xi^{(s)}) = \sum_{i=0}^{N_s} u_{s,i} H_i^{(s)}(\xi^{(s)}), \tag{13.3.12}$$

where $u_{s,i} = u_s^N(x_i^{(s)})$ are the unknown grid values of the approximate solution. If we fix s, and we take as test function v in (13.3.9) the function $H_j^{(s)}$, for $j = 1, \ldots, N_s - 1$ ($H_j^{(s)}$ is associated to a Chebyshev node internal to (a_s, a_{s+1}), extended by zero outside the interval (a_s, a_{s+1})), the equation arising from (13.3.9) takes the form

$$\sum_{k=1}^{N_s} C_{jk}^{(s)} u_{s,k} = \sum_{k=0}^{N_s} B_{jk}^{(s)} f_k^{(s)}. \tag{13.3.13}$$

We have set $f_k^{(s)} = f(x_k^{(s)})$, and $C^{(s)}$ and $B^{(s)}$ denote the elemental matrices. A straightforward calculation shows that $C^{(s)} = A^{(s)} + \lambda B^{(s)}$, where

$$\begin{aligned} A_{jk}^{(s)} &= \frac{8}{l_s(N_s)^2} \frac{1}{\bar{c}_k \bar{c}_j} \sum_{n,m=0}^{N_s} \frac{1}{\bar{c}_n \bar{c}_m} T_n(\xi_k^{(s)}) T_m(\xi_j^{(s)}) a_{nm}, \\ B_{jk}^{(s)} &= \frac{2l_s}{(N_s)^2} \frac{1}{\bar{c}_k \bar{c}_j} \sum_{n,m=0}^{N_s} \frac{1}{\bar{c}_n \bar{c}_m} T_n(\xi_k^{(s)}) T_m(\xi_j^{(s)}) b_{nm}. \end{aligned} \tag{13.3.14}$$

Here

$$a_{nm} = \int_{-1}^{1} \frac{dT_n}{dx} \frac{dT_m}{dx} dx = \begin{cases} 0 & n + m \text{ odd} \\ \dfrac{nm}{2} \{J_{|(n-m)/2|} - J_{(n+m)/2}\} & n + m \text{ even} \end{cases} \tag{13.3.15}$$

where $J_0 = 0$ and $J_k = -4\sum_{q=1}^{k} 1/(2q-1)$ if $k \geqslant 1$. Moreover,

$$b_{nm} = \int_{-1}^{1} T_n T_m \, dx = \begin{cases} 0 & n + m \text{ odd} \\ \dfrac{1}{1-(n+m)^2} + \dfrac{1}{1-(n-m)^2} & n + m \text{ even} \end{cases} \tag{13.3.16}$$

$A^{(s)}$ and $B^{(s)}$ are, respectively, the elemental stiffness and mass matrices. To complete their definition, it remains to fill their first and last rows. These correspond to the Lagrangian functions $H_0^{(s)}$ and $H_{N_s}^{(s)}$ respectively. In the former case, the test function to be taken in (13.3.9) is the function $v \in X_N$ such that $v_{s-1} = H_{N_{s-1}}^{(s-1)}$, $v_s = H_0^{(s)}$, and $v_r \equiv 0$ if $r \neq s-1, s$, for $s = 2, \ldots, S$. In the latter case, the test function $v \in X_N$ is such that $v_s = H_{N_s}^{(s)}$, $v_{s+1} = H_0^{(s+1)}$, and $v_r \equiv 0$ if $r \neq s, s+1$. In both cases, the support of v extends to two adjacent intervals. Hence, two integrals in (13.3.9) contribute to the entries of the elemental matrices. The modification of the formula (13.3.14) is straightforward. The Dirichlet boundary conditions $u^N(a) = u^N(b) = 0$ are imposed by matrix condensation, i.e., by eliminating from the system the rows and columns corresponding to the two boundary points. If Neumann boundary conditions were used instead, they would have been taken into account naturally by the variational principle.

After a global node numbering, and an assembly of the elemental equations, the algebraic representation of (13.3.9) takes the form

$$CU = BF. \tag{13.3.17}$$

The unknown vector U contains the values of the discrete solution u^N at all Chebyshev points $x_j^{(s)}$, for $j = 0, \ldots, N_s$ and for all $s = 1, \ldots, S$ except for $x_0^{(1)} = a$ and $x_{N_s}^{(S)} = b$. A similar notation is used for the source term F. The matrices C and B are positive-definite, symmetric and banded, the bandwidth being determined by the largest N_s.

When a two-dimensional problem is considered, the Cartesian products of the Chebyshev–Lobatto nodes are used on each subdomain Ω_s. Then Lagrangian interpolants at these points can be represented using tensor products of Chebyshev polynomials.

To date, direct methods have been favored for the solution of the implicit system represented by (13.3.17). The static condensation method (Przmieniecki (1963)) has been used successfully by Patera (1984) and Korczak and Patera (1986).

Let us now consider a spectral-element discretization of the non-linear hyperbolic equation

$$u_t + A(u)u_x = 0, \qquad \text{in } \Omega, \tag{13.3.18}$$

with appropriate initial and boundary conditions. A semi-discrete variational method reads

$$\sum_{s=1}^{S} \int_{\Omega_s} \frac{\partial u^N}{\partial t}(t) v \, dx \, dy = -\sum_{s=1}^{S} \langle A(u^N(t)) u_x^N(t), v \rangle_s, \quad \text{for all } v \in X_N. \tag{13.3.19}$$

As in (13.3.9), the symbol $\langle g, v \rangle_s$ is used to indicate a convenient approximation of the integral $\int_{\Omega_s} gv \, dx \, dy$. A pure Galerkin treatment of non-linear terms is rarely computationally efficient. Therefore, a collocation approach is recommended for the right-hand side of (13.3.19). This makes fast transform methods available for computing the non-linearities.

One can, of course, combine mapping techniques with the spectral-element domain decomposition strategy. This has been pursued extensively by Korczak and Patera (1986), who adopted the standard finite-element approach of isoparametric mappings (see, e.g., Ciarlet (1977)).

Most of the applications of the spectral-element method have been to incompressible Navier–Stokes flows. The incompressibility constraint has usually been handled by splitting methods as discussed in Sec. 7.3.2. These extend in a straightforward fashion to domain decomposition methods. The explicit treatment of the advection terms is handled in the manner of hyperbolic equations. The implicit Poisson and Helmholtz equations for the pressure and velocity components are solved as described above for the model problems.

Patera and his collaborators have produced spectral-element solutions of numerous fluid dynamical problems. Figure 13.11 illustrates a spectral-element solution to flow past a circular cylinder. Note how the spectral-element domain decomposition has been chosen to provide high resolution

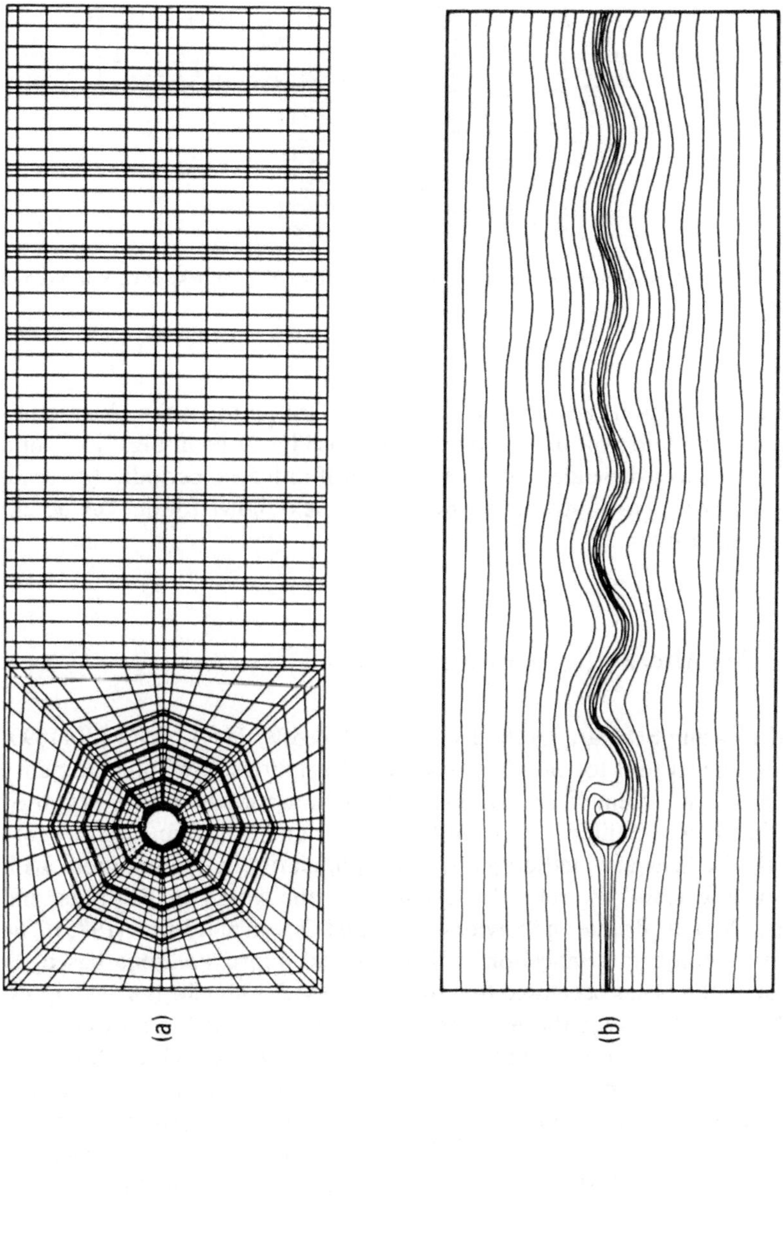

Figure 13.11. Spectral-element solution of flow past a circular cylinder: (a) computational grid, (b) streamlines of computed solution. (Courtesy of G. Karniadakis and A. Patera.)

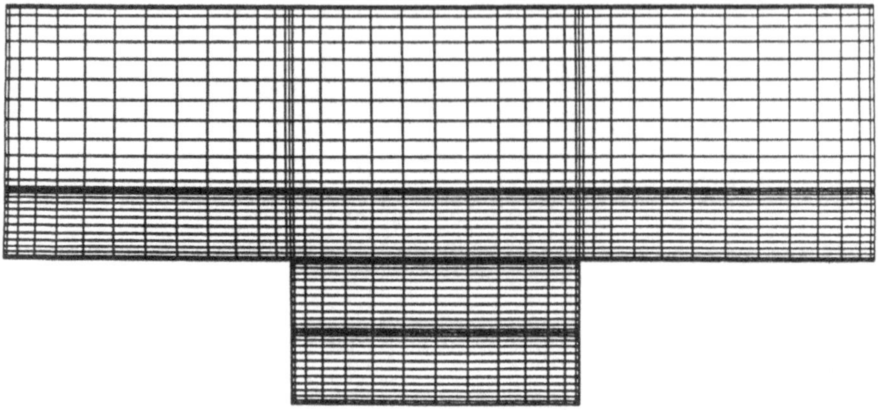

Figure 13.12. Spectral-element grid for flow in a grooved channel. (Courtesy of K. Korczak and A. Patera.)

in the critical regions adjacent to the cylinder and in the wake. A spectral-element solution to flow in a grooved channel is illustrated in Figs. 13.12 and 13.13. This time-dependent flow exhibits self-induced, periodic (in time) oscillations which are apparent in the pressure field shown for one full period T in the latter figure.

13.4. The Alternating Schwarz Method

The alternating Schwarz method was introduced by H. A. Schwarz (1890) as a theoretical tool for proving existence of solutions of elliptic boundary value problems. Numerical analogs to the Schwarz method were proposed some time ago by Miller (1965) and Kutznetsov and Matsokin (1972). More recently, the Schwarz method gained new popularity in computations with parallel processors. We refer, for example, to the works by Rodrigue and Simon (1984), Rodrigue and Saylor (1985), and Ortega and Voigt (1985). The Schwarz method in connection with numerical approximations based on spectral methods has been used first by Morchoisne (1984) for the incompressible Navier–Stokes equations. In order to illustrate the method, we consider here the two-dimensional elliptic boundary value problem (13.3.1), and we put for brevity $Lu \equiv -\Delta u + \lambda u$. What will be said, however, applies to a general second-order elliptic operator. For the sake of simplicity, as in Sec. 13.3.4, we consider a rectangular domain $\Omega = (a, b) \times (c, d)$. We suppose that Ω is partitioned, as illustrated in Fig. 13.14, into two overlapping subdomains

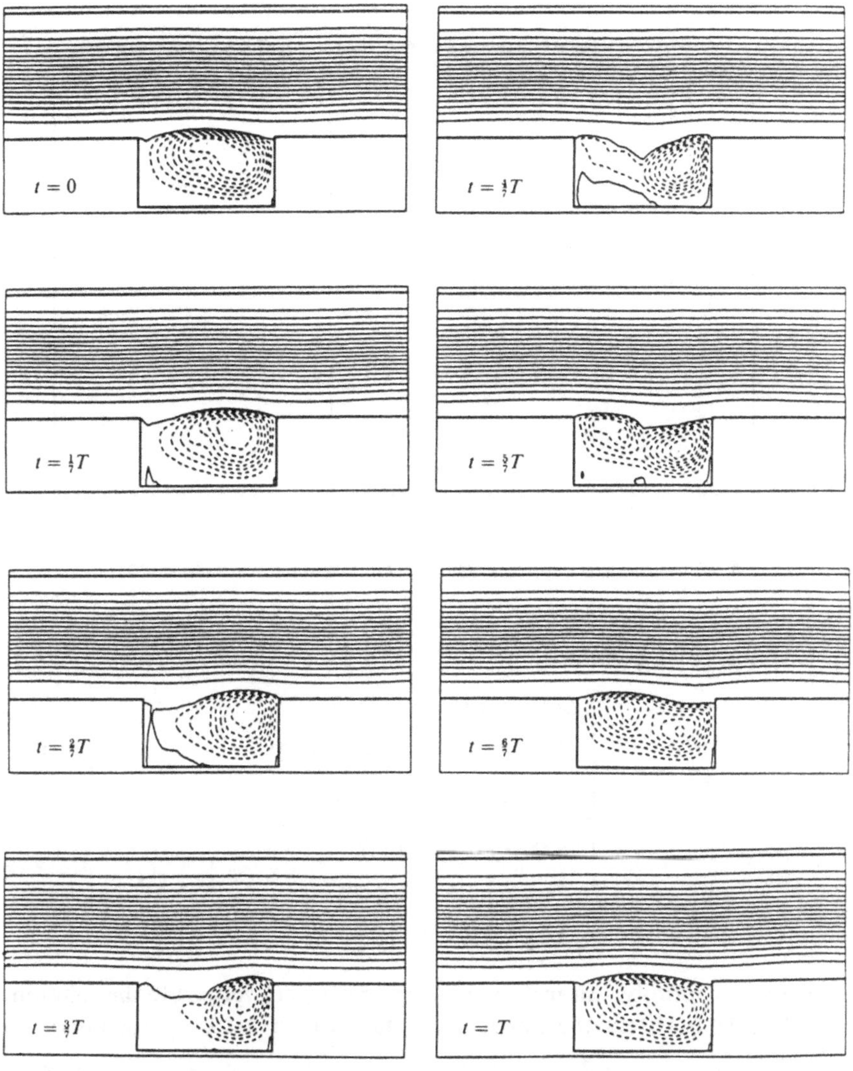

Figure 13.13. Computed pressure field for oscillatory flow in a grooved channel. (Courtesy of K. Korczak and A. Patera.)

Ω^- and Ω^+ of the form $\Omega^- = (a, \beta) \times (c, d)$, $\Omega^+ = (\alpha, b) \times (c, d)$, with $a < \alpha < \beta < b$. We also set $\Gamma^- = \{\beta\} \times (c, d)$, and $\Gamma^+ = \{\alpha\} \times (c, d)$.

Assuming the right-hand side f of (13.3.1) to be smooth enough, the Schwarz method for problem (13.3.1) consists of choosing an arbitrary initial function u^0 in Ω^+ such that $u^0|_{\partial\Omega^+/\Gamma^+} = 0$, and defining the two sequences

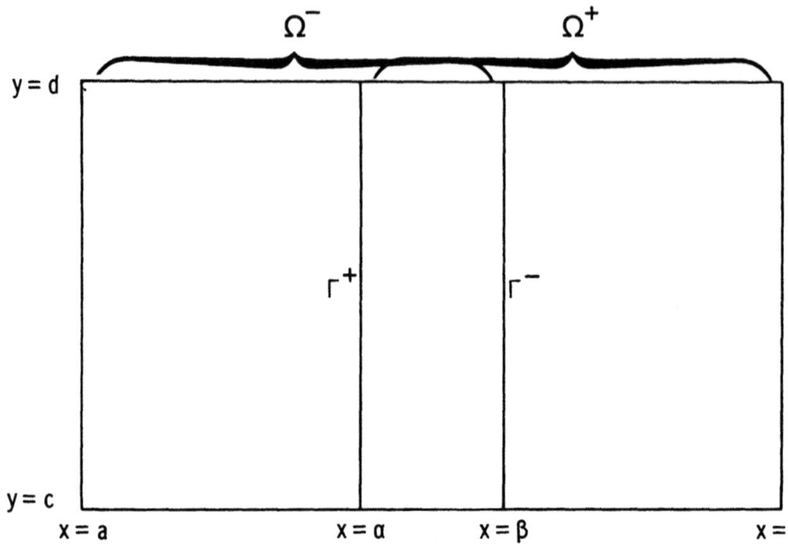

Figure 13.14. Overlapping domain decomposition in two dimensions.

$\{u^{2n-1}\}$ and $\{u^{2n}\}$, for $n \geqslant 1$, by

$$\begin{cases} Lu^{2n-1} = f & \text{in } \Omega^- \\ u^{2n-1} = 0 & \text{on } \partial\Omega^-/\Gamma^- \\ u^{2n-1} = u^{2n-2} & \text{on } \Gamma^-, \end{cases} \tag{13.4.1}$$

$$\begin{cases} Lu^{2n} = f & \text{in } \Omega^+ \\ u^{2n} = 0 & \text{on } \partial\Omega^+/\Gamma^+ \\ u^{2n} = u^{2n-1} & \text{on } \Gamma^+. \end{cases} \tag{13.4.2}$$

As n tends to infinity, this method is geometrically convergent to the solution u of (13.3.1). Precisely, there exists a constant k with $0 < k < 1$ such that:

$$\|u^{2n} - u|_{\Omega^+}\|_{H^1(\Omega^+)} + \|u^{2n-1} - u|_{\Omega^-}\|_{H^1(\Omega^-)} \leqslant Ck^n\|u^0 - u|_{\Omega^+}\|_{H^1(\Omega^+)}. \tag{13.4.3}$$

(See (A.11.2) for the definition of the space $H^1(\Omega)$.)

Now, we can apply a Chebyshev (or Legendre) collocation method in order to approximate both problems (13.4.1) and (13.4.2). The continuity conditions at Γ^- and Γ^+ are imposed at the collocation points on these boundaries respectively. Note that if the degrees of the polynomial expansions in the y-direction are different in Ω^- and Ω^+, then the transmission of data at the interfaces is done by resorting to the Chebyshev or (Legendre) expansion of the interpolating polynomials (see, e.g., (13.2.33) and (13.2.34)).

The discrete Schwarz method just described produces a sequence of polynomials $(u^{N_+})^{2n}$ on Ω^+ and $(u^{N_-})^{2n+1}$ on Ω^-, for $n = 1, 2, \ldots$. As n tends to ∞, the two sequences converge respectively to two polynomials u^{N_+} and u^{N_-} which match at both interfaces Γ^+ and Γ^-. Moreover, u^{N_-} satisfies the differential equation at the interior collocation points of the domain Ω^-, as well as the physical boundary conditions. Similar conditions are satisfied by u^{N_+}.

The convergence of the discrete Schwarz algorithm as $n \to \infty$ (and N_-, N_+ are fixed) has been proven for the model Laplace equation by Canuto and Funaro (1987). They proved that the rate of convergence obeys a law similar to (13.4.3) with k independent of N_- and N_+. Morever, at least for the one-dimensional version of the algorithm, they showed that the discrete solution (u^{N_-}, u^{N_+}) converges to the exact solution $(u|_{\Omega^-}, u|_{\Omega^+})$ with spectral accuracy as both N_- and N_+ tend to infinity.

The extension of the Schwarz algorithm to a general Cartesian or curvilinear overlapping partition of an arbitrarily shaped domain Ω is straightforward.

We now make some comments on the rate of convergence of the Schwarz algorithm. In the inequality (13.4.3), the constant k is called the reduction factor, since it is a measure of the reduction of the error over two successive iterates. As expected, the reduction factor is a monotonically decreasing function of the measure of the overlapping region $\Omega^- \cap \Omega^+$. In numerical computations, the following (more practical) form of the reduction factor may be used in order to estimate the rate of convergence of the Schwarz method. We set for each $n \geqslant 1$

$$e_+^n = u^{2n+2} - u^{2n} \qquad e_-^n = u^{2n+1} - u^{2n-1},$$

then we consider the maximum error at interfaces only, by setting

$$E_+^n = \max_{R \in J_{\Gamma+}} |e_+^n(R)| \qquad E_-^n = \max_{S \in J_{\Gamma-}} |e_-^n(S)|.$$

Here $J_{\Gamma+}$ and $J_{\Gamma-}$ denote the set of collocation points lying on Γ^+ and Γ^- respectively. We now define:

$$\kappa = \max\left\{\frac{E_+^n}{E_+^{n-1}}, \frac{E_-^n}{E_-^{n-1}}\right\}. \tag{13.4.4}$$

By an experimental analysis, Zanolli (1987) has found the value of κ for some model examples. In the case for which $\Omega = (0, 1) \times (0, 0.5)$ and $Lu = -\Delta u$, with $\Omega^- = (0, \beta) \times (0, 0.5)$, and $\Omega^+ = (\alpha, 1) \times (0, 0.5)$, with $0 < \alpha < \beta < 1$, the reduction factor (13.4.4) has the form

$$\kappa = \kappa(\delta) = \left(\frac{1 - \delta}{1 + \delta}\right)^2, \tag{13.4.5}$$

where $2\delta = \beta - \alpha$. For the one-dimensional domain $\Omega = (0, 1)$, and $Lu = -u_{xx}$ with $\Omega^- = (0, \beta)$, and $\Omega^+ = (\alpha, 1)$, it is

$$\kappa = \kappa(\delta) = \left(\frac{1 - 2\delta}{1 + 2\delta}\right)^2, \tag{13.4.6}$$

where $\delta = \beta - \alpha$ and now, of course, $\Gamma^- = \{\beta\}$ and $\Gamma^+ = \{\alpha\}$. In both one and two dimensions, the reduction factor does not depend on the degree N of the polynomial expansion. For operators of the form, $Lu = -\Delta u + \lambda u$ in two dimensions, and $Lu = -u_{xx} + \lambda u$ in one dimension, with $\lambda > 0$, it was observed by Metivet (1987) that the Schwarz algorithm has a convergence rate that is exponential rather than just geometric. This can be explained analytically by investigating the differential equations satisfied by the error terms e_+^n and e_-^n.

The same kind of experimental analysis was carried out for Cartesian subdivisions of Ω involving more than two subdomains. The behavior of the reduction factor is essentially the same as before, but now δ is the measure of the smallest overlapping region. A comparative experimental analysis between the Schwarz method and the patching methods presented in Sec. 13.2 was also carried out by Zanolli (1987).

Experimental results for Navier–Stokes approximations using the Schwarz algorithm for both Cartesian and curvilinear subdivisions of Ω can be found in Metivet (1987).

13.5. Mathematical Aspects of Domain Decomposition Methods

In this section, we discuss some mathematical aspects of the patching methods described in the previous sections. Moreover, we show that variational methods can be viewed as generalized patching methods and conversely.

13.5.1. Patching Methods

We begin by proving that the decomposed problem (13.2.2)—(13.2.5) is equivalent to the original problem (13.2.1). We shall actually show that (13.2.2)—(13.2.5) is equivalent to the variational form of (13.2.1), which reads as follows:

$$\begin{cases} u \in H_0^1(a, b) \\ \displaystyle\int_a^b [(\nu u_x - \alpha u)v_x + \lambda uv]\,dx = \int_a^b fv\,dx \quad \text{for all } v \in H_0^1(a, b). \end{cases} \tag{13.5.1}$$

(The space $H_0^1(a, b)$ is defined in (A.11.c).) We assume here that $f \in L^2(a, b)$. The formulation (13.5.1) is obtained from (13.2.2) by a simple integration-by-parts argument. By applying the Lax–Milgram lemma (A.5) one can easily show that problem (13.5.1) has a unique solution. Moreover, there exists a constant $C > 0$, such that $\|u\|_{H^2(a,b)} \leqslant C\|f\|_{L^2(a,b)}$. It follows (see (A.11.a)) that u is continuously differentiable in $[a,b]$, i.e., $u \in C^1([a,b])$. Clearly, u satisfies also (13.2.2) to (13.2.5). Conversely, let $u_s \in H^1(a_s, a_{s+1})$, $(s = 1,2)$ satisfy this set of equations. Denote by u the function such that $u|_{(a_s,a_{s+1})} \equiv u_s$. This function is continuous at $x = a_2$ by (13.2.4); thus, $u \in H_0^1(a,b)$. Moreover, since $f \in L^2(a_s, a_{s+1})$ for $s = 1, 2$, $u_s \in H^2(a_s, a_{s+1})$ by virtue of (13.2.2)—(13.2.3). Hence, $u \in C^1([a,b])$ because of (13.2.5). It remains to check that u satisfies the variational formulation (13.5.1). For every $v \in H_0^1(a,b)$, multiply (13.2.2) by $v|_{[a_1,a_2]}$ and (13.2.3) by $v|_{[a_2,a_3]}$ and then integrate. The result is

$$\sum_{s=1}^{2} \int_{a_s}^{a_{s+1}} Lu_s v \, dx = \sum_{s=1}^{2} \int_{a_s}^{a_{s+1}} fv \, dx.$$

Integration-by-parts on each interval yields

$$\int_a^b \{(\gamma u_x - \alpha u)v_x + \lambda uv\} \, dx + v(a_2)[vu_x - \alpha u](a_2) = \int_a^b fv \, dx.$$

(Recall that v vanishes at $x = a$ and $x = b$.) The jump $[vu_x - \alpha u]$ at $x = a_2$ is zero by (13.2.4) and (13.2.5). Therefore, u satisfies (13.5.1).

The equivalence between a second-order boundary value problem like (13.2.14) and its decomposed version with the interface condition (13.2.16) can be established by a suitable modification of the above argument.

13.5.2. Equivalence Between Patching and Variational Methods

Let us consider the patching collocation method (13.2.10)—(13.2.13). For the sake of simplicity we assume hereafter that the collocation nodes within each subdomain are relative to the Legendre rather than to the Chebyshev weight. We want to set the method in variational form. To this end, let us introduce the interpolation operator $\tilde{I}_N^{(s)}$: $C^0([a_s, a_{s+1}]) \to \mathbb{P}_{N_s-2}$ defined by

$$(\tilde{I}_N^{(s)}\phi)(x_j^{(s)}) = \phi(x_j^{(s)}) \qquad \text{for } 1 \leqslant j \leqslant N_s - 1, \tag{13.5.2}$$

i.e., $\tilde{I}_N^{(s)}\phi$ is the interpolant of ϕ of degree $N_s - 2$ at the internal Gauss–Lobatto nodes. Then the collocation equation (13.2.10) and (13.2.11) at these nodes are equivalent to the identities (between polynomials)

$$\tilde{I}_N^{(s)} Lu_s^N = \tilde{I}_N^{(s)} f \qquad (s = 1,2)$$

or

$$-vu_{s,xx}^N + \tilde{I}_N^{(s)}(\alpha u_{s,x}^N + \lambda u_s^N) = \tilde{I}_N^{(s)} f \qquad (s = 1,2). \tag{13.5.3}$$

Now, let $v \in X_N$ be an arbitrary test function. (The space X_N of the continuous, piecewise polynomial functions is defined in (13.3.3).) Let us multiply each side of (13.5.3) by $v|_{(a_s,a_{s+1})}$ and integrate over (a_s, a_{s+1}). Next, let us integrate by parts and take into account the interface condition (13.2.13) to eliminate the boundary term. We obtain

$$\sum_{s=1}^{2} \int_{a_s}^{a_{s+1}} [\nu u_{s,x}^N v_x + \tilde{I}_N^{(s)}(\alpha u_{s,x}^N + \lambda u_s^N)v]\,dx = \sum_{s=1}^{2} \int_{a_s}^{a_{s+1}} \tilde{I}_N^{(s)} f v\,dx \quad \text{for all } v \in X_N. \tag{13.5.4}$$

Conversely, let $u^N \in X_N$ be a solution of (13.5.4). Integrating by parts we get

$$\sum_{s=1}^{2} \int_{a_s}^{a_{s+1}} [-\nu u_{s,xx}^N + \tilde{I}_N^{(s)}(\alpha u_{s,x}^N + \lambda u_s^N - f)]v\,dx$$
$$- \nu(u_{1,x}^N - u_{2,x}^N)(a_2)v(a_2) = 0 \qquad \text{for all } v \in X_N.$$

Note that the term in square brackets is a polynomial of degree $N_s - 2$ in each subdomain. Choosing as test functions the Lagrange basis at the interior collocation points and using the exactness of the quadrature rule (see (2.2.25)), one gets (13.5.3) at the interior collocation points, and hence, everywhere in each subdomain. Finally, taking as v any test function with $v(a_2) \neq 0$ yields the interface condition (13.2.13). Thus, we have proved the following equivalence result.

Proposition 13.1. *The patching collocation method* (13.2.10)—(13.2.13) *is equivalent to the generalized variational method* (13.5.4).

We shall now show that, to some extent, a variational approximation method can be viewed as a generalized patching method for the same differential problem. Let us consider again the boundary value problem (13.2.1), in which for simplicity, we set $\lambda = 0$. Its variational spectral approximation, according to Sec. 13.3, reads as follows: find $u^N \in X_N$ such that

$$\sum_{s=1}^{2} \int_{a_s}^{a_{s+1}} [\nu u_{s,x}^N v_x + \alpha u_{s,x}^N v]\,dx = \sum_{j=1}^{2} (f, v)_{N_s} \quad \text{for all } v \in X_N. \tag{13.5.5}$$

Here $(f, v)_{N_s}$ denotes the discrete Legendre–Gauss–Lobatto inner product on the subdomain Ω_s (see (2.2.24)).

The following result can be established by a technique similar to that used in the proof of the previous proposition.

Proposition 13.2. *The variational method* (13.5.5) *is equivalent to the following generalized patching collocation method*:

$$\begin{aligned} &Lu_1^N - f|_{x=x_j^{(1)}} = 0 \qquad j = 1, \ldots, N_1 - 1 \\ &u_1^N(a_1) = 0 \end{aligned} \tag{13.5.6}$$

$$\begin{aligned} &Lu_2^N - f|_{x=x_j^{(2)}} = 0 \qquad j = 1, \ldots, N_2 - 1 \\ &u_2^N(a_3) = 0 \end{aligned} \tag{13.5.7}$$

$$u_1^N(a_2) = u_2^N(a_2) \tag{13.5.8}$$

$$\nu u_{1,x}^N(a_2) - \nu u_{2,x}^N(a_2) + w_0^1(Lu_N^1 - f)(a_2) + w_{N_2}^2(Lu_N^2 - f)(a_2) = 0. \tag{13.5.9}$$

In (13.5.9), w_0^1 and $w_{N_2}^2$ are, respectively, the first and last weights of the Legendre–Gauss–Lobatto quadrature formula on subdomains 1 and 2. The interface condition can be interpreted in two ways. One can think of it as a perturbation of the exact interface condition (13.2.13) through a small linear combination of the residuals at the interface (recall that $w_0^s = w_{N_s}^s = O(N_s^{-2})$, see (2.3.12)). Conversely, one can view it as a linear combination of the residuals perturbed by a penalization on the jump of the normal derivative. Note that a linear combination of the residuals alone would lead to a singular algebraic system. This approach is similar to that used by Deville and Mund (1985) (see Sec. 13.2.3).

The generalized patching method (13.5.6)—(13.5.9) has been proposed by Funaro (1986), who also pointed out the equivalences between variational and patching domain decomposition methods.

These equivalences give rise to a sort of duality principle for domain decomposition spectral approximations. It can be expressed by the following diagram:

Variational Approximation methods	$\Leftrightarrow$	Modified Patching Collocation Methods
Modified Variational Approximation Methods	$\Leftrightarrow$	Patching Collocation Methods

13.6. Some Stability and Convergence Results

13.6.1. Patching Methods

Let us consider the patching collocation method (13.2.10)—(13.2.13). In order to avoid an overly technical exposition let us assume that $\alpha = \lambda = 0$. We shall analyze this method using the more convenient equivalent form (13.5.4)

By choosing $v = u^N = (u_1^N, u_2^N)$ in (13.5.4), we verify that the patching collocation method satisfies the "energy" coercivity condition (10.4.51) with respect to the norm of the Hilbert space $E = H_0^1(a, b)$. Using the Cauchy–Schwarz inequality on the right-hand side of (13.5.4), we get the estimate

$$\|u^N\|_{H^1(a,b)} \leqslant C \sum_{s=1}^{2} \|\tilde{I}_N^{(s)} f\|_{L^2(a_s, a_{s+1})}, \tag{13.6.1}$$

which is a specific form of the general inequality (10.4.52). Let $u \in H_0^1(a, b) \cap$

$H^m(a,b)$, (for some $m > 1$) be the solution of the differential problem (13.2.1), and let R_N: $H_0^1(a,b) \to X_N$ be a suitable projection operator. The general convergence estimate (10.4.54) becomes

$$\|u - u^N\|_{H^1(a,b)} \leqslant C\left[\|u - R_N u\|_{H^1(a,b)} + \sum_{s=1}^{2} \|f - \tilde{I}_N^{(s)} f\|_{L^2(a_s,a_{s+1})}\right]. \quad (13.6.2)$$

Let us estimate the right-hand side. To this end, we define the operator R_N as follows. Let $r = r(x)$ be the piecewise linear polynomial such that $r(a_s) = u(a_s)$ for $s = 1, 2, 3$. Setting $u^0 = u - r$, one has $u^0|_{(a_s,a_{s+1})} \in H_0^1(a_s, a_{s+1})$ with $\|u^0\|_{H^m(a_s,a_{s+1})} \leqslant C\|u\|_{H^m(a_s,a_{s+1})}$. Now, let us define $R_N u \in X_N$ such that

$$R_N u|_{(a_s,a_{s+1})} = r + P_{N_s}^{1,0} u^0|_{[a_s,a_{s+1}]} \qquad s = 1, 2, \quad (13.6.3)$$

where $P_{N_s}^{1,0}$ is the orthogonal projection in $H_0^1(a_s, a_{s+1})$ upon the space of polynomials of degree N_s which vanish at $x = a_s$ and a_{s+1} (see (9.4.21)). Applying the error estimate (9.4.22) one obtains

$$\|u - R_N u\|_{H^1(a,b)} \leqslant C \sum_{s=1}^{2} N_s^{1-m} \|u\|_{H^m(a,b)} \qquad m \geqslant 1. \quad (13.6.4)$$

Let us now consider the interpolation error on the data f. To this end, let us denote by $\tilde{I}_N f$ the polynomial of degree $N-2$ which interpolates the function f at the interior Legendre–Gauss–Lobatto points (2.3.12), now on the reference interval $(-1, 1)$. The following error estimate

$$\|f - \tilde{I}_N f\|_{L^2(-1,1)} \leqslant CN^{1-\mu} \|f\|_{H^\mu(a,b)} \qquad \mu > 1 \quad (13.6.5)$$

holds. In particular, this estimate allows us to deduce that the right-hand side of (13.6.1) is bounded by a constant independent of N times the H^1-norm of the function f. A proof of this estimate is as follows. Note that $\tilde{I}_N f = \tilde{I}_N I_N f$, where I_N is the interpolation operator of degree N at all the Gauss–Lobatto points (see (2.2.19)). Thus, denoting as usual by $P_N f$ the truncation of the Legendre series, we get the identity

$$f - \tilde{I}_N f = f - P_{N-2} f + \tilde{I}_N(P_{N-2} f - I_N f).$$

The estimate (13.6.5) is now an easy consequence of (9.4.6), (9.4.24) and the following inequality

$$\|\tilde{I}_N \phi\|_{L^2(-1,1)} \leqslant C\sqrt{N}\,\|\phi\|_{L^2(-1,1)} \qquad \text{for all } \phi \in \mathbb{P}_N. \quad (13.6.6)$$

In turn, this inequality can be proven as follows. If $\phi \in \mathbb{P}_{N-2}$, then $\tilde{I}_N \phi = \phi$; hence (13.6.6) is trivial. By the orthogonality of the Legendre polynomials, it is therefore enough to check it for $\phi = L_{N-1}$ and $\phi = L_N$. This can be easily done by observing that indeed $\tilde{I}_N \phi$ is a polynomial which interpolates the interior values of ϕ and two unknown values at the endpoints of the interval $(-1, 1)$. These are determined by the condition that $\tilde{I}_N \phi \in \mathbb{P}_{N-2}$; hence, they can be computed using (2.2.20).

We obtain a convergence estimate for the patching collocation method (13.5.4) by using (13.6.4) and (13.6.5) in (13.6.2). Recalling that $f = -\nu u_{xx}$, one gets

$$\|u - u^N\|_{H^1(a,b)} \leqslant C \sum_{s=1}^{2} N_s^{3-m} \|u\|_{H_s^m(a,b)}, \qquad m \geqslant 3. \tag{13.6.7}$$

13.6.2. Variational Methods

We consider now spectral domain decomposition methods in variational form, as discussed in Sec. 13.3. For one-dimensional problems, the proof of stability and convergence essentially reproduces the one given in the previous section for the patching method, with the simplification that $I_{N_s} f$ replaces $\tilde{I}_N^{(s)} f$ in (13.6.1) and (13.6.2). Thus, using now the estimate (9.4.24), one obtains a convergence estimate like (13.6.7) with the exponent $2 - m$.

In two space dimensions, let us discuss the variational method (13.3.4) in which, for the sake of simplicity, we confine ourselves to the geometry of Fig. 13.5.

The stability of the approximation in the $H^1(\Omega)$-norm follows from the coercivity of the form on the left-hand side of (13.3.4), which allows us to choose $v = u^N$ as the proper test function. In the terminology of Chap. 10, this proves that (10.4.51) is fulfilled. The convergence estimate (10.4.54) reads as follows:

$$\|u - u^N\|_{H^1(\Omega)} \leqslant C\left[\|u - R_N u\|_{H^1(\Omega)} + \sum_{s=1}^{2} \|f - I_N f\|_{L^2(\Omega_s)}\right], \tag{13.6.8}$$

where $R_N u$ is a projection of the exact solution $u \in H^m(\Omega)$, which we shall now define. Let us denote by Π_x the one-dimensional projection operator introduced in (13.6.3) upon the continuous, piecewise polynomial functions of degree N_{x_1} on the left interval and N_{x_2} on the right interval. Moreover, let us denote by Π_y the orthogonal projection operator $P_{M_y}^{1,0}$ defined in (9.4.21), with $M_y = \min(N_{y_1}, N_{y_2})$. Then we set

$$R_N u = \Pi_y \Pi_x u. \tag{13.6.9}$$

Noting that $u - R_N u = (u - \Pi_x u) + \Pi_x(u - \Pi_y u)$, and using the triangle inequality and the estimates (13.6.4) and (9.4.22) we obtain

$$\|u - R_N u\|_{H^1(\Omega)} \leqslant C \sum_{s=1}^{2} (M_y^{1-m} + N_{x_s}^{1-m}) \|u\|_{H^m(\Omega_s)}, \quad m \geqslant 1. \tag{13.6.10}$$

Applying this estimate to (13.6.8) together with the usual interpolation error estimates (9.7.19) allows us to get an error bound for the variational approximation method. If, in order to avoid a cumbersome notation, we assume that $N_{x_1} = N_{y_1} (= N_1)$ and $N_{x_2} = N_{y_2} (= N_2)$, then from (13.6.8) we obtain

$$\|u - u_N\|_{H^1(\Omega)}$$

$$\leqslant C \sum_{s=1}^{2} \left\{ (M_y^{1-m} + N_s^{1-m}) \|u\|_{H^m(\Omega_s)} + N_s^{1/2-\mu} \|f\|_{H^\mu(\Omega_s)} \right\},$$

$$m \geqslant 1, \mu \geqslant 1. \qquad (13.6.11)$$

This is the error estimate for the spectral domain decomposition method in variational form.

13.7. Recent Developments

Spectral domain decomposition methods have been an extremely active area of research in the last few years. Indeed, the supplementary bibliography to this, the second printing of this book, contains over 50 papers on the subject that have appeared in just the last two years. Many of these represent significant advances over the techniques that were available at the time of the first printing. As general references we recommend: the proceedings of two recent conferences (edited by Chan, Glowinski, Periaux, and Widlund (1989) and by Canuto and Quarteroni (1990)), each of which contains a substantial number of papers on spectral domain decomposition methods; a recent special issue (Volume 6, edited by Macaraeg and Hussaini)) of **Appl. Numer. Math.** devoted entirely to the subject; and the lengthy review of the spectral element method by Maday and Patera (1989).

Some of the contributions to the spectral element method which have general applicability are: the construction of nonconforming spectral-element methods by Maday, Mavriplis, and Patera (1988); a method which uses finite-elements near a singularity and spectral methods elsewhere (Bernardi, Debit, and Maday 1989, Bernardi, Maday, and Sacchi-Landriani 1989); parallel implementations (Fischer 1989, Fischer, Ho, Karniadakis, Ronquist, and Patera 1988)); and spectral element multigrid (Ronquist and Patera 1987b, Maday and Munoz 1989).

Recently, Kopriva (1989a) presented a spectral domain decomposition method on a combination of patched and overlapped grids for hyperbolic equations. Kopriva (1989b) also coupled upwind finite-difference methods (to improve resolution in the vicinity of a shock wave) with spectral methods (in the smooth part of the flow).

Gastaldi and Quarteroni (1989) devised a well-posed mathematical formulation of a problem in which the equation itself differed on the subdomains, in particular, hyperbolic in one domain and parabolic in the other. They also analyzed a spectral domain decomposition method for this application.

Streett and Macaraeg (1989) made essential use of the domain decomposition approach in their development of a spectral method with a non-reflecting outflow boundary condition for incompressible Navier-Stokes simulations.

Although not confined to domain decomposition methods, the Uzawa scheme for solving steady and unsteady Stokes and Navier-Stokes problems has proven quite useful (Maday and Patera 1989, Streett and Hussaini 1990).

Appendix A

Basic Mathematical Concepts

A.1. Hilbert and Banach Spaces

(a) Hilbert Spaces

Let X be a real vector space. An *inner product* on X is a function $X \times X \to \mathbb{R}$ denoted by (u, v), which satisfies the following properties:

(i) $(u, v) = (v, u)$ for all $u, v \in X$;
(ii) $(\alpha u + \beta v, w) = \alpha(u, w) + \beta(v, w)$ for all $\alpha, \beta \in \mathbb{R}$ and all $u, v, w \in X$;
(iii) $(u, u) \geqslant 0$ for all $u \in X$;
(iv) $(u, u) = 0$ implies $u = 0$.

Two elements $u, v \in X$ are said to be *orthogonal* in X if $(u, v) = 0$. The inner product (u, v) defines a *norm* on X by the relation

$$\|u\| = (u, u)^{1/2} \qquad \text{for all } u \in X.$$

The *distance* between two elements u, $v \in X$ is the positive number $\|u - v\|$. A *Cauchy sequence* in X is a sequence $\{u_k | k = 0, 1, \ldots\}$ of elements of X which satisfies the following property:

> for each positive number $\varepsilon > 0$, there exists an integer $N = N(\varepsilon) > 0$ such that the distance $\|u_k - u_m\|$ between any two elements of the sequence is smaller than ε provided both k and m are larger than $N(\varepsilon)$.

A sequence in X is said to *converge* to an element $u \in X$ if the distance $\|u_k - u\|$ tends to 0 as k tends to ∞.

A *Hilbert space* is a vector space equipped with an inner product for which all the Cauchy sequences are convergent.

EXAMPLES. (i) $\mathbb{R}^n$ endowed with the Euclidean product

$$(\mathbf{u}, \mathbf{v}) = \sum_{i=1}^{n} u_i v_i$$

is a finite dimensional Hilbert space.
(ii) If $[a, b] \subset \mathbb{R}$ is an interval, the space $L^2(a, b)$ (see (A.9.f)) is an infinite dimensional Hilbert space for the inner product

$$(u, v) = \int_a^b u(x)v(x)\, dx.$$

If X is a complex vector space, the inner product on X will be a complex-valued function. Then condition (i) has to be replaced by

(i′) $$(u, v) = \overline{(v, u)} \qquad \text{for all } u,\ v \in X.$$

(b) Banach Spaces

The concept of Banach space extends that of Hilbert space. Given a vector space X, a *norm* on X is a function $X \to \mathbb{R}$ denoted by $\|u\|$ which satisfies the following properties

$$\|u + v\| \leqslant \|u\| + \|v\| \qquad \text{for all } u,\ v \in X;$$

$$\|\lambda u\| = |\lambda|\, \|u\| \qquad \text{for all } u \in X, \text{ and all } \lambda \in \mathbb{R};$$

$$\|u\| \geqslant 0 \qquad \text{for all } u \in X;$$

$$\|u\| = 0 \qquad \text{if and only if } u = 0.$$

A *Banach space* is a vector space equipped with a norm for which all the Cauchy sequences are convergent.

EXAMPLES. (i) $\mathbb{R}^n$ endowed with the norm

$$\|\mathbf{u}\| = \left(\sum_{i=1}^{n} |u_i|^p\right)^{1/p}$$

(with $1 \leqslant p < +\infty$) is a finite dimensional Banach space.
(ii) If $[a, b] \subset \mathbb{R}$ is an interval and $1 \leqslant p < +\infty$, the space $L^p(a, b)$ (see (A.9.f)) is an infinite dimensional Banach space for the norm

$$\|u\| = \left(\int_a^b |u(x)|^p \, dx\right)^{1/p}.$$

(c) Dual Spaces

Let X be a Hilbert or a Banach space. A linear form $F: X \to \mathbb{R}$ is said to be *continuous* if there exists a constant $C > 0$ such that

$$|F(u)| \leqslant C\|u\| \qquad \text{for all } u \in X.$$

The set of all the linear continuous forms on X is a vector space. We can define a norm on this space by setting

$$\|F\| = \sup_{\substack{u \in X \\ u \neq 0}} \frac{F(u)}{\|u\|}.$$

The vector space of all the linear continuous forms on X is called the *dual space* of X, and is denoted by X'. Endowed with the previous norm, it is itself a Banach space.

The bilinear form from $X' \times X$ into $\mathbb{R}$ defined by

$$\langle F, u \rangle = F(u)$$

is called the *duality pairing* between X and X'.

(d) The Riesz Representation Theorem

If X is a Hilbert space, the dual space X' can be canonically identified with X (hence, it is a Hilbert space). In fact, the Riesz representation theorem states that for each linear continuous form F on X, there exists a unique element $u \in X$ such that

$$\langle F, v \rangle = (u, v) \qquad \text{for all } v \in X.$$

Moreoever $\|F\|_{X'} = \|u\|_X$.

A.2. The Cauchy–Schwarz Inequality

Let X be a Hilbert space, endowed with the inner product (u, v) and the associated norm $\|u\|$ (see (A.1.a)). The Cauchy–Schwarz inequality states that

$$|(u, v)| \leqslant \|u\| \, \|v\| \qquad \text{for all } u, v \in X.$$

Of particular importance in the analysis of numerical methods for partial differential equations is the Cauchy–Schwarz inequality in the weighted Lebesgue spaces $L_w^2(\Omega)$, where Ω is a domain in $\mathbb{R}^n$ and $w = w(\mathbf{x})$ is a weight function (see (A.9.h)). The previous inequality becomes:

$$\left| \int_\Omega u(\mathbf{x}) v(\mathbf{x}) w(\mathbf{x}) \, d\mathbf{x} \right| \leqslant \left(\int_\Omega u^2(\mathbf{x}) w(\mathbf{x}) \, d\mathbf{x} \right)^{1/2} \left(\int_\Omega v^2(\mathbf{x}) w(\mathbf{x}) \, d\mathbf{x} \right)^{1/2}$$

for all functions $u, v \in L_w^2(\Omega)$.

A.3. Linear Operators Between Banach Spaces

Let X and Y be Banach spaces (see (A.1.b)). A linear operator L defined on X and taking values in Y, $L\colon X \to Y$, is said to be *bounded*, or *continuous*, if there exists a constant $C > 0$ such that

$$\|Lv\|_Y \leqslant C \|v\|_X \qquad \text{for all } v \in X.$$

The smallest constant C for which the inequality holds is denoted by $\|L\|$, i.e.,

$$\|L\| = \sup_{\substack{v \in X \\ v \neq 0}} \frac{\|Lv\|_Y}{\|v\|_X}.$$

The vector space of all the linear bounded operators between X and Y is denoted by $\mathscr{L}(X, Y)$. It is a Banach space for the norm $\|L\|$ just defined.

In the formulation of differential problems, it may be convenient to consider linear operators which are only defined on a subset of a Banach space X (say, with values in X). The *domain* $D(L)$ of a linear operator $L\colon X \to X$ is the largest subset of X on which L is defined, i.e., $v \in D(L)$ if and only if there exists $g \in X$ such that $Lv = g$. We say that L is an *unbounded* operator if

$$\sup_{\substack{v \in D(L) \\ v \neq 0}} \frac{\|Lv\|_X}{\|v\|_X} = +\infty.$$

EXAMPLE. Consider the linear differential operator $Lv = d^2v/dx^2$, where v is a function on the interval (a, b) of the real line. L can be considered as a *bounded* operator between the Banach spaces $X = C^2([a, b])$ and $Y = C^0([a, b])$ (see (A.7)), or as an *unbounded* operator in $X = C^0([a, b])$. In the former case, the numerator is $\|Lv\|_Y$, which measures the second derivative of v, and the denominator is $\|v\|_X$, which measures all the derivatives of v up to order 2. The ratio of these norms is bounded. In the latter case, the domain of L is $D(L) = C^2([a, b])$, considered now as a subspace of $C^0([a, b])$. Here the numerator is again the maximum norm of the second derivative, but the denominator is the weaker norm which measures only the function itself. Taking bounded, but rapidly oscillatory functions, this ratio can be arbitrarily large.

A linear continuous operator $L: X \to Y$ is said to be *compact* if for each sequence $\{v_n \in X | n = 0, 1, \ldots\}$ such that $\|v_n\|_X \leqslant C$, one can find a subsequence $\{v_{n_k} | k = 0, 1, \ldots\}$ and an element $v \in X$ such that

$$\|Lv_{n_k} - Lv\|_Y \to 0 \qquad \text{as } n_k \to \infty.$$

Finally an operator $L: X^l \to Y$ is said to be *multilinear* if it is linear in each its variables. A multilinear operator L is continuous if the quantity

$$\|L\| = \sup_{v_1, \ldots, v_l \in X} \frac{\|L(v_1, \ldots, v_l)\|_Y}{\|v_1\|_X \cdots \|v_l\|_X}$$

is finite. The space of the multilinear operators $L: X^l \to Y$ is denoted by $\mathscr{L}_l(X, Y)$ and is a Banach space for the norm just introduced.

A.4. The Fréchet Derivative of an Operator

Let A be a mapping between a Banach space X and a Banach space Y, i.e., $A: X \to Y$. We say that A is *Fréchet differentiable* at a point $u_0 \in X$ if there exists a linear continuous operator $L \in \mathscr{L}(X, Y)$ such that

$$\lim_{\substack{w \in X \\ \|w\|_X \to 0}} \frac{\|A(u_0 + w) - A(u_0) - Lw\|_Y}{\|w\|_X} = 0.$$

If this happens, the linear operator L is unique. It is termed the *Fréchet derivative of A* at the point u_0, and is denoted by $A'(u_0)$.

A.5. The Lax–Milgram Theorem

Let V be a real Hilbert space (see (A.1.a)). Let $a: V \times V \to \mathbb{R}$ be a bilinear continuous form on V, i.e., a satisfies:

(i) $a(\lambda u + \mu v, w) = \lambda a(u, w) + \mu a(v, w)$ and
$a(u, \lambda v + \mu w) = \lambda a(u, v) + \mu a(u, w)$
for all $u, v, w \in V$ and all $\lambda, \mu \in \mathbb{R}$;

(ii) there exists a constant $\beta > 0$ such that

$$|a(u, v)| \leqslant \beta \|u\|_V \|v\|_V \qquad \text{for all } u, v \in V.$$

Assume that the form a is *V-coercive*, or *V-elliptic*, i.e.,
(iii) there exists a constant $\alpha > 0$ such that

$$a(u, u) \geqslant \alpha \|u\|_V^2 \qquad \text{for all } u \in V.$$

Then for each form $F \in V'$ (the dual space of V, see (A.1.c)), there exists a unique solution $u \in V$ to the variational problem

$$a(u, v) = F(v) \qquad \text{for all } v \in V.$$

Moreover, the following inequality holds:

$$\|u\|_V \leqslant \frac{\beta}{\alpha} \|F\|_{V'}.$$

Note that the Riesz representation theorem (A.1.d) follows from the Lax–Milgram theorem applied to the inner product (u, v). This is indeed a symmetric bilinear form, for which (ii) is nothing but the Cauchy–Schwarz inequality (A.2) and (iii) follows from the definition of Hilbertian norm.

A.6. Dense Subspace of a Normed Space

Let X be a Hilbert or a Banach space, with norm $\|v\|$. Let $S \subset X$ be a subspace of X. S is said to be *dense* in X if for each element $v \in X$ there exists a sequence $\{v_n | n = 0, 1, \ldots\}$ of elements $v_n \in S$, such that

$$\|v - v_n\| \to 0 \qquad \text{as } n \to \infty.$$

Thus, each element of X can be approximated arbitrarily well by elements of S, in the distance induced by the norm of X.

For example, the subspace $C^0([a, b])$ of the continuous functions on a bounded, closed interval $[a, b]$ of the real line, is dense in $L^2(a, b)$, the space of the measurable square integrable functions on (a, b). Indeed, for each function $v \in L^2(a, b)$ and each $n > 0$, one can find a continuous function $v_n \in C^0([a, b])$ such that

$$\int_a^b |v(x) - v_n(x)|^2 \, dx \leqslant \frac{1}{n^2}.$$

A.7. The Spaces $C^m(\bar{\Omega})$, $m \geqslant 0$

Let $\Omega = (a, b)^d \subset \mathbb{R}^d$, with $d = 1, 2$ or 3. Let us denote by $\bar{\Omega}$ the closure of Ω, i.e., the closed poly-interval $[a, b]^d$. For each multi-index $\alpha = (\alpha_1, \ldots, \alpha_d)$ of non-negative integers, set $|\alpha| = \alpha_1 + \cdots + \alpha_d$ and $D^\alpha v = \partial^{|\alpha|} v / \partial x_1^{\alpha_1} \ldots \partial x_d^{\alpha_d}$.

We denote by $C^m(\bar{\Omega})$ the vector space of the functions $v: \bar{\Omega} \to \mathbb{R}$ such that for each multi-index α with $0 \leqslant |\alpha| \leqslant m$, $D^\alpha v$ exists and is continuous on $\bar{\Omega}$. Since a continuous function on a closed, bounded (poly)-interval is bounded there, one can set

$$\|v\|_{C^m(\bar{\Omega})} = \sup_{0 \leqslant |\alpha| \leqslant m} \sup_{\mathbf{x} \in \bar{\Omega}} |D^\alpha v(\mathbf{x})|.$$

This is a norm for which $C^m(\bar{\Omega})$ is a Banach space (see (A.1.b)).

The space $C^\infty(\bar{\Omega})$ is the space of the infinitely differentiable functions on $\bar{\Omega}$. Thus, a function v belongs $C^\infty(\bar{\Omega})$ if and only if it belongs to $C^m(\bar{\Omega})$ for all $m > 0$.

A.8. Functions of Bounded Variation and the Riemann (–Stieltjes) Integral

Let $[a, b] \subset \mathbb{R}$ be a bounded interval of the real line, and let $u: [a, b] \to \mathbb{R}$ be a given function. The *total variation* of u on $[a, b]$ is defined by

$$V(u) = \sup_{a = x_0 < x_1 < \cdots < x_n = b} \sum_{i=1}^{n} |u(x_i) - u(x_{i-1})|,$$

where the supremum is taken over all the partitions of $[a, b]$ by a finite number of points, i.e. over all the sets of $n + 1$ points such that $a = x_0 < x_1 < \cdots < x_n = b$, n being arbitrary.

A function is said to be of *bounded variation* in $[a, b]$ if $V(u)$ is finite. Note that a function of bounded variation is certainly bounded.

An absolutely continuous function in $[a, b]$, i.e., a continuous function which admits an integrable derivative in the sense of distributions (see (A.10.b)), is of bounded variation. In particular, so is a continuously differentiable function in $[a, b]$. However, a function of bounded variation need not be continuous. For instance, the step function

$$u(x) = \begin{cases} 0 & \text{if } x < 0 \\ 1 & \text{if } x \geqslant 0 \end{cases}$$

is of bounded variation on each interval $[a, b]$ of the real line. On the contrary, $u(x) = x \sin(1/x)$ is an example of a continuous function which is not of bounded variation in any interval containing the origin.

A function u of bounded variation can be split into the difference

$$u(x) = \alpha(x) - \beta(x),$$

where α and β are monotonically increasing functions. This property makes possible the definition of the *Riemann–Stieltjes* integral with respect to a function of bounded variation

$$\int_a^b f(x)\,du(x).$$

We start by defining the Riemann–Stieltjes integral of a bounded function on $[a, b]$, with respect to a monotonically increasing function $\alpha(x)$. Given a partition $\mathscr{P} = \{a = x_0 < x_1 < \cdots < x_n = b\}$, let us set $M_i = \sup\{f(x) | x_{i-1} \leqslant x \leqslant x_i\}$ and $m_i = \inf\{f(x) | x_{i-1} \leqslant x \leqslant x_i\}$. Next we define

$$\overline{\int_a^b} f(x)\,d\alpha = \inf_{\mathscr{P}} \sum_{i=1}^n M_i(\alpha(x_i) - \alpha(x_{i-1}))$$

and

$$\underline{\int_a^b} f(x)\,d\alpha = \sup_{\mathscr{P}} \sum_{i=1}^n m_i(\alpha(x_i) - \alpha(x_{i-1})),$$

the infimum and the supremum being taken over all the partitions $\mathscr{P}$ of $[a, b]$. If the two numbers just defined are equal, we denote their common value by

$$\int_a^b f(x)\,d\alpha$$

and we say that f is *Riemann–Stieltjes integrable* with respect to α.

If $\alpha(x) \equiv x$, the previous integral coincides with the classical *Riemann integral*.

The Riemann–Stieltjes integral of a bounded function on $[a, b]$ with respect to a function of bounded variation u is defined as

$$\int_a^b u(x)\,du = \int_a^b u(x)\,d\alpha - \int_a^b u(x)\,d\beta,$$

where $u = \alpha - \beta$ is any decomposition of u into the difference of two monotonically increasing functions. This definition is independent of the particular decomposition.

The following integration-by-parts rule for functions of bounded variation holds. Let u and v be continuous functions of bounded variation on $[a, b]$. Then,

$$\int_a^b u(x)\,dv = u(b)v(b) - u(a)v(a) - \int_a^b v(x)\,du.$$

A.9. The Lebesgue Integral and L^p-Spaces

Let us start with a schematic account of the Lebesgue measure on a bounded interval (a, b) of the real line. A complete introduction to the Lebesgue integration theory can be found, e.g., in Royden (1968) or Rudin (1966).

(a) The Lebesgue (Outer) Measure

Each set A contained in (a, b) can be covered by a countable union of open intervals I_N, i.e. $A \subset \bigcup_{n=0}^{\infty} I_n$. Taking into account this property, *the Lebesgue outer measure* $\mu(A)$ of the set A is defined as

$$\mu(A) = \inf \sum_n |I_n|,$$

where $|I_n|$ denotes the length of the interval I_n and the infimum is taken over all the coverings of A by open intervals. Note that the measure of an interval is its length. Each countable set has zero measure.

(b) Measurable Sets

For each set $A \subseteq (a, b)$, let $\tilde{A}$ denote the complementary set of A in (a, b), i.e. $\tilde{A} = \{x \in (a, b) : x \notin A\}$.

A set $A \subseteq (a, b)$ is said to be measurable if

$$\mu(A) + \mu(\tilde{A}) = \mu((a, b)) = b - a.$$

In Lebesgue's measure theory only measurable sets are of interest.

(c) Simple Measurable Functions

A functions s: $(a, b) \to [0, +\infty)$ is a *simple measurable function* if it assumes only a finite number of values $\{s_0, \ldots, s_n\}$ and if each set $A_i = \{x \in (a, b) : s(x) = s_i\}$ is measurable.

(d) Measurable Functions

A positive function u: $(a, b) \to [0, +\infty)$ is *measurable* if it is the pointwise limit of simple measurable functions, more precisely, if there exist simple measurable functions $s^{(k)}$ such that

(i) $0 \leqslant s^{(1)} \leqslant s^{(2)} \leqslant \cdots \leqslant u$

(ii) $s^{(k)}(x) \to u(x) \quad \text{as } k \to \infty, \quad \text{for all } x \in (a, b).$

A real function u: $(a, b) \to \mathbb{R}$ is measurable if both its positive and negative parts $u^+ = \max\{u, 0\}$ and $u^- = \max\{-u, 0\}$ are measurable.

(e) The Lebesgue Integral

If s is a simple measurable function on (a, b), we set

$$\int_a^b s\, d\mu = \sum_{i=0}^{n} s_i \mu(A_i).$$

If u is a positive measurable function on (a, b), we set

$$\int_a^b u\,d\mu = \sup \int_a^b s\,d\mu,$$

the supremum being taken over all the simple measurable functions such that $0 \leqslant s \leqslant u$. The value of the right-hand side is a non-negative number or $+\infty$. We call it the *Lebesgue integral* of u on (a, b).

A positive measurable function u is said to be *Lebesgue integrable on* (a, b) if

$$\int_a^b u\,d\mu < +\infty.$$

A real measurable function u on (a, b) is said to be Lebesgue integrable if both its positive and negative parts u^+ and u^- are Lebesgue integrable. In this case we define the Lebesgue integral of u on (a, b) as

$$\int_a^b u\,d\mu = \int_a^b u^+\,d\mu - \int_a^b u^-\,d\mu.$$

(f) The Spaces $L^p(a, b)$, $1 \leqslant p \leqslant \infty$

Let us now define several spaces of integrable functions in the sense of Lebesgue. Hereafter we will use the more conventional notation $\int_a^b u(x)\,dx$, $\int_\Omega u(\mathbf{x})\,d\mathbf{x}$, etc. to denote Lebesgue integrals. Since two integrable functions which differ on a set of zero measure have the same integral, they can be identified from the point of view of the Lebesgue integration theory, i.e., they belong to the same equivalence class. This identification is always presumed here and in the sequel.

Let (a, b) be a bounded interval of $\mathbb{R}$ and let $1 \leqslant p < +\infty$. We denote by $L^p(a, b)$ the space of the measurable functions $u\colon (a, b) \to \mathbb{R}$ such that $\int_a^b |u(x)|^p\,dx < +\infty$. Endowed with the norm

$$\|u\|_{L^p(a,b)} = \left(\int_a^b |u(x)|^p\,dx\right)^{1/p},$$

it is a Banach space (see (A.1.b)).

For $p = +\infty$, $L^\infty(a, b)$ is the space of the measurable functions $u\colon (a, b) \to \mathbb{R}$ such that $|u(x)|$ is bounded outside a set of measure zero. If M denotes the smallest real number such that $|u(x)| \leqslant M$ outside a set of measure zero, we define a norm on $L^\infty(a, b)$ by setting

$$\|u\|_{L^\infty(a,b)} = \operatorname*{ess\,sup}_{x \in (a,b)} |u(x)| = M.$$

(If u is continuous on $[a, b]$, then $\|u\|_{L^\infty(a,b)}$ is the maximum of the absolute value of u on $[a, b]$.) Again $L^\infty(a, b)$ is a Banach space.

The index $p = 2$ is of special interest, because $L^2(a, b)$ is not only a Banach space, but also a Hilbert space (see (A.1.a)). The inner product is

$$(u, v) = \int_a^b u(x)v(x)\,dx,$$

which induces the norm

$$\|u\|_{L^2(a,b)} = \left(\int_a^b |u(x)|^2\,dx\right)^{1/2}.$$

It is also possible to define L^p-spaces of complex measurable functions. The previous definitions and norms hold unchanged, provided the absolute value of u is replaced by the modulus of u. The inner product of the complex $L^2(a, b)$-space is

$$(u, v) = \int_a^b u(x)\overline{v(x)}\,dx.$$

(g) The Weighted Spaces $L^p_w(-1, 1)$, $1 \leqslant p \leqslant +\infty$

Let $w(x)$ be a weight function on the interval $(-1, 1)$, i.e., a continuous, strictly positive and integrable function on $(-1, 1)$. For $p < +\infty$, we denote by $L^p_w(-1, 1)$ the Banach space of the measurable functions u: $(a, b) \to \mathbb{R}$ such that $\int_a^b |u(x)|^p w(x)\,dx < +\infty$. It is endowed with the norm

$$\|u\|_{L^p_w(-1,1)} = \left(\int_a^b |u(x)|^p w(x)\,dx\right)^{1/p}.$$

For $p = \infty$ we set $L^\infty_w(-1, 1) = L^\infty(-1, 1)$.

The space $L^2_w(-1, 1)$ is a Hilbert space for the inner product

$$(u, v)_w = \int_a^b u(x)v(x)w(x)\,dx,$$

which induces the weighted norm

$$\|u\|_{L^2_w(a,b)} = \left(\int_a^b |u(x)|^2 w(x)\,dx\right)^{1/2}.$$

(h) The Spaces $L^p(\Omega)$ and $L^p_w(\Omega)$, $1 \leqslant p \leqslant +\infty$

The previous definitions can be extended in a straightforward way to more than one space dimension. Let Ω denote a bounded, open domain in $\mathbb{R}^d$, for $d = 2$ or 3 (for instance, $\Omega = (0, 2\pi)^d$ or $\Omega = (-1, 1)^d$), and let $d\mathbf{x}$ be the Lebesgue measure on $\mathbb{R}^d$.

For $p < +\infty$, we denote by $L^p(\Omega)$ the space of the measurable functions $u: \Omega \to \mathbb{R}$ such that $\int_\Omega |u(\mathbf{x})|^p \, d\mathbf{x} < +\infty$. It is a Banach space for the norm

$$\|u\|_{L^p(\Omega)} = \left(\int_\Omega |u(\mathbf{x})|^p \, d\mathbf{x} \right)^{1/p}.$$

$L^\infty(\Omega)$ is the Banach space of the measurable functions $u: \Omega \to \mathbb{R}$ which are bounded outside a set of measure zero, equipped with the norm

$$\|u\|_{L^\infty(\Omega)} = \operatorname*{ess\,sup}_{\mathbf{x} \in \Omega} |u(\mathbf{x})|.$$

The space $L^2(\Omega)$ is a Hilbert space for the inner product

$$(u, v) = \int_\Omega u(\mathbf{x}) v(\mathbf{x}) \, d\mathbf{x},$$

which induces the norm

$$\|u\|_{L^2(\Omega)} = \left(\int_\Omega |u(\mathbf{x})|^2 \, d\mathbf{x} \right)^{1/2}.$$

Again one can consider $L^p(\Omega)$ spaces of complex functions in a straightforward manner.

If $w(\mathbf{x})$ denotes a weight function on Ω, the weighted spaces $L^p_w(\Omega)$ can be defined, by analogy to $L^p_w(a, b)$, as the Banach spaces of the measurable functions $u: \Omega \to \mathbb{R}$ such that the function $\mathbf{x} \to |u(\mathbf{x})|^p w(\mathbf{x})$ is Lebesgue integrable on Ω. In particular, the space $L^2_w(\Omega)$ is a Hilbert space for the inner product

$$(u, v)_w = \int_\Omega u(\mathbf{x}) v(\mathbf{x}) w(\mathbf{x}) \, d\mathbf{x},$$

which induces the weighted norm

$$\|u\|_{L^2_w(\Omega)} = \left(\int_\Omega |u(\mathbf{x})|^2 \, w(\mathbf{x}) \, d\mathbf{x} \right)^{1/2}.$$

A.10. Infinitely Differentiable Functions and Distributions

Let Ω be a bounded, open domain in $\mathbb{R}^d$, for $d = 1, 2$ or 3. If $\alpha = (\alpha_1, \ldots, \alpha_d)$ is a multi-index of non-negative integers, let us set

$$D^\alpha v = \frac{\partial^{\alpha_1 + \cdots + \alpha_d} v}{\partial x_1^{\alpha_1} \ldots \partial x_d^{\alpha_d}}.$$

We denote by $\mathcal{D}(\Omega)$ the vector space of all the infinitely differentiable functions $\phi: \Omega \to \mathbb{R}$, for which there exists a closed set $K \subset \Omega$ such that $\phi \equiv 0$ outside K.

We say that a sequence of functions $\phi_n \in \mathscr{D}(\Omega)$ *converges in* $\mathscr{D}(\Omega)$ to a function $\phi \in \mathscr{D}(\Omega)$ as $n \to \infty$, if there exists a common closed set $K \subset \Omega$ such that all the ϕ_n vanish outside K, and $D^\alpha \phi_n \to D^\alpha \phi$ uniformly on K as $n \to \infty$, for all non-negative multi-indices α.

(a) Distributions

Let T be a linear form on $\mathscr{D}(\Omega)$, i.e., a linear mapping $T: \mathscr{D}(\Omega) \to \mathbb{R}$. We shall denote the value of T on the element $\phi \in \mathscr{D}(\Omega)$ by $\langle T, \phi \rangle$. T is said to be *continuous* if for each sequence $\phi_n \in \mathscr{D}(\Omega)$ which converges in $\mathscr{D}(\Omega)$ to a function $\phi \in \mathscr{D}(\Omega)$ as $n \to \infty$, one has

$$\langle T, \phi_n \rangle \to \langle T, \phi \rangle \qquad \text{as } n \to \infty.$$

A *distribution* is a linear continuous form on $\mathscr{D}(\Omega)$. The set of all the distributions on Ω is a vector space denoted by $\mathscr{D}'(\Omega)$.

EXAMPLES. (i) Each integrable function $f \in L^1(\Omega)$ (see (A.9.f)) can be identified with the distribution T_f defined by

$$\langle T_f, \phi \rangle = \int_\Omega f(\mathbf{x}) \phi(\mathbf{x}) \, d\mathbf{x} \qquad \text{for all } \phi \in \mathscr{D}(\Omega).$$

(ii) Let $\mathbf{x}_0 \in \Omega$. The linear form on $\mathscr{D}(\Omega)$

$$\langle \delta_{\mathbf{x}_0}, \phi \rangle = \phi(\mathbf{x}_0) \qquad \text{for all } \phi \in \mathscr{D}(\Omega)$$

is a distribution, which is commonly (but improperly) called the "Dirac function".

We notice that if T_1 and T_2 are two distributions, then they are "equal in the sense of distributions" if

$$\langle T_1, \phi \rangle = \langle T_2, \phi \rangle \qquad \text{for all } \phi \in \mathscr{D}(\Omega).$$

(b) Derivative of Distributions

Let α be a non-negative multi-index and set $m = \alpha_1 + \cdots + \alpha_d$. For each distribution $T \in \mathscr{D}'(\Omega)$ let us consider the linear form on $\mathscr{D}(\Omega)$

$$\langle D^\alpha T, \phi \rangle = (-1)^m \langle T, D^\alpha \phi \rangle \qquad \text{for all } \phi \in \mathscr{D}(\Omega).$$

This linear form is continuous on $\mathscr{D}(\Omega)$; hence, it is a distribution, which is called the α-distributional *derivative* of T.

It follows that each integrable function $u \in L^1(\Omega)$ is infinitely differentiable in the sense of distributions, and the following Green's formula holds

$$\langle D^\alpha u, \phi \rangle = (-1)^m \int_\Omega u(\mathbf{x}) D^\alpha \phi(\mathbf{x})\, d\mathbf{x} \qquad \text{for all } \phi \in \mathscr{D}(\Omega).$$

If u is m-times continuously differentiable in Ω, then the α-distributional derivative of u coincides with the classical derivative of index α. In general, a distributional derivative of an integrable function can be an integrable function or merely a distribution. We say that the α-distributional derivative of an integrable function $u \in L^1(\Omega)$ is an *integrable function* if there exists $g \in L^1(\Omega)$ such that

$$\langle D^\alpha u, \phi \rangle = \int_\Omega g(\mathbf{x}) \phi(\mathbf{x})\, d\mathbf{x} \qquad \text{for all } \phi \in \mathscr{D}(\Omega).$$

EXAMPLES. (i) Consider the function $u(x) = \frac{1}{2}|x|$ in the interval $(-1, 1)$. Note that u is not classically differentiable at the origin. The first derivative of u in the distributional sense is represented by the step function

$$v(x) = \begin{cases} 1/2 & \text{if } x > 0 \\ -1/2 & \text{if } x < 0. \end{cases}$$

(ii) Consider the function v now defined. Note that the classical derivative is zero at all the points $x \neq 0$. The first derivative of v in the sense of distributions is the "Dirac function" δ_0 at the origin. This distribution cannot be represented by an integrable function.

Functions having a certain number of distributional derivatives which can be represented by integrable functions play a fundamental role in the modern theory of partial differential equations. The spaces of these functions are named Sobolev spaces (see (A.11)).

(c) Periodic Distributions

Let $\Omega = (0, 2\pi)^d$, for $d = 1$, 2 or 3. We define the space $C_p^\infty(\bar{\Omega})$ as the vector space of the functions $u: \bar{\Omega} \to \mathbb{C}$ which have derivatives of any order continuous in the closure $\bar{\Omega}$ of Ω, and 2π-periodic in each space direction. A sequence $\phi_n \in C_p^\infty(\bar{\Omega})$ *converges in* $C_p^\infty(\bar{\Omega})$ to a function $\phi \in C_p^\infty(\bar{\Omega})$ if $D^\alpha \phi_n \to D^\alpha \phi$ uniformly on $\bar{\Omega}$, as $n \to \infty$ for all non-negative multi-indices α.

A *periodic distribution* is a linear form T: $C_p^\infty(\bar{\Omega}) \to \mathbb{C}$ which is continuous, i.e., such that

$$\langle T, \phi_n \rangle \to \langle T, \phi \rangle \qquad \text{as } n \to \infty,$$

whenever $\phi_n \to \phi$ in $C_p^\infty(\bar{\Omega})$.

The *derivative* of index α of a periodic distribution T is the periodic distribution $D^\alpha T$ defined by

$$\langle D^\alpha T, \phi \rangle = (-1)^m \langle T, D^\alpha \phi \rangle \qquad \text{for all } \phi \in C_p^\infty(\bar{\Omega})$$

(where $m = \alpha_1 + \cdots + \alpha_d$).

Note that each function in $\mathscr{D}(\Omega)$ also belongs to $C_p^\infty(\bar{\Omega})$. Thus, it is easily seen that each periodic distribution is indeed a distribution in the sense of (A.10.a).

A.11. Sobolev Spaces and Sobolev Norms

We introduce hereafter some relevant Hilbert spaces, which occur in the numerical analysis of boundary value problems. They are spaces of square integrable functions (see (A.9)), which possess a certain number of derivatives (in the sense of distributions, see (A.10.b)) representable as square integrable functions.

(a) The Spaces $H^m(a,b)$ and $H^m(\Omega)$, $m \geqslant 0$

Let (a, b) be a bounded interval of the real line, and let $m \geqslant 0$ be an integer.

We define $H^m(a,b)$ to be the vector space of the functions $v \in L^2(a,b)$ such that all the distributional derivatives of u of order up to m can be represented by functions in $L^2(a,b)$. In short,

$$H^m(a,b) = \left\{ v \in L^2(a,b)\text{: for } 0 \leqslant k \leqslant m, \frac{d^k u}{dx^k} \in L^2(a,b) \right\}.$$

$H^m(a,b)$ is endowed with the inner product

$$(u, v)_m = \sum_{k=0}^{m} \int_a^b \frac{d^k u}{dx^k}(x) \frac{d^k v}{dx^k}(x)\, dx$$

for which $H^m(a,b)$ is a Hilbert space. The associated norm is

$$\|v\|_{H^m(a,b)} = \left(\sum_{k=0}^{m} \left\| \frac{d^k v}{dx^k} \right\|^2_{L^2(a,b)} \right)^{1/2}.$$

The Sobolev spaces $H^m(a,b)$ form a hierarchy of Hilbert spaces, in the sense that $\ldots H^{m+1}(a,b) \subset H^m(a,b) \subset \cdots \subset H^0(a,b) \equiv L^2(a,b)$, each inclusion being continuous (see (A.3)). Clearly, if a function u has m classical continuous derivatives in $[a,b]$, then u belongs to $H^m(a,b)$: in other words $C^m([a,b]) \subset H^m(a,b)$ with continuous inclusion. Conversely, if u belongs to $H^m(a,b)$ for $m \geqslant 1$, then u has $m - 1$ classical continuous derivatives in $[a,b]$, i.e., $H^m(a,b) \subset C^{m-1}([a,b])$ with continuous inclusion. This is an example of the so-called "Sobolev Imbedding Theorems". As a matter of fact, $H^m(a,b)$ can be equivalently defined as

$$H^m(a,b) = \left\{ v \in C^{m-1}([a,b]) \colon \frac{d}{dx} v^{(m-1)} \in L^2(a,b) \right\},$$

where the last derivative is in the sense of distributions.

Functions in $H^m(a,b)$ can be approximated arbitrarily well by infinitely differentiable functions in $[a,b]$, in the distance induced by the norm of $H^m(a,b)$. In other words,

$$C^\infty([a,b]) \text{ is dense in } H^m(a,b)$$

(see (A.6) for the definition of density of a subspace).

Set now $\Omega = (a,b)^d$, for $d = 2$ or 3. Given a multi-index $\alpha = (\alpha_1, \ldots, \alpha_d)$ of non-negative integers, we set $|\alpha| = \alpha_1 + \cdots + \alpha_d$ and

$$D^\alpha v = \frac{\partial^{|\alpha|} v}{\partial x_1^{\alpha_1} \ldots \partial x_d^{\alpha_d}}.$$

The previous definition of Sobolev spaces can be extended to higher space dimensions as follows. We define

$$H^m(\Omega) = \{ v \in L^2(\Omega) \colon \text{for each non-negative multi-index } \alpha \text{ with } |\alpha| \leqslant m, \\ \text{the distributional derivative } D^\alpha v \text{ belongs to } L^2(\Omega) \}.$$

This is a Hilbert space for the inner product

$$(u, v)_m = \sum_{|\alpha| \leqslant m} \int D^\alpha u(\mathbf{x}) D^\alpha v(\mathbf{x}) \, d\mathbf{x}$$

which induces the norm

$$\|v\|_{H^m(\Omega)} = \left(\sum_{|\alpha| \leqslant m} \|D^\alpha v\|^2_{L^2(\Omega)} \right)^{1/2}.$$

Functions in $H^m(\Omega)$ for $m \geqslant 1$ need not have the derivatives of order $m-1$ continuous in Ω. However, the weaker Sobolev inclusion $H^m(\Omega) \subset C^{m-2}(\bar{\Omega})$ $(m \geqslant 2)$ holds. On the other hand, as in the one-dimensional case

$$C^\infty(\bar{\Omega}) \text{ is dense in } H^m(\Omega).$$

(b) The Spaces $H_w^m(-1,1)$ and $H_w^m(\Omega)$, $m \geqslant 0$

In the definition of a Sobolev space, one can require that the function as well as its distributional derivatives be square integrable with respect to a weight function w (see (A.9)). This is the most natural framework in dealing with Chebyshev methods.

Let now (a,b) be the interval $(-1,1)$. We choose the weight function w to be the Chebyshev weight $w(x) = (1-x^2)^{-1/2}$ (although the following definitions can be given for an arbitrary weight function). We set

$$H_w^m(-1,1) = \left\{ v \in L_w^2(-1,1)\text{: for } 0 \leqslant k \leqslant m, \text{ the distributional derivative } \frac{d^k u}{dx^k} \text{ belongs to } L_w^2(-1,1) \right\}.$$

$H_w^m(-1,1)$ is a Hilbert space for the inner product

$$(u, v)_{m,w} = \sum_{k=0}^{m} \int_{-1}^{1} \frac{d^k u}{dx^k}(x) \frac{d^k v}{dx^k}(x) \frac{dx}{\sqrt{1-x^2}},$$

which induces the norm

$$\|u\|_{H_w^m(-1,1)} = \left(\sum_{k=0}^{m} \left\| \frac{d^k v}{dx^k} \right\|^2_{L_w^2(-1,1)} \right)^{1/2}.$$

For $\Omega = (-1,1)^d$ ($d = 2$ or 3) and $w = w(\mathbf{x}) = \prod_{i=1}^{d} (1 - x_i^2)^{-1/2}$ (the d-dimensional Chebyshev weight), we define $H_w^m(\Omega)$ by analogy to $H^m(\Omega)$. Precisely we set

$$H_w^m(\Omega) = \{ v \in L_w^2(\Omega)\text{: for each non-negative multi-index } \alpha \text{ with } |\alpha| \leqslant m, \text{ the distributional derivative } D^\alpha v \text{ belongs to } L_w^2(\Omega) \}.$$

This space is endowed with the Hilbertian inner product

$$(u, v)_{m,w} = \sum_{|\alpha| \leqslant m} \int_\Omega D^\alpha u(\mathbf{x}) D^\alpha v(\mathbf{x}) w(\mathbf{x})\, d\mathbf{x}$$

and the associated norm

$$\|v\|_{H_w^m(\Omega)} = \left(\sum_{|\alpha| \leqslant m} \|D^\alpha v\|^2_{L_w^2(\Omega)} \right)^{1/2}.$$

The properties of inclusion and density previously recalled for $H^m(a, b)$ and $H^m(\Omega)$ hold for $H_w^m(-1,1)$ and $H_w^m(\Omega)$ as well. Moreover, we note that $H_w^m(\Omega) \subset H^m(\Omega)$ for all $m \geqslant 0$.

(c) The Spaces $H_0^1(a, b)$, $H_{w,0}^1(-1, 1)$ and $H_0^1(\Omega)$, $H_{w,0}^1(\Omega)$

Dirichlet conditions are among the simplest and most common boundary conditions to be associated with a differential operator. Therefore, the subspaces of the Sobolev spaces H^m spanned by the functions satisfying homogeneous Dirichlet boundary conditions play a fundamental role.

Since the functions of $H^1(a, b)$ are continuous up to the boundary by the Sobolev Imbedding Theorem, it is meaningful to introduce the following subspace of $H^1(a, b)$:

$$H_0^1(a, b) = \{ v \in H^1(a, b)\text{: } v(a) = v(b) = 0 \}.$$

This is a Hilbert space for the same inner product of $H^1(a,b)$. It is often preferable to endow $H_0^1(a,b)$ with a different, although equivalent, inner product. This is defined as

$$[u,v] = \int_a^b \frac{du}{dx} \cdot \frac{dv}{dx}\, dx.$$

By the Poincaré inequality (A.13), it is indeed an inner product on $H_0^1(a,b)$. The associated norm, denoted by

$$\|v\|_{H_0^1(a,b)} = \left(\int_a^b \left| \frac{dv}{dx} \right|^2 dx \right)^{1/2},$$

is equivalent to the $H^1(a,b)$-norm, in the sense that there exists a constant $C > 0$ such that for all $v \in H_0^1(a,b)$

$$C\|v\|_{H^1(a,b)} \leqslant \|v\|_{H_0^1(a,b)} \leqslant \|v\|_{H^1(a,b)}.$$

Again this follows from the Poincaré inequality.

The subspace $H_{w,0}^1(-1,1)$ of $H_w^1(-1,1)$ is defined similarly, namely, we set

$$H_{w,0}^1(-1,1) = \{v \in H_w^1(-1,1) \colon v(-1) = v(1) = 0\}.$$

Again, it can be endowed with the weighted inner product

$$[u,v]_w = \int_{-1}^1 \frac{du}{dx} \cdot \frac{dv}{dx} \frac{dx}{\sqrt{1-x^2}}.$$

The associated norm

$$\|v\|_{H_{w,0}^1(-1,1)} = \left(\int_{-1}^1 \left| \frac{dv}{dx} \right|^2 \frac{dx}{\sqrt{1-x^2}} \right)^{1/2}$$

is equivalent to the norm of $H_w^1(-1,1)$, due to the Poincaré inequality.

The functions of $H_0^1(a,b)$ can be approximated arbitrarily well in the norm of this space not only by infinitely differentiable functions on $[a,b]$, but also by infinitely differentiable functions which vanish identically in a neighborhood of $x = a$ and $x = b$. In other words

$$\mathscr{D}((a,b)) \text{ is dense in } H^1(a,b)$$

(see (A.10) and (A.6)). A similar result holds for $H_{w,0}^1(-1,1)$, i.e.,

$$\mathscr{D}((-1,1)) \text{ is dense in } H_{w,0}^1(-1,1).$$

We turn now to more space dimensions. If Ω is the Cartesian product of d intervals ($d = 2$ or 3), the functions of $H^1(\Omega)$ need not be continuous on the closure of Ω. Thus, their pointwise values on the boundary $\partial\Omega$ of Ω need not be defined. However, it is possible to extend the trace operator $v \mapsto v|_{\partial\Omega}$ (classically defined for functions $v \in C^0(\bar{\Omega})$) so as to be a linear continuous mapping between $H^1(\Omega)$ and $L^2(\partial\Omega)$, the space of the square integrable functions on $\partial\Omega$ (see Lions and Magenes (1972), Chapter 1, for the rigorous definition of the trace of a function $v \in H^1(\Omega)$). With this in mind, it is meaningful to define $H_0^1(\Omega)$ as the subspace of $H^1(\Omega)$ of the functions whose trace at the boundary is zero. Precisely we set

$$H_0^1(\Omega) = \{v \in H^1(\Omega) \colon v|_{\partial\Omega} = 0\}.$$

This is a Hilbert space for the inner product of $H^1(\Omega)$, or for the inner product

$$[u, v] = \int_\Omega \nabla u \cdot \nabla v \, d\mathbf{x}.$$

The associated norm is denoted by

$$\|v\|_{H_0^1(\Omega)} = \left(\int_\Omega |\nabla v|^2 \, d\mathbf{x} \right)^{1/2}$$

and is equivalent to the $H^1(\Omega)$-norm, by the Poincaré inequality (A.13).

In a completely similar manner we introduce the space

$$H_{w,0}^1(\Omega) = \{v \in H_w^1(\Omega) \colon v|_{\partial\Omega} \equiv 0\}$$

endowed with the inner product

$$[u, v]_w = \int_\Omega \nabla u \cdot \nabla v w(\mathbf{x}) \, d\mathbf{x}$$

and the norm

$$\|v\|_{H_{w,0}^1(\Omega)} = \left(\int_\Omega |\nabla v|^2 w(\mathbf{x}) \, d\mathbf{x} \right)^{1/2}.$$

Concerning the approximation of the functions of $H_0^1(\Omega)$ by infinitely smooth functions, the following result holds

$$\mathscr{D}(\Omega) \text{ is dense in } H_0^1(\Omega) \text{ (respectively in } H_{w,0}^1(\Omega)).$$

The dual spaces (see (A.1.c)) of the Hilbert spaces of type H_0^1 now defined are usually denoted by H^{-1}. Thus, $H^{-1}(a, b)$ is the dual space of $H_0^1(a, b)$, $H_w^{-1}(-1, 1)$ is the dual space of $H_{w,0}^1(-1, 1)$, and so on.

Finally let us mention that for $m \geqslant 2$, one can define the subspaces $H_0^m(a, b)$ of $H^m(a, b)$ (and similarly for $H_w^m(-1, 1)$, etc.) of the functions of $H^m(a, b)$ whose derivatives of order up to $m - 1$ vanish on the boundary of the domain of definition. Again, these spaces are Hilbert spaces for the inner product of $H^m(a, b)$, or for an equivalent inner product which only involves the derivatives of order m.

(d) The Spaces $H_p^m(0, 2\pi)$ and $H_p^m(\Omega)$, $m \geqslant 0$

In the analysis of Fourier methods, the natural Sobolev spaces are those of periodic functions. In this framework, functions are complex valued, and their derivatives are taken in the sense of the periodic distributions (see (A.10.c)). We set

$$H_p^m(0, 2\pi) = \left\{ v \in L^2(0, 2\pi) \text{: for } 0 \leqslant k \leqslant m \text{, the derivative } \frac{d^k v}{dx^k} \text{ in the sense of periodic distributions belongs to } L^2(0, 2\pi) \right\}.$$

$H_p^m(0, 2\pi)$ is a Hilbert space for the inner product

$$(u, v)_m = \sum_{k=0}^{m} \int_0^{2\pi} \frac{d^k u}{dx^k}(x) \overline{\frac{d^k v}{dx^k}}(x)\, dx,$$

whose associated norm is

$$\|v\|_{H_p^m(0, 2\pi)} = \left(\sum_{k=0}^{m} \left\| \frac{d^k v}{dx^k} \right\|^2_{L^2(0, 2\pi)} \right)^{1/2}.$$

The space $H_p^m(0, 2\pi)$ coincides with the space of the functions $v: [0, 2\pi] \to \mathbb{C}$ which have $m - 1$ continuously differentiable, 2π-periodic derivatives on $[0, 2\pi]$, and such that the periodic distributional derivative $(d/dx)v^{(m-1)}$ can be represented by a function of $L^2(0, 2\pi)$.

The space $C_p^\infty([0, 2\pi])$ introduced in (A.10.c) is dense in $H_p^m(0, 2\pi)$. If $\Omega = (0, 2\pi)$ for $d = 2$ or 3, we set

$$H_p^m(\Omega) = \{v \in L^2(\Omega) \text{: for each integral multi-index } \alpha \text{ with } |\alpha| \leqslant m, \text{ the derivative } D^\alpha v \text{ in the sense of periodic distributions belongs to } L^2(\Omega)\}.$$

This is a Hilbert space for the inner product

$$(u, v)_m = \sum_{|\alpha| \leqslant m} \int_\Omega D^\alpha u(\mathbf{x}) \overline{D^\alpha v(\mathbf{x})}\, d\mathbf{x},$$

with associated norm

$$\|v\|_{H_p^m(\Omega)} = \left(\sum_{|\alpha| \leqslant m} \|D^\alpha v\|^2_{L^2(\Omega)} \right)^{1/2}.$$

The space $C_p^\infty(\bar{\Omega})$ is dense in $H_p^m(\Omega)$. Note that since a periodic distribution is also a distribution (see (A.10.c)), each space $H_p^m(0, 2\pi)$ (resp. $H_p^m(\Omega)$) is a subspace of the space $H^m(0, 2\pi)$ (resp. $H^m(\Omega)$).

A.12. The Sobolev Inequality

Let $(a, b) \subset \mathbb{R}$ be a bounded interval of the real line. For each function $u \in H^1(a, b)$ (see (A.11.a)) the following inequality holds

$$\|u\|_{L^\infty(a, b)} \leqslant \left(\frac{1}{b - a} + 2 \right)^{1/2} \|u\|^{1/2}_{L^2(a, b)} \|u\|^{1/2}_{H^1(a, b)}.$$

A.13. The Poincaré Inequality

Let v be a function of $H^1(a,b)$ (see (A.11.a)). We know that v is continuous on $[a,b]$. Assume that at a point $x_0 \in [a,b]$, $v(x_0)=0$. The Poincaré inequality states that there exists a constant C (depending upon the interval length $b-a$) such that

$$\|v\|_{L^2(a,b)} \leqslant C\|v'\|_{L^2(a,b)}, \tag{A.13.1}$$

i.e., the L^2-norm of the function is bounded by the L^2-norm of the derivative. The Poincaré inequality applies to functions belonging to $H_0^1(a,b)$ (see (A.11.c)), for which $x_0 = a$ or b, and also to functions of $H^1(a,b)$ which have zero average on (a,b), since necessarily such functions change sign in the domain.

A similar inequality holds if we replace $H^1(a,b)$ with $H_w^1(a,b)$ (see (A.11.b)). Precisely, there exists a constant $C>0$ such that for all $v \in H_w^1(a,b)$ vanishing at a point $x_0 \in [a,b]$

$$\|v\|_{L_w^2(a,b)} \leqslant C\|v'\|_{L_w^2(a,b)}. \tag{A.13.2}$$

In space dimension $d \geqslant 2$, the functions to which the Poincaré inequality applies must vanish on a manifold of dimension $d-1$. Confining ourselves to the case of functions vanishing on the boundary $\partial\Omega$ of the domain of definition Ω, one has

$$\|v\|_{L^2(\Omega)} \leqslant C\|\nabla v\|_{(L^2(\Omega))^d} \qquad \text{for all } v \in H_0^1(\Omega), \tag{A.13.3}$$

and

$$\|v\|_{L_w^2(\Omega)} \leqslant C\|\nabla v\|_{(L_w^2(\Omega))^d} \qquad \text{for all } v \in H_{w,0}^1(\Omega). \tag{A.13.4}$$

(See (A.11.c) for the definition of the spaces $H_0^1(\Omega)$ and $H_{w,0}^1(\Omega)$.)

A.14. The Hardy Inequality

Let $a<b$ be two real numbers, and let $\alpha<1$ be a real constant. The following inequalities hold for all measurable functions ϕ on (a,b):

$$\int_a^b \left[\frac{1}{t-a}\int_a^t \phi(s)\,ds\right]^2 (t-a)^\alpha\,dt \leqslant \frac{4}{1-\alpha}\int_a^b \phi^2(t)(t-a)^\alpha\,dt,$$

and similarly

$$\int_a^b \left[\frac{1}{b-t}\int_t^b \phi(s)\,ds\right]^2 (b-t)^\alpha\,dt \leqslant \frac{4}{1-\alpha}\int_a^b \phi^2(t)(b-t)^\alpha\,dt.$$

A.15. The Gronwall Lemma

Let $\phi = \phi(t)$ be a continuous function in the interval $[0, t^*]$, which is differentiable on $(0, t^*)$. If there exists a constant $\alpha \in \mathbb{R}$ and a continuous function $g(t)$ such that for $0 < t < t^*$, ϕ satisfies the inequality

$$\phi'(t) \leqslant \alpha\phi(t) + g(t)$$

(or equivalently,

$$\phi(t) \leqslant \phi(0) + \int_0^t [\alpha\phi(s) + g(s)]\, ds),$$

then ϕ satisfies the inequality

$$\phi(t) \leqslant e^{\alpha t}\phi(0) + \int_0^t g(s)e^{\alpha(t-s)}\, ds.$$

Appendix B

Fast Fourier Transforms

Basics

The Fast Fourier Transform (FFT) is a recursive algorithm for evaluating the discrete Fourier transform and its inverse. The FFT is conventionally written for the evaluation of

$$\tilde{u}_k = \sum_{j=0}^{N-1} u_j e^{+2\pi ijk/N} \qquad k = 0, 1, \ldots, N-1 \tag{B.1.a}$$

$$\tilde{u}_k = \sum_{j=0}^{N-1} u_j e^{-2\pi ijk/N} \qquad k = 0, 1, \ldots, N-1, \tag{B.1.b}$$

where u_j, $j = 0, 1, \ldots, N-1$ are a set of complex data. The FFT quickly became a widely used tool in signal processing after its description by Cooley and Tukey (1965). (As noted later by Cooley, Lewis and Welch (1969), most essential components of the FFT date back to the 1920s.) The Cooley-Tukey algorithm enables the sums in (B.1) to be evaluated in $5N \log_2 N$ real operations (when N is a power of 2), instead of the $8N^2$ real operations required by the straightforward sum. Moreover, calculation of (B.1) via the FFT incurs less error due to round-off than the direct summation method (Cooley, Lewis and Welch (1969)).

Many versions of the FFT are now in existence. The review by Temperton (1983a) contains an especially clear description of a simple, yet efficient one. It allows N to be of the form

$$N = 2^p 3^q 4^r 5^s 6^t \tag{B.2}$$

and has the operation count

$$N(5p + 9\tfrac{1}{3}q + 8\tfrac{1}{2}r + 13\tfrac{3}{5}s + 13\tfrac{1}{3}t - 6). \tag{B.3}$$

No additional flexibility is gained by the inclusion of the factors 4 and 6. The algorithm is, however, more efficient when these factors are included. Not only is the operation count lower—for example, by 15% when $N = 64$—but, due to the higher ratio of arithmetic operations to memory accesses, most FORTRAN compilers generate more efficient code for the larger factors.

For the sake of simplicity, however, throughout this book we shall use $(5\log_2 N - 6)N$ as the operation count for the complex FFT; moreover, the lower order term linear in N will usually be omitted.

Temperton's article also describes how to take advantage of chaining (on Cray's) and linked triads (on Cyber 205's) to improve performance. We should also mention the book by Brigham (1974) which is devoted entirely to the Fast Fourier Transform.

Use in Spectral Methods

In applications of Fourier spectral methods, the sums that one must evaluate are

$$\tilde{u}_k = \frac{1}{N}\sum_{j=0}^{N-1} u_j e^{-2\pi ijk/N} \qquad k = -\frac{N}{2}, -\frac{N}{2}+1, \ldots, \frac{N}{2}-1 \quad \text{(B.4)}$$

and

$$u_j = \sum_{k=-N/2}^{N/2-1} \tilde{u}_k e^{2\pi ijk/N} \qquad j = 0, 1, \ldots, N-1 \quad \text{(B.5)}$$

(see (2.1.22) and (2.1.24)). From (B.4) it is apparent that, for integers p and k,

$$\tilde{u}_{k+pN} = \tilde{u}_k. \quad \text{(B.6)}$$

When the array $(u_0, u_1, \ldots, u_{N-1})$ is fed into a standard FFT for evaluating (B.1.b) it returns, in effect, the array

$$(N\tilde{u}_0, N\tilde{u}_1, \ldots, N\tilde{u}_{N/2-1}, N\tilde{u}_{-N/2}, N\tilde{u}_{-N/2+1}, \ldots, N\tilde{u}_{-1}).$$

Conversely, when this array (without the factor N) is fed into the standard FFT for evaluating (B.1.a) (with the plus sign), the array $(u_0, u_1, \ldots, u_{N-1})$ is returned.

In most applications of spectral methods the direct use of the complex FFT (B.1) is needlessly expensive. This is true, for example if the function u_j is real or if a cosine transform (for a Chebyshev spectral method) is desired. These issues have been addressed by Orszag (1971d, Appendix II) and by Brachet et al. (1983, Appendix C). A summary of some of the relevant transformations follows.

Real Transforms

The simplest case occurs when many real transforms are desired at once, as arises for multidimensional problems. They can be computed pairwise. Suppose that u_j^1 and u_j^2, $j = 0, 1, \ldots, N-1$ are two sets of real data. Then one can define

$$v_j = u_j^1 + iu_j^2 \quad \text{(B.7)}$$

and compute $\tilde{v}_k$ according to (B.4) by the standard N-point complex FFT.

Then the transforms $\tilde{u}_k^1$ and $\tilde{u}_k^2$ can be extracted according to

$$\left.\begin{aligned}\tilde{u}_k^1 &= \frac{1}{2}(\tilde{v}_k + \bar{\tilde{v}}_{-k})\\ \tilde{u}_k^2 &= -\frac{i}{2}(\tilde{v}_k - \bar{\tilde{v}}_{-k})\end{aligned}\right\} \quad k = 0, 1, \ldots, \frac{N}{2} - 1 \tag{B.8}$$

(The Fourier coefficients of real data for negative k are related to those for positive k by $\tilde{u}_{-k} = \bar{\tilde{u}}_k$.) This process is readily reversed. In fact, if one is performing a Fourier collocation derivative, one need not even bother with the separation (B.8) in Fourier space, since

$$\left.\frac{du^1}{dx}\right|_j + i\left.\frac{du^2}{dx}\right|_j = \sum_{k=-N/2}^{N/2-1} ik\tilde{v}_k e^{2\pi ijk/N}. \tag{B.9}$$

If only a single real transform is desired, then one may follow the prescription given by Orszag (1971d). Let $M = N/2$ and define

$$v_j = u_{2j} + iu_{2j+1} \qquad j = 0, 1, \ldots, M - 1. \tag{B.10}$$

Then take an M-point transform of v_j, set $\tilde{v}_M = \tilde{v}_0$, and extract the desired coefficients via

$$\tilde{u}_k = \frac{1}{2}(\tilde{v}_k + \bar{\tilde{v}}_{M-k}) - \frac{i}{2}e^{2\pi ik/N}(\tilde{v}_k - \bar{\tilde{v}}_{M-k}) \qquad k = 0, 1, \ldots, M - 1. \tag{B.11}$$

For both of these approaches the cost of a single, real to half-complex transform is essentially $(5/2)N\log_2 N$.

Chebyshev Transforms

The discrete Chebyshev transforms based on the Gauss-Lobatto points (2.4.14) are given by

$$\tilde{u}_k = \frac{2}{N\bar{c}_k}\sum_{j=0}^{N}\frac{1}{\bar{c}_j}u_j\cos\frac{\pi jk}{N} \qquad k = 0, 1, \ldots, N \tag{B.12}$$

(see (2.2.22) and (2.4.15)) and

$$u_j = \sum_{k=0}^{N}\tilde{u}_k\cos\frac{\pi jk}{N} \qquad j = 0, 1, \ldots, N \tag{B.13}$$

(see (2.2.21) and (2.4.17)). Suppose that the transform (B.12) is desired for two real sets of data u_j^1 and u_j^2. Then define the complex data v_j by

$$v_j = \begin{cases}u_j^1 + iu_j^2 & j = 0, 1, \ldots, N\\ v_{2N-j} & j = N + 1, N + 2, \ldots, 2N - 1\end{cases} \tag{B.14}$$

and by periodicity (with period $2N$) for other integers j. Next, define $\tilde{v}_k$, $k = 0, 1, \ldots, N$ by (B.12) and define $\tilde{V}_k$, $k = 0, 1, \ldots, 2N - 1$ by (B.1.a) with N

replaced by $2N$. It is readily shown that

$$\tilde{V}_k = \frac{1}{N\bar{c}_k}\tilde{v}_k \qquad k = 0, 1, \ldots, N \tag{B.15}$$

and that

$$\tilde{V}_k = \sum_{l=0}^{N-1} v_{2l} e^{2\pi ikl/N} + e^{\pi ik/N} \sum_{l=0}^{N-1} v_{2l+1} e^{2\pi ikl/N}. \tag{B.16}$$

Now, define w_j by

$$w_j = v_{2j} + i(v_{2j+1} - v_{2j-1}) \qquad j = 0, 1, \ldots, N-1 \tag{B.17}$$

and compute $\tilde{w}_k$ according to the complex FFT (B.1.a). We have

$$\begin{aligned} \tilde{w}_k &= \sum_{l=0}^{N-1} v_{2l} e^{2\pi ikl/N} + i(1 - e^{2\pi ik/N}) \sum_{l=0}^{N-1} v_{2l+1} e^{2\pi ikl/N} \\ \tilde{w}_{N-k} &= \sum_{l=0}^{N-1} v_{2l} e^{2\pi ikl/N} - i(1 - e^{2\pi ik/N}) \sum_{l=0}^{N-1} v_{2l+1} e^{2\pi ikl/N}. \end{aligned} \tag{B.18}$$

Consequently,

$$\begin{aligned} \tilde{v}_0 &= \frac{1}{N} \sum_{j=0}^{N} \frac{1}{\bar{c}_j} v_j \\ \tilde{v}_k &= \frac{1}{N}\left[\left(\frac{1}{2} + \frac{1}{4\sin\frac{\pi k}{N}}\right)\tilde{w}_k + \left(\frac{1}{2} - \frac{1}{4\sin\frac{\pi k}{N}}\right)\tilde{w}_{N-k}\right] \\ \tilde{v}_N &= \frac{1}{N} \sum_{j=0}^{N} (-1)^j \frac{1}{\bar{c}_j} v_j. \end{aligned} \tag{B.19}$$

The desired real coefficients $\tilde{u}_k^1$ and $\tilde{u}_k^2$ are the real and imaginary parts, respectively, of the $\tilde{v}_k$. Thus, the discrete Chebyshev transform (B.12) can be computed in $\frac{5}{2}N\log_2 N + 4N$ real operations per transform, assuming that a large number of such transforms are computed. The inverse discrete Chebyshev transform (B.13) can be evaluated with only minor modifications to the algorithm given by (B.14), (B.17) and (B.19).

Discrete sine transforms can be handled in a similar manner: (B.14) (with v_{2N-j} replaced by $-v_{2N-j}$) and (B.17) are retained as is the central equation in (B.19) with the coefficient of $\tilde{w}_{N-k}$ having the opposite sign; the entire $\tilde{v}_k$ term is multiplied by i and one sets $\tilde{v}_0 = \tilde{v}_N = 0$. Swarztrauber (1986) has recently described how real cosine and sine transforms can be computed without the pre- and post-processing costs incurred by (B.17) and (B.19).

Other Cosine Transforms

In some applications, such as the use of a staggered grid in Navier–Stokes calculations (see Sec. 5.6 and 7.3) and in simulations of flows with special symmetries (Brachet et al. (1983)), discrete Chebyshev transforms with respect to the Gauss points (see (2.4.12) but with $N-1$ in place of N) are required. Consider

$$\tilde{u}_k = \frac{2}{N}\sum_{j=0}^{N-1} u_j \cos\frac{(2j+1)\pi k}{2N} \qquad k = 0, 1, \ldots, N-1. \tag{B.20}$$

Brachet et al. (1983) have provided prescriptions for computing efficiently this and related sums. Put

$$v_j = \begin{cases} u_{2j} & j = 0, 1, \ldots, \dfrac{N}{2} - 1 \\[2ex] u_{2N-2j-1} & j = \dfrac{N}{2}, \dfrac{N}{2} + 1, \ldots, N-1 \end{cases} \tag{B.21}$$

and compute $\tilde{v}_k$ according to (B.1.a). Then $\tilde{u}_k$ may be extracted via

$$\tilde{u}_k = \frac{1}{N}[e^{2\pi i k/2N}\tilde{v}_k + e^{-2\pi i k/2N}\tilde{v}_{N-k}] \qquad k = 0, 1, \ldots, N-1. \tag{B.22}$$

The corresponding inverse Chebyshev transform

$$u_j = \sum_{k=0}^{N-1} \tilde{u}_k \cos\frac{(2j+1)\pi k}{2N} \tag{B.23}$$

can be evaluated by reversing these steps.

For some problems the Chebyshev expansion may be over the interval [0, 1] instead of [−1, 1]. Moreover, it may also be useful to use only the odd (or even) polynomials (Spalart (1984); see also Sec. 2.5.3). Spalart (1986, private communication) explained how to employ the FFT for an expansion over [0, 1] in terms of just the odd Chebyshev polynomials. The collocation points are

$$x_j = \cos\frac{(2j+1)\pi}{4N} \qquad j = 0, 1, \ldots, N-1, \tag{B.24}$$

the series expansion is

$$u^N(x) = \sum_{k=0}^{N-1} \tilde{u}_k T_{2k+1}(x), \tag{B.25}$$

and the discrete transforms are

$$\tilde{u}_k = \frac{2}{N}\sum_{j=0}^{N-1} u_j \cos\frac{(2k+1)(2j+1)\pi}{4N} \qquad k = 0, 1, \ldots, N-1 \tag{B.26}$$

and

$$u_j = \sum_{k=0}^{N-1} \tilde{u}_k \cos\frac{(2k+1)(2j+1)\pi}{4N} \qquad j = 0, 1, \ldots, N-1. \tag{B.27}$$

(In order for a half-interval Chebyshev expansion to be spectrally accurate, one needs $u(x)$ and all of its derivatives to vanish at $x = 0$.) Spalart's trick for evaluating (B.27) is to define

$$\tilde{v}_k = \frac{\tilde{u}_k + \tilde{u}_{k-1}}{2\cos\left(\dfrac{k\pi}{2N}\right)} \qquad k = 0, 1, \ldots, N, \tag{B.28}$$

where $\tilde{u}_{-1} = \tilde{u}_N = 0$, to compute v_j according to (B.13), and then to extract u_j via

$$u_j = \frac{\tilde{v}_j + \tilde{v}_{j+1}}{2\cos\dfrac{(2j+1)\pi}{4N}} \qquad j = 0, 1, \ldots, N-1. \tag{B.29}$$

(Note however, that this transform is not suitable for use with the Gauss–Lobatto points.)

Description of Sample Routines

The last section of this appendix consists of a listing of self-contained FORTRAN routines for computing Fourier and Chebyshev collocation derivatives. They are geared towards applications in multidimensional problems.

The dependent variable is stored in the array U, which is dimensioned $U(LEN, NP)$. The second dimension pertains to the y coordinate. If the approximation in this direction is a Fourier one, then N collocation points are used; if a Chebyshev approximation is employed, then $NP = N + 1$ points are taken. The first dimension of U pertains to the remaining coordinates—just x if the function is two-dimensional and the product of x and z if it is three-dimensional. It is assumed that the total number of points in the remaining directions, denoted here by the variable LEN, is even.

The routines illustrate collocation differentiation with respect to the y coordinate. Both transform and matrix-multiply differentiation procedures are illustrated for Chebyshev approximations.

The transform methods are based on the complex FFT. This is used to process real data as described above. The integer N, which specifies the number of collocation points, must have only prime factors 2 and 3 and must be even. Two versions of the FFT are included—FFT1 and FFT2. They differ only in the manner in which the data are stored. They are based on Temperton's (1983a) description and they take advantage of chaining on machines such as the Cray 1.

The cosine transforms—FCTRC and FCTCR—use the complex FFT (for multiple data) in conjunction with the pre- and post-processing stages described above. The Chebyshev matrix-multiply routine uses the first derivative matrix given by (2.4.31). It is computed in SETCHEB. The effects of round-off error have been minimized by using trigonometric identities to avoid subtraction of nearly equal numbers in the term $x_j - x_k$.

```
C
C  THIS IS A SET OF SUBROUTINES FOR PERFORMING THE DIFFERENTIATION
C  PROCESS USING FOURIER AND CHEBYSHEV COLLOCATION.  THE BASIC
C  ROUTINES ARE:
C
C
C    DY2DF:   PERFORMS A FOURIER COLLOCATION DIFFERENTIATION USING
C               TRANSFORM METHODS
C
C
C    DY3DC:   PERFORMS A CHEBYSHEV COLLOCATION DIFFERENTIATION USING
C               TRANSFORM METHODS
C
C
C    DY3DM:   PERFORMS A CHEBYSHEV COLLOCATION DIFFERENTIATION USING
C               MATRIX MULTIPLIES
C
C
C  ADDITIONAL ROUTINES WHICH THE USER MUST INVOKE ARE
C
C
C    PREFFT:  PERFORMS PRELIMINARY OPERATIONS FOR THE USE OF THE
C               FAST FOURIER TRANSFORMS
C
C
C    SETCHEB: COMPUTES THE CHEYBSHEV COLLOCATION GRID AND ALSO THE
C               MATRICES WHICH REPRESENT THE FIRST AND SECOND
C               DERIVATIVES
C
C
C  THE REMAINING ROUTINES ARE NEVER CALLED DIRECTLY BY THE USER.
C  THEY CONSIST PRIMARILY OF THE FAST FOURIER TRANSFORM ROUTINES.
C
C
C  THE TRANSFORM METHODS FOR DERIVATIVES PRESUME THAT MANY DERIVATIVES
C  ARE CALCULATED AT ONCE.  THIS IS THE CASE FOR MULTI-DIMENSIONAL
C  PROBLEMS.  IMAGINE THAT A FUNCTION  U  DEPENDS UPON BOTH  X  AND  Y.
C  THEN, WHEN COMPUTING DERIVATIVES WITH RESPECT TO  Y , RESULTS MAY
C  BE OBTAINED FOR 2 VALUES OF  X  AT ONCE BY USING THE STANDARD,
C  COMPLEX FFT WITH THE VALUES OF  Y  FOR THE FIRST  X  COMPRISING
C  THE REAL PART OF A FUNCTION AND THE VALUES FOR THE SECOND  X
C  COMPRISING THE IMAGINARY PART.  THE EXAMPLES GIVEN IN THIS SET
C  OF PROGRAMS PERFORM THE DERIVATIVES WITH RESPECT TO THE SECOND
C  COORDINATE.
C
C  THE KEY PARAMETERS THAT COMPRISE THE TEST CASE ARE
C
C  LEN:     DERIVATIVES FOR ALL THE VALUES OF  X  ARE CALCULATED
C           IN ONE CALL TO THE DERIVATIVE ROUTINES.  THE PARAMETER
C           LEN  IS THE NUMBER OF DISTINCT VALUES OF  X.  IT MUST
C           BE AN EVEN NUMBER.
C
C  N:       FOR A FOURIER APPROXIMATION, THERE ARE  N  POINTS IN  Y .
C           FOR A CHEBYSHEV APPROXIMATION, THERE ARE  N+1  POINTS IN Y
C
      PARAMETER (LEN=10)
      PARAMETER (LENH=LEN/2)
      PARAMETER (N=64)
      PARAMETER (NP=N+1)
C
```

```
C  TRIG TABLES AND FOURIER WAVENUMBERS
C
      COMMON /YFAC/NFAY,IFAY(20),TRIGY(2,N)
      COMMON /WVZ/WAVEY1(N),WAVEY2(N)
C
C  THE FOLLOWING ARRAYS ARE WORK ARRAYS.
C  THE FOURIER DERIVATIVE ROUTINE NEEDS  1  ARRAY AND THE
C  CHEBYSHEV DERIVATIVE ROUTINE NEEDS  2  WORK ARRAYS.
C
      REAL WC(LEN,NP),ZC(LEN,NP)
      COMMON /BIGWORK/WC,ZC
C
C  THE FOLLOWING  4  ARRAYS ARE USED FOR TEMPORARY STORAGE BY
C  THE CHEBYSHEV TRANSFORM ROUTINES (FCTRC  &  FCTCR).  STORAGE SHOULD
C  BE ALLOCATED IN THE MAIN PROGRAM AS IS DONE HERE.
C
      REAL C01(LENH)
      REAL C02(LENH)
      REAL CN1(LENH)
      REAL CN2(LENH)
      COMMON /DYN1/C01
      COMMON /DYN2/C02
      COMMON /DYN3/CN1
      COMMON /DYN4/CN2
C
C  ARRAYS FOR THE FUNCTION, ITS DERIVATIVES, AND THE CHEBYSHEV GRID
C
      REAL U(LEN,NP),UDEX(LEN,NP)
      REAL DUDX(LEN,NP)
      REAL Y(NP)
      DATA PI/3.1415926535898/
C
C  FACTOR THE LENGTH OF THE TRANSFORM (N) IN PRIME FACTORS AND SET UP
C  THE  TRIG  TABLES NEEDED BY THE FFT
C
      CALL PREFFT(N,NFAY,IFAY,TRIGY)
C
      KN = N
      WRITE(6,101) KN
C
C  SET UP THE WAVENUMBERS USED IN FOURIER DIFFERENTIATION
C
C  NOTE THE STORAGE ORDER OF THE FOURIER COEFFICIENTS:
C
C    K = 0, 1, ... , N/2-1, -N/2, -N/2+1, ... , -1
C
C  SET UP THE ARRAYS WHICH CONTAIN THE FACTORS  K  (FOR THE FIRST
C  DERIVATIVE AND  K**2  (FOR THE SECOND DERIVATIVE).  THESE ARE
C  USED IN THE ROUTINE  DY3DF .
C
      KHALFP=N/2+1
      DO 11 I=1,N
         MM=I/KHALFP
         M=MM*N+1
         WAVEY1(I)=(I-M)
         WAVEY2(I)=WAVEY1(I)*WAVEY1(I)
   11 CONTINUE
C
```

```
C   SET THE  N/2  FOURIER COEFFICIENT OF THE FIRST DERIVATIVE TO ZERO
C
       WAVEY1(KHALFP) = 0.0
C
C   SET UP THE FOURIER TEST CASE.  THE TEST CASE IS INDEPENDENT OF  X
C
       DY = 2. * PI / N
       DO 19 K=1,N
          YF = (K-1)*DY
          UY = COS(PI*SIN(YF))
          UDY = - PI * COS(YF) * SIN(PI*SIN(YF))
          DO 17 I=1,LEN
             U(I,K) = UY
             UDEX(I,K) = UDY
   17     CONTINUE
   19 CONTINUE
C
C   FOURIER TRANSFORM DERIVATIVE
C
       CALL DY2DF(U,1,DUDX)
       WRITE(6,110)
       WRITE(6,113)  (K,DUDX(1,K),UDEX(1,K),K=1,N)
C
C   SET UP THE MATRIX USED FOR MATRIX MULTIPLY CHEBYSHEV DIFFERENTIATION
C
       CALL SETCHEB(Y)
C
C   SET UP THE CHEBYSHEV DERIVATIVE TEST CASE (ALSO WITH A FUNCTION
C   WHICH IS INDEPENDENT OF  X)
C
       PIN=PI/N
       DO 29 K=1,NP
          UY = EXP(2.*Y(K))
          UDY = 2. * EXP(2.*Y(K))
          DO 27 I=1,LEN
             U(I,K) = UY
             UDEX(I,K) = UDY
   27     CONTINUE
   29 CONTINUE
C
C   CHEBYSHEV TRANSFORM DERIVATIVE
C
       CALL DY2DC(U,1,DUDX)
       WRITE(6,111)
       WRITE(6,113)  (K,DUDX(1,K),UDEX(1,K),K=1,NP)
C
C   MATRIX-MULTIPLY DERIVATIVE
C
       CALL DY2DM(U,1,DUDX)
       WRITE(6,112)
       WRITE(6,113)  (K,DUDX(1,K),UDEX(1,K),K=1,NP)
C
       STOP
  101 FORMAT(//25X,'N =',I4)
  110 FORMAT(//15X,'CHECK OF FOURIER TRANSFORM DERIVATIVE'//)
  111 FORMAT(//15X,'CHECK OF CHEBYSHEV TRANSFORM DERIVATIVE'//)
```

```
  112 FORMAT(//10X,'CHECK OF CHEBYSHEV MATRIX-MULTIPLY DERIVATIVE'//)
  113 FORMAT(5X,I5,5X,2F20.10)
      END
C ----------------------------------------------------
      SUBROUTINE DY2DF(U,NDERIV,UD)
C
C  COMPUTES THE  Y-DERIVATIVES  OF  U  STORING THE RESULT IN  UD
C
C    U:        INPUT ARRAY CONTAINING REAL SPACE VALUES
C                U  IS DIMENSIONED  U(LEN,0:N)  IN THE CALLING PROGRAM
C                THE FIRST INDEX LABELS DISTINCT DATA
C                THE SECOND INDEX LABELS THE REAL SPACE VALUES
C    NDERIV:   NUMBER OF DERIVATIVES DESIRED (MAY BE  1  OR  2)
C    UD:       OUTPUT ARRAY CONTAINING REAL SPACE VALUES OF THE DERIVATIVE
C              (UD  MAY BE THE SAME AS  U  --  U  IS THEN DESTROYED)
C                UD  IS DIMENSIONED  UD(LEN,0:N)  IN THE CALLING PROGRAM
C                THE FIRST INDEX LABELS DISTINCT DATA
C                THE SECOND INDEX LABELS THE REAL SPACE VALUES
C
C  PRESUMES FOURIER EXPANSION IN  Y  (THE LAST DIMENSION OF  U)
C
      IMPLICIT REAL(A-H,O-Z)
      PARAMETER (LEN=10)
      PARAMETER (LENH=LEN/2)
      PARAMETER (N=64)
      PARAMETER (NP=N+1)
      REAL U(LENH,2,NP),UD(LENH,2,NP)
      REAL WC(LENH,2,NP)
      COMMON /YFAC/NFAY,IFAY(20),TRIGY(2,N)
      COMMON /WVZ/WAVEY1(N),WAVEY2(N),WAVEY0(N)
      COMMON /BIGWORK/WC
C
C  COPY THE INPUT ARRAY INTO THE WORK ARRAY
C
      DO 40 K=1,NP
         DO 30 IJ=1,LENH
            WC(IJ,1,K)=U(IJ,1,K)
            WC(IJ,2,K)=U(IJ,2,K)
   30    CONTINUE
   40 CONTINUE
C
C  PERFORM  FFT  IN  Y
C
      CALL FFT1(WC,UD,N,NFAY,IFAY,-1,TRIGY,LENH)
C
C  MULTIPLY BY WAVENUMBER
C
      IF(NDERIV .EQ. 1) THEN
         DO 60 K=1,N
            DO 50 IJ=1,LENH
               WC(IJ,1,K)=-WAVEY1(K)*UD(IJ,2,K)
               WC(IJ,2,K)=WAVEY1(K)*UD(IJ,1,K)
   50       CONTINUE
   60    CONTINUE
      ELSEIF(NDERIV .EQ. 2) THEN
         DO 80 K=1,N
            DO 70 IJ=1,LENH
               WC(IJ,1,K)=-WAVEY2(K)*UD(IJ,1,K)
```

```
            WC(IJ,2,K)=-WAVEY2(K)*UD(IJ,2,K)
70       CONTINUE
80    CONTINUE
      ENDIF
C
C  PERFORM  INVERSE FFT  IN  Y
C
      CALL FFT1(WC,UD,N,NFAY,IFAY,+1,TRIGY,LENH)
C
      RETURN
      END
C -------------------------------------------------
      SUBROUTINE DY2DC(U,NDERIV,UD)
C
C  COMPUTES THE  Y-DERIVATIVES  OF  U  STORING THE RESULT IN  UD
C
C    U:         INPUT ARRAY CONTAINING REAL SPACE VALUES
C                 U  IS DIMENSIONED  U(LEN,0:N)
C                 THE FIRST INDEX LABELS DISTINCT DATA
C                 THE SECOND INDEX LABELS THE REAL SPACE VALUES
C    NDERIV:    NUMBER OF DERIVATIVES DESIRED (MAY BE  1  OR  2)
C    UD:        OUTPUT ARRAY CONTAINING REAL SPACE VALUES OF THE DERIVATIVE
C               (UD  MAY BE THE SAME AS  U  --  U  IS THEN DESTROYED)
C                 UD  IS DIMENSIONED  UD(LEN,0:N)
C                 THE FIRST INDEX LABELS DISTINCT DATA
C                 THE SECOND INDEX LABELS THE REAL SPACE VALUES
C
C  PRESUMES CHEBYSHEV EXPANSION  Y  (THE LAST DIMENSION OF  U)
C
C  PERFORMS A TRANSFORM METHOD DIFFERENTIATION
C
      IMPLICIT REAL(A-H,O-Z)
      PARAMETER (LEN=10)
      PARAMETER (N=64)
      PARAMETER (NP=N+1)
C
C  THE PARAMETER  SCALE = -1.0  BECAUSE THE USUAL CHEBYSHEV GRID
C  HAS BEEN REVERSED
C
      PARAMETER (SCALE=-1.0)
      REAL U(LEN,0:N),UD(LEN,0:N)
      REAL WC(LEN,0:N),ZC(LEN,0:N)
      COMMON /BIGWORK/WC,ZC
      COMMON /YFAC/NFAY,IFAY(20),TRIGY(2,N)
C
C  COMPUTE THE CHEBYSHEV COEFFICIENTS OF  U
C
      CALL FCTRC(U,ZC,WC,ZC,N,NFAY,IFAY,TRIGY,LEN,LEN)
C
C  APPLY THE RECURSION ONCE
C
      DO 10 IJ=1,LEN
         WC(IJ,N)=0.0E0
10    CONTINUE
      TWON = 2.E0 * N * SCALE
      DO 20 IJ=1,LEN
         WC(IJ,N-1)=TWON*ZC(IJ,N)
20    CONTINUE
```

```
      DO 40 K=N-2,0,-1
         TWON = 2.E0 * (K + 1) * SCALE
         DO 30 IJ=1,LEN
            WC(IJ,K)=WC(IJ,K+2)+TWON*ZC(IJ,K+1)
   30    CONTINUE
   40 CONTINUE
      DO 50 IJ=1,LEN
         WC(IJ,0)=0.5E0*WC(IJ,0)
   50 CONTINUE
C
C  IF ONLY  1  DERIVATIVE IS DESIRED RETURN TO PHYSICAL SPACE
C
      IF(NDERIV .EQ. 1) THEN
         CALL FCTCR(WC,UD,ZC,WC,N,NFAY,IFAY,TRIGY,LEN,LEN)
      ELSE
C
C  IF  2  DERIVATIVES ARE DESIRED REPEAT THE RECURSION
C
         DO 80 IJ=1,LEN
            ZC(IJ,N)=0.0E0
   80    CONTINUE
         TWON = 2.E0 * N * SCALE
         DO 90 IJ=1,LEN
            ZC(IJ,N-1)=TWON*WC(IJ,N)
   90    CONTINUE
         DO 110 K=N-2,0,-1
            TWON = 2.E0 * (K + 1) * SCALE
            DO 100 IJ=1,LEN
               ZC(IJ,K)=ZC(IJ,K+2)+TWON*WC(IJ,K+1)
  100       CONTINUE
  110    CONTINUE
         DO 120 IJ=1,LEN
            ZC(IJ,0)=0.5E0*ZC(IJ,0)
  120    CONTINUE
         CALL FCTCR(ZC,UD,UD,ZC,N,NFAY,IFAY,TRIGY,LEN,LEN)
      ENDIF
      RETURN
      END
C -------------------------------------------------
      SUBROUTINE DY2DM(U,NDERIV,UD)
C
C  COMPUTES THE  Y-DERIVATIVES  OF  U  STORING THE RESULT IN  UD
C
C    U:        INPUT ARRAY CONTAINING REAL SPACE VALUES
C                U  IS DIMENSIONED  U(LEN,0:N)
C                THE FIRST INDEX LABELS DISTINCT DATA
C                THE SECOND INDEX LABELS THE REAL SPACE VALUES
C    NDERIV:   NUMBER OF DERIVATIVES DESIRED (MAY BE  1  OR  2)
C    UD:       OUTPUT ARRAY CONTAINING REAL SPACE VALUES OF THE DERIVATIVE
C              (UD  MUST BE THE DISTINCT FROM  U)
C                UD  IS DIMENSIONED  UD(LEN,0:N)
C                THE FIRST INDEX LABELS DISTINCT DATA
C                THE SECOND INDEX LABELS THE REAL SPACE VALUES
C
C  PRESUMES CHEBYSHEV EXPANSION IN  Y  (THE LAST DIMENSION OF  U)
C
C  PERFORMS A MATRIX MULTIPLY DIFFERENTIATION
C
```

```
      IMPLICIT REAL(A-H,O-Z)
      PARAMETER (LEN=10)
      PARAMETER (LENH=LEN/2)
      PARAMETER (N=64)
      PARAMETER (NP=N+1)
C
C  THE PARAMETER  SCALE = -1.0  BECAUSE THE USUAL CHEBYSHEV GRID
C  HAS BEEN REVERSED
C
      PARAMETER (SCALE=-1.0)
      REAL U(LEN,NP),UD(LEN,NP)
      COMMON /DCHEB/DL1(NP,NP),DL2(NP,NP)
C
      DO 19 K=1,NP
         DO 11 IJ=1,LEN
            UD(IJ,K) = 0.0
 11      CONTINUE
         DO 15 L=1,NP
            CC = SCALE * DL1(L,K)
            DO 13 IJ=1,LEN
               UD(IJ,K) = UD(IJ,K) + CC * U(IJ,L)
 13         CONTINUE
 15      CONTINUE
 19   CONTINUE
      RETURN
      END
C -------------------------------------------------
      SUBROUTINE SETCHEB(Y)
C
C  SETS UP ARRAYS FOR CHEBYSHEV OPERATIONS
C
C  Y:   OUTPUT ARRAY CONTAINING THE CHEBYSHEV GRID
C
C  THE ARRAYS  DL1 & DL2  ARE THE FIRST AND SECOND DERIVATIVE MATRICES
C
      IMPLICIT REAL(A-H,O-Z)
      PARAMETER (N=64)
      PARAMETER (NP=N+1)
      REAL Y(*)
C
C  ARRAYS FOR CHEBYSHEV OPERATIONS
C
      COMMON /DCHEB/DL1(NP,NP),DL2(NP,NP)
      DATA PI/3.1415926535898E0/
C
C  COMPUTE THE CHEBYSHEV GAUSS-LOBATTO GRID, BUT REVERSE ITS ORIENTATION.
C
      PIN=PI/N
      DO 10 K=1,NP
         Y(K)=-COS((K-1)*PIN)
 10   CONTINUE
      Y(1) = -1.0
      Y(NP) = 1.0
C
C  COMPUTE ARRAY FOR CHEBYSHEV DIFFERENTIATION AT FACES, PRESUMING THE
C  STANDARD (NON-REVERSED) GAUSS-LOBATTO POINTS.
C
C  NOTE THAT THE USUAL ORDER OF INDICES IN  DL1 & DL2  HAS BEEN REVERSED.
```

```
C  THIS IS DONE TO IMPROVE THE EFFICIENCY OF MEMORY ACCESSES.
C
      PINR=3.1415926535898/N
      PINH=0.5*PIN
      DO 80 J=1,NP
         XJ=COS((J-1)*PINR)
         CBJ=1.0
         IF(J .EQ. 1 .OR. J .EQ. NP) CBJ=2.0
         DO 70 K=1,NP
            XK=COS((K-1)*PINR)
            CBK=1.0
            IF(K .EQ. 1 .OR. K .EQ. NP) CBK=2.0
            IF(J .NE. K) THEN
               XDIF=2.*SIN(PINH*(K+J-2))*SIN(PINH*(K-J))
               DL1(K,J)=(CBJ/CBK)*(-1.0)**(J+K)/XDIF
               ELSEIF(J .EQ. 1) THEN
               DL1(1,1)=(2*N**2+1)/6.
               ELSEIF(J .EQ. NP) THEN
               DL1(NP,NP)=-(2*N**2+1)/6.
            ELSE
               DL1(K,J)=-XJ/(2.*SIN((J-1)*PIN)**2)
            ENDIF
            DL1(K,J) = DL1(K,J)
   70    CONTINUE
   80 CONTINUE
C
C  COMPUTE ARRAY FOR CHEBYSHEV SECOND DERIVATIVE
C
      DO 110 J=1,NP
         DO 100 K=1,NP
            SUM=0.0
            DO 90 L=1,NP
               SUM=SUM+DL1(L,J)*DL1(K,L)
   90       CONTINUE
            DL2(K,J)=SUM
  100    CONTINUE
  110 CONTINUE
      RETURN
      END
C -----------------------------------------------
      SUBROUTINE FFT1(A,C,N,NFAX,IFAX,ISIGN,TRIG,LEN)
C
C  PERFORMS A COMPLEX FFT ON MULTIPLE DATA IN  A
C  AND RETURNS THE RESULT IN  C
C
C  THIS DIFFERS FROM  FFT2  IN THE STORAGE ORDER IN THE
C  ARRAYS  A  AND  C
C
C    A:     INPUT ARRAY (DESTROYED DURING CALCULATION)
C             A  IS DIMENSIONED  A(LEN,2,0:N-1)
C             THE FIRST INDEX LABELS DISTINCT DATA TO BE TRANSFORMED
C             THE SECOND INDEX LABELS REAL (1) OR IMAGINARY (2) PARTS
C             THE THIRD INDEX LABELS THE  N  POINTS TO BE TRANSFORMED
C    C:     OUTPUT ARRAY (MUST BE DISTINCT FROM INPUT ARRAY)
C             C  IS DIMENSIONED THE SAME AS  A
C    N:     NUMBER OF POINTS IN THE TRANSFORM DIRECTION
C             MUST HAVE ONLY PRIME FACTORS OF  2  AND  3
C    NFAX:  NUMBER OF PRIME FACTORS OF  N
```

```
C     IFAX:  INTEGER ARRAY CONTAINING THE PRIME FACTORS OF  N
C     ISIGN: SET  ISIGN = -1  TO COMPUTE FOURIER COEFFICIENTS
C            SET  ISIGN = +1  TO COMPUTE REAL SPACE VALUES
C     TRIG:  ARRAY CONTAINING TRIGONOMETRIC FACTORS
C               TRIG  IS DIMENSIONED  TRIG(2,0:N-1)
C                    TRIG(1,J)  = COS(2 * PI * J / N)
C                    TRIG(2,J)  = SIN(2 * PI * J / N)
C     LEN:   NUMBER OF DISTINCT TRANSFORMS TO BE PERFORMED
C
      REAL A(LEN,2,0:N-1),C(LEN,2,0:N-1)
      REAL TRIG(2,0:N-1)
      INTEGER IFAX(*)
      LOGICAL ODD
      DATA PI/3.1415926535898/
C
      LA=1
      ODD=.TRUE.
      DO 10 I=1,NFAX
         IFAC=IFAX(I)
         IF(ODD) THEN
            CALL PASS1(A,C,N,ISIGN,IFAC,LA,TRIG,LEN)
         ELSE
            CALL PASS1(C,A,N,ISIGN,IFAC,LA,TRIG,LEN)
         END IF
         ODD=.NOT. ODD
         LA=LA * IFAC
   10 CONTINUE
      IF(ODD) THEN
         DO 30 I=0,N-1
            DO 20 IJ=1,LEN
               C(IJ,1,I)=A(IJ,1,I)
               C(IJ,2,I)=A(IJ,2,I)
   20       CONTINUE
   30    CONTINUE
      END IF
      IF(ISIGN .EQ. -1) THEN
         XNI=1./N
         DO 50 I=0,N-1
            DO 40 IJ=1,LEN
               C(IJ,1,I)=XNI*C(IJ,1,I)
               C(IJ,2,I)=XNI*C(IJ,2,I)
   40       CONTINUE
   50    CONTINUE
      END IF
      RETURN
      END
C ------------------------------------------------
      SUBROUTINE PASS1(A,C,N,ISIGN,IFAC,LA,TRIG,LEN)
C
C  PERFORMS ONE PASS OF   FFT1
C
C  THIS ROUTINE IS NEVER CALLED DIRECTLY BY THE USER.
C  THE ARGUMENTS ARE SIMILAR TO THOSE OF  FFT1 .
C
C  THE FOLLOWING CONSTANTS PRESUME A  64-BIT  MACHINE.  THEY ARE THE
C  SINES OF  45 DEGREES  AND  60 DEGREES .
C
      PARAMETER (ROOT=.7071067811865)
```

```
      PARAMETER (ASN60=0.5 * 1.732050807569)
C
      REAL A(LEN,2,0:N-1),C(LEN,2,0:N-1),TRIG(2,0:N-1)
      INTEGER IND(0:20),JND(0:20)
      SN60 = ISIGN * ASN60
      M=N/IFAC
C
C  SET UP INDEXING
C
      DO 10 K=0,IFAC-1
         IND(K)=K * M
         JND(K)=K * LA
   10 CONTINUE
C
C  PERFORM THE ARITHMETIC
C
      I = 0
      J = 0
      JUMP = (IFAC-1) * LA
      DO 130 K=0,M-LA,LA
         DO 120 L=1,LA
         IF(IFAC .EQ. 2) THEN
            I0 = IND(0) + I
            I1 = IND(1) + I
            J0 = JND(0) + J
            J1 = JND(1) + J
            CC = TRIG(1,K)
            SS = ISIGN * TRIG(2,K)
            IF(K .EQ. 0) THEN
               DO 20 IJ=1,LEN
                  C(IJ,1,J0) = A(IJ,1,I0) + A(IJ,1,I1)
                  C(IJ,2,J0) = A(IJ,2,I0) + A(IJ,2,I1)
                  C(IJ,1,J1) = A(IJ,1,I0) - A(IJ,1,I1)
                  C(IJ,2,J1) = A(IJ,2,I0) - A(IJ,2,I1)
   20          CONTINUE
            ELSE
               DO 50 IJ=1,LEN
                  C(IJ,1,J0) = A(IJ,1,I0) + A(IJ,1,I1)
                  C(IJ,2,J0) = A(IJ,2,I0) + A(IJ,2,I1)
                  AM1 = A(IJ,1,I0) - A(IJ,1,I1)
                  AM2 = A(IJ,2,I0) - A(IJ,2,I1)
                  C(IJ,1,J1) = CC * AM1 - SS * AM2
                  C(IJ,2,J1) = SS * AM1 + CC * AM2
   50          CONTINUE
            END IF
         ELSEIF(IFAC .EQ. 3) THEN
            I0 = IND(0) + I
            I1 = IND(1) + I
            I2 = IND(2) + I
            J0 = JND(0) + J
            J1 = JND(1) + J
            J2 = JND(2) + J
            IF(K .EQ. 0) THEN
               DO 60 IJ=1,LEN
                  AP1 = A(IJ,1,I1) + A(IJ,1,I2)
                  AP2 = A(IJ,2,I1) + A(IJ,2,I2)
                  C(IJ,1,J0) = A(IJ,1,I0) + AP1
                  C(IJ,2,J0) = A(IJ,2,I0) + AP2
```

```
                TA1 = A(IJ,1,I0) - 0.5 * AP1
                TA2 = A(IJ,2,I0) - 0.5 * AP2
                AM1 = SN60 * (A(IJ,1,I1) - A(IJ,1,I2))
                AM2 = SN60 * (A(IJ,2,I1) - A(IJ,2,I2))
                C(IJ,1,J1) = TA1 - AM2
                C(IJ,2,J1) = TA2 + AM1
                C(IJ,1,J2) = TA1 + AM2
                C(IJ,2,J2) = TA2 - AM1
   60         CONTINUE
            ELSE
              C1 = TRIG(1,K)
              C2 = TRIG(1,2*K)
              S1 = ISIGN * TRIG(2,K)
              S2 = ISIGN * TRIG(2,2*K)
              DO 70 IJ=1,LEN
                AP1 = A(IJ,1,I1) + A(IJ,1,I2)
                AP2 = A(IJ,2,I1) + A(IJ,2,I2)
                C(IJ,1,J0) = A(IJ,1,I0) + AP1
                C(IJ,2,J0) = A(IJ,2,I0) + AP2
                TA1 = A(IJ,1,I0) - 0.5 * AP1
                TA2 = A(IJ,2,I0) - 0.5 * AP2
                AM1 = SN60 * (A(IJ,1,I1) - A(IJ,1,I2))
                AM2 = SN60 * (A(IJ,2,I1) - A(IJ,2,I2))
                T1 = TA1 - AM2
                T2 = TA2 + AM1
                C(IJ,1,J1)= C1 * T1 - S1 * T2
                C(IJ,2,J1)= S1 * T1 + C1 * T2
                T1 = TA1 + AM2
                T2 = TA2 - AM1
                C(IJ,1,J2)= C2 * T1 - S2 * T2
                C(IJ,2,J2)= S2 * T1 + C2 * T2
   70         CONTINUE
            END IF
          END IF
          I = I + 1
          J = J + 1
  120     CONTINUE
          J = J + JUMP
  130 CONTINUE
      RETURN
      END
C ---------------------------------------------------
      SUBROUTINE FFT2(A,C,N,NDIM,NFAX,IFAX,ISIGN,TRIG,LEN)
C
C  PERFORMS A COMPLEX FFT ON MULTIPLE DATA IN  A
C  AND RETURNS THE RESULT IN  C
C
C  THIS DIFFERS FROM  FFT1  IN THE STORAGE ORDER OF THE
C  ARRAYS  A  AND  C
C
C    A:     INPUT ARRAY (DESTROYED DURING CALCULATION)
C             A  IS DIMENSIONED  A(LEN,0:NDIM,2)
C             THE FIRST INDEX LABELS DISTINCT DATA TO BE TRANSFORMED
C             THE SECOND INDEX LABELS THE  N  POINTS TO BE TRANSFORMED
C             THE THIRD INDEX LABELS REAL (1) OR IMAGINARY (2) PARTS
C    C:     OUTPUT ARRAY (MUST BE DISTINCT FROM INPUT ARRAY)
C             C  IS DIMENSIONED THE SAME AS  A
C    N:     NUMBER OF POINTS IN THE TRANSFORM DIRECTION
```

```
C              MUST HAVE ONLY PRIME FACTORS OF  2  AND  3
C    NDIM:  SECOND DIMENSION OF  A  &  C
C    NFAX:  NUMBER OF PRIME FACTORS OF  N
C    IFAX:  INTEGER ARRAY CONTAINING THE PRIME FACTORS OF  N
C    ISIGN: SET  ISIGN = -1  TO COMPUTE FOURIER COEFFICIENTS
C           SET  ISIGN = +1  TO COMPUTE GRID VALUES
C    TRIG:  ARRAY CONTAINING TRIGONOMETRIC FACTORS
C              TRIG  IS DIMENSIONED  TRIG(2,0:N-1)
C                    TRIG(1,J) = COS(2 * PI * J / N)
C                    TRIG(2,J) = SIN(2 * PI * J / N)
C    LEN:   NUMBER OF DISTINCT TRANSFORMS TO BE PERFORMED
C
      REAL A(LEN,0:NDIM,2),C(LEN,0:NDIM,2)
      REAL TRIG(2,0:N-1)
      INTEGER IFAX(*)
      LOGICAL ODD
      DATA PI/3.1415926535898/
C
      LA=1
      ODD=.TRUE.
      DO 10 I=1,NFAX
         IFAC=IFAX(I)
         IF(ODD) THEN
            CALL PASS2(A,C,N,NDIM,ISIGN,IFAC,LA,TRIG,LEN)
         ELSE
            CALL PASS2(C,A,N,NDIM,ISIGN,IFAC,LA,TRIG,LEN)
         END IF
         ODD=.NOT. ODD
         LA=LA * IFAC
   10 CONTINUE
      IF(ODD) THEN
         DO 30 I=0,N-1
            DO 20 IJ=1,LEN
               C(IJ,I,1)=A(IJ,I,1)
               C(IJ,I,2)=A(IJ,I,2)
   20       CONTINUE
   30    CONTINUE
      END IF
      IF(ISIGN .EQ. -1) THEN
         XNI=1./N
         DO 50 I=0,N-1
            DO 40 IJ=1,LEN
               C(IJ,I,1)=XNI*C(IJ,I,1)
               C(IJ,I,2)=XNI*C(IJ,I,2)
   40       CONTINUE
   50    CONTINUE
      END IF
      RETURN
      END
C ---------------------------------------------
      SUBROUTINE PASS2(A,C,N,NDIM,ISIGN,IFAC,LA,TRIG,LEN)
C
C  PERFORMS ONE PASS OF   FFT2
C
C  THIS ROUTINE IS NEVER CALLED DIRECTLY BY THE USER.
C  THE ARGUMENTS ARE SIMILAR TO THOSE OF  FFT2 .
C
C  THE INNER LOOPS IN THIS ROUTINE (THOSE OVER THE INDEX  IJ)
```

```
C  EXTEND OVER PART OF THE SECOND DIMENSIONS OF THE ARRAYS
C  A  AND  C .  THIS PRODUCES LONGER VECTOR LENGTHS.
C
C  THE FOLLOWING CONSTANTS PRESUME A  64-BIT  MACHINE.  THEY ARE THE
C  SINES OF  45 DEGREES  AND  60 DEGREES .
C
      PARAMETER (ROOT=.7071067811865)
      PARAMETER (ASN60=0.5 * 1.732050807569)
C
      REAL A(LEN,0:NDIM,2),C(LEN,0:NDIM,2),TRIG(2,0:N-1)
      INTEGER IND(0:20),JND(0:20)
      SN60 = ISIGN * ASN60
      M=N/IFAC
C
C  SET UP INDEXING
C
      DO 10 K=0,IFAC-1
         IND(K)=K * M
         JND(K)=K * LA
   10 CONTINUE
      LLA = LA * LEN
C
C  PERFORM THE ARITHMETIC
C
      I = 0
      J = 0
      JUMP = (IFAC-1) * LA
      DO 130 K=0,M-LA,LA
         IF(IFAC .EQ. 2) THEN
            I0 = IND(0) + I
            I1 = IND(1) + I
            J0 = JND(0) + J
            J1 = JND(1) + J
            CC = TRIG(1,K)
            SS = ISIGN * TRIG(2,K)
            IF(K .EQ. 0) THEN
               DO 20 IJ=1,LLA
                  C(IJ,J0,1) = A(IJ,I0,1) + A(IJ,I1,1)
                  C(IJ,J0,2) = A(IJ,I0,2) + A(IJ,I1,2)
                  C(IJ,J1,1) = A(IJ,I0,1) - A(IJ,I1,1)
                  C(IJ,J1,2) = A(IJ,I0,2) - A(IJ,I1,2)
   20          CONTINUE
            ELSE
               DO 50 IJ=1,LLA
                  C(IJ,J0,1) = A(IJ,I0,1) + A(IJ,I1,1)
                  C(IJ,J0,2) = A(IJ,I0,2) + A(IJ,I1,2)
                  AM1 = A(IJ,I0,1) - A(IJ,I1,1)
                  AM2 = A(IJ,I0,2) - A(IJ,I1,2)
                  C(IJ,J1,1) = CC * AM1 - SS * AM2
                  C(IJ,J1,2) = SS * AM1 + CC * AM2
   50          CONTINUE
            END IF
         ELSEIF(IFAC .EQ. 3) THEN
            I0 = IND(0) + I
            I1 = IND(1) + I
            I2 = IND(2) + I
            J0 = JND(0) + J
            J1 = JND(1) + J
```

```
                J2 = JND(2) + J
                IF(K .EQ. 0) THEN
                   DO 60 IJ=1,LLA
                      AP1 = A(IJ,I1,1) + A(IJ,I2,1)
                      AP2 = A(IJ,I1,2) + A(IJ,I2,2)
                      C(IJ,J0,1) = A(IJ,I0,1) + AP1
                      C(IJ,J0,2) = A(IJ,I0,2) + AP2
                      TA1 = A(IJ,I0,1) - 0.5 * AP1
                      TA2 = A(IJ,I0,2) - 0.5 * AP2
                      AM1 = SN60 * (A(IJ,I1,1) - A(IJ,I2,1))
                      AM2 = SN60 * (A(IJ,I1,2) - A(IJ,I2,2))
                      C(IJ,J1,1) = TA1 - AM2
                      C(IJ,J1,2) = TA2 + AM1
                      C(IJ,J2,1) = TA1 + AM2
                      C(IJ,J2,2) = TA2 - AM1
 60                CONTINUE
                ELSE
                   C1 = TRIG(1,K)
                   C2 = TRIG(1,2*K)
                   S1 = ISIGN * TRIG(2,K)
                   S2 = ISIGN * TRIG(2,2*K)
                   DO 70 IJ=1,LLA
                      AP1 = A(IJ,I1,1) + A(IJ,I2,1)
                      AP2 = A(IJ,I1,2) + A(IJ,I2,2)
                      C(IJ,J0,1) = A(IJ,I0,1) + AP1
                      C(IJ,J0,2) = A(IJ,I0,2) + AP2
                      TA1 = A(IJ,I0,1) - 0.5 * AP1
                      TA2 = A(IJ,I0,2) - 0.5 * AP2
                      AM1 = SN60 * (A(IJ,I1,1) - A(IJ,I2,1))
                      AM2 = SN60 * (A(IJ,I1,2) - A(IJ,I2,2))
                      T1 = TA1 - AM2
                      T2 = TA2 + AM1
                      C(IJ,J1,1)= C1 * T1 - S1 * T2
                      C(IJ,J1,2)= S1 * T1 + C1 * T2
                      T1 = TA1 + AM2
                      T2 = TA2 - AM1
                      C(IJ,J2,1)= C2 * T1 - S2 * T2
                      C(IJ,J2,2)= S2 * T1 + C2 * T2
 70                CONTINUE
                END IF
             END IF
             I = I + LA
             J = J + LA
             J = J + JUMP
130 CONTINUE
    RETURN
    END
C ----------------------------------------------
    SUBROUTINE FCTRC(UR,UC,A,C,N,NFAX,IFAX,TRIG,LEN,LT)
C
C  PERFORMS MULTIPLE FAST CHEBYSHEV TRANSFORMS
C
C  THE CHEBYSHEV COEFFICIENTS OF  UR  ARE RETURNED IN  UC
C
C  THE ROUTINE  FCTCR  PERFORMS THE INVERSE OPERATION
C
C    UR:     INPUT ARRAY CONTAINING REAL SPACE VALUES
C              UR  IS DIMENSIONED  UR(LT,0:N)
```

```
C              THE FIRST INDEX LABELS DISTINCT DATA
C              THE SECOND INDEX LABELS THE REAL SPACE VALUES
C    UC:     OUTPUT ARRAY CONTAINING CHEBYSHEV COEFFICIENTS
C              UC  IS DIMENSIONED  UC(LT,0:N)
C              THE FIRST INDEX LABELS DISTINCT DATA
C              THE SECOND INDEX LABELS THE CHEBYSHEV COEFFICIENTS
C    A:      WORK ARRAY OF SAME LENGTH AS  UR
C              UR  AND  A  MAY BE THE SAME ARRAY (UR  IS THEN DESTROYED)
C    C:      THIS IS THE SAME ARRAY AS UC
C              IT OCCURS TWICE IN THE ARGUMENT LIST DUE TO THE NEED
C              TO INDEX DIFFERENTLY
C    N:      NUMBER OF POINTS IN THE TRANSFORM DIRECTION
C              N  MUST HAVE ONLY PRIME FACTORS OF  2  AND  3
C              N  MUST BE  EVEN
C    NFAX:   NUMBER OF PRIME FACTORS OF  N
C    IFAX:   INTEGER ARRAY CONTAINING THE PRIME FACTORS OF  N
C    TRIG:   ARRAY CONTAINING TRIGONOMETRIC FACTORS
C              TRIG  IS DIMENSIONED  TRIG(2,0:N-1)
C                   TRIG(1,J) = COS(2 * PI * J / N)
C                   TRIG(2,J) = SIN(2 * PI * J / N)
C    LEN:    NUMBER OF DISTINCT TRANSFORMS TO BE PERFORMED
C    LT:     FIRST DIMENSION OF  U
C
C
      COMMON /DYN1/C01
      COMMON /DYN2/C02
      COMMON /DYN3/CN1
      COMMON /DYN4/CN2
      REAL UR(LT,0:N),UC(LT,0:N)
      REAL A((LEN+1)/2,0:N,2),C((LEN+1)/2,0:N,2),TRIG(2,0:N-1)
      INTEGER IFAX(*)
      REAL C01(1),C02(1),CN1(1),CN2(1)
      DATA PI/3.1415926535898/
C
      ISIGN = 1
      LENH = (LEN + 1)/2
      LP = LENH + 1
      LPP = LP - 1
C
      XNI = 1.0/N
      N2 = 2 * N
      NH = N / 2
C
      DO 10 IJ=1,LENH
         C(IJ,0,1)=UR(IJ,0)
         C(IJ,NH,1)=UR(IJ,N)
         C01(IJ) = (0.5 * (UR(IJ,0) + UR(IJ,N))) + UR(IJ,N-1)
         CN1(IJ) = (0.5 * (UR(IJ,0) + UR(IJ,N))) - UR(IJ,N-1)
         C(IJ,0,2)=UR(LPP+IJ,0)
         C(IJ,NH,2)=UR(LPP+IJ,N)
         C02(IJ)=(0.5 * (UR(LPP+IJ,0) + UR(LPP+IJ,N))) + UR(LPP+IJ,N-1)
         CN2(IJ)=(0.5 * (UR(LPP+IJ,0) + UR(LPP+IJ,N))) - UR(LPP+IJ,N-1)
   10 CONTINUE
      L2N = 0
      DO 30 K=1,NH-1
         L2N = L2N + 2
         L2NM1 = L2N - 1
         L2NP1 = L2N + 1
```

```
         DO 20 IJ=1,LENH
            C(IJ,K,1)=UR(IJ,L2N)-UR(LPP+IJ,L2NP1)+UR(LPP+IJ,L2NM1)
            C(IJ,K,2)=UR(LPP+IJ,L2N)+UR(IJ,L2NP1)-UR(IJ,L2NM1)
   20    CONTINUE
   30 CONTINUE
C
      L2N = N
      DO 60 K=NH+1,N-1
         L2N = L2N - 2
         L2NM1 = L2N + 1
         L2NP1 = L2N - 1
         DO 50 IJ=1,LENH
            C(IJ,K,1)=UR(IJ,L2N)-UR(LPP+IJ,L2NP1)+UR(LPP+IJ,L2NM1)
            C(IJ,K,2)=UR(LPP+IJ,L2N)+UR(IJ,L2NP1)-UR(IJ,L2NM1)
   50    CONTINUE
   60 CONTINUE
C
      DO 110 K=1,N-2,2
         DO 100 IJ=1,LENH
            C01(IJ) = C01(IJ) + UR(IJ,K) + UR(IJ,K+1)
            C02(IJ) = C02(IJ) + UR(LPP+IJ,K) + UR(LPP+IJ,K+1)
            CN1(IJ) = CN1(IJ) - UR(IJ,K) + UR(IJ,K+1)
            CN2(IJ) = CN2(IJ) - UR(LPP+IJ,K) + UR(LPP+IJ,K+1)
  100    CONTINUE
  110 CONTINUE
C
      CALL FFT2(C,A,N,N,NFAX,IFAX,ISIGN,TRIG,LENH)
      PIN=PI/N
      DO 150 K=1,NH-1
         SSI = 0.25/SIN(K*PIN)
         ALJ = XNI * (0.5 + SSI)
         GLJ = XNI * (0.5 - SSI)
         DO 140 IJ=1,LENH
            UC(IJ,K)=ALJ*A(IJ,K,1)+GLJ*A(IJ,N-K,1)
            UC(IJ,N-K)=ALJ*A(IJ,N-K,1)+GLJ*A(IJ,K,1)
            UC(LPP+IJ,K)=ALJ*A(IJ,K,2)+GLJ*A(IJ,N-K,2)
            UC(LPP+IJ,N-K)=ALJ*A(IJ,N-K,2)+GLJ*A(IJ,K,2)
  140    CONTINUE
  150 CONTINUE
      DO 155 IJ=1,LENH
         UC(IJ,NH)=XNI*A(IJ,NH,1)
         UC(LPP+IJ,NH)=XNI*A(IJ,NH,2)
  155 CONTINUE
      DO 160 IJ=1,LENH
         UC(IJ,0)=XNI*C01(IJ)
         UC(LPP+IJ,0)=XNI*C02(IJ)
         UC(IJ,N)=XNI*CN1(IJ)
         UC(LPP+IJ,N)=XNI*CN2(IJ)
  160 CONTINUE
      RETURN
      END
C ----------------------------------------------------
      SUBROUTINE FCTCR(UC,UR,A,C,N,NFAX,IFAX,TRIG,LEN,LT)
C
C  PERFORMS MULTIPLE FAST CHEBYSHEV TRANSFORMS
C
C  THE CHEBYSHEV SERIES  UR  IS COMPUTED FROM THE CHEBYSHEV
C  COEFFICIENTS IN  UC
```

```
C
C  THE ROUTINE  FCTRC  PERFORMS THE INVERSE OPERATION
C
C    UC:     INPUT ARRAY CONTAINING CHEBYSHEV COEFFICIENTS
C              UC  IS DIMENSIONED  UC(LT,0:N)
C              THE FIRST INDEX LABELS DISTINCT DATA
C              THE SECOND INDEX LABELS THE CHEBYSHEV COEFFICIENTS
C    UR:     OUTPUT ARRAY CONTAINING REAL SPACE VALUES
C              UR  IS DIMENSIONED  UR(LT,0:N)
C              THE FIRST INDEX LABELS DISTINCT DATA
C              THE SECOND INDEX LABELS THE REAL SPACE VALUES
C    A:      WORK ARRAY OF SAME LENGTH AS  UC
C              UR  AND  A  MAY BE THE SAME ARRAY
C    C:      SAME ARRAY AS UC
C              IT IS INCLUDED SEPARATELY IN THE ARGUMENT LIST DUE TO
C              THE NEED TO INDEX IT DIFFERENTLY
C    N:      NUMBER OF POINTS IN THE TRANSFORM DIRECTION
C              N  MUST HAVE ONLY PRIME FACTORS OF  2  AND  3
C              N  MUST BE  EVEN
C    NFAX:   NUMBER OF PRIME FACTORS OF  N
C    IFAX:   INTEGER ARRAY CONTAINING THE PRIME FACTORS OF  N
C    TRIG:   ARRAY CONTAINING TRIGONOMETRIC FACTORS
C              TRIG  IS DIMENSIONED  TRIG(2,0:N-1)
C                   TRIG(1,J) = COS(2 * PI * J / N)
C                   TRIG(2,J) = SIN(2 * PI * J / N)
C    LEN:    NUMBER OF DISTINCT TRANSFORMS TO BE PERFORMED
C    LT:     FIRST DIMENSION OF  U
C
      COMMON /DYN1/U01
      COMMON /DYN2/U02
      COMMON /DYN3/UN1
      COMMON /DYN4/UN2
      REAL UR(LT,0:N),UC(LT,0:N)
      REAL A((LEN+1)/2,0:N,2),C((LEN+1)/2,0:N,2),TRIG(2,0:N-1)
      INTEGER IFAX(*)
      REAL U01(1),U02(1),UN1(1),UN2(1)
      DATA PI/3.1415926535898/
C
      ISIGN = 1
      LENH = (LEN + 1)/2
      LP = LENH + 1
      LPP = LP - 1
      N2 = 2 * N
      NH = N / 2
C
      DO 10 IJ=1,LENH
         U01(IJ) = (UC(IJ,0) + UC(IJ,N)) + UC(IJ,N-1)
         UN1(IJ) = (UC(IJ,0) + UC(IJ,N)) - UC(IJ,N-1)
         U02(IJ) = (UC(LPP+IJ,0) + UC(LPP+IJ,N)) + UC(LPP+IJ,N-1)
         UN2(IJ) = (UC(LPP+IJ,0) + UC(LPP+IJ,N)) - UC(LPP+IJ,N-1)
         A(IJ,0,1) = 2.*UC(IJ,0)
         A(IJ,0,2) = 2.*UC(LPP+IJ,0)
         A(IJ,NH,1) = 2.*UC(IJ,N)
         A(IJ,NH,2) = 2.*UC(LPP+IJ,N)
   10 CONTINUE
C
      DO 50 K=1,N-2,2
         DO 40 IJ=1,LENH
```

```
            U01(IJ) = U01(IJ) + UC(IJ,K) + UC(IJ,K+1)
            U02(IJ) = U02(IJ) + UC(LPP+IJ,K) + UC(LPP+IJ,K+1)
            UN1(IJ) = UN1(IJ) - UC(IJ,K) + UC(IJ,K+1)
            UN2(IJ) = UN2(IJ) - UC(LPP+IJ,K) + UC(LPP+IJ,K+1)
   40    CONTINUE
   50 CONTINUE
C
      L2N = 0
      DO 100 K=1,NH-1
         L2N = L2N + 2
         L2NM1 = L2N - 1
         L2NP1 = L2N + 1
         DO 90 IJ=1,LENH
            A(IJ,K,1)=UC(IJ,L2N)-UC(LPP+IJ,L2NP1)+UC(LPP+IJ,L2NM1)
            A(IJ,K,2)=UC(LPP+IJ,L2N)+UC(IJ,L2NP1)-UC(IJ,L2NM1)
   90    CONTINUE
  100 CONTINUE
C
      L2N = N
      DO 130 K=NH+1,N-1
         L2N = L2N - 2
         L2NM1 = L2N + 1
         L2NP1 = L2N - 1
         DO 120 IJ=1,LENH
            A(IJ,K,1)=UC(IJ,L2N)-UC(LPP+IJ,L2NP1)+UC(LPP+IJ,L2NM1)
            A(IJ,K,2)=UC(LPP+IJ,L2N)+UC(IJ,L2NP1)-UC(IJ,L2NM1)
  120    CONTINUE
  130 CONTINUE
      CALL FFT2(A,C,N,N,NFAX,IFAX,ISIGN,TRIG,LENH)
      PIN=PI/N
      DO 150 K=1,NH-1
         SSI = 0.125/SIN(K*PIN)
         ALJ = 0.25 + SSI
         GLJ = 0.25 - SSI
         DO 140 IJ=1,LENH
            UR(IJ,K)=ALJ*C(IJ,K,1)+GLJ*C(IJ,N-K,1)
            UR(IJ,N-K)=ALJ*C(IJ,N-K,1)+GLJ*C(IJ,K,1)
  140    CONTINUE
         DO 145 IJ=1,LENH
            UR(LPP+IJ,K)=ALJ*C(IJ,K,2)+GLJ*C(IJ,N-K,2)
            UR(LPP+IJ,N-K)=ALJ*C(IJ,N-K,2)+GLJ*C(IJ,K,2)
  145    CONTINUE
  150 CONTINUE
      DO 155 IJ=1,LENH
         UR(IJ,NH)=0.5*C(IJ,NH,1)
         UR(LPP+IJ,NH)=0.5*C(IJ,NH,2)
  155 CONTINUE
      DO 160 IJ=1,LENH
         UR(IJ,0)=U01(IJ)
         UR(IJ,N)=UN1(IJ)
  160 CONTINUE
      DO 165 IJ=1,LENH
         UR(LPP+IJ,0)=U02(IJ)
         UR(LPP+IJ,N)=UN2(IJ)
  165 CONTINUE

      RETURN
      END
```

```
C -------------------------------------------------
      SUBROUTINE PREFFT(N,NFAX,IFAX,TRIG)
C
C  COMPUTES PRELIMINARY QUANTITIES FOR  FFT  ROUTINES
C
C  N:      NUMBER OF POINTS IN THE DATA
C            N  MUST HAVE PRIME FACTORS  2  AND  3
C  NFAX:   NUMBER OF PRIME FACTORS OF  N  (OUTPUT)
C  IFAX:   ARRAY CONTAINING PRIME FACTORS OF  N  (OUTPUT)
C  TRIG:   ARRAY CONTAINING TRIGONOMETRIC FACTORS
C            TRIG  IS DIMENSIONED  TRIG(2,0:N-1)
C                 TRIG(1,J) = COS(2 * PI * J / N)
C                 TRIG(2,J) = SIN(2 * PI * J / N)
C
      REAL TRIG(2,0:N-1)
      INTEGER IFAX(*)
      DATA PI/3.1415926535898/
C
      CALL FACTOR(N,NFAX,IFAX)
      DO 10 K=0,N-1
         ARG=2.*PI*K/N
         TRIG(1,K)=COS(ARG)
         TRIG(2,K)=SIN(ARG)
   10 CONTINUE
      RETURN
      END
C -------------------------------------------------
      SUBROUTINE FACTOR(N,NFAX,IFAX)
C
C  COMPUTES THE FACTORS OF  N
C
C  THIS ROUTINE IS NOT CALLED DIRECTLY BY THE USER
C
      INTEGER IFAX(*)
      NFAX=0
      NN=N
C
C  EXTRACT FACTORS OF  3
C
      DO 10 II=1,20
         IF(NN .EQ. 3*(NN/3)) THEN
            NFAX=NFAX+1
            IFAX(NFAX)=3
            NN=NN/3
         ELSE
            GO TO 20
         END IF
   10 CONTINUE
   20 CONTINUE
C
C  EXTRACT FACTORS OF  2
C
      DO 30 II=NFAX+1,20
         IF(NN .EQ. 2*(NN/2)) THEN
            NFAX=NFAX+1
            IFAX(NFAX)=2
            NN=NN/2
         ELSE
```

```
            GO TO 40
         END IF
   30 CONTINUE
   40 CONTINUE
      IF(NN .NE. 1) THEN
         STOP
      END IF
      RETURN
      END
```

Appendix C

The Gauss–Lobatto points are the most frequently used collocation points for spectral methods based upon algebraic polynomials. For a Jacobi polynomial expansion (see Section 2.5.1) of degree N, these are the $N + 1$ roots of the polynomial $q(x)$ given by (2.2.15). The subroutines which follow compute the Gauss–Lobatto points for expansions in the polynomials $P_k^{(\alpha,\beta)}(x)$. They may be used for computing the Legendre roots by choosing $\alpha = \beta = 0$. The test program which accompanies the subroutines checks the roots for $\alpha = \beta = -\frac{1}{2}$. These are the Chebyshev–Gauss–Lobatto points which are given in closed form by (2.4.14). The convergence criterion EPS in the subroutine JACOBL should be set slightly higher than the precision of the machine.

```
C
C  TESTS THE ROUTINE TO COMPUTE JACOBI ROOTS
C
      REAL XJAC(65)
      REAL PI,PIN,XCHEB
      PI = ACOS(-1.0E0)
      ALPHA = -0.5
      BETA = -0.5
      DO 29 IN=2,6
         N = 2 ** IN
         NP = N + 1
         PIN = PI / N
         CALL JACOBL(N,ALPHA,BETA,XJAC)
         WRITE(6,100) N,ALPHA,BETA
         DO 19 I=1,NP
            XCHEB = COS((I-1)*PIN)
            WRITE(6,101) I,XJAC(I),XCHEB
   19    CONTINUE
   29 CONTINUE
      STOP
  100 FORMAT(//5X,'N =',I4,5X,'ALPHA =',F10.3,5X,'BETA =',F10.3/)
  101 FORMAT(I5,5X,F25.15,5X,F25.15)
      END
      SUBROUTINE JACOBL(N,ALPHA,BETA,XJAC)
C
C  COMPUTES THE GAUSS-LOBATTO COLLOCATION POINTS FOR JACOBI POLYNOMIALS
C
C  N:       DEGREE OF APPROXIMATION
C  ALPHA:   PARAMETER IN JACOBI WEIGHT
C  BETA:    PARAMETER IN JACOBI WEIGHT
C
C  XJAC:    OUTPUT ARRAY WITH THE GAUSS-LOBATTO ROOTS
C           THEY ARE ORDERED FROM LARGEST (+1.0) TO SMALLEST (-1.0)
C
      IMPLICIT REAL(A-H,O-Z)
      REAL XJAC(1)
      COMMON /JACPAR/ALP,BET,RV
      DATA KSTOP/10/
      DATA EPS/1.0E-12/
      ALP = ALPHA
      BET =BETA
      RV = 1 + ALP
      NP = N+1
C
C  COMPUTE THE PARAMETERS IN THE POLYNOMIAL WHOSE ROOTS ARE DESIRED
C
      CALL JACOBF(NP,PNP1P,PDNP1P,PNP,PDNP,PNM1P,PDNM1,1.0D0)
      CALL JACOBF(NP,PNP1M,PDNP1M,PNM,PDNM,PNM1M,PDNM1,-1.0D0)
      DET = PNP*PNM1M-PNM*PNM1P
      RP = -PNP1P
      RM = -PNP1M
      A = (RP*PNM1M-RM*PNM1P)/DET
      B = (RM*PNP-RP*PNM)/DET
C
      XJAC(1) = 1.0
      NH = (N+1)/2
C
```

```
C  SET-UP RECURSION RELATION FOR INITIAL GUESS FOR THE ROOTS
C
      DTH = 3.14159265/(2*N+1)
      CD = COS(2.*DTH)
      SD = SIN(2.*DTH)
      CS = COS(DTH)
      SS = SIN(DTH)
C
C  COMPUTE THE FIRST HALF OF THE ROOTS BY POLYNOMIAL DEFLATION
C
      DO 39 J=2,NH
         X = CS
         DO 29 K=1,KSTOP
            CALL JACOBF(NP,PNP1,PDNP1,PN,PDN,PNM1,PDNM1,X)
            POLY = PNP1+A*PN+B*PNM1
            PDER = PDNP1+A*PDN+B*PDNM1
            RECSUM = 0.0
            JM = J-1
            DO 27 I=1,JM
               RECSUM = RECSUM+1.0/(X-XJAC(I))
   27       CONTINUE
   28       CONTINUE
            DELX = -POLY/(PDER-RECSUM*POLY)
            X = X+DELX
            IF(ABS(DELX) .LT. EPS) GO TO 30
   29    CONTINUE
   30    CONTINUE
         XJAC(J) = X
         CSSAVE = CS*CD-SS*SD
         SS = CS*SD+SS*CD
         CS = CSSAVE
   39 CONTINUE
      XJAC(NP) = -1.0
      NPP = N+2
C
C  USE SYMMETRY FOR SECOND HALF OF THE ROOTS
C
      DO 49 I=2,NH
         XJAC(NPP-I) = -XJAC(I)
   49 CONTINUE
      IF(N .NE. 2*(N/2)) RETURN
      XJAC(NH+1) = 0.0
      RETURN
      END
      SUBROUTINE JACOBF(N,POLY,PDER,POLYM1,PDERM1,POLYM2,PDERM2,X)
C
C  COMPUTES THE JACOBI POLYNOMIAL (POLY) AND ITS DERIVATIVE
C  (PDER) OF DEGREE  N  AT  X
C
      IMPLICIT REAL(A-H,O-Z)
      COMMON /JACPAR/ALP,BET,RV
      APB = ALP+BET
      POLY = 1.0
      PDER = 0.0
      IF(N .EQ. 0) RETURN
      POLYLST = POLY
      PDERLST = PDER
      POLY = RV * X
      PDER = RV
      IF(N .EQ. 1) RETURN
```

```
      DO 19 K=2,N
         A1 = 2.*K*(K+APB)*(2.*K+APB-2.)
         A2 = (2.*K+APB-1.)*(ALP**2-BET**2)
         B3 = (2.*K+APB-2.)
         A3 = B3*(B3+1.)*(B3+2.)
         A4 = 2.*(K+ALP-1.)*(K+BET-1.)*(2.*K+APB)
         POLYN = ((A2+A3*X)*POLY-A4*POLYLST)/A1
         PDERN = ((A2+A3*X)*PDER-A4*PDERLST+A3*POLY)/A1
         PSAVE = POLYLST
         PDSAVE = PDERLST
         POLYLST = POLY
         POLY = POLYN
         PDERLST = PDER
         PDER = PDERN
   19 CONTINUE
      POLYM1 = POLYLST
      PDERM1 = PDERLST
      POLYM2 = PSAVE
      PDERM2 = PDSAVE
      RETURN
      END
```

References

Abarbanel, S., Gottlieb, D. Tadmor, E. (1986): "Spectral Methods for Discontinuous Problems", In *Numerical Methods for Fluid Dynamics. II.*, ed. by K. W. Morton, M. J. Baines (Oxford Univ. Press, London), pp. 129–153

Abramowitz, M., Stegun, I. A. (eds.) (1972): *Handbook of Mathematical Functions with Formulas, Graphs, and Mathematical Tables* (Gov. Printing Office, Washington, D.C.)

Ahlfors, L. V. (1966): *Complex Analysis* (McGraw-Hill, New York)

Ashby, S. F. (1985): "Chebycode, a Fortran Implementation of Manteuffel's Adaptive Chebyshev Algorithm", Rep. No. UIUCDCS-R-85-1203 (Dep. Comput. Sci., Univ. Illinois, Urbana, IL.)

Aydemir, A. Y., Barnes, D. C. (1984): Three-dimensional nonlinear incompressible MHD calculations. J. Comput. Phys. **53**, 100–123

Aydemir, A. Y., Barnes, D. C. (1985): An implicit algorithm for compressible three-dimensional magnetohydrodynamic calculations. J. Comput. Phys. **59**, 108–119

Babuška, I., Aziz A. K., (1972): "Survey Lectures on the Mathematical Foundations of the Finite Element Method", in *The Mathematical Foundations of the Finite Element Method with Application to Partial Differential Equations*, ed. by A. K. Aziz (Academic Press, London, New York) pp. 3–359

Babuška, I., Dorr, M. R. (1981): Error estimates for the combined h and p versions of the finite element method. Numer. Math. **37**, 257–277

Babuška, I., Szabo B. A., Katz I. N., (1981): The p-version of the finite element method. SIAM J. Numer. Anal. **18**, 515–545

Baer, F. (1964): Integration with the spectral vorticity equation. J. Atmosph. Sci. **21**, 260–276

Baer, F., Platzman G. W., (1961): A procedure for numerical integration of the spectral vorticity equation. J. Meteorol. **18**, 393–410

Bardina, J., Ferziger J. H., Reynolds W.C., (1983): "Improved Turbulence Models Based on Large Eddy Simulation of Homogeneous, Incompressible, Turbulent Flows", Rep. TF-19 (Dep. Mechanical Engineering, Stanford Univ.)

Bardina, J., Ferziger J. H., Rogallo R. S., (1985). Effect of rotation on isotropic turbulence: computation and modelling. J. Fluid Mech., **154**, 321–336.

Bartels, R. H., Stewart G. W., (1972): Solution of the matrix equation $AX + XB = C$. Comm. ACM **15**, 820–826

Basdevant, C. (1983): Technical improvements for direct numerical simulation of homogeneous three-dimensional turbulence. J. Comput. Phys. **50**, 209–214

Basdevant, C., Deville M., Haldenwang, P., Lacroix, J. M., Ouazzani, J., Peyret, R., Orlandi, P., Patera, A. T. (1986): Spectral and finite difference solutions of the Burgers equation. Comput. Fluids **14**, 23–41

Batchelor, G. K. (1967): *An Introduction to Fluid Dynamics* (Cambridge Univ. Press, Cambridge)

Bateman, H. (1915): Some recent researches on the motion of fluids. Mon. Weather Rev. **43**, 163–170

Battarra, V., Canuto, C., Quarteroni, A. (1985): A Chebyshev spectral method for gas transients in pipelines. Comput. Meth. Appl. Mech. Eng. **48**, 329–352

Bayliss, A., Matkowsky, B. J. (1987): Fronts, relaxation oscillations and period doubling in solid fuel combustion. J. Comput. Phys. **71**, 147–168

Belov, Y. Y., Yanenko, N. N. (1971): Influence of viscosity on the smoothness of solutions of incompletely parabolic systems. Math. Notes Acad. Sci. USSR **10**, 480–483

Bernal, L. P. (1981): "The Coherent Structure of Turbulent Mixing Layers. I. Similarity of the Primary Vortex Structure. II. Secondary Streamwise Vortex Structure", Ph.D. Thesis, Caltech, Pasadena, CA

Bernardi, C. (1982): Numerical approximation of a periodic linear parabolic equation, SIAM J. Numer. Anal. **19**, 1196–1207

Bernardi, C., Canuto, C. Maday, Y. (1986): "Generalized Inf-Sup Condition for Chebyshev Approximations to Navier–Stokes Equations", ICASE Rep. No. 86–61 (NASA Langley Research Center, Hampton, VA.) and *C. R. Acad. Sci. Paris* **303**, serie I, 971–974

Bernardi, C., Maday, Y. (1986a): "A Staggered Grid Spectral Method for the Stokes Problem," in *Proc. 6th Int. Symp. Finite Element Methods in Flow Problems*, ed. by M. O. Bristeau, R. Glowinski, A. Hausel, J. Periaux, pp. 33–37.

Bernardi, C., Maday, Y. (1986b): Propriétés de quelques espaces de Sobolev aved poids et application á la collocation de Tchebycheff. C. R. Acad. Sci. Paris **303**, serie I, 829–832

Bernardi, C., Maday, Y. Métivet, B. (1986): Une méthode directe de collocation pour le problème de Stokes. *C. R. Acad. Sci. Paris* **302**, serie I, 163–166

Bernardi, C., Maday, Y. Métivet, B. (1987a): Spectral approximation of the periodic/nonperiodic Navier–Stokes equations. Numer. Math. (in press)

Bernardi, C., Maday, Y. Métivet, B. (1987b): Calcul de la pression dans la résolution spectrale du problème de Stokes. Rech. Aerosp. (in press)

Biringen, S. (1985): Active control of transition by periodic suction-blowing. Phys. Fluids **27**, 1345–1347

Birkhoff, G., Rota, G. C. (1978): *Ordinary Differential Equations*, 3rd edn. (J. Wiley & Sons, New York)

Blackstock, D. T. (1966): Convergence of the Keck–Boyer perturbation solution for plane waves of finite amplitude in a viscous fluid. J. Acoust. Soc. Am. **39**, 411–413

Blinova, E. N. (1944): Hydrodynamic theory of pressure and temperature waves and center of action of the atmosphere. Trans. No. 113 (Regional Control Office, Second Weather Region, Patterson Field, OH)

Blum, E. K. (1962): A modification of the Runge–Kutta fourth-order method. Math. Comput. **16**, 176–187

Bontoux, P., Bondet De La Bernardie, B., Roux, B. (1981): "Spectral Methods for Natural Convection Problems", in *Numerical Methods for Coupled Problems*, ed. by E. Hinton, P. Bettess, R. W. Lewis (Pineridge, Swansea, UK) pp. 1018–1030

Bourke, W. (1972): An efficient, one-level, primitive-equation spectral model. Mon. Weather Rev. **100**, 683–689

Bourke, W. (1974): A multi-level spectral model. I. Formulation and hemispheric integrations. Mon. Weather Rev. **102**, 687–701

Bourke, W. McAvaney, B., Puri, K., Thurling, R. (1977): "Global Modelling of Atmospheric Flow by Spectral Methods", in *Methods in Computational Physics*, Vol. 17, ed. by Chang (Academic Press, London, New York) pp. 267–324

Boyd, J. P. (1978a): The choice of spectral functions on a sphere for boundary and eigenvalue problems: a comparison of Chebyshev, Fourier and associated Legendre expansions. Mon. Weather Rev. **106**, 1184–1191

Boyd, J. P. (1978b): Spectral and pseudospectral methods for eigenvalue and nonseparable boundary value problems. Mon. Weather Rev. **106**, 1192–1203

Boyd, J. P. (1982): The optimization of convergence for Chebyshev polynomial methods in unbounded domain, J. Comput. Phys. **45**, 43–79

Boyd, J. P. (1984): Asymptotic coefficients of Hermite function series. J. Comput. Phys. **54**, 382–410

Boyd, J. P. (1985): Complex coordinate methods for hydrodynamic instabilities and Sturm-Liouville eigenproblems with an interior singularity. J. Comput. Phys. **57**, 453–471

Boyd, J. P. (1987a): Spectral methods using rational basis functions on an infinite interval. J. Comput. Phys. **69**, 112–142

Boyd, J. P. (1987b): Orthogonal rational functions on a semi-infinite interval. J. Comput. Phys. **70**, 63–88

Brachet, M. E., Meiron, D. I., Orszag, S. A., Nickel, B. G., Morf, R. H., Frisch, U. (1983): Small-scale structure of the Taylor-Green vortex. J. Fluid Mech. **130**, 411–452

Bramley, J. S., Dennis, S. C. R. (1982): The calculation of eigenvalues for the stationary perturbations of Poiseuille flow. J. Comput. Phys. **47**, 179–198

Brandt, A., Fulton, S. R., Taylor, G. D. (1985): Improved spectral multigrid methods for periodic elliptic problems, J. Comput. Phys. **58**, 96–112

Bressan, N., Quarteroni, A. (1986a): An implicit/explicit spectral method for Burgers equation. Calcolo **23**, 265–284

Bressan, N., Quarteroni, A. (1986b): Analysis of Chebyshev collocation methods for parabolic equations. SIAM J. Numer. Anal. **23**, 1138–1154

Brezzi, F. (1974): On the existence, uniqueness and approximation of saddle point problems arising from Lagrangian multipliers. R.A.I.R.O. Numer. Anal. **8**, 129–151

Brezzi, F., Rappaz, J., Raviart, P. A. (1980): Finite dimensional approximation of nonlinear problems, Part I: Branches of nonsingular solutions. Numer. Math. **36**, 1–25

Brezzi, F., Rappaz, J., Raviart, P. A. (1981a): Finite dimensional approximation of nonlinear problems, Part II: Limit points. Numer. Math. **37**, 1–28

Brezzi, F., Rappaz, J., Raviart, P. A. (1981b): Finite dimensional approximation of nonlinear problems, Part III: Simple bifurcation points. Numer. Math. **38**, 1–30

Bridges, T. J., Morris, P. J. (1984a): Differential eigenvalue problems in which the parameter appears nonlinearly. J. Comput. Phys. **55**, 437–460

Bridges, T. J., Morris, P. J. (1984b): "Spectral Calculations of the Spatial Stability of Non-Parallel Boundary Layers," AIAA Pap. No. 84-0437

Brigham, E. O. (1974): *The Fast Fourier Transform* (Prentice-Hall, Englewood Cliffs, NJ.)

Bullister, E. T., Karniadakis, G. E., Ronquist, E. M., Patera, A. T. (1986): "Solution of the Unsteady Navier-Stokes Equations by Spectral Element Methods", in *Proc. 6th Int. Symp. Finite Element Methods in Flow Problems*, ed. by M. O. Bristeau, R. Glowinski, A. Haugel, J. Periaux, pp. 225–229

Burgers, J. M. (1948): A mathematical model illustrating the theory of turbulence. Adv. Appl. Mech. **1**, 171–199

Burgers, J. M. (1974): *The Nonlinear Diffusion Equation* (Reidel, Boston)

Butzer, P. L., Nessel, R. J. (1971): *Fourier Analysis and Approximation* (Birkhäuser, Basel)

Buzbee, B. L., Dorr, F. W., George, J. A., Golub, G. H. (1971): The direct solution of the discrete Poisson equation on irregular regions. SIAM J. Numer. Anal. **8**, 722–736

Cain, A. B., Ferziger, J. H., Reynolds, W. C. (1984): Discrete orthogonal function expansions for non-uniform grids using the fast Fourier transform. J. Comput. Phys. **56**, 272–286

Cain, A. B., Reynolds, W. C., Ferziger, J. H. (1981): "A Three-Dimensional Simulation of Transition and Early Turbulence in a Time-Developing Mixing Layer," Rep. No. TF-14 (Dep. Mechanical Engineering, Stanford Univ.)

Canuto, C. (1986): Boundary conditions in Legendre and Chebyshev methods. SIAM J. Numer. Anal. **23**, 815–831

Canuto, C. (1987): Spectral methods and a maximum principle. Math. Comput. (in press)

Canuto, C., Fujii, H., Quarteroni, A. (1983): Approximation of symmetry breaking bifurcations for the Rayleigh convection problem. SIAM J. Numer. Anal. **20**, 873–884

Canuto, C., Funaro, D. (1987): The Schwarz algorithm for spectral methods, SIAM J. Numer. Anal., in press

Canuto, C., Hariharan, S. I., Lustman, L. (1985): Spectral methods for exterior elliptic problems. Numer. Math. **46**, 505–520

Canuto, C., Maday, Y., Quarteroni, A. (1982): Analysis for the combined finite element and Fourier interpolation. Numer. Math. **39**, 205–220

Canuto, C., Maday, Y., Quarteroni, A. (1984): Combined finite element and spectral approximation of the Navier–Stokes equations. Numer. Math. **44**, 201–217

Canuto, C., Pietra, P. (1987): "Boundary and Interface Conditions within a Finite Element Preconditioner for Spectral Methods," Rep. No. 553 (I.A.N., Pavia Univ.)

Canuto, C., Quarteroni, A. (1981a): Spectral and pseudo-spectral methods for parabolic problems with nonperiodic boundary conditions. Calcolo **18**, 197–218

Canuto, C., Quarteroni, A. (1981b): Spectral methods for hyperbolic equations. R. Sem. Mat. Univ. Politec. Torino **39**, 21–31

Canuto, C., Quarteroni, A. (1982a): Approximation results for orthogonal polynomials in Sobolev spaces. Math. Comput. **38**, 67–86

Canuto, C., Quarteroni, A. (1982b): Error estimates for spectral and pseudo-spectral approximations of hyperbolic equations. SIAM J. Numer. Anal. **19**, 629–642

Canuto, C., Quarteroni, A. (1984): "Variational Methods in the Theoretical Analysis of Spectral Approximations," in *Spectral Methods for Partial Differential Equations*, ed. by R. G. Voigt, D. Gottlieb, M. Y. Hussaini (SIAM-CBMS, Philadelphia) pp. 55–78

Canuto, C., Quarteroni, A. (1985): Preconditioned minimal residual methods for Chebyshev spectral calculations. J. Comput. Phys. **60**, 315–337

Canuto, C., Quarteroni, A. (1987): On the boundary treatment in spectral methods for hyperbolic systems. J. Comput. Phys. **71**, 100–110

Canuto, C., Sacchi–Landriani, G. (1986): Analysis of the Kleiser–Schumann method. Numer. Math. **50**, 217–243

Carleson, L. (1966): On convergence and growth of partial sums of Fourier series. Acta Math. **116**, 135–157

Carrier, G. F., Krook, M., Pearson, C. E. (1966): *Functions of a Complex Variable* (McGraw-Hill, New York)

Cebeci, T., Bradshaw, P. (1977): *Momentum Transport in Boundary Layers* (McGraw-Hill, New York)

Cebeci, T., Stewartson, K. (1980): On stability and transition in three-dimensional flows. AIAA J. **18**, 398–405

Chan, T. F., Kerkhoven, T. (1985): Fourier methods with extended stability intervals for the Korteweg-de Vries equation. SIAM J. Numer. Anal. **22**, 441–454

Chandrasekhar, S. (1961): *Hydrodynamic and Hydromagnetic Stability* (Oxford Univ. Press, London)

Chaves, T., Ortiz, E. L. (1968): On the numerical solution of two-point boundary value problems for linear differential equations. Z. Angew. Math. Mech. **48**, 415–418

Cheney, E. W. (1966): *Introduction to Approximation Theory* (McGraw-Hill, New York)

Chorin, A. J. (1968): Numerical solution of the Navier–Stokes equations. Math. Comput. **22**, 745–762

Ciarlet, P. G. (1978): *The Finite Element Method for Elliptic Problems* (North-Holland, Amsterdam, New York)

Clark, R. A., Ferziger, J. H., Reynolds, W. C. (1979): Evaluation of subgrid-scale models using an accurately simulated turbulent flow. J. Fluid Mech. **91**, 1–16

Clenshaw, C. W. (1957): The numerical solution of linear differential equations in Chebyshev series, Proc. Cambridge Philos. Soc. **53**, 134–149

Clenshaw, C. W., Norton, H. J. (1963): The solution of nonlinear ordinary differential equations in Chebyshev series. Comput. J. **6**, 88–92

Cole, J. D. (1951): On a quasilinear parabolic equation occurring in aerodynamics. Q. Appl. Math. **9**, 225–236

Comte-Bellot, G. (1965): "Écoulement Turbulent Entre Deux Parois Parallèles". (P.U.B.L. Sci. Tech. M. Min. Air., 419 Paris)

Comte-Bellot, G., Corrsin, S. (1971): Simple Eulerian time correlation of full and narrow-band velocity signals in grid-generated, isotropic turbulence, J. Fluid Mech. **48**, 273–337

Concus, P., Golub, G. H., O'Leary, D. P. (1976): "A Generalized Conjugate Gradient Method for the Numerical Solution of Elliptic Partial Differential Equations," in *Sparse Matrix Computations*, ed. by J. R. Bunch, D. J. Rose (Academic Press, New York) pp. 309–332

Cooley, J. W., Lewis, P. A. W., Welch, P. D. (1969): The fast Fourier transform and its applications. IEEE Trans. Educ. **12**, 27–34

Cooley, J. W., Lewis, P. A. W., Welch, P. D. (1970): The fast Fourier transform algorithm: programming considerations in the calculation of sine, cosine, and Laplace transforms. J. Sound Vibr. **12**, 315–337

Cooley, J. W., Tukey, J. W. (1965): An algorithm for the machine calculation of complex Fourier series. Math. Comput. **19**, 297–301

Cornille, P. (1982): A pseudospectral scheme for the numerical calculation of shocks. J. Comput. Phys. **47**, 146–159

Courant, R., Friedrichs, K. O. (1948): *Supersonic Flow and Shock Waves* (Wiley-Interscience, New York)

Courant, R., Hilbert, D. (1953): *Methods of Mathematical Physics*, Vol. 1 (Wiley-Interscience, New York)

Crouzeix, M., Rappaz, J. (1987): *Approximation de Problèmes Faiblement non Lineaires* (in preparation)

Curry, J., Herring, J., Loncaric, J., Orszag, S. A. (1984): Order and disorder in two- and three-dimensional Benard convection. J. Fluid Mech. **147**, 1–38

Dahlburg, J. P. (1985): "Turbulent Disruptions from the Strauss Equations," Ph.D. Thesis, College of William and Mary, Williamsburg, VA.

Dahlburg, J. P., Montgomery, D., Doolen, G. D., Matthaeus, W. H. (1986): Large-scale disruptions in a current-carrying magnetofluid. J. Plasma Phys. **35**, 1–42

Dahlburg, J. P., Montgomery, D., Matthaeus, W. H. (1985): Turbulent disruptions from the Strauss equations. J. Plasma Phys. **34**, 1–46

Dahlburg, R. B., Zang, T. A., Montgomery, D. (1986): Unstable transition properties of the driven magnetohydrodynamic sheet pinch. J. Fluid Mech. **169**, 71–108

Dahlburg, R. B., Zang, T. A., Montgomery, D., Hussaini, M. Y. (1983): Viscous, resistive magnetohydrodynamic stability computed by spectral techniques. Proc. Natl. Acad. Sci. U.S.A. **80**, 5798–5802

Dang, K., Roy, P. (1985): "Direct and Large-Eddy Simulation of Homogeneous Turbulence Submitted to Solid Body Rotation," in *Proc. 5th Symp. Turbulent Shear Flows* (Cornell Univ. Ithaca, NY)

Davis, S. H. (1976): The stability of time-periodic flows. Ann. Rev. Fluid Mech. **8**, 57–74

Davis, P. J., Rabinowitz, P. (1984): *Methods of Numerical Integration*, 2nd edn. (Academic Press, London, New York)

Deardorff, J. W. (1970): A numerical study of three-dimensional turbulent channel flow at large Reynolds numbers. J. Fluid Mech. **41**. 453–480

Dekker, K., Verwer, J. G. (1984): *Stability of Runge–Kutta Methods for Stiff Nonlinear Differential Equations* (North-Holland, Amsterdam)

Delorme, P. (1984): Numerical simulation of homogeneous, isotropic, two-dimensional turbulence in compressible flow. Rech. Aerosp. **1984-1**, 1–13

Delprete, V. (1983): Convergence of the Fourier method for Euler's equation on the sphere. Boll. Un. Mat. It. **(6) 2-B**, 407–416

Delves, L. M., Hall, C. A. (1979): An implicit matching procedure for global element calculations. J. Inst. Math. Appl. **23**, 223–234

Delves, L. M., Phillips, C. (1980): A fast implementation of the global element method. J. Inst. Math. Appl. **25**, 177–197

Dennis, S. C. R., Ingham, D. B., Cook, R. N. (1979): Finite-difference methods for calculating steady incompressible flows in three dimensions. J. Comput. Phys. **33**, 325–339

Dennis, S. C. R., Quartapelle, L. (1983): Direct solution of the vorticity-stream function ordinary differential equations by a Chebyshev approximation. J. Comput. Phys. **45**, 448–463

Dennis, S. C. R., Quartapelle, L. (1985): Spectral algorithms for vector elliptic equations in a spherical gap. J. Comput. Phys. **61**, 218–241

Descloux, J., Rappaz, J. (1982): Approximation of solution branches of nonlinear equations. R.A.I.R.O. Anal. Numer. **16**, 319–342

Deville, M. (1984): "Recent Developments of Spectral and Pseudospectral Methods in Fluid Dynamics," von Karman Institute Lecture Series, Rhode-Saint Genese, Belgium.

Deville, M., Haldenwang, P., Labrosse, G. (1981): "Comparison of Time Integration (Finite Difference and Spectral) for the Nonlinear Burgers' Equation, in *Proc. 4th GAMM Conf. Numerical Methods in Fluid Mechanics* ed. by H. Viviand (Vieweg, Braunschweig)

Deville, M., Kleiser, L., Montigny-Rannou, F. (1984): Pressure and time treatment of Chebyshev spectral solution of a Stokes problem. Int. J. Numer. Meth. Fluids **4**, 1149–1163

Deville, M., Labrosse G. (1982): An algorithm for the evaluation of multidimensional (direct and inverse) discrete Chebyshev transform. J. Comput. Appl. Math. **8**, 293–304

Deville, M., Mund, E. (1985): Chebyshev pseudospectral solution of second-order elliptic equations with finite element preconditioning. J. Comput. Phys. **60**, 517–533

Dongarra, J., Bunch, J. R., Moler, C. B., Stewart, G. W. (1978): *LINPACK Users Guide* (SIAM, Philadelphia)

Doron, E., Hollingsworth, A., Hoskins, B. J., Simmonds, A. J. (1974): A comparison of grid-point and spectral methods in a meteorological problem. Q. J. R. Met. Soc. **100**, 371–383

Dorr, F. W. (1970): The direct solution of the discrete Poisson equation on a rectangle. SIAM Rev. **12**, 248–263

Dorr, M. R. (1984): The approximation theory for the *p*-version of the finite element method. I, SIAM J. Numer. Anal. **21**, 1180–1207

Douglas, J., Gunn, J. E. (1964): A general formulation of alternating direction methods. Numer. Math. **6**, 428–453

Drazin, P. G., Reid, W. M. (1981): *Hydrodynamic Stability* (Cambridge Univ. Press, Cambridge)

Drummond, J. P., Hussaini, M. Y., Zang, T. A. (1986): Spectral methods for modelling supersonic chemically reacting flow fields. AIAA J. **24**, 1461–1467

Dubiner, M. (1977): Asymptotic analysis of spectral methods (unpublished)

Duff, I. S. (1981): "MA32-A Package for Solving Sparse Unsymmetric Systems Using the Frontal Method," AERE Report R. 10079 (HMSO, London)

Ehrenstein, U., Peyret, R. (1986): "A Collocation-Chebyshev Method for Solving Stokes-Type Equations." in *Proc. 6th Int. Symp. Finite Element Methods in Flow Problems*, ed. by M. O. Bristeau, R. Glowinski, A. Hausel, J. Periaux, pp. 213–218

Eidson, T. M., Hussaini, M. Y., Zang, T. A. (1986): "Simulation of the Turbulent Rayleigh-Benard Problem Using a Spectral/finite Difference Technique," ICASE Rep. No. 86-6 pp. 188–209. (NASA Langley Research Center, Hampton, VA.)

Eisenstat, S. C., Elman, H. C., Schultz, M. H. (1983): Variational iterative methods for non-symmetric systems of linear equations. SIAM J. Numer. Anal. **20**, 345–357

Eisenstat, S. C., Gursky, M. C., Schultz. M. H., Sherman, A. H. (1982): Yale sparse matrix package I. The symmetric codes. Int. J. Numer. Meth. Eng. **18**, 1145–1152

Eliasen, E., Machenhauer, B., Rasmussen, E. (1970): "On a Numerical Method for Integration of the Hydrodynamical Equations with a Spectral Representation of the Horizontal Fields," Rep. No. 2 (Institut for Teoretisk Meteorologi, Univ. Copenhagen)

Ellsaesser, H. W. (1966): Evaluation of spectral versus grid methods of hemispheric numerical weather prediction. J. Appl. Meteorol. **5**, 246–262

Emmons, H. W. (1949): "The Numerical Solution of the Turbulence Problem", in *Proc. Symp. Applied Mathematics*, Vol. 1 (McGraw-Hill, New York) pp. 67–71

Erlebacher, G., (1986): "Incipient Transition Phenomena in Compressible Flows over a Flat Plate", in *Proc. 10th Inf. Conf. Numerical Methods in Fluid Dynamics*, ed. by F. G. Zhuang, Y. L. Zhu, (Springer, Berlin, Heidelberg, New York) pp. 264–269

Erlebacher, G. and Hussaini, M. Y. (1987): "Stability and Transition in Supersonic Boundary Layers," AIAA Paper No. 87-1416

Erlebacher, G., Hussaini, M. Y., Speziale, C., Zang, T. A. (1987): "Towards the Large-eddy Simulation of Compressible, Turbulent Flow," ICASE Rep. No. 87-20 (NASA Langley Research Center, Hampton, VA)

Erlebacher, G., Zang, T. A., Hussaini, M. Y. (1987): "Spectral Multigrid Methods for the Numerical Simulation of Turbulence," in *Multigrid Methods* ed. by S. McCormick, K. Stuben (Marcel Dekker, New York)

Evans, D. J. (ed.) (1983): *Preconditioning Methods: Analysis and Applications* (Gordon & Breach, New York)

Fasel, H.(1976): Investigation of stability of boundary layers by a finite-difference model of the Navier–Stokes equations J. Fluid Mech. **78**, 355–383

Feiereisen, W. J., Reynolds, W. J., Ferziger, J. H. (1981): "Numerical Simulation of Compressible, Homogeneous Turbulent Shear Flow." Rep. TF-13 (Dep. Mechanical Engineering, Stanford Univ.)

Feit, M. D., Fleck, J. A., Jr., Steiger, A. (1982): Solution of the Schroedinger equation by a spectral method. J. Comput. Phys. **47**, 412–433

Fenton, J. D., Rienecker, M. M. (1982): A Fourier method for solving nonlinear water-wave problems: Application to solitary-wave interactions. J. Fluid Mech. **118**, 411–443

Finlayson, B. A., Scriven, L. E. (1966): The method of weighted residuals—a review. Appl. Mech. Rev. **19**, 735–748

Fletcher, C. A. J. (1984): *Computational Galerkin Methods* (Springer, Berlin, Heidelberg, New York)

Fornberg, B. (1975): On a Fourier method for the integration of hyperbolic equations. SIAM J. Numer. Anal. **12**, 509–528

Fornberg, B. (1977): A numerical study of two-dimensional turbulence. J. Comput. Phys. **25**, 1–31

Fornberg, B. (1978): Pseudospectral calculations on 2-D turbulence and nonlinear waves. SIAM-AMS Proc. **11**, 1–18

Fornberg, B. (1980): A numerical method for conformal mappings. SIAM J. Sci. Stat. Comput. **1**, 386–400

Fornberg, B., Whitman, G. B. (1978): A numerical and theoretical study of certain nonlinear wave phenomena. Philos. Trans. Soc. London **289**, 373–404

Förster, K. (ed.) (1978): *Boundary Algorithms for Multidimensional Inviscid Hyperbolic Flows* (Vieweg, Braunschweig)

Fortin, M., Peyret, R., Temam, R. (1971): Résolution numérique des équations de Navier–Stokes pour un fluide incompressible. J. Mech **10**, 357–390

Fox, D. G., Orszag, S. A. (1973): Pseudospectral approximation to two-dimensional turbulence. J. Comput. Phys. **11**, 612–619

Fox, L., Parker, I. B. (1968): *Chebyshev Polynomials in Numerical Analysis* (Oxford Univ. Press, London)

Frazer, R. A., Jones, W. P., Skan, S. W. (1937): *Approximation to Functions and to the Solution of Differential Equations*, R&M 1799 (Aeronautical Research Council, London)

Frisch, U., Pouquet, A., Sulem, P.-L., Meneguzzi, M. (1983): The dynamics of two-dimensional MHD. J. Mec. Theor. Appl. (Numero special) 191–216

Fulton, S. R., Taylor, G. D. (1984): On the Gottlieb-Turkel time filter for Chebyshev spectral methods. J. Comput. Phys. **55**, 302–312

Funaro, D. (1981): "Approssimazione Numerica di Problemi Parabolici e Iperbolici con Metodi Spettrali", Thesis, Univ. Pavia

Funaro, D. (1983): Error estimates for spectral approximation of linear advection equation over an ipercube. Calcolo **20**, 335–353

Funaro, D. (1986): A multidomain spectral approximation of elliptic equations. Numer. Meth. PDEs **2**, 187–205

Funaro, D. (1987a): "Some Results About the Spectrum of the Chebyshev Differencing Operator", in *Numerical Approximation of P. D. E.*, ed. by E. L. Ortiz (North-Holland, Amsterdam), pp. 271–284

Funaro, D. (1987b): A preconditioning matrix for the Chebyshev differencing operator. SIAM J. Numer. Anal. (in press)

Funaro, D., Quarteroni, A., Zanolli, P. (1987): An iterative procedure with interface relaxation for domain decomposition methods SIAM J. Numer. Anal. (in press)

Fyfe, D., Joyce, G., Montgomery, D. (1977): Magnetic dynamo action in two-dimensional turbulent magnetofluids. J. Plasma Phys. **17**, 317–335

Fyfe, D., Montgomery, D., Joyce, G. (1977): Dissipative forced turbulence in two-dimensional magnetohydrodynamics. J. Plasma Phys. **17**, 369–398

Galerkin, B. (1915): Rods and plates: Series occurring in various questions concerning the elastic equilibrium of rods and plates. Vestn. Inzhen. **19**, 897–908

Gaster, M. (1962): A note on the relation between temporally-increasing and spatially-increasing disturbances in hydrodynamic stability. J. Fluid Mech. **14**, 222–224

Gazdag, J. (1976): Time-differencing schemes and transform methods. J. Comput. Phys. **20**, 196–207

Gear, C. W. (1971): *Numerical Initial Value Problems in Ordinary Differential Equations* (Prentice-Hall, Englewood Cliffs, NJ.)

Gear, C. W. (1981): Numerical solution of ordinary differential equations: is there anything left to do? SIAM Rev. **23**, 10–24

Ghaddar, N. K., Korczak, K. Z., Mikic, B. B., Patera, A. T. (1986a): Numerical investigation of incompressible flow in grooved channels, Part 1: Stability and self-sustained oscillations. J. Fluid Mech. **163**, 99–127

Ghaddar, N. K., Korczak, K. Z., Mikic, B. B., Patera A. T. (1986b): Numerical investigation of incompressible flow in grooved channels, Part 2: Resonance and oscillatory heat transfer. J. Fluid Mech. **168**, 541–567

Ghaddar, N. K., Patera, A. T., Mikic, B. B. (1984): "Heat Transfer Enhancement in Oscillatory Flow in a Grooved Channel," AIAA Pap. No. 84-0495

Girault, V., Raviart, P. A. (1986): *Finite Element Approximation of the Navier–Stokes Equations: Theory and Algorithms* (Springer, Berlin, Heidelberg, New York, Tokyo)

Glowinski, R. (1984): *Numerical Methods for Nonlinear Variational Problems* (Springer, Berlin, Heidelberg, New York, Tokyo)

Glowinski, R., Dinh, Q. V., Periaux, J. (1983): Domain decomposition methods for nonlinear problems in fluid dynamics. Comput. Meth. Appl. Mech. Eng. **40**, 27–109

Golub, G. H., Van Loan, C. F. (1983): *Matrix Computations* (John Hopkins Univ. Press, Baltimore)

Gordon, W. J., Hall, C. A. (1973a): Construction of curvilinear co-ordinate systems and their applications to mesh generation. Int. J. Numer. Meth. Eng. **7**, 461–477

Gordon, W. J., Hall, C. A. (1973b): Transfinite element methods: blending-function interpolation over arbitrary curved element domains. Numer. Math. **21**, 109–129

Gottlieb, D. (1981): The stability of pseudospectral Chebyshev methods. Math. Comput. **36**, 107–118

Gottlieb, D. (1985): "Spectral Methods for Compressible Flow Problems," in *Proc. 9th Int. Conf. Numerical Methods in Fluid Dynamics*, ed. by Soubbarameyer, J. P. Boujot (Springer, Berlin, Heidelberg, New York) pp. 48–61

Gottlieb, D., Gunzburger, M., Turkel, E. (1982): On numerical boundary treatment for hyperbolic systems. SIAM J. Numer. Anal. **19**, 671–697

Gottlieb, D., Gustafsson, B. (1976): Generalized DuFort–Frankel methods for parabolic initial-boundary value problems. SIAM J. Numer. Anal. **13**, 129–144

Gottlieb, D., Hussaini, M. Y., Orszag, S. A. (1984): "Theory and Applications of Spectral Methods," in *Spectral Methods for Partial Differential Equations*, ed. by R. G. Voigt, D. Gottlieb, M. Y. Hussaini (SIAM-CBMS, Philadelphia) pp. 1–54

Gottlieb, D., Lustman, L. (1983a): The Dufort–Frankel Chebyshev method for parabolic initial boundary value problems. Comput. Fluids **11**, 107–120

Gottlieb, D., Lustman, L. (1983b): The spectrum of the Chebyshev collocation operator for the heat equation. SIAM J. Numer. Anal. **20**, 909–921

Gottlieb, D., Lustman, L., Orszag, S. A. (1981): Spectral calculations of one-dimensional inviscid compressible flow. SIAM J. Sci. Stat. Comput. **2**, 296–310

Gottlieb, D., Lustman, L., Streett, C. S. (1984): "Spectral Methods for Two-Dimensional Shocks", in *Spectral Methods for Partial Differential Equations*, ed. by R. G. Voigt, D. Gottlieb, M. Y. Hussaini, (SIAM-CBMS, Philadelphia) pp.79–95

Gottlieb, D., Lustman, L., Tadmor, E. (1987a): "Stability analysis of spectral methods for hyperbolic initial-boundary value problems," SIAM J. Numer. Anal. **24**, 241–256

Gottlieb, D., Lustman, L., Tadmor, E. (1987b): "Convergence of spectral methods for hyperbolic initial-boundary value systems," SIAM J. Numer. Anal. **24**, 532–537

Gottlieb, D., Orszag, S. A. (1977): *Numerical Analysis of Spectral Methods: Theory and Applications* (SIAM-CBMS, Philadelphia.)

Gottlieb, D., Orszag, S. A., Turkel, E. (1981): Stability of pseudospectral and finite difference methods for variable coefficient problems. Math. Comput. **37**, 293–305

Gottlieb, D., Tadmor, E. (1985): "Recovering Pointwise Values of Discontinuous Data Within Spectral Accuracy," in *Progress and Supercomputing in Computational Fluid Dynamics*, ed. by E. M. Murman , S. S. Abarbanel (Birkhäuser, Boston) pp. 357–375

Gottlieb, D., Turkel, E. (1980): On time discretization for spectral methods. Stud. Appl. Math. **63**, 67–86

Gottlieb, D., Turkel, E. (1985): "Topics in Spectral Methods for Time Dependent Problems", in *Numerical Methods in Fluid Dynamics*, ed. by F. Brezzi (Springer, Berlin, Heidelberg, New York, Tokyo) pp. 115–155

Grant, H. L., Stewart, R. W., Moilliet, A. (1962): Turbulence spectra from a tidal channel. J. Fluid Mech. **12**, 241–268

Grimm, R. C. (1985): "Computational Fusion Magnetohydrodynamics", in *Large-Scale Computations in Fluid Mechanics*, Lectures in Applied Mathematics **22**, 241–269

Grosch, C. E., Orszag, S. A. (1977): Numerical solution of problems in unbounded regions: coordinate transformations. J. Comput. Phys. **25**, 273–296

Guo, Ben Yu, Babuška, I. (1985): "The h-p Version of the Finite Element Method", Tech. Note BN-1043, (Lab. Numerical Analysis, Univ. Maryland)

Guo, Ben Yu, Manozanjan, V. S. (1985): A spectral method for solving the RLW equation. IMA J. Numer. Anal. **5**, 307–318

Gustafsson, B., Kreiss, H.-O., Sundström, A. (1972): Stability theory of difference approximations for mixed initial boundary value problems. II. Math. Comput. **26**, 649–686

Hackbusch, W. (1985): *Multigrid Methods and Applications* (Springer, Berlin, Heidelberg, New York, Tokyo)

Hafez, M. M., South, J. C., Murman, E. M. (1979): Artificial compressibility methods for numerical solution of transonic full potential equation. AIAA J. **17**, 838–444

Hageman, A., Young, D. M. (1981): *Applied Iterative Methods* (Academic Press, London, New York)

Haidvogel, D. B. (1977): "Quasigeostrophic Regional and General Circulation Modelling: an Efficient Pseudospectral Approximation Technique," in *Computing Methods in Geophysical Mechanics*, Vol. 25, ed. by R. P. Shaw (ASME, New York)

Haidvogel, D. B., Zang, T. A. (1979): The accurate solution of Poisson's equation by expansion in Chebyshev polynomials. J. Comput. Phys. **30**, 167–180

Haj, A., Phillips, C., Delves, L. M. (1980): The global element method for stationary advective problems. Int. J. Numer. Meth. Eng. **15**, 167–175

Hald, O. H. (1981): Convergence of Fourier methods for Navier–Stokes equations, J. Comput. Phys. **40**, 305–317

Haldenwang, P. (1984): "Résolution Tridimensionnelle des Équations de Navier–Stokes par Méthodes Spectrales Tchébycheff: Application á la Convection Naturelle," Ph.D. Thèse, Université de Provence

Haldenwang, P., Labrosse, G. (1986): "2-D and 3-D Spectral Chebyshev Solutions for Free Convection at High Rayleigh Number," in *Proc. 6th Int. Symp. Finite Element Methods in Flow Problems*, ed. by M. O. Bristeau, R. Glowinski, A. Hausel, J. Periaux, pp. 261–266

Haldenwang, P., Labrosse, G., Abboudi, S., Deville, M. (1984): Chebyshev 3-D spectral and 2-D pseudospectral solvers for the Helmholtz equation. J. Comput. Phys. **55**, 115–128

Hall, M. G. (1981): "Computational Fluid Dynamics—a Revolutionary Force in Aerodynamics," AIAA Pap. No. 81-1014

Hall, P., Malik, M. R. (1986): On the instability of a three-dimensional attachment line boundary layer: weakly nonlinear theory and a numerical approach. J. Fluid Mech. **163**, 257–282

Haltiner, G. J., Williams, R. T. (1980): *Numerical Prediction and Dynamical Meteorology* (John Wiley & Sons, New York)

Hamming, R. W. (1977): *Digital Filters* (Prentice-Hall, Englewood Cliffs, NJ)

Hariharan, S. I. (1986): "Absorbing Boundary Conditions for Exterior Problems," in *Numerical Methods for Partial Differential Equations*, ed. by S. I. Hariharan, T. H. Moulden (Pitman)

Haurwitz, B., Craig, R. A. (1952): "Atmospheric Flow Patterns and Their Representation by Spherical Surface Harmonics, A.F.C.R.L. Geophysical Res. Pap. No. 14

Hendry, J. A., Delves L. M. (1979): The global element method applied to a harmonic mixed boundary value problem. J. Comput. Phys. **33**, 33–44

Herbert, T. (1977a): "Die Neutrale Fläche der Ebenen Poiseuille-Strömung," Habilitation, Univ. Stuttgart

Herbert, T. (1977b): "Finite Amplitude Stability of Plane Parallel Flows," in *Laminar-Turbulent Transition, AGARD Conference Proceedings No. 224*, Technical Editing and Reproduction Ltd, London

Herbert, T. (1983a): Stability of plane Poiseuille flow: theory, and experiment. Fluid Dyn. Trans. **11**, 77–126

Herbert, T. (1983b): Secondary instability of plane channel flow to subharmonic three-dimensional disturbances. Phys. Fluids **26**, 871–874

Herbert, T. (1984): "Analysis of the Subharmonic Route to Transition in Boundary Layers," AIAA Pap. No. 84-0009

Herbert, T. (1985): "Three-Dimensional Phenomena in the Transitional Flat-plate Boundary Layer," AIAA Pap. No. 85-0489

Herring, J. R., Orszag, S. A., Kraichnan, R. H., Fox, D. G. (1974): Decay of two-dimensional homogeneous turbulence. J. Fluid Mech. **66**, 417–444

Hestenes, M. R., Stiefel, E. (1952): Methods of conjugate gradients for solving linear systems. J. Res. Natl. Bur. Stand. **49**, 409–436

Heywood, J. C., Rannacher, R. (1986): Finite element approximation of the nonstationary Navier–Stokes problem, part II: stability of solutions and error estimates uniform in time. SIAM J. Numer. Anal. **23**, 750–777

Hicks, H. R., Carreras, B. A., Holmes, J. A., Lee, D. K., Waddell, B. V. (1981): 3-D nonlinear calculations of resistive tearing modes. J. Comput. Phys. **44**, 46–69

Hille, E., Phillips, R. S. (1957). *Functional Analysis and Semi-Groups*, Am. Math. Soc. (Providence, RI)

Hille, E., Tamarkin, J. D. (1935): On the absolute integrability of Fourier transforms. Fund. Math. **25**, 329–352

Hinze, J. O. (1975): *Turbulence* (McGraw-Hill, New York)

Hirsh, R. S., Taylor, T. D., Nadworny, M. M. (1983): An implicit predictor-corrector method for real space Chebyshev pseudospectral integration of parabolic equations. Comput. and Fluids **11**, 251–254

Hirsh, R. S., Taylor, T. D., Nadworny, M. M., Kerr, J. L. (1982): "Techniques for Efficient Implementation of Pseudo-Spectral Methods and Comparison with Finite Difference Solutions of the Navier–Stokes Equations, "in *Proc. 8th Int. Conf. Numerical Methods in Fluid*

Dynamics, ed. by E. Krause (Springer, Berlin, Heidelberg, New York), pp. 245–251
Holst, T. L. (1979): "A Fast, Conservative Algorithm for Solving the Transonic Full-Potential Equation," AIAA Pap. No. 79–1456
Holt, M. (1977): *Numerical Methods in Fluid Dynamics* (Springer, Berlin, Heidelberg, New York)
Hopf, E. (1950): The partial differential equation $u_t + uu_x = \mu u_{xx}$. Commun. Pure Appl. Math. **3**, 201–230
Huberson, S., Morchoisne, Y. (1983): "Large Eddy Simulation by Spectral Methods or by Multi-Level Particle Method," AIAA Pap. No. 83-1880
Hussaini, M. Y. (1986): "Some Recent Developments in Spectral Methods," in *Proc. 10th Int. Conf. Numerical Methods in Fluid Dynamics*, ed. by F. G. Zhuang and Y. L. Zhu, (Springer, Berlin, Heidelberg, New York) pp. 18–29
Hussaini, M. Y. (1987): "Stability, Transition and Turbulence," in *Supercomputing in Aerospace*, NASA Conf. Pub. 2454, pp. 211–220.
Hussaini, M. Y., Kopriva, D. A., Salas, M. D., Zang, T. A. (1985a): Spectral methods for the Euler equations, Part 1: Fourier methods and shock-capturing. AIAA J. **23**, 64–70
Hussaini, M. Y., Kopriva, D. A., Salas, M. D., Zang, T. A. (1985b): Spectral methods for the Euler equations: Part 2. Chebyshev methods and shock-fitting. AIAA J. **23**, 234–240
Hussaini, M. Y., Salas, M. D., Zang, T. A. (1985): "Spectral Methods for Inviscid, Compressible Flows," in *Advances in Computational Transonics*, ed. by W. G. Habashi (Pineridge, Swansea, UK) pp. 875–912
Hussaini, M. Y., Streett, C. L., Zang, T. A. (1983): "Spectral Methods for Partial Differential Equations," in *Trans. 1st Army Conf. Applied Mathematics and Computing*, pp. 883–925
Hussaini, M. Y., Zang, T. A. (1987): Spectral methods in fluid dynamics, Ann. Rev. Fl. Mech., **19**, 339–367
Ito, K., Teglas, R. (1986): Legendre–Tau approximations for functional differential equations. SIAM J. Control Optim. **24**, 737–759
Ito, K., Teglas, R. (1987): Legendre–Tau approximations for functional differential equations, part 2: The linear quadratic optimal control problem. SIAM J. Control Optim. (in press)
Jackson, D. (1930): *The Theory of Approximation*, Vol. 11 (AMS Colloquium Publications, New York)
Jameson, A. (1979): "Acceleration of Transonic Potential Flow Calculations on Arbitrary Meshes by the Multiple Grid Method," AIAA Pap. No. 79-1458
Jameson, A., Schmidt, H., Turkel, E. (1981): "Numerical Solutions of the Euler Equations by Finite Volume Methods Using Runge–Kutta Time Stepping Schemes," AIAA Pap. No. 81-1259
Jarraud, M., Baede, A. P. M. (1985): "The Use of Spectral Techniques in Numerical Weather Prediction," in *Large-scale Computations in Fluid Mechanics*, Lectures in Applied Mathematics **22**, 1–41
Kantorovic, L. V. (1934): On a new method of approximate solution of partial differential equations. Dokl. Akad. Nauk SSSR **4**, 532–536 (in Russian)
Karamzin, Y. N., Tsvetkova, I. L. (1982): On the convergence of the spectral method of solving a problem of nonlinear optics. USSR Comput. Math. Math. Phys. **22**, 251–257
Karniadakis, G. E., Bullister, E. T., Patera, A. T. (1986): "A Spectral Element Method for Solution of the Two- and Three-Dimensional Time-Dependent Incompressible Navier–Stokes Equations," in *Proc. Europe-U.S. Conf. Finite Element Methods for Nonlinear Problems*, ed. by P. Bergan, K. J. Bathe, Wunderlich (Springer, Berlin, Heidelberg, New York), pp. 803–817
Kasahara, A. (1978): Further studies on a spectral model of the global barotropic primitive equations with Hough harmonic expansions. J. Atmosph. Sci. **35**, 2043–2051
Kerr, R. M. (1985): Higher-order derivative correlations and the alignment of small-scale structures in isotropic numerical turbulence. J. Fluid Mech. **153**, 31–58
Kerr, R. M., Nakano, T., Nelkin, M. (1985): "Decay of a Coherent Scalar Disturbance in a Turbulent Flow," NASA TM-86700
Kim, J. (1985): "Evolution of a Vortical Structure Associated with the Bursting Event in a Channel Flow," in *Proc. 5th Symp. Turbulent Shear Flows* (Cornell Univ., Ithaca, New York)
Kim, J., Moin, P. (1985): Application of a fractional-step method to incompressible Navier–Stokes equations. J. Comput. Phys. **59**, 308–323
Kim, J., Moin, P. (1986): The structure of the vorticity field in turbulent channel flow, Part 2: Study of ensemble-averaged fields. J. Fluid Mech. **162**, 339–363
King, G. P., Li, Y., Lee, W., Swinney, H. L., Marcus, P. S. (1984): Wave speeds in wavy Taylor-vortex flow. J. Fluid Mech. **141**, 365–390

Kleiser, L. (1982a): "Spectral Simulation of Laminar-Turbulent Transition in Plane Poiseuille Flow and Comparison with Experiments," in *Proc. 8th Int. Conf. Numerical Methods in Fluid Dynamics*, ed. by E. Krause (Springer, Berlin, Heidelberg, New York) pp. 280–285
Kleiser, L. (1982b): "Numerische Simulationen zum Laminar-Turbulenten Umschlagsprozeß der Abenen Poiseuille-Strömung," Ph.D. Thesis, Univ. Karlsruhe, Kernforschungszentrum Karlsruhe, KLK 3271
Kleiser, L., Schumann, U. (1980): "Treatment of Incompressibility and Boundary Conditions in 3-D Numerical Spectral Simulations of Plane Channel Flows," in *Proc. 3rd GAMM Conf. Numerical Methods in Fluid Mechanics*, ed. by E. H. Hirschel (Vieweg, Braunschweig) pp. 165–173
Kleiser, L., Schumann, U. (1984): "Spectral Simulation of the Laminar-Turbulent Transition Process in Plane Poiseuille Flow," in *Spectral Methods for Partial Differential Equations*, ed. by R. G. Voigt, D. Gottlieb, M. Y. Hussaini (SIAM-CBMS, Philadelphia) pp. 141–163
Kopriva, D. A. (1986a): A spectral multidomain method for the solution of hyperbolic systems Appl. Numer. Math. **2**, 221–241.
Kopriva, D. A. (1986b): "A Multidomain Spectral Collocation Computation of the Sound Generated by a Shock-Vortex Interaction," in *Computational Acoustics and Wave Propagation*, ed. by M. Schultz, D. Lee, R. Sternberg (North-Holland, Amsterdam)
Kopriva, D. A. (1987): A practical assessment of spectral accuracy for hyperbolic problems with discontinuities. J. Sci. Comput. (in press)
Kopriva, D. A., Zang, T. A., Salas, M. D., Hussaini, M. Y. (1984): "Pseudospectral Solution of Two-dimensional Gas-Dynamics Problems," in *Proc. 5th GAMM Conf. Numerical Methods on Fluid Mechanics*, ed. by M. Pandolfi, R. Piva (Vieweg, Braunschweig) pp. 185–192
Korczak, K. Z., Patera, A. T. (1986): Isoparametric spectral element method for solution of the Navier–Stokes equations in complex geometry. J. Comput. Phys. **62**, 361–382
Kovasznay, L. S., Komoda, H., Vasudeva, B. R. (1962): "Detailed Flow Field in Transition," in *Proc. 1962 Heat Transfer and Fluid Mechanics Institute*, pp. 1–26
Kreiss, H.-O. (1970): Initial boundary value problems for hyperbolic systems. Commun. Pure Appl. Math. **23**, 277–298
Kreiss, H.-O., Oliger, J. (1972): Comparison of accurate methods for the integration of hyperbolic equations. Tellus, **24**, 199–215
Kreiss, H.-O. Oliger, J. (1979): Stability of the Fourier method. SIAM J. Numer. Anal. **16**, 421–433
Krist, S., Zang, T. A. (1987): "Numerical Simulation of Channel Flow Transition: Resolution Requirements and the Structure of the Hairpin Vortex," NASA TP-2667
Ku, H. C., Taylor, T. D., Hirsh, R. S. (1987): Pseudospectral methods for solution of the incompressible Navier–Stokes equations. Comput. Fluids **15**, 195–214
Kubota, S. (1959): Surface spherical harmonic representations of the system of equations for analysis. Pap. Meteorol. Geophys. **10**, 145–166
Kubota, S., Hirose, M., Kikuchi, Y., Kurihara, Y. (1961): Barotropic forecasting with the use of surface spherical harmonic representation. Pap. Meteorol. Geophys. **12**, 199–215
Kuznetsov, J. A., Matsokin, A. M. (1972): Solution of the Helmholtz equation by the method of fictitious domains, Computational methods of linear algebra. Akad. Nauk SSSR **36**, 127–145
Ladyzhenskaya, O. A. (1969): *The Mathematical Theory of Viscous Incompressible Flow* (Gordon & Breach, New York)
Lambert, J. D. (1973): *Computational Methods in Ordinary Differential Equations* (John Wiley & Sons, New York)
Lambiotte, J., Bokhari, S., Hussaini, M. Y., Orszag, S. A. (1982): "Navier–Stokes Solutions on the Cyber-203 by a Pseudospectral Technique," in *10th IMACS World Congr. System Simulation and Scientific Computation* (Montreal, Can.)
Lanczos, C. (1938): Trigonometric interpolation of empirical and analytical functions. J. Math. Phys. **17**, 123–199
Lanczos, C. (1973): "Legendre Versus Chebyshev Polynomials," in *Topics in Numerical Analysis*, ed. by J. H. Miller (Academic Press, London, New York) pp. 191–201
Laurien, E. (1986): "Numerische Simulation zur Aktiven Beeinflussung des Laminar-Turbulenten Übergangs in der Plattengrenzschichtströmung", DFVLR-FB 86-05
Lax, P. D. (1978): "Accuracy and Resolution in the Computation of Solutions of Linear and

Nonlinear Equations," in *Recent Advances in Numerical Analysis* (Academic Press, London, New York) pp. 107–117

Lax, P. D., Wendroff, B. (1960): Systems of conservation laws. Commun. Pure Appl. Math. **13**, 217–237

Lee, M. J., Reynolds, W. C. (1985): "On the Structure of Homogeneous Turbulence," in *Proc. 5th Symp. Turbulent Shear Flows* (Cornell Univ., Ithaca, New York)

Leonard, A. (1984): "Numerical Simulation of Turbulent Fluid Flows," NASA TM-84320

Leonard, A., Wray, A. (1982): "New Numerical Method for the Simulation of Three-Dimensional Flow in a Pipe," in *Proc. 8th Int. Conf. Numerical Methods in Fluid Dynamics*, ed. by E. Krause (Springer, Berlin, Heidelberg, New York), pp. 335–342

Leorat, J., Pouquet, A., Poyet, J. P., Passot, T. (1985): "Spectral Simulations of 2D Compressible Flows," in *Proc. 9th Int. Conf. Numerical Methods in Fluid Dynamics*, ed. by Soubbarameyer, J. P. Boujet (Springer, Berlin, Heidelberg, New York), pp. 369–374

LeQuere, P., Alziary De Roquefort, T. (1985): Computation of natural convection in two-dimensional cavities with Chebyshev polynomials. J. Comput. Phys. **57**, 210–228

LeQuere, P., Pecheux, J. (1986): "Simulation numérique des écoulements de convection thermique dans un tore de section rectangulaine en rotation," in *Proc. 6th Int. Symp. Finite Element* Methods in Flow Problems, ed. by M. O. Bristeau, R. Glowinski, A. Haugel, J. Periaux, pp. 219–223

Lerat, A. (1979): Une classe de schémas implicites pour les systèmes hyperboliques de lois de conservation C. R. Acad. Sci. Paris **288**, serie A, 1033–1036

Lerat, A. (1983): "Implicit Methods of Second-Order Accuracy for the Euler Equations," AIAA Pap. No. 83-1925

Leray, J. (1933): Étude de diverse équations, intégrales non linéaires et de quelques problèmes que pose l'hydrodynamique. J. Math. Pure Appl. **12**, 1–82

Liebovitch, L. S. (1978): "Two-Dimensional Calculations of Gas Flow in Barred Spiral Galaxies," Ph.D. Thesis, Harvard Univ., Cambridge, MA.

Liepmann, H. W., Roshko, A. (1957): *Elements of Gas Dynamics* (John Wiley & Sons, New York)

Lighthill (1956): "Viscosity Effects in Sound Waves of Finite Amplitude," in *Surveys in Mechanics*, ed. by G. K. Batchelor, R. Davies (Cambridge Univ. Press, Cambridge)

Lin, C. C. (1955): *The Theory of Hydrodynamics Stability* (Cambridge Univ. Press, Cambridge)

Lions, J. L. (1969): *Quelques Methodes de Résolution des Problèmes aux Limites non Linéaires* (Dunod, Paris).

Lions, J. L., Magenes, E. (1972): *Nonhomogeneous Boundary Value Problems and Applications*, Vol. 1 (Springer, Berlin, Heidelberg, New York)

Lorenz, E. N. (1960): Maximum simplification of the dynamic equations. Tellus **12**, 243–254

Lustman, L. (1986): The time evolution of spectral discretization of hyperbolic equation. SIAM J. Numer. Anal. **23**, 1193–1198

Lynch, R. E., Rice, J. R., Thomas, D. H. (1964): Direct solution of partial difference equations by tensor product methods. Numer. Math. **6**, 185–199

MacCormack (1969): "The Effect of Viscosity in Hypervelocity Impact Cratering," AIAA Pap. No. 69-354

Mac Giolla Mhuiris, N. (1986): "Numerical Calculations of the Stability of Some Axisymmetric Flows Proposed as a Model of Vortex Breakdown," Ph.D. Thesis, Cornell Univ., Ithaca, NY.

Macaraeg, M. C. (1983): "Numerical Solution of the Axisymmetric Flow in a Heated Rotating Spherical Shell," Ph.D. Thesis, Atmospheric Sciences Division, UTSI, Tullahoma, TN.

Macaraeg, M. C. (1986): "A Spectral Multi-Domain Technique with Application to Generalized Curvilinear Coordinates," in *Proc. 6th Int. Symp. Finite Element Methods in Flow Problems*, ed. by M. O. Bristeau, R. Glowinski, A. Haugel, J. Periaux pp. 231–238

Macaraeg, M., Streett, C. L. (1986): Improvements in spectral collocation through a multiple domain technique. Appl. Numer. Math. **2**, 95–108

Machenhauer, B. (1973): "On the Use of the Spectral Method in Numerical Integrations of Atmospheric Models," in *Proc. Symp. Difference and Spectral Methods for Atmosphere and Ocean Dynamics Problems* (USSR Acad. Sci. Siberian Branch, Novosibirsk)

Machenhauer, B., Rasmussen, E. (1972): "On the Integration of the Spectral Hydrodynamical Equations by a Transform Method," Rep. No. 3 (Institute for Theoretical Meteorology, Univ. Copenhagen)

Maday, Y. (1981): "Sur Quelques Propriétes des Approximations par des Méthodes Spectrales

dans les Espaces de Sobolev à Poids, Applications à la Résolution de Problèmes non Linéaires," Thése de Troisieme cycle, Univ. Paris VI
Maday, Y. (1987a): Analysis of spectral operators in one dimensional domains. Math. Comput., in press.
Maday, Y. (1987b): Contributions à l'Analyse Numérique des Méthodes Spectrales, Thése de Doctorate, Univ. Paris 6
Maday, Y., Métivet, B. (1983): Error estimates for spectral approximation of Stokes equations. Rech. Aerosp. **1983–4**, 21–28
Maday, Y., Métivet, B. (1986): "Chebyshev spectral approximation of Navier–Stokes equation in a two dimensional domain (Lab. d'Anal. Numer. Paris 6)
Maday, Y., Pernaud-Thomas, B., Vandeven, H. (1985): Reappraisal of Laguerre type spectral methods. Rech. Aerosp. **1985–6**, 13–35
Maday, Y., Quarteroni, A. (1981): Legendre and Chebyshev spectral approximations of Burgers' equation. Numer. Math. **37**, 321–332
Maday, Y., Quarteroni, A. (1982a): Approximation of Burgers' equation by pseudospectral methods R.A.I.R.O. Anal. Numer. **16**, 375–404
Maday, Y., Quarteroni, A. (1982b): Spectral and pseudospectral approximations of Navier–Stokes equations. SIAM J. Numer. Anal. **19**, 761–780
Majda, A., McDonough, J., Osher, S. (1978): The Fourier method for nonsmooth initial data Math. Comput. **32**, 1041–1081
Malik, M. R., Zang, T. A., Hussaini, M. Y. (1985): A spectral collocation method for the Navier–Stokes equations. J. Comput. Phys. **61**, 64–88
Manteuffel, T. A. (1977): The Tchebychev iteration for nonsymmetric linear systems. Numer. Math. **28**, 307–327
Manteuffel, T. A. (1978): Adaptive procedure for estimating parameters for the nonsymmetric Tchebychev iteration. Numer. Math. **31**, 183–208
Marchuk, G. I. (1975): *Methods of Numerical Mathematics* (Springer, Berlin, Heidelberg, New York)
Marcus, P. S. (1981): Effects of truncation on modal representations of thermal convection. J. Fluid Mech. **103**, 241–255
Marcus, P. S. (1984a): Simulation of Taylor–Couette flow. Part 1. Numerical methods and comparison with experiment. J. Fluid Mech. **146**, 45–64
Marcus, P. S. (1984b): Simulation of Taylor–Couette flow. Part 2. Numerical results for wavy-vortex flow with one traveling wave. J. Fluid Mech. **146**, 65–113
Marcus, P. S., Orszag, S. A., Patera, A. T. (1982): "Simulation of Cylindrical Couette Flow", in *Proc. 8th Int. Conf. Numerical Methods in Fluid Dynamics*, ed. by E. Krause (Springer, Berlin, Heidelberg, New York), pp. 371–376
Marcus, P. S., Tuckerman, L. S. (1987a): Simulation of flow between concentric rotating spheres. Part 1. Steady states. J. Fluid Mech. (in press)
Marcus, P. S., Tuckerman, L. S. (1987b): Simulation of flow between concentric rotating spheres, Part 2: Transitions. J. Fluid Mech. (in press)
Marion. Y., Gay, B. G. (1986): "Résolution des équations de Navier–Stokes par méthode pseudo-spectrale via une technique de coordination," in *Proc. 6th Int. Symp. Finite Element Methods in Flow Problems*, ed. by M. O. Bristeau, R. Glowinski, A. Haugel, J. Periaux, pp. 239–243
Matthaeus, W. M., Montgomery, D. (1981): Nonlinear evolution of the sheet pinch. J. Plasma Phys. **25**, 11–41
McCrory, R. L., Orszag, S. A. (1980): Spectral methods for multidimensional diffusion problems. J. Comput. Phys. **37**, 93–112
McKerrell, A., Phillips, C., Delves, L. M. (1981): Chebyshev expansion methods for the solution of elliptic partial differential equations. J. Comput. Phys. **40**, 444–452
McLaughlin, J. B., Orszag, S. A. (1982): Transition from periodic to chaotic thermal convection. J. Fluid Mech. **122**, 123–142
McMillan, O. J., Ferziger, J. H. (1979): Direct testing of subgrid-scale models. AIAA J. **17**, 1340–1346
McMurtry, P. A., Jou, W.-H., Riley, J. J., Metcalfe, R. W. (1986): Direct numerical simulations of a reacting mixing layer with chemical heat release. AIAA J., **24**, 962–970
Meijerink, J. A., Van der Vorst, H. A. (1981): Guidelines for the usage of incomplete decompositions in solving sets of linear equations as they occur in practical problems. J. Comput. Phys. **44**, 134–155.

Meiron, D. I., Orszag, S. A., Israeli, M. (1981): Applications of numerical conformal mapping. J. Comput. Phys. **40**, 345–360

Mercier, B. (1981): "Analyse Numérique des Méthodes Spectrales," Note CEA-N-2278 (Commissariat a l'Énergie Atomique Centre d'Études de Limeil, 94190 Villeneuve-Saint Georges)

Mercier, B. (1982): Stabilité et convergence des méthodes spéctrales polynômiales: application a l'équation d'advection. R.A.I.R.O. Anal. Numer. **16**, 67–100

Mercier, B., Raugel, G. (1982): Resolution d'un problème aux limites dans un ouvert axisymétrique par éléments finis en r, z et séries de Fourier en théta. R.A.I.R.O. Anal. Numer. **16**, 405–461

Merilees, P. E. (1966): Harmonic representation applied to large scale atmospheric waves A.M.R.G. Publ. Meteorol. **83** (McGill Univ.)

Merilees, P. E. (1968): The equations of motion in spectral form. J. Atmosph. Sci. **25**, 736–743

Merilees, P. E. (1972): Truncation error in a spectral model. Atmosphere **10**, 1–9

Merilees, P. E. (1973): An alternative scheme for the summation of a series of spherical harmonics. J. Appl. Meteorol. **132**, 224–227

Merilees, P. E. (1974): Numerical experiments with the pseudospectral method in spherical coordinates. Atmosphere **12**, 77–96

Merilees, P. E., Ducharme, P., Jacques, G. (1977): Experiments with a polar filter and a one-dimensional semi-implicit algorithm. Atmosphere **15**, 19–32

Mestayer, P. G., Gibson, C. H., Coantic, M. F., Patel, A. S. (1970): Local anisotropy in heated and cooled turbulent boundary layers. Phys. Fluid **19**, 1279–1287

Metcalfe, R. W., Orszag, S. A., Brachet, M. E., Menon, S., Riley J. J. (1987): Secondary instability of a temporally growing mixing layer. J. Fluid Mech. (in press)

Metcalfe, R. W., Rutland, C. J., Duncan, J. H., Riley, J. J. (1986): Numerical simulations of active stabilizations of laminar boundary layers. AIAA J. **24**, 1494–1501

Métivet, B. (1987): "Résolution Spectrale des Équations de Navier–Stokes par une Méthodes de Sous-Domaines Courbes," Thèse de Doctorate, Univ. Paris 6

Métivet, B. Morchoisne, Y. (1982): "Multi-Domain Spectral Technique for Viscous Flow Calculation," in *Proc. 4th GAMM Conf. Numerial Methods in Fluid Mechanics*, ed. by H. Viviand (Vieweg, Braunschweig) pp. 207–219

Miketinac, M. J., Parter, S. V. (1981): Numerical computation of certain three-dimensional stellar structures using a semidiscrete pseudospectral method. J. Appl. Math. Phys. (ZAMP) **32**, 204–228

Miller, K. (1965): Numerical analogs of the Schwarz alternating procedure. Numer. Math. **7**, 91–103

Milne-Thomson, L. M. (1966): *Theoretical Aerodynamics* (MacMillan, New York)

Moin, P., Kim, J. (1980): On the numerical solution of time-dependent viscous incompressible fluid flows involving solid boundaries. J. Comput. Phys. **35**, 381–392

Moin, P., Kim, J. (1982): Numerical investigation of turbulent channel flow. J. Fluid Mech. **118**, 341–377

Moin P., Kim, J. (1985): The structure of the vorticity field in turbulent channel flow. Part 1. Analysis of instantaneous fields and statistical correlations. J. Fluid Mech. **155**, 441–464

Moin, P., Rogers, M. M., Moser, R. D. (1985): "Structure of Turbulence in the Presence of Uniform Shear," in *Proc. 5th Symp. Turbulent Shear Flows* (Cornell Univ., Ithaca, NY.)

Montigny–Rannou, F. (1982): Effect of "aliasing" on spectral method solution of Navier–Stokes equations. Rech. Aerosp. **1982-2**, 36–41

Montigny–Rannou, F. (1984): "Influence of Compatibility Conditions in Numerical Simulation of Inhomogeneous Incompressible Flows," in *Proc. 5th GAMM Conf. Numerical Methods in Fluid Mechanics*, ed. by M. Pandolfi, R. Piva (Vieweg, Braunschweig) pp. 234–242

Montigny–Rannou, F., Morchoisne, Y. (1987): A spectral method with staggered grid for incompressible Navier–Stokes equations. Int. J. Numer. Meth. Fluids **7**, 175–189

Morchoisne, Y. (1979): Resolution of Navier–Stokes equations by a space-time pseudospectral method. Rech. Aerosp. **1979-5**, 293–306

Morchoisne, Y. (1981a): "Pseudo-spectral Space-Time Calculations of Incompressible Viscous Flows," AIAA Pap. No. 81-0109

Morchoisne, Y. (1981b): "Pseudo-Spectral Methods for Homogeneous or Inhomogeneous Flows," in *Proc. 3rd Symp. Turbulent Shear Flows* (Davis, CA)

Morchoisne, Y. (1983): "Résolution des Équations de Navier–Stokes par une Méthode Spectrale de Sous-Domaines," in *Proc. 3rd Int. Conf. Num. Meth. Sci. Eng.* (Gamni, Paris)

Morchoisne, Y. (1984): "Inhomogeneous Flow Calculations by Spectral Methods: Mono-Domain and Multi-Domain Techniques," in *Spectral Methods for Partial Differential Equations*, ed. by R. G. Voigt, D. Gottlieb, M. Y. Hussaini (SIAM-CBMS) pp. 181–208
Moretti, G. (1968): "Inviscid Blunt Body Shock Layers," PIBAL Rep. No. 68-15 (Polytech. Inst. Brooklyn, New York)
Moretti, G. (1972): "Thoughts and Afterthoughts About Shock Computations," PIBAL Rep. No. 72-37 (Polytech. Inst. Brooklyn, New York)
Morf, R. H., Orszag, S. A., Frisch, U. (1980): Spontaneous singularity in three-dimensional, inviscid, incompressible flow. Phys. Rev. Lett. **44**, 572–575
Moser, R. D., Moin P. (1987): The effects of curvature in wall-bounded turbulent flows. J. Fluid Mech. **175**, 479–510
Moser, R. D., Moin, P., Leonard, A. (1983): A spectral numerical method for the Navier–Stokes equations with applications to Taylor–Couette flow. J. Comput. Phys. **52**, 524–544
Murdock, J. W. (1977): A numerical study of nonlinear effects on boundary-layer stability. AIAA J. **15**, 1167–1173
Murdock, J. W. (1986): "Three-Dimensional Numerical Study of Boundary-Layer Stability," AIAA Pap. No. 86-0434
Murman, E. M., Cole, J. D. (1971): Calculation of plane steady transonic flow. AIAA J. **9**, 114–121
NASA conference Publication 2201 (1982)
Nayfeh, A. H. (1980): Stability of three-dimensional boundary layers. AIAA J. **18**, 406–416
Necas, J. (1962): Sur une méthode pour resoudre les equations aux derivées partielles du type elliptique, voisine de la variationnelle. Ann. Sc. Norm. Sup. Pisa **16**, 305–326
Nikolskii, S. M. (1951): Inequalities for entire functions of finite degree and their application to the theory of differentiable functions of several variables. Dokl. Akad. Nauk. SSSR **58**, 244–278 (in Russian)
Nikolskii, S. M. (1975): *Approximation of Functions of Several Variables and Imbedding Theorems* (Springer, Berlin, Heidelberg, New York)
Nishioka, M., Asai, M., Iida, S. (1980): "An Experimental Investigation of the Secondary Instability in Laminar-Turbulent Transition," in *Laminar-Turbulent Transition*, ed. by R. Eppler, H. Fasel (Springer, Berlin, Heidelberg, New York), pp. 37–46
Oliger, J., Sundström, A. (1978): Theoretical and practical aspects of some initial boundary value problems in fluid dynamics. SIAM J. Appl. Math. **35**, 419–446
Orszag, S. A. (1969): Numerical methods for the simulation of turbulence. Phys. Fluids, Suppl. II, **12**, 250–257
Orszag, S. A. (1970): Transform method for calculation of vector coupled sums: Application to the spectral form of the vorticity equation. J. Atmosph. Sci. **27**, 890–895
Orszag, S. A. (1971a): On the elimination of aliasing in finite-difference schemes by filtering high-wavenumber components. J. Atmosph. Sci. **28**, 1074
Orszag, S. A. (1971b): Numerical simulations of incompressible flows within simple boundaries: accuracy. J. Fluid Mech. **49**, 75–112
Orszag, S. A. (1971c): Galerkin approximations to flows within slabs, spheres, and cylinders. Phys. Rev. Lett. **26**, 1100–1103
Orszag, S. A. (1971d): Numerical simulation of incompressible flows within simple boundaries: I. Galerkin (spectral) representations. Stud. Appl. Math. **50**, 293–327
Orszag, S. A. (1971e): Accurate solution of the Orr-Sommerfeld stability equation. J. Fluid Mech. **50**, 689–703
Orszag, S. A. (1972): Comparison of pseudospectral and spectral approximations. Stud. Appl. Math. **51**, 253–259
Orszag, S. A. (1974): Fourier series on spheres. Mon. Weather Rev. **102**, 56–75
Orszag, S. A. (1976): "Turbulence and Transition: A Progress Report," in *Proc. 5th Int. Conf. Numerical Fluid Dynamics Proceedings*, ed. by A. I. Van der Vooren, P. J. Zandbergen (Springer, Berlin, Heidelberg, New York), pp. 39–51
Orszag, S. A. (1980): Spectral methods for problems in complex geometries. J. Comput. Phys. **37**, 70–92
Orszag, S. A. (1986): Fast Eigenfunction Transforms," in *Science and Computers*, Advances in Mathematics Supplementary Series, ed. by G. C. Rota (Academic Press, London, New York), pp. 23–30

Orszag, S. A., Gottlieb, D. (1980) "High Resolution Spectral Calculations of Inviscid Compressible Flows," in *Approximation Methods for Navier–Stokes Problems* (Springer, Berlin, Heidelberg, New York), pp. 381–398

Orszag, S. A., Israeli, M. (1974): Numerical simulation of viscous incompressible flows. Ann. Rev. Fluid Mech. **6**, 281–318

Orszag, S. A., Israeli, M., Deville, M. O. (1986): Boundary conditions for incompressible flows. J. Sci. Comput. **1**, 75–111

Orszag, S. A., Kells, L. C. (1980): Transition to turbulence in plane Poiseuille flow and plane Couette flow. J. Fluid Mech. **96**, 159–205

Orszag, S. A., Pao, Y. H. (1974): "Numerical Computation of Turbulent Shear Flows," in *Proc. Symp. Turbulent Diffusion in Environmental Pollution*, ed. by F. N. Frenkiel, R. E. Mann (Academic Press, London, New York)

Orszag, S. A., Patera, A. T. (1980): Subcritical transition to turbulence in plane channel flows. Phys. Rev. Lett. **45**, 989–993

Orszag, S. A., Patera, A. T. (1981a): "Three-Dimensional Instability of Plane Channel Flows at Subcritical Reynolds Numbers," in *Proc. Symp. Numerical and Physical Aspects of Aerodynamic Flows* (Long Beach, CA)

Orszag, S. A., Patera, A. T. (1981b): "Subcritical Transition to Turbulence in Planar Shear Flows," in *Transition and Turbulence*, ed. by R. E. Meyer (Academic Press, London, New York), pp. 127–146

Orszag, S. A., Patera, A. T. (1981c): Calculation of Von Karman's constant for turbulent channel flow. Phys. Rev. Lett. **47**, 832–835

Orszag, S. A., Patera, A. T. (1983): Secondary instability of wall-bounded shear flows. J. Fluid Mech. **128**, 347–385

Orszag, S. A., Patera, A. T., Balasubramanian, R. (1983): "Spectral Methods for Flows in Complex Geometries," AIAA Pap. No. 83–0229

Orszag, S. A., Patterson, G. S., Jr. (1972a): "Numerical Simulation of Turbulence," in *Statistical Models and Turbulence* (Springer, Berlin, Heidelberg, New York), pp. 127–147

Orszag, S. A., Patterson, G. S., Jr. (1972b): Numerical simulation of three dimensional homogeneous isotropic turbulence. Phys. Rev. Lett. **28**, 76–79

Orszag, S. A., Tang, C. M. (1979): Small-scale structure of two-dimensional magnetohydrodynamic turbulence. J. Fluid Mech. **90**, 129–143

Ortega, J. M., Voigt, R. G. (1985): Solution of partial differential equations on vector and parallel computers. SIAM Rev. **27**, 149–240

Ortiz, E. L., Samara, H. (1983): Numerical solution of differential eigenvalue problems with an operational approach to the tau method. Computing **31**, 95–103

Osher, S. (1969): Systems of difference equations with general homogeneous boundary conditions. Trans. Am. Math. Soc. **137**, 177–201

Osher, S. (1984): "Smoothing for Spectral Methods," in *Spectral Methods for Partial Differential Equations*, ed. by R. G. Voigt, D. Gottlieb, M. Y. Hussaini (SIAM-CBMS Philadelphia), pp. 209–216

Ouazzani, J., Peyret, R. (1984): "A Pseudo-Spectral Solution of Binary Gas Mixture Flows," in *Proc. 5th GAMM Conf. Numerical Methods in Fluid Mechanics*, ed. by M. Pandolfi, R. Piva (Vieweg, Braunschweig), pp. 225–282

Ouazzani, J., Peyret, R., Zakaria, A. (1985): "Stability of Collocation-Chebyshev Schemes with Application to the Navier–Stokes Equations," in *Proc. 6th GAMM Conf. Numerical Methods in Fluid Mechanics*, ed. by D. Rues, W. Kordulla (Vieweg, Braunschweig) pp. 287–294

Pasciak, J. E. (1980): Spectral and pseudospectral methods for advection equations. Math. Comput. **35**, 1081–1092

Pasciak, J. E. (1982): Spectral methods for a nonlinear initial value problem involving pseudodifferential operators. SIAM J. Numer. Anal. **19**, 142–154

Pasciak, J. E. (1984): "Spectral Methods for Atmospheric Laser Calculations," in *Spectral Methods for Partial Differential Equations*, ed. by R. G. Voigt, D. Gottlieb, M. Y. Hussaini (SIAM-CBMS, Philadelphia), pp. 217–238

Pasquarelli, F., Quarteroni, A., Sacchi–Landriani, G. (1987): "Spectral Approximations of the Stokes Problem by Divergence Free Functions," IAN-CNR Rep. No. 559 (I. A. N., Pavia Univ.)

Patera, A. T. (1984): A spectral element method for fluid dynamics: laminar flow in a channel expansion. J. Comput. Phys. **54**, 468–488

Patera, A. T. (1986): Fast direct Poisson solvers for high-order finite element discretizations in rectangularly decomposable domains. J. Comput. Phys. **65**, 474–480

Patera, A. T., Orszag, S. A. (1981): Finite-amplitude stability of axisymmetric pipe flow. J. Fluid Mech. **112**, 467–474

Patera, A. T., Orszag, S. A. (1980): "Transition and Turbulence in Plane Channel Flows," in *Proc. 7th Int. Conf. Numerical Methods in Fluid Dynamics*, ed. by R. W. MacCormack, W. C. Reynolds (Springer, Berlin, Heidelberg, New York), pp. 329–335

Patterson, G. S., Jr., Orszag, S. A. (1971): Spectral calculations of isotropic turbulence: Efficient removal of aliasing interaction. Phys. Fluids **14**, 2538–2541

Peyret, R. (1986): "Introduction to Spectral Methods," von Karman Institute Lecture Series 1986-04, Rhode–Saint Genese, Belgium

Phillips, T. N., Zang, T. A., Hussaini, M. Y. (1986): Preconditioners for the spectral multigrid method. IMA J. Numer. Anal. **6**, 273–292

Platzman, G. W. (1960): The spectral form of the vorticity equation. J. Meteorol. **17**, 635–644

Platzman, G. W. (1961): An approximation to the product of discrete functions. J. Meteorol. **18**, 31–37

Pouquet, A., Patterson, G. S., Jr. (1978): Numerical simulation of helical magnetohydrodynamic turbulence. J. Fluid Mech. **85**, 305–323

Prandtl, L. (1904): "Über Flüssigkeitsbewegung bei Sehr Kleiner Reibung," in *Proc. 3rd Int. Mathematics Congr.* (Heidelberg)

Przmieniecki, J. S. (1963): Matrix structural analysis of sub-structures. AIAA J. **1**, 138–147

Quarteroni, A. (1982): "Theoretical and Computational Aspects of Spectral Methods," in *Computing Methods in Applied Sciences and Engineering V*, ed. by R. Glowinski, J. L. Lions (North-Holland, Amsterdam) pp. 325–345

Quarteroni, A. (1983): "Theoretical Motivations Underlying Spectral Methods," in *Numerical Solutions of Nonlinear Problems* (Inria, Rocquencourt), pp. 79–92

Quarteroni, A. (1984): Some results of Bernstein and Jackson type for polynomial approximation in L_p-spaces. Jpn. J. Appl. Math. **1**, 173–181

Quarteroni, A. (1985): "Approximation Theory and Analysis of Spectral Methods," in *Multivariate Approximation Theory III*, ed. by W. Schempp, K. Zeller (Birkhäuser, Basel), pp. 322–331

Quarteroni, A. (1986a): "Gas Transient Simulations with Spectral Methods," in *Computing Methods in Applied Science and Engineering*, ed. by R. Glowinski, J. L. Lions (North-Holland, Amsterdam), pp. 123–136

Quarteroni, A. (1986b): "Semi-Implicit Time Advancing Schemes for Spectral Methods," in *Proc. INRIA School on Spectral Methods*, ed. by R. Temam

Quarteroni, A. (1987a): Blending Fourier and Chebyshev interpolation. J. Approx. Theor. (in press)

Quarteroni, A. (1987b): Spectral methods for pseudo-parabolic equations. SIAM J. Numer. Anal. **24**, 323–335

Reid, J. K. (1971): "On the Method of Conjugate Gradients for the Solution of Large Sparse Systems of Linear Equations," in *Large Sparse Sets of Linear Equations*, ed. by J. K. Reid (Academic Press, London, New York), pp. 231–254

Reif, J., Barakat, R. (1977): Numerical solution of the Fokker–Planck equation via Chebyshev polynomial approximations with reference to first passage time probability density functions. J. Comput. Phys. **23**, 425–445

Reyna, L. G. M. (1982): "Stability of Chebyshev Collocation," Ph.D. Thesis, Caltech, Pasadena, CA

Richardson, L. F. (1910): The approximate solution by finite differences of physical problems involving differential equations, with an application to the stresses in a masonry dam. Philos. Trans. R. Soc. London Ser. A **210**, 307–357

Richtmyer, R. D. (1978): *Principles of Advanced Mathematical Physics*, Vol. 1 (Springer, Berlin, Heidelberg, New York)

Rienecker, M. M., Fenton, J. D. (1981): A Fourier approximation method for steady water waves. J. Fluid Mech. **104**, 119–137

Riley, J. J., Metcalfe, R. W. (1980): "Direct Numerical Simulation of a Perturbed, Turbulent Mixing Layer," AIAA Pap. No. 80-0274

Riley, J. J., Metcalfe, R. W., Orszag, S. A. (1986): Direct numerical simulations of chemically reacting turbulent mixing layers. Phys. Fluids **29**, 406–422

Ringleb, F. (1940): Exakte Lösungen der Differentialgleichungen einer adiabatischen Gasströmung. Z. Angew. Math. Mech. **20**, 185–198

Rivlin, T. J. (1974): *The Chebyshev Polynomials* (John Wiley & Sons, New York)

Robert, A. J. (1966): The integration of a low order spectral form of the primitive meteorological equations. J. Meteorol. Soc. Jpn. **44**, 237–244

Rodrigue, G., Saylor, P. (1985): "Inner/Outer Iterative Methods and Numerical Schwarz Algorithm II," in *Proc. IBM Conf. Vector and Parallel Processors for Scientific Computations* (Rome)

Rodrigue, G., Simon, J. (1984): "A Generalization of the Numerical Schwarz Algorithm," in *Computing Methods in Applied Sciences and Engineering VI*, ed. by R. Glowinski, J. L. Lions (North-Holland, Amsterdam)

Rogallo, R. S. (1977): "An ILLIAC Program for the Numerical Simulation of Homogeneous, Incompressible Turbulence," NASA TM-73203

Rogallo, R. S. (1981): "Numerical Experiments in Homogeneous Turbulence," NASA TM-81315

Rogallo, R. S. Moin, P. (1984): Numerical simulation of turbulent flows. Ann. Rev. Fluid Mech. **16**, 99–137

Rosenhead, L. (ed.) (1963): *Laminar Boundary Layers* (Clarendon, Oxford)

Roy, P. (1980): Solution of Navier-Stokes equations by a method of high accuracy in space and time. Rech. Aerosp. **1980–6**, 3–15

Roy, P. (1982): "Numerical Simulation of Homogeneous Anisotropic Turbulence," in *Proc. 8th Int. Conf. Numerical Methods in Fluid Dynamics*, ed. by E. Krause (Springer, Berlin, Heidelberg, New York) pp. 440–447

Roy, P. (1984): "Numerical Simulation of Homogeneous Turbulence Submitted to Two Successive Plane Strains and to Solid Body Rotation," ONERA Rep. No. 1984-4, Chatillon, France

Royden, H. L. (1968): *Real Analysis* (McMillan, New York)

Rudin, W. (1966): *Real and Complex Analysis*, (McGraw–Hill, New York)

Saad, Y. Schultz, M. H. (1983): "GMRES: A Generalized Minimum Residual Algorithm for Solving Nonsymmetric Linear Systems," Res. Rep. No. YALEU/DCS/RR-254 (Yale Univ., New Haven, CT)

Sacchi–Landriani, G. (1986): "Spectral Tau Approximation of the Two-Dimensional Stokes Problem," IAN-CNR Rep. No. 528 (I. A. N., Pavia Univ.)

Sacchi–Landriani, G. (1987): Convergence of the Kleiser–Schumann Method for the Navier–Stokes equations. Calcolo (in press)

Sacchi–Landriani, G., Vandeven, H. (1987): Approximation polynômiale de functions à divergence nulle. C. R. Acad. Sci. Paris **304**, Serie I, 87–90.

Sakell, L. (1982): "Solution to the Euler Equation of Motion, Pseudospectral Techniques," in *Proc. 10th IMACS World Congr. System, Simul. and Sci. Comput.*, ed. by R. Stepleman (North-Holland, Amsterdam)

Sakell, L. (1984): Pseudospectral solutions of one- and two-dimensional inviscid flows with shock waves. AIAA J. **22**, 929–934

Salas, M. D., Zang, T. A., Hussaini, M. Y. (1982): "Shock-Fitted Euler Solutions to Shock-Vortex Interactions," in *Proc. 8th Int. Conf. Numerical Methods in Fluid Dynamics*, ed. by E. Krause (Springer, Berlin, Heidelberg, New York), pp. 461–467

Saric, W. S., Kozlov, V. V., Levchenko V. Y. (1984): "Forced and Unforced Subharmonic Resonance in Boundary-layer Transition." AIAA Pap. No. 84-0007

Schamel, H., Elsässer, K. (1976): The application of the spectral methods to nonlinear wave propagation. J. Comput. Phys. **22**, 501–516

Schlichting (1979): *Boundary Layer Theory* (McGraw-Hill, New York)

Schnack, D. D., Baxter, D. C., Caramana, E. J. (1984): A pseudospectral algorithm for three-dimensional magnetohydrodynamic simulation. J. Comput. Phys. **55**, 485–514

Schnack, D., Killeen, J. (1980): Nonlinear two-dimensional magnetohydrodynamic calculations. J. Comput. Phys. **35**, 110–145

Schumann, U. (1975): Subgrid scale model for finite difference simulations of turbulent flows in plane channels and annuli. J. Comput. Phys. **18**, 376–404

Schumann, U. (1976): Numerical simulation of the transition from three- to two-dimensional turbulence under a uniform magnetic field. J. Fluid Mech. **74**, 31–58

Schumann, U. (1980): "Fast Elliptic Solvers and Their Application in Fluid Dynamics," in *Computational Fluid Dynamics*, ed. by W. Kollmann (Hemisphere, Washington) pp. 401–430

Schumann, U. (1985): "Algorithms for Direct Numerical Solution of Shear-Periodic Turbulence," in *Proc. 9th Int. Conf. Numerical Methods in Fluid Dynamics*, ed. by Soubbarameyer, J. Boujot (Springer, Berlin, Heidelberg, New York), pp. 492–496
Schumann, U., Grötzbach, G., Kleiser, L. (1980): "Direct Numerical Simulation of Turbulence," in *Prediction Methods for Turbulent Flows*, ed. by W. Kollman (Hemisphere, Washington), pp. 123–258
Schwarz, H. A. (1890): *Gesammelte Mathematische Abhandlungen*, Vol. 2 (Springer, Berlin), pp. 133–134
Sela, J. (1980): Spectral modelling at the National Meteorological Center. Mon. Weather Rev. **108**, 1279–1292
Shampine, L. F., Gordon, M. K. (1975): *Computer Solution of Ordinary Differential Equations: The Initial Value Problem* (Freeman, San Francisco)
Shampine, L. F., Watts, H. A., Davenport, S. M. (1976): Solving nonstiff ordinary differential equations—the state of the art. SIAM Rev. **18**, 376–411
Siggia, E. D. (1981): Numerical study of small-scale intermittency in three-dimensional turbulence. J. Fluid Mech. **107**, 375–406
Siggia, E. D., Patterson, G. S., Jr. (1978): Intermittancy effects in a numerical simulation of stationary three-dimensional turbulence. J. Fluid Mech. **86**, 567–592
Siggia, E. D., Zippelius, A. (1981): Dynamics of defects in Rayleigh-Benard convection. Phys. Rev. A **24**, 1036–1049
Silberman, I. (1954): Planetary waves in the atmosphere. J. Meteorol. **11**, 27–34
Simmonds, I. (1975): The spectral representation of moisture. J. Appl. Meteorol. **14**, 175–179
Simmons, A. J., Hoskins, B. J. (1975): A comparison of spectral and finite-difference simulations of a growing baroclinic wave. Q. J. R. Meteorol. Soc. **101**, 551–565
Singer, B., Reed, H. L., Ferziger, J. H. (1986): "Investigation of the Effects of Initial Disturbances on Plane Channel Transition," AIAA Pap. No. 86-0433
Slater, J. C. (1934): Electronic energy bands in metal. Phys. Rev. **45**, 794–801
Smagorinsky, J. (1963): General circulation experiments with the primitive equations. Mon. Weather Rev. **91**, 99–164
Solomonoff A., Turkel, E. (1986): "Global Collocation Methods for Approximation and the Solution of Partial Differential Equations," ICASE Rep. No. 86-60 (NASA Langley Research Center, Hampton, VA)
Spalart, P. R. (1984): A spectral method for external viscous flows. Contemp. Math. **28**, 315–335
Spalart, P. R. (1985): "Numerical Simulation of Boundary-Layer Transition," in *Proc. 9th Int. Conf. Numerical Methods in Fluid Dynamics*, ed. by Soubbarameyer, J. P. Boujot (Springer, Berlin, Heidelberg), pp. 531–535
Spalart, P. R. (1986): "Numerical Simulation of Boundary Layers, Part 1: Weak Formulation and Numerical Method," NASA TM-88222
Spalart, P. R. (1987): Numerical study of sink-flow boundary layers J. Fluid Mech. **172**, 307–328
Spalart, P. R., Leonard, A. (1985): "Direct Numerical Simulation of Equilibrium Turbulent Boundary Layers," in *Proc. 5th Symp. Turbulent Shear Flows* (Cornell University, Ithaca, NY)
Spalart, P. R., Yang K.-S. (1987): Numerical simulation of ribbon-induced transition in Blasius flow. J. Fluid Mech. **178**, 345–365
Squire, H. B. (1933): On the stability of the three-dimensional disturbances of viscous flow between parallel walls. Proc. R. Soc. London Ser. A **142**, 621–628
Stewart, G. W. (1985): A Jacobi-like algorithm for computing the Schur decomposition of a non Hermitian matrix. SIAM J. Sci. Stat. Comput. **6**, 853–864
Strang, W. G., Fix, G. J. (1973): *An Analysis of the Finite Element Method* (Prentice-Hall, Englewood Cliffs, NJ.)
Streett, C. L. (1983): "A Spectral Method for the Solution of Transonic Potential Flow about an Arbitrary Airfoil," AIAA Pap. No. 83-1949
Streett, C. L., Hussaini, M. Y. (1987): "Finite Length Taylor–Couette Flow", in *The Stability of Time-Dependent/Spatially Varying Flows*, ed. by D. L. Dwoyer, M. Y. Hussaini (Springer, Berlin, Heidelberg, New York), pp. 312–334
Streett, C. L., Zang, T. A., Hussaini, M. Y. (1984): "Spectral Methods for Solution of the Boundary–Layer Equations," AIAA Pap. No. 84-0170
Streett, C. L., Zang, T. A., Hussaini, M. Y. (1985): Spectral multigrid methods with applications to transonic potential flow. J. Comput. Phys. **57**, 43–76

Stuben, K., Trottenberg, U. (1982): "Multigrid Methods: Fundamental Algorithms, Model Problem Analysis and Applications," in *Multigrid Methods*, ed. by W. Hackbusch, U. Trottenberg (Springer, Berlin, Heidelberg, New York), pp. 1–176

Sulem, C., Sulem, P. L., Frisch, H. (1983): Tracing complex singularities with spectral methods, J. Comput. Phys. **50**, 138–161

Swarztrauber, P. N. (1977): The methods of cyclic reduction, Fourier analysis and the FACR algorithm for the discrete solution of Poisson's equation on a rectangle. SIAM Rev. **19**, 490–501

Swarztrauber, P. N. (1986): Symmetric FFTs. Math. Comput. **47**, 323–346

Swarztrauber, P. N., Sweet, R. (1975): "Efficient FORTRAN Subprograms for the Solution of Elliptic Partial Differential Equations," NCAR-TN/IA-109

Szegö, G. (1939): *Orthogonal Polynomials*, Vol. 23 (AMS Coll. Publ., New York)

Tadmor, E. (1984): Skew-selfadjoint form for systems of conservation laws. J. Math. Anal. Appl. **103**, 428–442

Tadmor, E. (1986): The exponential accuracy of Fourier and Chebyshev differencing methods. SIAM J. Numer. Anal. **23**, 1–10

Tadmor, E. (1987): Stability analysis of finite difference, pseudospectral and Fourier–Galerkin approximations for time-dependent problems. SIAM Rev. (in press)

Tal–Ezer, H. (1986a): Spectral methods in time for hyperbolic equations. SIAM J. Numer. Anal. **23**, 11–26

Tal–Ezer, H. (1986b): A pseudospectral Legendre method for hyperbolic equations with an improved stability condition. J. Comput. Phys. **67**, 145–172

Tal–Ezer, H. (1987a): The eigenvalues of the pseudospectral Fourier approximation to the operator $\sin(2x)(\partial/\partial x)$. Math. Comput. (in press)

Tal–Ezer, H. (1987b): Spectral methods in time for parabolic problems. SIAM J. Numer. Anal. (in press)

Tan, C. S. (1985): Accurate solution of three-dimensional Poisson's equation in cylindrical coordinates by expansion in Chebyshev polynomials. J. Comput. Phys. **59**, 81–95

Tang, C. M. (1979): Comparison of spectral methods for flow on spheres. J. Comput. Phys. **32**, 80–88

Taylor, A. E. (1958): *Introduction to Functional Analysis* (John Wiley & Sons, New York)

Taylor, M. (1981): *Pseudodifferential Operators* (Princeton Univ. Press, Princeton, NJ)

Taylor, T. D., Hirsh, R. S., Nadworny, M. M. (1981): "FFT Versus Conjugate Gradient Method for Solutions of Flow Equations by Pseudospectral Methods," *in Proc. 4th GAMM Conf. Numerical Methods in Fluid Mechanics*, ed. by H. Viviand (Vieweg, Braunschweig), pp. 311–325

Taylor, T. D., Hirsh, R. S., Nadworny, M. M. (1984): Comparison of FFT, direct inversion and conjugate gradient methods for use in pseudo-spectral methods. Comput. Fluids **12**, 1–9

Taylor, T. D., Murdock, J. W. (1981): Application of spectral methods to the solution of Navier–Stokes equations. Comput. Fluids **9**, 255–263

Taylor, T. D., Myers, R. B., Albert, J. H. (1981): Pseudospectral calculations of shock waves, rarefaction waves and contact surfaces. Comput. Fluids **9**, 469–473

Temam, R. (1968): Une méthode d'approximation de la solution des équations de Navier–Stokes. Bull. Soc. Math. Fr. **96**, 115–152

Temam, R. (1977): *Navier–Stokes Equations* (North-Holland, Amsterdam)

Temam, R. (1983): *Navier–Stokes Equations and Nonlinear Functional Analysis*, (SIAM-CBMS, Philadelphia)

Temperton, C. (1983a): Self-sorting mixed-radix fast Fourier transforms. J. Comput. Phys. **52**, 1–23

Temperton, C. (1983b): Fast mixed-radix real Fourier transforms. J. Comput. Phys. **52**, 340–350

Temperton, C. (1985): Implementation of a self-sorting in-place prime factor FFT algorithm. J. Comput. Phys. **58**, 283–299

Tennekes, H., Lumley, J. L. (1972): *A First Course in Turbulence* (M.I.T., Cambridge)

Timan, A. F. (1963): *Theory of Approximation of Functions of a Real Variable* (Pergamon, Oxford)

Titchmarsch, E. C. (1962): *Eigenfunction Expansions*, Part 1 (Oxford Univ. Press, London)

Toomre, J., Gough, D. O., Spiegel, E. A. (1977): Numerical solution of single-mode convection equations. J. Fluid Mech. **79**, 1–31

Trefethen, L. N. (1980): Numerical computation of the Schwarz–Christoffel transformation. SIAM J. Sci. Stat. Comput. **1**, 82–102

Trefethen, L. N. (1983): Group velocity interpretation of the stability theory of Gustafsson, Kreiss, and Sundstrom. J. Comput. Phys. **49**, 199–217
Trefethen, L. N., Trummer, M. R. (1987): An instability phenomenon in spectral methods. SIAM J. Numer. Anal. (in press)
Tuckerman, L. (1988): "Divergence-free Velocity Field, in Nonperiodic Geometries," J. Comput. Phys., submitted.
Turkel, E. (1980): "Numerical Methods for Large-Scale Time-Dependent Partial Differential Equations," in *Computational Fluid Dynamics*, Vol. 2 (Hemisphere, Washington), pp. 127–262
Van Albada, G. D., Van Leer, B., Roberts, W. W. Jr., (1982): A comparative study of computational methods in cosmic gas dynamics. Astron. Astrophys. **108**, 76–84
Van Dyke, M. D., Guttmann A. J. (1983): Subsonic potential flow past a circle and the transonic controversy. J. Aust. Math. Soc. **B 24**, 243–261
Vanel, J. M., Peyret, R., Bontoux, P. (1985): "A Pseudo-Spectral Solution of Vorticity-Stream Function Equations Using the Influence Matrix Technique," Preprint No. 74 (Univ. Nice)
Varga, R. S. (1962): *Matrix Iterative Analysis* (Prentice-Hall, Englewood Cliffs, NJ)
Villadsen, J. V., Stewart, W. E. (1967): Solution of boundary value problems by orthogonal collocation. Chem. Eng. Sci. **22**, 1483–1501
Vinsome, P. K. W. (1976): "ORTHOMIN, an Iterative Method for Solving Sparse Sets of Simultaneous Linear Equations", in *Proc. Symp. Numerical Simulation of Reservoir Performance* (Los Angeles, CA)
Voigt, R. G., Gottlieb, D., Hussaini, M. Y. (eds.) (1984): *Spectral Methods for Partial Differential Equations* (SIAM-CBMS, Philadephia)
Voke, P. R., Collins, M. W. (1983): "Fluid Dynamic Simulations in General Coordinates," in *Proc. 3rd Int. Conf. Numerical Methods in Laminar and Turbulent Flow*, ed. by C. Taylor, J. A. Johnson, W. R. Smith (Pineridge, Swansea, UK), pp. 1204–1213
Von Kerczek, C. H. (1982): The instability of oscillatory plane Poiseuille flow. J. Fluid Mech. **116**, 91–114
Von Neumann, J. (1963): "Recent Theories of Turbulence," 1949 Report to ONR, in *John von Neumann Collected Works*, Vol. 6, ed. by A. H. Taub (Macmillan, New York) pp. 437–472
Wambecq, A. (1978): Rational Runge–Kutta methods for solving systems of ordinary differential equations. Computing **20**, 333–342
Whitham, G. B. (1974): *Linear and Nonlinear Waves* (John Wiley & Sons, New York)
Wilkinson, J. H. (1965): *The Algebraic Eigenvalue Problem* (Clarendon, Oxford)
Williamson, J. H. (1980): Low-storage Runge–Kutta schemes. J. Comput. Phys. **35**, 48–56
Wimmer, M. (1976): Experiments on a viscous fluid flow between concentric rotating spheres. J. Fluid Mech. **78**, 317–335
Woodward, P. R. (1975): On the nonlinear time development of gas flow in spiral density waves. Astrophys. J. **195**, 61–73
Wong, Y. S., Zang, T. A., Hussaini, M. Y. (1986): Efficient iterative techniques for the solution of spectral equations. Comput. Fluids **14**, 85–95
Wornom, S. F. (1978): "Critical Study of Higher Order Numerical Methods for Solving the Boundary-Layer Equations," NASA TP-1302
Wray, A., Hussaini, M. Y. (1984): Numerical experiments in boundary-layer stability. Proc. R. Soc. London Ser. A **392**, 373–389
Wright, K. (1964): Chebyshev collocation methods for ordinary differential equations. Computer J. **6**, 358–365
Wu, C.-T., Ferziger, J. H., Chapman, D. R. (1985): "Simulation and Modeling of Homogeneous Compressed Turbulence," Rep. TF-21 (Dep. Mechanical Engineering, Stanford Univ.)
Yanenko, N. N. (1971): *The Method of Fractional Steps for Solving Multi-dimensional Problems of Mathematical Physics in Several Variables* (English translation), ed. by M. Holt (Springer, Berlin, Heidelberg, New York)
Young, D. M. (1954): On Richardson's method for solving linear systems with positive definite matrices. J. Math. Phys. **22**, 243–255
Young, D. M. (1971): *Iterative Solution of Large Linear Systems* (Academic Press, London, New York)
Young, D. M., Jea, K. C. (1980): Generalized conjugate gradient acceleration of nonsymmetrizable iterative methods. Lin. Alg. Appl. **34**, 159–194
Zakaria, A. (1985): "Étude de divers schémas pseudo-spectraux de type collocation pour la

résolution des équations aux dérivées partielles application aux équations de Navier–Stokes," Thèse, Univ. Nice
Zang, T. A. (1988): "On the rotation and skew-symmetric forms for incompressible flow simulations." (submitted).
Zang, T. A., Drummond, J. P., Erlebacher, G., Speziale, C., Hussaini, M. Y. (1987): "Numerical Simulation of Transition, Compressible Turbulence and Reacting Flows," AIAA Pap. No. 87-0130
Zang, T. A., Krist, S. E., Erlebacher, G., Hussaini, M. Y. (1987): "Nonlinear Structures in the Later Stages of Transition," AIAA Pap. No. 87-1204
Zang, T. A., Hussaini, M. Y. (1981): "Mixed Spectral/Finite Difference Approximations for Slightly Viscous Flows," in *Proc. 7th Int. Conf. Numerical Methods in Fluid Dynamics*, ed. by W. C. Reynolds, R. W. MacCormack (Springer, Berlin, Heidelberg, New York), pp. 461–466
Zang, T. A., Hussaini, M. Y. (1985a): "Numerical Experiments on Subcritical Transition Mechanisms," AIAA Pap. No. 85-0296
Zang, T. A., Hussaini, M. Y. (1985b): "Numerical Experiments on the Stability of Controlled Shear Flows,", AIAA Pap. No. 85-1698
Zang, T. A., Hussaini, M. Y. (1985c): "Recent Applications of Spectral Methods in Fluid Dynamics," in *Large-scale Computations in Fluid Mechanics*, Lectures in Applied Mathematics **22**, 379–409
Zang, T. A., Hussaini, M. Y. (1985d): "Fourier–Legendre Spectral Methods for Incompressible Channel Flow," in *Proc. 9th Int. Conf. Numerical Methods in Fluid Dynamics*, Soubbarameyer, J. P. Boujet (Springer, Berlin, Heidelberg, New York), pp. 603–607
Zang, T. A., Hussaini, M. Y. (1986): On spectral multigrid methods for the time-dependent Navier–Stokes equations. Appl. Math. Comput. **19**, 359–372
Zang, T. A., Hussaini, M. Y., Bushnell, D. M. (1984): Numerical computations of turbulence amplification in shock-wave interactions. AIAA J. **22**, 12–21
Zang, T. A., Kopriva, D. A., Hussaini, M. Y. (1983): "Pseudospectral Calculation of Shock Turbulence Interactions," in *Proc. 3rd Int. Conf. Numerical Methods in Laminar and Turbulent Flow*, ed. by C. Taylor, J. A. Johnson, W. R. Smith (Pineridge, Swansea, UK), pp. 210–220
Zang, T. A., Wong, Y.-S., Hussaini, M. Y. (1982): Spectral multigrid methods for elliptic equations. J. Comput. Phys. **48**, 485–501
Zang, T. A., Wong, Y.-S., Hussaini, M. Y. (1984): Spectral multigrid methods for elliptic equations II. J. Comput. Phys. **54**, 489–507
Zanolli, P. (1987): Domain decomposition algorithms for spectral methods. Calcolo (in press)
Zebib, A. (1984): A Chebyshev method for the solution of boundary value problems. J. Comput. Phys. **53**, 443–455
Zlatev, Z., Berkowicz, R., Prahm, L. P. (1984): Implementation of a variable stepsize variable formula method in the time-integration part of a code for treatment of long-range transport of air pollutants. J. Comput. Phys. **55**, 278–301
Zygmund, A. (1968). *Trigonometric Series* (Cambridge Univ. Press, London)

Supplementary Bibliography

Amon, C. H. (1988): "Heat Transfer Enhancement and Three-dimensional Transitional Flows by a Spectral Element-Fourier Method", Ph.D. Thesis, M.I.T., Cambridge, MA.

Augenbaum, J. M. (1989): An adaptive pseudospectral method for discontinuous problems. Appl. Numer. Math. **5**, 459–480.

Azaiez, M., LaBrosse, G., Vandeven, H. (1989): "A Pressure Field Pseudo-spectral Evaluation for 3-D Numerical Experiments in Incompressible Fluid Dynamics," in *Proc. 11th Int. Conf. Numerical Methods in Fluid Dynamics*, ed. by D. L. Dwoyer, M. Y. Hussaini, R. G. Voigt (Springer, Berlin, Heidelberg, New York), pp. 121–125.

Babuska, I., Guo, B. Q. (1988): The *h-p* version of the finite element method for domains with curved boundaries. SIAM J. Numer. Anal. **25**, 837–861.

Babuska, I., Suri, M. (1987): The optimal convergence rate of the *p*-version of the finite element method. SIAM J. Numer. Anal. **24**, 750–776.

Balachandar, S., Maxey, M. R. (1989): Methods for evaluating fluid velocities in spectral simulations of turbulence. J. Comput. Phys. **83**, 96–125.

Bayliss, A., Gottlieb, D., Matkowsky, B. J., Minkoff, M. (1989): An adaptive pseudo-spectral method for reaction diffusion problems. J. Comput. Phys. **81**, 421–443.

Bayly, B. J., Orszag, S. A., Herbert, T. (1988): Instability mechanisms in shear-flow transition. Ann. Rev. Fluid Mech. **20**, 359–391.

Bernardi, C., Canuto, C., Maday, Y. (1988): Generalized inf-sup conditions for Chebyshev spectral approximation of the Stokes problem. SIAM J. Numer. Anal. **25**, 1237–1271.

Bernardi, C., Canuto, C., Maday, Y., Metivet, B. (1990): Single-grid spectral collocation for the Navier-Stokes equations. IMA J. Numer. Anal. **10**, 253–297.

Bernardi, C., Debit, N., Maday, Y. (1990): Coupling finite element and spectral methods: first results. Math. Comp. **54**, 21–39.

Bernardi, C., Maday, Y. (1988): Spectral methods for the approximation of fourth-order problems: application to the Stokes and Navier-Stokes equations. Computers & Structures **30**, 205–216.

Bernardi, C., Maday, Y. (1988): Analysis of a staggered grid algorithm for the Stokes equation. Int. J. Num. Meth. Fluids **8**, 537–557.

Bernardi, C., Maday, Y. (1989): Properties of some weighted Sobolev spaces, and application to spectral approximations. SIAM J. Numer. Anal. **26**, 769–829.

Bernardi, C., Maday, Y. (1989): Approximation results for spectral methods with domain decomposition. Appl. Numer. Math. **6**, 33–52.

Bernardi, C., Maday, Y. (1990): Relevement polynomial de traces et applications. MMAN **24**(5), 557–611.

Bernardi, C., Maday, Y. (1991): Some spectral approximations of monodimensional fourth-order problems. J. Approx. Theory (in press).

Bernardi, C., Maday, Y., Metivet, B. (1987): Spectral approximation of the periodic/nonperiodic Navier-Stokes equations. Numer. Math. **51**, 655–700.

Bernardi, C., Maday, Y., Metivet, B. (1987): Calcul de la pression dans la résolution spectrale du problème de Stokes. La Rech. Aéro. 1–21.

Bernardi, C., Maday, Y., Patera, A. T. (1989): "A New Nonconforming Approach to Domain Decomposition: the Mortar Element Method", in *Nonlinear Partial Differential Equations and Their Applications*, ed. by H. Brezis, J. L. Lions (Pitman).

Bernardi, C., Maday, Y., Sacchi-Landriani, G. (1989): Nonconforming matching conditions for

coupling spectral and finite element methods. Appl. Numer. Math. **6**, 65–84.
Bisseling, R. H., Kosloff, R. (1988): Optimal choice of grid points in multidimensional pseudospectral Fourier methods. J. Comput. Phys. **76**, 243–262.
Boyd, J. P. (1987): Exponentially convergent Fourier-Chebyshev quadrature schemes on bounded and infinite intervals. J. Sci. Comput. **2**, 99–121.
Brenier, B., Roux, B., Bontoux, P. (1986): Comparasion des méthodes tau-Chebyshev et Galerkin dans l'étude de stabilité des mouvements de convection naturelle. Probléme des valeurs propres parasites. J. Theor. Appl. Mech. **5**, 95–119.
Briggs, W. L., Hart, L. B., Sweet, R. A., O'Gallagher, A. (1987): Multiprocessor FFT methods. SIAM J. Sci. Stat. Comput. **8**, s27–s42.
Bullister, E. T. (1986): "Development and Application of High Order Numerical Methods for Solution of the Three Dimensional Navier-Stokes Equations,", Ph.D. Thesis, M.I.T., Cambridge, MA.
Cai, W., Gottlieb, D., Shu, C.-W. (1989): Non-oscillatory spectral Fourier methods for shock wave calculations. Math. Comp. **52**, 389–410.
Canuto, C. (1988): Spectral methods and a maximum principle. Math. Comput. **51**, 615–629.
Canuto, C., Funaro, D. (1988): The Schwarz algorithm for spectral methods. SIAM J. Numer. Anal. **25**, 24–40.
Canuto, C., Pietra, P. (1990): Boundary and interface conditions within a finite element preconditioner for spectral methods, Report No. 553, Pavia University. J. Comput. Phys. **91**(2), 310–343.
Canuto, C., Quarteroni, A. (eds.) (1990): *Spectral and High Order Methods for Partial Differential Equations*, (North Holland, Amsterdam).
Cartage, T., Deville, M. (1986): "Improved Time-marching Schemes for Navier-Stokes Equations Using Chebyshev-tau Approximation", in *Proc. 6th GAMN Conf. Numerical Methods in Fluid Mechanics*, ed. by D. Rues, W. Kordulla (Vieweg, Braunschweig), pp. 47–54.
Chan, T. F., Glowinski, R., Periaux, J., Widlund, O. B. (eds.) (1989): *Domain Decomposition Methods*, (SIAM, Philadelphia).
Chan, T. F., Goovaerts, D. (1989): Schur complement domain decomposition algorithms for spectral methods. Appl. Numer. Math. **6**, 53–64.
Crespo del Arco, E., Bontoux, P., Sani, R. L., Hardin, G., Extrémet, G. P. (1988): Steady and oscillatory convection in vertical cylinders heated from below. Numerical simulation of asymmetric flow regimes, Adv. Space Res. **8**, 281–292.
Crespo del Arco, E., Peyret, R., Pulicani, J. P., Randriamampianina, A. (1989): "Spectral Calculations of Oscillatory Convective Flows", in *Proc. 7th Int. Conf. Finite Element Methods in Flow Problems*, ed. by T. J. Chung, G. R. Karr (UAH Press, Huntsville), pp. 1464–1472.
Crespo del Arco, E., Randriamampianina, A., Bontoux, P. (1987): "Two-dimensional Simulation of Time Dependent Convective Flow of a Pr $\rightarrow$ 0 Fluid. A Period Doubling Transition to Chaotic Motion", in *Proc. Conf. Synergetics, Order and Chaos*, ed. by M. G. Velarde (World Scientific, Singapore), pp. 244–257.
Dang, K., Loisel, P. (1987): Direct Simulation of Viscous Compressible Transitional Flows, ONERA Report No. 1987-23.
Davies, A. R., Anderssen, R. S. (1986): Improved estimates of statistical regularization parameters in Fourier differentiation and smoothing. Numer. Math. **48**, 671–697.
Davies, A. R., Karageorghis, A., Phillips, T. N. (1988): Spectral Galerkin methods for the primary two-point boundary value problem in modelling viscoelastic flows. Int. J. Numer. Methods Eng. **26**, 647–662.
Demaret, P., Deville, M. (1988): Chebyshev FEPS Solution of the Stokes Equations Using Subdomain Decomposition", *Computational Fluid Dynamics*, ed. by G. de Vahl Davis, C. Fletcher (North-Holland, Amsterdam), pp. 359–368.
Demaret, P., Deville, M. (1988): "Navier-Stokes Solutions by Spectral Multidomain Decomposition and Finite Element Preconditioning", in *Proc. 10th IMACS Con.*, **4**, 372–374.
Demaret, P., Deville, M. (1989): "Chebyshev Pseudospectral Solutions of the Navier-Stokes Equations by Domain Decomposition and Finite Element Preconditioning", in *Proc. 7th Int. Conf. Finite Element Methods in Flow Problems*, ed. by T. J. Chung, G. R. Karr (UAH Press, Huntsville), pp. 1409–1415.
Demaret, P., Deville, M. (1989): Chebyshev pseudospectral solution of the Stokes equations using finite element preconditioning. J. Comput. Phys. **83**, 463–484.

Demaret, P., Deville, M. O., Schneidesch, C. (1989): Thermal convection solutions by Chebyshev pseudospectral multi-domain decomposition and finite element preconditioning. Appl. Numer. Math. **6**, 107–121.

Demay, Y., Lacroix, J. M., Peyret, R. (1987): "Étude Numérique des mouvements d'un fluide stratifié soumis á une source de chaleur", Prepublication No. 208, Université de Nice.

Deville, M., Mund, E. (1985): "On a Mixed One Step/Chebyshev Pseudospectral Technique for the Integration of Parabolic Problems Using Finite Element Preconditions", in *Polynomes Orthogonaux et Applications*, ed. by C. Brezinski et al (Springer, Berlin, Heidelberg, New York), pp. 399–407.

Deville, M., Mund, E. (1990): Finite element preconditioning for pseudospectral solutions of elliptic problems. SIAM J. Sci. Stat. Comput. **11**, 311–342.

Deville, M., Mund, E., Patera, A. T. (1987): Iterative solution of isoparametric spectral element equations by low-order finite element preconditioning. J. Comput. Appl. Math. **20**, 189–197.

Dubiner, M. (1987): Asymptotic analysis of spectral methods. J. Sci. Comput. **2**, 3–31.

Ehrenstein, U. (1988): "Stability of a Chebyshev-Fourier Approximation for the Stokes Equations", Prepublication No. 198, Universite de Nice.

Ehrenstein, U., Peyret, R. (1987): "A Chebyshev-collocation Method for the Navier-Stokes Equations with Application to Double-Diffusive Convection", Prepublication No. 169, Universite de Nice.

Elia, F., Chikhaoui, A., Randriamampianina, A., Bontoux, P., Roux, B. (1983): "Spectral Approximation for Boussinesq Double Diffusion", in *Proc. 5th GAMM Conf. Numerical Methods in Fluid Mechanics*, ed. by M. Pandolfi, R. Piva (Vieweg, Braunschweig), pp. 57–63.

Elowitz, M., Hill, F., Duvall, T. L., Jr. (1989): A test of a modified algorithm for computing spherical harmonic coefficients using an FFT. J. Comput. Phys. **80**, 506–511.

Erlebacher, G., Bokhari, S. H., Hussaini, M. Y. (1990): Parallelization of a three-dimensional compressible transition code. AIAA J. **28**, 83–90.

Erlebacher, G., Hussaini, M. Y. (1990): Numerical experiments in supersonic boundary-layer stability. Phys. Fluids A **2**, 94–104.

Erlebacher, G., Zang, T. A., Hussaini, M. Y. (1987): "Spectral Multigrid Methods for the Numerical Simulation of Turbulence", in *Multigrid Methods, Theory, Applications, and Supercomputing*, ed. by S. F. McCormick (Marcel Dekker, New York), pp. 177–194.

Fishelov, D. (1988): Spectral methods for the small disturbance equation of transonic flows. SIAM J. Sci. Stat. Comput. **9**, 232–251.

Fishelov, D. (1988): The spectrum and the stability of the Chebyshev collocation operator for transonic flow. Math. Comp. **51**, 559–579.

Fischer, P. (1989): "Spectral Element Solution of the Navier-Stokes Equations on High Performance Distributed-Memory Parallel Processors", Ph.D. Thesis, M.I.T., Cambridge, MA.

Fischer, P. F., Ho., L. W., Karniadakis, G. E., Ronquist, E. M., Patera, A. T. (1988): Recent advances in parallel spectral element simulation of unsteady incompressible flows. Computers & Structures **30**, 217–231.

Fischer, P., Ronquist, E. M., Dewey, D., Patera, A. T. (1988): "Spectral element methods: Algorithms and architectures", in *Proc. First Int. Symp. Domain Decomposition Methods for PDEs*, ed. by Glowinski et al (SIAM, Philadelphia), pp. 173–197.

Fischer, P., Ronquist, E. M., Patera, A. T. (1989): "Parallel spectral element methods for viscous flow" in *Parallel Supercomputing Methods, Algorithms and Applications*, ed. by G. F. Carey (Chichester, Wiley), pp. 223–232.

Foias, C., Manley, O., Temam, R. (1988): Modelling of the interaction of small and large eddies in two-dimensional turbulent flows. MMAN **22**, 93–114.

Fontaine, J. P., Crespo del Arco, E., Randriamampianina, A., Extrémet, G. P., Bontoux, P. (1988): Convective motions in liquid metals for material processings. Numerical simulation of oscillation regimes and effect of rotation. Adv. Space Res. **8**, 265–279.

Fontaine, J. P., Randriamampianina, A., Extrémet, G. P., Bontoux, P. (1989): "Buoyancy-driven and Forced Convection in Crystal Growth from Melts. Effects of Rotation and Simulations of Oscillation Regimes", in *Proc. 7th Int. Conf. Finite Element Methods in Flow Problems*, ed. by T. J. Chung, G. R. Karr (UAH Press, Huntsville), pp. 434–442.

Fornberg, B. (1987): The pseudospectral method: Comparisions with finite differences for the elastic wave equation. Geophysics **52**, 483–501.

Frölich, J. (1988): "Comparison of Solution Algorithms for the Variable Coefficient Heat Equa-

tion Applying a Spectral Collocation Method", Prepublication No. 212, Université de Nice.
Francken, P., Deville, M., Mund, E. (1990): On the spectrum of the iteration operator associated to the finite element preconditioning of the Chebyshev collocation calculations. Comput. Meth. Appl. Mech. Eng. **80**(1–3), 295–304.
Funaro, D. (1987): A preconditioning matrix for the Chebyshev differencing operator. SIAM J. Numer. Anal. **24**, 1024–1031.
Funaro, D. (1988): Computing the inverse of the Chebyshev collocation derivative. SIAM J. Sci. Stat. Comput. **9**, 1050–1057.
Funaro, D. (1988): Domain decomposition methods for pseudospectral approximations. I. Second order equations in one dimension. Numer. Math. **52**, 329–344.
Funaro, D., Gottlieb, D. (1988): A new method of imposing boundary conditions in pseudospectral approximations of hyperbolic equations. Math. Comp. **51**, 519–613.
Funaro, D., Quarteroni, A., Zanolli, P. (1988): An iterative procedure with interface relaxation for domain decomposition methods. SIAM J. Numer. Anal. **25**, 1213–1236.
Gardner, D. R., Trogdon, S. A., Douglass, R. W. (1989): A modified tau spectral method that eliminates spurious eigenvalues. J. Comput. Phys. **80**, 137–167.
Gastaldi, F., Quarteroni, A. (1989): On the coupling of hyperbolic and parabolic systems: analytical and numerical approach. Appl. Numer. Math. **6**, 3–31.
Gastaldi, F., Quarteroni, A., Sacchi-Landriani, G. (1989): "Effective Methods for the Treatment of Interfaces Separating Equations of Different Character," in *Computers and Experiments in Fluid Flow*, ed. by G. M. Carlomagno, C. A. Brebbia (Springer, Berlin, Heidelberg, New York), pp. 65–74.
Gauthier, S. (1988): A spectral collocation method for two-dimensional compressible convection. J. Comput. Phys. **75**, 217–235.
Gilbert, N. (1988): "Numerische Simulation der Transition von der laminaren in die turbulente Kanalströmung", DFVLR-FD 88-55.
Gilbert, N., Kleiser, L. (1986): "Subcritical Transition to Turbulence in Channel Flow, in *Direct and Large Eddy Simulation of Turbulence*, ed. by U. Schumann, R. Friedrich (Vieweg, Braunschweig), pp. 1–18.
Gilbert, N., Kleiser, L. (1990): "Near-wall Phenomena in Transition to Turbulence", in *Near-Wall Turbulence: 1988 Zoran Zaric Memorial Conference*, ed. by S. J. Kline, N. H. Afgan (Hemisphere, Washington), pp. 7–27.
Gottlieb, D., Hirsh, R. S. (1989): Parallel pseudospectral domain decomposition techniques. J. Sci. Comput. **6**(4), 309–325.
Guillard, H., Peyret, R. (1986): "On the Use of Spectral Methods for the Numerical Solution of Stiff Problems", Prepublication No. 108, Université de Nice.
Guo, B. (1983): The convergence of the spectral scheme for solving two-dimensional vorticity equation. J. Comp. Math. **1**, 353–362.
Guo, B. (1985): Error estimations of the spectral method for solving K.D.V.-Burgers equation. Acta Math. Sin. **28**, 1–15.
Guo, B. (1985): Spectral method for solving Navier-Stokes equations. Scientia Sin. **28**, 1139–1153.
Guo, B., Cao, W. (1988): The Fourier pseudospectral method with a restrain operator for the RLW equation. J. Comput. Phys. **74**, 110–126.
Guo, B., Ma, H. (1987): The Fourier pseudo-spectral method for three-dimensional vorticity equations. Acta Math. Appl. Sin. **3**, 296–309.
Guo, B., Ma, H. (1989): Fourier pseudo-spectral method for Burgers equation. Acta Math. Appl. Sin. (in press).
Heinrichs, W. (1988a): Line relaxation for spectral multigrid methods. J. Comput. Phys. **77**, 166–182.
Heinrichs, W. (1988b): Multigrid methods for combined finite difference and Fourier problems. J. Comput. Phys. **78**, 424–436.
Herbert, T. (1988): Secondary instability of boundary layers. Ann. Rev. Fluid Mech. **20**, 487–526.
Ho, L. W. (1989): "A Legendre Spectral Element Method for Simulation of Incompressible Unsteady Viscous Free-Surface Flows", Ph.D. Thesis, M.I.T., Cambridge, MA.
Horiuti, K. (1987): Comparison of conservative and rotational forms in large eddy simulation of turbulent channel flow. J. Comput. Phys. **71**, 343–370.
Hussaini, M. Y. (1988): "Computational Fluid Dynamics: a Personal View", in *Proc. 11th Int. Conf. Numerical Methods in Fluid Dynamics*, ed. by D. L. Dwoyer, M. Y. Hussaini, R. G. Voigt (Springer, Berlin, Heidelberg, New York), pp. 3–17.

Hussaini, M. Y., Kopriva, D. A., Patera, A. T. (1989): Spectral collocation methods. Appl. Numer. Math. **5**, 177–208.

Jauberteau, F., Rosier, C., Temam, R. (1990): Nonlinear Galerkin methods in computational fluid dynamics. Appl. Num. Math. **6**, 361–370.

Karageorghis, A., Phillips, T. N. (1989): Spectral collocation methods for Stokes flow in contraction geometries and unbounded domains. J. Comput. Phys. **80**, 314–330.

Karageorghis, A., Phillips, T. N. (1989): Chebyshev spectral element methods for laminar flow through a channel contraction. J. Comput. Phys. **84**, 114–133.

Karageorghis, A., Phillips, T. N., Davies, A. R. (1988): Spectral collocation methods for the primary two-point boundary value problem in modelling viscoelastic flows. Int. J. Numer. Methods Eng. **26**, 805–813.

Karniadakis, G. E. (1987): "The Spectral Element Method Applied to Heat Transport Enhancement by Flow Destabilization", Ph.D. Thesis, M.I.T., Cambridge, MA.

Karniadakis, G. E. (1989): Spectral element simulations of laminar and turbulent flows in complex geometries. Appl. Numer. Math. **6**, 85–105.

Khorrami, M. R., Malik, M. R., Ash, R. L. (1989): Application of spectral collocation techniques to the stability of swirling flows. J. Comput. Phys. **81**, 206–229.

Kopriva, D. A. (1987): A practical assessment of spectral accuracy for hyperbolic problems with discontinuities. J. Sci. Comput. **10**, 249–261.

Kopriva, D. A. (1989a): Computation of hyperbolic equations on complicated domains with patched and overset Chebyshev grids. SIAM J. Sci. Stat. Comput. **10**, 120–132.

Kopriva, D. A. (1989a): Domain decomposition with both spectral and finite difference methods for the accurate computation of flows with shocks. Appl. Numer. Math. **6**, 141–151.

Kopriva, D. A., Hussaini, M. Y. (1989): "Multidomain Spectral Solution of Shock-Turbulence Interactions", in *Domain Decomposition Methods*, ed. by T. F. Chan et al (SIAM, Philadelphia), pp. 340–350.

Ku, H. C., Hirsh, R. S., Taylor, T. D., Rosenberg, A. P. (1988): "A Domain Decomposition Approach for the Computation of Incompressible Flow by the Pseudospectral Matrix Element Method and Its Parallel Implementation", in *Domain Decomposition Methods*, ed. by T. F. Chan et al (SIAM, Philadelphia), pp. 351–360.

Lacroix, J. M., Peyret, R., Pulicani, J. P. (1987): "A Pseudospectral Multi-domain Method for the Navier-Stokes Equations with Application to Double-diffusive Convection", in *Proc. 7th GAMM Conf. Numerical Methods in Fluid Mechanics*, (Vieweg, Braunschweig).

Laurien, E., Kleiser, L. (1989): Numerical simulation of boundary-layer transition and transition control. J. Fluid Mech. **199**, 403–440.

Le Quere, P. (1989): "A Pseudo-spectral Chebyshev Algorithm on a Staggered Grid", in *Proc. 8th GAMN Conf. Numerical Methods in Fluid Mechanics.*

Le Quere, P. (1989): "Mono and Multidomain Chebyshev Algorithm on a Staggered Grid", in *Proc. 7th Int. Conf. Finite Element Methods in Flow Problems*, ed. by T. J. Chung, G. R. Karr (UAH Press, Huntsville).

Loisel, P. (1986): "Résolution des Équations de Navier-Stokes Compressibles Stationnaires par Méthodes Spectrales de Tchebysheff", Ph. D. Thesis, Université Pierre et Marie Curie, Paris.

Liu, K. M., Ortiz, E. L. (1988): Tau method approximation of differential eigenvalue problems where the spectral parameter enters nonlinearly. J. Comput. Phys. **67**, 299–310.

Ma, H., Guo, B. (1986): The Fourier pseudospectral method with a restrain operator for the Korteweg-de Vries equation. J. Comput. Phys. **65**, 120–137.

Ma, H., Guo, B. (1987): The Fourier pseudospectral method for two-dimensional vorticity equations. IMA J. Numer. Anal. **7**, 47–60.

Macaraeg, M. G. (1989): "A Spectral Multi-domain Technique Applied to Stability and Transition of Incompressible and Compressible Flows", in *Proc. 5th Int. Symp. Numerical Methods in Engineering.*

Macaraeg, M. G., Streett, C. L. (1988): "An Analysis of Artificial Viscosity Effects on Reacting Flows Using a Spectral Multi-domain technique", in *Computational Fluid Dynamics*, ed. by G. de Vahl Davis, C. Fletcher (North-Holland, Amsterdam), pp. 503–514.

Macaraeg, M. G., Streett, C. L. (1988): A spectral multi-domain technique for viscous compressible reacting flows, Int. J. Numer. Meth. Fluids **8**, 1121–1134.

Macaraeg, M. G., Streett, C. L. (1989): New instability modes for bounded, free shear flows. Phys. Fluids A, **1**, 1305–1307.

Macaraeg, M. G., Streett, C. L. (1991): Linear stability of high-speed mixing layers, Appl. Numer. Math. 7(1), 93–127.

Macaraeg, M. G., Streett, C. L., Hussaini, M. Y. (1988): "A Spectral Collocation Solution to the Compressible Stability Eigenvalue Problem", NASA TP-2858.

Macaraeg, M. G., Streett, C. L., Hussaini, M. Y. (1989): "A Spectral Multi-Domain Technique Applied to High-Speed Chemically Reacting Flows", in *Domain Decomposition Methods*, ed. by T. F. Chan et al (SIAM, Philadelphia), pp. 361–391.

Maday, Y. (1990): Analysis of spectral operators in one dimensional domains. Math. Comput. **55**, 537.

Maday, Y. (1989): Relèvements de traces polynomiales et interpolation hilbertiennes entre espaces de polynomes. C. R. Acad Sci. Paris I **309**, 463–468.

Maday, Y., Metivet, B. (1986): Chebyshev spectral approximation of Navier-Stokes' equation in a two dimentional domain. MMAN **21**, 93–123.

Maday, Y., Mavriplis, C., Patera, A. T. (1989): "Nonconforming Mortar Element Methods: Application to Spectral Discretizations". in *Domain Decomposition Methods*, ed. by T. F. Chan et al (SIAM, Philadelphia), pp. 392–418.

Maday, Y., Meiron, D., Patera, A. T., Ronquist, E. M. (1990): Analysis of iterative methods for the steady and unsteady Stokes problem: Application to spectral element discretizations. SIAM J. Sci. Stat. Comput. (in press)

Maday, Y., Munoz, R. (1988): Spectral element multigrid. II. Theoretical justification. J. Sci. Comput. **3**, 323–354.

Maday, Y., Munoz, R. (1989): "Numerical Analysis of a Multigrid Method for Spectral Approximations", in *Proc. 11th Int. Conf. Numerical Methods in Fluid Dynamics*, ed. by D. L. Dwoyer, M. Y. Hussaini, R. G. Voigt (Springer, Berlin, Heidelberg, New York), pp. 389–394.

Maday, Y., Patera, A. T. (1989): "Spectral Element Methods for the Incompressible Navier-Stokes Equations", in *State of the Art Surveys in Computational Mechanics*, ed. by A. Noor, J. T. Oden (ASME, New York), pp. 71–143

Maday, Y., Patera, A. T., Ronquist, E. M. (1991): A well-posed optimal spectral element approximation for the Stokes problem. SIAM J. Numer. Anal. **28**, 333.

Maday, Y., Quarteroni, A. (1988): Error analysis for spectral approximations to the Korteweg de Vries equation. MMAN **22**, 539–569.

Maday, Y., Tadmor, E. (1989): Analysis of the spectral vanishing viscosity method for periodic conservation laws. SIAM J. Numer. Anal. **26**, 854–870.

Malik, M. R. (1990): Numerical methods for hypersonic boundary layer stability. J. Comput. Phys. **86**, 376–413.

Marcus, P. S., Tuckerman, L. S. (1987): Simulation of flow between concentric rotating spheres. Part 1. Steady states. J. Fluid Mech. **185**, 1–30.

Marcus, P. S., Tuckerman, L. S. (1987): Simulation of flow between concentric rotating spheres. Part 2. Transitions. J. Fluid Mech. **185**, 31–66.

Marion, M., Temam, R. (1989): Nonlinear Galerkin methods. SIAM J. Numer. Anal. **26**, 1139–1157.

Mavriplis, C. (1989): "Nonconforming Discretizations and A Posteriori Error Estimators for Adaptive Spectral Element Techniques". Ph.D. Thesis, M.I.T., Cambridge, MA.

Mavriplis, C. (1989): Laguerre polynomials for infinite-domain spectral elements. J. Comput. Phys. **80**, 480–488.

Mercier, B. (1981): Analyse Numérique des Méthodes Spectrales, Note CEA-N-2278, Commissariat a l'Energie Atomique Centre d'Étude de Limeil (English translation to be published as *An Introduction to the Numerical Analysis of Spectral Methods* (Springer, Berlin, Heidelberg, New York).

Metcalfe, R. W., Orszag, S. A., Brachet, M. E., Menon, S., Riley, J. (1987): Secondary instability of a temporally growing mixing layer. J. Fluid Mech. **184**, 207–243.

Munoz, R. (1989): "Algunas Contribuciones al Estudio Teórico de los Métodos "Multimalla" en el Contexto de las Discretizaciones de Tipo Espectral, Ph.D. Thesis, Universidad de Santiago de Compostela (Spain).

Ng, L. L. (1990): "Resonances and Instabilities in Subsonic Heated Round Jets", Ph.D. Thesis, Univ. of Arizona, Tucson.

Otto, J. S. (1989): Symmetric prime factor fast Fourier transform algorithms. SIAM J. Sci. Stat. Comput. **10**, 419–431.

Ouazzani, J., Bontoux, P., Elia, F., Peyret, R., Roux, B. (1987): "Numerical Solutions for Binary Mixture Flows. Applications to Crystal Growth Processings", in *Proc. 6th European Symp.*

Material Sciences Under Microgravity Conditions, ESA SP-256, pp. 291–301.
Pasquarelli, F., Quarteroni, A., Sacchi-Landriani, G. (1987): Spectral approximations of the Stokes problem by divergence-free functions, J. Sci. Comput. **2**, 195–226.
Passot, T., Pouquet, A. (1988): Hyperviscosity for compressible flows using spectral methods. J. Comput. Phys. **75**, 300–313.
Pelz, R. B. (1991): The parallel Fourier pseudospectral method. J. Comput. Phys. **92**(2), 296–312.
Pernaud-Thomas, B. (1988): "Méthode d'Ordre Élevé Appliquées au Calcul d'Écoulements de Fluides Compressibles", Ph.D. Thesis, Université Pierre et Marie Curie, Paris.
Phillips, T. N. (1989): Singular matched eigenfunction expansions for Stokes flow around a corner. IMA J. Appl. Math. **42**, 13–26.
Phillips, T. N. (1989): On the potential of spectral methods to solve problems in non-Newtonian fluid mechanics. Numer. Meth. PDE's. **5**, 36–43.
Phillips, T. N. (1989): On the Legendre coefficients of a general order derivative of an infinitely differentiable function. IMA J. Numer. Anal. **8**, 455–459.
Phillips, T. N. (1989): Fourier series solutions to Poisson's equation in rectangularly decomposable regions. IMA J. Numer. Anal. **9**, 337–352.
Phillips, T. N., Davies, A. R. (1988): On semi-infinite spectral elements for Poisson problems with reentrant boundary singularities, J. Comput. Appl. Math. **21**, 173–188.
Phillips, T. N., Karageorghis, A. (1989): Efficient direct methods for solving the spectral collocation equations for Stokes flow in rectangularly decomposable domains. SIAM J. Sci. Stat. Comput. **10**, 89–130.
Phillips, T. N., Karageorghis, A. (1990): On the coefficients of integrated expansions of ultraspherical polynomials. SIAM J. Numer. Anal. **27**, 823–830.
Phillips, T. N., Solinan, I. M. (1991): The influence matrix technique for the numerical spectral simulation of viscous incompressible flows. Numer. Meth. PDE's **7**, 9–24.
Pulicani, J. P. (1986): "A Spectral Multi-method for the Solution of 1-D Helmholtz and Stokes Type Equations", Prepublication No. 122, University de Nice.
Pulicani, J. P., Crespo del Arco, E., Randriamampianina, A., Bontoux, P., Peyret, R. (1990): Spectral simulations of oscillatory convection at low Prandtl Number. Int. J. Numer. Meth. Fluids **10**(5), 481–517.
Quarteroni, A. (1987): Domain decomposition techniques using spectral methods. Calcolo **24**, 141–177.
Quarteroni, A. (1989): "Domain Decomposition Algorithms for the Stokes Equations," in *Domain Decomposition Methods*, ed. by T. Chan et al (SIAM, Philadelphia), pp. 431–442.
Quarteroni, A. (1989): "Iteration by Subdomain Algorithms for Systems of Hyperbolic Equations", in *Parallel Computing: Methods, Algorithms, Applications*, ed. by D. J. Evans, C. Sutti (Adam Hilger, Bristol).
Quarteroni, A. (1990): Domain decomposition methods for systems of conservation laws: spectral collocation approximations. SIAM J. Sci. Stat. Comput. **11**(6), 1029–1052.
Quarteroni, A., Sacchi-Landriani, G. (1988): Domain decomposition preconditioners for the spectral collocation method. J. Sci. Comput. **3**, 45–75.
Quarteroni, A., Sacchi-Landriani, G. (1988): Parallel algorithms for the capacitance matrix method in domain decompositions. Calcolo **25**, 75–102.
Quarteroni, A., Valli, A. (1989): "Domain Decomposition for a Generalized Stokes Problem," in *Proc. 2nd ECMI Conference.*
Randriamampianina, A., Bontoux, P., Roux, B. (1987): Écoulements induits par la force gravique dans une cavité cylindrique en rotation. Int. J. Heat Mass Transfer **30**, 1275–1292.
Randriamampianina, A., Bontoux, P., Roux, B., Argoul, P. (1986): "Multistep Methods for Spectral Tau-Chebyshev Approximation: Application to Rotating and Buoyancy Driven Internal Flows", in *Proc. 6th GAMM Conf. Numerical Methods in Fluid Mechanics*, ed. by D. Rues, W. Kordulla (Vieweg, Braunschweig), pp. 302–309.
Randriamampianina, A., Crespo del Arco, E., Fontaini, J. P., Bontoux, P. (1988): "Spectral Method for Two-dimensional Time-dependent Pr $\rightarrow 0$ Convection", in *Proc. GAMM Workshop on Low Prandtl Number Convection* (Vieweg, Braunschweig).
Ronquist, E. M. (1988): Optimal Spectral element methods for the unsteady three-dimensional incompressible Navier-Stokes equations, Ph.D. Thesis, M.I.T., Cambridge, MA.
Ronquist, E.M., Patera, A.T. (1987a): A Legendre spectral element method for the Stefan problem. Int. J. Numer. Method Eng. **24**, 2273–2299.

Ronquist, E. M., Patera, A. T. (1987b): Spectral element multigrid. I. Formulation and numerical results. J. Sci. Comput. **2**, 389–406.

Ronquist, E. M., Patera, A. T. (1988): "A Legendre Spectral Element Method for the Incompressible Navier-Stokes Equations, in *Proc 7th GAMM Conf. Numerical Methods in Fluid Mechanics* (Vieweg, Braunschweig), pp. 318–326.

Sacchi Landriani, G., Vandeven, H. (1988): Polynomial approximation of divergence free functions. Applications to the Stokes problem. Math. Comput. **52**, 103–130.

Sacchi Landriani, G., Vandeven, H. (1989): "A Multidomain Spectral Collocation Method for the Stokes Problem", Preprint No 642 of the Istituto di Analisi Numerica del C.N.R. (Pavia, Italy).

Sacchi Landriani, G., Vandeven, H. (1990): Error estimates for the spectral approximation of the nonstationary Stokes problem. SIAM J. Numer Anal. **27**(5), 1160–1186.

Schumann, U., Sweet, R. A. (1988): Fast Fourier transforms for direct solution of Poisson's equation with staggered boundary conditions, J. Comput. Phys. **75**, 123–137.

Shen, J. (1987): "Résolution Numérique des Équations de Stokes et de Navier-Stokes par les Méthodes Spectrales", Ph.D. Thesis, Université de Paris-Sud, Paris.

Solomonoff, A., Turkel, E. (1989): Global properties of pseudospectral methods. J. Comput. Phys. **81**, 239–276.

Streett, C. L. (1987): Spectral methods and their implementation to solution of aerodynamic and fluid mechanic problems. Int. J. Numer. Meth. Fluids **7**, 1159–1189.

Streett, C. L., Hussaini, M. Y. (1991): A numerical simulation of the appearance of chaos in finite-length Taylor-Couette flow. Appl. Numer. Math. **7**(1), 41–71.

Streett, C. L., Macaraeg, M. G. (1986): "Preconditioning for First-order Spectral Discretizations", NASA TM 87619.

Streett, C. L., Macaraeg, M. G. (1989): Spectral multi-domain for large-scale fluid dynamic simulations. Appl. Numer. Math. **6**, 123–139.

Tadmor, E. (1989): The convergence of spectral methods for nonlinear conservation laws. SIAM J. Numer. Anal. **26**, 30–44.

Tal-Ezer, H. (1989): Spectral methods in time for parabolic problems. SIAM J. Numer. Anal. **26**, 1–11.

Tal-Ezer, H. (1991): High degree interpolation polynomial in Newton form. SIAM J. Sci. Stat. Comput. **12**, 648.

Temam, R. (1989): "Dynamical Systems, Turbulence and the Numerical Solution of the Navier Stokes Equations", in *Proc. 11th Int. Conf. Numerical Methods in Fluid Dynamics*, ed. by D. L. Dwoyer, M. Y. Hussaini, R. G. Voigt (Springer, Berlin, Heidelberg, New York), pp. 79–98.

Temperton, C. (1988): A new set of minimum-add small-n rotated DFT modules. J. Comput. Phys. **75**, 190–198.

Temperton, C. (1988): A self-sorting in-place prime factor real/half-complex FFT algorithm. J. Comput. Phys. **75**, 199–216.

Tomboulian, S., Streett, C. L., Macaraeg, M. G. (1989): "Spectral Solution of the Incompressible Navier-Stokes Equations on the Connection Machine-2", in *Super-computing 88: Volume II*, ed. by J. L. Martin, S. F. Lundstrom (IEEE Computer Society Press, Washington), pp. 45–50.

Trefethen, L. N. (1988): "Lax-stability vs. Eigenstability of Spectral Methods", in *Numerical methods for Fluid Dynamics III*, ed. by K. W. Morton, M. J. Baines (Clarendon Press).

Trefethen, L. N., Trummer, M. R. (1987): An instability phenomenon in spectral methods. SIAM J. Numer. Anal. **24**, 1008–1023.

Tuckerman, L. (1989): Divergence-free velocity fields in nonperiodic geometries. J. Comput. Phys. **80**, 403–441.

Tuckerman, L. (1989): Transformations of matrices into banded form. J. Comput. Phys. **84**, 360–376.

Tuckerman, L. S. (1989): "Steady-state Solving via Stokes Preconditioning; Recursion Relations for Elliptic Operators", in *Proc. 11th International Conference on Numerical Methods in Fluid Dynamics*, ed. by D. L. Dwoyer, M. Y. Hussaini, R. G. Voigt (Springer, Berlin, Heidelberg, New York), pp. 573–577.

Vandeven, H. (1989): Family of spectral filters for discontinuous problems. J. Approx. Theory (in press).

Vandeven, H. (1990): Compatibilité des espaces discrets pour l'approximation par méthodes spectales du problème de Stokes périodique-nonpériodique. MMAN **23**, 649–688.

Van Dooren, R., Janssen, H. (1989): Period doubling solutions in the Duffing oscillator: a Galerkin approach. J. Comput. Phys. **82**, 161–171.
Weideman, J. A. C., Trefethen, L. N. (1988): The eigenvalues of second-order spectral differentiation matrices. SIAM J. Numer. Anal. **25**, 1279–1298.
Yeung, P. K., Pope, S. B. (1988): An algorithm for tracking fluid particles in numerical simulations of homogeneous turbulence. J. Comput. Phys. **79**, 373–416.
Zang, T. A. (1991): On the rotation and skew-symmetric forms for incompressible flow simulations. Appl. Numer. Math. **7**, 27–40.
Zang, T. A., Hussaini, M. Y. (1987): "Numerical Simulation of Nonlinear Interactions in Channel and Boundary-layer Transition", in *Nonlinear Wave Interactions in Fluids*, AMD-87, ed. by R. W. Miksad, T. R. Akylas, T. Herbert, (ASME, New York), pp. 131–145.
Zang, T. A., Krist, S. E. (1989): Numerical experiments on stability and transition in plane channel flow. Theor. Comput. Fluid Dynamics **1**, 41–64.
Zang, T. A., Krist, S. E., Hussaini, M. Y. (1990): Resolution requirements for numerical simulations of transition. J. Sci. Comput. **4**, 197–217.
Zang, T. A., Streett, C. L., Hussaini, M. Y. (1989): Spectral methods for CFD, ICASE Rep. No. 89-13 (NASA Langley Research Center, Hampton, VA).

Index

GPSR Compliance

The European Union's (EU) General Product Safety Regulation (GPSR) is a set of rules that requires consumer products to be safe and our obligations to ensure this.

If you have any concerns about our products, you can contact us on ProductSafety@springernature.com

In case Publisher is established outside the EU, the EU authorized representative is:

Springer Nature Customer Service Center GmbH
Europaplatz 3
69115 Heidelberg, Germany

Batch number: 10704711

Printed by Printforce, the Netherlands